Table of Atomic Masses Based on Carbon-12

Name	Symbol	Atomic Number	Atomic Mass	Name	Symbol	Atomic Number	Atomic Mass
Actinium	Ac	89	227.028	Mendelevium	Md	101	(258)
Aluminum	Al	13	26.9815	Mercury	Hg	80	200.59
Americium	Am	95	(243)	Molybdenum	Mo	42	95.94
Antimony	Sb	51	121.75	Neodymium	Nd	60	144.24
Argon	Ar	18	39.948	Neon	Ne	10	20.1797
Arsenic	As	33	74.9216	Neptunium	Np	93	237.048
Astatine	At	85	(210)	Nickel	Ni	28	58.69
Barium	Ba	56	137.327	Niobium	Nb	41	92.9064
Berkelium	Bk	97	(247)	Nitrogen	N	7	14.0067
Beryllium	Be	4	9.01218	Nobelium	No	102	(259)
Bismuth	Bi	83	208.980	Osmium	Os	76	190.23
Bohrium	Bh	107	(264)	Oxygen	O	8	15.9994
Boron	B	5	10.811	Palladium	Pd	46	106.42
Bromine	Br	35	79.904	Phosphorus	P	15	30.9738
Cadmium	Cd	48	112.411	Platinum	Pt	78	195.08
Calcium	Ca	20	40.078	Plutonium	Pu	94	(244)
Californium	Cf	98	(251)	Polonium	Po	84	(209)
Carbon	C	6	12.011	Potassium	K	19	39.0983
Cerium	Ce	58	140.115	Praseodymium	Pr	59	140.908
Cesium	Cs	55	132.905	Promethium	Pm	61	(145)
Chlorine	Cl	17	35.4527	Protactinium	Pa	91	231.036
Chromium	Cr	24	51.9961	Radium	Ra	88	226.025
Cobalt	Co	27	58.9332	Radon	Rn	86	(222)
Copper	Cu	29	63.546	Rhenium	Re	75	186.207
Curium	Cm	96	(247)	Rhodium	Rh	45	102.906
Dubnium	Db	105	(262)	Rubidium	Rb	37	85.4678
Dysprosium	Dy	66	162.50	Ruthenium	Ru	44	101.07
Einsteinium	Es	99	(252)	Rutherfordium	Rf	104	(261)
Erbium	Er	68	167.26	Samarium	Sm	62	150.36
Europium	Eu	63	151.965	Scandium	Sc	21	44.9559
Fermium	Fm	100	(257)	Seaborgium	Sg	106	(263)
Fluorine	F	9	18.9984	Selenium	Se	34	78.96
Francium	Fr	87	(223)	Silicon	Si	14	28.0855
Gadolinium	Gd	64	157.25	Silver	Ag	47	107.868
Gallium	Ga	31	69.723	Sodium	Na	11	22.9898
Germanium	Ge	32	72.61	Strontium	Sr	38	87.62
Gold	Au	79	196.967	Sulfur	S	16	32.066
Hafnium	Hf	72	178.49	Tantalum	Ta	73	180.948
Hassium	Hs	108	(265)	Technetium	Tc	43	(98)
Helium	He	2	4.00260	Tellurium	Te	52	127.60
Holmium	Ho	67	164.930	Terbium	Tb	65	158.925
Hydrogen	H	1	1.00794	Thallium	Tl	81	204.383
Indium	In	49	114.818	Thorium	Th	90	232.038
Iodine	I	53	126.904	Thulium	Tm	69	168.934
Iridium	Ir	77	192.22	Tin	Sn	50	118.710
Iron	Fe	26	55.847	Titanium	Ti	22	47.88
Krypton	Kr	36	83.80	Tungsten	W	74	183.85
Lanthanum	La	57	138.906	Uranium	U	92	238.029
Lawrencium	Lr	103	(260)	Vanadium	V	23	50.9415
Lead	Pb	82	207.2	Xenon	Xe	54	131.29
Lithium	Li	3	6.941	Ytterbium	Yb	70	173.04
Lutetium	Lu	71	174.967	Yttrium	Y	39	88.9059
Magnesium	Mg	12	24.3050	Zinc	Zn	30	65.39
Manganese	Mn	25	54.9381	Zirconium	Zr	40	91.224
Meitnerium	Mt	109	(266)				

Atomic masses in this table are relative to carbon-12 and limited to six significant figures, although some atomic masses are known more precisely. For certain radioactive elements the numbers listed (in parentheses) are the mass numbers of the most stable isotopes.

Chemistry
for Changing Times

Chemistry
for Changing Times

NINTH EDITION

John W. Hill

University of Wisconsin–River Falls

Doris K. Kolb

Bradley University

PRENTICE HALL
Upper Saddle River, NJ 07458

Library of Congress Cataloging-in-Publication Data

Hill, John William.
    Chemistry for changing times.—9th ed. / John W. Hill, Doris K. Kolb.
      p. cm.
    Includes bibliographical references and index.
    ISBN 0-13-087489-2
    1. Chemistry.   I. Kolb, Doris K.   II. Title

  QD33.H65   2001
  540—dc21   00-034683

Senior Editor: *Kent Porter Hamann*
Editorial Director: *Paul F. Corey*
Development Editor: *Mary Ginsburg*
Editor in Chief, Development: *Carol Trueheart*
Production Editor: *Nicole Bush*
Assistant Vice President of Production and Manufacturing: *David W. Riccardi*
Executive Managing Editor: *Kathleen Schiaparelli*
Marketing Manager: *Steve Sartori*
Manufacturing Manager: *Trudy Pisciotti*
Manufacturing Buyer: *Michael Bell*
Creative Director: *Paul Belfanti*
Art Director: *Joseph Sengotta*
Interior Designer: *Judith A. Matz-Coniglio*
Cover Photo: *Peter Beck/The Stock Market*
Cover Designer: *Stacey Abraham*
Cover Molecular Art: © *Kenneth Eward/BioGrafx, 2000*
Art Manager: *Gus Vibal*
Art Editor: *Karen Branson*
Illustrations: *Academy Artworks*
Editorial Assistant: *Richard Moriarty*
Photo Editor: *Beth Boyd*
Photo Researcher: *Stuart Kenter Associates*
Photo Coordinator: *Michelina Viscusi*
Copy Editor: *Carol Dean*
Composition: *Preparé Inc.*

Prentice Hall

© 2001, 1998 by Prentice-Hall, Inc.
Upper Saddle River, NJ 07458

Printed in the United States of America

10 9 8 7 6 5 4 3 2

ISBN 0-13-087489-2

Prentice-Hall International (UK) Limited, London
Prentice-Hall of Australia Pty. Limited, Sydney
Prentice-Hall Canada Inc., Toronto
Prentice-Hall Hispanoamericana, S.A., Mexico
Prentice-Hall of India Private Limited, New Delhi
Prentice-Hall of Japan, Inc., Tokyo
Pearson Education Asia Pte., Ltd.
Editora Prentice-Hall do Brasil, Ltda., Rio de Janeiro

Brief Contents

MediaLabs

Each of the twelve MediaLabs is introduced to the student in a two-page spread after appropriate chapters throughout the textbook. The topics have been carefully chosen not only to interest students but also to focus on information that is timely and relevant in daily life. All of the necessary background information is included in the text. Once the student accesses the Companion Website, the static investigations and exercises become a dynamic and exciting discovery activity for students to actively participate in and report on individually or in groups.

Contents

4   Nuclear Chemistry: The Heart of Matter   79

5   Chemical Bonds: The Ties That Bind   112

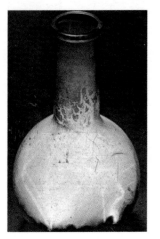

10 Polymers: Giants Among Molecules 266

11 Chemistry of Earth: Metals and Minerals 294

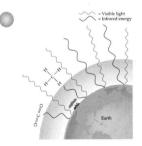

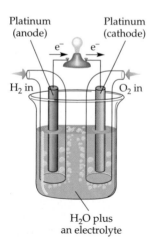

Platinum (anode)   Platinum (cathode)

e^- e^-

H_2 in O_2 in

H_2O plus an electrolyte

15   Biochemistry: A Molecular View of Life   408

16   Food: Those Incredible Edible Chemicals   446

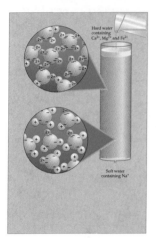

20 Poisons: Chemical Toxicology 600

Appendix A

Appendix B

Appendix C

Photo Credits

Index

Preface

Chemistry for Changing Times is now in its ninth edition. Times have indeed changed since the first edition appeared in 1972, and the book has changed accordingly. Our knowledge base has expanded enormously since that first edition, yet we have resisted the pressure to increase the size of the book. This has forced us to make some tough choices in deciding what to include and what to leave out. We live in what has been called the "information age." Our main focus, therefore, is not so much on providing information as it is on helping students evaluate that information.

We believe that a chemistry course for students who are not majoring in science should be quite different from the course we offer our science majors. It should present basic chemical concepts with intellectual honesty, but it should not focus on esoteric theories or rigorous mathematics. It should include lots of modern everyday applications. The textbook should be appealing to look at, easy to understand, and interesting to read.

Three-fourths of the legislation considered by the U. S. Congress involves questions having to do with science or technology, yet only rarely does a scientist or engineer enter politics. Most of the people who make important decisions regarding our health and our environment are not trained in science, but it is critical that these decision makers are scientifically literate. A chemistry course for students who are not science majors should emphasize practical applications of chemistry to problems involving such things as environmental pollution, radioactivity, energy sources, and human health. The students who take our liberal arts chemistry courses include future teachers, lawyers, accountants, journalists, and judges. There are probably some future legislators, too.

Objectives

Our main objectives in a chemistry course for students who are not majoring in science are as follows:

- To attract as many students as possible. If students are not enrolled in the course, we can't teach them.
- To use topics of current interest to illustrate chemical principles. We want students to appreciate the importance of chemistry in the real world.
- To relate chemical problems to the everyday lives of our students. Chemical problems seem more significant to students if they can see a personal connection.
- To instill in students an appreciation for chemistry as an open-ended learning experience. We hope that our students will want to continue learning throughout their lives.
- To acquaint students with scientific methods. We want students to be able to read about science and technology with some degree of critical judgment.
- To help students become literate in science. We want our students to develop a comfortable knowledge of science so that they find news articles relating to science interesting rather than intimidating.

New Features in the Ninth Edition

In response to suggestions from users and reviewers of the eighth edition, as well as using our own writing and teaching experience, we have thoroughly updated all the text to reflect the latest scientific knowledge. The organization of the 20 chapters remains much the same as in the eighth edition. The major exception lies within Chapters 5 and 6.

Changes in Content

Some of the more important changes are as follows:

- Chapter 6 was substantially revised and has a new title, "Chemical Accounting: Mass and Volume Relationships." We have concentrated much of the quantitative material in this one chapter.
- We moved the sections on naming chemical compounds from Chapter 6 to Chapter 5.
- We moved the kinetic-molecular theory from Chapter 5 to Chapter 6.
- At the request of several users and reviewers, we have added a new section (Section 6.7) on solutions. This addition includes new worked-out examples, exercises, and end-of-chapter problems.
- We revised much of the early treatment of acid-base chemistry (Chapter 7), making the Brønsted–Lowry treatment more explicit.
- We added a discussion of half-reactions in Chapter 7, with new worked-out examples, exercises, and end-of-chapter problems.
- We added a brief discussion of IUPAC naming in Chapter 9.
- In Chapter 13 we added a brief subsection, "Calculations of Parts per Billion," including a new worked-out example and end-of-chapter problems.
- We have added new tables and reorganized others. New tables include Physical Properties, Chemical Properties, Potential Energy, and Kinetic Energy (Chapter 1); Mendeleev's Original Periodic Table (Chapter 2); Types of Radiation, Differences Between Chemical and Nuclear Reactions, and Symbols for Subatomic Particles (Chapter 4).

Additions to Pedagogy

The following changes have been made to strengthen and improve the pedagogy in this edition.

- We follow worked-out examples by A and B exercises in some cases. The B exercise is intended to be a bit more challenging, often requiring a knowledge of material from earlier in the book.
- To improve the organization of the text, we use superheads in some chapters.
- We use voice balloons in problem solving to carefully guide the student through the process and thus improve the pedagogy.
- Focusing on the importance of providing interesting, relevant applications, we have added several new box features: Cost-Benefit Analysis and Health Care and Body Temperature, Hypothermia, and Hyperthermia (Fever) (Chapter 1); What a Difference an O Makes, and Recycling (Chapter 2); What Makes for Nuclear Stablility? (Chapter 4); Who's Number Is It Anyway? (Chapter 6); Conducting Polymers: Polyacetylene (Chapter 10); Asbestos: Risks and Benefits (Chapter 11); Air Pollution in China, An Air Pollution Episode: London, England, and Wood Smoke (Chapter 12); and Entropy (Chapter 14).

- We have updated the References and Readings at the end of each chapter.
- We continue to include Critical Thinking exercises at the end of each chapter.
- We have chosen several new photographs and produced new diagrams to improve the pedagogy and the visual appeal of the book.

Web-Related Activities

- For this ninth edition, we have added 12 MediaLabs, which are spread throughout the text and placed appropriately after the chapter to which they relate. Examples include: Chapter 9 MediaLab—Fragrances: Stop and Smell the Roses, Chapter 15 MediaLab—Genetic Recombination—Promise or Peril?, and Chapter 18 MediaLab—Keeping Fit or Overexerting? These MediaLabs help tie the chapter topics to current events and use the power of the internet to explore those events.
- For this ninth edition, a major change is that we have added web references with brief descriptions as margin notes within each chapter. They can be accessed through the *Chemistry for Changing Times World Wide Web Center* (http://www.prenhall.com/hillkolb).
- We have also added interesting Online Projects at the end of each chapter that may be assigned as a collaborative group or individual activity.

Use of Color

New color photographs and diagrams have been added. Visual material adds greatly to the general appeal of a textbook. Color diagrams can also be highly instructive, and colorful photographs relating to descriptive chemistry do much to enhance the learning process.

Readability

Over the years, students have told us that they have found this textbook easy to read. The language is simple, and the style is conversational. Explanations are clear and easy to understand. The friendly tone of the book has been maintained in this edition.

Units of Measurement

The United States continues to cling to the traditional English system for many kinds of measurement even though the metric system has long been used internationally. A modern version of the metric system, the Système International (SI), is now widely used, especially by scientists. So what units should be used in a text for liberal arts students? In presenting chemical principles, we use primarily metric units. In other parts of the book we use those units that the students are most likely to encounter elsewhere in the same context.

Chemical Structures

The structures of many complicated molecules are presented in the text, especially in the later chapters. These structures are presented mainly to emphasize that they are actually known and to illustrate the fact that substances with similar properties often have similar structures. Students should not feel that they must learn all these structures, but they should take the time to look at them. We hope that they will come to recognize familiar features in these molecules.

Glossary

The Glossary (Appendix B) gives definitions of terms that appear in boldface throughout the text. These terms include all key terms listed at the end of each chapter.

Questions and Problems

The end-of-chapter exercises include review questions, a set of matched-pair problems, and suggested projects and online projects. Answers to many review questions and to all the odd-numbered problems are given in Appendix C. Problems are given within some of the chapters, with worked-out examples followed by similar exercises. Answers to all the in-chapter exercises are also given in Appendix C.

References and Suggested Readings

An updated list of recommended books and articles appears at the end of each chapter. A student whose interest has been sparked by a topic can delve more deeply into the subject in the library. Instructors might also find these lists useful.

Supplementary Materials

The most important learning aid is the teacher. In order to make the instructor's job easier and enrich the education of students, we have provided a variety of supplementary materials.

Print Resources for Students

Student Study Guide (0-13-087497-3) by John W. Hill of University of Wisconsin–River Falls and Richard Jones of Sinclair Community College. This book assists students through the text material and contains learning objectives, chapter outlines, key terms, and additional problems along with self-tests and answers.

Chemical Investigations for Changing Times, Ninth Edition (0-13-087499-X) by Alton C. Hassell and Paula Marshall. Contains 44 laboratory experiments and is specifically referenced to *Chemistry for Changing Times*. An *Instructor's Manual* (0-13-087491-4) prepared by Paula Marshall is also available.

New York Times Themes of the Times. This newspaper-format offprint uses current chemistry-related articles to emphasize the importance and relevance of chemistry in our lives. (Free in quantity to qualified adopters.)

Print Resources for Instructors

Instructor's Resource Manual (0-13-087480-9) by Paul Karr of Wayne State College. This useful guide describes all the different resources available to instructors and shows how to integrate them into your course. Organized by chapter, this manual offers lecture outlines, answers and solutions to all questions and problems that are not answered by the authors in the answer appendix, suggested in-class demonstrations recommended by Doris Kolb, and other suggested resources. The lecture outline is also available in an electronic format.

Instructor's Manual for Chemical Investigations for Changing Times (0-13-087491-4) by Paula Marshall. This laboratory manual reference includes notes for experiments, safety regulations, procedural instructions, and specifications for equipment and supplies.

Transparencies (0-13-087492-2) This set contains 150 full-color acetates.

Test Item File (0-13-089605-5) by William D. Scott, III, University of Mississippi. Contains over 1800 multiple-choice questions that are referenced to the text.

Media Resources for Students

Chemistry for Changing Times World Wide Web Center (http://www.prenhall.com/hillkolb). This user-friendly site emphasizes that chemistry is an open-ended learning experience, with both interesting applications in our daily lives and exciting new developments. The site features the online versions of MediaLabs in this book, where students are encouraged to do web-based investigations on a topic and then communicate their results in a report. In addition, the links featured in the margins of the text are maintained on this site. Finally, the site also features interactive quizzes with carefully crafted hints and specific feedback that can be assigned in an online syllabus and submitted directly to the student's professor or a teaching assistant via e-mail.

Media Resources for Instructors

Matter 2001: Instructor Presentation CD (coming late fall 2000). This dual-platform presentation CD includes art from the textbook, as well as animations and video clips selected to be appropriate for each chapter. Build a presentation in your office, carry the CD and a small presentation file to your classroom, and use our high-quality images to enhance your lecture.

Course Management Options. Prentice Hall offers a variety of online course management options to suit your needs. The Companion Website comes also in a *Companion Website Plus* version, which includes online tracking of the performance by the students in your class on their online assignments, plus broadcast messaging, all tied to a calendar-driven syllabus. For schools interested in using *WebCT* or *Blackboard*, we offer complete, enhanced website content for use in these powerful course management tools (available late fall 2000).

Computerized Prentice Hall Custom Test. This computerized version of the *Test Item File*, available for both Windows and Macintosh, includes electronic versions of all 1800 test questions as well as the latest Prentice Hall Test Manager Software. Test Manager allows you to create and tailor exams to your own needs and includes tools for course management, algorithmic question generation, and offering tests over a local network. *Windows* (0-13-087498-1); Macintosh (0-13-087496-5)

Acknowledgments

Through the last quarter century we have greatly benefited from hundreds of helpful reviews. It would take far too many pages to list all of those reviewers here. Many of you have contributed to the flavor of the book and helped us minimize our errors. Please know that your contributions are deeply appreciated. For the eighth and ninth editions, we are grateful for challenging reviews from:

Wanda Bailey, *John Jay College of Criminal Justice*
Kerry Bruns, *Southwestern University*
Richard Cahill, *DeAnza College*
Agnes Chartier, *Utah State University*
Frank Darrow, *Ithaca College*
Ronald Frystak, *Hawaii Pacific University*
Greg Gellene, *Texas Tech University*
Hal Harris, *University of Missouri*
Paul Karr, *Wayne State College*

Dominick A. Labianca, *Brooklyn College of the City University of New York*

Maria Longas, *Purdue University – Calumet*

Richard Peterson, *Montana State University*

Jeffrey A. Rahn, *Eastern Michigan University*

Mitchell J. Robertson, *Southeastern Louisiana University*

Ruth Russo, *Whitman College*

Salvatore Russo, *Western Washington University*

James Rybarczyk, *Ball State University*

Wayne Stalick, *George Mason University*

Lawrence Stephens, *Elmira College*

Darrell Watson, *University of Mary Hardin Baylor*

Sandra Wheeler, *Furman University*

Alan S. Wingrove, *Towson State University*

Gary Wolf, *Spokane Falls Community College*

We also appreciate the many people who have called or written or e-mailed with corrections and other helpful suggestions.

Cynthia S. Hill prepared much of the original material on biochemistry, food, and health and fitness.

Four of the verses that appear in this volume were first published in the *Journal of Chemical Education.* We acknowledge, with thanks, the permission to reprint them here. Doris Kolb wrote those verses plus all of the others, including the chapter openers.

We also want to thank our colleagues at the University of Wisconsin–River Falls and Bradley University for all their help and support.

We have been blessed with a team of careful and considerate editors. We especially appreciate all the help we have received from our senior editor, Kent Porter-Hamann. Mary Ginsburg, our development editor, has been a marvelous help in so many ways. Our outstanding production editor, Nicole Bush, has been especially helpful in guiding the book through the putting-it-together process.

We owe a very special kind of thanks to our wonderful spouses, Ina and Ken. Ina has done typing, library research, and so many other things. Ken has done chapter reviews, made suggestions, and given invaluable help with this ninth edition. Most of all, we are grateful to both of them for their enduring love and their boundless patience.

Finally, we also thank all those many students whose enthusiasm has made teaching such a joy. It is gratifying to have students learn what you are trying to teach them, but it is a supreme pleasure to find that they want to learn even more. Finally, we want to thank all of you who have made so many helpful suggestions. We welcome and appreciate all your comments, corrections, and criticisms.

J. W. H.
D. K. K.

To the Student

Welcome to Our Chemical World!

Chemistry is fun. Through this book, we would like to share with you some of the excitement of chemistry and some of the joy of learning about it. We hope to convince you that chemistry does not need to be excluded from your learning experiences. Learning chemistry will enrich your life—now and long after this course is over—through a better understanding of the natural world, the technological questions now confronting us, and the choices we must face as citizens within a scientific and technological society.

Chemistry Directly Affects Our Lives

How does the human body work? How does aspirin cure our headaches? Do steroids enhance athletic ability? Is table salt poisonous? Can scientists cure genetic diseases? Why do most weight-loss diets seem to work in the short run but fail in the long run? Does fasting cleanse the body? Why do our moods swing from happy to sad? Can a chemical test on urine predict possible suicide attempts? How does penicillin kill bacteria without harming our healthy body cells? Chemists have found answers to questions such as these and continue to seek the knowledge that will unlock still other secrets of our universe. As these mysteries are resolved, the direction of our lives often changes—sometimes dramatically. We live in a chemical world—a world of drugs, biocides, food additives, fertilizers, detergents, cosmetics, and plastics. We live in a world with toxic wastes, polluted air and water, and dwindling petroleum reserves. Knowledge of chemistry will help to better understand the benefits and hazards of this world and will enable you to make intelligent decisions in the future.

Chemical Dependency

We are all chemically dependent. Even in the womb we depend on a constant supply of oxygen, water, glucose, and a multitude of other chemicals.

Our bodies are intricate chemical factories. They are durable but delicate systems. Innumerable chemical reactions that allow our bodies to function properly are constantly taking place within us. Thinking, learning, exercising, feeling happy or sad, putting on too much weight or not gaining enough, and virtually all life processes are made possible by these chemical reactions. Everything that we ingest is part of a complex process that determines whether our bodies work effectively or not. The consumption of some substances can initiate chemical reactions that will stop body functions altogether. Other substances, if consumed, can cause permanent handicaps, and still others can make living less comfortable. A proper balance of the right foods provides the chemicals and generates the reactions we need in order to function at our best. The knowledge of chemistry that you will soon be gaining will help you better understand how your body works so that you will be able to take proper care of it.

Changing Times

We live in a world of increasingly rapid change. It has been said that the only constant is change itself. At present, we are facing some of the greatest problems that humans have ever encountered, and the dilemmas with which we are now confronted seem to have no perfect solutions. We are sometimes forced to make a best choice among only bad alternatives, and our decisions often provide only temporary solutions to our problems. Nevertheless, if we are to choose properly, we must understand what our choices are. Mistakes can be costly, and they cannot always be rectified. It is easy to pollute, but cleaning up pollution once it is there is enormously expensive. We can best avoid mistakes by collecting as much information as possible before making critical decisions. Science is a means of gathering and evaluating information, and chemistry is central to all the sciences.

Chemistry and the Human Condition

Above all else, our hope is that you will learn that the study of chemistry need not be dull and difficult. Rather, it can enrich your life in so many ways—through a better understanding of your body, your mind, your environment, and the world in which you live. After all, the search to understand the universe is an essential part of what it means to be human.

Student Media Resources

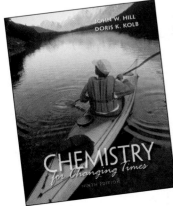

John Hill and Doris Kolb have written *Chemistry for Changing Times* with an emphasis on connecting our world to the chemistry that is all around us. In doing so, they have made chemistry exciting and interesting for thousands of students.

The advantages of media present innumerable educational opportunities. By using Prentice Hall's *Companion Website* technology and in-text MediaLabs, the authors link key chapter concepts to websites and other Internet resources that will expand your study. The next two pages explain how to use these integrated, innovative resources to enhance your understanding of chemistry.

How to Access the Companion Website

You can access the Companion Website (CW) for *Chemistry for Changing Times*, ninth edition by using a standard web browser such as Internet Explorer or Netscape, by typing in the URL **www.prenhall.com/hillkolb**. No password is needed and you may use this site for free with the purchase of the ninth edition.

◀ The MediaLabs in the ninth edition are live on the **Companion Website** and provide current and maintained links for your investigations. Web References and Online Projects are also made live in the Hot Sites sections. Also, you can test your understanding by using the Practice Quiz for each chapter.

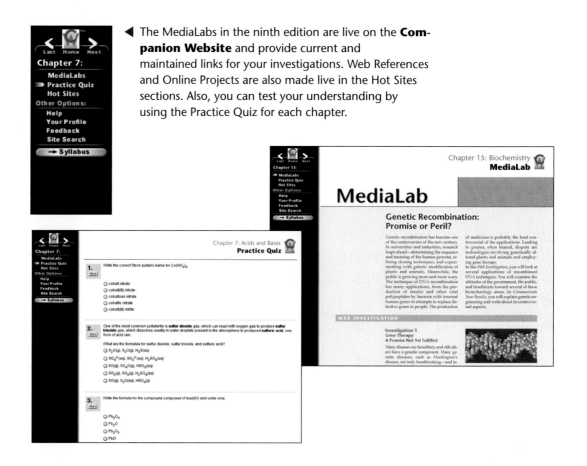

MediaLab

Genetic Recombination: Promise or Peril?

Genetic recombination has become one of the controversies of the new century. In universities and industries, research leaps ahead—determining the sequence and meaning of the human genome, refining cloning techniques, and experimenting with genetic modification of plants and animals. Meanwhile, the public is growing more and more wary. The technique of DNA recombination has many applications, from the production of insulin and other vital polypeptides by bacteria with inserted human genes to attempts to replace defective genes in people. The production

of medicines is probably the least controversial of the applications. Leading to greater, often heated, dispute are technologies involving genetically altered plants and animals and employing gene therapy.

In this *Web Investigation*, you will look at several applications of recombinant DNA techniques. You will examine the attitudes of the government, the public, and bioethicists toward several of these biotechnology areas. In *Communicate Your Results*, you will explain genetic engineering and write about its controversial aspects.

WEB INVESTIGATION

Investigation 1
Gene Therapy:
A Promise Not Yet Fulfilled

Many diseases are hereditary, and still others have a genetic component. Many genetic diseases, such as Huntington's disease, are truly heartbreaking—and inescapable for the person inheriting the gene. Thus, great excitement has accompanied each report of apparently successful gene therapy trials. However, though the concept of injecting replacement genes seems straightforward, many obstacles lie in the way, including the body's own defenses against intruders. Select Keyword **THERAPY** for a look at early successes and disappointments. This site also has a nice animated illustration of the technique. Select Keyword **NEWS** for current news and bioethical debates. (Read what seems interesting and relevant at this site.)

Investigation 2
A New Gene in My Food?

Some of the fiercest reactions from the public, particularly in Europe, have

arisen in response to genetic engineering of various foods. For centuries, people have improved plants and animals by selective breeding. Sometimes, however, breeding for one trait has diminished another desirable attribute. Genetic engineering, however, speeds up the process and enables the insertion of totally new genes, such as genes for selective pest resistance. Select Keyword **FARMING** to read about the benefits of biotechnology in farming and to explore the concerns some people have about it. For an interesting account of the use of gene transfer to protect plants against pests, select Keyword **PLANTS**.

Investigation 3
Cloning: Techniques and Ethics

Among the most controversial topics in genetic engineering is that of cloning. Since Dolly the sheep was created in 1997 by the nuclear replacement technique, interest has focused on the possibility of cloning human beings, and governments around the world have attempted to legislate limits. For a clear but lengthy description of the cloning technique, select Keyword **CLONING**. This British document discusses potential benefits and perils of cloning and related genetic research and technology. It also includes a glossary of genetic engineering terms and a brief overview of national laws concerning

Dolly and Her Surrogate Mother

human cloning. If you wish to see a list of U.S. federal bills, with links to transcripts, select Keyword **BILLS**.

COMMUNICATE YOUR RESULTS

Exercise 1
Risk–Benefit Ratio

Little controversy exists over the use of recombinant methods to produce human insulin, growth hormone, and other essential medications. Is this because people realize that in these cases the benefits of the drugs are worth any perceived risk from genetic engineering? List some benefits of current recombinant DNA techniques, primarily as used in human medicine, and discuss any accompanying risks in terms of these benefits.

Exercise 2
Cloning: Benefits and Pitfalls

Looking at *Web Investigation 3*, discuss the potential benefits and ethical pitfalls of genetic engineering in general and vertebrate cloning in particular. For this discussion, concentrate on techniques used in animals. What are some potential benefits to agriculture? Are these worth less risk than medical applications? Should the risks dictate a "go slow" approach or lead to an outright ban of some applications?

Exercise 3
Bioethics and the Law

In the not-so-distant past, scientists generally regulated themselves, enforcing intellectual integrity by the peer review process. Sometimes they seem to have given little thought to the impact of their findings on society. With an accelerated pace of research and development, including genetic recombination, is this sufficient? Drawing on the *Web Investigations* and any other resources you wish to use, discuss the role of law in regulating the course of biotechnical endeavors. (You might evaluate the degree to which scientific research has moved from academic laboratories into the realm of big business. Does this make federal controls more essential?)

Exercise 4
It's in the Genes

To what degree is the fear of genetic engineering due to the fact that it involves the *genes*—the "stuff of life"—and the concern that any changes will be passed on to future generations? How much concern is due to the fact that many technological advances in the past (DDT, for example) had hidden costs? Write an op-ed article on this topic.

◀ MediaLabs

Throughout the ninth edition, the authors have included **MediaLabs.** These pages cover interesting topics and tie the text discussion to the CW, where a more detailed discussion of the topic can be found. Students are prompted when to access a given website and are given questions to answers. By reading the MediaLabs and completing the exercises provided, students learn chemistry by discovering the answers for themselves, not by just reading about them.

MediaLab Features

- The MediaLabs start with an introduction to the topic under discussion.

- The *Web Investigation* sections provide thought-provoking issues and questions that you can explore by using the text's Companion Website.

- The *Communicate Your Results* sections give you the opportunity to answer questions about the topics you have investigated.

[Web Reference 1]
www Friedrich Wöhler's 1828 report on his synthesis of urea from inorganic materials can be read.

◀ Web References in the margin of the book prompt you to research specific key concepts using the Companion Website.

Online Projects at the end of ▶ the chapters provide you the opportunity to further investigate and answer questions on a high-interest topic using the web address provided.

Online Projects

71. The theory that organic chemistry was the chemistry of living organisms, which was overturned by Wöhler's discovery in 1828, was known as vitalism. What was the effect of this philosophy on areas outside chemistry. You might begin your research at http://www.rit.edu/~flw-stv/biology.html, which has a long but interesting history of vitalism, biology, and the origin of life.

Chemistry

A Science for All Seasons

The vast and fascinating field of chemistry,
Which we shall be discussing in these pages,
Extends from tiny molecules to galaxies.
A Science for All Seasons—and all Ages!

Chemistry is everywhere! Spectacular chemical changes occur in autumn as the chlorophyll in green leaves breaks down, revealing the brilliant colors of other pigments.

"I never buy this brand because of the chemicals."

"I don't use chemicals in *my* garden."

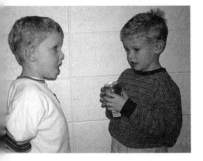

"Mom won't let me drink that stuff. It has chemicals in it."

Technology is the practical application of knowledge.

Look around you. Everything you see is made of chemicals. Chemistry is everywhere. It is involved with the clothes we wear, the food we eat, the air we breathe, and the buildings we live and work in.

Everything we *do* involves chemistry, too. Whenever we wash our hands, watch television, drive a car, surf the internet, eat an apple, or simply take a walk, we are using chemistry. Even when we are asleep, chemical reactions are constantly going on throughout our bodies.

The news media often carry reports about harmful chemicals. They tell us about dangerous chemical spills, toxic chemicals polluting our air and water, and common chemicals that have been found to cause cancer. Several decades ago we rarely heard about such health risks. The news media used to emphasize the wonders of chemical technology. Has chemistry changed over the years, or just our perception of it?

It is true that certain chemicals do cause problems, but there are many others that are extremely helpful. Some chemicals kill bacteria that cause dreadful diseases; some greatly relieve our pain and suffering; some increase our food production; some provide fuel for our heating, cooling, lighting, and transportation; and some provide materials for building our machines, making our clothing, or constructing our houses. Chemistry has provided ordinary people with many luxuries that were not available even to the mightiest of kings in ages past. Chemicals are highly important to our lives. In fact, life itself would be impossible without chemicals.

So, what is chemistry anyway? According to the usual definition, *chemistry* is the study of matter and the changes it undergoes. *Matter* is anything that has *mass*, which means that if you can weigh it, it is matter. Even air, which you cannot see, has weight; therefore, it is matter. Matter is the stuff of which all material things are made. The science of chemistry deals with every kind of matter, from the tiniest parts of atoms to the most complex materials in living plants and animals.

Then, there are those *changes* that matter undergoes. Sometimes matter changes on its own, as when a tool left out in the wet grass gets rusty. But often we change matter to make it more useful, as when we light a candle or cook an egg. Most changes in matter are accompanied by changes in energy. For example, when we burn gasoline, the reaction gives off energy that we can use to propel an automobile or a lawn mower.

Your own body is an incredibly marvelous chemical factory. It takes the food you eat and turns it into skin, bones, blood, and muscle, while also generating energy for all your many activities. Your body is an amazing chemical plant that operates continuously 24 hours a day for as long as you live. Not only does chemistry affect your own individual life every moment, but it also affects society as a whole. Chemistry in a very real way shapes our civilization.

1.1 Science and Technology: The Roots of Knowledge

Chemistry is a **science**, but what is science? Let's examine the roots of science. Our study of the material universe has two facets: the *technological* (or *factual*) and the *philosophical* (or *theoretical*).

Technology arose long before science, having its origins in antiquity. The ancients used fire to bring about chemical changes. For example, they cooked food, baked pottery, and smelted ores to produce metals such as copper. They made beer and wine by fermentation, and obtained dyes and drugs from plant materials. These things—and many others—were accomplished without an understanding of the scientific principles involved.

The Greek philosophers, about 2500 years ago, were perhaps the first to formulate theories explaining the behavior of matter. They generally did not test their the-

The Alchemist, a painting done by the Dutch artist Cornelis Vega around 1660, depicts a laboratory of the seventeenth century.

Aristotle (384–322 B.C.E.), Greek philosopher and tutor of Alexander the Great, believed that we could understand nature through logic. The idea of experimental science did not triumph over Aristotelian logic until about C.E. 1500.

ories by experimentation, however. Nevertheless, their view of nature—attributed mainly to Aristotle—dominated natural philosophy for 2000 years.

The experimental roots of chemistry are planted in **alchemy**, a mystical chemistry that flourished in Europe during the Middle Ages, about C.E. 500 to 1500. Modern chemists inherited from the alchemists an abiding interest in human health and the quality of life. Alchemists not only searched for a philosophers' stone that would turn cheaper metals into gold, but also sought an elixir that would confer immortality on those exposed to it. Alchemists never achieved these goals, but they discovered many new chemical substances and perfected techniques such as distillation and extraction that are still used today.

Technology also developed rapidly during the Middle Ages in Europe, in spite of the generally nonproductive Aristotelian philosophy that prevailed. The beginnings of modern science were more recent, however, coming with the emergence of the experimental method. What we now call science grew out of **natural philosophy**—that is, out of philosophical speculation about nature. Science had its true beginnings in the seventeenth century, when astronomers, physicists, and physiologists began to rely on experimentation.

In this book, we use the more universal designations C.E. and B.C.E., instead of A.D. and B.C., with which you may be more familiar. C.E. means Common Era and is used when referring to the past 2000 years. B.C.E. means Before the Common Era and is used for more ancient times.

Modern Alchemy

Modern chemists can transform matter in ways that would astound the alchemists. They can change ordinary salt into lye and laundry bleach. They can convert sand into transistors and computer chips. And they can turn crude oil into plastics, fibers, pesticides, drugs, detergents, and a host of other products. Perhaps the ultimate example of this modern alchemy was achieved by James E. Butler of the Naval Research Laboratory in Washington, DC, who made diamonds from sewer gas. Many products of modern chemistry are much more valuable than the golden metal that the alchemists vainly sought.

Francis Bacon (1561–1626), English philosopher and Lord Chancellor to King James I.

Rachel Carson (1907–1964) at Woods Hole, Massachussets, in 1951.

 [Web Reference 1] This is a source of information about the history of chemistry.*

 [Web Reference 2] More on the history of chemistry.

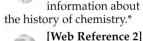 **[Web Reference 3]** Monsanto Chemical Company.

[Web Reference 4] United States Government.

 [Web Reference 5] U.S. Environmental Protection Agency (EPA).

 [Web Reference 6] Greenpeace, an international environmental agency.

1.2 The Baconian Dream and the Carsonian Nightmare

Francis Bacon was a philosopher who practiced law and served as a judge. Although not a scientist, he was very interested in science and argued that it should be experimental. It was his dream that science could solve the world's problems and enrich human life with new inventions, thereby increasing happiness and prosperity.

By the middle of the twentieth century, science and technology appeared to have made the Baconian dream come true. Many dread diseases, such as smallpox, polio, and plague, had been virtually eliminated. The use of fertilizers, pesticides, and scientific animal breeding had increased and enriched our food supply. New materials had been developed to improve our clothing and shelter. Transportation was swift, and communication nearly instantaneous. In addition, nuclear energy seemed to promise unlimited power for our every need. Science and technology had done much toward creating our "modern" world.

The Baconian dream has lost much of its luster in recent decades. People have learned that the products of science are not an unmitigated good. Some people have predicted that science might bring not wealth and happiness, but death and destruction.

Perhaps most noteworthy among these critics of modern technology was Rachel Carson, a biologist. Her poetic and polemic book *Silent Spring* was published in 1962. The book's main theme is that through our use of chemicals to control insects, we are threatening the destruction of all life, including ourselves. People in the pesticide industry (and their allies) roundly denounced Carson as a "propagandist," while other scientists rallied to her support. By the late 1960s, however, we had experienced massive fish kills, that threatened extinction of several species of birds, and the disappearance of fish from rivers, lakes, and areas of the ocean that had long been productive. Most scientists had moved into Carson's camp. Popular support for Carson's views was overwhelming.

Carson was not the first prophet of doom. In 1798 Thomas Malthus, in his "Essay upon the Principles of Population," had predicted that the rapid increase in world population would outpace the increase in food supply and result in widespread famine. During the nineteenth century and the first half of the twentieth century, however, technological developments enabled food production to keep up with population growth, at least in developed countries.

Today the picture has changed. In spite of serious efforts at birth control over the past few decades, population growth threatens to overtake even the most optimistic projections of food production. Some scientists project a dismal future for our world; others confidently predict that science and technology, properly applied, will save us from disaster.

1.3 Science: Testable, Reproducible, Explanatory, Predictive, and Tentative

What *is* science? If scientists disagree about what is and what will be, is science merely a guessing game in which one guess is as good as another? Science is difficult to define precisely, but we will try to describe it.

Scientific Hypotheses

Science is an accumulation of knowledge about nature and our physical world, based on observations. By making careful observations, scientists collect data. Data report-

---

* For information on using Web References, please refer to page xxv.

ed by a scientist must also be observable by other scientists; it must be *reproducible*. Scientists develop **hypotheses** (guesses) to try to explain the observed data, and they test these hypotheses by designing and performing experiments. This is the main thing that distinguishes science from the arts and humanities: The tenets of science are *testable*. In the humanities, people often still argue about some of the same questions that were being debated thousands of years ago: What is truth? What is beauty? However, experiments can be devised to answer most scientific questions. Ideas can be tested and thereby either verified or rejected. As a result, scientists have established a firm foundation of knowledge so that each new generation can build on the past.

> Scientists must make careful observations and measurements, but their work is not really accepted in the scientific community until it has been verified by other scientists.

Scientific Laws

Large amounts of scientific data can sometimes be summarized in brief statements called **scientific laws**. For example, Robert Boyle (1627–1691), an Englishman, conducted many experiments on gases. In each experiment, he found the volume of the gas to decrease when the pressure applied to the gas was increased. Many scientific laws can be stated mathematically. For example, Boyle's law can be written as $PV = k$, where P is the pressure on a gas, V is its volume, and k is a constant number. If P is doubled, V will be cut in half. These laws are *universal*; under the stated conditions they hold everywhere in the observable universe.

Experimental observations are just the beginning of the intellectual processes of science. There are many different paths to scientific discovery, and there is no general set of rules. Science is not just a straightforward logical process for cranking out discoveries.

> A Possible Scientific Process
>
> Observation reported
> ↓
> Observation confirmed by others
> ↓
> Hypothesis suggested
> ↓
> Experiments designed to test hypothesis
> ↓
> Experiments unsuccessful
> ↓
> Hypothesis rejected
> ↓
> New hypothesis offered
> ↓
> New experiments tried
> ↓
> Experiments successful
> ↓
> Experiments repeated and results confirmed
> ↓
> Theory formulated
> ↓
> Many further experiments
> ↓
> and so on

Scientific Theories

Science is a body of knowledge. Scientists organize this knowledge on a framework of detailed explanations called **theories**. The theories represent the best explanations for various phenomena as of today, but they are always *tentative*. Tomorrow, a theory may have to be modified or even discarded in the light of new observations. Science is a body of knowledge that is rapidly growing and always changing.

Theories are useful mainly for their *predictive* value. Predictions based on theories are then tested by further experiments. Theories that make successful predictions are generally widely accepted by the scientific community. A theory developed in one area is often found to apply in others.

> Example of a discarded "fact": It was long thought that people die with the same collection of brain cells that they are born with. However, recent findings show that new brain cells are continually being created, at least in certain regions of the brain.

Scientific Models

Scientists often use models to help explain complicated phenomena. **Scientific models** use tangible items or pictures to represent invisible processes. For example, the invisible particles of a gas can be visualized as billiard balls or marbles, or as dots or circles on paper. We know that when a glass of water is left standing for a period of time, the water disappears through a process called evaporation (Figure 1.1). Scientists use a theory, termed the kinetic–molecular theory, based on tiny, invisible particles called *molecules*, which are in constant motion. To explain evaporation, the kinetic–molecular model pictures the liquid water as molecules. In the bulk of the liquid, these molecules are held together by forces of attraction. The molecules collide with one another like billiard balls on a playing table. Through collisions, some of the molecules gain sufficient energy to break the attraction to their neighbors, escape from the liquid, and disperse among the widely spaced air molecules. The water in the glass gradually disappears. It is much more satisfying to understand evaporation through the use of a model than merely to have a name for it.

> Five characteristics of science are that it is
> (1) *testable*,
> (2) *reproducible*,
> (3) *explanatory*, and
> (4) *predictive*, but always
> (5) *tentative*.

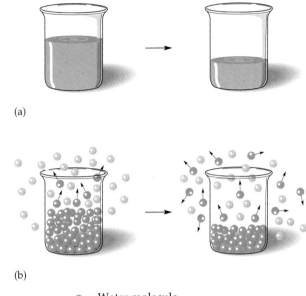

Figure 1.1　The evaporation of water. (a) When a container of water is left standing open to the air, the water slowly disappears. (b) Scientists explain evaporation in terms of the motion of molecules.

　= Water molecule

　= Air (nitrogen or oxygen) molecule

In assessing data, it is important to note that a *correlation* between two items does not prove that one *causes* the other. For example, many people have allergies in the fall when goldenrod is in bloom. However, research has shown that the main cause of these allergies is ragweed pollen. There is a correlation between the goldenrod blooming and autumnal allergies, but goldenrod pollen is not the cause. Ragweed happens to bloom at the same time.

Molecular Modeling

Molecular models are three-dimensional representations of molecules. They can be put together with simple balls and sticks or with various space-filling model kits, or they can be computer-generated. Using a computer, one can draw various kinds of molecules and then rotate them on the screen to get a three-dimensional perspective.

　Some molecular modeling programs can compute the structures of quite complicated molecules, even indicating the distribution of electrons. These computational programs can be used not only to look at individual molecules, but also to study their interactions with each other. Computer modeling is especially useful in the study of such complex systems as biological molecules and pharmaceuticals.

1.4　The Limitations of Science

We sometimes hear scientists and nonscientists alike state that we could solve all our problems if we would only attack them using the scientific method. We have seen already that there is no single scientific method. But why can't the procedures of the scientist be applied to social, political, ethical, and economic problems? Why do scientists disagree over environmental, social, and political issues?

　Often the disagreement results from the inability to control **variables**. If, for example, we wanted to study in the laboratory how the volume of gases varies with changes in pressure, we would hold constant such factors as temperature and the amount and kind of matter. If, on the other hand, we wanted to determine the health effect of low levels of a particular pollutant on a human population, we would find

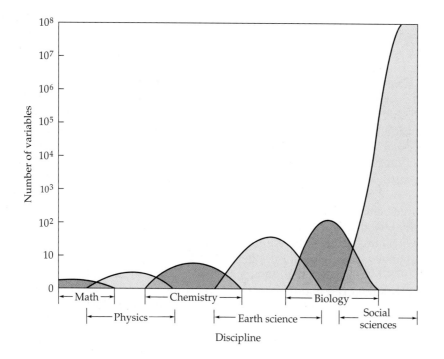

Figure 1.2   A rough estimate of the number of variables involved in scientific disciplines.

it difficult, if not impossible, to control such variables as the people's diets, habits, and exposure to other substances (Figure 1.2).

We can make observations, formulate hypotheses, and conduct experiments, even if most of the variables are not subject to control. Interpretation of the results, however, is much more difficult and more subject to disagreement. Nonscientists sometimes use some of the methods and language of scientists. Artists experiment with new techniques and new materials. Playwrights and novelists observe life as it is before trying to express its essence in their writings. Scientific ideas and methods pervade nearly every aspect of society.

1.5 Science and Technology: Risks and Benefits

Science and technology are interrelated. In everyday life, people often fail to distinguish between the two. A distinction can be useful, however. **Technology** is the sum total of the processes by which humans modify the materials of nature to better satisfy their needs and wants. These processes need not be based on scientific principles. For example, ancient peoples were able to smelt ores to produce metals, such as copper and iron, without having any understanding of the chemistry involved.

Few people question that society has benefited from science and technology, but there are risks associated with technological advances. How can we determine when the benefits outweigh the risks? One approach, called **risk–benefit analysis**, involves the calculation of a desirability quotient (DQ).

$$DQ = \frac{\text{Benefits}}{\text{Risks}}$$

A *benefit* is anything that promotes well-being or has a positive effect. Benefits may be economic, social, or psychological. A *risk* is any hazard that leads to loss or injury. Some of the risks in modern technology have led to disease, death, economic

A **variable** is something that can change over the course of an experiment. Historically, scientists have tried to vary only one quantity at a time, holding most factors constant. Today, with computers to do the complicated mathematics, scientists can often vary several factors at the same time and still determine the effect of each variable.

For most people in Canada, northern Europe, and the United States, milk is a wholesome food; its benefits far outweigh its risks. In other countries, with high rates of lactose intolerance among adults, the desirability quotient for milk is much smaller.

Milk
(northern Europeans)

$$\frac{\text{Large benefits}}{\text{Small risks}} = \text{Large DQ}$$

Thalidomide
(for morning sickness)

$$\frac{\text{Small benefits}}{\text{Large risks}} = \text{Small DQ}$$

Aspartame

$$\frac{\text{Small benefits}}{\text{Small risks}} = \text{Uncertain DQ}$$

Coal conversion

$$\frac{\text{Large benefits}}{\text{Large risks}} = \text{Uncertain DQ}$$

---

loss, and environmental deterioration. Risks may involve one individual, a group, or society as a whole.

The automobile is a technological development that certainly benefits us greatly. But driving a car involves risk, including the risk of possible death in a traffic accident. Clearly, most people seem to feel that the benefits of driving a car far outweigh the risks.

Weighing the benefits and risks connected with a product is more difficult when one considers a group of people. For most people, for example, pasteurized milk is a safe, clean, nutritious beverage. Some people, however, can't tolerate lactose, the sugar in milk. For these people, and others who are allergic to milk, drinking milk can be harmful. Fortunately, milk allergies and lactose intolerance are relatively rare in North America and northern Europe. In these societies milk is generally beneficial, with large benefits and small risks, resulting in a large DQ. However, this analysis generally fits only people of northern European descent. Adults in much of the rest of the world are lactose-intolerant and would find that milk has a small DQ. Thus, milk is not always suitable for use in programs to relieve malnutrition.

In 1958, a drug called thalidomide was introduced in Germany. Many pregnant women took the drug in order to prevent morning sickness. But soon thalidomide was shown to involve an enormous risk. Many malformed infants were born to women who had taken the drug during pregnancy. For thalidomide, then, there are small benefits, very large risks, and a very small DQ. Thalidomide was eventually judged to present unacceptable risks and was banned. But more recently there has been renewed interest in thalidomide. Its use in treating a debilitating skin condition of leprosy was approved in September 1998. Current research is evaluating the use of thalidomide in treating Kaposi's sarcoma (a complication of AIDS), as well as for graft-versus-host reactions in organ transplants and autoimmune diseases, for ulcers in AIDS, and for tuberculosis.

The artificial sweetener aspartame is the most highly studied of all food additives. There is little evidence that use of an artificial sweetener helps one to lose weight. There are anecdotal reports of problems with the sweetener, but these have not been confirmed in controlled studies. To most people, the risk involved in using aspartame is small. This leads to an uncertain DQ—and, it seems, to endless debate over the safety of aspartame. (Aspartame may provide some benefit to diabetics, however, because sugar consumption presents a large risk to them.)

Other technologies provide large benefits and present large risks. For these technologies, too, the DQ is uncertain. An example is the conversion of coal to liquid fuels. Liquid fuels provide large benefits to society in the areas of transportation, home heating, and industry. The risks associated with coal conversion are also large, however. These include air and water pollution and the exposure of workers to toxic chemicals. The result, again, is an uncertain DQ and political controversy.

There are yet other problems in risk–benefit analysis. Some technologies benefit one group of people while presenting a risk to another. For example, it may be economically advantageous to a community to spend as little as possible on a sewage treatment plant and to dump raw wastes into a nearby stream. These wastes might present a hazard to downstream communities, however. Difficult political decisions are needed in such cases.

Other technologies provide current benefits but present future risks. For example, although nuclear power now provides useful electricity, wastes from nuclear power plants, if improperly stored, might present hazards for centuries. Thus, the use of nuclear power is controversial.

Science and technology obviously involve *both* risks and benefits. The determination of benefits is almost entirely a social judgment; risk assessment also involves social decisions, but scientific investigation can help considerably in risk evaluation.

Example 1.1

Some doctors think heroin is more effective for the relief of severe pain than other medicines. However, heroin is highly addictive and often renders the user unable to function in society. Do a risk–benefit analysis for the use of heroin in treating the pain of **(a)** a young athlete's broken leg and **(b)** a terminally ill cancer patient.

Solution

a. The heroin would provide the benefit of pain relief, but its use for such purposes has been judged to be too risky by the U.S. Food and Drug Administration.
b. The heroin would provide the benefit of pain relief. The risk of addiction in a dying person is irrelevant. Heroin is used this way in Great Britain, but it is banned for any purpose in the United States. (Both answers involve judgments that are not clearly scientific; people can differ in their assessments of each.)

Exercise 1.1

A. Chloramphenicol is a powerful antibacterial drug that often destroys bacteria unaffected by other drugs. It is highly dangerous to some individuals, however, causing fatal aplastic anemia in about 1 in 30,000 people. Do a risk–benefit analysis for the use of chloramphenicol in **(a)** sick farm animals, from which people might consume milk or meat with residues of the drug, and **(b)** a person with Rocky Mountain spotted fever faced with a high probability of death or permanent disability.
B. Coal is a fuel widely used for generating electricity. It is abundant and relatively inexpensive. But burning coal produces large amounts of soot, which pollutes the atmosphere and generates much carbon dioxide, contributing to global warming. Burning high-sulfur coal also leads to acid rain, which can kill the fish in a lake or the trees in a forest. Do a risk–benefit analysis for using coal to produce electricity.

Cost–Benefit Analysis and Health Care

When buying a new home, car, or major appliance, people usually consider the cost of the purchase against the benefits it will provide. They are much less likely to make such a cost–benefit analysis when entering a hospital or scheduling a surgical procedure. Health care insurers, however, do this for their clients all the time.

Some people complain because they must get approval from their health insurance company before they check into a hospital or schedule a medical treatment. Insurance companies may argue that they must have such approval rights because some treatments are still experimental and certain procedures are too unreliable to justify their very high cost. Since the insurers bear the financial risks, but their clients receive any health benefits that result, these two groups have quite different viewpoints. It is not surprising that they might disagree about the DQ for a particular type of health care.

1.6 Chemistry: Its Central Role

Chemistry is not only useful in itself, but it is also fundamental to other scientific disciplines. A knowledge of chemistry is applied in much of biology, medicine, agriculture, communication, and many other fields. Science is a unified whole. The various areas of science interact and support one another. Science provides a powerful

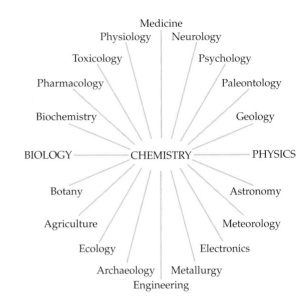

Figure 1.3 Chemistry has a central role among the sciences.

According to the National Safety Council, the chemical industry ranks at or near the top each year in worker safety among 42 basic industries. The number of incidents of occupational illness and injury in the chemical industry is only one-third of the average rate for all U.S. industries. This is in sharp contrast to the popular belief that chemicals are so dangerous. (Some are, of course, but even they can be used safely with proper precautions.)

explanatory scheme. The application of chemical principles has revolutionized biology, provided materials for computers and other electronics equipment, and profoundly influenced other fields such as psychology. The social goals of better health and more and better food, housing, and clothing are dependent to a large extent on the knowledge and techniques of chemists. The recycling of basic materials—paper, glass, and metals—is primarily a matter of chemical processes. Chemistry is indeed a central science (Figure 1.3). There is scarcely a single area of our daily lives that is not affected by chemistry.

Chemistry is also important to the *economy* of industrial nations. In the United States, chemical process industries make about 70,000 consumer products, including pharmaceutical and personal care products, agricultural products, plastics, coatings, soaps, and detergents. The U.S. chemical industry employs over a million people in more than 10,000 plants. It is the nation's fifth largest industry with sales of about $200 billion per year. The exports of the chemical industry help to keep the U.S. international trade deficit from being even worse than it now is. The U.S. chemical industry has had a trade surplus every year since 1981.

1.7 Solving Society's Problems: Scientific Research

Chemistry is one of the most powerful forces shaping society today. Chemical research not only plays a pivotal role in other sciences, but it also has a profound influence on society as a whole. Chemists, like other scientists, do research in one of two categories, *applied research* or *basic research.* The two often overlap, and it isn't always possible to label a particular project as one or the other.

Applied Research

Most chemists work in the area of applied research. They analyze polluted soil, air, and water. They synthesize new chemical compounds for use as drugs or pesticides. They formulate plastics for new applications. They analyze foods, fuels, cosmetics, detergents, and drugs. These are examples of **applied research**—work oriented toward the solution of a particular problem in industry or the environment.

Among the most monumental accomplishments in the realm of applied research were those of George Washington Carver. Born in slavery, Carver attended Simpson

College and later graduated from Iowa State University. A botanist and agricultural chemist, Carver taught and did research at Tuskegee Institute. He developed over 300 products from peanuts, including peanut butter. He made other new products from sweet potatoes, pecans, and clay. Carver also taught southern farmers to rotate crops and to use legumes to replenish the nitrogen removed from the soil by cotton crops. Carver's work helped to revitalize the economy of the South.

Another example of applied research is the synthesis of an antifungal drug. Many drugs are effective against bacteria, but few are useful and safe against fungal infections in humans. Seeking an effective fungal antibiotic, Rachel Brown and Elizabeth Hazen of the New York State Department of Health discovered nystatin, a compound effective against such fungal infections as thrush and vaginitis. Nystatin also has been used in veterinary medicine, in agriculture (to protect fruit), and in the art world (to prevent the growth of mold on art treasures in Florence, Italy, after the objects were damaged by floods). Brown and Hazen set out to find a fungicide and did so—an excellent example of applied research.

George Washington Carver (1860–1943), research scientist, in his laboratory at Tuskegee Institute.

Basic Research: The Search for Knowledge

Many chemists are involved in **basic research**, the search for knowledge for its own sake. Some chemists work out the fine points of atomic and molecular structure. Others measure the intricate energy changes that accompany complex chemical reactions. Chemists synthesize new compounds and determine their properties. This type of investigation is called basic research. Done for the sheer joy of unraveling the secrets of nature and discovering order in our universe, basic research is characterized by the absence of any predictable, marketable product.

Findings from basic research often *are* applied at some point, but this is not the primary goal of the researcher. In fact, most of our modern technology is based on results obtained in basic research. Without this base of factual information, technological innovation would be haphazard and slow.

Applied research is carried out mainly by industries seeking a competitive edge with a novel, better, or more salable product. Its ultimate aim is usually profit for the stockholders. Basic research is conducted mainly at universities and research institutes. Most of its support comes from federal and state governments and foundations, although some larger industries also support it.

An example of basic research later applied to improving human welfare is the work of Gertrude Elion and George Hitchings, who studied compounds called purines in order to try to understand their role in the chemistry of the cell. Their basic research at Burroughs Wellcome Research Laboratories in North Carolina led to the discovery

Rachel Brown (left) and Elizabeth Hazen, who discovered nystatin, a safe fungal antibiotic, in 1950.

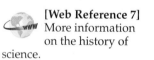

[Web Reference 7] More information on the history of science.

Importance of Basic Research

There are two compelling reasons why society must support basic science. One is substantial: The theoretical physics of yesterday is the nuclear defense of today; the obscure synthetic chemistry of yesterday is curing disease today. The other reason is cultural. The essence of our civilization is to explore and analyze the nature of man and his surroundings. As proclaimed in the Bible in the Book of Proverbs: "Where there is no vision, the people perish."

Arthur Kornberg (1918–), *American Biochemist*
Nobel Prize in Physiology and Medicine, 1959

Gertrude Elion, a basic research chemist, won the Nobel prize in 1988 because of a chance discovery.

Spinning Nuclei

An example of basic research later applied to improving human welfare is the work of two physicists, Otto Stern (1888–1969) and Isidor Isaac Rabi (1898–1988). In 1930 they determined that certain atomic nuclei have a property called *spin*, and that in spinning, the nuclei act as tiny magnets. For their basic research in measuring these minuscule magnetic fields and thoroughly studying this phenomenon, Stern, a German, and Rabi, an American, won the Nobel prize in physics in 1943 and 1944, respectively. In the 1940s teams led by Edward M. Purcell (1912–1997) and Felix Bloch (1905–1983) used the fact that nuclear spin is influenced by its chemical environment (that is, by the atoms around the nucleus) to work out the structure of complicated molecules. This technique, called *nuclear magnetic resonance* (NMR) has become a major tool of chemists in determining molecular structure. Purcell and Bloch shared the Nobel prize in physics in 1952 for this research. In the 1960s scientists applied the principles of NMR to scans of the human body. This technique, called *magnetic resonance imaging* (MRI), has replaced many exploratory surgical operations with a noninvasive procedure and made other surgery processes much more precise.

of a number of valuable new drugs for carrying out successful organ transplants and for treating various diseases such as gout, malaria, herpes, and cancer. Elion and Hitchings shared the 1988 Nobel prize in physiology and medicine, and in 1991 she became the first woman to be inducted into the National Inventors Hall of Fame.

1.8 Chemistry: A Study of Matter and Its Changes

Earlier we defined **chemistry** as the study of matter and the changes it undergoes. Because the entire physical universe is made up of nothing more than matter and energy, the field of chemistry extends from atoms to stars, from rocks to living organisms. Now let's look at matter a little more closely.

Matter is the stuff that makes up all material things. Matter has *mass*; you can weigh it. Wood, sand, water, air, and people have mass and are therefore matter. **Mass** is a measure of the quantity of matter that an object contains. The greater the mass of an object, the more difficult it is to change its velocity. You can easily deflect a tennis ball coming toward you at 30 meters per second (m/s), but you would have difficulty stopping a cannonball of the same size moving at the same speed. A cannonball has more mass than a tennis ball of equal size.

The mass of an object does not vary with location. An astronaut has the same mass on the moon as on Earth. **Weight**, on the other hand, measures a force. On Earth, it measures the force of attraction between our planet and the mass in question. On the moon, where gravity is one-sixth that on Earth, an astronaut weighs only one-sixth as much as on Earth. Weight varies with gravity; mass does not.

Example 1.2

On Mars gravity is one-third that on Earth. **(a)** What would be the mass on Mars of a person who has a mass of 60 kilograms (kg) on Earth? **(b)** What would be the weight on Mars of a person who weighs 120 pounds (lb) on Earth?

Solution
a. The person's mass would be the same (60 kg) as on Earth; the quantity of matter has not changed.
b. The person would weigh only 40 lb; the force of attraction between planet and person is only one-third that on Earth.

Astronaut John W. Young leaps from the lunar surface, where gravity pulls at him with only one-sixth the force on Earth.

Exercise 1.2

A. At the surface of Planet X, the force of gravity is only half that on Earth's surface. **(a)** What would be the mass of a standard 1.00-kg object on Planet X? **(b)** A man who weighs 198 lb on Earth would weigh how much on Planet X?

B. On Jupiter, at the boundary between the gaseous atmosphere and the liquid that makes up the bulk of the planet, the force of gravity is 2.4 times that on Earth. **(a)** What would be the mass of a 52-kg woman at that location on Jupiter? **(b)** A man who weighs 200 lb on Earth would weigh how much on Jupiter?

Physical and Chemical Properties

We can use our knowledge of chemistry to change matter to make it more useful. Chemists can change crude oil into gasoline, plastics, pesticides, drugs, detergents, and thousands of other products. Changes in matter are accompanied by changes in energy. Often we change matter to extract part of its energy. For example, we burn gasoline to get energy to propel our automobiles.

To distinguish between samples of matter, we can compare their properties (Figure 1.4). The **physical properties** of a substance are its physical characteristics and

Figure 1.4 A comparison of the physical properties of two elements. Copper, obtained as pellets, can be hammered into thin foil or drawn into wire. When hammered, lumps of sulfur crumble into a fine powder.

Table 1.1 ∎ Some Examples of Physical Properties

Property	Examples
Temperature	0°C for ice water, 100°C for boiling water.
Mass	A nickel weighs 5 g; a penny weighs 2.5 g.
Structure	Ice is crystalline; glass is amorphous.
Color	Sulfur is yellow; bromine is reddish brown.
Taste	Acids are sour; bases are bitter.
Odor	Benzyl acetate smells like jasmine; hydrogen sulfide smells like rotten eggs.
Boiling point	Water boils at 100°C; alcohol boils at 78.5°C.
Freezing point	Water freezes at 0°C; methane freezes at −182°C.
Heat capacity	Water has a high heat capacity; iron has a low heat capacity.
Hardness	Diamond is exceptionally hard; sodium metal is soft.
Conductivity	Copper conducts electricity; diamond does not. Aluminum is a good conductor of heat; glass is a poor heat conductor.
Solubility	Ethyl alcohol dissolves in water; gasoline does not.
Density	For water it is 1.0 g/mL; for gold it is 19.3 g/mL.

Table 1.2 ∎ Some Examples of Chemical Properties

Substance	Typical Chemical Property
Iron	gets rusty (combines with oxygen to form iron oxide).
Carbon	burns (combines with oxygen to form carbon dioxide).
Silver	tarnishes (combines with sulfur to form silver sulfide).
Nitroglycerin	explodes (decomposes to produce a mixture of gases).
Neon	is inert (does not react with anything).

behavior, such as color, odor, and hardness (Table 1.1). Its **chemical properties** describe how it reacts with other types of matter—how its atomic building blocks can change (Table 1.2).

A **physical change** is one that does not entail any change in chemical composition: solid iron melting in a blast furnace or an ice cube changing to liquid water in a glass. The process of melting is a physical change, and the temperature at which it occurs—the melting point—is a physical property. A **chemical change** involves a change in chemical composition. In exhibiting a chemical property, matter undergoes a chemical change; the original substance is replaced by one or more new substances. Iron metal reacts with oxygen from the air to form rust (iron oxide); when sulfur burns in air, sulfur, which is made up of one type of atom, and oxygen (from air), which is made up of another type of atom, combine to form sulfur dioxide, which is comprised of molecules that have sulfur and oxygen atoms in the ratio 1:2. (A *molecule* is a group of atoms bound together as a single unit. More about atoms and molecules later.)

It is difficult at times to determine whether a change is physical or chemical, but we can decide on the basis of what happens to the composition or structure of the matter involved. *Composition* refers to the types of atoms that are present and their relative proportions, and *structure* to the arrangement of those atoms in particular assemblages or in space. A chemical change results in a change in composition or structure, whereas a physical change does not.

Chemical properties are inherent in a substance; they are manifested through chemical change.

Example 1.3

Which of the following events involve chemical changes and which involve physical changes?

a. You trim your fingernails.
b. Lemon juice converts milk to curds and whey.
c. Water boils.
d. Water is broken down into hydrogen gas and oxygen gas.

Solution

a. Physical change: The composition of the nail is not changed by clipping.
b. Chemical change: The compositions of curds and whey are different from the composition of the milk.
c. Physical change: Liquid water and invisible water vapor formed when liquid water boils have the same composition; the water merely changes from a liquid to a gas.
d. Chemical change: New substances, hydrogen and oxygen, are formed.

Exercise 1.3

A. Which of the following events involve chemical changes and which involve physical changes?
    a. Liquid alcohol vaporizes from an open container.
    b. A piece of lithium metal burns in air to form a white powder called *lithium oxide*.
    c. A dull saw is sharpened with a file.
B. Which of the following events involve chemical changes and which involve physical changes?
    a. A car fender gets rusty.
    b. A stick of butter melts.
    c. A pile of leaves is burned.
    d. A piece of wood is ground up into saw dust.

1.9 Classification of Matter

Matter can be classified in many ways. We will examine three ways of doing so in this section. First we look at the physical forms or *states of matter*. Then we examine classification as a single substance or a mixture of substances. Finally, we classify substances as elements or compounds.

The States of Matter

There are three familiar states of matter: solid, liquid, and gas (Figure 1.5). **Solid** objects ordinarily maintain their shape and volume regardless of their location. A **liquid** occupies a definite volume but assumes the shape of the occupied portion of its container. If you have a 355-milliliter (mL) soft drink, you have 355 mL whether the soft drink is in a can, in a bottle, or, through some mishap, on the floor—which demonstrates another property of liquids. Unlike solids, liquids flow readily. A **gas** maintains neither shape nor volume. It expands to fill completely whatever container it occupies. Gases can be easily compressed. For example, enough air for many minutes of breathing can be compressed into a steel tank for underwater diving. We discuss the states of matter in more detail in Chapter 5.

Substances and Mixtures

Figure 1.6 shows several other ways that chemists classify matter. A **substance** has a definite, or fixed, composition that does not vary from one sample to another. The composition of a **mixture** is variable. Pure gold (24-carat gold) consists entirely of gold atoms; it is a substance. All samples of pure water are comprised of molecules consisting of two hydrogen atoms and one oxygen atom; water is a substance. On the other hand, a saline solution—a solution of salt in water—is a mixture. The proportions of salt and water can vary from one solution to another.

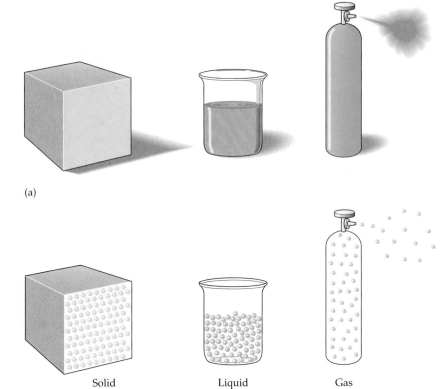

Figure 1.5 Solids, liquids, and gases. (a) Bulk properties. (b) Interpretation of bulk properties in terms of the kinetic–molecular theory. In liquids, the particles are close together, but they are free to move about. In gases, the particles are far apart and are in rapid random motion.

(a)

(b)

Solid   Liquid   Gas

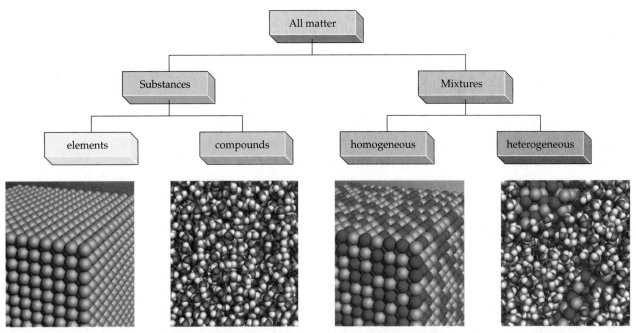

All matter

Substances   Mixtures

elements   compounds   homogeneous   heterogeneous

Figure 1.6 A scheme for classifying matter. The "molecular-level" views are of gold—an *element*; water—a *compound*; 12-carat gold—a *homogeneous mixture* of silver and gold; and a *heterogeneous mixture* of particles of 12-carat gold in water.

Elements and Compounds

Substances are either elements or compounds. **Elements** are the fundamental substances from which all material things are constructed. **Compounds** are substances made up of two or more elements chemically combined. Our ideas about elements have changed over the years. Fire was once considered an element, but it is now regarded as energy rather than matter. Water was once thought to be an element, but we now know it to be a compound composed of two elements, hydrogen and oxygen. We presently regard as elements 115 substances that cannot be broken down by chemical means into simpler substances. Sulfur, oxygen, carbon, and iron are elements. Sulfur dioxide, carbon disulfide, and iron sulfate are compounds.

Because elements are so fundamental to our study of chemistry, we find it useful to refer to them in a shorthand form. Each element can be represented by a **chemical symbol** made up of one or two letters derived from the name of the element (or, sometimes, from the Latin name of the element). The first letter of the symbol is always capitalized; the second is always lowercase. (It makes a difference. For example, Co is the symbol for cobalt, an element, but CO is the formula for carbon monoxide, a compound. Similarly, Hf stands for hafnium, an element, whereas HF stands for hydrogen fluoride, a compound.)

Symbols are the alphabet of chemistry. Symbols for all the elements studied are listed in the Table of Atomic Masses (inside front cover). It is well worth your time to memorize the names and symbols listed in red in that table. They are also listed in Table 1.3. (The symbols based on Latin names are a bit more difficult to learn.)

Of the 115 elements now known, 25 are too unstable to exist in nature. We deal with only about one-third of the known elements in any detail in this book.

The monetary value of the elements contained in the human body is about $2.00. But the compounds in which they occur are much more valuable. The biochemical compounds in the human body are worth hundreds of thousands of dollars.

A chemical symbol in a formula stands for one atom of the element. If more than one atom is to be indicated in a formula, a subscripted number is used after the symbol. For example, the formula Cl_2 represents two atoms of chlorine, and the formula CCl_4 stands for one atom of carbon and four atoms of chlorine. Table 1.3 gives actual formulas (Br_2, P_4, S_8, and so on) for some of the elements as they occur in the elemental form. You need not concern yourself with such formulas now; they are included in the table for your future reference.

Example 1.4

Which of the following represent elements and which represent compounds?

Hg   HI   BN   In

Solution
Hg and In represent elements (each is a single symbol). HI and BN are composed of two symbols each and represent compounds.

Exercise 1.4
Which of the following represent elements and which represent compounds?

He   CuO   No   NO   KI

Atoms and Molecules

An **atom** is the smallest characteristic part of an element. Each element is composed of atoms of a particular kind. For example, the element copper is made up of copper atoms, and gold is made up of gold atoms. All copper atoms are alike in a fundamental way and are different from gold atoms. The smallest characteristic part of most compounds is a molecule. A **molecule** is a group of atoms bound together as a unit. Each molecule of a given compound has the same atoms in the same proportions and in the same structural arrangement as all the other molecules of the compound. We will discuss atoms in some detail in Chapter 2, and much of the focus of many of the remaining chapters is on molecules.

1.10 The Measurement of Matter

Accurate measurements of such quantities as mass, volume, time, and temperature are essential to the compilation of dependable scientific data. This data may be used by a chemist interested in basic research, but similar information is also of critical importance in every other science-related field. Measurements of both temperature and

Table 1.3 ▌ Names, Symbols, and Physical Characteristics of Some Common Elements

Name (Latin name)	Symbol	Selected Properties
Aluminum	Al	Light, silvery metal
Argon	Ar	Colorless gas
Arsenic	As	Grayish-white solid
Barium	Ba	Silvery-white metal
Beryllium	Be	Steel gray, hard, light solid
Boron	B	Black or brown powder; several crystal forms
Bromine	Br	Reddish-brown liquid (Br_2)
Calcium	Ca	Silvery-white metal
Carbon	C	Soft, black solid (graphite) or hard, brilliant crystal (diamond)
Chlorine	Cl	Greenish-yellow gas (Cl_2)
Copper (Cuprum)	Cu	Light reddish-brown metal
Fluorine	F	Pale-yellow gas (F_2)
Gold (Aurum)	Au	Yellow, malleable metal
Helium	He	Colorless gas
Hydrogen	H	Colorless gas (H_2)
Iodine	I	Lustrous black solid (I_2)
Iron (Ferrum)	Fe	Silvery, ductile, malleable metal
Lead (Plumbum)	Pb	Bluish-gray, soft, heavy metal
Lithium	Li	Silvery-white, soft, light metal
Magnesium	Mg	Silvery-white, ductile, light metal
Mercury (Hydrargyum)	Hg	Silvery-white, liquid, heavy metal
Neon	Ne	Colorless gas
Nickel	Ni	Silvery-white, ductile, malleable metal
Nitrogen	N	Colorless gas (N_2)
Oxygen	O	Colorless gas (O_2)
Phosphorus	P	Yellowish-white, waxy solid (P_4) or red powder
Plutonium	Pu	Silvery-white, radioactive metal
Potassium (Kalium)	K	Silvery-white, soft metal
Silicon	Si	Lustrous-gray solid
Silver (Argentum)	Ag	Silvery-white metal
Sodium (Natrium)	Na	Silvery-white, soft metal
Sulfur	S	Yellow solid (S_8)
Tin (Stannum)	Sn	Silvery-white, soft metal
Uranium	U	Silvery, radioactive metal
Zinc	Zn	Bluish-white metal

blood pressure are routinely made in medicine, and modern medical diagnosis depends on a whole battery of other measurements, including careful chemical analyses of blood and urine.

The measurement system agreed on since 1960 is the *International System of Units* or *SI units* (from the French *Système Internationale*), a modernized version of the metric system established in France in 1791. Most countries use metric measures in everyday life, but in the United States SI units are used mainly in science laboratories.

Figure 1.7 Comparisons of metric and customary units of measure. The left beaker contains 1 kg of candy, and the right beaker contains 1 lb (1 kg = 2.2 lb). The red ribbon is 1 in. wide and is tied around a stick 1 yd long. The green ribbon is 1 cm wide and is tied around a stick 1 m long (1 m = 1.0936 yd; 1 cm = 0.3937 in.). The flask on the right and the bottle behind it each contain 1 L of orange juice. The flask on the left and the carton behind it each contain 1 qt of milk (1 L = 1.06 qt).

Table 1.4 ▌ The Seven SI Base Units

Physical Quantity	Name of Unit	Symbol of Unit
Length	meter*	m
Mass	kilogram	kg
Time	second	s
Temperature	kelvin	K
Amount of substance	mole	mol
Electric current	ampere	A
Luminous intensity	candela	cd

*Spelled *metre* in most countries other than the United States.

However, metric measures are increasingly being used in commerce, especially in businesses with an international component. The contents of most bottled beverages are now given in metric units, and metric measurements are also common in sporting events. Figure 1.7 compares some customary and metric units.

Because SI units are based on the decimal system, it is easy to convert from one unit to another. All measured quantities can be expressed in terms of the seven base units listed in Table 1.4. We will use the first five in this text.

Because the basic SI units are often of awkward magnitude, we use prefixes (Table 1.5) to indicate units larger and smaller than the base unit.

Scientists and others are often faced with the task of converting from one unit to another. Unit conversions are discussed in some detail in Appendix A, where you will also find tables of common conversion factors.

Example 1.5

Convert each of the following measurements to a unit that replaces the power of ten by a prefix.
a. 2.89×10^{-3} g   **b.** 4.30×10^{3} m

Solution
Our goal is to replace each power of ten with the appropriate prefix from Table 1.5. For example, $10^{-3} = 0.001$, leading to milli(unit).
a. 10^{-3} corresponds to the prefix milli; 2.89 mg
b. 10^{3} corresponds to the prefix kilo; 4.30 km

Table 1.5 ■ Approved Numerical Prefixes*

Exponential Expression	Decimal Equivalent	Prefix	Pronounced	Symbol
10^{12}	1,000,000,000,000.	tera-	TER-uh	T
10^{9}	1,000,000,000	giga-	GIG-uh	G
10^{6}	1,000,000	mega-	MEG-uh	M
10^{3}	1,000	kilo-	KIL-oh	k
10^{2}	100	hecto-	HEK-toe	h
10	10	deka-	DEK-uh	da
10^{-1}	0.1	deci-	DES-ee	d
10^{-2}	0.01	centi-	SEN-tee	c
10^{-3}	0.001	milli-	MIL-ee	m
10^{-6}	0.000,001	micro-	MY-kro	μ
10^{-9}	0.000,000,001	nano-	NAN-oh	n
10^{-12}	0.000,000,000,001	pico-	PEE-koh	p
10^{-15}	0.000,000,000,000,001	femto-	FEM-toe	f
10^{-18}	0.000,000,000,000,000,001	atto-	AT-toe	a
10^{-21}	0.000,000,000,000,000,000,001	zepto-	ZEP-toe	z

*The most commonly used prefixes are shown in color.

Exercise 1.5

Convert each of the following measurements to a unit that replaces the power of ten by a prefix.

a. 7.24×10^{-3} g   **b.** 5.14×10^{-6} m   **c.** 1.91×10^{-9} s   **d.** 5.58×10^{3} m

Mass

The SI base quantity of mass is the **kilogram (kg)**, about 2.2 pounds (lb). This base quantity is unusual in that it already has a prefix. A more convenient mass unit for most laboratory work is the gram (g).

$$1 \text{ kg} = 10^3 \text{ g} = 1000 \text{ g}$$

The milligram (mg) is a suitable unit for small quantities of materials, such as some drug dosages.

$$1 \text{ mg} = 10^{-3} \text{ g}$$

Chemists can now detect masses in the microgram (μg), nanogram (ng), picogram (pg), and even smaller ranges (Table 1.6).

Length, Area, and Volume

The SI base unit of length is the **meter (m)**, a unit about 10% longer than 1 yard (yd). To measure distances along a highway, we often use the kilometer (km). In the laboratory, we usually find lengths smaller than the meter most convenient. For exam-

Table 1.6 ■ Some Metric Units of Mass

$$1 \text{ kilogram (kg)} = 1000 \text{ grams (g)} = 10^3 \text{ g}$$
$$1 \text{ gram (g)} = 1000 \text{ milligrams (mg)}$$
$$1 \text{ milligram (mg)} = 1000 \text{ micrograms (μg)} = 0.001 \text{ g} = 10^{-3} \text{ g}$$
$$1 \text{ microgram (μg)} = 1000 \text{ nanograms (ng)} = 10^{-6} \text{ g}$$
$$1 \text{ nanogram (ng)} = 1000 \text{ picograms (pg)} = 10^{-9} \text{ g}$$
$$1 \text{ picogram (pg)} = 1000 \text{ femtograms (fg)} = 10^{-12} \text{ g}$$

ple, we use the centimeter (cm), which is about 0.4 inch (in.)—and the millimeter (mm), which is about the thickness of the cardboard backing in a note pad. For measurements at the atomic and molecular level, we use the micrometer (μm), the nanometer (nm), and the picometer (pm). For example, a chlorophyll molecule is about 0.1 μm or 100 nm long, and the diameter of a sodium atom is 372 pm.

The units for area and volume are derived from the base unit of length. The SI unit of area is the square meter (m^2), but we often find square centimeters (cm^2) or square millimeters (mm^2) more convenient for laboratory work.

$$1 \text{ cm}^2 = (10^{-2}\text{ m})^2 = 10^{-4}\text{ m}^2 \qquad 1 \text{ mm}^2 = (10^{-3}\text{ m})^2 = 10^{-6}\text{ m}^2$$

Similarly, the SI unit of volume is the cubic meter (m^3), but two units more likely to be used in the laboratory are the cubic centimeter (cm^3 or cc) and the cubic decimeter (dm^3). A cubic centimeter is about the volume of a sugar cube, and a cubic decimeter is slightly larger than 1 quart (qt).

$$1 \text{ cm}^3 = (10^{-2}\text{ m})^3 = 10^{-6}\text{ m}^3 \qquad 1 \text{ dm}^3 = (10^{-1}\text{ m})^3 = 10^{-3}\text{ m}^3$$

The cubic decimeter is commonly called a liter. A **liter (L)** is 1 cubic decimeter or 1000 cubic centimeters.

$$1 \text{ L} = 1 \text{ dm}^3 = 1000 \text{ cm}^3$$

The milliliter (mL) and/or cubic centimeter are frequently used in laboratories.

$$1 \text{ mL} = 1 \text{ cm}^3$$

The units liter, deciliter, and milliliter are commonly used for volumes of liquids and gases; the units cubic centimeter and cubic decimeter are more often used for solids.

Time

The SI base unit for measuring intervals of time is the second (s). Extremely short time periods are expressed through the usual SI prefixes: milliseconds (ms), microseconds (μs), nanoseconds (ns), and picoseconds (ps). Long time intervals, in contrast, are usually expressed in traditional, non-SI units: minutes (min), hours (h), days (d), and years (y).

Exponential Numbers: Powers of Ten

Scientists deal with objects smaller than atoms and as large as the universe. We usually use exponential numbers to describe the sizes of such objects. The diameter of an electron is about 10^{-15} m, and its mass is about 10^{-30} kg. At the other extreme, a galaxy typically measures about 10^{23} m across and has a mass of about 10^{41} kg. It is difficult even to imagine numbers so small or so large. The following data offer some perspectives on size.

An essential part of the SI system is the use of exponential (powers of ten) notation for numbers. If you are not already familiar with this notation, you will find a discussion of this topic in Appendix A.

Object	Diameter (m)	Mass (kg)
Atom	10^{-10}	10^{-25}
Molecule (hemoglobin)	10^{-9}	6×10^{-22}
Virus (tobacco mosaic)	10^{-7}	4×10^{-19}
Red blood cell	6×10^{-6}	6×10^{-9}
Grain of sand	10^{-3}	2×10^{-6}
Baseball	8×10^{-2}	1.5×10^{-1}
Earth	6×10^{6}	6×10^{24}
Sun	1.4×10^{9}	2×10^{30}

Example 1.6

Use exponential notation to express each of the following measurements in terms of an SI base unit.
a. 4.12 cm **b.** 947 μs

Solution

a. Our goal is to find the power of ten that relates the given unit to the SI base unit. That is, centi(base unit) $= 10^{-2} \times$ (base unit)

$$4.12 \text{ centimeter} = 4.12 \times 10^{-2} \text{ m}$$

b. To change microsecond to the base unit second, we need to replace the prefix micro- by 10^{-6}. To obtain an answer in the conventional exponential form, we also need to replace the coefficient 947 by 9.47×10^2. The result of these two changes is

$$947 \text{ } \mu s = 947 \times 10^{-6} \text{ s} = 9.47 \times 10^2 \times 10^{-6} \text{ s} = 9.47 \times 10^{-4} \text{ s}$$

Exercise 1.6

Use exponential notation to express each of the following measurements in terms of an SI base unit.

a. 745 nm   b. 525 ns   c. 1415 km   d. 2.06×10^6 g

1.11 Density

In everyday life, we might speak of lead as "heavy" or aluminum as "light," but such descriptions are imprecise at best. Scientists use the term density to describe this important property. The **density**, d, of a substance is the quantity of mass, m, per unit of volume, V.

$$d = \frac{m}{V}$$

For substances that don't mix, such as oil and water, the concept of density allows us to predict which will float on the other. It isn't just the mass of the materials, but the *mass per unit volume*—the density—that determines the result. Density is one of the properties that helps us to know that in an oil-and-vinegar salad dressing, oil (lower density) floats on water (higher density). Figure 1.8 gives more examples. We can rearrange the equation for density to give

$$m = d \times V \quad \text{and} \quad V = \frac{m}{d}$$

These equations are useful for calculations. Densities (Table 1.4) are usually reported in grams per milliliter (g/mL) for liquids or grams per cubic centimeter (g/cm³) for solids.

Example 1.7

What is the density of iron if 156 g of iron occupies a volume of 20.0 cm³?

Solution

The given quantities are

$$m = 156 \text{ g} \quad \text{and} \quad V = 20.0 \text{ cm}^3$$

We can use the equation that defines density.

$$d = \frac{M}{V} = \frac{156 \text{ g}}{20.0 \text{ cm}^3} = 7.80 \text{ g/cm}^3$$

Figure 1.8 Since the density of water is 1 g/mL, we can tell that cork, which floats on water, has a density less than 1. A coin, however, sinks in water because its density is greater than 1.

Exercise 1.7

A. What is the density, in grams per milliliter, of a salt solution if 50.0 mL has a mass of 55.5 g?

B. What is the density, in grams per cubic centimeter, of a metal alloy if a cube that measures 2.00 cm on an edge has a mass of 74.4 g?

Example 1.8

What is the mass of 1.00 L of gasoline if its density is 0.660 g/mL?

Solution

We can express density as a ratio, 0.660 g/1.00 mL, and use it as a conversion factor. We also need the factor 1000 mL/1 L to convert liters to milliliters.

$$m = d \times V = \frac{0.660 \text{ g}}{1 \text{ mL}} \times 1.00 \text{ L} \times \frac{1000 \text{ mL}}{1 \text{ L}} = 660 \text{ g}$$

Mass does not vary with temperature, but volume does, and so does density. We can assume the density of water to be 1.00 g/mL near room temperature.

Exercise 1.8

A. What is the mass of 50.0 mL of glycerol, which has a density of 1.264 g/mL at 20°C?

B. What is the mass of a cubic block of gold (Table 1.7) that is 4.50 cm on an edge?

Example 1.9

What volume is occupied by 461 g of mercury? (See Table 1.7.)

Solution

Here we can express the inverse of density as a ratio, 1.00 mL/13.6 g, and use it as a conversion factor.

$$V = 461 \text{ g} \times \frac{1 \text{ mL}}{13.6 \text{ g}} = 33.9 \text{ mL}$$

Ethylene glycol (the principal ingredient in many antifreeze solutions), water, and ethanol are all colorless liquids. They can be distinguished by measuring their densities. At 20°C, the density of ethylene glycol is 1.114 g/mL, that of water is 0.998 g/mL, and that of ethanol is 0.789 g/mL.

Exercise 1.9

A. What volume is occupied by a 10.0-g piece of magnesium? (See Table 1.7.)

B. What volume is occupied by 10.0 g of hexane? (See Table 1.7.)

Table 1.7 ▌ Densities of Some Common Substances at Specified Temperatures

Substance*	Density	Temperature
Solids		
Copper (Cu)	8.94 g/cm^3	25°C
Gold (Au)	19.3 g/cm^3	25°C
Magnesium (Mg)	1.738 g/cm^3	20°C
Water (ice) (H$_2$O)	0.917 g/cm^3	0°C
Liquids		
Ethyl alcohol (C$_2$H$_5$OH)	0.789 g/mL	20°C
Hexane (C$_6$H$_{14}$)	0.660 g/mL	20°C
Mercury (Hg)	13.534 g/mL	25°C
Urine (a mixture)	1.003–1.030 g/mL	25°C
Water (H$_2$O)	0.997 g/mL	0°C
	1.000 g/mL	4°C

*Formulas are provided for possible future reference.

1.12 Energy: A Matter of Moving Matter

The physical and chemical changes that matter undergoes are almost always accompanied by changes in energy. **Energy** is the ability to change matter, either physically or chemically. In order to make something happen that wouldn't happen by itself, energy is required.

Energy exists in two main forms. Energy due to position or arrangement is called **potential energy**. The water at the top of a dam has potential energy due to gravitational attraction. When the water is allowed to flow through a turbine to a lower level, the potential energy is converted to **kinetic energy** (the energy of motion). In Figure 1.8, the water at the top of the dam, as it flows through the hydroelectric plant, moves the blades of a turbine. As the water falls, it moves faster. Its kinetic energy becomes greater as its potential energy decreases. The turbine can convert part of the kinetic energy of the water into electrical energy. The electricity thus produced can be carried by wires to homes and factories where it can be converted to light energy, to heat, or to mechanical energy. Table 1.8 gives several examples each of kinetic and potential energy.

Nearly all the energy available on Earth comes ultimately from the sun. Much of the energy we use today was captured long ago by green plants through photosynthesis. It is available to us today in the form of fossil fuels—coal, petroleum, and natural gas. Chemistry is concerned not only with matter and its changes, but also with the energy involved in those changes.

Temperature and Heat

Two important concepts in science that are related, and sometimes confused, are *temperature* and *heat*. When two objects at different temperatures are brought together, heat flows from the warmer to the cooler object until both are at the same temperature. **Heat** is the energy that flows from a warmer object to a cooler one. **Temperature** is a measure of how hot or cold an object is. Temperature, therefore, tells us in which direction heat will flow. Heat (a form of kinetic energy) flows from more energetic (higher-temperature) to less energetic (lower-temperature) atoms or molecules. For example, if you touch a hot test tube, heat will flow from the tube to your hand. If the tube is hot enough, your hand will be burned.

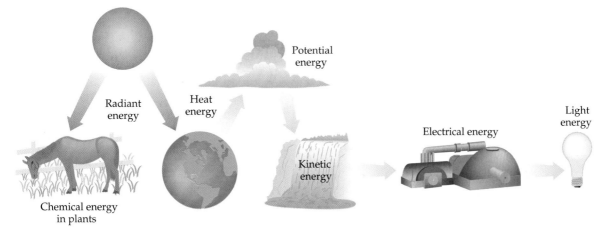

Figure 1.9   Energy can be changed from one form to another.

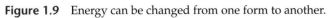

Table 1.8 ∎ Some Examples of Potential and Kinetic Energies	
Potential Energy	
Energy stored by position	Water at the top of a waterfall
	A child at the top of a sliding board
	A skier poised at the top of a mountain slope
	A ball ready to roll down a hill
	A swimmer ready to dive
	A baseball player poised to swing his bat
	A golfer at the top of her backswing ready to tee off
Energy stored in chemical bonds	Fuel (coal, gasoline, natural gas)
	Food (carbohydrates, fats, proteins)
	Explosives (nitroglycerin)
Energy stored in bound nuclear particles	Nuclear energy (power plants, bombs)
Energy stored by compression	A compressed spring
	A squeezed rubber ball
Kinetic Energy	
Energy of any moving object	A rolling freight train
	A spinning waterwheel
	A rolling bowling ball
	A moving molecule
	A sailboat skimming across a lake
	An eagle soaring above a mountain
	An exploding firecracker
	A pair of figure skaters
	A baseball hurtling toward home plate
	A football passing through the goal posts
	Horses racing along a track
	The explosion of a volcano
	The explosion of a building being demolished
	A nuclear explosion

Temperature

The SI base unit of temperature is the **kelvin (K)**. For routine laboratory work, we often use the more familiar **Celsius scale**. On this temperature scale, the freezing point of water is 0 degrees Celsius (°C) and the boiling point is 100°C. The interval between these two reference points is divided into 100 equal parts, each a degree Celsius. The Kelvin scale is called an *absolute scale* because its zero point is the cold-est temperature possible, or absolute zero. (This fact was determined by theoretical considerations and has been confirmed by experiment, as we will see in Chapter 6.) The zero point on the Kelvin scale, 0 K, is equal to −273.15°C, often rounded to −273°C. A kelvin is the same size as a degree Celsius, and so the freezing point of water on the Kelvin scale is 273 K. The Kelvin scale has no negative temperatures, and we don't use a degree sign with the K. To convert from degrees Celsius to kelvins, simply add 273 to the Celsius temperature.

$$K = °C + 273$$

Example 1.10

Ether boils at 36°C. What is the boiling point of ether on the Kelvin scale?

Solution

$$36°C + 273 = 309 \text{ K}$$

Exercise 1.10
A. How is the boiling point of water (100°C) expressed in kelvins?
B. Express a temperature of −78°C in kelvins.

The Fahrenheit temperature scale is widely used in the United States. Figure 1.10 compares the three temperature scales. Note that on the Fahrenheit scale, the freezing point of water is 32°F and the boiling point is 212°F, so that a 10-degree temperature interval on the Celsius scale equals an 18-degree interval on the Fahren-heit scale. Conversion between Fahrenheit and Celsius temperatures is discussed in Appendix A.

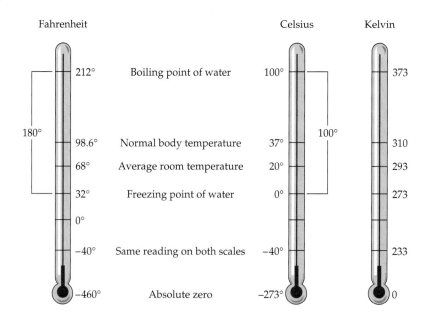

Figure 1.10 A compari-son of the Fahrenheit, Cel-sius, and Kelvin temperature scales.

Heat

The SI unit of heat is the **joule (J)**, but we will use the more familiar **calorie (cal)**.

$$1 \text{ cal} = 4.184 \text{ J}$$

$$1000 \text{ cal} = 1 \text{ kcal} = 4184 \text{ J}$$

A calorie is the amount of heat required to raise the temperature of 1 g of water 1°C.

The "calorie" used for measuring the energy content of foods is actually a **kilocalorie (kcal)**. A dieter might be aware that a banana split contains 1500 "calories." If the same dieter realized that this was really 1,500,000 calories, giving up a banana split might be easier. The concept of heat is discussed further in Appendix A.

> Since it varies slightly with temperature, a calorie is defined more precisely as the amount of heat needed to raise the temperature of 1 g of water from 14.5 to 15.5°C.

Body Temperature, Hypothermia, and Hyperthermia (Fever)

Carl Wunderlich, a German physician, first recognized fever as a symptom of disease. During the 1840s and 1850s he averaged a large number of human temperature measurements and rounded the value to 37°C. When this value was converted to the Fahrenheit scale, it somehow (improperly) acquired an extra significant figure, and 98.6°F became widely (and incorrectly) known as the *normal body temperature*. In recent years millions of measurements have revealed that *average* normal body temperature is actually 98.2°F and that body temperatures in healthy people range from 97.7 to 99.5°F.

The healthy human body maintains a fairly constant temperature. Heat is a by-product of metabolism, and we must constantly get rid of some of it. Because heat always flows spontaneously from a hot object to a cold one, we can get rid of excess heat by simple conduction only if the surrounding temperature is less than 37°C. Above that temperature, the body depends on air movement, blood vessel dila-tion, and perspiration to keep its temperature normal. If the outside temperature is above 40°C, the National Weather Service usually issues a heat advisory because the body can *gain* heat from the environment. The body temperature then rises above the normal range, a condition known as *hyperthermia*. Certain diseases also cause abnormally high body temperatures or *fevers*. Mild fevers (up to 102°F) are not dangerous and may even help the body fight off an infection. Prolonged high fevers (104°F or more in adults) can be fatal.

When exposed to prolonged cold, the body can lose too much heat to the environment. The body temperature drops below the normal range, a condition known as *hypothermia*. A drop of only 2 or 3°F leads to shivering, a condition in which muscles contract in an attempt to generate more heat. Prolonged or severe hypothermia (body temperature below 93°F) can lead to unconsciousness and death.

1.13 Critical Thinking

One of the hallmarks of science is the ability to think critically. This ability can be learned, and it will serve you well in everyday life as well as in science courses. We will use an approach adapted from one developed by James Lett,[1] first outlining the approach and then working through some simple examples.

You can use the acronym FLaReS (ignore the vowels) to remember four rules used to test a claim: *f*alsifiability, *l*ogic, *r*eplicability, and *s*ufficiency.

> "Falsify" is one of those words that can have several different meanings. One definition, the one intended here, is "to prove false." A second definition, nearly the opposite of the first, is "to make false."

---

[1] Lett used the acronym FiLCHeRS (ignore the vowels) as a mnemonic for six rules: *f*alsifiability, *l*ogic, *c*omprehensiveness, *h*onesty, *r*eplicability, and *s*ufficiency. See Reference 10 for a full description.

- **Falsifiability:** It must be possible to conceive of evidence that would prove the claim false. Falsifiability is an essential component of the scientific method. A hypothesis that cannot be falsified is of no value. Science cannot prove anything true; it can only prove something false. Things not proven false may be tentatively accepted as true, but always with the possibility they may later be shown to be false.

- **Logic:** Any argument offered as evidence in support of any claim must be sound. An argument is sound if its conclusion follows inevitably from its premises and if its premises are true. It is unsound if a premise is false, and it is unsound if there is a single exception in which the conclusion does not necessarily follow from the premises.

- **Replicability:** If the evidence for any claim is based on an experimental result, or if the evidence offered in support of any claim could logically be explained as coincidental, it is necessary for the evidence to be replicable in subsequent experiments or trials. Scientific research is almost always reviewed by other qualified scientists before publication and published in a form that enables others to repeat the experiment. Sometimes bad science slips through the peer-review process, but such science would fail the test of replicability.

- **Sufficiency:** The evidence offered in support of any claim must be adequate to establish the truth of that claim, with these stipulations: (1) the burden of proof for any claim rests on the claimant, (2) extraordinary claims demand extraordinary evidence, and (3) evidence based on authority and/or testimony is never adequate.

If a claim passes all four FLaReS tests, then it *might* be true. On the other hand, it could still be proven false. If a claim fails even one of the FLaReS tests, however, it is likely to be false. Let's work through some simple examples.

Critical Thinking Examples

1.1 A psychic claims he can bend a spoon using only the powers of his mind. However, he says he can do so only when the conditions are right; there must be no one with negative energy present. Apply the FLaReS tests to evaluate the psychic's claim.

Solution
1. Is the claim falsifiable? No. If the psychic fails, he can always claim that someone present had negative energy.
2. Is the claim logical? No. What kind of matter or energy could move from the psychic's mind to the spoon with enough force to bend it? Just what is "negative energy?"
3. Is the claim reproducible? No. He can do it only when "conditions are right." (Actually, one such psychic was caught cheating; he was seen bending the spoon with his hands. Now his proponents claim he cheats only sometimes!)
4. Is the claim sufficient? No. Any such claim is quite extraordinary; it would require extraordinary evidence. The burden of proof for any such claim rests on the claimant, and only flimsy evidence is provided.

1.2 Some people claim that biorhythm cycles that date from a person's birth date can predict airplane crashes. They say that more crashes occur when the pilot, copilot, and/or navigator are experiencing critically low points in their intellectual, emotional, and/or physical cycles. Several studies show no such correlation. Apply the FLaReS tests to evaluate this claim.

Solution

1. Is the claim falsifiable? Yes. Crash data and birth dates of the crew members can be analyzed to see whether or not there is a correlation.

2. Is the claim logical? No. Hundreds of thousands of people are born each day. It is highly unlikely that all of them would share the same up-and-down cycles day by day for a lifetime.

3. Is the claim reproducible? No. It has not held up in other studies.

4. Is the claim sufficient? No. It does not address the real causes of airplane crashes.

1.3 Psychics such as the late Jeanne Dixon claim to be able to predict the future. These predictions are often made near the end of one year for the next year. Look up several such predictions and apply the FLaReS tests to evaluate the claim.

Solution

1. Is the claim falsifiable? Yes. You can write a claim down and then check it yourself at the end of the year.

2. Is the claim logical? No. If psychics could really predict the future, they could readily win lotteries, ward off all kinds of calamities by providing advance warning, and so on.

3. Is the claim reproducible? No. Psychics usually make broad, general claims. They later make claim-specific references to "hits" while ignoring "misses." (Some of Jeanne Dixon's past predictions: World War III will begin in 1958; Kennedy will *not* be elected in 1960; Fidel Castro will die in 1969; George Bush will be elected in 1992.)

4. Is the claim sufficient? No. Any such claim is quite extraordinary; it would require extraordinary evidence. The burden of proof for any such claim rests on the claimant, and only flimsy evidence is provided.

Critical Thinking Exercises

Apply knowledge that you've gained in this chapter and one or more of the FLaReS principles to evaluate the following statements or claims.

1.1 Some people claim that Elvis is alive and well, working at a McDonald's restaurant and living in Michigan.

1.2 A doctor claims that she can cure a patient of arthritis by simply massaging the affected joints. If she has the patient's complete trust, she can cure the arthritis within a year. Several of her patients have testified that the doctor has cured their arthritis.

1.3 Some people claim that UFOs are actually alien spacecraft. Some even claim to have had personal encounters with aliens on their spaceships.

1.4 Sarah, who is in sixth grade, has a beautiful tiger-eye stone that her grandfather gave her. When she holds it in her hand, she can think more clearly. Sarah knows that the stone really works because one day when she left the stone at home, she made the worst grade she had ever received on an exam. She believes that her stone has magic power. What do you think?

1.5 A woman claims that she has memorized the New Testament. She offers to quote any chapter of any book entirely from memory. Apply the FLaReS tests to evaluate her claim.

1.6 John, an eighth-grade spelling bee champion, claims that he can spell any word in *Webster's Collegiate Dictionary*. You can ask him to spell any word if you simply define the word or use it in a sentence. Is John's claim believable?

Summary

1. *Chemistry* is the study of matter and the changes it undergoes. The changes may be physical or chemical. Matter is anything that has mass.

2. With its roots in natural philosophy and alchemy, chemistry finally became a science in the seventeenth century, when chemists began to rely on experimentation.

3. Much older than science is technology, which consists of all the processes we use to modify nature, processes that need not be based on science.

4. It was suggested by Francis Bacon that science and technology might solve all the world's problems and bring about universal prosperity, but Rachel Carson pointed out that science and technology can also create some serious problems.

5. When a scientist reports an observation, it is not really accepted unless it can be reproduced by other scientists. An observation about nature may lead to an explanation called a hypothesis. If a hypothesis can stand up to testing and further experimentation, it may become a theory.

6. Although scientific theories have been tested many times, they are always tentative and can be rejected if they do not continue to stand up to further testing. A valid theory can be used to predict new scientific facts.

7. Science and technology have provided many benefits for our world, but they have also introduced new risks. A risk–benefit analysis can help us decide which outweighs the other.

8. Because it is so important to other sciences, chemistry plays a central role in science.

9. Chemists are usually involved in some kind of research. The purpose of applied research is to make particular kinds of useful products. Although it may happen to result in a useful product, basic research is carried out simply to obtain new knowledge or to answer some fundamental question.

10. Whenever matter undergoes a physical or chemical change, there is also a change in energy. Energy is either given off or absorbed in each process.

11. Calories of heat are often used to measure the quantity of energy. The intensity of energy is measured in terms of temperature.

12. One way to classify matter is according to physical state: solids, liquids, and gases. Heat can convert a solid to a liquid or a liquid to a gas.

13. Another way to classify matter is according to purity: pure substances or mixtures. Pure substances are either elements or compounds.

14. There are only about a hundred elements, each represented by a chemical symbol. A compound contains two or more elements that are chemically combined.

15. Scientific measurements are made using *SI* metric units: the meter (m) for length, the kilogram (kg) for mass, the liter (L) for volume, and the kelvin (K) for temperature. Prefixes make basic units larger or smaller by factors of ten.

16. Density is a measure of heaviness. It is the amount of mass per unit volume. The density of water is 1 g/mL.

17. In doing critical thinking exercises the acronym FLaReS can be useful. The FLRS letters represent the following rules: falsifiability, logic, replicability, and sufficiency.

Key Terms

alchemy   1.1
applied research   1.7
atom   1.9
basic research   1.7
calorie   1.12
Celsius scale   1.12
chemical change   1.8
chemical property   1.8
chemical symbol   1.9
chemistry   1.8
compound   1.9

density   1.11
element   1.9
energy   1.12
gas   1.9
heat   1.12
hypothesis   1.3
kelvin   1.12
kilocalorie   1.12
kilogram   1.10
kinetic energy   1.12
liquid   1.9

liter   1.10
mass   1.8
matter   1.8
meter   1.10
mixture   1.9
molecule   1.9
natural philosophy   1.1
physical change   1.8
physical property   1.8
potential energy   1.12
risk–benefit analysis   1.5

science   1.3
scientific law   1.3
scientific model   1.3
SI units   1.10
solid   1.9
substance   1.9
technology   1.5
temperature   1.12
theory   1.3
variable   1.4
weight   1.8

Review Questions

1. Define chemistry. What is a chemical?

2. What is matter? How is it related to mass?

3. State five distinguishing characteristics of science. Which characteristic best serves to distinguish science from other disciplines?

4. Why were the ancient Greek philosophers, such as Aristotle, not successful as scientists?

5. What is alchemy?

6. What is natural philosophy?

7. What did Francis Bacon envision for us as a result of science?

8. What is the main theme of Rachel Carson's *Silent Spring*?
9. Why have Thomas Malthus's predictions not been fulfilled in developed countries?
10. What is a scientific hypothesis? How are hypotheses tested?
11. What is a scientific law?
12. What is a theory?
13. Why can't scientific methods always be used to solve social, political, ethical, and economic problems?
14. How does technology differ from science?
15. What is risk–benefit analysis?
16. What sort of judgments go into the evaluation of benefits?
17. What sort of judgments go into the evaluation of risks?
18. What is a DQ?
19. Why is it often difficult to estimate a DQ?

20. Explain the difference between mass and weight.
21. Distinguish between chemical and physical properties.
22. What is energy?
23. What is the difference between kinetic energy and potential energy?
24. How do substances and mixtures differ?
25. How do gases, liquids, and solids differ in their properties?
26. What are the names and symbols of the basic SI units for mass, length, and temperature? What derived units are more often used in the laboratory?
27. What is the SI-derived unit for volume? What volume units are more often used in the laboratory?
28. What is applied research? What is basic research? Give an example of each.

Problems

(*Hint:* You must work a large proportion of these practice problems. You cannot learn to solve problems by just reading them or watching your teacher work them any more than you can become a good pianist solely by reading about piano skills or watching a performance.)

Risk–Benefit Analysis

29. Synthetic food colors make food more attractive and increase sales. Some such dyes are suspected carcinogens (cancer inducers). Who derives most of the benefits from the use of food colors? Who assumes most of the risk associated with use of these dyes?
30. Penicillin kills bacteria, thus saving the lives of thousands of people who otherwise might die of infectious diseases. Penicillin causes allergic reactions in some people; in extreme cases the allergic reaction can lead to death if the resulting condition is not treated. Do a risk–benefit analysis of the use of penicillin for society as a whole.
31. Do a risk–benefit analysis of the use of penicillin for a person who is allergic to it. (See Problem 30.)
32. An artificial sweetener is 4000 times as sweet as table sugar but exhibits possible toxic side effects. Do a risk–benefit analysis of the sweetener.
33. Nitrogen mustard is extremely toxic but is an effective anticancer drug. Do a risk–benefit analysis of nitrogen mustard.
34. Do a risk–benefit analysis for the smoking of cigarettes.

Matter and Energy

35. Which of the following are examples of matter?
    a. natural gas   b. love   c. iron
36. Which of the following are examples of matter?
    a. the human body   b. air   c. an idea

Mass and Weight

37. Two samples are weighed under identical conditions in a laboratory. Sample A weighs 1 lb and Sample B weighs 2 lb. Does Sample B have twice the mass of Sample A?
38. Sample A, which is on the moon, has exactly the same mass as Sample B, which is on Earth. Do the two samples weigh the same? Explain your answer.

Physical and Chemical Change

39. Which of the following describes a physical change, and which describes a chemical change?
    a. Sheep are sheared and the wool is spun into yarn.
    b. Silkworms feed on mulberry leaves and produce silk.
    c. Milk that has been left outside a refrigerator for many hours turns sour.
40. Which of the following describes a physical change, and which describes a chemical change?
    a. Because a lawn is watered and fertilized, it grows thicker.
    b. An overgrown lawn is manicured by mowing it with a lawn mower.
    c. Ice cubes form when a tray filled with water is placed in a freezer.

Substances and Mixtures

41. Identify each of the following as a substance or a mixture.
    a. carbon dioxide   b. oxygen
    c. smog   d. a carrot
42. Identify each of the following as a substance or a mixture.
    a. gasoline   b. mercury
    c. soup   d. 24-carat gold
43. Every sample of the sugar glucose (collected anywhere on Earth) consists of 8 parts (by mass) oxygen, 6 parts carbon, and 1 part hydrogen. Is glucose a substance or a mixture? Explain.

44. A sample of saline has 5.0% salt. Another sample of saline has 7.0% salt. Is saline a substance or a mixture? Explain.

Elements and Compounds

45. Which of the following represent elements and which represent compounds?
  a. H   **b.** He   **c.** HF   **d.** Ca

46. Which of the following represent elements and which represent compounds?
  a. C   **b.** CO   **c.** Cl   **d.** $CaCl_2$

47. Without consulting tables, write symbols for each of the following.
  a. carbon   **b.** chlorine   **c.** iron

48. Without consulting tables, write symbols for each of the following.
  a. calcium   **b.** potassium   **c.** plutonium

49. Without consulting tables, name each of the following.
  a. H   **b.** O   **c.** Na

50. Without consulting tables, name each of the following.
  a. N   **b.** Pb   **c.** U

The Metric System

51. How many millimeters and how many centimeters are there in 1 m?

52. How many meters are there in each of the following?
  a. 10 km   **b.** 75 cm

53. How many millimeters are there in each of the following?
  a. 7.5 m   **b.** 46 cm

54. How many liters are there in each of the following?
  a. 2056 mL   **b.** 47 kL

55. For each of the following, indicate which is the larger unit.
  a. mm or cm   **b.** kg or g   **c.** dL or μL

56. For each of the following, indicate which is the larger unit.
  a. L or cm^3 **b.** cm^3 or mL

57. How many milliliters are there in 1 cm^3? In 15 cm^3?

58. How many millimeters are there in 1 cm? In 1.83 m?

Density

(You may need data from Table 1.7 for some of these problems.)

59. What is the density, in grams per milliliter, of **(a)** a salt solution if 75.0 mL has a mass of 87.5 g? **(b)** 2.75 L of the liquid glycerol, which has a density of 1.26 g/mL?

60. What is the density of **(a)** a sulfuric acid solution if 5.00 mL has a mass of 7.52 g (in grams per milliliter)? **(b)** a 10.0-cm^3 block of plastic with a mass of 9.23 g (in grams per cubic centimeter)?

61. What is the mass, in grams, of **(a)** 125 mL of castor oil, a laxative, which has a density of 0.962 g/mL? **(b)** a 33.0-mL sample of the liquid hexane (C_6H_{14}), a solvent used to extract oil from soybeans?

62. What is the mass, in grams, of **(a)** 30.0 mL of the liquid propylene glycol, a moisturizing agent for foods, which has a density of 1.036 g/mL at 25°C? **(b)** 30.0 mL of grenadine, which has a density of 1.32 g/mL?

63. What is the volume of **(a)** a 475-g piece of copper (in cubic centimeters)? **(b)** a 253-g sample of mercury (in milliliters)?

64. What is the volume of **(a)** 227 g of hexane (in milliliters)? **(b)** a 454-g block of ice (in cubic centimeters)?

Additional Problems

65. The brass weight and the pillow (below) have the same mass. Which has the greater density? Explain.

The brass weight has the same mass as the pillow, but their densities are quite different.

66. What has changed when a formerly chubby person completes a successful diet—the person's weight or the person's mass?

67. Arrange the following in order of increasing length (shortest first): (1) a 1.21-m chain, (2) a 75-in. board, (3) a 3-ft 5-in. rattlesnake, (4) a yardstick.

68. Arrange the following in order of increasing mass (lightest first): (1) a 5-lb bag of potatoes, (2) a 1.65-kg cabbage, (3) 2500 g sugar.

69. One of the women pictured below has a mass of 38.5 kg and a height of 1.51 m. Which one is it likely to be?

70. Some metal chips having a volume of 3.29 cm^3 are placed on a piece of paper and weighed. The combined mass is found to be 18.43 g. The paper itself weighs 1.21 g. Calculate the density of the metal.

71. A glass container weighs 48.462 g. A sample of 4.00 mL of antifreeze solution is added, and the container plus the antifreeze weigh 54.51 g. Calculate the density of the antifreeze solution.

72. The sign on the steam room of the fitness center reads, Do not enter if the temperature is above 118°F. The digital thermometer on the door indicates that the temperature is 29.7°C. Is it safe to enter?

Projects

73. List five chemical activities you have engaged in today.

74. Prepare a brief biographical report on one of the following.
    a. Francis Bacon
    b. Rachel Carson
    c. Thomas Malthus
    d. George Washington Carver

Online Projects

75. Using a popular search engine, such as Alta Vista, Yahoo, or Webcrawler, make a search of the Internet for information about alchemy.

76. Search the world wide web for information about MRI.

77. Compare the information on fertilizers found at several different websites, such as
    http://www.farmsource.com and
    http://www.greenpeace.org.

References and Readings

1. Bate, Roger. *What Risk?* Stoneham, MA: Butterworth-Heinemann, 1999.

2. Breslow, Ronald. *Chemistry Today and Tomorrow: The Central, Useful and Creative Science*. Boston: Jones and Bartlett, 1997.

3. Callery, Michael L., and Helen G. Koritz. "What Science Is and Is Not: Ten Myths." *Journal of College Science Teaching*, December 1992–January 1993, pp. 154–157.

4. Carson, Rachel. *Silent Spring*. Boston: Houghton Mifflin, 1962. The classic book on dangers to the environment.

5. Derry, Gregory N. *What Science Is and How It Works*. Princeton, NJ: Princeton University Press, 1999.

6. Emsley, John. *The Elements*, 3rd ed. Oxford, UK: Oxford University Press, 1998.

7. Giere, Ronald N. *Understanding Scientific Reasoning*, 4th ed. New York: Harcourt Brace, 1997.

8. lhde, A. J. *The Development of Modern Chemistry*. New York: Harper and Row, 1964.

9. Jones, A. V., M. Clemmet, A. Higton, and E. Golding. *Access to Chemistry*. New York: Springer, 1999.

10. Laidler, Keith J. *To Light Such a Candle: Chapters in the History of Science and Technology*. Oxford, UK: Oxford University Press, 1998.

11. Lett, James. "A Field Guide to Critical Thinking." *The Skeptical Inquirer*, Winter 1990, pp. 153–160.

12. LeMay, H. Eugene, Herbert Beall, Karen M. Robblee, and Douglas C. Brower. *Chemistry: Connections to Our Changing World*. Upper Saddle River, NJ: Prentice Hall, 1996.

13. Pacey, Arnold. *Technology in World Civilization: A Thousand-Year History*. Cambridge, MA: MIT Press, 1990.

14. Paul, Robin. "Chemicals and the Quality of Living." *Chemistry and Industry*, 20 May 1996, pp. 370–373.

15. Principe, Lawrence M. *The Aspiring Adept: Robert Boyle and His Alchemical Quest*. Princeton, NJ: Princeton University Press, 1998.

16. Randal, Judith. "Going Metric: American Foods and Drugs Measure Up." *FDA Consumer*, September 1994, pp. 23–26.

17. Ross, John F. "Risk: Where Do Real Dangers Lie?" *Smithsonian*, November 1995, pp. 42–53.

18. Sagan, Carl. *The Demon-Haunted World: Science as a Candle in the Dark*. New York: Random House, 1996.

19. Segal, Marian. "Determining Risk." *FDA Consumer*, June 1990, pp. 7–11.

20. Tyson, Neil de Grasse. "On Being Dense." *Natural History*, January 1996, pp. 66–67.

21. Varmus, Harold E. "Science as a Way of Thinking." *Journal of College Science Teaching*, November 1996, pp. 119–122.

MediaLab

Green Chemistry: Chemical Technology in the Service of the Environment

Since the publication of Rachel Carson's *Silent Spring* in 1962, people have become increasingly aware that our advances in chemical technology have not led to unalloyed benefits. Chemicals that increase the food supply run off into rivers and damage the habitats of fish. Pesticides that control insects threaten beneficial insects and birds. The energy that keeps us warm and powers our civilization pollutes our air and warms our atmosphere. The changing times in which we live have seen changes in the attitudes of some scientists and industrialists. Many who formerly forged ahead without regard for the ecological effects of research and development now look at the side effects of progress. An important advance in this regard has been the development of the concept of "green chemistry," the use of chemistry in environmentally benign or beneficial ways. The green chemistry initiative began with the Office of Pollution Prevention and Toxics (OPPT) of the Environmental Protection Agency (EPA) in 1991.

In the *Web Investigation* of this MediaLab, you will explore the history and current status of green chemistry. In *Communicate Your Results*, you will evaluate the success of this initiative as seen in the websites of some corporations, assess how well it has been communicated to the general public, and attempt to play a role in promulgating it. Green chemistry is an important concept, and you will want to return to this MediaLab throughout the course as you learn about the benefits and risks of chemistry!

WEB INVESTIGATION

Investigation 1
The EPA Green Chemistry Program

Select the Keyword **EPA** on your website. Read the goals and history of the EPA's Green Chemistry Program.

Investigation 2
Partnerships with Industry

Select the Keyword **INDUSTRY** on your Website. Select an industry, such as screen printing, from the buttons on the left (or links at the bottom). Follow the links to learn what this industry is doing to reduce the impact of its technology on the environment. Look at such aspects as: Does this industry have an association targeted toward environmental concerns? What is it doing? Are there links to specific companies? What are individual companies doing? Do you think that their efforts are making a difference?

Investigation 3
Green Chemistry Internationally

Select the Keyword **NETWORK** on your website. Select "About GCN" and compare the principles of green chemistry found there with those on the EPA site. Choose an area of interest—awards, current issues, or industry, for example—and explore what's happening in other countries to encourage ecologically responsible chemistry. (Don't be discouraged by technical terminology—you should be able to read most of it.)

Investigation 4
A Matter of Terminology

Another term for green chemistry is "sustainable chemistry." (A related, broader, term is "sustainable development.") Select the Keyword **SEARCH**. Using the "meta" search engine Metacrawler (or your own favorite), search first on "green chemistry" and then on "sustainable chemistry." (Be sure to specify your search by phrase.) How do your results compare? Do you find any corporations not mentioneds in *Investigations 2* or *3*?

COMMUNICATE YOUR RESULTS

Exercise 1
Voluntary Partnerships or Regulations?

The EPA's Design for the Environment (DfE) Program acts through voluntary partnerships with industry, universities, and public interest groups. Other branches of the government, and of the EPA itself, accomplish their ecological and health-related goals through regulations. Review what you learned about these approaches in Web Investigations 1 and 2. Write a one-page essay (about 250 words) about the advantages and disadvantages of a voluntary versus a regulatory approach, or about when each should be used.

Exercise 2
Getting the Word Out

Based on how much—or how little—you have noticed in the media, do you think that coverage of the green chemistry initiative is adequate? Return to **NETWORK** on your website and select "Publications." Take a quick look at what's been published recently. Is any of it in the mass media? If not, search the website of a prominent newspaper or television network using the term "green chemistry" or "sustainable chemistry." Compose a letter to a newspaper or to a television or radio network, explaining what green chemistry is and urging expanded coverage.

COOPERATIVE EXERCISE

The worldwide web is vast, with a rapidly increasing abundance of information. Working in groups can help! After you have become acquainted with the concepts of green chemistry, form small groups of "reporters." Using your results from *Investigations 2* or *3*, look closely at a winner of a green chemistry award. Write a brief report on the company and on why it won the award. Compose a letter of commendation to the company.

2

Atoms

Are They for Real?

An atom is invisible, much tinier than a minute;
Yet atoms make up all the world and everything that's in it!

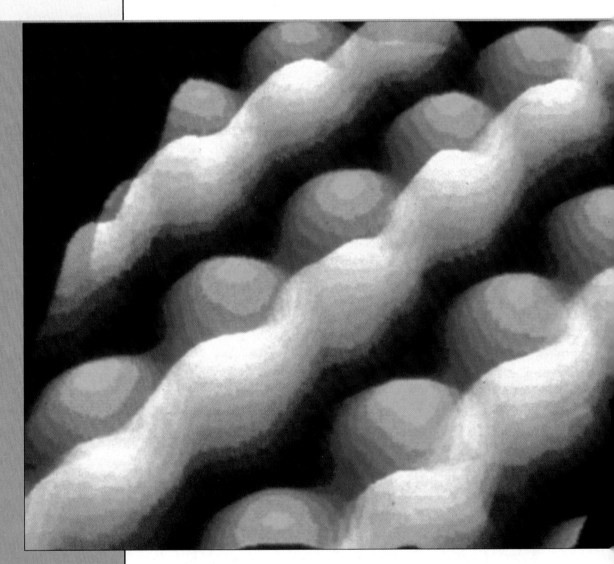

False color image of the atoms in gallium arsenide, a semiconductor used in some solar cells. Gallium (Ga) atoms are shown in red and arsenic (As) atoms in blue. The image was made by a technique called scanning tunneling microscopy. The hypothesis that all matter is composed of atoms is over 2000 years old, but it was only in the latter part of the twentieth century that scientists were able to obtain images of atoms.

The twentieth century has been called the Atomic Age. During the latter half of the century, terms such as "atomic power," "atomic energy," and "atomic bomb" entered our common vocabulary. People also became aware that certain kinds of atoms could be "split," and that the splitting of atoms in a bomb could be devastating. But just what are these things called atoms?

Every material thing in the world is made up of atoms, tiny particles that are much too small to see. The tiniest speck of matter that can barely be detected by the human eye is made up of many billions of atoms bound together. The number of atoms that make up a penny, for example, is enormous. Imagine the atoms in a penny being enlarged until they were barely visible, like tiny grains of sand. The atoms in a single penny would make enough "sand" to cover the entire state of Texas to a depth of several feet. Comparing an atom to a penny is like comparing a grain of sand to a huge sandbox as big as Texas.

Why should we care about something so insignificant in size as an atom? Because our entire world is made up of atoms, and they constantly affect whatever we happen to be doing. *Everything* is made of atoms, including you.

Atoms are not all alike. Each element has its own kind of atoms. On Earth there are about 90 elements that occur in nature, and, as far as we know, the entire universe is made up of these same elements. An atom is the smallest particle that is characteristic of a given element.

Democritus appeared on a 1983 postage stamp from Greece.

In addition to the 90 elements found in nature, about two dozen additional elements have been synthesized by scientists.

2.1 Atoms: The Greek Idea

A pool of water can be separated into drops, and then each drop can be split into smaller and smaller drops. Suppose you could keep splitting these drops into still smaller ones even after they became much too small to see. Would you ever reach a point at which the tiny drop could no longer be separated into smaller droplets of water?

Leucippus (ca. 450 B.C.E.), the Greek philosopher, and his pupil Democritus might well have discussed this question as they strolled along the beach of the Aegean Sea back in the fifth century B.C.E. Based only on his intuition, Leucippus thought that there must ultimately be tiny particles of water that could not be subdivided. After all, from a distance the sand on the beach looked continuous, but closer inspection showed it to be made up of tiny grains.

Democritus expanded on this idea of Leucippus and referred to the particles as *atomos* (cannot be cut). We still refer to the tiny unit particles of elements as **atoms**. Democritus thought that each kind of atom was distinct in shape and size (Figure 2.1).

Figure 2.1 Democritus imagined that "atoms" of water might be smooth, round balls and that atoms of fire could have sharp edges.

The sand looks continuous when you look at a beach from a distance.

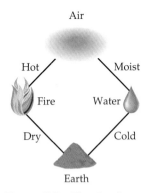

Air

Hot   Moist

Fire   Water

Dry   Cold

Earth

Figure 2.2 The Greek view of matter was that there were only four elements connected by four "principles."

[Web Reference 1] More about the history of chemistry, including contributions of the ancient Greeks.

Antoine Lavoisier and his wife, painted by Jacques Louis David in 1788 (detail).

Real substances were thought to be mixtures of various kinds of atoms. The Greeks at that time believed that there were four basic elements: earth, air, fire, and water. The relationships among these elements and the four "principles"—hot, moist, dry, and cold—are shown in Figure 2.2.

Four centuries later, the Roman poet Lucretius (ca. 95–ca. 55 B.C.E.) wrote a long didactic poem (a poem meant to teach), "On the Nature of Things," in which he presented strong arguments for the atomic nature of matter. Unfortunately, a few centuries earlier, Aristotle (ca. 384–ca. 322 B.C.E.) had declared that matter was continuous, not atomistic, and the ancient Greeks and Romans had no way to determine which of these two views of matter was correct. The continuous view of matter seemed more logical and reasonable to most of them, and so the view of Aristotle prevailed for 2000 years even though it was wrong.

2.2 Lavoisier: The Law of Conservation of Mass

By the eighteenth century scientists were observing more carefully and measuring more accurately. Antoine Laurent Lavoisier perhaps did more than anyone else to establish chemistry as a quantitative science. He found that when a chemical reaction was carried out in a closed system, the total mass of the system was not changed. Perhaps the most important chemical reaction that Lavoisier performed was decomposition of the red oxide of mercury to form metallic mercury and a gas he named oxygen. Karl Wilhelm Scheele (1742–1786), a Swedish apothecary, and Joseph Priestley, a Unitarian minister who later fled England and eventually settled in America, had carried out the same reaction earlier, but Lavoisier was the first to weigh all the substances present before and after the reaction. He was also the first to interpret the reaction correctly.

Lavoisier carried out many quantitative experiments. He found that when coal was burned, it united with oxygen to form carbon dioxide. He also experimented with animals, observing that when a guinea pig breathed, oxygen was consumed and carbon dioxide was formed. Lavoisier therefore concluded that respiration was related to combustion. In each of these reactions, he found that matter was conserved.

Lavoisier summarized his findings in a scientific law. The **law of conservation of mass** states that matter is neither created nor destroyed during a chemical change (Figure 2.3). The total mass of the reaction products is always equal to the total mass of the reactants (starting materials).

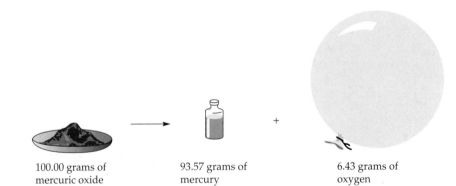

100.00 grams of mercuric oxide

93.57 grams of mercury

+

6.43 grams of oxygen

Figure 2.3 Although mercuric oxide (a red solid) has none of the properties of mercury (a silver liquid) or oxygen (a colorless gas), when 100.00 g of mercuric oxide is decomposed by heating, the products are 93.57 g of mercury and 6.43 g of oxygen. Properties are completely changed in this reaction, but there is no change in mass.

Scientists had by this time abandoned the Greek idea of the four elements and were almost universally using Robert Boyle's operational definition put forth over a century before. In his book *The Sceptical Chymist* (published in 1661), Boyle said that a supposed *element* must be tested to see if it really was simple. If a substance could be broken down into simpler substances, it was not an element. The simpler substances might be elements and would be so regarded until such time (if it ever came) as they in turn could be broken down into still simpler substances. On the other hand, two or more elements might combine to form a complex substance called a *compound*.

Using Boyle's definition, Lavoisier included a table of elements in his book *Elementary Treatise on Chemistry*. The table included some substances we now know to be compounds. (It also included light and heat, which he called "caloric".) Lavoisier was the first to use systematic names for chemical elements. He is often called the "father of modern chemistry," and his book is usually regarded as the first chemistry textbook.

The law of conservation of mass is the basis for many chemical calculations. We can calculate the mass of a product that can be made from a given mass of reactant or the mass of reactant that must be used to yield a certain mass of product (Chapter 6). This law is not just a matter of academic interest. It states that we cannot create materials from nothing; we therefore deduce that we can make new materials only by changing the way atoms are combined. Nor can we get rid of wastes by the destruction of matter. We must put wastes somewhere. Chemistry offers alternatives, however. Through chemical reactions, we can change some kinds of potentially hazardous wastes to less harmful forms. Such transformations of matter from one form to another are what chemistry is all about.

Robert Boyle.

[Web Reference 2] 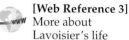 Some of Lavoisier's writings (in English translation).

[Web Reference 3] More about Lavoisier's life and his contributions to chemistry.

[Web Reference 4] Biographical information about several of the chemists mentioned here.

2.3 Proust: The Law of Definite Proportions

By the end of the eighteenth century, Lavoisier and other scientists noted that many substances were composed of two or more elements. Each compound had the same elements in the same proportions, regardless of where it came from or who prepared it. The painstaking work of Joseph Louis Proust (1754–1826) convinced most chemists of the general validity of these observations. In one set of experiments, for example, Proust found that basic copper carbonate, whether prepared in the laboratory or obtained from natural sources, was always composed by mass of 5.3 parts copper, 1.0 part carbon, and 4.0 parts oxygen (Figure 2.4). To summarize these and many other

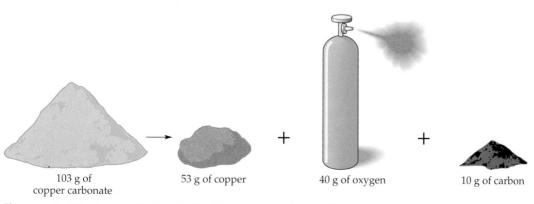

| 103 g of copper carbonate | 53 g of copper | 40 g of oxygen | 10 g of carbon |

Figure 2.4 Whether synthesized in the laboratory or obtained from various natural sources, copper carbonate always has the same composition. Analysis of this compound led Proust to formulate the law of definite proportions.

Figure 2.5 Berzelius's experiment illustrating the law of definite proportions.

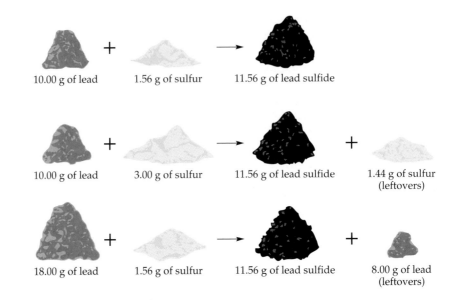

10.00 g of lead + 1.56 g of sulfur → 11.56 g of lead sulfide

10.00 g of lead + 3.00 g of sulfur → 11.56 g of lead sulfide + 1.44 g of sulfur (leftovers)

18.00 g of lead + 1.56 g of sulfur → 11.56 g of lead sulfide + 8.00 g of lead (leftovers)

Proust, like Lavoisier, was a member of the French nobility. He was working in Spain, temporarily safe from the ravages of the French revolution. His laboratory was destroyed and he was reduced to poverty, however, when the French troops of Napoleon Bonaparte occupied Madrid in 1808.

The idea of definite proportions (or constant composition) had been first expressed by Hieronymus Richter (Richter's rule) in 1792, but the work of Proust, published in 1799, was more extensive and more convincing.

Jöns Jakob Berzelius was the first person to prepare an extensive list of atomic weights. Published in 1828, it agrees remarkably well with most of our accepted values today.

experiments, Proust in 1799 formulated a new scientific law. A compound, he said, always contains the same elements in certain definite proportions and in no other combinations. He called this generalization the **law of definite proportions**. (It is also sometimes called the *law of constant composition*.)

An early illustration of the law of definite proportions is found in the work of the noted Swedish chemist J. J. Berzelius (1779–1848). Berzelius heated 10.00 g of lead with various amounts of sulfur to form lead sulfide. Lead is a soft, grayish metal. Sulfur is a yellow solid. Lead sulfide is a shiny, black solid. It was easy, therefore, to tell when all the lead had reacted. Excess sulfur was washed away with carbon disulfide, a solvent that dissolves sulfur but not lead sulfide. As long as he used at least 1.56 g of sulfur with 10.00 g of lead, Berzelius got exactly 11.56 g of lead sulfide. Any sulfur in excess of 1.56 g was left over, unreacted. If he used more than 10.00 g of lead with 1.56 g of sulfur, he got 11.56 g of lead sulfide, with lead left over. These reactions are illustrated in Figure 2.5. (An explanation is given later in Figure 2.7.)

The law of definite proportions is further illustrated by the electrolysis of water. In 1783 Henry Cavendish (1731–1810), a wealthy, eccentric English nobleman, found that water forms when hydrogen burns in oxygen. (It was Lavoisier, however, who correctly interpreted the experiment and who first used the names "hydrogen" and "oxygen"). Later, in 1800, two English chemists, William Nicholson and Anthony Carlisle, decomposed water into hydrogen and oxygen gases by passing an electric current through the water (Figure 2.6). (The Italian scientist Alessandro Volta had invented the chemical battery only 6 weeks earlier.) The two gases are always produced in a 2:1 volume ratio. This scientific breakthrough led to rapid developments in chemistry and dealt a death blow to the ancient Greek idea of water as an element.

The law of definite proportions is the basis for chemical formulas, such as H_2O (Chapter 6). It also has wider meaning. Not only do compounds have constant composition, but they also have constant properties. Pure water always dissolves salt or sugar, and at normal pressure it always freezes at 0°C and boils at 100°C.

2.4 John Dalton and the Atomic Theory of Matter

Lavoiser's law of conservation of mass and Proust's law of definite proportions were repeatedly verified by experiment. This work led to attempts to develop theories to explain these laws.

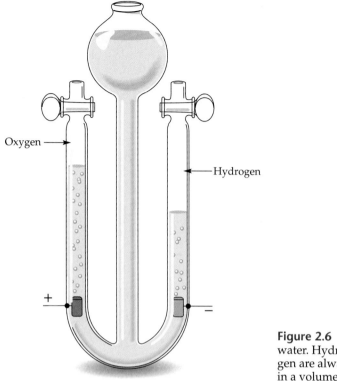

Figure 2.6 Electrolysis of water. Hydrogen and oxygen are always produced in a volume ratio of 2:1.

John Dalton (1766–1848).

In 1803 John Dalton, an English schoolteacher, proposed a model to explain the accumulating experimental data. As he refined his model, he discovered another law that his theory would have to explain. Proust had stated that a compound contains elements in certain proportions and only those proportions. Dalton's new law, called the **law of multiple proportions**, stated that elements might combine in *more* than one set of proportions, with each set corresponding to a different compound. For example, carbon combines with oxygen in a mass ratio of 3.0:8.0 to form carbon dioxide, a gas familiar as a product of respiration and of the burning of coal and wood. But Dalton found that carbon also combines with oxygen in a mass ratio of 3.0:4.0 to form carbon monoxide, a poisonous gas formed when a fuel is burned in the presence of a limited air supply.

Dalton then proposed his **atomic theory**, a model he used to explain the various laws. Following are the important points of Dalton's atomic theory, with some modern modifications that we will consider later.

Dalton's Atomic Theory

1. All matter is composed of extremely small particles called atoms.
2. All atoms of a given element are alike, but atoms of one element differ from the atoms of any other element.
3. Compounds are formed when atoms of different elements combine in fixed proportions.
4. A chemical reaction involves a *rearrangement* of atoms. No atoms are created, destroyed, or broken apart in a chemical reaction.

Modern Modifications

1. Dalton assumed atoms to be indivisible. This isn't quite true, as we will see in the next chapter.

The ancient Greeks' ideas of atoms were mainly intuitive. However, by 1800 scientists had accumulated considerable evidence for their existence.

2. Dalton assumed that all the atoms of a given element were identical in all respects, including mass. This is now known to be incorrect, as we will see in Chapter 3.

3. Unmodified. The numbers of each kind of atom in a compound usually form a simple ratio. For example, the ratio of carbon atoms to oxygen atoms is 1 : 1 in carbon monoxide and 1 : 2 in carbon dioxide.

4. Unmodified for *chemical* reactions. Atoms are broken apart in *nuclear* reactions.

Explanations Using Atomic Theory

Dalton's theory clearly explains the difference between elements and compounds. *Elements* are composed of only one kind of atom. For example, a sample of the element phosphorus has only phosphorus atoms in it. (We will explain more precisely what we mean by "kind" in Section 3.6.) *Compounds* are made up of two or more kinds of atoms chemically combined in definite proportions.

To explain the law of definite proportions, Dalton's reasoning went something like this. Why should 1.0 g of hydrogen always combine with 19 g of fluorine? Why shouldn't 1.0 g of hydrogen also combine with 18 g of fluorine? Or 20 g of fluorine? Or any other mass of fluorine? If an atom of fluorine has a mass 19 times that of a hydrogen atom, the compound formed by the union of one atom of each element would have to consist of 1 part by mass of hydrogen and 19 parts by mass of fluorine. Matter must be atomic for the law of definite proportions to be valid. Figure 2.7 shows Berzelius's experiment interpreted in terms of Dalton's atomic theory.

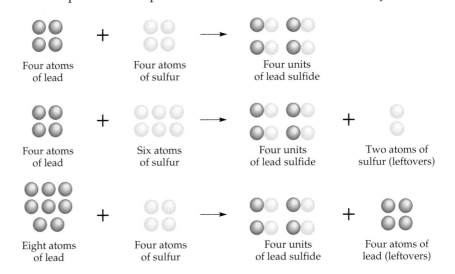

Figure 2.7   The law of definite proportions: Berzelius's experiment interpreted in terms of Dalton's atomic theory.

The atomic theory also explains the law of multiple proportions. For example, 1.000 g of carbon combines with 1.332 g of oxygen to form carbon monoxide, and with 2.664 g of oxygen to form carbon dioxide. Carbon dioxide has twice the mass of oxygen per gram of carbon as does carbon monoxide. This is explained by the fact that one atom of carbon combines with *one* atom of oxygen to form carbon monoxide and one atom of carbon combines with *two* atoms of oxygen to form carbon dioxide. Figure 2.8 uses the oxides of carbon to illustrate the law of multiple proportions, and Table 2.1 shows some of the proportions in which nitrogen and oxygen combine.

Figure 2.8   The law of multiple proportions. A carbon atom can combine with either one or two atoms of oxygen.

Table 2.1 ∎ The Law of Multiple Proportions

Compound	Representation[a]	Mass of N per 1.000 g of O	Ratio of the Masses of N[b]
Nitrous oxide		1.750 g	$(1.750 \div 0.4375) = 4.000$
Nitric oxide		0.8750 g	$(0.8750 \div 0.4375) = 2.000$
Nitrogen dioxide		0.4375 g	$(0.4375 \div 0.4375) = 1.000$

[a] ● = nitrogen atom and ○ = oxygen atom

[b] We obtain the ratio of the masses of N by dividing each quantity in the third column by the smallest (0.4375 g).

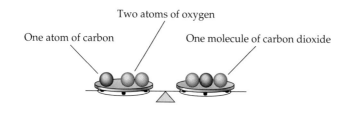

One atom of carbon — Two atoms of oxygen — One molecule of carbon dioxide

Figure 2.9 The law of conservation of mass. When one atom of carbon reacts with two atoms of oxygen to form carbon dioxide, the atoms are merely rearranged (not created or destroyed). Mass is therefore conserved.

Finally, the law of conservation of mass is also explained by atomic theory. When carbon atoms combine with oxygen atoms to form carbon dioxide, the atoms are merely rearranged (Figure 2.9). Matter is neither lost nor gained; the mass does not change.

As part of his theory, Dalton set up a table of relative atomic masses based on hydrogen as 1. Many of Dalton's atomic masses were incorrect, as we might expect because of the equipment available at that time. Using modern values, we assign oxygen a mass of 16.0 and carbon a mass of 12.0. On this scale carbon monoxide is seen to be made up of one atom of carbon combined with one atom of oxygen to give a mass ratio of 12.0 parts carbon to 16.0 parts oxygen (or 3.00 : 4.00), and carbon dioxide to be made up of one atom of carbon combined with two atoms of oxygen to give a mass ratio of 12.0 parts carbon to $2 \times 16.0 = 32.0$ parts oxygen (or 3.00 : 8.00). Dalton also invented a set of symbols to represent the different kinds of atoms. In fact, symbols similar to Dalton's are used in Table 2.1 and Figure 2.8. These symbols have since been replaced by modern symbols of one or two letters (see Table 1.2).

Some of Dalton's Atomic Symbols

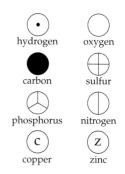

hydrogen   oxygen

carbon   sulfur

phosphorus   nitrogen

copper   zinc

Some of Dalton's symbols for the elements.

| **What A Difference an O Makes!**

An example of the law of multiple proportions is the fact that carbon forms two oxides, CO and CO_2. Both are colorless, odorless gases, but how very different they are! CO_2 (carbon dioxide) is always present in your body. It is collected from your cells and carried by your bloodstream to your lungs and then exhaled. The bubbles in "soda"-type drinks are also carbon dioxide. Most fuels produce CO_2, along with H_2O, when they burn. At temperatures around $-80°C$ carbon dioxide solidifies and becomes dry ice, which we often use to keep ice cream from melting. CO (carbon monoxide) is quite another thing. It is so highly poisonous that only 0.2% of CO in the air is enough to kill. Unfortunately, most fuels also produce some CO when they burn. This is why an indoor kerosene heater should always be well vented and a car engine should never be left running in a closed garage.

Relative atomic masses were determined by comparison with a standard mass, a technique called *weighing*. For this historical reason, many chemists refer to relative masses as *atomic weights*.

We can use proportions, such as those determined by Dalton, to calculate the amount of one substance needed to combine with a given amount of another substance. To learn how to do this, let's took at some examples.

Example 2.1

Carbon combines with hydrogen in a ratio of 3.00 parts by mass of carbon to 1.00 part by mass of hydrogen to form a gas called *methane*. How much hydrogen is needed to combine with 90.0 g of carbon to form methane?

Solution

We can express the parts ratio in any units we choose as long as it is the same for both elements.

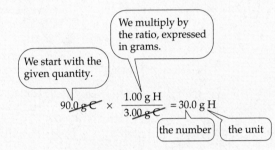

Exercise 2.1

A. The gas arsine can be decomposed to give 24.8 parts by mass of arsenic and 1.00 part by mass of hydrogen. What mass of arsenic is obtained if 1.27 g of arsine is decomposed?

B. Nitrous oxide, sometimes called "laughing gas," can be decomposed to give 7.00 parts by mass of nitrogen and 4.00 parts by mass of oxygen. What mass of nitrogen is obtained if enough nitrous oxide is decomposed to yield 36.0 g of oxygen?

Example 2.2

Hydrogen sulfide gas can be decomposed to give sulfur and hydrogen in a mass ratio of 16.0:1.00. If the relative mass of sulfur is 32.0 when the mass of hydrogen is taken to be 1.00, how many hydrogen atoms are combined with each sulfur atom in the gas?

Solution

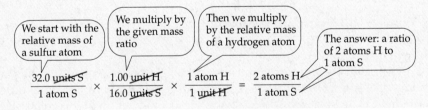

Exercise 2.2

Methane gas can be decomposed to give carbon and hydrogen in a mass ratio of 3.00:1.00. If the relative mass of carbon is 12.0 when the mass of hydrogen is taken to be 1.00, how many hydrogen atoms are combined with each carbon atom in the gas?

Despite some inaccuracies, Dalton's atomic theory was a great success. Why? Because it served—and still serves—to explain a large amount of experimental data. It also successfully predicted how matter would behave under a wide variety of circumstances. Dalton arrived at his atomic theory by reasoning based on a growing body of experimental facts, and with modest modification it has stood the test of time and modern, highly sophisticated instrumentation. Formulation of so successful a theory was quite a triumph for a Quaker schoolteacher in 1803.

2.5 Out of Chaos: The Periodic Table

Before moving on, let's take a look at a remarkable parallel development. New elements were being discovered with surprising frequency, and by 1830, there were 55 known elements, all with different properties and with no apparent order in these properties. John Dalton had set up a table of relative atomic masses in his book *A New System of Chemical Philosophy* in 1808. Many of his values were wrong, but they were improved in subsequent years, notably by Berzelius, who published a table of atomic weights in 1828 containing 54 elements. Except for a few cases, Berzelius's values are in quite good agreement with modern values.

Relative Atomic Masses

Although it was impossible to determine actual masses of atoms in the 1800s, chemists were able to determine relative atomic weights by measuring the amounts of various elements that combined with a given mass of another element. Dalton's atomic weights were based on an atomic mass of 1 for hydrogen. As more accurate atomic weights were determined, this standard was replaced by one in which oxygen was assigned a value of 16.0000. The oxygen standard was used until 1961 when it was replaced by a more logical one based on an isotope of carbon (Section 3.6). Adoption of this new standard caused little change in atomic weights. These relative atomic weights are usually expressed in **atomic mass units (amu)**, commonly referred to today simply as *units (u)*.

Mendeleev's Periodic Table

Various attempts were made to arrange the elements in some sort of systematic fashion. The most successful arrangement, and one that soon became widely accepted by chemists, was published in 1869 by Dmitri Ivanovich Mendeleev (1834–1907), a Russian chemist. Mendeleev's **periodic table** arranged the elements primarily in order of increasing atomic mass, although in a few cases he put a slightly heavier element before a lighter one in order to place elements with similar chemical properties in the same column (Table 2.2). For example, he put tellurium, with an atomic mass of 127.6 u, ahead of iodine, which has an atomic mass of 126.9 u. He did this in order to place tellurium in the same column as sulfur and selenium, which it resembles in chemical properties. This rearrangement also put iodine in the same column as chlorine and bromine, which it resembles.

Mendeleev left gaps in his table. This was also necessary in order to place elements in groups with similar properties. Instead of considering these blank spaces defect, he boldly predicted the existence of elements yet undiscovered. Further, he even predicted the properties of some of the missing elements. For example, the missing elements he called eka-boron, eka-aluminum, and eka-silicon were soon discovered and named scandium, gallium, and germanium.[1] From their positions in the periodic table, Mendeleev predicted the properties of these elements with amazing success (Table 2.3). This remarkable predictive value led to wide acceptance of Mendeleev's table.

Dmitri Mendeleev, the Russian chemist who invented the periodic table of the elements, continues to be honored in his native land.

---

[1]Gallium was discovered in 1875 by P. E. Lecoq de Boisbaudran, who named it for his native land, Gaul (France). Swedish chemist Lars F. Nilson discovered scandium in 1879 and named it for Scandinavia. Finally, in 1886 C. A. Winkler discovered germanium and named it for his country, Germany.

Table 2.2 ■ The Periodic Table of Mendeleev*

Reihen	Gruppe 1 — R²O	Gruppe 2 — RO	Gruppe 3 — R²O³	Gruppe 4 RH⁴ RO²	Gruppe 5 RH³ R²O⁵	Gruppe 6 RH² RO³	Gruppe 7 RH R²O⁷	Gruppe 8 — RO⁴
1	H = 1							
2	Li = 7	Be = 9,4	B = 11	C = 12	N = 14	O = 16	F = 19	
3	Na = 23	Mg = 24	Al = 27,3	Si = 28	P = 31	S = 32	Cl = 35,5	
4	K = 39	Ca = 40	— = 44	Ti = 48	V = 51	Cr = 52	Mn = 55	Fe = 56, Co = 59, Ni = 59, Cu = 63.
5	(Cu = 63)	Zn = 65	— = 68	— = 72	As = 75	Se = 78	Br = 80	
6	Rb = 85	Sr = 87	?Yt = 88	Zr = 90	Nb = 94	Mo = 96	— = 100	Ru = 104, Rh = 104, Pd = 106, Ag = 108.
7	(Ag = 108)	Cd = 112	In = 113	Sn = 118	Sb = 122	Te = 125	J = 127	
8	Cs = 133	Ba = 137	?Di = 138	?Ce = 140	—	—	—	—————
9	(—)	—	—	—	—	—	—	
10	—	—	?Er = 178	?La = 180	Ta = 182	W = 184	—	Os = 195, Ir = 197, Pt = 198, Au = 199.
11	(Au = 199)	Hg = 200	Ti = 204	Pb = 207	Bi = 208	—	—	
12	—	—	—	Th = 231	—	U = 240	—	—————

*Spaces are left for the unknown elements with atomic masses 44, 68, 72, and 100, as well as some others.

Table 2.3 ▮ Properties of Germanium: Predicted and Observed

Property	Predicted by Mendeleev for Eka-silikon (1871)	Observed by Winkler for Germanium (1886)
Atomic mass	72	72.6
Density (g/cm^3)	5.5	5.47
Color	Dirty gray	Grayish white
Density of oxide (g/cm^3)	EsO_2: 4.7	GeO_2: 4.703
Boiling point of chloride	$EsCl_4$: below 100°C	$GeCl_4$: 86°C
Density of chloride (g/cm^3)	$EsCl_4$: 1.9	$GeCl_4$: 1.887

Precursors of the Periodic Table

"Triads" of Dobereiner

As early as 1816 Johann Dobereiner, a German chemist, noticed that there were several groups of three elements that were very similar (lithium, sodium, potassium; calcium, strontium, barium; sulfur, selenium, tellurium; chlorine, bromine, iodine). In each case, the middle element seemed to be halfway between the other two in atomic mass, reactivity, and other properties. He published this in 1829.

"Telluric Helix" of de Chancourtois

In 1862 Beguyer de Chancourtois, a French geologist, arranged the elements in order of atomic mass. When he wound the list spirally around a cylinder, he found that similar elements fell along the same vertical lines.

Newlands' "Law of Octaves"

In 1863 John Newlands, an English chemist, noted that when elements were listed in order of atomic mass, every eighth element had similar properties; however, the rule seemed to break down for elements past calcium.

Meyer's System of Elements

In 1868 Lothar Meyer in Germany came up independently with an arrangement of elements similar to that of Mendeleev. Many believe that he should share the credit for the periodic table. Unfortunately, Meyer did not write up his table until December 1869, and it was not published until March 1870. Mendeleev had published his table in 1869.

The modern periodic table (inside front cover) contains more than 110 elements. Each element is represented by a "box"' in the periodic table, and the data typically shown are

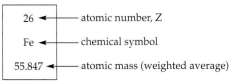

We will discuss the periodic table and its theoretical basis in Chapter 3.

2.6 Atoms: Real and Relevant

Are atoms real? Certainly they are real as a concept, a highly useful concept at that. Scientists can even observe computer-enhanced *images* of individual atoms. These portraits reveal little detail of atoms, but they provide powerful (though still indirect) evidence that atoms exist.

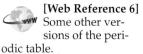

[Web Reference 5] A very useful periodic table that includes information about each element.

[Web Reference 6] Some other versions of the periodic table.

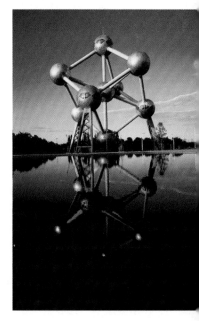

The Atomium in Brussels was built for the 1958 World's Fair. The spheres represent atoms in an iron crystal. Each "atom" is 18 meters in diameter and houses an exhibit that demonstrates some kind of progress in science and technology (except for the top sphere, which is a restaurant). The "bonds" connecting the atoms are giant tubes containing escalators so that visitors can move easily from one sphere to another.

Are atoms relevant? Much of modern science and technology—including the production of new materials and the technology of pollution control—is ultimately based on the concept of atoms. We have seen that atoms are conserved in chemical reactions. Thus, material things—things made of atoms—can be recycled, for the atoms are not destroyed no matter how we use them. The one way we might lose a material from a practical standpoint is to spread the atoms so thinly that it would take too much time and energy to put them back together again.

2.7 Leucippus Revisited: Molecules

Now back to Leucippus and his musings by the seashore. We now know that if we keep dividing drops of water into smaller drops, we will ultimately obtain a small particle—called a *molecule*—that is still water. If we divide this particle still further, we will obtain *two atoms* of hydrogen and *one atom* of oxygen. And if we divide these …, but that is a story for another time.

Dalton regarded the atom as indivisible, as did his successors up until the discovery of radioactivity in 1895. We examine the changing concept of the atom in the next chapter.

A **molecule** is a group of atoms that are chemically bonded together. Molecules are represented by chemical formulas. The symbol H represents an *atom* of hydrogen; the formula H_2 represents a *molecule* of hydrogen, which is composed of two hydrogen atoms. The formula H_2O represents a molecule of water, which is composed of two hydrogen atoms and one oxygen atom.

Recycling

Consider the following two different pathways for the recycling of iron.

1. Iron ore (hematite) is mined from the ground and converted to pig iron and then into steel (an alloy of iron with carbon). The steel is used in making an automobile, which is driven for a decade and then sent to the junkyard. The junkyard compresses the automobile and sends it to a recycling plant where the steel is recovered and ultimately used again in a new automobile. Once the iron was removed from its ore, it has been conserved in its elemental metallic form.

2. Pig iron (obtained from iron ore) is used to make wire for a fence. Over the years the wire gets rusty. Eventually it is sent to a junkyard and ends up in a vat of sulfuric acid where it is completely dissolved as iron sulfate. The iron sulfate is later poured down the drain so that it flows into the river and eventually winds up in the ocean. The original iron atoms have become dissolved ions that are usable by plants. Marine plants absorb and incorporate the iron, so that it is available to any nearby marine creatures that eat plants. The original iron atoms are now widely separated in space. Perhaps a few of them might even become part of the hemoglobin in your own bloodstream. The iron has been recycled, but it will never again resemble the original pig iron.

Critical Thinking Exercises

Apply knowledge that you have gained in this chapter and one or more of the FlaReS principles (Chapter 1) to evaluate the following statements or claims.

2.1 A doctor claims that by giving patients small amounts of selenium, he can cure some types of cancer. He says that selenium atoms have the ability to get inside certain kinds of cancer cells and kill them. He has a list of patients that he says he has cured with this treatment.

2.2 A health food store has a large display of bracelets made of copper metal. Some people claim that wearing a copper bracelet will protect the wearer against arthritis or rheumatoid diseases.

2.3 Mary has just learned that her red blood cell count is low, and her doctor has given her some pills that contain iron. The doctor says that the pills should raise the level of hemoglobin in her bloodstream and keep Mary from becoming anemic. Should Mary take the pills or should she seek another opinion?

Summary

1. An atom is the smallest unit particle of an element. Although the concept of atoms was first suggested in ancient Greece by Leucippus and Democritus, the atomic theory was not proposed until 1803 by John Dalton.

2. The law of conservation of mass resulted from careful experiments by Lavoisier and others who weighed all the reactants and all the products for a number of chemical reactions and found that no change in mass occurred.

3. The law of definite proportions (or the law of constant composition) was formulated by Proust, based on the work of Berzelius plus his own experiments. A given compound always contains the same elements in exactly the same proportions.

4. The law of multiple proportions states that the same group of elements might combine in several different sets of proportions, each set corresponding to a different compound. John Dalton discovered this law while working on his atomic theory.

5. The atomic theory states that:
   a. All matter is made up of tiny particles called atoms.
   b. All atoms of the same element are alike.
   c. Compounds are formed when atoms of different elements combine in certain proportions.
   d. During chemical reactions atoms are only rearranged, not destroyed or broken apart.

6. The most successful systematic arrangement of the elements is the periodic table, published in 1869 by a Russian chemist named Mendeleev.

7. Since atoms are conserved in chemical reactions, matter (which is made of atoms) can be recycled. If we hope to recycle a particular kind of matter, however, we should take care not to let it spread too thinly throughout nature.

8. Just as atoms are the smallest unit particles of elements, molecules are the smallest unit particles of most compounds. A molecule is a group of atoms chemically bonded together.

Key Terms

atom   2.1
atomic mass
   units   2.5
atomic theory   2.4

law of conservation of
   mass   2.2
law of definite
   proportions   2.3

law of multiple
   proportions   2.4

molecule   2.7
periodic table   2.5

Review Questions

1. What is the distinction between the atomistic view and the continuous view of matter? What was Democritus' contribution to atomic theory?

2. Why did the idea that matter was continuous (rather than atomic) prevail for so long? What discoveries finally refuted the idea?

3. If we applied the idea of divisible (atomic) or continuous to foods at the macroscopic (visible to the unaided eye) level, which designation would you use for each of the following?
   a. milk
   b. mashed potatoes
   c. peas
   d. olives
   e. hard-boiled eggs
   f. scrambled eggs

4. Describe Lavoisier's contribution to the development of modern chemistry.

5. State the law of conservation of mass.

6. How does the ancient Greek definition of an element differ from the modern one? How did Robert Boyle define an element?

7. State the law of definite proportions.

8. State the law of multiple proportions.

9. Outline the main points of Dalton's atomic theory.

10. Sulfur and oxygen form two compounds. One is 50.0% sulfur and 50.0% oxygen; the other is 40.0% sulfur and 60.0% oxygen. What law does this illustrate?

11. A photographic flashbulb weighing 0.750 g contains magnesium and air. The flash produces magnesium oxide. After cooling, the bulb weighs 0.750 g. What law does this illustrate?

12. Consider the following set of compounds. What principle does this group illustrate?

$$N_2O \quad NO \quad NO_2 \quad N_2O_4$$

13. Consider the following set of compounds. What principle does this group illustrate?

$$FeO \quad Fe_2O_3 \quad Fe_3O_4$$

14. Use Dalton's atomic theory to explain the law of conservation of mass. Give an example that illustrates this law.

15. Use Dalton's atomic theory to explain the law of definite proportions. Give an example that illustrates this law.

16. Use Dalton's atomic theory to explain the law of multiple proportions. Give an example that illustrates this law.

17. When 3.00 g of carbon is burned in 8.00 g of oxygen, 11.00 g of carbon dioxide is formed. What mass of carbon dioxide is formed when 3.00 g of carbon is burned in 50.00 g of oxygen? What law does this illustrate?

18. Heptane is always composed of 84.0% carbon and 16.0% hydrogen. What law does this illustrate?

19. The ancient Greeks thought that water was an element. In 1800 Nicholson and Carlisle decomposed water into hydrogen and oxygen. What did their experiment prove?

20. What did each of the following contribute to the development of modern chemistry?
    a. J. J. Berzelius
    b. Henry Cavendish
    c. Joseph Proust

Problems

Conservation of Mass

21. A balloon filled with helium floats near the ceiling. After several days, the balloon is deflated and lying on the floor. Have the helium atoms been destroyed? If so, how? If not, where are they?

22. A copper coin is dissolved in a solution of nitric acid. Have the copper atoms been destroyed? If so, how? If not, where are they?

23. Sugar consists of carbon, hydrogen, and oxygen atoms. When a sugar cube is burned in a crucible, nothing remains in the vessel. Have the carbon, hydrogen, and oxygen atoms of the sugar been destroyed? If so, how? If not, where are they?

24. Aspirin consists of carbon, hydrogen, and oxygen atoms. A student in an organic chemistry laboratory dissolves some aspirin in hot water in order to purify it by recrystallization. No crystals form on cooling, and so the student pours out the solution and starts the experiment over. Have the carbon, hydrogen, and oxygen atoms of the aspirin been destroyed? If so, how? If not, where are they?

25. A student heats 1.0000 g of zinc powder with 0.2000 g of sulfur. He reports that he obtains 0.6080 g of zinc sulfide and recovers 0.5920 g of unreacted zinc. Show by calculation whether or not his results obey the law of conservation of mass.

26. A student heats 0.5585 g of iron with 0.3550 g of sulfur. She reports that she obtains 0.8792 g of iron sulfide and recovers 0.0433 g of unreacted sulfur. Show by calculation whether or not her results obey the law of conservation of mass.

27. A city has to come up with a plan to solve its solid waste problem. The solid wastes consist of many different kinds of materials, and the materials are comprised of many different kinds of atoms. The options for disposal include burying the wastes in a landfill, incinerating them, and dumping them at sea. Which method, if any, will get rid of the atoms that make up the waste? Which method, if any, will change the chemical form of the waste?

28. Polychlorinated biphenyls (PCBs), frequently found as environmental contaminants, are composed of carbon, hydrogen, and chlorine atoms. A news report of a new method for disposal of PCBs describes the process as converting the compounds completely to carbon dioxide (made of carbon and oxygen atoms) and water (composed of hydrogen and oxygen atoms). The oxygen atoms can come from atmospheric oxygen. Examine the report in light of the law of conservation of mass.

Dalton's Atomic Theory

29. To the nearest atomic mass unit, an atom of calcium has a mass of 40 u and an atom of vanadium has a mass of 50 u. Do these findings contradict Dalton's atomic theory? Explain.

30. To the nearest atomic mass unit, an atom of calcium has a mass of 40 u and an atom of potassium has a mass of

40 u. Do these findings contradict Dalton's atomic theory? Explain.

31. To the nearest atomic mass unit, one atom of calcium has a mass of 40 u and another calcium atom has a mass of 44 u. Do these findings contradict Dalton's atomic theory? Explain.

32. An atom of uranium-235 is struck by a neutron and splits into two, smaller atoms. Do these findings contradict Dalton's atomic theory? Explain.

Chemical Compounds

33. A colorless liquid is thought to be a pure compound. Analyses of three samples of the material yield the following results.

	Mass of Sample	Mass of Carbon	Mass of Hydrogen
Sample 1	1.000 g	0.862 g	0.164 g
Sample 2	1.549 g	1.335 g	0.254 g
Sample 3	0.988 g	0.852 g	0.162 g

Could the material be a pure compound?

34. A blue substance called azulene is thought to be a pure compound. Analyses of three samples of the material yield the following results.

	Mass of Sample	Mass of Carbon	Mass of Hydrogen
Sample 1	1.000 g	0.937 g	0.0629 g
Sample 2	0.244 g	0.229 g	0.0153 g
Sample 3	0.100 g	0.094 g	0.0063 g

Could the material be a pure compound?

Definite Proportions

35. When 18.0 g of water is decomposed by electrolysis, 16.0 g of oxygen and 2.0 g of hydrogen are formed. According to the law of definite proportions, how much hydrogen is formed by the electrolysis of 360 g of water?

36. Hydrogen from the decomposition of water has been promoted as the fuel of the future (Chapter 14). How much water would have to be electrolyzed to produce 100 kg of hydrogen? (See Problem 35.)

37. With a plentiful supply of air, 3.0 parts carbon react with 8.0 parts oxygen to produce carbon dioxide. Use this mass ratio to calculate how much carbon is required to produce 990 g of carbon dioxide.

38. With a limited supply of air, 3.0 parts carbon react with 4.0 parts oxygen to produce carbon monoxide. Use this mass ratio to calculate how much carbon monoxide can be formed from 42 g of carbon.

Additional Problems

39. When we burn a 10-kg piece of wood, only 0.05 kg of ash remains. Explain this apparent contradiction of the law of conservation of mass.

40. The gas silane can be decomposed, to yield silicon and hydrogen in a ratio of 7 parts by mass of silicon to 1 part by mass of hydrogen. If the relative mass of silicon atoms is 28 and the mass of hydrogen atoms is taken to be 1, how many hydrogen atoms are combined with each silicon atom?

41. Jan Baptista van Helmont (1579–1644), a Flemish alchemist, performed an experiment in which he planted a young willow tree in a weighed bucket of soil. After 5 years, he found that the tree had gained 75 kg, yet the soil had lost only 0.057 kg. He had added only water to the system, and so he concluded that the substance of the tree had come from water. Criticize his conclusion.

Projects

42. Write a brief report on any one of the following scientists:
   a. Henry Cavendish   **c.** John Newlands
   b. Joseph Proust   **d.** Lothar Meyer

43. Color blindness is sometimes called *daltonism*. What can you find out about daltonism? (Does this refer to the same Dalton who proposed the atomic theory?)

44. Mendeleev is probably the most famous of all Russian scientists. Examine the literature (or the internet) to find out more about this interesting man.

Online Projects

45. Search the web and make a list of what you can learn there about the contributions and ideas of the early Greeks in the field of science. (See Web Reference **1**.)

46. Lavoisier has been called the father of modern chemistry. Write a brief report on the too short life of this talented scientist. (See Web Reference **3**.)

47. Write a brief essay on recycling, contrasting a recycling method that maintains the properties of an element with one that changes them. Search the internet for information on the recycling of metals.

References and Readings

1. Brock, William H. *The Norton History of Chemistry*. New York: Norton, 1993.

2. Cobb, Cathy, and Harold Goldwhite. *Creations of Fire: Chemistry's Lively History from Alchemy to the Atomic Age*. New York: Plenum, 1996.

3. Cole, K. C. "On Imagining the Unseeable." *Discover*, December 1982, pp. 70–72.

4. Gribbin, John. *Almost Everyone's Guide to Science*. London: Phoenix (Orion Books), 1998.

5. Hellman, Hal. *Great Feuds in Science: Ten of the Liveliest Disputes Ever*. New York: John Wiley, 1998.

6. Jaffe, Bernard. *Crucibles: The Story of Chemistry*. New York: Fawcett World Library, 1957.

7. Kolb, Doris. "Chemical Principles Revisited: But if Atoms Are So Tiny…" *Journal of Chemical Education*, September 1977, pp. 543–547.

8. Lloyd, G. E. R. *Methods and Problems in Greek Science*. New York: Cambridge University Press, 1991.

9. Quinn, Susan. *Madame Curie: A Life*. New York: Simon and Schuster, 1995.

10. Salzberg, Hugh W. *From Caveman to Chemist: Circumstances and Achievements*, 2nd edition. Washington: American Chemical Society, 1991.

11. Stillman, John Maxon. *The Story of Alchemy and Early Chemistry*. New York: Dover Publications, 1960.

12. Young, Louise B. (Ed.). *The Mystery of Matter*. New York: Oxford University Press, 1965.

Atomic Structure
Images of the Invisible

A cloud of electrons, so swift and intense,
Surrounding a nucleus tiny and dense.

The brilliant colors of a fireworks display result from changes in electron energy levels of atoms. These changes occur because the atoms absorb energy from the combustion of materials in the fireworks. In fact, it was from the colors imparted to flames that scientists first worked out the arrangement of electrons within atoms.

toms are exceedingly tiny particles that are much too small to see, and yet we are about to discuss their inner structure. If atoms are so small that we cannot see them, even with a microscope, how can we possibly know what their structures are like?

In order to see even the roughest images of atoms, we must use special kinds of microscopes. Atoms are much too small to be seen with an ordinary light microscope.

It turns out that we really know quite a lot about the structure of atoms, even though all our information has been gathered indirectly. Although scientists have never examined an atom directly, by simply exercising their powers of deduction and designing some clever experiments, they have been able to put together an amazingly detailed model of what an atom must be like.

It is not quite true to say that no one has ever seen an atom. In 1970, at the University of Chicago, Albert Crewe used an electron microscope to make the photograph shown in Figure 3.1. The seven spots are images of individual uranium atoms. Similar photographs have been taken by others. More recently, scientists have used scanning tunneling microscopy (STM) to obtain pictures such as the one in Figure 3.2, showing a pattern of atoms lying along the surface of a silicon crystal. We can see outlines of atoms in such photographs, and we can tell quite a bit about how they are arranged, but these pictures tell us nothing about the inner structure of the atoms. No one has ever seen inside an atom.

Why do we care about the structure of particles as tiny as atoms? It is the arrangement of various parts of their atoms that determines the properties of different kinds of matter. Only by understanding atomic structure can we learn how atoms combine to make the many different substances in nature. With such knowledge, we can modify and synthesize materials to meet our needs more precisely. A knowledge of atomic structure is even essential to our health. Many medical diagnoses are based on chemical analyses that have been developed from our understanding of atomic structure.

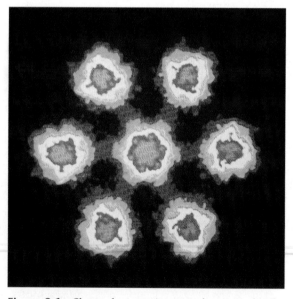

Figure 3.1 Shown here are images of seven individual uranium atoms (the colored spots with red-orange centers). The images are made with an electron microscope. The atoms are in the form of a compound called uranyl acetate and are pictured on an extremely thin carbon layer that appears black in this photograph. The atoms are 0.34 nm apart.

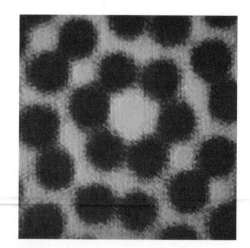

Figure 3.2 Image of the surface of silicon, produced by a scanning tunneling microscope. The red spots are individual silicon atoms arranged in a regular pattern that repeats itself across the surface. Images such as this aid our understanding of the surface structures of many different materials; silicon, in particular, is a material of vital significance to the semiconductor and computer industries.

Perhaps of greater interest to you is the fact that your understanding of chemistry (as well as much of biology and other sciences) depends, at least in part, on your knowledge of atomic structure. Let's start our study of atomic structure by going back to the time of John Dalton.

3.1 Electricity and the Atom

Dalton, who set forth his atomic theory in 1803, regarded the atom as hard and indivisible. It wasn't long, however, before evidence accumulated to show that matter has electrical properties. Indeed, the electrolytic decomposition of water by Nicholson and Carlisle in 1800 (Section 2.3) had already indicated this. Electricity played an important role in unraveling the structure of the atom.

Static electricity has been known since ancient times, but the notion of continuous electric current was born with the nineteenth century. In 1800 Alessandro Volta invented what is now known as the voltaic pile, an electrochemical cell much like a modern battery. If the poles of a pile are connected by a wire, current flows through the wire. The current is sustained by chemical reactions inside the pile. Volta's invention soon was applied in many areas of science and everyday life.

Electrolysis

Soon after Volta's invention, Humphry Davy (1778–1829), a British chemist, built a powerful battery that he used to pass electricity through molten (melted) salts. Davy quickly discovered several new elements. In 1807 he liberated highly reactive potassium metal from molten potassium hydroxide. Shortly thereafter he produced sodium metal by passing electricity through molten sodium hydroxide. Within a year, Davy had also produced magnesium, strontium, barium, and calcium metals for the first time. The science of electrochemistry was born.

Davy's protégé Michael Faraday (1791–1867) greatly extended this new science and defined many of the terms we still use today. **Electrolysis** is the splitting of compounds by electricity (Figure 3.3). An **electrolyte** is a compound that conducts electricity when melted or dissolved in water. **Electrodes** are carbon rods or metal strips inserted into a molten compound or solution to carry the electric current. The **anode** is the electrode that bears a positive charge, and the **cathode** is the electrode that is negatively charged. Faraday hypothesized that electric current is carried through a melted compound or solution by charged atoms—later named **ions**. An **anion** is an ion with a negative charge; anions travel toward the anode. A **cation** is a positively charged ion; cations move toward the cathode.

Faraday's electrochemical work established that atoms are electrical in nature, but further details of atomic structure had to wait several decades for the development of gas discharge tubes and for more powerful sources of electrical voltage. Actually, Faraday tried and failed to pass electricity through a tube that had part of the air pumped out. His vacuum was not sufficient for the voltage he had available.

[Web Reference 1]
A biography of Volta.

Electrochemistry, which includes the study of electrochemical cells and electrolysis, is discussed in Chapter 8.

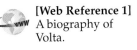

Figure 3.3 Electrolysis apparatus.

[Web Reference 2]
The paper by Faraday in which he named the two electrodes.

Ions are simply atoms, or unit groups of atoms, that have an electric charge.

This 20-pound English bank note bears a picture of Michael Faraday.

Figure 3.4   A simple gas discharge tube.

Electrode (cathode)

Cathode ray   Electrode (anode)

High voltage

Connection to vacuum pump

Cathode rays travel in straight lines unless some kind of external field is applied.

[Web Reference 3]
A multimedia exhibit celebrating the 100th anniversary of the discovery of the electron.

The greater the charge on a particle, the more it is deflected in an electric (or magnetic) field. (The greater the charge, the greater the attraction or repulsion.) The greater the mass of the particle, the less it is deflected by a force. (Consider two spheres coming toward you. It is much easier to deflect a ping-pong ball than a cannonball.)

For many purposes, the electron is considered the unit of electrical charge. The charge is shown as a superscript minus sign (meaning 1^-). In indicating charges on ions, we use $^-$ to indicate a net charge of one electron, 2^- to indicate a charge of two electrons, and so on.

[Web Reference 4]
A discussion of the history of chemistry, emphasizing the work of J. J. Thomson.

[Web Reference 5]
The story of the discovery of the electron.

Cathode Ray Tubes

By 1875 tubes with better vacuum than Faraday could achieve were available. William Crookes (1832–1919), an English chemist, passed an electric current through such a tube containing air at low pressure (Figure 3.4). His experiment can be repeated today. The tube has metal electrodes sealed in it. It is connected to a vacuum pump, and most of the air is removed. A beam of current is seen as green fluorescence, which is observed when the beam strikes a screen coated with zinc sulfide. This beam seems to leave the cathode and travel to the anode (Figure 3.5). The beam is called a **cathode ray**.

Thomson's Experiment: Mass-to-Charge Ratio

Considerable speculation arose as to the nature of cathode rays. In general, British scientists believed these rays to be beams of particles. Most German scientists held that they were more likely to consist of a form of energy much like visible light. The answer came (as scientific answers should) from an experiment performed by the English physicist Joseph John Thomson (1856–1940) in 1897. Thomson showed that cathode rays were deflected in an electric field (Figure 3.5). The beam was attracted to the positive plate and repelled by the negative plate. Thomson therefore concluded that cathode rays consist of negatively charged particles. His experiments also showed that the particles were the same regardless of the materials from which the electrodes were made or the type of gas in the tube. He concluded that these negative particles are constituents of all kinds of atoms. Thomson named these negatively charged units **electrons**. Cathode rays, then, are beams of electrons.

Cathode rays are deflected in magnetic fields as well as in electric fields. By measuring the amount of deflection in fields of known strength, Thomson was able to calculate the ratio of the mass of the electron to its charge. He could not measure either the mass or the charge separately. This is like knowing that you weigh 25 lb for each foot of height but not knowing either your weight or height. Once either of these measurements is determined, it is easy to calculate the other from the known value and the 25 lb/ft ratio. Thomson was awarded the Nobel prize in physics in 1906.

Goldstein's Experiment: Positive Particles

In 1886 a German scientist named Eugen Goldstein performed experiments with gas discharge tubes that had perforated cathodes (Figure 3.6). He found that although

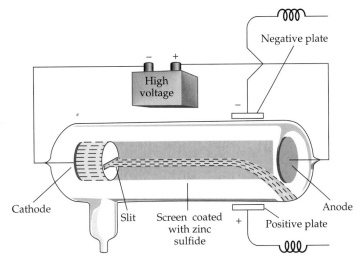

(a)

(b)

Figure 3.5 Thomson's apparatus, showing deflection of cathode rays (a beam of electrons). Cathode rays are themselves invisible but are observed through the green fluorescence they produce when they strike a zinc sulfide–coated screen. The diagram (a) shows deflection of the beam in an electric field. The photograph (b) shows the deflection in a magnetic field. The magnetic field is created by the magnet to the right of and slightly behind the screen.

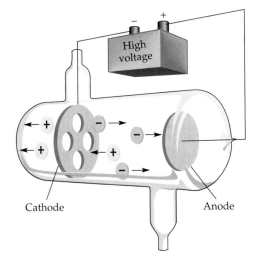

Figure 3.6 Goldstein's apparatus for the study of positive particles.

Figure 3.7   The Millikan oil-drop experiment. Oil drops irradiated with X-rays pick up electrons and become negatively charged. Their fall due to gravity can be balanced by adjusting the voltage of the electric field. From the applied voltage and the mass of the oil drop, the charge on the oil drop can be determined. Each drop carries a charge corresponding to some whole number of electrons.

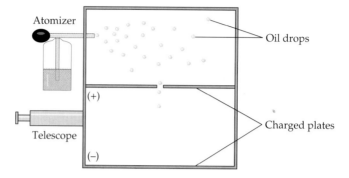

electrons were formed and sped off toward the anode as usual, positive ions were also formed and shot in the opposite direction toward the cathode. Some of these positive particles went through the holes in the cathode. In 1907 a study of the deflection of these particles in a magnetic field indicated that they were of varying mass. The lightest particles, formed when there was a little hydrogen gas in the tube, were later shown to have a mass 1837 times that of an electron.

Millikan's Oil-Drop Experiment: Electron Charge

Thomson measured the mass-to-charge ratio for an electron, but he could not measure the mass or the charge. The charge on the electron was not determined until 1909. An American physicist, Robert A. Millikan, experimented with electrically charged oil drops in an electric field. A diagram of Millikan's apparatus is shown in Figure 3.7. A spray bottle is used to inject tiny droplets of oil. Some of the oil droplets pass into a chamber where they can be viewed through a microscope. Some acquire negative charges by picking up electrons from the friction generated as the particles rub against the opening of the spray nozzle and against each other. (The charge is static electricity, just like the charge you get from walking across a nylon carpet.) The negative droplets can acquire one or more extra electrons by this process. Charges can also be produced by irradiation with X-rays.

The mass of the droplets can be determined by the rate at which they fall in air under the influence of gravity. As they fall between the charged plates, those that bear a negative charge are attracted upward toward the positive plate. The voltage differential between the plates can be adjusted to cause the particle to remain in place, with the upward attraction of the positive plate just balanced by the downward pull of gravity. From the voltage necessary to establish this stalemate, the mass-to-charge ratio of the particle can be calculated. Because the mass has already been determined from the free fall of the droplet, its charge can be calculated. Millikan took the smallest possible difference in charge between two droplets to be the charge of an individual electron. For his research, he received the Nobel prize in physics in 1923.

From Millikan's value for the charge and Thomson's value for the mass-to-charge ratio, the mass of the electron was readily calculated. Electrons are extremely light particles, having a mass of only 9.1×10^{-28} g.

If you do not understand numbers such as 9.1×10^{-28}, you can find out about them in Appendix A.

3.2 Serendipity in Science: X-Rays and Radioactivity

Let's return now to the structure of the atom and look at a little scientific serendipity. Often scientific discoveries are described as happy accidents. Have you ever wondered why these accidents always seem to happen to scientists? It is probably because scientists are trained observers. The same accident could happen right before the eyes of an untrained person and go unnoticed. Or, if it were noticed, its significance might not be grasped.

Roentgen: The Discovery of X-Rays

Two such happy accidents occurred in the last years of the nineteenth century. In 1895 a German scientist named Wilhelm Conrad Roentgen was working in a dark room, studying the glow produced in certain substances by cathode rays. To his surprise, he noted this glow on a chemically treated piece of paper some distance from the cathode ray tube. The paper even glowed when taken into the next room. Roentgen had discovered a new type of ray that could travel through walls. When he waved his hand between the radiation source and the glowing paper, he suddenly was able to see through the paper the bones of his own hand. He called these mysterious rays, which seemed to make his flesh disappear, **X-rays**.

Today X-rays are one of the most widely used tools in the world for medical diagnosis (Figure 3.8). Not only are they employed for examining decayed teeth, broken bones, and diseased lungs, but they are also the basis for such procedures as mammography and computerized tomography (Chapter 4). In the United States alone, payment for various radiological procedures totals more than $20 billion each year. How ironic that Roentgen himself made no profit at all from his discovery. He considered X-rays a "gift to humanity" and refused to patent any part of the discovery. He did, however, receive much popular acclaim and in 1901 was awarded the first Nobel prize in physics.

The Discovery of Radioactivity

Certain chemicals exhibit *fluorescence* after exposure to strong sunlight; they continue to glow even when taken into a dark room. In 1895 Antoine Henri Becquerel (1852–1908), a French physicist, was studying fluorescence by wrapping photographic film in black paper, placing a few crystals of the fluorescing chemical on top of the paper, and then placing the paper in strong sunlight. If the glow was like ordinary light, it would not pass through the paper. On the other hand, if it was similar to X-rays, it would pass through the black paper and fog the film.

Before Becquerel learned much more about fluorescence, he made an important accidental discovery when working with a uranium compound. When placed in sunlight (which caused the compound to fluoresce), the film was fogged. On several cloudy days when exposure to sunlight was not possible, he prepared samples and placed them in a drawer. To his great surprise, the photographic film was fogged even though the uranium compound had not been exposed to sunlight. Further experiments showed that the radiation coming from the uranium compound had nothing to do with fluorescence but was a characteristic of the element uranium.

Other scientists immediately began to study this new radiation. Becquerel had a graduate student from Poland, Marie Sklodowska, who gave it a name: radioactivity. **Radioactivity** is the spontaneous emission of radiation from certain unstable elements. Marie later married Pierre Curie, a French physicist. Together they discovered the radioactive elements polonium and radium, and with Becquerel they shared the 1903 Nobel prize in physics.

After her husband's death in 1906, Marie Curie continued to work with radioactive substances, winning the Nobel prize for chemistry in 1911. For more than 50 years she was the only person ever to have received two Nobel prizes.

3.3 Three Types of Radioactivity

Scientists soon showed that three types of radiation emanated from various radioactive elements. Ernest Rutherford (1871–1937), a New Zealander who spent his career in Canada and Great Britain, chose the names alpha, beta, and gamma for the three types of radiation. When passed through a strong magnetic or electric field, alpha rays were deflected in a manner indicating that they were beams of positive particles (Figure 3.9). Later experiments showed that an **alpha particle** has a mass

Serendipity is an aptitude for making fortunate discoveries by accident. Horace Walpole coined the term, alluding to the Persian fairy tale about *The Three Princes of Serendip*, who made many such discoveries.

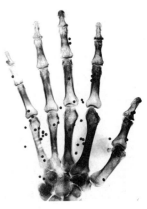

Figure 3.8 X-rays were used in medicine shortly after they were discovered by Wilhelm Roentgen (1845–1923) in 1895. Michael Purpin of Columbia University made this X-ray in 1896 to aid in the surgical removal of gunshot pellets (the dark spots) from the hand of a patient.

Marie Curie died in 1934 of leukemia, probably brought on by her long exposure to radioactive materials.

 [Web Reference 6] An interesting account of the life of Marie Curie, presented as a series of simulated articles. This site also has biographical information on other scientists.

Alpha particles are identical to helium ions (He^{2+}) or helium nuclei. They are helium atoms with both electrons removed.

Figure 3.9   Behavior of radioactive rays in an electric field.

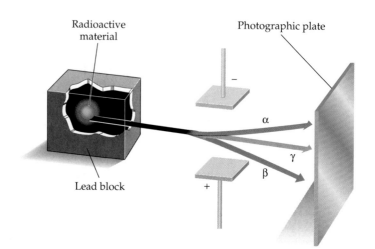

Figure 3.9 *(labels: Radioactive material, Photographic plate, Lead block, α, γ, β, − , +)*

Marie Sklodowska Curie in her laboratory and Marie and Pierre Curie on a French postage stamp.

Table 3.1 ▌ Types of Radioactivity			
Name	**Greek Letter**	**Mass (u)**	**Charge**
Alpha	α	4	2+
Beta	β	$\frac{1}{1837}$	1−
Gamma	γ	0	0

four times that of a hydrogen atom and a charge twice the magnitude of, but opposite in sign to, that of an electron.

Beta rays were shown to be made up of negatively charged particles identical to those of cathode rays. Therefore, **beta particles** are electrons.

Gamma rays are not deflected by a magnetic field. They are a form of energy, much like the X-rays used in medical work but even more penetrating. The three types of radioactivity are summarized in Table 3.1.

The discoveries of the late nineteenth century paved the way for an entirely new picture of the atom, which developed rapidly during the early years of the twentieth century.

3.4 Rutherford's Experiment: The Nuclear Model of the Atom

At Rutherford's suggestion, two of his coworkers, Hans Geiger (1882–1945) a German physicist, and Ernest Marsden (1889–1970), an English undergraduate student, bombarded very thin metal foils with alpha particles from a radioactive source (Figure 3.10).

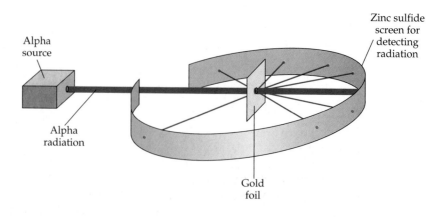

Figure 3.10   Rutherford's gold-foil experiment. Most alpha rays passed right through the gold foil, but now and then a ray was deflected.

(labels: Alpha source, Alpha radiation, Zinc sulfide screen for detecting radiation, Gold foil)

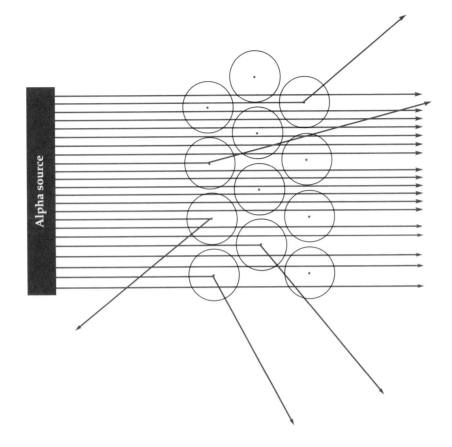

Figure 3.11 Model explaining the results of Rutherford's gold-foil experiment. Most of the alpha particles pass right through the foil because it is mainly empty space. But some alpha particles are deflected as they pass close to an atomic nucleus (which is very dense and positive in charge). Once in a while an alpha particle approaches an atomic nucleus head-on and is knocked back in the direction from which it came.

One target was a thin piece of gold foil. Most of the particles behaved as Rutherford expected, going right through the foil with little or no scattering. However, a few particles were deflected sharply, and occasionally one was sent right back in the direction from which it had come. Rutherford had assumed the positive charge to be spread smoothly over all the space occupied by the atom, but obviously it was not. To explain the experiment, Rutherford concluded that all the positive charge and nearly all the mass of an atom are concentrated at the center of the atom in a tiny core called the **nucleus**.

When an alpha particle, which is positively charged, approached the positively charged nucleus, it was strongly repelled and therefore sharply deflected (Figure 3.11). Since only a few alpha particles were deflected, Rutherford concluded that the nucleus must occupy only a tiny fraction of the volume of an atom. Most of the particles passed right through because most of an atom is empty space. The space outside the nucleus isn't completely empty, however. It is here that Rutherford placed the negatively charged electrons. He concluded that the electrons had so little mass that they were no match for the alpha bullets. It would be analogous to a mouse trying to stop the charge of a bull elephant.

Rutherford's nuclear theory of the atom, set forth in 1911, was revolutionary. He postulated that all the positive charge and nearly all the mass of an atom are concentrated in a tiny, tiny nucleus. The negatively charged electrons have almost no mass, yet they occupy nearly all the volume of an atom. To picture Rutherford's model, visualize a sphere as big as a giant indoor football stadium. The nucleus at the middle of the sphere is as small as a pea but weighs several million tons. A few flies flitting here and there throughout the sphere represent the electrons.

[Web Reference 7] An abstract of Rutherford's 1911 paper proposing the nuclear model of the atom.

[Web Reference 8] The following website presents the full text of a lecture that Rutherford gave in 1920. Although it is quite long and technical, it is worth a look.

3.5 The Structure of the Nucleus

In 1914 Rutherford suggested that the smallest positive-ray particle (the one formed when there is hydrogen gas in the Goldstein apparatus—see Section 3.1) is the unit of positive charge in the nucleus. This particle, called a **proton**, has a charge equal in magnitude to that of the electron and has nearly the same mass as a hydrogen atom. Rutherford's suggestion was that protons constitute the positively charged matter in all atoms. The nucleus of a hydrogen atom consists of one proton, and the nuclei of larger atoms contain greater numbers of protons.

Except for hydrogen atoms, though, atomic nuclei were found to be heavier than would be indicated by the number of positive charges (number of protons). For example, the helium nucleus was found to have a charge of 2+ (and therefore it contained two protons, according to Rutherford's theory), but its mass was four times that of hydrogen. This excess mass puzzled scientists at first. But in 1932 an English physicist, James Chadwick, discovered a particle with about the same mass as a proton but with no electric charge. It was called a **neutron**, and its existence made possible an explanation of the unexpectedly high mass of the helium nucleus. Whereas the hydrogen nucleus contained only one proton of mass 1 u, the helium nucleus contained not only two protons (2 u) but also two neutrons (2 u), giving the nucleus a total mass of 4 u.

With the discovery of the neutron, the list of "building blocks" we will need for "constructing" atoms is complete. The properties of these particles are summarized in Table 3.2.

The number of protons in the nucleus of an atom of any element is equal to the **atomic number** of that element. This number determines the kind of atom—that is, the identity of the element. Dalton had said that the mass of an atom determines the element. We now know it is not the mass but the number of protons that determines the identity of an element. For example, an atom with 26 protons (one whose atomic number is 26) is an atom of iron (Fe). An atom with 50 protons is an atom of tin (Sn).

In a neutral (nonionized) atom the positive charge of the protons is exactly neutralized by the negative charge of the electrons. The attractive forces between the unlike charges play an important role in holding the atom together.

A proton and a neutron have virtually the same mass, 1.0073 u and 1.0087 u, respectively. This is equivalent to saying that two different people weigh 100.7 kg and 100.9 kg. The difference is so small that it usually can be ignored. Thus, for many purposes, we assume the masses of the proton and the neutron to be the same, 1 u. The proton has a charge equal in magnitude but opposite in sign to that of an electron. This charge on a proton is written as 1+. The electron has a charge of 1− and a mass of 0.00055 u. The electrons in an atom contribute so little to its total mass that their mass is usually disregarded and treated as if it were 0.

> An *element* is a substance in which all the atoms have the same atomic number. The **atomic number** of an element is the number of protons in each of its atoms. For neutral atoms (those without an electric charge), the atomic number is also the number of electrons in each atom.

Isotopes

The number of neutrons in the nuclei of atoms of a given element may vary. For example, most hydrogen (H) atoms have a nucleus consisting of a single proton and no neutrons (and therefore a mass of 1 u.) About 1 hydrogen atom in 5000, however,

Table 3.2 ▌ Subatomic Particles

Particle	Symbol	Mass (u)	Charge	Location in Atom
Proton	p^+	1	1+	Nucleus
Neutron	n	1	0	Nucleus
Electron	e^-	$\frac{1}{1837}$	1−	Outside nucleus

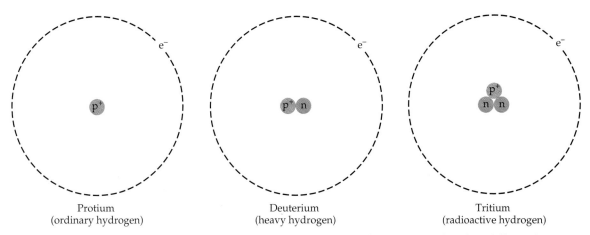

| Protium (ordinary hydrogen) | Deuterium (heavy hydrogen) | Tritium (radioactive hydrogen) |

Figure 3.12 The three isotopes of hydrogen. Each has one proton and one electron, but they differ in the number of neutrons in the nucleus.

does have a neutron as well as a proton in the nucleus. This heavier hydrogen atom is called **deuterium**, and it has a mass of 2 u. Both kinds are hydrogen atoms (any atom with atomic number 1—that is, with one proton—is a hydrogen atom). Atoms that have this sort of relationship—the same number of protons but different numbers of neutrons—are called **isotopes** (Figure 3.12). A third, rare isotope of hydrogen is **tritium**, which has two neutrons and one proton in the nucleus (and thus a mass of 3 u).

Most, but not all, elements exist in nature in isotopic forms. For example, tin (Sn) is present in nature in ten different isotopic forms. It also has 15 radioactive isotopes that do not occur in nature. This fact also requires a modification of Dalton's original theory. He said that all atoms of the same element are alike. We now say that all atoms of the same element have the same number of protons. Different isotopes of the same element have atoms with the same number of protons but with different numbers of neutrons (and therefore different masses).

Isotopes usually are of little importance in ordinary chemical reactions. All three hydrogen isotopes react with oxygen to form water. Since the isotopes differ in mass, compounds formed with different hydrogen isotopes have different physical properties, but such differences are usually slight. In nuclear reactions, however, isotopes are of utmost importance, as we shall see in the next chapter.

> Water in which both hydrogen atoms are deuterium is called *heavy water*, often written D_2O. Heavy water boils at 101.4°C and freezes at 3.8°C. Its density is 1.108 g/cm^3. (The density of ordinary water is 1.000 g/cm^3.)

Symbols for Isotopes

Collectively, the two principal nuclear particles, protons and neutrons, are called **nucleons**. Isotopes are represented by symbols with subscripts and superscripts. In the general symbol

$$^A_Z X$$

Z is the nuclear charge, or the atomic number (the number of protons), and A is the **mass number**, or the **nucleon number** (the number of protons plus the number of neutrons). As an example, the isotope with the symbol

$$^{35}_{17}Cl$$

has 17 protons and 35 nucleons. The number of neutrons is therefore $35 - 17 = 18$.

Isotopes also are identified by placing the nucleon number as a suffix to the name of the element. The three hydrogen isotopes are therefore represented as

$$^1_1H \quad ^2_1H \quad ^3_1H$$

or as hydrogen-1, hydrogen-2, and hydrogen-3.

> We will refer to A as the nucleon number in order to stress the fact that the mass number of an atom is equal to the total number of nucleons.

Example 3.1

How many neutrons are there in the $^{235}_{92}$U nucleus?

Solution

Simply subtract the atomic number (number of protons) from the nucleon number (number of protons plus neutrons).

$$\text{Nucleon number} - \text{atomic number} = \text{number of neutrons}$$

$$235 - 92 = 143$$

There are 143 neutrons in the nucleus.

Exercise 3.1

How many neutrons are there in the $^{90}_{38}$Sr nucleus?

Example 3.2

How many neutrons are there in the bromine-81 nucleus?

Solution

The atomic number of bromine is 35. The nucleon number is given as 81. The number of neutrons is therefore

$$81 - 35 = 46$$

Exercise 3.2

How many neutrons are there in the molybdenum-90 nucleus?

Example 3.3

(a) Which of the following are isotopes of the same element? (We are using the letter X as the symbol for all elements so that the symbol will not identify the elements.) **(b)** Which of the five isotopes have the same number of neutrons?

$$^{16}_{8}X \qquad ^{16}_{7}X \qquad ^{14}_{7}X \qquad ^{14}_{6}X \qquad ^{12}_{6}X$$

Solution

a. $^{16}_{7}X$ and $^{14}_{7}X$ are isotopes of nitrogen (N). $^{14}_{6}X$ and $^{12}_{6}X$ are isotopes of carbon (C). $^{16}_{8}X$ and $^{16}_{7}X$ have the same nucleon number. The first is an isotope of oxygen, and the second an isotope of nitrogen. $^{14}_{7}X$ and $^{14}_{6}X$ have the same nucleon number. The first is an isotope of nitrogen, and the second an isotope of carbon.
b. $^{16}_{8}X$ and $^{14}_{6}X$ each have eight neutrons ($16 - 8 = 8$ and $14 - 6 = 8$, respectively).

Exercise 3.3

Which of the following are isotopes of the same element?

$$^{90}_{37}X \qquad ^{90}_{35}X \qquad ^{88}_{37}X \qquad ^{88}_{38}X \qquad ^{93}_{38}X$$

Although protons, neutrons, and electrons are the primary components of an atom, there are dozens of other subatomic particles that have been shown to exist within the atom, at least on a temporary basis.

 [Web Reference 9] The story of nuclear structure does not stop with the nucleons. There are quarks and leptons and .... To get an idea of this complexity, look at The Particle Adventure Home Page, which manages to be entertaining as well as informative.

3.6 Electron Arrangement: The Bohr Model

Let us now turn our attention once more to electrons. Rutherford demonstrated that atoms have a tiny, positively charged nucleus with electrons outside the nucleus. Evidence soon accumulated that the electrons were not randomly distributed but were arranged in a quite structured fashion. We will examine the evidence soon. But first let's take a side trip into some colorful chemistry and physics to provide a background for our study of the electron structure of atoms.

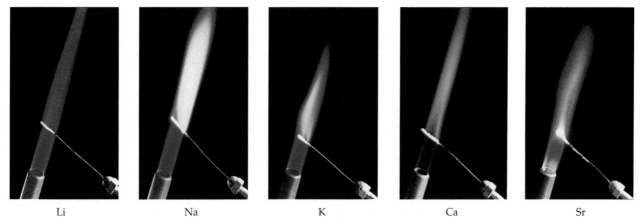

| Li | Na | K | Ca | Sr |

Figure 3.13   Certain chemical elements can be identified by the characteristic colors their compounds impart to flames. Five examples are shown here.

Fireworks and Flame Tests

Chemists of the eighteenth and nineteenth centuries developed flame tests that used the colors of flames to identify several elements (Figure 3.13). Sodium salts give a persistent yellow flame, potassium salts a fleeting lavender flame, and lithium salts a brilliant red flame. Like those of fireworks, these flame colors result from the electron structures of atoms of the specific elements.

Fireworks originated earlier than flame tests, in ancient China. The brilliant colors of aerial displays still mark our celebrations of patriotic holidays (Figure 3.14). The colors of fireworks are attributable to specific elements. Brilliant reds are produced by strontium compounds, whereas barium compounds give yellow-green, sodium compounds yield yellow, and copper salts produce a greenish blue.

The colors of fireworks and flame tests are not what they seem to the unaided eye. If the light from the flame is passed through a prism, it is separated into light of several different colors.

Figure 3.14   Flame colors (see Figure 3.13) also are the basis for the brilliant colors of fireworks displays. Strontium compounds produce red, copper compounds produce blue, and sodium compounds produce yellow.

Continuous and Line Spectra

When white light from an incandescent lamp is passed through a prism, it produces a continuous spectrum, or rainbow of colors (Figure 3.15). When sunlight passes through a raindrop, the same thing occurs (Figure 3.16). The different colors of light

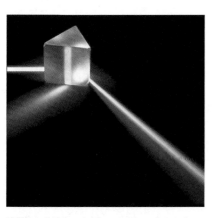

Figure 3.15   A glass prism separates white light into a continuous spectrum or rainbow of colors.

Figure 3.16   A rainbow is an example of a continuous spectrum.

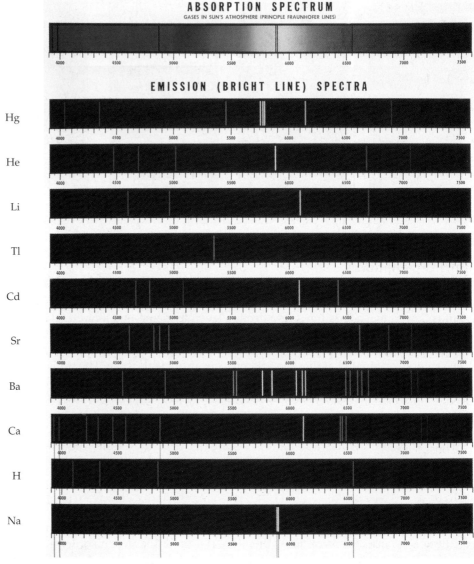

ABSORPTION SPECTRUM
GASES IN SUN'S ATMOSPHERE (PRINCIPLE FRAUNHOFER LINES)

EMISSION (BRIGHT LINE) SPECTRA

Hg

He

Li

Tl

Cd

Sr

Ba

Ca

H

Na

Figure 3.17 Line spectra of selected elements are shown here. Some of the components of the light emitted by excited atoms appear as colored lines. A continuous spectrum is shown at the top for comparison. The numbers are wavelengths of light given in Angstrom units (1 Å = 10^{-10} m). Each element has its own characteristic line spectrum that is different from all others. In addition to their practical use in analyzing matter, atomic spectra are the basis for many of the concepts of atomic structure.

represent different wavelengths. Blue light has shorter wavelengths than red light, but there is no sharp transition as one moves from one color to the next. All wavelengths are present in a continuous spectrum. White light is simply a combination of all the various colors.

If the light from a gas discharge tube containing a particular element is passed through a prism, only narrow colored lines are observed (Figure 3.17). Each line corresponds to light of a particular wavelength. The pattern of lines emitted by an element is called its *line spectrum*. The line spectrum of an element is characteristic of that element and can be used to identify it. Not all the lines in the spectrum of an atom are visible; some lines appear as infrared or ultraviolet radiation.

The line spectrum of hydrogen is fairly simple; it consists of four lines in the visible portion of the electromagnetic spectrum. It was to explain this spectrum that Niels Bohr worked out his model for the electron structure of the hydrogen atom.

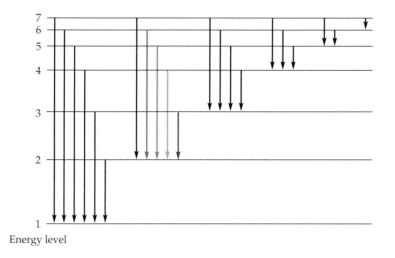

Figure 3.18 Possible electron shifts between energy levels in atoms to produce the lines found in spectra. Not all the lines are in the visible portion of the spectrum. The four colored lines correspond to the colored lines in the hydrogen spectrum.

Bohr's Explanation of Line Spectra

Niels Bohr presented his explanation of line spectra in 1913. He suggested that electrons cannot have just any amount of energy but can have only certain specified amounts; that is to say, the energy of an electron is quantized. The specified energy values for an electron are called its **energy levels**.

An electron, by absorbing a quantum of energy (for example, when atoms of the element are heated), is elevated to a higher energy level (Figure 3.18). By giving up a quantum of energy, the electron can return to a lower energy level. The energy released or absorbed in these transitions shows up as a line spectrum. Each line has a specific wavelength corresponding to a **quantum** of energy. An electron moves instantaneously from one energy level to another, and there are no intermediate stages.

Consider as an analogy a person on a ladder. The person can stand on the first rung, the second rung, the third rung, and so on, but is unable to stand between rungs. As the person goes from one rung to another, the potential energy (energy due to position) changes by definite amounts or quanta. As an electron moves from one energy level to another, its total energy (both potential and kinetic) changes.

Bohr's model of the atom was based on the laws of planetary motion that had been set down by the German astronomer Johannes Kepler (1571–1630) three centuries before. Bohr imagined the electrons to be orbiting about the nucleus much as planets orbit the sun (Figure 3.19). Different energy levels were pictured as different orbits. The modern picture is different, as we shall see in subsequent sections.

Ground States and Excited States

The electron in a hydrogen atom is usually in the first energy level. Given the choice, electrons usually remain in their lowest possible energy levels (those nearest the nucleus); atoms whose electrons are in this situation are said to be in their **ground state**. When a flame or other source supplies energy to an atom (hydrogen, for example) and an electron jumps from the lowest possible level to a higher level, the atom is said to be in an **excited state**. An atom in an excited state eventually emits a photon of energy as the electron jumps back down to one of the lower levels and ultimately reaches the ground state.

Bohr's theory was a spectacular success in explaining the line spectrum of hydrogen. It established the important idea of energy levels in atoms. Bohr was awarded the Nobel prize in physics in 1922 for this work.

Atoms larger than hydrogen have more than one electron, and Bohr was also able to deduce that the various energy levels of an atom could handle only a certain

A **quantum** is a tiny unit of energy, the value of which depends on the frequency.

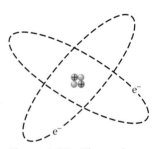

Figure 3.19 The nuclear atom, as envisioned by Bohr, has most of its mass in an extremely small nucleus. Electrons orbit about the nucleus, occupying most of the volume of the atom but contributing little to its mass.

[Web Reference 10] An interactive explanation of spectral lines in relation to Bohr's theory.

Niels Bohr in 1922 received the Nobel prize in physics for his planetary model of the atom with its quantized electron energy levels.

 [Web Reference 11] An interesting site called Hands-on Hydrogen Atom.

A **photon** is a quantum or "particle" of light.

number of electrons at one time. We shall simply state Bohr's findings in this regard. The maximum number of electrons that can be in a given level is indicated by the formula $2n^2$, where n is equal to the energy level being considered. For the first energy level ($n = 1$), the maximum population is 2×1^2, or 2. For the second energy level ($n = 2$), the maximum number of electrons is 2×2^2, or 8. For the third level, the maximum is 2×3^2, or 18. (However, the outermost level usually has no more than 8 electrons.)

Example 3.4

What is the maximum number of electrons in the fifth energy level?
Solution
For the fifth level, $n = 5$, and so we have

$$2 \times 5^2 = 2 \times 25 = 50$$

Exercise 3.4
What is the maximum number of electrons in the fourth energy level?

Building Atoms

Imagine building up atoms by adding one electron to the proper energy level as *each* proton is added to the nucleus, keeping in mind that electrons will go to the lowest energy level available. For hydrogen (H), with a nucleus of only one proton, the one electron goes into the first energy level. For helium (He), with a nucleus of two protons (and two neutrons), both electrons go into the first energy level. According to Bohr, two electrons is the maximum population of the first energy level; that level is filled in the helium atom.

With lithium (Li), which has three electrons, two electrons go into the first level; the other must go into the second energy level. This process of adding electrons is continued until the second energy level is filled with eight electrons, as in a neon (Ne) atom, which has two of its 10 electrons in the first energy level and the remaining eight in the second energy level.

A sodium (Na) atom has 11 electrons. Two are in the first energy level, the second level is filled with 8 electrons, and the remaining electron is in the third energy level. We can use a modified Bohr diagram to indicate this **electron configuration** (or arrangement).

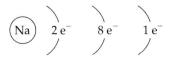

The circle with the symbol indicates the sodium nucleus. The arcs represent the energy levels, the one closest to the nucleus being the first energy level, the next the second, and so on.

We could now continue to add electrons to the third energy level until we get to argon (Ar).

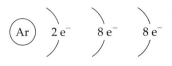

Even though the third energy level can have up to 18 electrons, it is temporarily filled with only 8. The next element, potassium (K), has the following electron configuration.

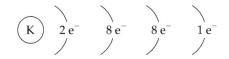

Sometimes the Bohr configuration is written without the arcs to indicate energy levels. The following designations illustrate this method.

Na	2, 8, 1
Ar	2, 8, 8
K	2, 8, 8, 1
F	2, 7

Similarly, calcium (Ca) has 2 electrons in the fourth energy level.

After calcium, things get a bit more complicated. The next 10 electrons resume filling the third energy level. The topic of atomic structure is discussed in greater detail in Section 3.8. Meanwhile, let us try to draw a few Bohr diagrams.

Example 3.5

Draw Bohr diagrams for fluorine (F) and aluminum (Al).

Solution

Fluorine is element number 9; it has 9 electrons. Two of these electrons go into the first energy level, and the remaining seven go into the second level.

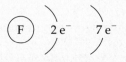

Aluminum is element number 13; it has 13 eletrons. Two go into the first energy level, 8 go into the second, and the remaining three go into the third energy level.

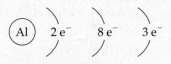

Notice that fluorine is in Group 7A of the periodic table and that it has 7 outer shell electrons. Aluminum is in Group 3A, and it has 3 electrons in its outermost shell.

Exercise 3.5

Draw Bohr diagrams for beryllium (Be), magnesium (Mg), and calcium (Ca). These elements are all in the same group of the periodic table. What do you notice about the number of electrons in their outermost shells?

3.7 The Quantum Mechanical Atom

The simple planetary Bohr model of the atom has been replaced for many purposes by more sophisticated models in which electrons are treated as waves and their locations are indicated as probabilities. The theory that the electron should have

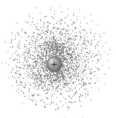

An *s* orbital

A *p* orbital

Figure 3.20 Charge-cloud representations of atomic orbitals.

Fortunately, we need not understand the elaborate mathematics used by Schrödinger in order to make use of some of his results.

A **sublevel** (or *subshell*) is a subdivision of an electron energy level. For example, the third main energy level in an atom has three sublevels—the 3*s*, 3*p*, and 3*d*.

wavelike properties was first suggested (in 1924) by Louis de Broglie, a young French physicist. Although it was hard to accept because of Thomson's proof that electrons are particles, de Broglie's theory was experimentally verified within a few years.

Erwin Schrödinger, an Austrian physicist, used highly mathematical quantum mechanics in the 1920s to develop equations that describe the properties of electrons in atoms. The solutions to these equations measure the probability of finding an electron in a given volume of space. These shaped volumes of space, called **orbitals**, replace the planetary orbits of the Bohr model.

Suppose you had a camera that could photograph electrons and you left the shutter open while an electron zipped about the nucleus. The developed picture would give a record of where the electron had been. (Doing the same thing with an electric fan would give a blurred image of the rapidly moving blades, a picture resembling a disk.) The electrons in the first energy level would appear as a fuzzy ball (often referred to as a *charge cloud* or an *electron cloud* (Figure 3.20).

Building Atoms by Orbital Filling

Schrödinger concluded that each electron orbital could contain a maximum of two electrons; thus a given energy level could contain more than one orbital. The first energy level contains a single spherical orbital named 1*s*. The second contains four orbitals: 2*s* is spherical, and the other three orbitals, called 2*p*, are dumbbell-shaped (Figures 3.20 and 3.21). The third energy level contains nine orbitals: a spherical 3*s*, three dumbbell-shaped 3*p*, and five 3*d* orbitals (resembling four-leaf clovers).

In building up the electron configuration of atoms of the various elements, the lower **sublevels** are filled first.

Hydrogen (H), with only one electron in the *s* orbital of the first energy level, has an electron configuration of $1s^1$; that of helium (He) is $1s^2$. Lithium (Li) has three electrons—two in the first energy level and one in the *s* orbital of the second energy level: $1s^2 2s^1$.

Let's look at argon (Ar), which has 18 electrons. Its configuration is $1s^2 2s^2 2p^6 3s^2 3p^6$. Note that the highest occupied energy level, 3*p*, is filled. When we move to potassium (K) with 19 electrons, we find that the 4*s* sublevel fills before the 3*d* sublevel. (The order of filling the various electron sublevels is shown in Figure 3.22.)

Figure 3.21 Electron orbitals. In these drawings, the nucleus of the atom is located at the intersection of the axes. The eight electrons that would be placed in the second energy level of Bohr's model are distributed among these four orbitals in the current model of the atom, with two electrons per orbital.

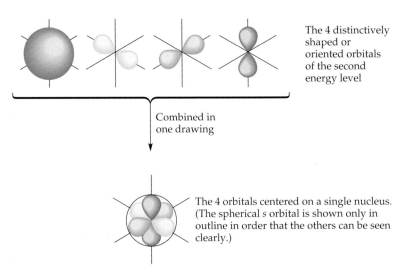

The 4 distinctively shaped or oriented orbitals of the second energy level

Combined in one drawing

The 4 orbitals centered on a single nucleus. (The spherical *s* orbital is shown only in outline in order that the others can be seen clearly.)

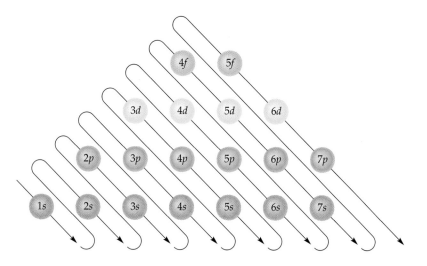

Figure 3.22 An order-of-filling chart for determining the electron configurations of atoms.

Example 3.6

Write out the electron configuration for nitrogen (N) using sublevel notation.

Solution

Nitrogen has seven electrons. Place them in the lowest unfilled energy sublevels. Two go into the 1s orbital and two into the 2s orbital. That leaves three electrons to be placed in the p sublevel. The electron configuration is $1s^2 2s^2 2p^3$.

Exercise 3.6

Write out the electron configuration for fluorine (F).

Example 3.7

Write out the electron configuration for sulfur (S).

Solution

Sulfur atoms have 16 electrons each. The electron configuration is $1s^2 2s^2 2p^6 3s^2 3p^4$. Note that the total of the superscripts is 16 and that we have not exceeded the maximum capacity for any sublevel.

Exercise 3.7

Write out the electron configuration for chlorine (Cl).

3.8 Electron Configurations and the Periodic Table

In general, the properties of all elements can be correlated with their electron configurations. (We explore bond formation using electron configurations in Chapter 5.) Because the number of electrons equals the number of protons, the periodic table tells us about electron configuration as well as atomic number.

The modern periodic table (inside front cover) has vertical columns called **groups** or (sometimes) families. The horizontal rows of the periodic table are called **periods**. Elements in a group have similar chemical properties. The properties of elements vary periodically across a period.

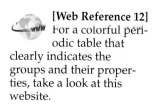

[Web Reference 12] For a colorful periodic table that clearly indicates the groups and their properties, take a look at this website.

Table 3.3 ▮ Electron Structures for Atoms of the First 20 Elements

Name	Atomic Number	Electron Structure
Hydrogen	1	$1s^1$
Helium	2	$1s^2$
Lithium	3	$1s^2 2s^1$
Beryllium	4	$1s^2 2s^2$
Boron	5	$1s^2 2s^2 2p^1$
Carbon	6	$1s^2 2s^2 2p^2$
Nitrogen	7	$1s^2 2s^2 2p^3$
Oxygen	8	$1s^2 2s^2 2p^4$
Fluorine	9	$1s^2 2s^2 2p^5$
Neon	10	$1s^2 2s^2 2p^6$
Sodium	11	$1s^2 2s^2 2p^6 3s^1$
Magnesium	12	$1s^2 2s^2 2p^6 3s^2$
Aluminum	13	$1s^2 2s^2 2p^6 3s^2 3p^1$
Silicon	14	$1s^2 2s^2 2p^6 3s^2 3p^2$
Phosphorus	15	$1s^2 2s^2 2p^6 3s^2 3p^3$
Sulfur	16	$1s^2 2s^2 2p^6 3s^2 3p^4$
Chlorine	17	$1s^2 2s^2 2p^6 3s^2 3p^5$
Argon	18	$1s^2 2s^2 2p^6 3s^2 3p^6$
Potassium	19	$1s^2 2s^2 2p^6 3s^2 3p^6 4s^1$
Calcium	20	$1s^2 2s^2 2p^6 3s^2 3p^6 4s^2$

In the United States, the groups are often indicated by a numeral followed by the letter A **(main group elements)** or B **(transition elements)**. The International Union of Pure and Applied Chemistry (IUPAC) recommends numbering the groups from 1 to 18. Both systems are indicated on the periodic table on the inside front cover, but we follow the traditional U.S. method in this book.

Family Features: Outer Electron Configurations

The period in which an element appears in the periodic table tells us how many main electron energy levels there are in that atom. Sulfur, for example, is in the third period, and so the sulfur atom has three main energy levels. The group number (for main group elements) tells us how many electrons are in the outermost energy level. These are called **valence electrons**. Because sulfur is in Group 6A, it has six valence electrons. Two of these are in an s orbital, and the other four are in p orbitals. We can indicate the outer electron configuration of the sulfur atom as follows: $3s^2 3p^4$.

It is these outermost electrons that mainly determine the chemistry of an atom. Since all the elements in the same group of the periodic table have the same number of valence electrons, they should have very similar chemistry, and they do. Figure 3.23 relates the sublevel configurations to the groups in the periodic table.

Family Groups

Elements within a group have similar properties. All the elements in Group 1A have one valence electron, and all are very active metals (except for hydrogen, which is a unique nonmetal). All have the outer electron structure ns^1, where n denotes the number of the outermost main energy level. Often called **alkali metals**, they react vigorously with water to evolve hydrogen gas. There are important trends within a family. For example, lithium is the hardest metal in the group. Sodium is softer than lithium; potassium is softer still, and so on down the group. Lithium is also the least reactive toward water. Sodium, potassium, rubidium, and cesium are progressively more reactive. (Francium is highly radioactive and extremely rare; few of its properties have been measured.) Hydrogen is the odd one in Group 1A. It is not an alkali

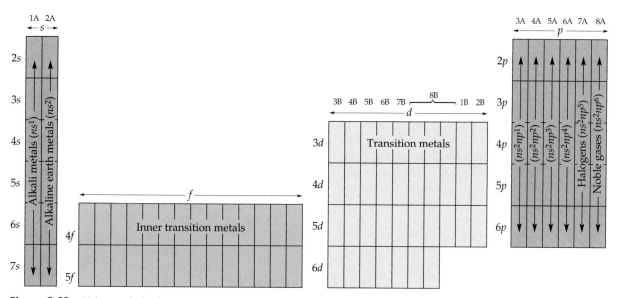

Figure 3.23   Valence shell electron configurations and the periodic table.

metal. Indeed, it is a rather characteristic nonmetal. As far as its properties are concerned, hydrogen probably should be put in a group of its own.

Group 2A elements are sometimes known as **alkaline earth metals**. The metals in this group are fairly soft and moderately reactive with water. Beryllium is an odd member of the group in that it is rather hard and does not react with water. There are trends in properties within the group—as there are in the other families. For example, magnesium, calcium, strontium, barium, and radium are progressively more reactive toward water.

Group 7A elements, often called **halogens**, also consist of reactive elements. All have seven valence electrons in the configuration ns^2np^5. Halogens react vigorously with alkali metals to form crystalline solids (this is discussed further in Chapter 5). There are trends in the halogen family. Fluorine is most reactive toward alkali metals, chlorine is the next most reactive, and so on. Fluorine and chlorine are gases at room temperature, bromine is a liquid, and iodine is a solid. (Astatine, like francium, is highly radioactive and extremely rare; few of its properties have been determined.)

Members of Group 8A, to the far right of the periodic table, have a complete set of valence electrons and therefore undergo few, if any, chemical reactions. They are called **noble gases**, known for their lack of chemical reactivity.

Example 3.8

Write out the sublevel notation for the electrons in the highest main energy level for strontium (Sr) and arsenic (As).

Solution

Strontium is in Group 2A and the fifth period of the periodic chart. Its outer electron configuration is $5s^2$. (Each period of the periodic table corresponds to the filling, or at least partial filling, of a main energy level.) Arsenic is in Group 5A and the fourth period. Its five outer (valence) electrons have the configuration denoted by $4s^24p^3$.

Exercise 3.8

Write out the sublevel notation for the electrons in the highest main energy level for rubidium (Rb) and selenium (Se).

Metals, Nonmetals, and Metalloids

Elements in the periodic table also are divided into two classes by a heavy, stepped diagonal line. Those to the left of the line are **metals**, elements that have a characteristic luster and generally are good conductors of heat and electricity. Except for mercury, which is a liquid, all metals are solids at room temperature. Metals generally are malleable; that is, they can be hammered into thin sheets. Most also are ductile; they can be drawn into wires.

Elements to the right of the stepped line are **nonmetals**. These elements lack metallic properties. Several are gases (oxygen, nitrogen, fluorine, chlorine). Others are solids (carbon, sulfur, phosphorus, iodine). Bromine is the only nonmetal that is a liquid at room temperature.

Some of the elements bordering the stepped line are **metalloids**, elements that have intermediate properties. Metalloids have properties that resemble those of both metals and nonmetals.

3.9 Which Model to Choose?

For some purposes in this text, Bohr diagrams of atoms are used to picture the distribution of electrons among the main energy levels. At other times, the electron clouds of the quantum mechanical model are more useful. Even Dalton's model sometimes proves to be the best way to describe certain phenomena (the behavior of gases, for example). The choice of model is always based on which one is most helpful in understanding a particular concept. This, after all, is the whole purpose of scientific models.

Critical Thinking Exercises

Apply knowledge that you have gained in this chapter and one or more of the FLaReS principles (Chapter 1) to evaluate the following statements or claims.

3.1 Suppose you read in the newspaper that a chemist in South America claims to have discovered a new element with an atomic mass of 34. An extremely rare element, it was found in a sample recovered from the Andes Mountains. Unfortunately, he has used up all of the sample in his analyses.

3.2 Some aboriginal tribes have rain-making ceremonies in which they toss pebbles of gypsum up into the air. (Gypsum is the material used to make plaster of Paris by heating the rock to remove some of its water.) Sometimes it really does rain several days after these rain-making ceremonies.

Summary

1. The fact that matter is electrical in nature was established during the eighteenth century by Davy, Faraday, and others when they split compounds into their component elements by electrolysis.

2. Experiments with cathode ray tubes showed that matter contained negatively charged particles, which were called electrons. By deflecting cathode rays with a magnet, Thomson was able to determine the mass-to-charge ratio for the electron. Later, Millikan's oil drop experiment measured the charge on the electron, so its mass could then be calculated.

3. Goldstein's experiment, using gas discharge tubes with perforated cathodes, showed that matter also contained positively charged particles, but different gases produced particles with different masses. The positive particles of least mass were formed when the gas in the tube was hydrogen. These smallest positive particles were later called protons.

4. Other important chemical developments of the late nineteenth century included the discovery of X-rays by Roentgen and the discovery of radioactivity by Becquerel.

5. Rutherford's gold-foil experiment indicated that an atom must have a tiny, very dense nucleus, with electrons occupying most of the atom's space. It was not until 1932 that Chadwick discovered the neutron, a particle in the nucleus as massive as a proton but bearing no charge.

6. The number of protons determines the positive charge on a particular nucleus, and this is known as the atomic number of that element. When atoms have the same atomic number but differ in mass, they are called isotopes. Isotopes of the same element differ only in number of neutrons.

7. A study of the spectral lines of hydrogen led Bohr to propose an atom with concentric shells of electrons surrounding the positively charged nucleus. Each shell represents an electron energy level. The farther a shell is from the nucleus, the higher the electron energy level and the greater the electron capacity. The first four energy levels have capacities of 2, 8, 18, and 32 electrons, respectively.

8. Although the simple Bohr model is very useful, our modern view of the atom is much more complicated. Electrons act as waves as well as particles, and they can be investigated by the mathematical methods of quantum mechanics. Bohr's energy levels can be further split into sublevels that differ slightly in energy because of the differing shapes of their electron orbitals. (An orbital is an electron cloud that can hold one pair of electrons.)

9. The shape of *s* orbitals is spherical, *p* orbitals are dumbbell-shaped, and *d* orbitals are shaped like four-leaf clovers. The first main energy level can hold only one *s* orbital; the second level can hold one *s* and three *p* orbitals; and the third level can hold one *s*, three *p*, and five *d* orbitals.

10. The periodic table is made up of horizontal rows called periods and vertical rows called groups (or families). A stepped diagonal line near the right end of the table (from B to A) divides the metals on the left from the nonmetals on the right. Most of the elements are metals. Some elements along the dividing line are metalloids.

11. An element's location in the periodic table can tell you a lot about its atoms. The period number tells you the number of electron energy levels in an atom. The group number (for main group elements) tells you how many valence electrons the atom has.

12. The first group in the periodic table is made up of alkali metals (most active of all metals), and the second group contains the alkaline earth metals. The last group is made up of the noble gases (least reactive of all the elements), and the next to last group (the most reactive group of nonmetals) are known as the halogens. The ten short groups across the middle of the table are the transition metals.

Key Terms

alkali metals 3.8
alkaline earth metals 3.8
alpha particle 3.3
anion 3.1
anode 3.1
atomic number 3.5
beta particles 3.3
cathode 3.1
cathode ray 3.1
cation 3.1
deuterium 3.5
electrodes 3.1

electrolysis 3.1
electrolyte 3.1
electron 3.1
electron configuration 3.6
energy levels 3.6
excited state 3.6
gamma rays 3.3
ground state 3.6
groups 3.8
halogens 3.8
ions 3.1
isotopes 3.5

main group elements 3.8
mass number 3.5
metalloids 3.8
metals 3.8
neutron 3.5
noble gases 3.8
nonmetals 3.8
nucleus 3.4
nucleon 3.5
nucleon number 3.5
orbital 3.7
periods 3.8

photon 3.6
proton 3.5
quantum 3.6
radioactivity 3.2
sublevel 3.7
transition elements 3.8
tritium 3.5
valence electrons 3.8
X-rays 3.2

Review Questions

1. What did each of the following scientists contribute to our knowledge of the atom?
   a. Crookes   b. Thomson
   c. Goldstein   d. Millikan
   e. Roentgen   f. Becquerel

2. What evidence is there that electrons are particles?

3. What is radioactivity? How did the discovery of radioactivity contradict Dalton's atomic theory?

4. Define or identify each of the following.
   a. alpha particle   b. beta particle
   c. gamma ray   d. deuterium
   e. tritium

5. What are isotopes?

6. How are X-rays and gamma rays similar? How are they different?

7. The table below describes four atoms.

	Atom A	Atom B	Atom C	Atom D
Number of protons	10	11	11	10
Number of neutrons	11	10	11	10
Number of electrons	10	11	11	10

Are atoms A and B isotopes? A and C? A and D? B and C?

8. What are the masses of the atoms in Question 7?

9. Discuss Rutherford's gold-foil experiment. What did it tell us about the structure of the atom?

10. What is the atomic nucleus?

11. Give the distinguishing characteristics of the proton, the neutron, and the electron.

12. Should the proton and electron attract or repel one another?

13. Should the neutron and proton attract or repel one another?

14. Which subatomic particles are found in the nuclei of atoms?

15. In an atom, what are the extranuclear (outside the nucleus) subatomic particles?

16. Compare Dalton's model of the atom with the nuclear model of the atom.

17. If the nucleus of an atom contains ten protons, how many electrons are there in the neutral atom?

18. What is the symbol of the element with atomic number 98?

19. What is the name of the element that has 18 protons in the nucleus of its atoms?

20. Explain what is meant by the term "atomic weight."

21. What is meant by the nucleon number of an isotope?

22. Give the nuclear symbols for protium, deuterium, and tritium (which are hydrogen-1, hydrogen-2, and hydrogen-3, respectively).

23. How did Bohr refine the nuclear model of the atom?

24. What particles travel in the orbits of the Bohr model of the atom?

25. According to Bohr, what is the maximum number of electrons in the fourth energy level ($n = 4$)?

26. If the third energy level of an atom contains two electrons, what is the total number of electrons in the atom?

27. Define the following terms.
    a. ground state
    b. excited state

28. When light is emitted by an atom, what change has occurred within the atom?

29. When an electron moves from the fourth energy level to the second, is energy being emitted or absorbed?

30. What is the electron capacity of the 3d sublevel in an atom?

31. What is the electron capacity of the 4p sublevel in an atom?

32. If a neutral atom in its ground state contains only five electrons in its outermost p sublevel, it is an atom of what group of elements?

33. If a neutral atom in its ground state has a d sublevel that contains five electrons, to what group of elements does the atom belong?

34. If an atom contains only s electrons in its outermost energy level, is the element a metal or a nonmetal?

35. Which atom absorbs more energy, one in which an electron moves from the second energy level to the third level or an otherwise identical atom in which an electron moves from the first to the third energy level?

36. Use the periodic table to determine the numbers of protons in atoms of the following elements.
    a. helium (He)
    b. sodium (Na)
    c. chlorine (Cl)
    d. oxygen (O)
    e. magnesium (Mg)
    f. sulfur (S)

37. How many electrons are there in the neutral atoms of the elements listed in Question 36?

38. Draw Bohr diagrams for the elements listed in Question 36.

39. The following Bohr diagram is supposed to represent the neutral atoms of an element. The diagram is incorrectly drawn. Identify the error.

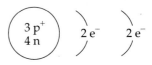

40. Elements are defined on a theoretical basis as being composed of atoms that share the same atomic number. On the basis of this theory, would you think it possible for someone to discover a new element that would fit between magnesium (atomic number 12) and aluminum (atomic number 13)?

41. What is the difference between a Bohr orbit and the orbital of wave mechanics?

42. Where are metals and nonmetals located in the periodic table?

43. List some characteristic properties of metals.

44. List some characteristic properties of nonmetals.

45. Identify the following elements as metals or nonmetals. You may refer to the periodic table. The numbers in parentheses are the atomic numbers of the elements.
    a. sulfur (16)
    b. chromium (24)
    c. iodine (53)

46. Indicate the group numbers of the following families.
    a. alkali metals
    b. halogens
    c. alkaline earth metals

47. Identify the periods of the following elements. You may refer to the periodic table. The numbers in parentheses are the atomic numbers of the elements.
    a. chlorine (17)
    b. osmium (76)
    c. hydrogen (1)

48. List some of the properties of alkali metals.

49. What is the most distinguishing property of noble gases?

50. Which of the following elements are halogens?
    a. Ag   b. At   c. As

51. Which of the following elements are alkali metals?
    a. K   b. Y   c. W

52. Which of the following elements are noble gases?
    a. Fe   b. Ne   c. Ge
    d. He   e. Xe

53. Which of the following elements are transition metals?
    a. Ti (22)   b. Tc (43)   c. Te (52)

54. Which of the following elements are alkaline earth metals?
    a. Bi   b. Ba   c. Be   d. Br

55. How many electrons are in the outermost energy levels of Group 2A elements?

56. In what period of elements are electrons first introduced into the fourth energy level?

Problems

Nuclear Symbols and Isotopes

57. Give the nuclear symbol for an isotope with a nucleon number of 8 and an atomic number of 5.

58. Give the nuclear symbol for an isotope with $Z = 35$ and $A = 83$.

59. Give the nuclear symbol for an isotope with 53 protons and 72 neutrons.

60. Give the nuclear symbols for the following isotopes. You may refer to the periodic table.
    a. gallium-69   b. molybdenum-98
    c. molybdenum-99   d. technetium-98

61. Indicate the number of protons and the number of neutrons in atoms of the following isotopes:
    a. $^{62}_{30}Zn$ b. $^{241}_{94}Pu$ c. $^{99m}_{43}Tc$ d. $^{81m}_{36}Kr$

62. Which of the following pairs represent isotopes?
    a. $^{70}_{34}X$ and $^{70}_{33}X$ b. $^{57}_{28}X$ and $^{66}_{28}X$
    c. $^{186}_{74}X$ and $^{186}_{74}X$ d. $^{8}_{2}X$ and $^{6}_{4}X$
    e. $^{22}_{11}X$ and $^{44}_{22}X$

Quantum Mechanical Notation for Electron Structure

63. In the quantum mechanical notation $2s^2$, how many electrons are described? What is the general shape of the orbitals described in the notation? How many orbitals are included in the notation?

64. In the quantum mechanical notation $2p^6$, how many electrons are described? What is the general shape of the orbitals described in the notation? How many orbitals are included in the notation?

65. Use quantum mechanical notation to describe the electron configuration of the atom represented in the following Bohr diagram.

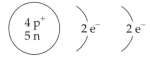

66. Give the electron configurations (using quantum mechanical notation) for the elements in Question 36. You may refer to the periodic table.

67. Identify the elements from their electron configurations. You may refer to the periodic table.
    a. $1s^2 2s^2$ b. $1s^2 2s^2 2p^3$ c. $1s^2 2s^2 2p^6 3s^2 3p^1$

68. Without referring to the periodic table, give the atomic numbers of the elements described in Problem 67.

69. None of the following electron configurations is reasonable. In each case, explain why.
    a. $1s^2 2s^2 3s^2$ b. $1s^2 2s^2 2p^3 3s^1$ c. $1s^2 2s^2 2p^6 2d^5$

70. None of the following electron configurations is reasonable. In each case, explain why.
    a. $1s^1 2s^1$ b. $1s^2 2s^2 2p^7$ c. $1s^2 2p^2$

71. Referring only to the periodic table, indicate what similarity in electron structure is shared by fluorine (F) and chlorine (Cl). What is the difference between their electron structures?

72. Referring only to the periodic table, indicate similarity in electron structure is shared by oxygen (O) and sulfur (S). What is the difference between their electron structures?

73. What is the difference between the electron configurations of sulfur (S) and chlorine (Cl)?

74. What is the difference between the electron configurations of fluorine (F) and sulfur (S)?

75. If three electrons were added to the outermost energy level of a phosphorus atom, the new electron configuration would resemble that of what element?

76. If two electrons were removed from the outermost energy level of a magnesium atom, the new electron configuration would resemble that of what element?

Projects

77. Write a brief essay about one of the following:
   a. Humphry Davy   d. Marie Curie
   b. William Crookes   e. Robert Millikan
   c. Wilhelm Roentgen   f. Niels Bohr

78. Draw a grid containing 16 squares, in two rows of 8, representing the 16 elements in the periodic table from lithium to argon. Draw Bohr diagrams for each of the elements, showing numbers of protons in the nucleus and numbers of electrons in each shell.

79. Choose three alkali metals and three alkaline earth metals. List commercial uses for each one.

80. Using this book and other references, design a time line of important milestones in the elucidation of atomic structure. Begin with the ancient Greeks and conclude with the present.

81. Draw Bohr diagrams for the first 20 elements in the periodic table.

Online Projects

82. Search the web for information about one of the following:
   a. Alessandro Volta   c. Ernest Rutherford
   b. Michael Faraday   d. J. J. Thomson

83. What information can you find on the web about fireworks?

84. Search the web for information about commercial uses of halogens.

References and Readings

1. Abrahams, Marc (Ed). *The Best Annals of Improbable Research*. New York: Freeman, 1998. A collection of anecdotes about scientific research showing that scientists are real people and that some of them, at least, can laugh at themselves.

2. Andrade, E. N. da C. *Rutherford and the Nature of the Atom*. Garden City, NY: Doubleday, 1964.

3. Asimov, Isaac. *Atom: Journey Across the Subatomic Cosmos*. New York: Dutton, 1991.

4. Bowden, Mary Ellen. *Chemical Achievers: The Human Face of the Chemical Sciences*. Philadelphia: Chemical Heritage Foundation, 1997.

5. Conkling, John A. "Pyrotechnics." *Scientific American*, July 1990, pp. 96–102. Discusses the science of fireworks.

6. Farmelo, Graham. "The Discovery of X-rays." *Scientific American*, November 1995, pp. 86–91.

7. Freedman, David H. "Weird Science." *Discover*, November 1990, pp. 62–68. Reveals some of the latest experiments in quantum mechanics.

8. Houwink, R. *Data: Mirrors of Science*. New York: American Elsevier, 1970. Chapter 3 includes marvelous illustrations showing the minute size of atoms.

9. Jaffe, Bernard. *Moseley and the Numbering of the Elements*. London: Heinemann Educational Books, 1971. Tells how atomic numbers were determined.

10. Knight, David M. *Humphry Davy: Science and Power*. Cambridge, MA: Blackwell, 1992.

11. Krebs, Robert E. *The History and Use of Our Earth's Chemical Elements: A Reference Guide*. Westport, CT: Greenwood Press, 1998.

12. Quinn, Susan. *Madame Curie: A Life*. New York: Simon and Schuster, 1995.

13. Radke, Neil. "Atomic 'Cities.'" *The Science Teacher*, January 1978, p. 35. Relates the filling of electron shells to the placement of people in houses and neighborhoods.

14. Walker, Jearl. "The Amateur Scientist: The Spectra of Street Lights Illuminate Basic Principles of Quantum Mechanics." *Scientific American*, January 1984, pp. 138–143.

15. Wolff, Peter. *Breakthroughs in Chemistry*. New York: Signet Science Library, 1967. Chapters 6–9 relate the research of Faraday, Mendeleev, Marie Curie, and Bohr.

Nuclear Chemistry
The Heart of Matter

Deep down within the atom, at the center of it all,
*A **nucleus**—extremely dense and very, very small!*

Enormous amounts of energy are constantly being generated in the sun by nuclear reactions that cause small atomic nuclei to combine and form larger ones.

Nuclear energy—the very term conjures up images of a mighty, often fearsome force! Prominent danger signs around radioactive areas and news reports about power plant disasters come to mind, along with visions of giant mushroom clouds. We know that nuclear energy can destroy a city with a single blow, but it can also destroy tiny cancer cells. Today there are many applications of nuclear energy apart from weapons and power plants. They range from medical diagnosis and treatment to dozens of uses in chemistry, physics, biology, industry, agriculture, and archaeology.

In Chapter 3 we discussed the structure of an atom. Although we mentioned the nucleus, we mainly discussed the electrons in the atom, the particles that determine its chemistry. Let us now take a closer look at that tiny speck in the center of the atom—the atomic nucleus.

If an atom is incomprehensibly small, the infinitesimal size of the atomic nucleus is completely beyond our imagination. The diameter of an atom is 100,000 times greater than the diameter of its nucleus. If an atom could be blown up in size until it was as large as your classroom, the nucleus would be about as big as the period at the end of this sentence.

Yet this tiny nucleus contains almost all the atom's mass. How incredibly dense the atomic nucleus must be! A cubic centimeter of water weighs 1 g, a cubic centimeter of lead about 11 g, and a cubic centimeter of gold about 19 g. A cubic centimeter of pure atomic nuclei would weigh more than 100 million metric tons!

Even more amazing than its density is the enormous amount of energy contained within the atomic nucleus. Some atomic nuclei undergo reactions that can fuel the most powerful bombs ever built or they can provide electricity for millions of people.

Much of what we hear today about nuclear processes is negative, but there is a positive side, too. Nuclear medicine saves lives, and many applications of nuclear chemistry in science and industry have improved the human condition significantly. For example, the use of radioisotopes in agricultural research has led to increased crop production, which provides more food for our hungry world.

[Web Reference 1] A discussion of the impact of nuclear science on society.

4.1 Natural Radioactivity: Nuclear Equations

Recall from Section 3.5 that most elements occur in nature in several isotopic forms, with nuclei differing in their number of neutrons. Some of these nuclei are unstable and undergo **radioactive decay**. Writing balanced equations for such nuclear processes is quite different from writing chemical equations. First, chemical equations must have the same elements on both sides of the arrow, whereas nuclear equations rarely do. Second, it is *atoms* that we must balance in ordinary chemical equations; in nuclear equations we must balance the *nucleons*, protons and neutrons. What this really means is that we must balance the atomic numbers (number of protons) and mass numbers (number of nucleons) of the starting materials and products. For this reason, we must always specify the isotope of each element appearing in a nuclear equation. We normally use *nuclear symbols* (Section 3.5) in writing nuclear equations because this makes the equations easier to balance.

Radium-226 atoms break down spontaneously, giving off alpha (α) particles. The process is called *alpha decay*. Because alpha particles are identical to helium nuclei, this reaction ran be summarized by the equation

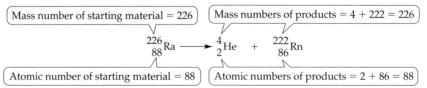

$$\underset{88}{\overset{226}{}}\text{Ra} \longrightarrow \underset{2}{\overset{4}{}}\text{He} + \underset{86}{\overset{222}{}}\text{Rn}$$

Mass number of starting material = 226   Mass numbers of products = 4 + 222 = 226

Atomic number of starting material = 88   Atomic numbers of products = 2 + 86 = 88

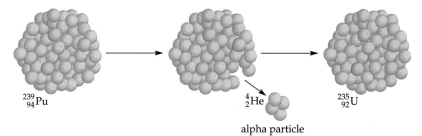

(a) Nuclear changes accompanying alpha decay.

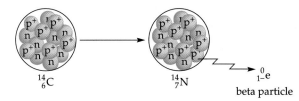

(b) Nuclear changes accompanying beta decay.

Figure 4.1   Nuclear emission of (a) an alpha particle and (b) a beta particle.

The atomic number (86) identifies the new element as radon (Rn). Note that the mass number of the starting material must equal the total of the mass numbers of the products. The same is true for atomic numbers. We use the symbol $^4_2$He for the alpha particle (rather than α) because it allows us to check the balance of mass and atomic numbers more readily. Another example of alpha decay is shown in Figure 4.1 (a).

Hydrogen-3, often called tritium, decomposes by *beta decay*. Because a beta (β) particle is identical to an electron, this process can be written as

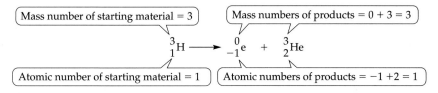

The atomic number (2) identifies the product isotope as helium. A second example of beta decay is shown in Figure 4.1 (b).

When gamma radiation occurs, the emitted radiation has no charge and no appreciable mass. Thus, the emitting atoms are changed in neither mass number nor atomic number. They simply become less energetic. Table 4.1 compares the properties of alpha, beta, and gamma radiation.

Two other types of radioactive decay, discovered later, are *positron emission* and *electron capture*. An interesting thing about these two processes is that they have the same effect on the atomic nucleus. Both result in a decrease of 1 in atomic number, with no change in mass number, but they accomplish the change by different pathways. Positron-emitting isotopes and those that undergo electron capture both have important medical applications (Section 4.7).

Table 4.1 ▮ Types of Radiation

Radiation	Mass (u)	Charge	Identity	Velocity*	Penetrating Power
Alpha (α)	4	2+	He^{2+}	$0.1c$	Very low
Beta (β)	0.00055	1−	e^-	$< 0.9c$	Moderate
Gamma (γ)	0	0	High-energy photon	c	Extremely high

*c is the speed of light.

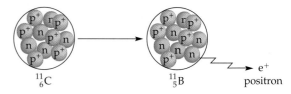

(a) Nuclear change accompanying positron emission.

Figure 4.2 Nuclear change accompanying (a) positron emission. (b) Nuclear change accompanying electron capture.

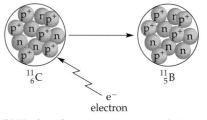

(b) Nuclear change accompanying electron capture.

The **positron** (β^+) is a particle equal in mass but opposite in charge to the electron. It is represented as $_{+1}^{0}e$. Fluorine-18 decays by positron emission.

$$^{18}_{9}\text{F} \longrightarrow {}^{0}_{+1}e + {}^{18}_{8}\text{O}$$

We can envision a proton in the nucleus changing into a neutron and a positron (Figure 4.2).

$$^{1}_{1}\text{p} \longrightarrow {}^{1}_{0}\text{n} + {}^{0}_{+1}e$$

After the positron is emitted, the original radioactive nucleus has one fewer proton and one more neutron than it had before. Therefore, the mass number of the product nucleus is the same, but its atomic number has been reduced by 1. The emitted positron quickly encounters an electron (there are numerous electrons in all kinds of matter), and both particles are annihilated, with the production of two gamma photons.

$$^{0}_{+1}e + {}^{0}_{-1}e \longrightarrow 2{}^{0}_{0}\gamma$$

Electron capture (EC) is a process in which a nucleus absorbs an electron from an inner electron shell, usually the first or second. When an electron from a higher shell drops to the level vacated by the captured electron, an X-ray is released; EC is always accompanied by X-radiation. Once inside the nucleus, the captured electron combines with a proton to form a neutron.

$$^{1}_{1}\text{p} + {}^{0}_{-1}e \longrightarrow {}^{1}_{0}\text{n}$$

Iodine-125, used in medicine to diagnose pancreatic function and intestinal fat absorption, decays by EC.

$$^{125}_{53}\text{I} + {}^{0}_{-1}e \longrightarrow {}^{125}_{52}\text{Te}$$

Conversion of a proton to a neutron (by the absorbed electron) yields a nucleus lower by 1 in atomic number but unchanged in atomic mass. Emission of a positron and absorption of an electron have the same effect on an atomic nucleus (lowering the atomic number by 1), except that positron emission is accompanied by gamma radiation and electron capture by X-radiation.

Alpha and beta, the major types of particulate radioactive decay, are pictured in Figure 4.1, positron emission and electron capture are pictured in Figure 4.2, and all five types of decay are summarized in Table 4.2.

A *photon* is a tiny bundle of energy, a "particle" of insignificant mass. Visible light and all other kinds of electromagnetic radiation are made up of photons.

The air surrounding us is about 1% argon. Most of this argon is believed to have come from the radioactive decay of potassium-40, which decays by electron capture.

Table 4.2 ▮ Radioactive Decay and Nuclear Change

Type of Decay	Decay Particle	Particle Mass (u)	Particle Charge	Change in Nucleon Number	Change in Atomic Number
alpha decay	α	4	2+	Decreases by 4	Decreases by 2
beta decay	β	0	1−	No change	Increases by 1
gamma radiation	γ	0	0	No change	No change
positron emission	β^+	0	1+	No change	Decreases by 1
electron capture (EC)	e- absorbed	0	1−	No change	Decreases by 1

Example 4.1

Write balanced nuclear equations for each of the following processes. In each case, indicate what new element is formed.

a. Plutonium-239 emits an alpha particle when it decays.
b. Protactinium-234 undergoes beta decay.
c. Carbon-10 emits a positron when it decays.
d. Iridium-192 undergoes electron capture.

Solution

a. We start by writing the symbol for plutonium-239 and a partial equation showing that one of the products is an alpha particle (helium nucleus).

$$^{239}_{94}\text{Pu} \longrightarrow {}^{4}_{2}\text{He} + ?$$

Mass and charge are conserved. The new element must have a mass of $239 - 4 = 235$ and a charge of $94 - 2 = 92$. The nuclear charge (atomic number) of 92 identifies the element as uranium (U).

$$^{239}_{94}\text{Pu} \longrightarrow {}^{4}_{2}\text{He} + {}^{235}_{92}\text{U}$$

b. Write the symbol for protactinium-234 and a partial equation showing that one of the products is a beta particle (electron).

$$^{234}_{91}\text{Pr} \longrightarrow {}^{0}_{-1}\text{e} + ?$$

The new element still has a mass number of 234. It must have a nuclear charge of 92 in order for the total charge to be the same on each side of the equation. The nuclear charge identifies the new atom as another isotope of uranium (U).

$$^{234}_{91}\text{Pr} \longrightarrow {}^{0}_{-1}\text{e} + {}^{234}_{92}\text{U}$$

c. Write the symbol for carbon-10 and a partial equation showing that one of the products is a positron.

$$^{10}_{6}\text{C} \longrightarrow {}^{0}_{+1}\text{e} + ?$$

To balance the equation, a particle with a mass number of 10 and an atomic number of 5 (B) is required.

$$^{10}_{6}\text{C} \longrightarrow {}^{0}_{+1}\text{e} + {}^{10}_{5}\text{B}$$

d. We write the symbol for indium-192 and a partial equation showing it capturing an electron.

$$^{192}_{77}\text{Ir} + {}^{0}_{-1}\text{e} \longrightarrow ?$$

To balance the equation, the product must have a mass number of 192 and an atomic number of 76 (O).

$$^{192}_{77}\text{Ir} + {}^{0}_{-1}\text{e} \longrightarrow {}^{192}_{76}\text{Os}$$

Table 4.3 ▌ Some Differences Between Chemical Reactions and Nuclear Reactions

Chemical Reactions	Nuclear Reactions
Atoms retain their identity.	Atoms usually change from one element to another.
Reactions involve only electrons, and usually only outermost electrons.	Reactions involve mainly protons and neutrons. It does not matter what the valence electrons are doing.
Reaction rates can be speeded up by raising the temperature.	Reaction rates are unaffected by changes in temperature.
Energy absorbed or given off in reactions is comparatively small.	Reactions sometimes involve enormous changes in energy.
Mass is conserved. The mass of products equals the mass of starting materials.	Huge changes in energy are accompanied by measurable changes in mass $(E = mc^2)$.

Exercise 4.1

Write balanced nuclear equations for each of the following processes. In each case, indicate what new element is formed.
 a. Fermium-250 undergoes alpha decay.
 b. Selenium-85 undergoes beta decay.
 c. Gold-188 decays by positron emission.
 d. Argon-37 undergoes electron capture.

 [Web Reference 2] A tutorial on radiation and radioactivity.

Nuclear reactions are completely different from ordinary chemical reactions (Table 4.3). It is not surprising then that nuclear equations differ so much from ordinary chemical equations. In a nuclear equation, it is only the mass numbers and the atomic numbers that must be balanced. The sum of the mass numbers on the left of the arrow must equal the sum of the mass numbers on the right. Likewise, the sum of the atomic numbers on the left must equal the sum of the atomic numbers on the right. When the unknown product has an atomic number that does not correspond to an atom, perhaps the particle is subatomic. A list of nuclear symbols for subatomic particles is given in Table 4.4.

4.2 Half-Life

Thus far we have discussed radioactivity as applied to single atoms. In the laboratory, we generally deal with great numbers of atoms—numbers far larger than the number of all the people on Earth. If we could see the nucleus of an individual atom, we could tell whether or not it would undergo radioactive decay by noting its composition. Certain combinations of protons and neutrons are unstable. We could not,

Table 4.4 ▌ Nuclear Symbols for Subatomic Particles

Particles	Symbols	Nuclear Symbols
Proton	p	${}^1_1p$ or ${}^1_1H$
Neutron	n	${}^1_0n$
Electron	e^- or β	${}^0_{-1}e$ or ${}^0_{-1}\beta$
Positron	e^+ or β^+	${}^0_{+1}e$ or ${}^0_{+1}\beta$
Alpha particle	α	${}^4_2He$ or ${}^4_2\alpha$
Beta particle	β or β^-	${}^0_{-1}e$ or ${}^0_{-1}\beta$
Gamma ray	γ	${}^0_0\gamma$

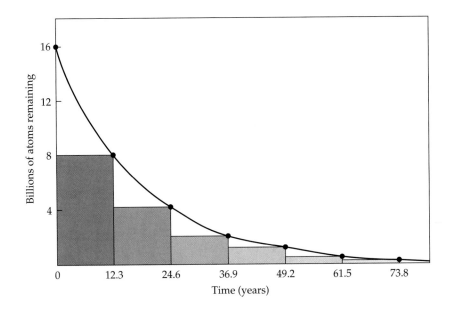

Figure 4.3 The radioactive decay of tritium (hydrogen-3). The top of each successive colored block represents one half-life.

however, determine when the atom would undergo a change. Radioactivity is a random process, generally independent of outside influences.

With large numbers of atoms, the process of radioactive decay becomes more predictable. We can measure the half-life, a property characteristic of each radioisotope. The **half-life** of a radioactive isotope is the period in which one-half of the original number of atoms undergo radioactive decay to form a new element. Suppose, for example, we had 16 billion atoms of tritium, the radioactive isotope of hydrogen. The half-life of tritium is 12.3 years. This means that in 12.3 years, 8 billion atoms of the tritium will have decayed, and there will be 8 billion atoms left. In another 12.3 years, half of the remaining 8 billion atoms will have decayed. After two half-lives, then, one-quarter of the original tritium atoms will remain. Two half-lives, then, do not make a whole. The concept of half-life is illustrated by the graph in Figure 4.3.

We can calculate the fraction of the original isotope that remains after a given number of half-lives from the relationship

$$\text{Fraction remaining} = \frac{1}{2^n}$$

where n is the number of half-lives.

Example 4.2

You obtain a new sample of cobalt-60, half-life 5.25 years, with a mass of 400 mg. How much cobalt-60 remains after 15.75 years (three half-lives)?

Solution
The fraction remaining after three half-lives is

$$\frac{1}{2^n} = \frac{1}{2^3} = \frac{1}{2 \times 2 \times 2} = \frac{1}{8}$$

The amount of cobalt-60 remaining is $\left(\frac{1}{8}\right)(400 \text{ mg}) = 50 \text{ mg}$.

Exercise 4.2
You have 1.224 mg of freshly prepared gold-189, half-life 30 min. How much of the gold-189 sample remains after five half-lives?

The half-life of an element can be very long (millions of years) or extremely short (tiny fractions of a second). The half-life of uranium-238 is 4.5 billion years; that of boron-9 is 8×10^{-19} s.

It is impossible to say when *all* the atoms of a radioactive isotope will have decayed. For most samples, we can assume that the activity is essentially gone after about ten half-lives. (After ten half-lives, the activity is $\frac{1}{2} = \frac{1}{1024}$ of the original value.) Generally, we can say that $\frac{1}{1000}$ of the original activity remains.

Example 4.3

You obtain a 20.0-mg sample of mercury-190, half-life 20 min. How much of the mercury-190 sample remains after 2 hr?

Solution

There are 120 min in 2 hr. There are $\left(\frac{120}{20}\right) = 6$ half-lives in 2 hr. The fraction remaining after six half-lives is

$$\frac{1}{2^n} = \frac{1}{2^6} = \frac{1}{2 \times 2 \times 2 \times 2 \times 2 \times 2} = \frac{1}{64}$$

The amount of mercury-190 remaining is $\left(\frac{1}{64}\right)(20.0)$ mg $= 0.313$ mg.

Exercise 4.3

A sample of 16.0 mg of nickel-57, half-life 36.0 hr, is produced in a nuclear reactor. How much of the nickel-57 sample remains after 7.5 days?

What Makes for Nuclear Stability?

Most isotopes are radioactive.

What are the factors that tend to make an atomic nucleus stable?

1. *Even* numbers of either protons or neutrons. (Of the 264 stable isotopes, 157 have *even* numbers of both protons and neutrons. Only 4 have odd numbers of both protons and neutrons.) Elements of even atomic number have more stable isotopes than those of odd atomic number.

2. "Magic" numbers of either protons or neutrons. (Magic numbers are 2, 8, 20, 50, 82, and 126.)

3. An atomic number of 83 or less. (All isotopes with atomic numbers greater than 83 are radioactive.)

4. There should be no more protons than neutrons in the nucleus, and the ratio of neutrons to protons should be close to 1 if the atomic number is 20 or below. (As atomic numbers get larger, the stable n/p ratio also increases, up to about 1.5. There is a zone of stability within which the n/p ratio should lie for an atom of a given atomic number.)

4.3 Radioisotopic Dating

The half-lives of certain isotopes can be used to estimate the ages of rocks and archaeological artifacts. Uranium-238 decays with a half-life of 4.5 billion years. The initial products of this decay are also radioactive, and breakdown continues until an isotope of lead (lead-206) is formed. By measuring the relative amounts of uranium-238 and lead-206, chemists can estimate the age of a rock. Some of the older rocks on Earth have been found to be 3.0–4.5 billion years old. Moon rocks and meteorites have been dated at a maximum age of about 4.5 billion years. Thus, the age of Earth is generally estimated to be about 4.5 billion years.

Carbon-14 Dating

The dating of artifacts derived from plants or animals usually involves a radioactive isotope of carbon. Carbon-14 is formed in the upper atmosphere by the bombardment of ordinary nitrogen by neutrons from cosmic rays.

$$^{14}_{7}\text{N} + ^{1}_{0}\text{n} \longrightarrow ^{12}_{6}\text{C} + ^{1}_{1}\text{H}$$

This process leads to a steady-state concentration of carbon-14 in Earth's CO_2. Living plants and animals incorporate this isotope into their own cells. When they die, however, the incorporation of carbon-14 ceases, and the carbon-14 in the organisms decays—with a half-life of 5730 years—to nitrogen-14. Thus, we need merely to measure the carbon-14 activity remaining in an artifact of plant or animal origin to determine its age. For instance, a sample that has half the carbon-14 activity of new plant material is 5730 years old; it has been dead for one half-life. Similarly, an artifact with 25% of the carbon-14 activity of new plant material is 11,460 years old; it has been dead for two half-lives.

Carbon-14 dating, as outlined here, assumes that the formation of the isotope was constant over the years. This is not quite the case. However, for the most recent 7000 years or so, carbon-14 dates have been correlated with those obtained from the annual growth rings of trees. Calibration curves have been constructed from which accurate dates can be determined. Generally, carbon-14 is reasonably accurate for dating objects up to about 50,000 years old. Objects older than 50,000 years have too little of the isotope left for accurate measurement.

Charcoal from the fires of an ancient people, dated by determining the carbon-14 activity, is used to estimate the age of other artifacts found at the same archaeological site. Carbon-14 dating also has been used to detect forgeries of supposedly ancient artifacts.

The Shroud of Turin

The Shroud of Turin is a very old piece of linen cloth, about 4 m long, bearing a faint human likeness. Since about C.E. 1350 it had been alleged to be part of the burial shroud of Christ. However, carbon-14 dating studies in 1988 by three different nuclear laboratories indicated that the flax used in making the cloth was not grown until sometime between C.E. 1260 and 1390. Therefore, the cloth could not possibly have existed at the time of Christ. Unlike the Dead Sea Scrolls, which were shown by carbon-14 dating to be authentic records from a civilization that existed about 2000 years ago, the Shroud of Turin has been shown to be less than 800 years old.

Tritium Dating

Tritium, the radioactive isotope of hydrogen, also is useful for dating. Its half-life of 12.3 years makes it useful for dating items up to about 100 years old. An interesting application is the dating of brandies. These alcoholic beverages are quite expensive when aged from 10 to 50 years. Tritium dating can be used to check the veracity of advertising claims about the aging process of the most expensive kinds.

Many other isotopes are useful for estimating the ages of objects and materials. Several of the more important ones are listed in Table 4.5.

Table 4.5 ▌ Several Isotopes Useful in Radioactive Dating

Isotope	Half-Life (years)	Useful Range	Dating Applications
Carbon-14	5730	500 to 50,000 years	Charcoal, organic material
Tritium ($^{3}_{1}\text{H}$)	12.3	1 to 100 years	Aged wines
Potassium-40	1.3×10^{9}	10,000 years to the oldest Earth samples	Rocks, the Earth's crust, the moon's crust
Rhenium-187	4.3×10^{10}	4×10^{7} years to the oldest samples in the universe	Meteorites
Uranium-238	4.5×10^{9}	10^{7} years to the oldest Earth samples	Rocks, the Earth's crust

Example 4.4

A piece of fossilized wood has carbon-14 activity one-eighth that of new wood. How old is the artifact? The half-life of carbon-14 is 5730 years.

Solution

The carbon-14 has gone through three half-lives.

$$\frac{1}{8} = \left(\frac{1}{2}\right)^3 = \frac{1}{2} \times \frac{1}{2} \times \frac{1}{2}$$

It is therefore about $3 \times 5730 = 17,190$ years old.

Exercise 4.4

How old is a piece of cloth that has carbon-14 activity $\frac{1}{16}$ that of new cloth fibers? The half-life of carbon-14 is 5730 years.

4.4 Artificial Transmutation

During the Middle Ages, alchemists had tried to turn base metals, such as lead, into gold. But they were trying to do it chemically, and so they were doomed to failure. In order to accomplish *transmutation* (changing one element into another), one must alter the *nucleus*. Today, by using nuclear bombardment, scientists achieve transmutation every day.

The forms of radioactivity encountered thus far occur in nature. Other nuclear reactions may be brought about by bombardment of stable nuclei with alpha particles, neutrons, or other subatomic particles. These particles, given sufficient energy, penetrate the formerly stable nucleus and bring about some form of radioactive emission. Like natural radioactivity, this sort of nuclear change brings about a **transmutation**: one element is changed into another. Because the change would not have occurred naturally, the process is called *artificial transmutation*.

In 1919, a few years after his famous gold-foil experiment (Chapter 3), Ernest Rutherford reported on the bombardment of a variety of light elements with alpha particles. One such experiment, in which he bombarded nitrogen, resulted in the production of protons.

$$^{14}_{7}\text{N} + {}^{4}_{2}\text{He} \longrightarrow {}^{17}_{8}\text{O} + {}^{1}_{1}\text{H}$$

(The hydrogen nucleus is simply a proton; hence the alternative symbol ${}^{1}_{1}\text{H}$ for the proton. Notice that the nucleon numbers and the atomic numbers are both balanced.) This provided the first empirical verification of the existence of protons in atomic nuclei, which Rutherford had first postulated in 1914.

Recall that Eugen Goldstein had produced protons in his gas discharge tube experiments in 1886 (Chapter 3). The significance of Rutherford's experiment lay in the fact that he obtained protons from the nucleus of an atom other than hydrogen, thus establishing their nature as **fundamental particles**. By fundamental particles we mean basic units comprising more complicated structures such as nuclei. Rutherford's experiment was the first induced nuclear reaction.

A great many transmutations were carried out during the 1920s. In the 1930s, one such reaction led to the discovery of another fundamental particle. James Chadwick, in 1932, bombarded beryllium with alpha particles.

$$^{9}_{4}\text{Be} + {}^{4}_{2}\text{He} \longrightarrow {}^{12}_{6}\text{C} + {}^{1}_{0}\text{n}$$

Among the products was the neutron.

Ernest Rutherford carried out the first nuclear bombardment experiment.

Example 4.5

When potassium-39 is bombarded with neutrons, chlorine-36 is produced. What other particle is emitted?

$$^{39}_{19}\text{K} + {}^{1}_{0}\text{n} \longrightarrow {}^{36}_{17}\text{Cl} + ?$$

Solution

Write a balanced nuclear equation. To balance the equation, we need four mass units and two charge units (that is, a particle with a nucleon number of 4 and an atomic number of 2). That's an alpha particle.

$$^{39}_{19}K + ^{1}_{0}n \longrightarrow ^{36}_{17}Cl + ^{4}_{2}He$$

Exercise 4.5

Technetium-97 is produced by bombarding molybdenum-96 with a deuteron (hydrogen-2 nucleus). What other particle is emitted?

$$^{96}_{42}Mo + ^{2}_{1}H \longrightarrow ^{97}_{43}Tc + ?$$

4.5 Induced Radioactivity

The first artificial nuclear reactions produced isotopes already known to occur in nature. This was perhaps fortuitous, because it was inevitable that an unstable nucleus would be produced sooner or later. Irène Curie (daughter of the 1903 Nobel prize winners) and her husband, Frederic Joliot, were studying the bombardment of aluminum with alpha particles. Neutrons were produced, leaving behind an isotope of phosphorus.

$$^{27}_{13}Al + ^{4}_{2}He \longrightarrow ^{30}_{15}P + ^{1}_{0}n$$

Much to their surprise, the target continued to emit particles after the bombardment had been halted. The isotope of phosphorus was radioactive, emitting particles equal in mass to the electron but opposite in charge. These particles are called **positrons**. The reaction they observed is written

$$^{30}_{15}P \longrightarrow ^{0}_{+1}e + ^{30}_{14}Si$$

They had discovered the process of positron emission (discussed in Section 4.1). For this work the Joliot-Curies were awarded a Nobel prize of their own in 1935.

4.6 Uses of Radioisotopes

Most of the 3000 known **radioisotopes** (radioactive isotopes) are produced by artificial transmutation from stable isotopes. The value of both naturally occurring and artificial radioisotopes goes far beyond their contributions to our knowledge of chemistry. Scientists in a wide variety of fields use radioisotopes as **tracers** in physical, chemical, and biological systems. Isotopes of a given element, whether radioactive or not, behave nearly identically in chemical and physical processes. Because radioactive isotopes are easily detected, it is relatively easy to trace their movement, even through a complicated system.

As a simple example, let's consider the flow of a liquid through a pipe. Suppose there is a leak in the pipe, which is buried beneath a concrete floor. We could locate the leak by digging up extensive areas of the floor, or we could add a small amount of radioactive material to liquid poured into the drain and trace the flow of the liquid with a Geiger counter (an instrument that detects radioactivity). Once we locate the leak, only a small area of the floor would have to be dug up to repair the leak. Short-lived isotopes—which disappear soon after doing their job—usually are employed for such purposes.

Radioactive tracers are also put to good use in agricultural research and development. Uses include study of the effectiveness of fertilizers and weed killers, comparison of nutritional value of various feeds, determination of optimal insect control methods, and improvement of commercial crop strains by purposeful mutation. For

Frédéric and Irène Joliot-Curie discovered artificially **induced radioactivity** in 1934. They are shown here working in their laboratory.

The Joliot-Curies adopted a combined surname to perpetuate the Curie name. Marie and Pierre Curie had two daughters, but no son.

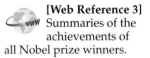 **[Web Reference 3]** Summaries of the achievements of all Nobel prize winners.

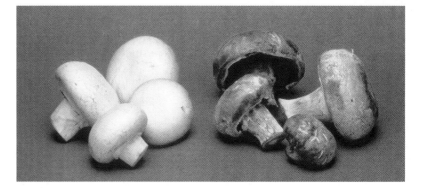

Figure 4.4 Gamma radiation delays the decay of mushrooms. Those on the left were irradiated; the ones on the right were not.

example, to trace the uptake of phosphorus by a green plant, a scientist feeds it fertilizer containing radioactive phosphorus. When the plant is later placed on a photographic film, radiation from the phosphorus isotopes exposes the film, much as light does. This type of exposure, called a *radiograph*, shows the distribution of phosphorus in the plant.

Radioisotopes are also used as sources to irradiate foodstuffs as a method of preservation (Figure 4.4). The radiation destroys microorganisms that cause food spoilage. Irradiated food shows little change in taste or appearance. Some people are concerned about possible harmful effects of chemical substances produced by the radiation, but there is no good evidence of harm to laboratory animals fed irradiated food, nor are there any known adverse effects in humans in countries where irradiation has been used for years. There is no residual radiation in the food after the sterilization process.

An important example of the extensive use of radioisotopes in basic scientific research was determination of the mechanism of photosynthesis, in which plants make the sugar glucose from carbon dioxide and water. Plants were exposed to carbon dioxide containing radioactive carbon-14 ($^{14}CO_2$) as a tracer. Researchers identified the compounds formed from these starting materials and their order of formation, and then determined which new compounds became radioactive and in what sequence. Using data from radioactive tracer experiments, scientists determine metabolic pathways in plants, animals, and humans.

Figure 4.5 shows another application of radiography—in art.

Figure 4.5 (a) *Saint Rosalie Interceding for the Plague-Stricken of Palermo*, by Anthony van Dyck (1599–1641). (b) A radiograph reveals the presence of an earlier painting on the same canvas.

(a)

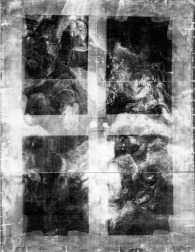

(b)

4.7 Nuclear Medicine

Nuclear medicine involves two distinct uses of radioisotopes: therapeutic and diagnostic. In radiation therapy, an attempt is made to treat or cure disease with radiation. The diagnostic use of radioisotopes is aimed at obtaining information about the state of a patient's health.

Radiation Therapy

Cancer is not one disease but many. Some forms are particularly susceptible to radiation therapy. The aim of radiation therapy is to destroy cancerous cells before too much damage is done to healthy tissue. Radiation is most lethal to rapidly reproducing cells, and this is precisely the characteristic of cancer cells that allows radiation therapy to be successful. Radiation is carefully aimed at cancerous tissue, while minimizing the exposure of normal cells. If the cancer cells are killed by the destructive effects of the radiation, the malignancy is halted.

Patients undergoing radiation therapy often get sick from the treatment. Nausea and vomiting are the usual early symptoms of radiation sickness. Radiation therapy can also interfere with white blood cell replenishment and increase susceptibility to infection.

Diagnostic Uses of Radioisotopes

Radioisotopes are used for diagnostic purposes to provide information about the type or extent of an illness. Table 4.6 lists some radioisotopes in common use in medicine. The list is necessarily incomplete. Even this abbreviated discussion should give you an idea of the importance of radioisotopes in medicine. The claim that nuclear science has saved many more lives than nuclear bombs have destroyed is not an idle one.

Table 4.6 ▌ Some Radioisotopes and Their Medical Applications

Isotope	Name	Half-Life*	Use
^{11}C	Carbon-11	20.3 m	Brain scans
^{51}Cr	Chromium-51	27.8 d	Blood volume determination
^{57}Co	Cobalt-57	270 d	Measuring vitamin B_{12} uptake
^{60}Co	Cobalt-60	5.26 y	Radiation cancer therapy
^{153}Gd	Gadolinium-153	242 d	Determining bone density
^{67}Ga	Gallium-67	78.1 h	Scan for lung tumors
^{131}I	Iodine-131	8.07 d	Thyroid therapy
^{192}Ir	Iridium-192	74 d	Breast cancer therapy
^{59}Fe	Iron-59	45 d	Detection of anemia
^{32}P	Phosphorus-32	14.3 d	Detection of skin cancer or eye tumors
^{238}Pu	Plutonium-238	86 y	Provides power for pacemakers
^{226}Ra	Radium-226	1600 y	Radiation therapy for cancer
^{75}Se	Selenium-75	120 d	Pancreas scans
^{24}Na	Sodium-24	15.0 h	Locating obstructions in blood flow
^{99m}Tc	Technetium-99^m	6.0 h	Imaging of brain, liver, bone marrow, kidney, lung, or heart
^{201}Tl	Thallium-201	73 h	Detecting heart problems with treadmill stress test
^{3}H	Tritium	12.3 y	Determining total body water
^{133}Xe	Xenon-133	5.27 d	Lung imaging

*Abbreviations: y, years; d, days; h, hours; m, minutes.

PET scans using carbon-11 have shown that a schizophrenic brain metabolizes only about one-fifth as much glucose as a normal brain. These scans can also reveal metabolic changes that occur in the brain during tactile learning (learning by the sense of touch).

Radioactive iodine-131 is used to determine the size, shape, and activity of the thyroid gland, as well as to treat cancer located in this gland and to control a hyperactive thyroid. Small doses are used for diagnostic purposes, and large doses for treatment of thyroid cancer. After the patient drinks a solution of potassium iodide incorporating iodine-131, the body concentrates iodide in the thyroid. A detector showing the differential uptake of the isotope is used in diagnosis. The resulting **photoscan** can pinpoint the location of tumors or other abnormalities in the thyroid. In cancer treatment, radiation from therapeutic (large) doses of iodine-131 kills the thyroid cells in which the radioisotope has concentrated. This occurs even in thyroid-type cells that have spread to other parts of the body.

The radioisotope most widely used in medicine is gadolinium-153. This isotope is used to determine bone mineralization. Its widespread use is an indication of the large number of people, mostly women, who suffer from osteoporosis (reduction in the quantity of bone) as they grow older. Gadolinium-153 gives off two characteristic radiations, a gamma ray and an X-ray. A scanning device compares these radiations after they pass through bone. Bone densities are then determined by differences in absorption of the rays.

Technetium-99^m is used in a variety of diagnostic tests (Figure 4.6). The m stands for "metastable," which means that this isotope gives up some energy to become a more stable version of the same isotope (same atomic number, same atomic mass). The energy it gives up is the gamma ray needed to detect the isotope.

$$^{99m}_{43}\text{Tc} \longrightarrow ^{99}_{43}\text{Tc} + \gamma$$

Notice that the decay of technetium-99m produces no alpha or beta particles, which could cause unnecessary damage to the body. Technetium-99m also has a short half-life (about 6 hr), which means that the radioactivity does not linger in the body long after the scan has been completed. With so short a half-life, use of the isotope must be carefully planned. In fact, the isotope itself is not what is purchased. Technetium-99m is formed by the decay of molybdenum-99.

$$^{99}_{42}\text{Mo} \longrightarrow ^{99m}_{43}\text{Tc} + ^{0}_{-1}\text{e} + \gamma$$

A container of this molybdenum isotope is obtained, and the decay product, technetium-99m, is "milked" from the container as needed.

Using modern computer technology, positron emission tomography (PET), can measure dynamic processes occurring in the body, such as blood flow or the rate at which oxygen or glucose is being metabolized. PET scans can pinpoint the area of brain damage that triggers severe epileptic seizures. Compounds incorporating positron-emitting isotopes, such as carbon-11, are inhaled or injected prior to the

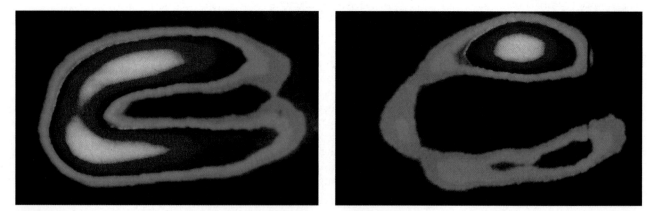

Figure 4.6 Blood flow patterns in a healthy heart (left) and in a damaged heart (right). The highlighted images from a technetium-99^m compound indicate regions receiving adequate blood flow.

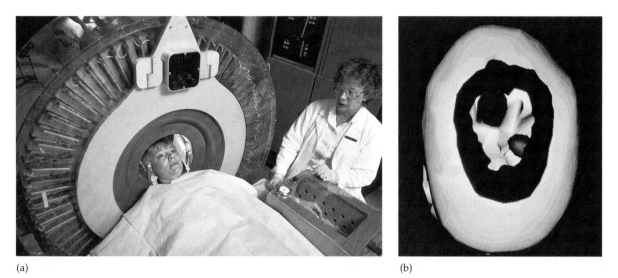

(a)   (b)

Figure 4.7 Modern computer technology used for medical diagnosis. (a) Patient in position for positron emission tomography (PET), a technique that uses radioisotopes to scan internal organs. (b) Image created by computerized tomography (CT), a scanning technique that uses X-rays rather than radioisotopes, looking throug the skull at a pituitary brain tumor.

scan. Before the emitted positron can travel very far in the body, it encounters an electron (numerous in any ordinary matter), and two gamma rays are produced, exiting from the body in exactly opposite directions.

$$^{11}_{6}C \longrightarrow ^{11}_{5}B + ^{0}_{+1}e$$

$$^{0}_{+1}e + ^{0}_{-1}e \longrightarrow 2\gamma$$

Detectors, positioned on opposite sides of the patient, record the gamma rays. Computerized calculations of the points within the body at which annihilation of the positrons and electrons occurred result in an image of that area (Figure 4.7).

Example 4.6

One of the isotopes used for PET scans is oxygen-15, a positron emitter. What new element is formed when oxygen-15 decays?

Solution

First write the nuclear equation

$$^{15}_{8}O \longrightarrow ^{0}_{+1}e + ?$$

The nucleon number does not change, but the atomic number becomes 821, or 7; and so the new product is nitrogen-15.

$$^{15}_{8}O \longrightarrow ^{0}_{+1}e + ^{15}_{7}N$$

Exercise 4.6

Phosphorus-30 is a positron-emitting radioisotope suitable for use in PET scans. What new element is formed when phosphorus-30 decays?

4.8 Radiation and Us

We have seen how radioisotopes are used in beneficial ways, and we are all familiar with the use of X-rays in medical diagnosis. But radiation also presents a hazard to living things. Radiation with enough energy to knock electrons from atoms and molecules, converting them into ions (electrically charged atoms or groups of atoms), is called **ionizing radiation**. Nuclear radiation and X-rays are examples. Because radiation is invisible and because it has such great potential for harm, we are very much concerned with our exposure to it.

Radiation Damage to Cells

Radiation-caused chemical changes in living cells can be highly disruptive. Ionizing radiation can devastate living cells by interfering with their normal chemical processes. Water in cells can be transformed to highly reactive hydrogen peroxide (H_2O_2), which can disrupt the delicate chemical balance in the cells. Particularly vulnerable are white blood cells, the body's first line of defense against bacterial infection. Radiation also affects bone marrow, causing a drop in the production of red blood cells, which results in anemia. Radiation also has been shown to induce leukemia, a cancerlike disease of the blood-forming organs.

Radiation also causes changes in the molecules of heredity (DNA) in reproductive cells. Such changes show up as mutations in the offspring of exposed parents. Little is known of the effects of such exposure on humans. However, many of the mutations that occurred during the evolution of present species may have been caused by background radiation.

Background Radiation

Humans have always been exposed to radiation. Some of it, called **cosmic rays**, comes from the sun and outer space. Other radiation reaches us from natural radioactive isotopes in air, water, soil, and rocks. This ever-present radiation is called **background radiation**.

Human activities have added to our exposure to radiation. Figure 4.8 shows that about two-thirds of the average radiation exposure comes from background radiation. Most of the remaining one-third comes from medical irradiation such as X-rays. Other sources, such as fallout, releases from the nuclear industry, and occupational exposure, account for only a minute fraction of our total exposure. Nevertheless, accidents such as those at the Three Mile Island plant near Harrisburg, Pennsylvania, in 1979 and at Chernobyl in Ukraine, USSR, in 1986 did much to increase public apprehension about nuclear power (Table 4.7). Even though no one was hurt at Three Mile Island, some people are still concerned about the long-term effects of exposure to the small amounts of radioactivity released. The accident at Chernobyl was much more severe. Some people were killed outright; many others died later of severe ra-

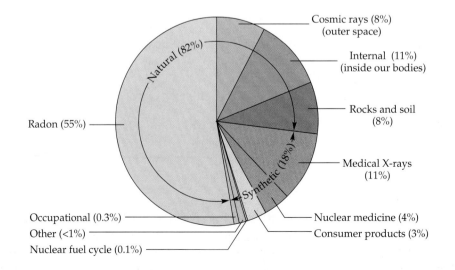

Figure 4.8 Most of our exposure to ionizing radiation comes from natural sources, including radon that seeps into our homes from the underlying rocks.

Cosmic rays (8%) (outer space)

Internal (11%) (inside our bodies)

Natural (82%)

Rocks and soil (8%)

Radon (55%)

Medical X-rays (11%)

Synthetic (18%)

Occupational (0.3%)

Other (<1%)

Nuclear fuel cycle (0.1%)

Nuclear medicine (4%)

Consumer products (3%)

Table 4.7 ▌ Most Serious of Reported Nuclear Accidents

Date	Country	Location	Cause	Extent of Damage	Casualties
1952	Canada	Chalk River, Ottawa	Four control rods accidentally removed.	Partial meltdown. Millions of gallons of radioactive water in reactor.	No one injured
1957	England	Windscale Pile No. 1 near Liverpool	Fire in graphite-cooled reactor.	Radiation spewed over 200 sq. mi. of countryside.	Probably caused 39 cancer deaths later
1957	USSR	Ural Mountains near Kyshtym	Explosion of nuclear waste tank.	Much release of radioactivity.	None reported
1976	Germany	Lubin near Greifswald	Safety systems failed during a fire.	Near meltdown of reactor core.	None reported
1979	USA	Three Mile Island Middletown, PA near Harrisburg	One reactor lost coolant and overheated.	Partial meltdown. Some radioactivity released.	No lives lost
1986	Ukraine	Chernobyl near Kiev	Poor design. Safety features lacking. Human error.	Reactor meltdown. Enormous release of radioactivity (70% fell on Belarus).	31 reported dead. Thousands died soon after. Many thousands remain ill.
1999	Japan	Tokaimura	Too much U-235 added in uranium dioxide plant.	Accidental fission reaction caused release of radioactivity.	49 workers gravely overexposed

diation poisoning. Increased levels of radioactivity throughout Europe created great concern about possible harmful effects.

Penetrating Power of Radiation

Radioactive materials can be dangerous because the radiation emitted can damage living tissue. The ability to inflict injury depends in part on the power of the radiation to penetrate the tissue.

All other things being equal, the more massive the particle, the less its penetrating power. Alpha particles, which are helium nuclei with a mass of 4 u, are the least penetrating of the three main types of radioactivity. Beta particles, which are identical to the almost massless electrons, are somewhat more penetrating. Gamma rays, like X-rays, truly have no mass; they are considerably more penetrating than the other two types.

But all other things are not always equal. The faster a particle moves or the more energetic the radiation is, the more penetrating power it has.

It may seem contrary to common sense that the biggest particles make the least headway. Consider that penetrating power reflects the ability of radiation to make its way through a sample of matter. It is as if you were trying to roll some rocks through a field of boulders. The alpha particle acts as if it were a boulder itself. Because of its size, it cannot get very far before it bumps into and is stopped by other boulders. The beta particle acts as if it were a small stone. It can sneak between and perhaps ricochet off boulders until it makes its way farther into the field (Figure 4.9). The gamma ray can be compared with a grain of sand that can get through the smallest openings.

In 1993 a Ukrainian Academy of Science study group led by Vladimir Chernousenko investigated the aftereffects of the Chernobyl accident. Chernousenko claims that 15,000 of those who helped clean up the accident site have died and that another 250,000 have been left as invalids. He also says that 200,000 children have experienced radiation-induced illnesses and that half of the children in Ukraine and Belarus have symptoms. The incident at Chernobyl greatly influenced the desire of Ukrainians for independence after the breakup of the Soviet Union in 1991.

[Web Reference 4] A discussion of the radiation all around us.

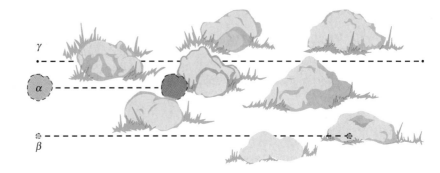

Figure 4.9 Shooting radioactive particles through matter is like rolling rocks through a field of boulders—the larger rocks stop more quickly.

The danger of a specific type of radiation to human tissue depends on the location of the source as well as on the penetrating power. If a radioactive substance is outside the body, alpha particles of low penetrating power are the least dangerous; they are stopped by the outer layer of skin. Beta particles also usually are stopped before they reach vital organs. Gamma rays readily pass through tissues, and so an external gamma source can be quite dangerous. When the radioactive source is inside the body, the situation is reversed. The nonpenetrating alpha particles can do great damage. All such particles are trapped within the body, which must then absorb all the energy released by the particle. Alpha particles inflict all their damage in a very small area because they do not travel far. Beta particles distribute their damage over a somewhat larger area because they travel farther. Tissue may recover from limited damage spread over a large area; it is less likely to survive concentrated damage.

People working with radioactive materials can do several things to protect themselves. The simplest is to move away from the source, because intensity of radiation decreases with distance from the source. Workers can also be protected by shielding. A sheet of paper can stop most alpha particles, a block of wood or a thin sheet of aluminum can stop beta particles, but it takes several meters of concrete or several centimeters of lead to stop gamma rays (Figure 4.10).

4.9 Energy from the Nucleus

We have seen that radioactivity—quiet and invisible—can be beneficial or dangerous. A much more dramatic—and equally paradoxical—aspect of nuclear chemistry is the release of nuclear energy by either fission (splitting of heavy nuclei into smaller nuclei) or fusion (combining of light nuclei to form heavier ones).

Figure 4.10 The relative penetrating powers of alpha, beta, and gamma radiation. Alpha particles are stopped by a sheet of paper (a). Beta particles will not penetrate a sheet of aluminum foil (b). It takes several centimeters of lead to block gamma rays (c).

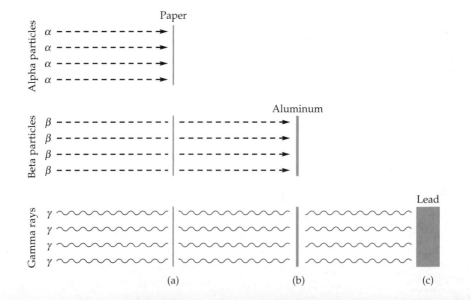

Einstein and the Equivalence of Mass and Energy

The potential power in the nucleus was worked out by Albert Einstein, one of the most famous and most unusual scientists. Whereas most scientists work with glassware and instruments in laboratories, Einstein worked with a pencil and a note pad. By 1905, at the age of 26, he had already worked out his special theory of relativity and had developed his famous **mass–energy equation**

$$E = mc^2$$

where E represents energy, m is mass, and c is the speed of light. This equation suggests that mass and energy are just two different aspects of the same thing.

According to Einstein's equation, a chemical reaction that gives off heat must lose mass in the process, although the change in mass is much too small to measure. Reaction energy must be enormous in order for the mass loss to be measurable—for example, the energy given off by nuclear explosions. (Converting a single gram of matter to energy would provide enough heat to warm an average home for 1000 years, if conversion were complete; but even when a hydrogen bomb explodes, less than 1% of the matter is converted to energy.)

Albert Einstein on a Chinese postage stamp.

The word "nucleons" refers to both protons and neutrons.

Binding Energy

As we see in the atomic bomb and in nuclear power plants, nuclear fission involves a tremendous discharge of energy. Where does all this energy come from? It is locked inside the atomic nucleus. When protons and neutrons are combined to form atomic nuclei, a small amount of mass is converted to energy. This is the **binding energy** that holds the nucleons together in the nucleus. For example, the helium nucleus contains two protons and two neutrons. The mass of these four particles is 4.0320 u (Figure 4.11). However, the actual mass of the helium nucleus is only 4.0015 u, and the missing mass amounts to 0.0305 u. Using Einstein's equation $E = mc^2$, we can calculate a value of 28.3 million electron volts (MeV) for the binding energy of the helium nucleus. This is the amount of energy it would take to separate one helium nucleus into two protons and two neutrons. If we divide by 4 (the number of nucleons), the binding energy per nucleon is 7.1 MeV.

When binding energy per nucleon is calculated for all the elements and plotted against mass number, a graph such as that in Figure 4.12 is obtained. The elements with the highest binding energies per nucleon have the most stable nuclei. They include iron and nearby elements. When uranium atoms undergo nuclear fission, they split into fragments with higher binding energies; in other words, the fission reaction converts large atoms into smaller ones with greater nuclear stability.

We can also see from Figure 4.12 that even more energy can be obtained by combining small atoms, such as hydrogen or deuterium, to form larger atoms with more stable nuclei. This kind of reaction is called **nuclear fusion**. It is what happens when a hydrogen bomb explodes, and it is also the source of the sun's energy.

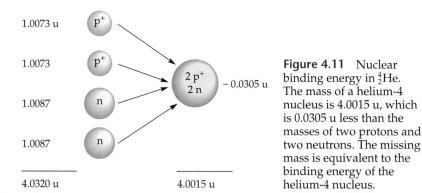

Figure 4.11 Nuclear binding energy in $^4_2$He. The mass of a helium-4 nucleus is 4.0015 u, which is 0.0305 u less than the masses of two protons and two neutrons. The missing mass is equivalent to the binding energy of the helium-4 nucleus.

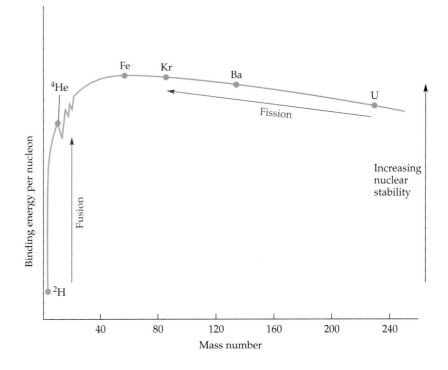

Figure 4.12 Nuclear stability is greatest near iron in the periodic table. Fission of very large atoms or fusion of very small ones results in greater nuclear stability.

4.10 The Building of the Bomb

In 1934 the Italian scientists Enrico Fermi and Emilio Segrè bombarded uranium atoms with neutrons. They were trying to make elements higher in atomic number than uranium, which then had the highest known number. To their surprise, they found four radioactive species among the products. One presumably was element 93, formed by the initial conversion of uranium-238 to uranium-239.

$$^{238}_{92}U + {}^{1}_{0}n \longrightarrow {}^{239}_{92}U$$

The latter then underwent beta decay.

$$^{239}_{92}U \longrightarrow {}^{0}_{-1}e + {}^{239}_{93}Np$$

The two scientists were unable to explain the remaining radioactivity.

Nuclear Fission

In repeating the Fermi–Segrè experiment in 1938, German chemists Otto Hahn and Fritz Strassman were perplexed to find isotopes of barium (Ba) among the many reaction products. Hahn wrote to Lise Meitner, his former long-time colleague, to ask what she thought about these strange results.

Lise Meitner was an Austrian physicist who had worked with Hahn in Berlin. Because she was Jewish she had recently fled to Sweden when the Nazis took over Austria in 1938. On hearing about Hahn's work, she noted that barium atoms were only about half the size of uranium atoms. Was it possible that the uranium nucleus might be splitting into fragments? She made some calculations that convinced her that the uranium nuclei had indeed been split apart. Her nephew, Otto Frisch, was visiting for the winter holidays, and they discussed this new discovery with great excitement. It was Frisch who later coined the term **nuclear fission** (Figure 4.13).

Frisch was working with Niels Bohr at the University of Copenhagen, and when he returned to Denmark he took the news about the fission reaction to Bohr, who

Enrico Fermi.

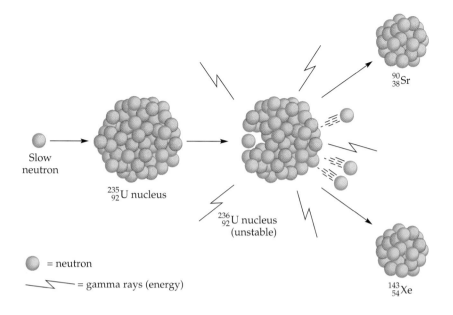

Figure 4.13 The splitting of a uranium atom. The neutrons produced in the fission can split other uranium atoms, thus sustaining a chain reaction. The splitting of one uranium-235 atom yields 8.9×10^{-18} kWh of energy. Fission of a mole of uranium-235 (6.02×10^{23} atoms) produces 5.3 million kWh of energy.

Lise Meitner portrayed on an Austrian postage stamp.

happened to be going to the United States to attend a physics conference. The discussions in the corridors about this new reaction would be the most important talks to take place at that meeting.

Meanwhile, Enrico Fermi had just received the 1938 Nobel prize in physics. Because Fermi's wife Laura was Jewish, and the fascist Italian dictator Mussolini was an ally of Hitler, Fermi accepted the award in Stockholm and then immediately fled with his wife and child to the United States. Thus, by 1939 the United States had received news about the German discovery of nuclear fission and had also acquired from Italy one of the world's foremost nuclear scientists.

Nuclear Chain Reaction

Leo Szilard was one of the first scientists to realize that nuclear fission could be a practical chain reaction. Szilard had been born in Hungary and educated in Germany, but he came to the United States in 1937 as another Jewish refugee. He saw that neutrons released in the fission of one atom could trigger the fission of other uranium atoms, thus setting off a **chain reaction** (Figure 4.14). Because massive amounts of energy could be obtained from the fission of uranium, he saw that the fission process might produce a bomb with tremendous explosive force.

Aware of the destructive forces that could be produced and concerned that Germany might develop such a bomb, Szilard prevailed on Einstein to sign a letter to President Franklin D. Roosevelt indicating the importance of the discovery. It was critical that the U.S. government act quickly.

The Manhattan Project

President Roosevelt launched a highly secret research project for the study of atomic energy. Called the Manhattan Project, it eventually became a massive research effort involving more scientific brainpower than has ever been devoted to a single project. Amazingly, it was conducted under such extreme secrecy that even Vice-President Harry Truman did not know about its existence until after Roosevelt's death.

The Manhattan Project included four separate research teams working on the following problems:

Leo Szilard persuaded Einstein to send the letter to President Franklin Roosevelt that resulted in the Manhattan Project. But later he was one of the scientists who begged that the bomb not be used.

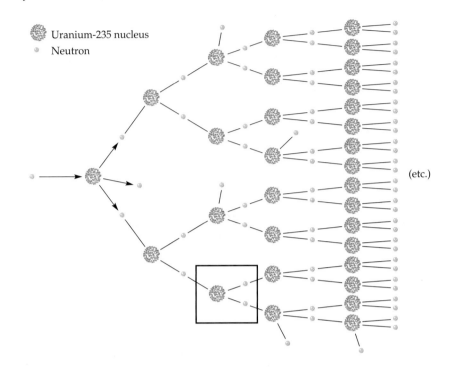

○ Uranium-235 nucleus

· Neutron

(etc.)

Figure 4.14 Schematic representation of a nuclear chain reaction. Neutrons released in the fission of one uranium-235 nucleus can strike other nuclei, causing them to split and release more neutrons. For simplicity, fission fragments are not shown. An area equivalent to that inside the box is shown in detail in Figure 4.13.

A *fissile* isotope is one that undergoes fission (splitting). A *nonfissile* isotope does not undergo fission.

1. how to sustain the nuclear fission chain reaction,
2. how to enrich uranium-235 (the rare fissile isotope),
3. how to make plutonium-239 (another fissile isotope), and
4. how to build a bomb based on nuclear fission.

Sustaining the Nuclear Chain Reaction   The Manhattan Project was created in 1939. By that time it had been established that neutron bombardment could initiate the fission reaction, but there were many things about fission that were unknown. Enrico Fermi and his group, working in a lab under the bleachers at Stagg Field on the campus of the University of Chicago, worked on the fission reaction itself and how to sustain it.

It was found that the neutrons used to trigger the reaction had to be slowed down in order to increase the probability that they would hit a uranium nucleus. Because graphite slows down neutrons, a large "pile" of graphite was built to house the reaction. Then the uranium "fuel" was gradually increased. The major question was how large a **critical mass** of uranium-235 was needed to sustain the fission reaction. There had to be enough fissile nuclei for the neutrons released in one fission process to have a good chance of being captured by another fissile nucleus before escaping from the pile.

On 2 December 1942, Fermi and his group achieved the first sustained nuclear fission reaction. The critical mass for uranium-235 turned out to be about 4 kg, an amount about the size of a baseball.

Isotopic Enrichment

Natural uranium is more than 99% uranium-238, which does not undergo fission. Uranium-235, the fissile isotope, makes up only 0.7% of natural uranium. Because making a bomb required that the uranium be enriched to about 90% uranium-235, it was necessary to find a way to separate the uranium isotopes. This was the job of a top-secret research team in Oak Ridge, Tennessee.

Chemical separation was almost impossible. The separation method eventually used involved the conversion of uranium to uranium hexafluoride, UF_6, which can be vaporized. Molecules containing uranium-235 are slightly lighter in weight and therefore move slightly faster than molecules containing the uranium-238 isotope. Vapors of uranium hexafluoride were allowed to pass through a series of thousands of pinholes, and the molecules containing uranium-235 gradually outdistanced the others. About 15 kg of enriched uranium-235 was finally obtained, enough to make a small explosive device.

The Synthesis of Plutonium

While the tedious work of separating uranium isotopes was under way at Oak Ridge, other workers, led by Glenn T. Seaborg, approached the problem of obtaining fissionable material by another route. It was known that uranium-238 would not fission when bombarded by neutrons. However, it had been determined that when uranium-238 was bombarded by neutrons, a new element, named neptunium (Np), was formed and that this product quickly decayed to another new element, plutonium (Pu).

$$^{238}_{92}U + {}^{1}_{0}n \longrightarrow {}^{239}_{92}U$$

$$^{239}_{92}U \longrightarrow {}^{0}_{-1}e + {}^{239}_{93}Np$$

$$^{239}_{93}Np \longrightarrow {}^{239}_{94}Pu + {}^{0}_{-1}e$$

The isotope plutonium-239 was found to be fissile and thus was suitable material for the making of a bomb. A series of large reactors were built near Hanford, Washington, to produce plutonium.

Seaborg was involved in the discovery of several other transuranium elements. In 1994 he was honored by having element 106 named for him: seaborgium (Sg). Seaborg was the first person ever to be so honored while he was still alive.

Building the Bomb

The actual building of the nuclear bomb was carried out at Los Alamos, New Mexico, under the direction of J. Robert Oppenheimer. In a top-secret laboratory at a remote site, a group of scientists worked at planning and then constructing what would become known as the atomic bomb.

The critical mass of uranium-235 had been determined to be about 4 kg, and so it was important that no single piece of fissionable material be that large. The bomb would contain a number of pieces of subcritical mass, plus a neutron source to initiate the fission reaction. Then, at the chosen time, all the pieces would be forced together by setting off a charge of TNT (trinitrotoluene) thus triggering a runaway nuclear chain reaction.

By July 1945, enough plutonium had been made for a bomb to be assembled. The first atomic bomb was tested in the desert near Alamogordo, New Mexico, on 16 July 1945. The heat from the explosion vaporized the 30-m steel tower on which the bomb was placed and melted the sand for several hectares around the site. The light produced was the brightest anyone had ever seen.

Some of the scientists were so awed by the force of the blast that they argued against its use against Japan. A few, led by Leo Szilard, suggested a demonstration of its power at an uninhabited site. But fear of a well-publicized "dud" and the desire to avoid millions of casualties in an invasion of Japan led President Harry S. Truman to order the dropping of bombs on Japanese cities. A uranium bomb called Little Boy was dropped on Hiroshima on 6 August 1945, causing over 100,000 casualties (Figures 4.15 and 4.16). Three days later, a plutonium bomb called Fat Man was dropped on Nagasaki with comparable results. World War II ended with the surrender of Japan on 14 August 1945.

A *neutron bomb*, a more recent weapon, produces large amounts of neutrons. Neutrons are a serious radiation problem for people because water absorbs neutrons and the human body is about two-thirds water. Therefore, neutron bombs are highly effective at killing people without doing much damage to buildings or creating much radioactive fallout.

Figure 4.15 A nuclear bomb of the type exploded over Hiroshima. The bomb is 71 cm in diameter and 305 cm long; it weighs 4000 kg and has an explosive power equivalent to about 18 million kg (20,000 tons) of high explosive.

Figure 4.16 The now familiar mushroom cloud that follows a nuclear explosion.

4.11 Radioactive Fallout

When a nuclear explosion occurs in the open atmosphere, radioactive materials can rain down on parts of Earth thousands of miles away, days and weeks later. This is called *radioactive fallout*.

The uranium atom can split in many different ways. Some examples are

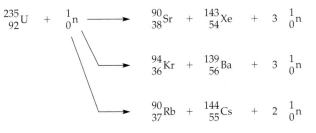

$$^{235}_{92}U + {}^{1}_{0}n \longrightarrow {}^{90}_{38}Sr + {}^{143}_{54}Xe + 3\ {}^{1}_{0}n$$

$$^{94}_{36}Kr + {}^{139}_{56}Ba + 3\ {}^{1}_{0}n$$

$$^{90}_{37}Rb + {}^{144}_{55}Cs + 2\ {}^{1}_{0}n$$

The primary fission products are radioactive. They decay to daughter isotopes, many of which are also radioactive. In all, over 200 different fission products are produced, with half-lives varying from less than a second to more than a billion years. In addition, the neutrons produced in the explosion act on molecules in the atmosphere to produce carbon-14, tritium, and other radioisotopes. Fallout is therefore exceedingly complex. We consider only three of the more worrisome isotopes here.

Of all the isotopes, strontium-90 presents the greatest hazard to people. The isotope has a half-life of 28 years. Strontium-90 reaches us primarily through milk and vegetables (Figure 4.17). Because of its similarity to calcium (both are Group 2A elements), strontium-90 is incorporated into bone. There it remains a source of internal radiation for many years.

Iodine-131 may present a greater threat immediately after a nuclear explosion. Its half-life is only 8 days, but it is produced in relatively large amounts. Iodine-131 is efficiently carried through the food chain. In the body it is concentrated in the thyroid gland, and it is precisely this characteristic that makes it so useful for diagnostic scanning. However, for a healthy individual, the incorporation of radioactive iodine offers no useful information, only damaging side effects.

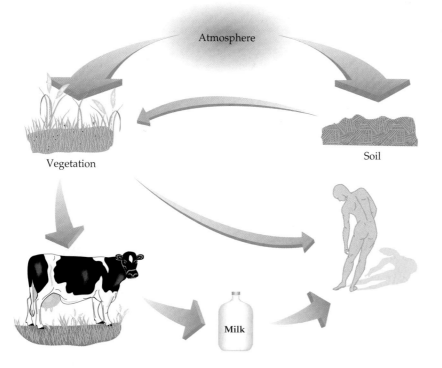

Figure 4.17 Pathways of strontium-90 from fallout.

Another important isotope in fallout is cesium-137. Cesium is similar to potassium (both are Group 1A elements), and it mimics potassium in the body. Cesium-137 is a gamma emitter and has a half-life of 30 years. It is less of a threat than strontium-90, however, because it is removed from the body more readily. We ingest cesium-137 through vegetables, milk, and meat.

By the late 1950s, radioactive isotopes from atmospheric testing of nuclear weapons were detected in the environment. Concern over radiation damage from nuclear fallout led to a movement to ban atmospheric testing. Many scientists were leaders in the movement. Linus Pauling, who won the Nobel prize in chemistry in 1954 for his bonding theories and for his work in determining the structure of proteins, was a particularly articulate advocate of banning atmospheric nuclear testing. In 1963 a nuclear test ban treaty was signed by the major powers—with the exception of France and the People's Republic of China, who continued aboveground tests. Since the signing of the treaty, other countries have joined the nuclear club. Pauling, who had endured being called a Communist and a traitor because of his outspoken position, was awarded the Nobel prize for peace in 1962.

4.12 Nuclear Winter

Radioactive fallout isn't the only concern in the aftermath of nuclear explosions. The nations of planet Earth have acquired nuclear weapons with an explosive power equal to more than that of a million Hiroshima bombs. Studies suggest that the explosion of only half of these weapons would produce enough soot, smoke, and dust to blanket Earth, block out the sun, and bring on a **nuclear winter** that would threaten survival of the human race. Computer simulations suggest that as much as 90% of the sun's light could be prevented from reaching the surface of Earth, dropping the temperature to -25°C for several months. Crops would freeze, farm animals would die of thirst and hunger, and many species would be extinguished, including perhaps our own.

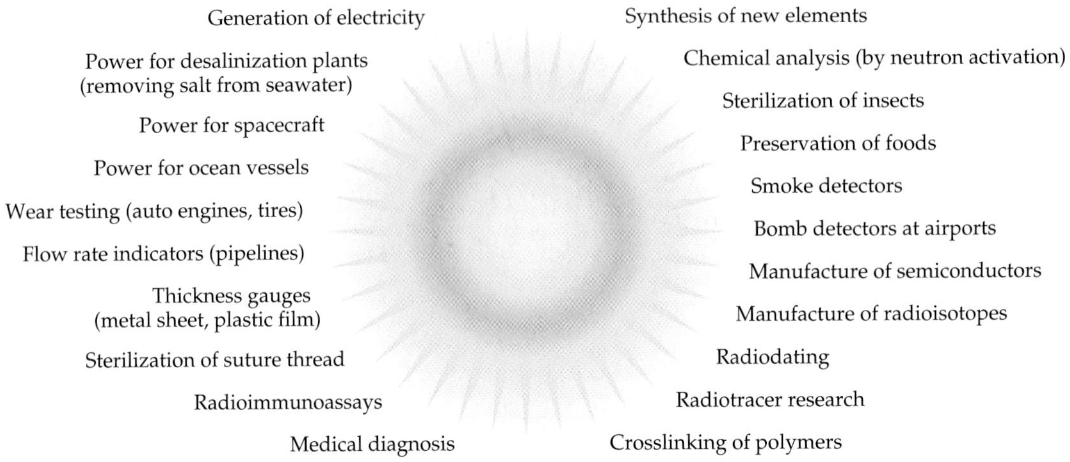

Generation of electricity

Power for desalinization plants
(removing salt from seawater)

Power for spacecraft

Power for ocean vessels

Wear testing (auto engines, tires)

Flow rate indicators (pipelines)

Thickness gauges
(metal sheet, plastic film)

Sterilization of suture thread

Radioimmunoassays

Medical diagnosis

Synthesis of new elements

Chemical analysis (by neutron activation)

Sterilization of insects

Preservation of foods

Smoke detectors

Bomb detectors at airports

Manufacture of semiconductors

Manufacture of radioisotopes

Radiodating

Radiotracer research

Crosslinking of polymers

Figure 4.18   Some peacetime uses of nuclear energy.

4.13 Nuclear Power Plants

Nuclear energy has been used for highly destructive bombs and has led to harmful radioactive fallout, and it might even bring on a devastating nuclear winter. Yet, many more people have been helped by nuclear energy than have been hurt by it. Many applications of nuclear energy are extremely beneficial. We have already mentioned the use of radioactive isotopes in medical diagnosis (Section 4.9) and in dating objects of unknown age (Section 4.3). Figure 4.18 suggests a number of other peacetime uses of nuclear energy. Notice that the first one is generation of electricity. Much of the electric power being used today is generated by nuclear power plants.

At present, one-fifth of all the electricity produced in the United States comes from nuclear power plants. In Europe, there is even greater reliance on nuclear energy for the production of electricity. In France, for example, more than 70% of the country's electric power comes from nuclear plants.

4.14 Thermonuclear Reactions

Almost all of our energy here on Earth comes from thermonuclear reactions taking place on the sun. They are called **thermonuclear reactions** because very high temperatures (millions of degrees) are required in order to initiate them. The intense temperatures and pressures on the sun cause nuclei to fuse and release enormous amounts of energy. The principal reaction is believed to be the fusion of four hydrogen nuclei to produce one helium nucleus and two positrons.

$$4\,_1^1\text{H} \longrightarrow \,_2^4\text{He} + 2\,_{+1}^{\,0}\text{e}$$

The fusion of only 1 g of hydrogen releases an amount of energy equivalent to the burning of nearly 20 tons of coal.

Although controlled nuclear fusion is not yet a reality on Earth, we can carry out uncontrolled fusion in the form of the **hydrogen bomb** (Figure 4.19). In this case a small fission bomb containing uranium or plutonium must be used as the core in order to provide enough heat to achieve the million-degree temperatures needed to start the nuclear fusion reaction. The outer layer of a hydrogen bomb is mainly lithium deuteride ($_3^6\text{Li}\,_1^2\text{H}$). The lithium produces tritium by neutron bombardment from the fission reaction

Figure 4.19 A cloud and fireball produced by explosion of a hydrogen bomb.

$$^6_3\text{Li} + ^1_0\text{n} \longrightarrow ^4_2\text{He} + ^3_1\text{H}$$

and then the deuterium and tritium undergo nuclear fusion.

$$^2_1\text{H} + ^3_1\text{H} \longrightarrow ^4_2\text{He} + ^1_0\text{n}$$

4.15 The Nuclear Age

We live in exciting times. The goal of the alchemists, to change one element into another, has been achieved through the application of scientific principles. New elements have been formed, and the periodic chart has been extended beyond uranium ($Z = 92$) to element 118. This modem alchemy produces plutonium by the ton; neptunium ($Z = 93$), americium ($Z = 95$), and curium ($Z = 96$) by the kilogram; and berkelium ($Z = 97$) and einsteinium ($Z = 99$) by the milligram. These new elements have been used in medicine and to power spacecraft and build bombs. We live in an age in which extraordinary forces have been unleashed. The threat of nuclear war has been a constant specter for the last five decades. Still, it is hard to believe that the world would be a better place if we had not discovered the secrets of the atomic nucleus.

[Web Reference 5]
A discussion of nuclear fusion.

[Web Reference 6]
Risk from radiation from various sources is compared with risk from other factors.

[Web Reference 7]
A discussion of the synthesis and the nature of the heaviest elements.

[Web Reference 8]
More about the work of Glenn Seaborg and his colleagues at Berkeley.

Critical Thinking Exercises

Apply knowledge that you have gained in this chapter and one or more of the FLaReS principles (Chapter 1) to evaluate the following statements or claims.

4.1 A cave in Mexico is claimed to have wonderful healing powers. People who are ill sometimes sit for hours on the benches inside this cave. Studies have shown that the cave really is unusual. Its walls are radioactive, and they constantly give off a low level of radiation. Do you think the cave really has healing power?

4.2 There are some who claim that people who live near high-voltage cables, especially children, run a higher than normal risk of getting leukemia. It is true that electromagnetic fields are associated with high-voltage wires. There have been hundreds of scientific studies on this subject. In only a few of these studies has a slightly higher incidence of leukemia been observed among families living near high-voltage wires. What do you think about the possible danger of living near high-voltage cables?

4.3 For more than 600 years it has been alleged that the Shroud of Turin was the burial shroud of Jesus Christ. In 1988 several laboratories carried out carbon-14 analyses indicating that the flax from which the shroud was made was grown during the period between C.E. 1260 and 1390. Recently there have been new claims that the 1988 analyses are unreliable because the shroud is coated with pollen from plants grown in the fourteenth century, and it is the age of this pollen that the scientists were actually measuring in 1988. At least one of the scientists has offered to repeat the analysis on a very carefully cleaned sample of the cloth, but further access to the shroud has been refused.

Summary

1. The atomic nucleus is a tiny speck in the middle of an atom, containing all of its protons and neutrons and therefore almost all of its mass.

2. A nuclear symbol includes the atomic number as a subscript and the mass number (or nucleon number) as a superscript, both placed in front of the symbol.

3. Nuclear equations are written using nuclear symbols, with superscript mass numbers and subscript atomic numbers, each in balance on both sides of the arrow.

4. Radioactive decay may involve alpha emission, beta emission, or gamma radiation. There are also other modes of decay, such as positron emission and electron capture.

5. The half-life of a radioactive isotope is the time it takes for half of a sample to undergo radioactive decay.

6. Artificial transmutation can be achieved by nuclear bombardment of stable atomic nuclei with protons, neutrons, alpha particles, or other subatomic particles.

7. Alpha particles (helium nuclei) are relatively slow and low in penetrating power. Beta particles (electrons) are much faster and more penetrating. Gamma rays (high-energy photons) travel at the speed of light and have great penetrating power.

8. Scientists often use radioactive isotopes as tracers. A radioactive atom in a molecule labels it so that it can be followed by a radiation detector.

9. In medicine, radioisotopes are used both for diagnosis and for therapy. Diagnostic procedures (such as PET scans) use radioisotopes as tracers. Radiation therapy uses radiation to kill cancer cells.

10. Half-lives of radioisotopes can be used to measure the ages of certain objects. The ages of rocks containing uranium can be estimated by measuring their uranium to lead ratios. Carbon-14 dating can be used to estimate the ages of artifacts containing carbon.

11. Einstein's mass-energy equation, $E = mc^2$, was derived in 1905, but it was not fully appreciated until the explosion of the atom bomb 40 years later.

12. The nuclear fission reaction, although discovered in Germany, became the center of the massive but secret U.S. Manhattan Project. Its goals were (a) to achieve sustained nuclear fission and determine the critical mass required; (b) to enrich the amount of fissile uranium-235 in ordinary uranium; (c) to synthesize plutonium-239, which is also fissile; and (d) to construct a nuclear fission bomb before the Germans were able to do so.

13. In addition to the devastation at the site of a nuclear explosion, much radioactive debris is produced, resulting in radioactive fallout all over the planet.

14. It is believed that a full-scale nuclear war would produce enough smoke and soot to block out the sunlight and bring on a nuclear winter.

15. Radiation is hazardous because it can knock electrons from molecules, turning the molecules into ions. When this happens to molecules in human cells, the cells often die.

16. The total mass of the individual nucleons in an atom is greater than the actual mass of the nucleus. The missing mass is present as binding energy holding the nucleons together.

17. Whereas fission reactions break up large atomic nuclei into smaller ones, nuclear fusion reactions combine small nuclei to form larger ones. Nuclear fusion produces even more energy than nuclear fission and is the basis of the hydrogen bomb. It is also the source of the sun's energy.

Key Terms

background radiation 4.8	electron capture 4.1	mass-energy equation 4.9	radioactive decay 4.1
binding energy 4.9	fundamental particles 4.4	nuclear fission 4.10	radioisotopes 4.6
carbon-14 dating 4.3	half-life 4.2	nuclear fusion 4.9	thermonuclear
chain reaction 4.10	hydrogen bomb 4.14	nuclear winter 4.12	reactions 4.14
cosmic rays 4.8	induced radioactivity 4.5	photoscan 4.7	tracers 4.6
critical mass 4.1	ionizing radiation 4.8	positron 4.1	transmutation 4.4

Review Questions

1. Define or identify each of the following.
   a. half-life   b. positron
   c. background radiation   d. radioisotope
   e. fission   f. fusion
   g. artificial transmutation h. binding energy

2. How does the size of the nucleus compare with that of an atom as a whole?

3. Why are isotopes important in nuclear reactions?

4. What is meant by the nucleon number of an isotope?

5. Give the nuclear symbols for protium, deuterium, and tritium (which are hydrogen-1, hydrogen-2, and hydrogen-3, respectively).

6. Give the nuclear symbol for an isotope with a nucleon number of 8 and an atomic number of 5.

7. Give the nuclear symbol for an isotope with $Z = 35$ and $A = 83$.

8. Give the nuclear symbol for an isotope with 53 protons and 72 neutrons.

9. Give the nuclear symbols for the following isotopes. You may refer to the periodic table.
   a. gallium-69
   b. molybdenum-98
   c. molybdenum-99
   d. technetium-98

10. Indicate the number of protons and the number of neutrons in atoms of the following isotopes:
   a. $^{62}_{30}Zn$
   b. $^{241}_{94}Pu$
   c. $^{99m}_{43}Tc$
   d. $^{81m}_{36}Kr$

11. Which of the following pairs represent isotopes?
   a. $^{70}_{34}X$ and $^{70}_{33}X$
   b. $^{57}_{28}X$ and $^{66}_{28}X$
   c. $^{186}_{74}X$ and $^{186}_{74}X$
   d. $^{8}_{2}X$ and $^{6}_{4}X$
   e. $^{22}_{11}X$ and $^{44}_{22}X$

12. The longest-lived isotope of fermium (Fm) has a nucleon number of 257. How many neutrons are there in the nucleus of this isotope?

13. The longest-lived isotope of technetium (Tc) has 54 neutrons. What is the nucleon number of this isotope?

14. The two principal isotopes of lithium are lithium-6 and lithium-7. The atomic mass of lithium is 6.9 u. Which is the predominant isotope of lithium?

15. Out of every five atoms of boron, one has a mass of 10 u and four have a mass of 11 u. What is the atomic mass of boron? Use the periodic table only to check your answer.

16. Give the nuclear symbols for the following subatomic particles.
   a. alpha particle
   b. beta particle
   c. neutron
   d. positron

17. When a nucleus emits a beta particle, what changes occur in the nucleon number and atomic number of the nucleus?

18. When a nucleus emits a neutron, what changes occur in the nucleon number and the atomic number of the nucleus?

19. When a nucleus emits a proton, what changes occur in the nucleon number and the atomic number of the nucleus?

20. When a nucleus emits an alpha particle, what changes occur in the nucleon number and atomic number of the nucleus?

21. When a nucleus emits a gamma ray, what changes occur in the nucleon number and the atomic number of the nucleus?

22. When a nucleus emits a positron, what changes occur in the nucleon number and the atomic number of the nucleus?

23. Explain how radioisotopes can be used for therapeutic purposes.

24. Which radioisotope has been used extensively for the treatment of overactive or cancerous thyroid glands?

25. Describe the use of a radioisotope as a diagnostic tool in medicine.

26. What are some of the characteristics that make technetium-99m such a useful radioisotope for diagnostic purposes?

27. A pair of gloves would be sufficient to shield the hands from which type of radiation: the heavy alpha particles or the massless gamma rays?

28. Heavy lead shielding is necessary as protection from which type of radiation: alpha, beta, or gamma?

29. Plutonium is especially hazardous when inhaled or ingested because it emits alpha particles. Why would alpha particles cause more damage to tissue than beta particles?

30. What form of radiation is detected in PET scans?

31. List two ways in which workers can protect themselves from the radioactive materials with which they work.

32. Which subatomic particles are responsible for carrying on the chain reactions characteristic of nuclear fission?

33. Compare nuclear fission and nuclear fusion. Why is energy liberated in each case?

34. Discuss nuclear winter.

35. What is the source of the greatest proportion of our exposure to artificial radiation?

36. What is the source of the energy that triggers the thermonuclear reactions of the hydrogen bomb?

37. Did President Harry S. Truman make the right decision when he decided to drop nuclear bombs on Japanese cities? Would your answer be the same if you were living in 1945 and had relatives among the troops preparing for the invasion of Japan? If you were an inhabitant of one of the cities bombed?

38. Discuss the impact of nuclear science on the following topics. (See References and Readings for resource material.)
   a. war and peace
   b. industrial progress
   c. medicine
   d. agriculture
   e. human, animal, and plant genetics

Problems

Nuclear Equations

Write a balanced equation for each of the following.

39. Emission of a positron by sulfur-31.

40. Emission of a neutron by bromine-87.

41. Emission of a proton by magnesium-21.

42. Beta decay of lead-209.

43. Alpha decay of thorium-225.

44. Emission of a gamma ray by gold-186.

45. Complete the following equations.
   a. $^{179}_{79}Au \longrightarrow {}^{175}_{77}Ir + ?$
   b. $^{23}_{10}Ne \longrightarrow {}^{23}_{11}Na + ?$

46. Complete the following equations.
   a. $^{10}_{5}B + {}^{1}_{0}n \longrightarrow {}^{4}_{2}He + ?$
   b. $^{12}_{6}C + {}^{2}_{1}H \longrightarrow {}^{13}_{6}C + ?$
   c. $^{121}_{51}Sb + ? \longrightarrow {}^{121}_{52}Te + {}^{1}_{0}n$
   d. $^{154}_{62}Sm + {}^{1}_{0}n \longrightarrow 2\,{}^{1}_{0}n + ?$

47. When magnesium-24 is bombarded with a neutron, a proton is ejected. What new element is formed? (*Hint:* Write a balanced nuclear equation.)

48. When clorine-37 is bombarded with a neutron, a proton is ejected. What new element is formed?

49. A radioactive isotope decays to give an alpha particle and bismuth-211. What was the original element?

50. A radioisotope decays to give an alpha particle and protactinium-233. What was the original element?

Half-Life

51. C. E. Bemis and colleagues at Oak Ridge National Laboratory confirmed the synthesis of element 104, the half-life of which was only 4.5 s. Only 3000 atoms of the element were created in the tests. How many atoms were left after 4.5 s? After a total of 9.0 s?

52. Krypton-81m is used for lung ventilation studies. Its half-life is 13 s. How long does it take the activity of this isotope to reach one-quarter of its original value?

53. A 100-mg technetium-99m sample is used in a medical study. How much of the technetium-99m sample remains after 24 hr? The half-life of technetium-99m is 6.0 hr.

54. The half-life of molybdenum-99 is 67 hr. How much time passes before a sample with an activity of 160 counts/min has decreases to 5.0 counts/min?

Radioisotopic Dating

55. Living matter has a carbon-14 activity of 16 counts/min per gram of carbon. What is the age of an artifact for which the carbon-14 activity is 8 counts/min per gram of carbon?

56. A piece of wood from an Egyptian tomb has carbon-14 activity of 980 counts per hour. A piece of new wood of the same size gave 3920 counts/hr. What is the age of the wood from the tomb?

57. The ratio of carbon-14 to carbon-12 in a piece of charcoal from an archaeological excavation is found to be one-half the ratio in a sample of modern wood. Approximately how old is the site? How old would it be if the ratio were 25% of the ratio in a sample of modern wood?

58. How old is a bottle of wine if the tritium activity is 25% that of new wine? The half-life of tritium is 12.3 years.

Additional Problems

59. To make element 106, a 0.25-mg sample of californium-249 was used as the target. Four neutrons were emitted to yield a nucleus with 106 protons and a mass of 263 u. What was the bombarding particle?

60. One atom of element 109 with a nucleon number of 266 was produced in 1982 by bombarding a target of bismuth-209 with iron-58 nuclei for 1 week. How many neutrons were released in the process?

61. Element 109 undergoes alpha emission to form element 107, which in turn also emits an alpha particle. What are the atomic number and nucleon number of the isotope formed by these two steps? Write balanced nuclear equations for the two reactions.

62. Radium-223 nuclei usually decay by alpha emission. For every billion alpha decays, one atom emits a carbon-14 nucleus. Write a balanced nuclear equation for each type of emission.

63. Uranium has a density of 19 g/cm^3. What volume is occupied by a critical mass of 8 kg of uranium?

64. Neptunium-237 undergoes a series of seven alpha and four beta decays. What stable isotope results from this radioactive decay series?

65. Write an equation to represent each of the following nuclear processes.

    a. The reaction of two deuterons to produce helium-3.

    b. The production of $^{243}_{97}$Bk by the alpha particle bombardment of $^{241}_{95}$Am.

    c. The bombardment of $^{121}_{51}$Sb by alpha particles to produce $^{124}_{53}$I, followed by its radioactive decay by positron emission.

66. What is the new nucleus formed in each of the following processes?

    a. Lead-196 goes through two successive EC processes.

    b. Bismuth-215 decays through two successive beta emissions.

    c. Protactinium-231 decays through four successive alpha emissions.

67. Complete the following nuclear equations.

    a. $^{10}_{5}$B + $^{1}_{0}$n $\longrightarrow$? + $^{1}_{1}$H

    b. $^{121}_{51}$Sb + ? $\longrightarrow$ $^{121}_{52}$Te + $^{1}_{0}$n

    c. $^{59}_{27}$Co + $^{1}_{0}$n $\longrightarrow$ $^{56}_{25}$Mn + ?

68. Complete the following nuclear equations.

    a. $^{154}_{62}$Sm + $^{1}_{0}$n $\longrightarrow$? + 2$^{1}_{0}$n

    b. ? + $^{4}_{2}$He $\longrightarrow$ $^{133}_{57}$La + 4$^{1}_{0}$n

    c. $^{246}_{96}$Cm + $^{13}_{6}$C $\longrightarrow$ $^{254}_{102}$No + ?

Projects

69. Write a brief report on one of the following.

    a. Radiodating of archaeological objects

    b. Use of radioisotopes in medicine

    c. The many and varied uses of nuclear energy

70. Write a brief biography of one of the following scientists.

    a. Otto Hahn   b. Enrico Fermi

    c. Glenn T. Seaborg   d. J. Robert Oppenheimer

    e. Lise Meitner   f. Albert Einstein

71. Make a list of Nobel prize winners whose work involved nuclear science.

72. Write a brief essay on the discovery of radioactivity. (You might want to take a look at the *Journal of Chemical Education*, January 1992, p. 10.)

Online Projects

73. The positron is a particle of antimatter. Search the web for information about antimatter, and the positron in particular.

74. Find a website that is strongly in favor of nuclear power plants and one that is strongly opposed. Note their sponsors and analyze their viewpoints. Try to find a website with a balanced approach.

75. Go to Web Reference 3 in the chemistry and physics areas. Make a list of all the Nobel prize recipients we have encountered thus far.

76. Glenn T. Seaborg was the only person ever to have an element named for him while he was alive. See how much you can find out about Seaborg by searching the web.

77. In July 1999 the discovery of a new element was reported. Search for details about this new element on the internet.

References and Readings

1. Armbruster, Paul, and Fritz Hessberger. "Making New Elements." *Scientific American*, September 1998, pp. 72–77. Discusses synthesis of the transuranium elements.

2. Borman, Stu. "Scientists Honor the Centennial of the Discovery of Radioactivity." *Chemical and Engineering News*, 29 April 1996, pp. 55–65.

3. Brennan, Mairin B. "Positron Emission Tomography Merges Chemistry with Biological Imaging." *Chemical and Engineering News*, 19 February 1996, pp. 26–33.

4. Budavari, Susan (Ed.). *The Merck Index*, 11th edition. Rahway, NJ: Merck and Co., 1989. Contains extensive tables of radioisotopes (pp. MISC 31–45) and radioisotopes used in medical diagnosis and therapy (pp. MISC 46–52).

5. Cobb, Cathy, and Harold Goldwhite. *Creations of Fire: Chemistry's Lively History from Alchemy to the Atomic Age*. New York: Plenum, 1995.

6. Ehrlich, Paul R., et al. "Long-Term Biological Consequences of Nuclear War." *Science*, 23 December 1983, pp. 1293–1300.

7. Hoffman, Darleane C., and Diana M. Lee. "Chemistry of the Heaviest Elements—One Atom at a Time." *Journal of Chemical Education*, March 1999, pp. 331–347.

8. Jaworowski, Zbigniew. "Radiation Risk and Ethics." *Physics Today*, September 1999, pp. 24–29.

9. Keen, Judy. "Chernobyl: Coping with Nuclear Disaster." *USA Today*, 17 September 1991, p. 6A.

10. Lenihan, Daniel J. "Ground Zero Revisited." *Natural History*, July 1995, pp. 42–50. The first nuclear explosion.

11. Rhodes, Richard. *The Making of the Atomic Bomb*. New York: Simon & Schuster, 1986. Part 1, "Profound and Necessary Truth."

12. Rigden, John S. "J. Robert Oppenheimer: Before the War." *Scientific American*, July 1995, pp. 76–81.

13. Seaborg, Glenn T. "The Positive Power of Radioisotopes." *The Skeptical Inquirer*, January–February 1995, pp. 39–40, 62.

14. Seaborg, Glenn T., and Walter D. Loveland. *The Elements Beyond Uranium*. New York: Wiley, 1990.

15. Seaborg, Glenn T., and Walter D. Loveland. *The New Chemistry*. New York: Cambridge University Press, 1999.

16. Sime, Ruth Lewin. *Lise Meitner: A Life in Physics*. Berkeley, CA: University of California Press, 1996.

17. Sime, Ruth Lewin, "Meitner and the Discovery of Nuclear Fission. *Scientific American*, January 1998, p. 80.

18. Sparberg, Esther B. "Hindsight and the History of Chemistry." *Journal of Chemical Education*, March 1996, pp. 199–202. Good discussion about the discovery of nuclear fission.

19. Wetherill, George W. "Dating Very Old Objects." *Natural History*, September 1982, pp. 14–20.

20. Wheelwright, Jeff. "For Our Nuclear Wastes, There's Gridlock on the Way to the Dump." *Smithsonian*, May 1995, pp. 40–51.

21. Yalow, Rosalyn S. "Radioactivity in the Service of Man." *Journal of Chemical Education*, September 1982, pp. 735–738.

MediaLab

Radiation and You: Risks and Benefits

Radiation and nuclear power arouse misgivings or anxiety in many people. Whether it is a nuclear power plant, a diagnostic procedure using radioisotopes, or radiation-treated food, many persons oppose even the most peaceful uses of nuclear energy.

Radiation *is* dangerous. In sufficient doses, it can cause mutations, induce cancer formation, or even directly cause death. However, many uses are so beneficial that few knowledgeable persons would turn back the clock to the days before we began to exploit "the power of the atom."

In the *Web Investigation* of this MediaLab, you will explore the risk of radiation exposure and try to determine your personal radiation dosage. You will also look at food irradiation and nuclear power—and read arguments both for and against these technologies. In *Communicate Your Results*, you will use what you've learned to write about radiation and its uses for an uninformed audience.

WEB INVESTIGATION

Investigation 1
Radiation Dosage

The measurement of radiation and of radiation exposure is extremely complex, with many different units, both metric and nonmetric, for measuring various aspects. In reading news items or scientific articles about radiation, it is important to have an understanding of what is being measured—energy emitted from a source or dose absorbed by a body. Select Keyword **UNITS** for an explanation of the various units. The definitions of terms at Keyword **TERMS** may also be useful.

Investigation 2
Personal Radiation Dosage Inventory

In the United States an average person receives about 360 mrem of whole-body radiation annually. Select Keyword **PERSONAL DOSAGE** to learn the sources of this dose and to calculate how much you're receiving. Then go to **DOSE** and read the effects of various doses on the human body.

Investigation 3
Communicating Risk

Many articles have been written about radiation and its dangers. Much of this information is valid, but some overstate the benefits, others exaggerate the risk, and some verge on hysteria. Select the Keyword **COMMUNICATE** and read the advice on communicating with the community about nuclear risk. Although this site is geared toward government agencies preparing for accidents involving radiation, many of the guidelines can have universal application.

Investigation 4
Nuclear Power: From Enthusiasm to Fear

Many countries have embraced nuclear power as a route to independence from the oil cartel. In the United States, however, public opposition has stopped the growth of the nuclear power industry.

Select Keyword **PBS** for a variety of links about nuclear power and American fears. Explore one or more of these links. ("Searching for Safety" gives a good historical overview.)

Investigation 5
Food Irradiation: A Controversial Topic

Another area of increasing controversy is the irradiation of food to kill bacteria and lengthen shelf life. Arguments both for and against this technology often cast more heat than light. Select Keyword **BALANCED** for a fairly balanced view of the topic, which lists disadvantages as well as advantages. Select Keyword **FOOD** for a good discussion that attempts to counter arguments against the technology. (This site also has numerous links.) For a look at some vehement opposition to food irradiation, select Keyword **OPPOSE**.

COMMUNICATE YOUR RESULTS

Exercise 1
A Matter of Numbers

Using the definitions and tables in *Web Investigation 1*, write an explanation of radioactivity and absorbed radiation, defining key units. In "What Effects Do Different Doses of Radiation Have on People?" (Keyword **UNITS**), convert the radiation doses from sieverts to rems. (Remember that your audience may not understand such terms as "milli.")

Exercise 2
How Much Radiation Is Safe?

As with many drugs and poisons, the effect of radiation on the human body depends on the dose per unit of time and on the total accumulated dose. Expand your *Exercise 1* article with a discussion of radiation dose and its effects.

Exercise 3
Explaining the Risk

Refer back to the discussion of benefit and risk in Chapter 1. Write an analysis of the benefits versus the risks of food irradiation *or* nuclear power plants.

Exercise 4
Critical Reading

Return to the website at Keyword **OPPOSE**. Write a critical evaluation of the article, pointing out aspects that should arouse a reader's skepticism.

Chemical Bonds
The Ties That Bind

Whether it is hard or soft, or prone to liquefaction,
Depends upon its bonding and its forces of attraction.

Just as these skydivers bond together to form a pattern, atoms bond together in molecules and crystals, giving structure to matter.

Zinc is a silvery metal that conducts electricity. Sulfur is a yellow solid that melts to become a reddish-brown liquid. When these two elements react with each other, they produce zinc sulfide, which is white. Zinc sulfide is nothing like zinc or sulfur.

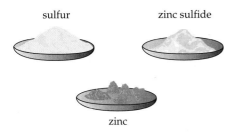

sulfur

zinc sulfide

zinc

White zinc sulfide powder is nothing like the zinc and sulfur from which it was made.

Mercury is a silver liquid, and oxygen is a colorless gas. These two elements combine to form mercuric oxide, which is a bright orange-red solid, quite different from mercury or oxygen. When elements combine to form a compound, they do not keep their original properties. They are transformed into a completely different substance with its own properties.

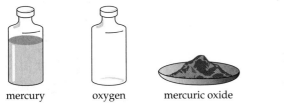

mercury

oxygen

mercuric oxide

Mercuric oxide, a red solid, is made from a silvery liquid (mercury) and a colorless gas (oxygen).

Why should compounds be so unlike their component elements? The answer lies in the bonds that tie the various atoms together. We have learned a bit about the structure of the atom and have looked briefly at the atomic nucleus. Now we are ready to consider chemical bonds, the ties that bind atoms together.

Chemical bonds are the forces that hold together atoms in molecules and ions in crystals. Let us consider some of the reasons why chemical bonds are so important.

- The great number and variety of compounds in the world result from the fact that atoms can form bonds with certain other atoms.
- Whether a substance is a solid, a liquid, or a gas depends mainly on its bonding.
- Whether a solid is hard and strong or soft and waxy depends on its bonding.
- Whether a liquid is light and volatile or heavy and viscous depends on its bonding.
- The melting points and boiling points of substances depend largely on their bonding.
- The strength of materials used for building bridges, houses, and thousands of other structures results from their chemical bonding.
- Molecular bonding determines the shapes of molecules. Taste, odor, and drug action are some of the many properties that depend on molecular shape.

- Ionic bonding plays some important biological roles. For example, bones and teeth depend on the bonding of calcium and phosphate ions.
- Because it forms stronger bonds with hemoglobin than oxygen does, carbon monoxide gas can kill.
- Knowledge of molecular structure and bonding enables chemists to design drugs, perfumes, plastics, synthetic fibers, pesticides, and many other chemical products.

In this chapter we look at several kinds of chemical bonds, each associated with unique properties.

5.1 The Art of Deduction: Stable Electron Configurations

In our discussion of the atom and its structure (Chapters 2 and 3), we followed the historical development of some of the more important atomic concepts. Some of the nuclear concepts (Chapter 4) were approached in the same way. We could continue to look at chemistry in this manner, but that would require several volumes of print—and perhaps more of your time than you care to spend. We won't abandon the historical approach entirely, but we will emphasize another important aspect of scientific endeavor: deduction.

The art of deduction works something like this.

> The outermost energy level is "filled" when it contains eight electrons, two *s* and six *p* (except for the first level, which holds only two *s* electrons).

1. *Fact* Noble gases, such as helium, neon, and argon, are inert (that is, they undergo few, if any, chemical reactions).
2. *Theory* The inertness of noble gases is a result of their electron structures; each (except helium) has an octet of electrons in its outermost energy level.
3. *Deduction* If other elements could alter their electron structures to become more like those of noble gases, they would become less reactive.

> Diagrams indicating electron structure are often called *Bohr diagrams.*

To illustrate, let's look at an atom of the element sodium (Na). It has 11 electrons, 2 in the first energy level, 8 in the second, and 1 in the third. If the atom could get rid of an electron, it would have the same electronic structure as an atom of the noble gas neon (Ne).

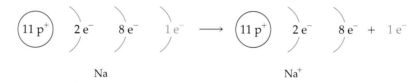

Recall that neon has the structure

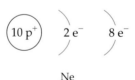

If a chlorine atom (Cl) could gain an electron, it would have the same electronic structure as argon (Ar).

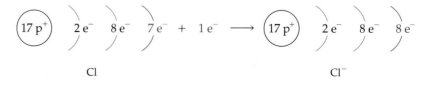

The structure of the argon atom is

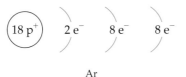

Ar

The sodium atom, having lost an electron, becomes positively charged. It has 11 protons (11+) and only 10 electrons (10−). It is written Na^+ and is called a *sodium ion*. The chlorine atom, having gained an electron, becomes negatively charged. It has 17 protons (17+) and 18 electrons (18−). It is written Cl^- and is called a *chloride ion*. Note that a positive charge, as in Na^+, indicates that one electron has been lost. Similarly, a negative charge, as in Cl^-, indicates that one electron has been gained.

5.2 Electron-Dot Structures

In forming ions, the cores of sodium atoms and chlorine atoms do not change. It is convenient therefore to let the symbol represent the *core* of the atom (nucleus plus inner electrons). The **valence (outer) electrons** are then represented by dots. The equations of the preceding section then can be written as follows.

$$Na\cdot \longrightarrow Na^+ + 1e^-$$

and

$$:\overset{..}{\underset{..}{Cl}}\cdot + 1e^- \longrightarrow :\overset{..}{\underset{..}{Cl}}:^-$$

Such forms, in which the symbol of the element represents the core and dots stand for valence electrons, are called **electron-dot symbols**.

Electron-Dot Symbols and the Periodic Table

It is especially easy to write electron-dot symbols for most of the main group elements. The number of valence electrons for most of these elements is equal to the group number (Table 5.1).

Example 5.1

Without referring to Table 5.1, give electron-dot symbols for magnesium, oxygen, and phosphorus. You may use the periodic table.

Solution

Magnesium is in Group 2A, oxygen is in Group 6A, and phosphorus is in Group 5A. The electron-dot symbols are therefore

$$\cdot Mg\cdot \qquad :\overset{\cdot}{\underset{\cdot}{O}}: \qquad :\overset{\cdot}{P}\cdot$$

Exercise 5.1

Without referring to Table 5.1, give electron-dot symbols for each of the following elements. You may use the periodic table.
a. Ar   **b.** Ca   **c.** F   **d.** N   **e.** K   **f.** S

Electron-dot symbols are often called *Lewis symbols* after G. N. Lewis, the famous American chemist (1875–1946) who invented this symbolism. He also made important contributions in the fields of thermodynamics, acids and bases, and spectroscopy. He might have been awarded a Nobel prize for any *one* of his major achievements, yet he never received a Nobel prize.

Let us emphasize that chlorine does not *become* argon. The chloride ion and the argon atom have the same stable electron configuration, but the chloride ion has only 17 protons in its nucleus, while the argon atom has 18. Neither does sodium *become* neon when it gives up an electron. The sodium ion and the neon atom simply have the same stable electron configuration.

In writing a Lewis symbol, it is only the *number* of dots that is important. The dots need not be drawn in any specific positions, except that there should be no more than two dots on any given side of the chemical symbol (right, left, top, or bottom).

Table 5.1 ▌ Electron-Dot Symbols for Selected Main Group Elements

Group 1A	Group 2A	Group 3A	Group 4A	Group 5A	Group 6A	Group 7A	Noble Gases
H·							He :
Li·	·Be·	·Ḃ·	·Ċ·	:Ṅ·	:Ö·	:Ḟ:	:N̈e:
Na·	·Mg·	·Al·	·Si·	:Ṗ·	:S̈·	:C̈l·	:Ä̈r:
K·	·Ca·				:Se·	:B̈r·	:K̈r:
Rb·	·Sr·				:Te·	:Ï·	:Ẍe:
Cs·	·Ba·						

If you have not yet learned the chemical symbols listed in Table 1.1, you should take the time to memorize them now. You will save time in the long run.

The use of symbolism is not unique to chemistry, of course. Symbols are used in almost every area of human endeavor, from traffic signs to musical notation.

Chemical Symbolism

The reason why chemistry often seems mysterious to the nonchemist is probably because of chemical symbolism. Symbolism is just a convenient shorthand way to convey a lot of information quickly and easily. Chemists find it convenient and much easier to represent sodium and chlorine atoms as Na· and :Cl· instead of writing

Learning chemical symbolism is much like learning a foreign language. Once you have learned a basic "vocabulary," the rest is a lot easier. In Chapter 1 we introduced symbols for the chemical elements. Now you also know how to write Lewis symbols using dots to represent valence electrons.

5.3 Sodium Reacts with Chlorine: The Facts

Sodium is a highly reactive metal. It is soft enough to be cut with a knife. When freshly cut, it is bright and silvery, but it dulls rapidly because it reacts with oxygen in the air. In fact, it reacts so readily in air that it is usually stored under oil or kerosene. Sodium reacts violently with water also, becoming so hot that it melts. A small piece forms a spherical bead after melting and races around on the surface of the water as it reacts.

Chlorine is a greenish-yellow gas. It is familiar as a disinfectant for swimming pools and city water supplies. (The actual substance added is usually a compound that reacts with water to form chlorine.) Chlorine is an extremely irritating gas. In fact, it was used as a poison gas in World War I.

If a piece of sodium is dropped into a flask containing chlorine gas, a violent reaction ensues, producing sodium chloride, beautiful white crystals that you might sprinkle on your food at the dinner table. It is ordinary table salt. Sodium chloride has none of the properties of sodium and none of the properties of chlorine (Figure 5.1).

Sodium metal and chlorine gas react violently with each other, producing a great deal of heat and light. The product is sodium chloride, ordinary table salt.

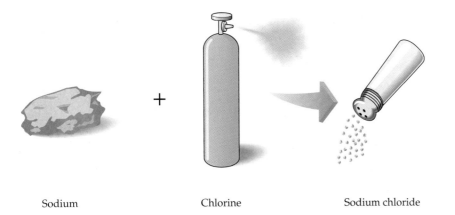

Figure 5.1 Sodium, a soft, silvery metal, reacts with chlorine, a greenish gas, to form sodium chloride (ordinary table salt), a white crystalline solid.

Sodium Chlorine Sodium chloride

5.4 Sodium Reacts with Chlorine: The Theory

A sodium atom becomes less reactive by *losing* an electron. A chlorine atom becomes less reactive by *gaining* an electron. What happens when sodium atoms come into contact with chlorine atoms? The obvious: Chlorine extracts an electron from a sodium atom.

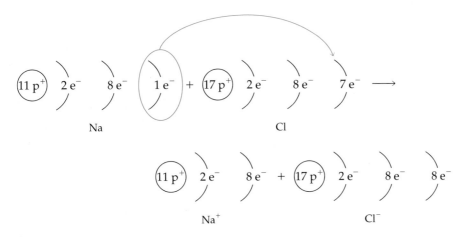

In the abbreviated electron-dot form, this reaction is written

$$Na\cdot \; + \; :\overset{..}{\underset{..}{Cl}}\cdot \; \longrightarrow \; Na^+ \; + \; :\overset{..}{\underset{..}{Cl}}:^-$$

Chlorine gas is actually composed of Cl_2 molecules. Each atom of the molecule takes an electron from a sodium atom. Two sodium ions and two chloride ions are formed.

$$Cl_2 + 2\,Na \longrightarrow 2\,Cl^- + 2\,Na^+$$

Electron loss is known as *oxidation*, and gain of electrons is called *reduction*. We shall explore this concept in Chapter 8.

Ionic Bonds

The two ions formed from sodium and chlorine atoms have opposite charges and are strongly attracted to one another. Remember, however, that for even a tiny grain of salt, there are billions and billions of each kind of ion. These ions arrange themselves in an orderly fashion. The arrangements are repeated many times in all directions—above and below, left and right, top and bottom—to make a crystal of sodium chloride (Figure 5.2). Each sodium ion attracts (and is attracted by) six chloride ions (the ones to its front and back, its top and bottom, and its two sides). Each chloride ion attracts (and is attracted by) six sodium ions. The forces holding the crystal together (the attractive forces between positive and negative ions) are called **ionic bonds**.

A **crystal** is a solid substance with a regular arrangement of its constituent particles. The solid has a well-defined regular shape.

(a)   (b)   (c)

Figure 5.2 Structure of a sodium chloride crystal. (a) Each Na^+ ion (small sphere) is surrounded by six Cl^- ions (large spheres), and each Cl^- ion by six Na^+ ions. (b) This arrangement repeats itself many, many times. (c) The highly ordered pattern of alternating Na^+ and Cl^- ions is observed in the macro world as a crystal of sodium chloride.

We cannot emphasize too strongly the difference between ions and the atoms from which they are made. They are as different as a whole peach (an atom) and a peach pit (a positive ion). The names and symbols may look a lot alike, but the substances themselves are quite different (Figure 5.3). Unfortunately, the situation is confusing because people talk about needing iron to perk up "tired blood" and calcium for healthy teeth and bones. What they really mean is iron(II) *ions* (Fe^{2+}) and calcium *ions* (Ca^{2+}). You wouldn't think of eating iron nails to get iron. Nor would you eat highly reactive calcium metal. Although careful distinctions are not always made by persons who are not chemists, we try to use precise terminology here.

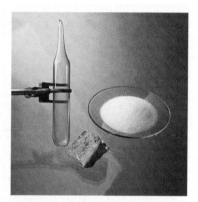

Figure 5.3 Ions differ greatly from the atoms from which they are made. Sodium atoms are the constituents of a soft, highly reactive metal. Chlorine atoms—paired in chlorine molecules—make up a corrosive, greenish-yellow gas. Sodium ions and chloride ions make up ordinary table salt.

5.5 Using Electron-Dot Symbols: More Ionic Compounds

As we might expect, potassium, a metal in the same family as sodium, also reacts with chlorine. The reaction yields potassium chloride (KCl).

$$K\cdot \ + \ \cdot \overset{..}{\underset{..}{Cl}}: \ \longrightarrow \ K^+ \ + \ :\overset{..}{\underset{..}{Cl}}:^-$$

Potassium also reacts with bromine, a reddish-brown liquid in the same family as chlorine, to form stable white, crystalline potassium bromide (KBr).

$$K\cdot \ + \ \cdot \overset{..}{\underset{..}{Br}}: \ \longrightarrow \ K^+ \ + \ :\overset{..}{\underset{..}{Br}}:^-$$

Example 5.2

Use electron-dot symbols to show the transfer of electrons from sodium atoms to bromine atoms to form ions with noble gas configurations.

Solution

Sodium has one valence electron, and bromine has seven. Transfer of the single electron from sodium to bromine leaves each with a noble gas configuration.

$$\text{Na} \cdot \; + \; \cdot \ddot{\text{Br}} \!: \; \longrightarrow \; \text{Na}^+ \; + \; : \ddot{\text{Br}} \!: ^-$$

Exercise 5.2

Use electron-dot symbols to show the transfer of electrons from lithium atoms to fluorine atoms to form ions with noble gas configurations.

Magnesium, a Group 2A metal, is harder and less reactive than sodium. Magnesium reacts with oxygen, a Group 6A element (a colorless gas), to form another stable white, crystalline solid called magnesium oxide (MgO).

$$\cdot \text{Mg} \cdot \; + \; \cdot \ddot{\text{O}} \!: \; \longrightarrow \; \text{Mg}^{2+} \; + \; : \ddot{\text{O}} \!: ^{2-}$$

Magnesium must give up two electrons and oxygen must gain two electrons for each to have the same configuration as the noble gas neon.

An atom such as oxygen, which needs two electrons to complete a noble gas configuration, may react with potassium atoms, which have only one electron each to give. In this case, two atoms of potassium are needed for each oxygen atom. The product is potassium oxide (K_2O).

$$
\begin{array}{ccc}
\text{K} \cdot & & \text{K}^+ \\
& + \; \cdot \ddot{\text{O}} \!: \; \longrightarrow & + \; : \ddot{\text{O}} \!: ^{2-} \\
\text{K} \cdot & & \text{K}^+
\end{array}
$$

By this process, each potassium atom achieves the argon configuration. Oxygen again assumes the neon configuration.

Example 5.3

Use electron-dot symbols to show the transfer of electrons from magnesium atoms to nitrogen atoms to form ions with noble gas configurations.

Solution

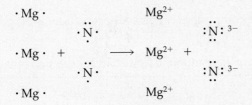

Each of three magnesium atoms gives up two electrons (a total of six), and each of the two nitrogen atoms acquires three (a total of six). Notice that the total positive and negative charges on the products are equal (6+ and 6−). Magnesium reacts with nitrogen to yield magnesium nitride (Mg_3N_2).

Exercise 5.3

Use electron-dot symbols to show the transfer of electrons from aluminum atoms to oxygen atoms to form ions with noble gas configurations.

Generally speaking, metallic elements in Groups 1A and 2A (those from the left side of the periodic table) react with nonmetallic elements in Groups 6A and 7A (those from the right side) to form ionic compounds. These are stable crystalline solids.

When $\cdot$ Mg $\cdot$ becomes Mg^{2+}, its second electron shell becomes its outermost shell. Since the second shell has a full octet of electrons, the Mg^{2+} ion, like the Ne atom, has a very stable 2, 8 electron structure.

The K atom loses one electron to become a K^+ ion, which has a full octet of electrons in its outermost (third) shell. The K^+ ion and the Ar atom both have 2, 8, 8 electron structures.

The Octet Rule

Each atom of metal tends to give up the electrons in its outer shell, and each atom of nonmetal takes on enough electrons to complete its valence shell. The ions so formed have noble gas configurations. An octet of valence electrons is the characteristic arrangement of all noble gases except helium. When atoms react with each other, they often appear to be trying to attain this stable electron configuration. Thus they are said to be following the **octet rule**, or the "rule of eight." (In the case of helium, only two electrons can exist in its single energy level, and so hydrogen follows the "rule of two.")

In following the octet rule, atoms of Group 1A metals give up one electron to form 1+ ions, those of Group 2A metals give up two electrons to form 2+ ions, and Group 3A metals give up three electrons to form 3+ ions. Group 7A nonmetal atoms take on one electron to form 1− ions, and Group 6A atoms tend to pick up two electrons to form 2− ions. Atoms of B group metals can give up various numbers of electrons to form positive ions with various charges. These periodic relationships are summarized in Figure 5.4.

Table 5.2 lists symbols and names for some ions formed by the gain or loss of electrons. You can calculate the charge on the negative ions in the table by subtracting 8 from the group number. For example, the charge on the oxide ion (since oxygen is in Group 6A) is $6 - 8 = -2$. The nitride ion (nitrogen is in Group 5A) has a charge of $5 - 8 = -3$.

Example 5.4

What is the formula of the compound formed by the reaction of sodium and sulfur?

Solution

Sodium is in Group 1A; the sodium atom has one valence electron. Sulfur is in Group 6A; the sulfur atom has six valence electrons

$$\text{Na}\cdot \quad \cdot\overset{\cdot\cdot}{\underset{\cdot\cdot}{\text{S}}}\cdot$$

Sulfur needs two electrons to gain an argon configuration, but sodium has only one to give. The sulfur atom must react with two sodium atoms.

$$\begin{array}{c} \text{Na}\cdot \\ + \quad \cdot\overset{\cdot\cdot}{\underset{\cdot\cdot}{\text{S}}}\cdot \quad \longrightarrow \\ \text{Na}\cdot \end{array} \qquad \begin{array}{c} \text{Na}^+ \\ + \quad :\overset{\cdot\cdot}{\underset{\cdot\cdot}{\text{S}}}:^{2-} \\ \text{Na}^+ \end{array}$$

The formula of the compound, called sodium sulfide, is Na_2S.

Exercise 5.4

What is the formula of the compound formed by the reaction of calcium with fluorine?

1A	2A	3B	4B	5B	6B	7B	8B			1B	2B	3A	4A	5A	6A	7A	Noble gases
Li⁺														N³⁻	O²⁻	F⁻	
Na⁺	Mg²⁺											Al³⁺		P³⁻	S²⁻	Cl⁻	
K⁺	Ca²⁺						Fe²⁺ Fe³⁺			Cu⁺ Cu²⁺	Zn²⁺					Br⁻	
Rb⁺	Sr²⁺									Ag⁺						I⁻	
Cs⁺	Ba²⁺																

Figure 5.4 Periodic relationships of some simple ions. B group elements often form ions of more than one kind, with different charges.

Table 5.2 ∎ Symbols and Names for Some Simple (Monatomic) Ions

Group	Element	Name of Ion	Symbol for Ion
1A	Hydrogen	Hydrogen ion*	H^+
	Lithium	Lithium ion	Li^+
	Sodium	Sodium ion	Na^+
	Potassium	Potassium ion	K^+
2A	Magnesium	Magnesium ion	Mg^{2+}
	Calcium	Calcium ion	Ca^{2+}
3A	Aluminum	Aluminum ion	Al^{3+}
5A	Nitrogen	Nitride ion	N^{3-}
6A	Oxygen	Oxide ion	O^{2-}
	Sulfur	Sulfide ion	S^{2-}
7A	Fluorine	Fluoride ion	F^-
	Chlorine	Chloride ion	Cl^-
	Bromine	Bromide ion	Br^-
	Iodine	Iodide ion	I^-
1B	Copper	Copper(I) ion (cuprous ion)	Cu^+
		Copper(II) ion (cupric ion)	Cu^{2+}
	Silver	Silver ion	Ag^+
2B	Zinc	Zinc ion	Zn^{2+}
8B	Iron	Iron(II) ion (ferrous ion)	Fe^{2+}
		Iron(III) ion (ferric ion)	Fe^{3+}

*Does not exist independently in aqueous solution.

5.6 Formulas and Names of Binary Ionic Compounds

Names of simple positive ions (*cations*) are derived from those of their parent elements by the addition of the word "ion". A sodium atom (Na), on losing an electron, becomes a *sodium ion* (Na^+). A magnesium atom (Mg), on losing two electrons, becomes a *magnesium ion* (Mg^{2+}). Names of simple negative ions (*anions*) are derived from those of their parent elements by changing the usual ending to *-ide* and adding the word "ion". A chlor*ine* atom (Cl), on gaining an electron, becomes a chlor*ide ion* (Cl^-). A sul*fur* atom (S), on gaining two electrons, becomes a sul*fide ion* (S^{2-}) (Table 5.2).

Simple ions of opposite charge can be combined to form **binary** (two-component) **compounds**. To get the correct **formula** for a binary compound, simply write each ion with its charge (positive ion to the left), then cross over the numbers (but not the plus and minus signs) and write them as subscripts. The process is best learned by practice. Work through Examples 5.5–5.9 and Exercises 5.5–5.9. There are also problems at the end of the chapter for further practice.

Note that the crossover method works because it is based on the transfer of electrons and the conservation of charge. Two aluminum atoms lose three electrons each (a total of six electrons lost), and three oxygen atoms gain two electrons each (a total of six electrons gained). Electrons lost equal electrons gained, and all is well. Similarly, two aluminum ions have six positive charges (three each), and three oxide ions have six negative charges (two each). The net charge on aluminum oxide is 0, just as it should be.

Example 5.5

Give the formula for calcium chloride.

Solution

First, write the symbols for the ions.

$$Ca^{2+} \quad Cl^{1-}$$

Then cross over the numbers as subscripts.

Then rewrite the formula, dropping the charges. The formula for calcium chloride is

$$Ca_1Cl_2 \quad \text{or} \quad CaCl_2$$

Exercise 5.5
Give the formula for potassium oxide.

Example 5.6
Give the formula for aluminum oxide.

Solution
Write the symbols for the ions.

$$Al^{3+} \quad O^{2-}$$

Cross over the numbers as subscripts

Then rewrite the formula, dropping the charges. The formula for aluminum oxide is

$$Al_2O_3$$

Exercise 5.6
Give the formula for calcium nitride.

Example 5.7
Give the formula for magnesium oxide.

Solution
The ions are

$$Mg^{2+} \quad O^{2-}$$

Crossing over,

and we have

$$Mg_2^{2+}O_2^{2-}$$

Dropping the charges, we get

$$Mg_2O_2$$

Such formulas usually are reduced to the simplest whole-number ratio, and we write magnesium oxide as MgO.

Exercise 5.7
Give the formula for calcium sulfide.

Now you are able to translate the "English," such as aluminum oxide, into the "chemistry," Al_2O_3. You also can translate in the other direction.

Example 5.8

What is the name of MgS?

Solution

Table 5.2 tells us that MgS is made up of Mg^{2+} (magnesium ions) and S^{2-} (sulfide ions). The name is simply magnesium sulfide.

Exercise 5.8

What is the name of CaF_2?

Example 5.9

What is the name of $FeCl_3$?

Solution

The ions are

$$Fe^{3+} \qquad Cl^-$$

(How do we know the iron is Fe^{3+} and not Fe^{2+}? Since there are three Cl^- ions, each 1−, the one Fe ion must be 3+ because the compound $FeCl_3$ is neutral.) The names of these ions are iron(III) ion (or ferric ion) and chloride ion. Therefore, the compound is iron(III) chloride (or, by the older system ferric cloride).

Exercise 5.9

What is the name of $CuBr_2$?

5.7 Covalent Bonds: Shared Electron Pairs

One might expect a hydrogen atom, with its one electron, to acquire another electron and assume the more stable helium configuration. Indeed, hydrogen atoms do just that in the presence of atoms of a reactive metal such as lithium—that is, a metal that finds it easy to give up an electron.

$$Li\cdot \; + \; H\cdot \; \longrightarrow \; Li^+ \; + \; H\!:^-$$

But what if there are no other kinds of atoms around? What if there are only hydrogen atoms? One atom can't gain an electron from another, for among hydrogen atoms all have an equal attraction for electrons. They can compromise, however, by *sharing a pair* of electrons.

$$H\cdot \; + \; \cdot H \; \longrightarrow \; H\!:\!H$$

By sharing electrons, the two hydrogen atoms form a hydrogen molecule. The bond formed by a shared pair of electrons is called a **covalent bond**.

$$H\!:\!H$$

covalent bond (shared pair of electrons)

Consider next the case of chlorine. A chlorine atom picks up an extra electron from anything willing to give one up. But again, what if the only thing around is another chlorine atom? Chlorine atoms also can attain a more stable arrangement by sharing a pair of electrons.

$$:\!\ddot{C}l\cdot \; + \; \cdot\ddot{C}l\!: \; \longrightarrow \; :\!\ddot{C}l\!:\!\ddot{C}l\!:$$

When two atoms that are combining are very different (for example, a metal and a nonmetal from opposite ends of the periodic table), they form an *ionic bond*. When two atoms that are combining are very similar (for example, two nonmetal atoms), they form a *covalent bond*.

Nonbonding pairs of electrons are often called *lone pairs*.

The shared pair of electrons in the chlorine molecule is another example of a covalent bond; they are called a **bonding pair**. The other electrons that stay on one atom and are not shared are called **nonbonding pairs**.

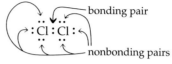

For simplicity, the hydrogen molecule is often represented as H_2, and the chlorine molecule as Cl_2. In each case, the covalent bond between the atoms is understood. Sometimes the covalent bond is indicated by a dash, $H-H$ and $Cl-Cl$. Nonbonding pairs of electrons often are not shown.

Each chlorine atom in the chlorine molecule has eight electrons around it, an arrangement like that of the noble gas argon. Thus, the atoms in a covalent bond follow the octet rule by sharing electrons, even as those in an ionic bond follow it by giving up or taking on electrons.

Multiple Bonds

Atoms can share more than one pair of electrons. In carbon dioxide, for example, the carbon atom shares two pairs of electrons with each of the two oxygen atoms.

$$:\overset{..}{O}::C::\overset{..}{O}:$$

A single pair of shared electrons is called a **single bond**.

Note that each atom has an octet of electrons about it as a result of this sharing. We say that the atoms are joined by a **double bond**, a covalent linkage in which the two atoms share two pairs of electrons.

Atoms also can share three pairs of electrons. In the nitrogen (N_2) molecule, for example, each nitrogen atom shares three pairs of electrons with the other.

Covalent bonds are usually represented as dashes. The three kinds of covalent bonds are simply written as follows:

$$H-Cl \qquad O=C=O$$
$$N\equiv N$$

$$:N:::N:$$

The atoms are joined by a **triple bond**, a covalent linkage in which two atoms share three pairs of electrons. Note that each of the nitrogen atoms has an octet of electrons around it.

Names of Covalent Compounds

Many molecular compounds have common and widely used names. Examples are water (H_2O), methane (CH_4), and ammonia (NH_3). For other compounds, the prefixes *mono-, di-, tri-*, and so on, are used to indicate the number of atoms of each element in the molecule. A list of these prefixes for up to ten atoms is given in Table 5.3.

Simply use the prefixes to indicate the number of each kind of atom. For example, the compound N_2O_4 is called *dinitrogen tetroxide*. (The *a* often is dropped from *tetra-* and other prefixes when they precede another vowel.) We often leave off the *mono-* prefix (NO_2 is nitrogen dioxide) but do include it to distinguish between two compounds of the same pair of elements (CO is carbon monoxide; CO_2 is carbon dioxide).

Table 5.3 ▌ Prefixes that Indicate the Number of Atoms of an Element in a Compound

Prefix	Number of Atoms
Mono-	1
Di-	2
Tri-	3
Tetra-	4
Penta-	5
Hexa-	6
Hepta-	7
Octa-	8
Nona-	9
Deca-	10

Example 5.10

What are the names of SCl_2 and SF_6?

Solution

With one sulfur atom and two chlorine atoms, SCl_2 is sulfur dichloride. With one sulfur atom and six fluorine atoms, SF_6 is sulfur hexafluoride.

Exercise 5.10

What are the names of BrF_3 and BrF_5?

Example 5.11

Give the formula for carbon tetrachloride.

Solution

The name indicates one carbon atom and four chlorine atoms. The formula is CCl_4.

Exercise 5.11

Give the formula for dinitrogen pentoxide.

Example 5.12

Give the formula for tetraphosphorus hexoxide.

Solution

The name indicates four phosphorus atoms and six oxygen atoms. The formula is P_4O_6.

Exercise 5.12

Give the formula for tetraphosphorus triselenide. (The symbol for selenium is Se.)

5.8 Unequal Sharing: Polar Covalent Bonds

So far we have seen that atoms combine in two different ways. Some that are quite different in electron structure (from opposite sides of the periodic table) react by the complete transfer of an electron from one atom to another to form an ionic bond. Atoms that are identical combine by sharing a pair of electrons to form a covalent bond. Now let's consider bond formation between atoms that are different, but not different enough to form ionic bonds.

Hydrogen Chloride

Hydrogen and chlorine react to form a colorless gas called hydrogen chloride. This reaction may be represented as

$$H\cdot\ +\ \cdot\ddot{\underset{\cdot\cdot}{Cl}}:\ \longrightarrow\ H:\ddot{\underset{\cdot\cdot}{Cl}}:\quad\text{(or H—Cl)}$$

Both hydrogen and chlorine need an electron to achieve a noble gas configuration, and so they share a pair and form a covalent bond.

Both hydrogen and chlorine actually consist of diatomic molecules; the reaction is more accurately represented by the scheme

$$H:H\ +\ :\ddot{\underset{\cdot\cdot}{Cl}}:\ddot{\underset{\cdot\cdot}{Cl}}:\ \longrightarrow$$
$$2\,H:\ddot{\underset{\cdot\cdot}{Cl}}:$$

We use the individual atoms in order to focus on the sharing of electrons to form a covalent bond.

Example 5.13

Use electron-dot structures to show the formation of a covalent bond between
a. two fluorine atoms.
b. a fluorine atom and a hydrogen atom.

Solution

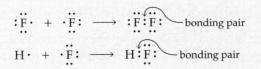

Exercise 5.13

Use electron-dot structures to show the formation of a covalent bond between
a. two bromine atoms
b. a hydrogen atom and a bromine atom
c. an iodine atom and a chlorine atom

When two identical nonmetal atoms combine, they form a *nonpolar* covalent bond. When two atoms of different nonmetals combine, they form a *polar* covalent bond.

One might reasonably ask why a hydrogen molecule and a chlorine molecule react at all. Have we not just explained that they themselves were formed to provide a more stable arrangement of electrons? Yes, indeed, we did say that. But there is stable, and there is more stable. The chlorine molecule represents a more stable arrangement than two separate chlorine atoms. Given the opportunity, a chlorine atom would rather form a bond with a hydrogen atom than with another chlorine atom.

For convenience and simplicity, the reaction of hydrogen (molecule) and chlorine (molecule) to form hydrogen chloride often is represented as

$$H_2 + Cl_2 \longrightarrow 2\ HCl$$

Molecules of hydrogen chloride consist of one atom of hydrogen and one atom of chlorine. These unlike atoms share a pair of electrons. Sharing, however, does not mean sharing equally. Chlorine atoms have a greater attraction for a shared pair of electrons than hydrogen atoms do; chlorine is said to be more *electronegative* than hydrogen.

Electronegativity

The **electronegativity** of an element is a measure of the attraction of an atom in a molecule for a pair of shared electrons. The atoms to the right in the periodic table are, in general, more electronegative than those to the left. The ones on the right are precisely the atoms that, in forming ions, tend to gain electrons and form negative ions. The ones on the left, metals, tend to give up electrons and become positive ions. The more electronegative an atom, the greater its tendency to pull the electrons in the bond toward its end of the bond when it is involved in covalent bonding.

Chlorine is more electronegative than hydrogen. In the hydrogen chloride molecule, the shared electrons are held more tightly by the chlorine atom, and this results in the chlorine end of the molecule being more negative than the hydrogen end (Figure 5.5). When the electrons in a covalent bond are not equally shared, the bond is said to be *polar*. Thus, the bonds in hydrogen chloride are described as **polar covalent bonds**, whereas the bonds in the hydrogen molecule or in the chlorine molecule are **nonpolar covalent bonds**. A polar covalent bond is not an ionic bond. In an ionic bond, one atom completely loses an electron. In a polar covalent bond, the atom at the positive end of the bond (hydrogen in HCl) still has some share in the bonding pair of electrons (Figure 5.6). To distinguish this arrangement from that in an ionic bond, the following notation is used.

$$\overset{\delta+}{H} - \overset{\delta-}{Cl}$$

The line between the atoms represents the covalent bond, a pair of shared electrons. The $\delta+$ and $\delta-$ (read "delta plus" and "delta minus") signify which end is partially positive and which is partially negative. (The word "partially" is used to distinguish this charge from the full charge on an ion).

5.9 Polyatomic Molecules: Water, Ammonia, and Methane

To obtain an octet of electrons, an oxygen atom must share electrons with two hydrogen atoms, a nitrogen atom must share electrons with three hydrogen atoms, and a carbon atom must share electrons with four hydrogen atoms. In general, many nonmetals often form a number of covalent bonds equal to eight minus the group number. Oxygen, which is in Group 6A, forms $8 - 6 = 2$ covalent bonds in most compounds. Nitrogen, in Group 5A, forms $8 - 5 = 3$ covalent bonds in most of its compounds. Carbon, in Group 4A, forms $8 - 4 = 4$ covalent bonds in most carbon compounds, including the great host of organic compounds (Chapter 9). These simple rules will enable you to write formulas for many molecules.

Figure 5.5 Chlorine hogs the electron blanket, leaving hydrogen partially, but positively, exposed.

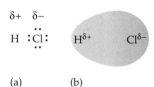

Figure 5.6 Representation of the polar hydrogen chloride molecule. (a) The electron-dot formula, with the shared electron pair shown nearer the chlorine atom. The symbols $\delta+$ and $\delta-$ indicate partial positive and partial negative charges, respectively. (b) A diagram depicting the unequal distribution of electron density in the hydrogen chloride molecule.

Carbon	forms 4 bonds.
Nitrogen	forms 3 bonds.
Oxygen	forms 2 bonds.
Hydrogen	forms 1 bond.

Water

Water is one of the most familiar chemical substances. The electrolysis experiment of Nicholson and Carlisle (Section 2.3) and the fact that both hydrogen and oxygen are diatomic gases indicate that the molecular formula for water is H_2O. In order to be surrounded by an octet, oxygen shares two pairs of electrons. But a hydrogen atom shares only one pair of electrons. An oxygen atom must therefore bond with two hydrogen atoms.

$$\cdot \ddot{O}\!: \;+\; 2\,H\cdot \;\longrightarrow\; H\!:\!\ddot{\underset{H}{O}}\!:$$

Water is discussed in considerable detail in Chapter 13.

This arrangement completes the octet in the valence energy level of oxygen, giving it the neon structure. It also completes the outer energy level of the hydrogen atoms, each of which now has the helium structure. The H—O bonds formed are *covalent* and *polar*.

Ammonia

An atom of the element nitrogen has five electrons in its valence energy level. It can assume the neon configuration by sharing three pairs of electrons with *three* hydrogen atoms. The result is the compound ammonia.

$$\cdot \overset{\cdot\cdot}{\underset{\cdot}{N}}\cdot \;+\; 3\,H\cdot \;\longrightarrow\; H\!:\!\overset{\cdot\cdot}{\underset{H}{N}}\!:\!H$$

Ammonia (NH_3) is a gas at room temperature. Vast quantities of it are compressed into tanks and used as fertilizer (Chapter 16).

In ammonia, the bond arrangement is that of a tripod with a hydrogen atom at the end of each "leg" and the nitrogen atom with its unshared pair of electrons sitting at the top. All three N—H bonds are *polar covalent*.

Methane

An atom of carbon has four electrons in its valence energy level. It can assume the neon configuration by sharing pairs of electrons with four hydrogen atoms, forming the compound methane.

$$\cdot \overset{\cdot}{\underset{\cdot}{C}}\cdot \;+\; 4\,H\cdot \;\longrightarrow\; H\!:\!\overset{H}{\underset{H}{\overset{\cdot\cdot}{\underset{\cdot\cdot}{C}}}}\!:\!H$$

Methane (CH_4) is the simplest of the hydrocarbons, a group of organic compounds discussed in detail in Chapter 9. It is the principal component of natural gas, which is used as a fuel.

The four bonds in the methane molecule, as written above, appear to be planar but actually are not (Section 5.14).

5.10 Polyatomic Ions

Many compounds contain both ionic and covalent bonds. Sodium hydroxide, commonly known as lye, consists of sodium ions (Na^+) and hydroxide ions (OH^-). The hydroxide ion contains an oxygen atom covalently bonded to a hydrogen atom, plus an "extra" electron. Whereas the sodium atom becomes a cation by giving up an electron, the hydroxide group becomes an anion by gaining an electron.

$$e^- \;+\; \cdot \ddot{O}\cdot \;+\; \cdot H \;\longrightarrow\; [\!:\!\ddot{O}\!:\!H]^-$$

Table 5.4 ▮ Some Common Polyatomic Ions

Charge	Name	Formula
1+	Ammonium ion	NH_4^+
	Hydronium ion	H_3O^+
1−	Hydrogen carbonate (bicarbonate) ion	HCO_3^-
	Hydrogen sulfate (bisulfate) ion	HSO_4^-
	Acetate ion	$CH_3CO_2^-$ (or $C_2H_3O_2^-$)
	Nitrite ion	NO_2^-
	Nitrate ion	NO_3^-
	Cyanide ion	CN^-
	Hydroxide ion	OH^-
	Dihydrogen phosphate ion	$H_2PO_4^-$
	Permanganate ion	MnO_4^-
2−	Carbonate ion	CO_3^{2-}
	Sulfate ion	SO_4^{2-}
	Chromate ion	CrO_4^{2-}
	Monohydrogen phosphate ion	HPO_4^{2-}
	Oxalate ion	$C_2O_4^{2-}$
	Dichromate ion	$Cr_2O_7^{2-}$
3−	Phosphate ion	PO_4^{3-}

The formula for sodium hydroxide is NaOH; for each sodium ion there is one hydroxide ion.

There are many groups of atoms that (like hydroxide ion) remain together through most chemical reactions. **Polyatomic ions** are charged particles containing two or more covalently bonded atoms. A list of common polyatomic ions is given in Table 5.4. You can use these ions, in combination with the simple ions in Table 5.2 to determine formulas for compounds that contain polyatomic ions.

Acetate ion

Ammonium ion

Hydrogen carbonate ion
(Bicarbonate ion)

Carbonate ion   Nitrite ion

Polyatomic ions have both covalent bonds (dashes) and ionic charges (superscript + or −).

Example 5.14

What is the formula for ammonium sulfide?

Solution
The ions are

$$NH_4^+ \qquad S^{2-}$$

Crossing over,

$$NH_4^+ \quad \text{⟡} \quad S^{2-}$$

we get

$$(NH_4^+)_2S_1^{2-}$$

Dropping the charges gives

$$(NH_4)_2S$$

The parentheses with a subscript 2 indicate that the entire ammonium unit is taken twice; there are two nitrogen atoms and eight ($4 \times 2 = 8$) hydrogen atoms.

Exercise 5.14

What is the formula for calcium acetate?

Example 5.15

What is the formula for ammonium nitrate?

Solution

The ions are

$$NH_4^+ \qquad NO_3^-$$

Both superscripts are 1 (understood). The formula for ammonium nitrate is NH_4NO_3.

Exercise 5.15

What is the formula for calcium monohydrogen phosphate?

Example 5.16

What is the name of the compound NaCN?

Solution

The ions are

$$Na^+ \qquad CN^-$$

The name is sodium cyanide.

Exercise 5.16

What is the name of the compound $CaCO_3$?

Example 5.17

What is the name of KH_2PO_4?

Solution

The ions are

$$K^+ \qquad H_2PO_4^-$$

The name is potassium dihydrogen phosphate.

Exercise 5.17

What is the name of $K_2Cr_2O_7$?

Example 5.18

What is the name of $Cu_3(PO_4)_2$?

Solution

The ions are

$$Cu^{2+} \qquad PO_4^{3-}$$

The name is copper(II) phosphate.

Exercise 5.18

What is the name of $Fe_2(CO_3)_3$?

5.11 Rules for Writing Electron-Dot Formulas

Recall that electrons are transferred (Section 5.4) or shared (Section 5.7) in ways that leave most atoms with octets of electrons in their outermost energy levels. In this section we describe how to write **electron-dot formulas** for molecules. First, we must put the atoms of the molecules in their proper places. The only way to do this accurately is by experimentation, but there are some generalizations that will help us to get many structures right.

The *skeletal structure* of a molecule tells us the order in which the atoms are attached to one another. In the absence of experimental evidence, the following rules help us to devise likely skeletal structures.

1. Hydrogen atoms form only single bonds; they are always at the end of a sequence of atoms. Hydrogen is often bonded to carbon, nitrogen, or oxygen.
2. Polyatomic molecules and ions often consist of a central atom surrounded by more electronegative atoms. (Hydrogen is an exception; it is always on the outside, even when bonded to a more electronegative element.)

After a skeletal formula for a polyatomic molecule or ion has been chosen, we can use the following steps to write an electron-dot formula.

1. Calculate the total number of valence electrons. The total for a molecule is the sum of the valence electrons for all the atoms. For a polyatomic anion, add the number of negative charges. For a polyatomic cation, subtract the number of positive charges.

   Examples:

   $$N_2O_4 \text{ has } (2 \times 5) + (4 \times 6) = 34 \text{ valence electrons.}$$

   $$NO_3^- \text{ has } 5 + (3 \times 6) + 1 = 24 \text{ valence electrons.}$$

   $$NH_4^+ \text{ has } 5 + (4 \times 1) - 1 = 8 \text{ valence electrons.}$$

2. Write the skeletal structure and connect bonded pairs of atoms by a dash (one electron pair).
3. Place electrons around outer atoms so that each (except hydrogen) has an octet.
4. Subtract the number of electrons assigned so far from the total calculated in Step 1. Any electrons that remain are assigned in pairs to the central atom(s).
5. If a central atom has fewer than eight electrons after Step 4, a multiple bond is likely. Move one or more nonbonding pairs from an outer atom to the space between the atoms to form a double or triple bond. A deficiency of two electrons suggests a double bond, and a shortage of four electrons indicates a triple bond or two double bonds to the central atom.

Example 5.19

Give the electron-dot formula for methanol, CH_4O.

Solution

1. The total number of valence electrons is $4 + (4 \times 1) + 6 = 14$.
2. The skeletal structure in which all the hydrogen atoms are on the outside and the least electronegative carbon atom is most central is

$$
\begin{array}{c}
\quad\;\; H \\
\quad\;\; | \\
H - C - O - H \\
\quad\;\; | \\
\quad\;\; H
\end{array}
$$

3. Now, counting each bond as two electrons gives ten electrons. The four remaining electrons are placed (as two nonbonding pairs) on the oxygen atom.

(The remaining steps are not necessary; both carbon and oxygen have octets of electrons.)

Exercise 5.19

Give the electron-dot formula for ethyl chloride, C_2H_5Cl.

Example 5.20

Give the electron-dot formula for nitrogen trifluoride, NF_3.

Solution

1. There are $5 + (3 \times 7) = 26$ valence electrons.
2. The skeletal structure is

$$F\!-\!N\!-\!F$$
$$\vert$$
$$F$$

3. Place three nonbonding pairs on each fluorine atom.

$$:\!\ddot{F}\!-\!N\!-\!\ddot{F}\!:$$
$$\vert$$
$$:\!\ddot{F}\!:$$

4. We have assigned 24 electrons. Place the remaining two as a nonbonding pair on the nitrogen atom.

$$:\!\ddot{F}\!-\!\ddot{N}\!-\!\ddot{F}\!:$$
$$\vert$$
$$:\!\ddot{F}\!:$$

(Each atom has an octet; Step 5 is not needed.)

Exercise 5.20

Give the electron-dot structure for oxygen difluoride, OF_2.

Example 5.21

Give the electron-dot structure for the BF_4^- ion.

Solution

1. There are $3 + (4 \times 7) + 1 = 32$ electrons.
2. The skeletal structure is

$$F$$
$$\vert$$
$$F\!-\!B\!-\!F$$
$$\vert$$
$$F$$

3. Place three nonbonding pairs on each fluorine atom.

$$:\!\ddot{F}\!:$$
$$\vert$$
$$:\!\ddot{F}\!-\!B\!-\!\ddot{F}\!:$$
$$\vert$$
$$:\!\ddot{F}\!:$$

4. We have assigned 32 electrons. None remains to be assigned.

Exercise 5.21

Give the electron-dot structure for the PH_4^+ ion.

132 Chapter 5 Chemical Bonds

Example 5.22

Give the electron-dot structure for carbon dioxide, CO_2.

Solution

1. There are $4 + (2 \times 6) = 16$ valence electrons.
2. The skeletal structure is

$$O-C-O$$

3. Place three nonbonding pairs on each oxygen atom.

$$:\overset{..}{\underset{..}{O}}-C-\overset{..}{\underset{..}{O}}:$$

4. We have assigned 16 electrons. None remains to be placed.
5. The carbon atom has only four electrons. It needs to form two double bonds in order to have an octet. (It is not reasonable to expect that carbon would form a triple bond to one of the oxygen atoms. Why?) Move a nonbonding pair from each oxygen atom to the space between the atoms to form a double bond on each side of the carbon atom.

$$:\overset{..}{O}=C=\overset{..}{O}:$$

Exercise 5.22

Give the electron-dot formula for nitryl fluoride, NO_2F (O — N — O — F skeleton).

The rules we have used here lead to results that are summarized and illustrated for selected elements in Table 5.5. Figure 5.7 relates the number of covalent bonds to the periodic table.

Every atom or molecule with an odd number of electrons must have one unpaired electron. Filled energy levels and sublevels have all their electrons paired, with two electrons in each orbital (Section 3.8). We need only consider valence electrons to determine whether or not an atom or molecule is a free radical. Electron-dot structures of NO, NO_2, and ClO_2 show that one atom of each has an unpaired electron; this atom obviously does not have an octet of electrons in its outer energy level.

$$:\overset{.}{N}::\overset{..}{O}: \quad :\overset{..}{O}:\overset{..}{N}::\overset{..}{O}:$$

$$:\overset{..}{O}:\overset{..}{Cl}:\overset{..}{O}:$$

5.12 Exceptions to the Octet Rule

Many molecules made of atoms of the main group elements have electron structures that follow the octet rule. There are numerous exceptions, however. The exceptions fall into three main groups. Each type is readily identified by some structural characteristic.

Odd-Electron Molecules: Free Radicals

Molecules with odd numbers of valence electrons obviously cannot satisfy the octet rule. Examples of such molecules are nitrogen monoxide (NO, also called nitric oxide), with $5 + 6 = 11$ valence electrons; nitrogen dioxide (NO_2), with 17 valence electrons; and chlorine dioxide (ClO_2), which has 19 outer electrons. Obviously, one of the atoms in each of these molecules has an odd number of electrons and therefore cannot have an octet.

Atoms and molecules with unpaired electrons are called **free radicals**. Most free radicals are highly reactive and have only transitory existence as intermediates in chemical reactions. An example is the chlorine atom that is formed from the breakdown of chlorofluorocarbons in the stratosphere and leads to depletion of the ozone shield (Chapter 12). Some free radicals are quite stable, however. Nitrogen oxides are major components of smog. Chlorine dioxide is made in multiton quantities and is used for bleaching flour.

Table 5.5 ▌ Number of Bonds Formed by Selected Elements

Electron-Dot Picture	Bond Picture	Number of Bonds	Representative Molecules	Ball-and-Stick Models
H·	H—	1	H—H H—Cl	HCl
He:		0	He	He
·Ċ·	—Ċ—	4	$\overset{H}{\underset{H}{H-\overset{\mid}{\underset{\mid}{C}}-H}}$ $H-\overset{O}{\overset{\parallel}{C}}-F$	CH₄
·N̈·	—N̈—	3	$H-\overset{\mid}{\underset{H}{N}}-H$ $\overset{N-O-H}{\underset{}{H-\overset{\parallel}{C}-H}}$	NH₃
·Ö:	—Ö	2	$H-\overset{\mid}{\underset{H}{O}}$ $H-\overset{O}{\overset{\parallel}{C}}-H$	H₂O
·F̈:	—F	1	H—F F—F	F₂
·C̈l:	—Cl	1	Cl—Cl $\overset{H}{\underset{H}{H-\overset{\mid}{\underset{\mid}{C}}-Cl}}$	CH₃Cl

Molecules with Incomplete Octets: Too Few Electrons

Boron atoms have three valence electrons; fluorine atoms have seven. When boron reacts with fluorine, it shares these electrons with three fluorine atoms to form boron trifluoride.

$$\overset{\cdot\cdot}{:}\overset{\cdot\cdot}{F}\overset{\cdot\cdot}{:}$$
$$\overset{\mid}{B}-\overset{\cdot\cdot}{F}\overset{\cdot\cdot}{:}$$
$$\overset{\mid}{:}\overset{\cdot\cdot}{F}\overset{\cdot\cdot}{:}$$

Beryllium reacts with halogens to form compounds with the simplest formula BeX_2, where X = F, Cl, Br, or I. In these molecules, beryllium shares its pair of valence electrons with each of two bromine atoms, giving it only four electrons in its valence shell.

$$: \overset{\cdot\cdot}{Br} : Be : \overset{\cdot\cdot}{Br} :$$

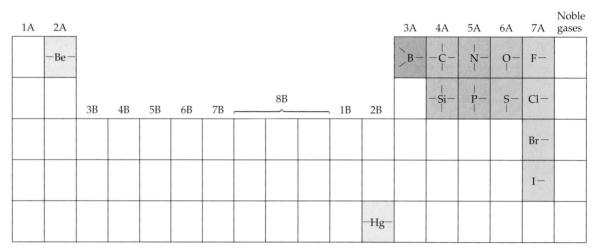

Figure 5.7   Covalent bonding of representative elements of the periodic table.

Molecules with Expanded Octets: Too Many Electrons

It is a good idea to use the octet rule except in cases where it obviously doesn't apply: when there is an odd number of electrons, when there are too few electrons to make an octet, and (third period and beyond) when there are obviously more than eight electrons that must be in the valence level.

The second-period elements—carbon, nitrogen, oxygen, and fluorine—nearly always obey the octet rule. (Odd-electron compounds are obvious exceptions.) The valence electron level of second-period elements holds a maximum of eight electrons ($2s^2 2p^6$).

The third main energy level can hold up to 18 electrons ($3s^2 3p^6 3d^{10}$). Elements in the third period and beyond therefore can have more than eight electrons in their valence level, violating the octet rule. These so-called expanded octets are evident in the following compounds.

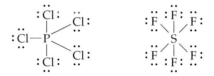

Phosphorus pentachloride   Sulfur hexafluoride

5.13 Molecular Shapes: The VSEPR Theory

We have represented molecules in two dimensions on paper, but molecules have three-dimensional shapes. We can use electron-dot structures as part of the process of predicting molecular shapes. The shapes that we consider in this book are shown in Figure 5.8.

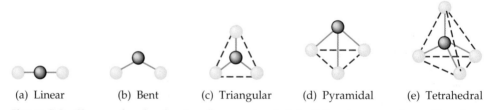

(a) Linear   (b) Bent   (c) Triangular   (d) Pyramidal   (e) Tetrahedral

Figure 5.8   Shapes of molecules. In a *linear* molecule (a), all the atoms are along a line; the bond angle is 180°. A *bent* molecule (b) has an angle less than 180°. Connecting the three outer atoms of a *triangular* molecule (c) with imaginary lines produces a triangle with an atom at the center. Imaginary lines connecting all four atoms of a *pyramidal* molecule (d) form a three-sided pyramid. Connecting the four outer atoms of a *tetrahedral* molecule (e) with imaginary lines produces a tetrahedron (a four-sided figure in which each side is a triangle) with an atom at the center.

Table 5.6 ▌ Bonding and the Shape of Molecules

Number of Bonded Atoms	Number of NBP*	Number of Sets	Molecular Shape	Examples	Ball-and-Stick Models
2	0	2	Linear	$BeCl_2$ $HgCl_2$ CO_2 HCN	$BeCl_2$
3	0	3	Triangular	BF_3 $AlBr_3$ CH_2O	BF_3
4	0	4	Tetrahedral	CH_4 CBr_4 $SiCl_4$	CH_4
3	1	4	Pyramidal	NH_3 PCl_3	NH_3
2	2	4	Bent	H_2O H_2S SCl_2	H_2O
2	1	3	Bent	SO_2 O_3	SO_2

*NBP, Nonbonding pair(s) of electrons.

The **valence shell electron pair repulsion (VSEPR) theory** is often used to predict the arrangement of atoms about a central atom. The basis of the VSEPR theory is that electron pairs arrange themselves about a central atom in a way that minimizes repulsion between like-charged particles. This means that they will get as far apart as possible. Table 5.6 gives the geometric shapes associated with the arrangement of two, three, or four entities about a central atom.

The farthest apart two substituent atoms can get are the opposite sides of the central atom at an angle of 180°.

Three groups assume a triangular arrangement about the central atom, forming angles of separation of 120°. Four groups form a tetrahedral array around the central atom, giving a separation of about 109.5°.

You can determine the shapes of many molecules (and polyatomic ions, Chapter 6) by following these simple rules.

1. Draw an electron-dot structure in which a shared electron pair (bonding pair, BP) is indicated by a line. Use dots to show any nonbonding pairs (NBPs) of electrons.

2. To determine shape, count the number of atoms *and NBPs attached to the central atom*. Note that a multiple bond counts only as one *set*. Examples are

The shapes of molecules are of considerable importance. We are interested in the shapes of biologically active molecules (Chapters 15, 16, 18, and 19). These molecules must have the right groups in the right places in order to function properly. For now, we limit our discussion to a few simple molecules for which we can use the VSEPR theory to predict shapes.

[Web Reference 1]
www This is a good
place to look for a
helpful VSEPR tutorial.

[Web Reference 2]
www Animated mole-
cules that may
help you better under-
stand the VSEPR theory.

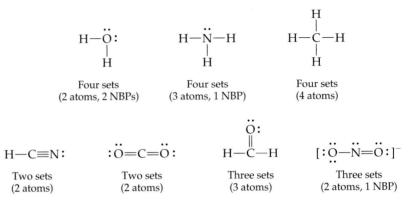

H—O̤:
|
H

Four sets
(2 atoms, 2 NBPs)

H—N̈—H
|
H

Four sets
(3 atoms, 1 NBP)

H
|
H—C—H
|
H

Four sets
(4 atoms)

H—C≡N:

Two sets
(2 atoms)

:Ö=C=Ö:

Two sets
(2 atoms)

Ö:
‖
H—C—H

Three sets
(3 atoms)

[:Ö—N̈=Ö:]⁻

Three sets
(2 atoms, 1 NBP)

3. Determine the number of electron sets and draw a shape *as if* all were bonding pairs.

4. Sketch this shape, placing the electron pairs as far apart as possible (Table 5.6). If there is *no* NBP, this is the shape of the molecule. If there *are* NBPs, remove them, leaving the BPs exactly as they were. (This may seem strange, but it stems from the fact that *all* the sets determine the geometry but only the arrangement of bonded atoms is considered in the shape of the molecule.)

Example 5.23

What is the shape of the BH_3 molecule?

Solution

1. The electron-dot structure is

H
|
B—H
|
H

2. There are three sets to consider

H
|
B—H
|
H

3. The three sets get as far apart as possible, giving a triangular arrangement of the sets.

H⟋
  \
   B⟍H
  /
H   120°

4. All the sets are bonding pairs; the molecular shape is triangular, the same as the arrangement of the electrons.

Exercise 5.23

What is the shape of the BeH_2 molecule?

Example 5.24

What is the shape of the SCl_2 molecule?

Solution

1. The electron-dot structure is

$$:\ddot{S}-\ddot{\underset{..}{Cl}}:$$
$$|$$
$$:\ddot{\underset{..}{Cl}}:$$

2. There are four sets on the sulfur atom to consider.

$$:\ddot{S}-\ddot{\underset{..}{Cl}}:$$
$$|$$
$$:\ddot{\underset{..}{Cl}}:$$

3. The four sets get as far apart as possible, giving a tetrahedral arrangement of the sets about the central atom.

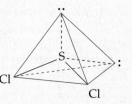

4. Two of the sets are bonding pairs and two are NBP. Remove (ignore) the NBP; the molecular shape is *bent*, with a bond angle of about 109.5°.

Exercise 5.24

What is the shape of the PH_3 molecule?

5.14 Shapes and Properties: Polar and Nonpolar Molecules

In Section 5.8 we discussed polar and nonpolar bonds. Diatomic molecules are polar if their bonds are polar, and nonpolar if their bonds are nonpolar.

$$\overset{\delta+}{H}-\overset{\delta-}{Cl} \qquad Cl-Cl$$
$$\text{Polar} \qquad\quad \text{Nonpolar}$$

For molecules with three or more atoms, we must also consider the orientation of the bonds to determine whether or not the molecule as a whole is polar.

Water: A Bent Molecule

We should expect the bonds in water to be polar because oxygen is more electronegative than hydrogen. (Like chlorine, oxygen is to the right in the periodic table, and electronegativity increases as one moves to the right in the table.) Just because a molecule contains polar bonds, however, does not mean that the molecule as a whole is polar. If the atoms in the water molecule were in a straight row (that is, in a linear arrangement), the two polar bonds would cancel one another out.

$$\overset{\delta+}{H}-\overset{\delta-}{O}-\overset{\delta+}{H} \qquad \text{(incorrect geometry)}$$

Instead of having one end of the molecule positive and the other end negative, the electrons would be pulled toward the right in one bond and toward the left in the

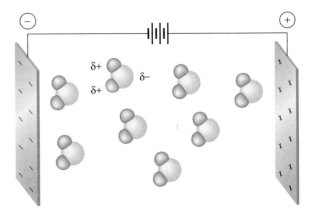

Figure 5.9 Polar molecules are aligned in an electric field, with the positive end of each molecule preferentially pointing toward the negative plate and the negative end of the molecule directed toward the positive plate.

A **polar molecule** has separate centers of positive and negative charge, just as a magnet has north and south poles.

other. Overall there would be no net dipole. By a **dipole**, we mean a molecule that has a positive end and a negative end.

But water *does* act like a dipole. If you place a sample of water between two electrically charged plates, the water molecules align themselves with one end attracted toward the positive plate and the other end toward the negative plate. To act like a dipole, the molecule must be bent so that the dipoles do not cancel one another out.

Dipoles are often represented by an arrow with a plus at the tail end ($+\!\!\longrightarrow$).

$$\delta^+ H : \overset{..}{\underset{..}{O}} : \delta 2- \quad \text{or} \quad \delta^+ H \diagdown O \delta 2- \\ \qquad\quad \overset{..}{H} \delta^+ \qquad\qquad\qquad \underset{H \delta^+}{|}$$

$$\overset{+\longrightarrow}{H - Cl}$$

The positive end of the arrow is obvious; the head of the arrow indicates the negative end of the dipole. Using these arrows, we see that carbon dioxide is nonpolar despite its polar bonds.

$$\overset{\longleftarrow +\quad +\longrightarrow}{O = C = O}$$

The two cancel one another out. In water, both dipoles point toward the oxygen, giving a net dipole toward that end of the molecule.

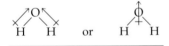

Such molecules would align themselves between charged plates as shown in Figure 5.9.

The shape of the water molecule can be accounted for by an extension of the VSEPR theory. According to VSEPR, the two bonds and two NBP should form a tetrahedral arrangement.

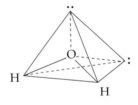

Ignoring the NBP, the molecular shape has the atoms in a bent arrangement, with a bond angle of about 109.5° (the tetrahedral angle).

The predicted bond angle of 109.5° for water is a bit larger than the measured angle of 104.5°. The difference is explained by the fact that the NBPs occupy a greater volume than the bonding pairs. These larger orbitals push the smaller BPs closer together.

Ammonia: A Pyramidal Molecule

 [Web Reference 3] This website has a hypermedia tutorial on chemical bonds, molecular shapes, and molecular models.

There are three BPs and one NBP about the nitrogen atom of ammonia. The VSEPR theory predicts a tetrahedral arrangement with bond angles of 109.5°. The actual bond angles are 107°, close to the theoretical value. Again, the unshared pair of electrons occupies a greater volume than a shared pair, pushing the latter slightly closer together. The arrangement is therefore that of a tripod with a hydrogen atom at the end of each leg and the nitrogen atom with its unshared pair sitting at the top (Figure 5.10). Each nitrogen-to-hydrogen bond is somewhat polar, making the ammonia molecule polar.

Methane: A Tetrahedral Molecule

There are four pairs of electrons on the central carbon atom in methane. Using the VSEPR theory, we would expect a tetrahedral arrangement and bond angles of 109.5°. The actual bond angles are 109.5°, as predicted by the theory (Figure 5.11). All four electron pairs are shared with hydrogen atoms; thus, all four pairs occupy identical volumes. Each carbon-to-hydrogen bond is slightly polar, but the methane molecule as a whole is symmetric. The slight bond polarities cancel out, leaving the methane molecule, as a whole, nonpolar.

Many of the properties of compounds—such as melting point, boiling point, and solubility—depend on the polarity of the molecules of the compounds.

5.15 Intermolecular Forces and the States of Matter

The shape and polarity of molecules determine how they interact with each other. The physical state of a material (whether it is a gas, liquid, or solid) depends on the presence or absence of *intermolecular* forces that hold molecules together in a liquid or solid. Gas molecules, which are so far apart that they do not undergo significant attractive forces, interact only during collisions in most cases.

Solids and Liquids

Solids are highly ordered assemblies of particles in close contact with one another. Their motion is primarily vibration within a solid lattice. In liquids, the particles are still in close contact, but they are more loosely organized and are freer to move about. In gases the molecules are no longer in close contact with one another but are separated by relatively great distances. They are also moving quite rapidly in random directions. We shall look more closely at gases in the next chapter.

Table salt (sodium chloride) is a typical crystalline solid. In this solid, ionic bonds hold the ions in position and maintain the orderly arrangement. Not all solids are held together by ionic bonds, but some attractive force is necessary to maintain the characteristic orderly array of particles.

To get a better image of a liquid at the molecular level, think of a box of marbles being shaken continuously. The marbles move back and forth, rolling over one another. The particles of a liquid (like the marbles) are not so rigidly held in place as are particles in a solid. The particles in a liquid are held close together, however, and this means that there must be some force attracting them.

Solids can be changed to liquids; that is, they can be *melted*. The solid is heated, and the heat energy is absorbed by the particles of the solid. The energy causes the particles to vibrate in place with more and more vigor until, finally, the forces holding the particles in a particular arrangement are overcome. The solid has become a liquid. The temperature at which this happens is called the **melting point** of the solid. A high melting point is one indication that the forces holding a solid together are very strong.

A liquid can change to a gas or vapor in a process called **vaporization**. Again, one need only supply sufficient heat to achieve this change. Energy is absorbed by the liquid particles, which move faster and faster as a result. Finally, the attractive forces holding the liquid particles in contact are overcome by this increasingly violent motion and the particles fly away from one another. The liquid has become a gas.

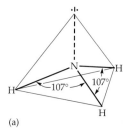

(a)

(b)

Figure 5.10 The ammonia molecule. In the drawing (a), solid lines indicate covalent bonds; the colored lines outline a tetrahedron. The photograph (b) shows a space-filling model of an ammonia molecule.

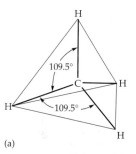

(a)

(b)

Figure 5.11 The methane molecule. In the drawing (a) all bond angles are 109.5°. The photograph (b) shows a space-filling model of the methane molecule.

Hydrocarbons generally are, like methane, nonpolar molecules.

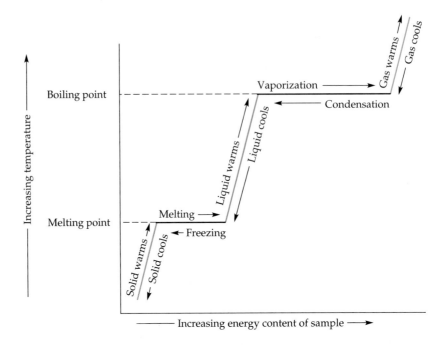

Figure 5.12 Diagram of changes in state of matter on heating or cooling.

Some substances go directly from the solid to the gaseous state, a process called *sublimation*. For example, dry ice (solid CO_2) sublimes; the solid "disappears" without first melting to a liquid.

 [Web Reference 4] This website provides an animated explanation of the behavior of atoms in sodium chloride as a solid, a liquid, and a gas.

The entire sequence of changes can be reversed by removing energy from the sample and slowing down the particles. Vapor changes to liquid in a process referred to as **condensation**; liquid changes to solid in a process called **freezing**. Figure 5.12 presents a diagram of the changes in state that occur as energy is added to or removed from a sample.

The amount of energy required to accomplish these changes depends on the type of forces responsible for maintaining the solid or the liquid state. The ionic bonds found in salt crystals are very strong. Sodium chloride must be heated to about 800°C before it melts. Generally, interionic forces are the strongest of all the forces that hold solids and liquids together. We will now consider some other interactions that hold the particles of solids and liquids together.

Dipole Forces

Hydrogen chloride melts at −112°C and boils at −85°C (it is a gas at room temperature). The attractive forces between molecules are not nearly as strong as the interionic forces in salt crystals. We know that covalent bonds hold the hydrogen and chlorine *atoms* together to form the hydrogen chloride *molecule*, but what makes one molecule interact with another in the solid or liquid state?

Remember that the hydrogen chloride molecule is a dipole. It has a positive end and a negative end. Two dipoles brought close enough together attract one another. The positive end of one molecule attracts the negative end of another. Such forces may exist throughout the structure of a liquid or solid (Figure 5.13). In general, attractive forces between dipoles are much weaker than attractive forces between ions. **Dipole interactions** are, however, stronger than the forces between nonpolar molecules of comparable size.

Hydrogen Bonds

Hydrogen bonds may be as much as 5% to 10% as strong as covalent bonds.

Certain polar molecules exhibit stronger attractive forces than expected on the basis of ordinary dipolar interactions. These forces are strong enough to be given a special name, *hydrogen bonds*. Note that hydrogen bond is a somewhat misleading name,

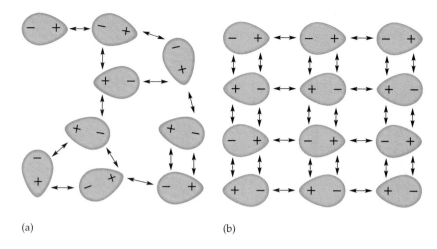

(a)   (b)

Figure 5.13 An idealized representation of dipole forces in (a) a liquid and (b) a solid. In a real liquid or solid, interactions are more complex.

because it emphasizes only one component of the interaction. Not all compounds containing hydrogen exhibit this strong attractive force; the hydrogen *must* be attached to a small electronegative atom such as fluorine, oxygen, or nitrogen. These atoms permit us to offer an explanation for the extra strength of hydrogen bonds as compared with other dipolar forces. Fluorine, oxygen, and nitrogen are all highly electronegative, and they are small (they are at the top of the periodic table). A hydrogen—fluorine bond, for example, is strongly polarized, with a negative fluorine end and a positive hydrogen end. Both hydrogen and fluorine are small atoms, and so the negative end of one dipole can approach very closely the positive end of a second dipole. This results in an unusually strong interaction between the hydrogen atom of one molecule and an unshared pair of electrons on another. This special force between two molecules is called a **hydrogen bond**. Hydrogen bonds are often explicitly represented by *dotted* lines to emphasize their exceptional strength compared with ordinary dipolar interactions. A dotted line is used to distinguish a hydrogen bond from the much stronger covalent bond, which is represented by a *solid* line (Figure 5.14).

Water has both an unusually high melting point and an unusually high boiling point for a compound with molecules of such small size. These abnormal values are attributed to water's ability to form hydrogen bonds. Water is discussed in detail in Chapter 13.

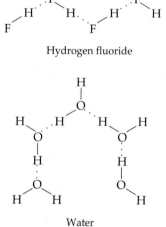

Hydrogen fluoride

Water

Figure 5.14 Hydrogen bonding in hydrogen fluoride and in water.

Dispersion Forces

If one understands that positive attracts negative, it is easy enough to understand how ions or polar molecules maintain contact with one another. But how can we explain the fact that nonpolar compounds can exist in the liquid and solid states? Even hydrogen can exist as a liquid or a solid if the temperature is low enough (its melting point is −259°C). Some force must be holding these molecules in contact with one another in the liquid and solid states.

Up to this point we have pictured the electrons in a covalent bond as being held in place between the two atoms sharing the bond. But the electrons are *not* really static; they actually move about in the bonds. On average, the two electrons in the hydrogen molecule (or any nonpolar bond) are between and equidistant from the two nuclei. At any given instant, however, the electrons may be at one end of the molecule. At some other time, a moment later, the electrons may be at the other end of the

Dispersion forces are weak for small, nonpolar molecules such as H_2, N_2, and CH_4. These forces can be substantial between large molecules. The properties of polymers such as polyethylene (Chapter 10) are determined to a large degree by dispersion forces between long chains of repeating $\text{(CH}_2\text{CH}_2\text{)}_n$ units. (In this formula, n is several hundred or even several thousand.)

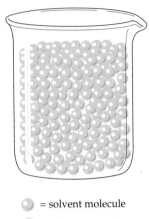

○ = solvent molecule

○ = solute molecule

Figure 5.15 In a solution, solute molecules are randomly distributed among solvent molecules.

(a)

(b)   (c)

Figure 5.16 (a) Lawn mowers with two-cycle engines are fueled and lubricated with a solution of nonpolar lubricating oil in nonpolar gasoline. (b) In Italian dressing, polar vinegar and nonpolar olive oil are mixed. However, the two liquids do not form a solution and separates on standing. (c) Wine is a solution of polar ethyl alcohol in polar water.

molecule. At the instant the electrons of one molecule are at one end, the electrons in the next molecule move away from its adjacent end. Thus, at this instant there is an attractive force between the electron-rich end of one molecule and the electron-poor end of the next. These momentary, usually weak, attractive forces between molecules are called **dispersion forces**. To a large extent, dispersion forces determine the physical properties of nonpolar compounds.

Forces in Solutions

To complete our look at chemical bonding, we shall briefly examine the interactions that occur in solutions. A **solution** is an intimate, homogeneous mixture of two or more substances. By intimate we mean that mixing occurs down to the level of individual ions and molecules. In a salt–water solution, for example, there are no clumps of ions floating around, but single ions randomly distributed among the water molecules. **Homogeneous** means that the mixing is thorough. All parts of the solution have the same distribution of components. For a salt solution, the saltiness is the same at the top, bottom, and middle of the solution. The substance being dissolved, and usually present in a lesser amount (salt in a salt solution), is called the **solute**. The substance doing the dissolving, and usually present in a greater amount, is the **solvent** (water in the salt–water solution) (Figure 5.15).

Ordinarily solutions form most readily when the substances involved have *similar* bonding characteristics. An old chemical adage is "Like dissolves like." Nonpolar solutes dissolve best in nonpolar solvents. For example, oil and gasoline, both nonpolar, mix (Figure 5.16 a), but oil and water do not (Figure 5.16 b). Ethyl alcohol is a liquid held together by hydrogen bonds. So is water. Ethyl alcohol readily dissolves in water because the two substances can hydrogen-bond with one another (Figure 5.16 c). In general, a solute dissolves when attractive forces between it and the solvent overcome the attractive forces operating in the pure solute and in the pure solvent.

Why, then, does salt dissolve in water? Ionic solids are held together by strong ionic bonds. We have already indicated that very high temperatures are required to melt ionic solids and break these bonds. Yet, by simply placing sodium chloride in water at room temperature we can dissolve the salt (or, rather, the water can). And when such a solid dissolves, its bonds *are* broken. The difference between the two processes is the difference between brute force and persuasion. In the melting process, we are simply pouring in enough energy (as heat) to pull the crystal apart. In the dissolving process, we offer the ions an attractive alternative to the ionic interactions in the crystal.

It works this way. Water molecules surround the crystal. Those that approach a negative ion align themselves so that the positive ends of their dipoles point toward the ion. With a positive ion, the process is reversed, and the negative end of the water dipole points toward the ion. Still, the attraction between a dipole and an ion is not as strong as that between two ions. To compensate for their weaker attractive power, several molecules surround each ion, and in this way the many *ion–dipole* interactions overcome the *ion–ion* interactions (Figure 5.17).

In an ionic solid, the positive and negative ions are strongly bonded together in an orderly crystalline arrangement. In solution, cations and anions move about more or less independently, each surrounded by a cage of solvent molecules. Water—including the water in our bodies—is an excellent solvent for many ionic compounds. Water also dissolves many molecules that are polar covalent like itself. These solubility principles explain how nutrients reach the cells of our bodies (dissolved in blood, which is mostly water) and how pollutants get into our water supplies.

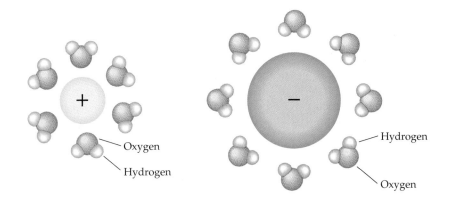

Figure 5.17   The interaction of polar water molecules with ions.

Oxygen
Hydrogen
Hydrogen
Oxygen

Critical Thinking Exercises

Apply knowledge that you have gained in this chapter and one or more of the FLaReS principles (Chapter 1) to evaluate the following statements or claims.

5.1 Some people believe that crystals have special powers. Crystal therapists claim that they can use quartz crystals to restore balance and harmony to a person's spiritual energy. Do you think their claim is credible?

5.2 Sodium chloride, NaCl, is a metal–nonmetal compound held together by ionic bonds. A scientist has studied mercury(II) chloride, $HgCl_2$, and says that the atoms are held together by covalent bonds. The scientist bases this contention largely on the fact that a water solution of the substance does not conduct an electric current, indicating that it does not contain an appreciable amount of ions. In your opinion, is the scientist correct?

5.3 Another scientist, noting that the noble gas xenon does not contain any ions, states that xenon atoms must be held together by covalent bonds. Is this statement plausible?

Summary

1. The most unreactive of all the elements are noble gases. They all (except helium) have an octet of electrons in their outer shells, an electron configuration that confers stability.

2. Electron-dot symbols are chemical symbols surrounded by dots representing valence electrons.

3. When a very active metal reacts with a very active nonmetal, the metal gives up electrons and the nonmetal takes them. The metal atoms become cations and the nonmetal atoms become anions, all being held together by ionic bonds.

4. Simple binary compounds are named by naming the more positive (or more metallic) element first and then putting an *-ide* ending on the name of the more negative element.

5. Names of polyatomic ions often end in *-ate*, but if there are two anions containing the same elements (including oxygen), the one with less oxygen has a name ending in *-ite*.

6. When a metal forms several different compounds with the same nonmetal, each compound is named by indicating the charge on the metal ion as a parenthetical Roman numeral.

7. When two similar atoms react, they combine by sharing electrons. A pair of shared electrons is called a covalent bond.

8. When two nonmetals form several different compounds, prefixes such as *mono-, di-, tri-*, and so on, are used to indicate numbers of atoms of a particular element.

9. The valence shell of a noble gas atom contains eight electrons (except for helium), and other atoms seem to be striving for that same stable octet of electrons when they react. They are said to be following the octet rule or the rule of eight. In the case of helium, only two electrons can exist in its only electron shell, and so small atoms such as hydrogen follow the rule of two.

10. Two similar but different atoms combine by sharing electrons, but one atom has a stronger attraction for electrons than the other, and so the electrons are unequally shared. The attraction of an atom for a pair of shared electrons is called electronegativity, and an unequally shared pair of electrons is a polar covalent bond. An equally shared pair of electrons between two identical atoms is a nonpolar covalent bond.

11. When two atoms share two pairs of electrons, the covalent bond is a double bond. Three shared electron pairs constitute a triple bond.

12. Using electron-dot symbols for each atom, you can write electron-dot formulas for molecules.

13. There are exceptions to the octet rule. In some molecules there are incomplete octets of electrons, and in others expanded octets. Some molecules have unpaired electrons, which make them free radicals.

14. According to the VSEPR theory, if you assume that sets of electrons around a central atom will get as far away from each other as possible, you can predict the shapes of simple molecules.

15. If you know the shape of a molecule, you can decide whether or not it is polar. The water molecule (H_2O) has polar O—H bonds and is bent, and so it is polar.

16. Small nonpolar molecules are held together only by weak dispersion forces, which makes for low melting points and boiling points. Polar molecules are held together by dipole forces as well as dispersion forces. Molecules that contain H—O—H—N, or H—F bonds have special attractive forces called hydrogen bonds.

Key Terms

binary compounds 5.6
bonding pair 5.7
condensation 5.15
covalent bond 5.7
crystal 5.4
dipole 5.14
dipole interactions 5.15
dispersion forces 5.15
double bond 5.7

electron-dot
    formulas 5.11
electron-dot symbols 5.2
electronegativity 5.8
formula 5.6
free radicals 5.12
freezing 5.15
homogeneous 5.15
hydrogen bond 5.15

ionic bonds 5.4
melting point 5.15
nonbonding pairs 5.7
nonpolar covalent
    bonds 5.8
octet rule 5.5
polar covalent bonds 5.8
polar molecule 5.9
polyatomic ions 5.10

single bond 5.7
solute 5.15
solution 5.15
solvent 5.15
triple bond 5.7
valence electrons 5.2
vaporization 5.15
VSEPR theory 5.13

Review Questions

1. Which group of elements in the periodic table is characterized by stable electron arrangements?

2. What is the structural difference between a sodium atom and a sodium ion?

3. How does sodium metal differ from sodium ions (in sodium chloride, for example) in properties?

4. What is the structural difference between a sodium ion and a neon atom? What is similar?

5. What are the structural differences among chlorine atoms, chlorine molecules, and chloride ions? How do their properties differ?

6. Give electron-dot symbols for each of the following elements. You may use the periodic table.
  a. sodium   **b.** oxygen
  c. fluorine   **d.** aluminum

7. Give electron-dot symbols for each of the following elements. You may use the periodic table.
  a. carbon   **b.** potassium
  c. magnesium   **d.** chlorine
  e. nitrogen

8. Give electron structures for each of the following.
  a. K^+ **b.** S^{2-} **c.** F^- **d.** Al^{3+}

9. Give electron structures for each of the following.
  a. Mg^{2+} **b.** Cl^- **c.** Li^+ **d.** N^{3-}

10. Using electron-dot formulas, show the formation of an ion from an atom for each of the following.

  a. barium   **b.** bromine
  c. aluminum   **d.** sulfur

11. Use electron-dot symbols to show the transfer of electrons from calcium atoms to bromine atoms to form ions with noble gas configurations.

12. Use electron-dot symbols to show the transfer of electrons from magnesium atoms to sulfur atoms to form ions with noble gas configurations.

13. Use electron-dot symbols to show the transfer of electrons from aluminum atoms to sulfur atoms to form ions with noble gas configurations.

14. Use electron-dot symbols to show the transfer of electrons from magnesium atoms to phosphorus atoms to form ions with noble gas configurations.

15. Use electron-dot symbols to show the sharing of electrons between two iodine atoms to form an iodine (I_2) molecule. Label all electron pairs as bonding or nonbonding.

16. Use electron-dot symbols to show the sharing of electrons between a hydrogen atom and a fluorine atom. Label the ends of the molecule with symbols that indicate polarity.

17. Indicate charges on simple ions formed from the following elements.
  a. Group 3A   **b.** Group 6A
  c. Group 1A   **d.** Group 7A

18. In what group of the periodic table would elements that form ions with the following charges likely be found?
  a. 2+   **b.** 3−   **c.** 1−

19. Chlorine dioxide is used to bleach flour. Give the formula for chlorine dioxide.

20. Tetraphosphorus trisulfide is used in the tips of "strike anywhere" matches. Give the formula for tetraphosphorus trisulfide.

21. The gas phosphine (PH_3) is used as a fumigant to protect stored grain and other durable produce from pests. Phosphine is generated in situ by adding water to aluminum phosphide or magnesium phosphide. Give formulas for the two phosphides.

22. How many hydrogen atoms are indicated in one formula unit of each of the following?
   a. NH_4NO_3 **b.** CH_3OH
   c. $CH_3CH_2CH_3$ **d.** C_6H_5COOH

23. How many hydrogen atoms are indicated in one formula unit of each of the following?
   a. $(NH_4)_2HPO_4$ **b.** $Al(C_2H_3O_2)_3$

24. How many oxygen atoms are indicated in one formula unit of each of the following?
   a. $Al_2(C_2O_4)_3$ **b.** $Ca_3(PO_4)_2$

25. How many atoms of each kind (Al, C, H, and O) are indicated by the notation $2\ Al(C_2H_3O_2)_3$?

26. How many atoms of each kind (N, P, H, and O) are indicated by the notation $6\ (NH_4)_2HPO_4$?

27. Classify the following bonds as ionic or covalent. For bonds that are covalent, indicate whether they are polar or nonpolar.
   a. KF   **b.** IBr   **c.** MgO

28. Classify the following bonds as ionic or covalent. For bonds that are covalent, indicate whether they are polar or nonpolar.
   a. NO   **b.** CaO   **c.** NaBr

29. Classify the following bonds as ionic or covalent. For bonds that are covalent, indicate whether they are polar or nonpolar.
   a. Br_2 **b.** F_2 **c.** HCl

30. Classify the following covalent bonds as polar or nonpolar.
   a. H—O   **b.** N—Cl   **c.** B—F

31. Classify the following covalent bonds as polar or nonpolar.
   a. H—N   **b.** Be—F   **c.** P—Cl

32. Use the symbol $\longmapsto$ to indicate the direction of the dipole in each of the bonds in Question 30.

33. Use the symbol $\longmapsto$ to indicate the direction of the dipole in each of the bonds in Question 31.

34. The molecule BeF_2 is linear. Is it polar or nonpolar? Use the symbol $\longmapsto$ to explain your answer.

35. The molecule SF_2 is bent. Is it polar or nonpolar? Use the symbol $\longmapsto$ to explain your answer.

36. How many covalent bonds do each of the following usually form? You may refer to the periodic table.
   a. H   **b.** C   **c.** O
   d. F   **e.** N   **f.** Br

37. Of the elements H, O, N, and C, which can readily form double bonds?

38. Of the elements H, O, N, and C, which can readily form triple bonds?

39. List three main types of compounds that are exceptions to the octet rule. Give an example of each.

40. In what ways are liquids and solids similar? In what ways are they different?

41. List four types of interactions between particles in the liquid and solid states. Give an example of each type.

42. Define each of the following terms.
   a. melting   **b.** vaporization
   c. condensation   **d.** freezing

43. In which process is energy absorbed by the material undergoing the change of state?
   a. melting or freezing
   b. condensation or vaporization

44. Label each arrow with the term listed in Question 42 that correctly identifies the process presented.

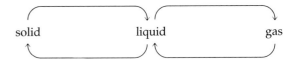

45. Define each of the following terms.
   a. solution   **b.** solute   **c.** solvent

Problems

Names and Symbols for Simple Ions

46. Name the following ions.
   a. K^+ **b.** Ca^{2+} **c.** Zn^{2+}
   d. Br^- **e.** Li^+ **f.** S^{2-}

47. Name the following ions.
   a. Na^+ **b.** Mg^{2+} **c.** Al^{3+}
   d. Cl^- **e.** O^{2-} **f.** N^{3-}

48. Name the following ions.
   a. Fe^{2+} **b.** Cu^+ **c.** I^-

49. Name the following ions.
   a. Fe^{3+} **b.** Cu^{2+} **c.** Ag^+

50. Give symbols for the following ions.
   a. sodium ion   **b.** aluminum ion
   c. oxide ion   **d.** copper(II) ion

51. Give symbols for the following ions.
   a. bromide ion   **b.** calcium ion
   c. potassium ion   **d.** iron(II) ion

Names and Formulas for Polyatomic Ions

52. Name the following ions.
  a. CO_3^{2-} **b.** HPO_4^{2-} **c.** MnO_4^- **d.** OH^-

53. Name the following ions.
  a. NO_3^- **b.** SO_4^{2-} **c.** $H_2PO_4^-$ **d.** HCO_3^-

54. Give formulas for the following ions.
  a. ammonium ion   **b.** hydrogen sulfate ion
  c. cyanide ion   **d.** nitrite ion

55. Give formulas for the following ions.
  a. phosphate ion   **b.** hydrogen carbonate ion
  c. dichromate ion   **d.** oxalate ion

Ions and Ionic Bonding

56. Give electron-dot symbols for the following.
  a. Al and Al^{3+} **b.** Br and Br^-
  c. Mg and Mg^{2+} **d.** O^{2-} and Ne

57. Give electron-dot symbols for the following.
  a. Ca and Ca^{2+} **b.** S and S^{2-}
  c. Rb and Rb^+ **d.** P and P^{3-}

58. Draw electronic structures for the elements and ions in Problem 56.

59. Draw electronic structures for the elements and ions in Problem 57.

60. Give electron-dot formulas for each of the following ionic compounds.
  a. magnesium fluoride (MgF_2)
  b. calcium chloride ($CaCl_2$)
  c. sodium oxide (Na_2O)
  d. potassium sulfide (K_2S)

61. Give electron-dot formulas for each of the following ionic compounds.
  a. sodium fluoride (NaF)
  b. potassium chloride (KCl)
  c. potassium fluoride (KF)

62. Give electron-dot formulas for each of the following ionic compounds.
  a. magnesium oxide (MgO)
  b. aluminum nitride (AlN)
  c. aluminum sulfide (Al_2S_3)

63. Give electron-dot formulas for each of the following ionic compounds.
  a. sodium nitride (Na_3N)
  b. aluminum chloride ($AlCl_3$)
  c. magnesium nitride (Mg_3N_2)

Molecules: Covalent Bonds

64. Use electron-dot symbols to show the sharing of electrons between a phosphorus atom and hydrogen atoms to form a molecule in which phosphorus has an octet of electrons.

65. Use electron-dot symbols to show the sharing of electrons between a silicon atom and hydrogen atoms to form a molecule in which silicon has an octet of electrons.

66. Use electron-dot symbols to show the sharing of electrons between a carbon atom and fluorine atoms to form a molecule in which each atom has an octet of electrons.

67. Use electron-dot symbols to show the sharing of electrons between a nitrogen atom and chlorine atoms to form a molecule in which each atom has an octet of electrons.

Electronegativity: Polar Covalent Bonds

68. Use the symbols $\delta+$ and $\delta-$ to indicate partial charges, if any, on the following bonds.
  a. Si — Cl   **b.** Cl — Cl   **c.** O — F

69. Use the symbols $\delta+$ and $\delta-$ to indicate partial charges, if any, on the following bonds.
  a. N — H   **b.** C — F   **c.** C — C

Names and Formulas for Ionic Compounds

70. Name the following ionic compounds.
  a. NaBr   **b.** $CaCl_2$ **c.** $FeCl_3$
  d. LiI   **e.** K_2S **f.** CuBr

71. Name the following ionic compounds.
  a. KCl   **b.** $MgBr_2$ **c.** CuI_2
  d. CaS   **e.** $FeCl_2$ **f.** Al_2O_3

72. Give formulas for the following ionic compounds.
  a. magnesium sulfate
  b. sodium hydrogen carbonate
  c. potassium nitrate
  d. calcium monohydrogen phosphate

73. Give formulas for the following ionic compounds.
  a. calcium carbonate
  b. potassium dihydrogen phosphate
  c. magnesium cyanide
  d. lithium hydrogen sulfate

74. Give formulas for the following ionic compounds.
  a. iron(II) phosphate
  b. potassium dichromate
  c. copper(I) iodide
  d. ammonium nitrite

75. Give formulas for the following ionic compounds.
  a. iron(III) oxalate
  b. sodium permanganate
  c. copper(II) bromide
  d. zinc monohydrogen phosphate

76. Name the following ionic compounds.
  a. KNO_2 **b.** LiCN   **c.** NH_4I
  d. $NaNO_3$ **e.** $KMnO_4$ **f.** $CaSO_4$

77. Name the following ionic compounds.
  a. $NaHSO_4$ **b.** $Al(OH)_3$ **c.** Na_2CO_3
  d. $KHCO_3$ **e.** NH_4NO_2 **f.** $Ca(HSO_4)_2$

78. Name the following ionic compounds.
  a. Na_2HPO_4 **b.** $(NH_4)_3PO_4$
  c. $Al(NO_3)_3$ **d.** NH_4NO_3

79. Name the following ionic compounds.
  a. Li_2CO_3 **b.** $Na_2Cr_2O_7$
  c. $Ca(H_2PO_4)_2$ **d.** $(NH_4)_2C_2O_4$

Names and Formulas for Covalent Compounds

80. Give formulas for the following covalent compounds.
  a. dinitrogen monoxide
  b. tetraphosphorus trisulfide
  c. phosphorus pentachloride
  d. sulfur hexafluoride

81. Give formulas for the following covalent compounds.
    a. oxygen difluoride   **b.** dinitrogen pentoxide
    c. phosphorus tribromide   **d.** tetrasulfur tetranitride
82. Name the following covalent compounds.
    a. CS_2 **b.** N_2S_4 **c.** PF_5 **d.** S_2F_{10}
83. Name the following covalent compounds.
    a. CBr_4 **b.** Cl_2O_7 **c.** P_4S_{10} **d.** I_2O_5

Intermolecular Forces

84. With which of the following would hydrogen bonding be an important intermolecular force?

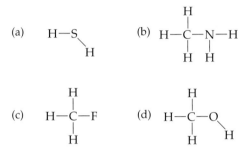

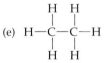

85. Alcohol that is used as a disinfectant to clean the skin prior to an injection is actually a solution of 3 parts water and 7 parts alcohol. Which component is the solvent and which is the solute?
86. Explain why a salt dissolves in water.
87. Benzene (C_6H_6) is a nonpolar solvent. Would you expect NaCl to dissolve in benzene? Explain.
88. Motor oil is nonpolar. Would you expect it to dissolve in water? In benzene? Explain.
89. A 250-mL can of motor oil is poured into a can that contains 2.0 L of gasoline. Which component is the solute and which is the solvent?

VSEPR Theory: The Shapes of Molecules

90. Use VSEPR theory to predict the shape of each of the following molecules.
    a. silane (SiH_4)   **b.** hydrogen selenide (H_2Se)
91. Use the VSEPR theory to predict the shape of each of the following molecules.
    a. beryllium chloride ($BeCl_2$)
    b. boron chloride (BCl_3)

92. Use the VSEPR theory to predict the shape of each of the following molecules.
    a. arsine (AsH_3)   **b.** carbon tetrafluoride (CF_4)
93. Use the VSEPR theory to predict the shape of each of the following molecules.
    a. oxygen difluoride (OF_2)
    b. silicon tetrachloride ($SiCl_4$)
94. Use the VSEPR theory to predict the shape of each of the following molecules.
    a. nitrogen trichloride (NCl_3)
    b. sulfur dichloride (SCl_2)
95. Use the VSEPR theory to predict the shape of each of the following molecules.
    a. phosphorus trifluoride (PF_3)
    b. dichlorodifluoromethane (CCl_2F_2)

Molecules That Are Exceptions to the Octet Rule

96. Give an electron-dot formula for each of the following.
    a. NO   **b.** BeI_2 **c.** BCl_3
97. Give an electron-dot formula for each of the following.
    a. PF_5 **b.** $AlBr_3$

Electron-Dot Formulas

98. Give electron-dot formulas that follow the octet rule, showing all valence electrons as dots, for the following covalent molecules.
    a. CH_4O **b.** NOH_3 **c.** CH_5N **d.** N_2H_4
99. Give electron-dot formulas that follow the octet rule, showing all valence electrons as dots, for the following covalent molecules.
    a. NF_3 **b.** C_2H_4 **c.** C_2H_2 **d.** CH_2O
100. Shared pairs of electrons can be represented by dashes. Give dash line formulas for the molecules in Problem 98.
101. Shared pairs of electrons can be represented by dashes. Give dash line formulas for the molecules in Problem 99.
102. Give electron-dot formulas that follow the octet rule for the following covalent molecules.
    a. COF_2 **b.** PCl_3 **c.** H_3PO_3 **d.** HCN
103. Give electron-dot formulas that follow the octet rule for the following covalent molecules.
    a. SCl_2 **b.** H_2SO_4 **c.** XeO_3 **d.** $HClO_4$
104. Give electron-dot formulas that follow the octet rule for the following ions.
    a. ClO^- **b.** HPO_4^{2-} **c.** ClO_2^- **d.** BrO_3^-
105. Give electron-dot formulas that follow the octet rule for the following ions.
    a. CN^- **b.** IO_4^- **c.** HSO_4^- **d.** PO_4^{3-}

Additional Problems

106. Why does neon tend not to form chemical bonds?
107. Draw a charge-cloud picture for the HF molecule. Use the symbols $\delta+$ and $\delta-$ to indicate the polarity of the molecule.

108. There are two different covalent molecules with the formula C_2H_6O. Give electron-dot formulas for the two molecules.

109. Fill in the following table, assuming that elements X, Y, and Z are all in A subgroups in the periodic table.

	Element X	Element Y	Element Z
Group number	1A		
Electron-dot formula		$\cdot\ddot{Y}\cdot$	
Charge on ion			2–

110. Consider the hypothetical elements X, Y, and Z with the following electron-dot formulas.

$$:\overset{\cdot\cdot}{\underset{\cdot\cdot}{X}}\cdot \qquad :\overset{\cdot\cdot}{Y}\cdot \qquad :\overset{\cdot}{Z}\cdot$$

a. To which group in the periodic table would each element belong?
b. Give the electron-dot formula for the simplest compound of each element with hydrogen.

c. Give electron-dot formulas for the ions formed when X and Y react with sodium.

111. Solutions of iodine chloride (ICl) are used as disinfectants. Are the molecules of ICl ionic, polar covalent, or nonpolar covalent?

112. Potassium is a soft, silvery metal that reacts violently with water and ignites spontaneously in air. Your doctor recommends you take a potassium supplement. Would you take potassium metal? What would you take?

113. Is there any such thing as a sodium chloride molecule? Explain.

114. Use sublevel notations to give electron configurations for the most stable simple ion formed by each of the following elements.
a. Ba   b. K   c. Se

115. Use sublevel notations to give electron configurations for the most stable simple ion formed by each of the following elements.
a. I   b. N   c. Te

Projects

116. Starting with 20 balloons, blown up to about the same size and tied, tie 2 of the balloons together. Then tie 3 balloons together. Repeat this with 4, 5, and 6 balloons. (*Suggestion*: 3 sets of 2 balloons can be twisted together to form a 6-balloon set, and so on). Show how these balloon sets can be used to explain the VSEPR theory.

117. Write a brief essay on Gilbert N. Lewis, a famous American chemist.

118. Exceptions to the octet rule include atoms with too few electrons, atoms with too many electrons, and atoms with an odd number of electrons. List examples of each.

Online Projects

119. See how many websites you can find (in about 15 min) on the subject of the VSEPR theory.

120. Search the web for information on the hydrogen bond.

121. Look for information about G. N. Lewis by searching the web.

References and Readings

1. Atkins, P. W. *Molecules*. New York: W. H. Freeman, 1987. A friendly and colorful discussion about some common molecules and their chemical bonds.

2. Asimov, Isaac. *A Short History of Chemistry*. Garden City, NY: Doubleday, 1965. Chapter 7, "Molecular Structure."

3. Benfey, O. T. *Classics in the Theory of Chemical Combination*. New York: Dover Publications, 1963.

4. Companion, Audrey. *Chemical Bonding*, 2nd edition. New York: McGraw-Hill, 1979.

5. Dahl, Peter. "The Valence-Shell Electron-Pair Repulsion Theory." *Chemistry*, March 1973, pp. 17–19.

6. Ebbing, Darrell D., and Steven D. Gammon. *General Chemistry*, 6th edition. Boston, MA: Houghton Mifflin, 1999.

7. Emsley, John. *Molecules at an Exhibition: Portraits of Interesting Molecules in Everyday Life*. Oxford, UK: Oxford University Press, 1998. A witty "tour" through a museum" of various elements and compounds.

8. Gillespie, Ronald J. "Covalent and Ionic Molecules: Why Are BeF_2 and AlF_3 High Melting Point Molecules Whereas BF_3 and SiF_4 Are Gases?" *Journal of Chemical Education*, July 1998, pp. 923–925.

9. Gillespie, Ronald J. *Molecular Geometry*, London: Van Nostrand Reinhold, 1972.

10. Hill, John W., and Ralph H. Petrucci. *General Chemistry*, 2nd edition. Upper Saddle River, NJ: Prentice Hall, 1999. Chapters 9 and 10.

11. Normile, Dennis. "Search for Better Crystals Explores Inner, Outer Space." *Science*, 22 December 1995, pp. 1921–1922.

12. Pauling, L., and R. Hayward. *The Architecture of Molecules*. San Francisco: W. H. Freeman, 1964. Easy reading, great artwork.

MediaLab

Compounds and Their Elements: What a Difference a Bond Makes!

We've seen in this chapter that the properties of a compound typically differ greatly from those of its component elements. In this MediaLab, we'll take a closer look at two ionic compounds and the elements of which they're composed: lithium fluoride, a compound that is important in optics, and copper(II) sulfate, the crystals of which are sometimes collected as decorative items.

Recall what happens during the formation of an ionic bond between a metal on the far left side of the periodic table and a nonmetal on the right. The metal tends to lose its valence electron(s), acquire a noble gas electron configuration, and become less reactive. At the same time, the

nonmetal gains one or more electrons to complete an octet. The resulting ions have gained stable electron configurations and are held together in a crystal lattice by the attractions between unlike charges. This seemingly simple change results in markedly different properties. In the *Web Investigation* of this MediaLab, you will look at the physical and chemical properties of these compounds and the elements in them. You will have an opportunity to view some virtual experiments. In *Communicate Your Results*, you will compare elements and ions and use your ingenuity to explain what happens to the atoms when they form ionic compounds.

Investigation 1
Lithium and Fluorine: An Alkali Metal and a Halogen

Let's begin by looking at the elements lithium and fluorine themselves. The periodic table is a good "jumping–off point." Select Keyword **TABLE1**. There's a lot of information here about each element, much of it very technical. (*Hint*: Many terms have hyperlinks to their definitions.) This site has many graphs, but most refer to the elements by atomic number. Be sure you have a periodic table handy before you start investigating. Look at the basic information about each element and at Electronic Configuration, Compounds, and Crystal Struc-

ture. Also glance at Uses, Geology, and Biology. An alternate table, with more narrative background information, is found at Keyword **TABLE2**.

Investigation 2
Binary Ionic Solids

As we saw in Section 5.3, sodium chloride consists of sodium ions (Na^+) and chloride ions (Cl^-) held together by electrostatic forces. The same is true of every ionic compound composed of an alkali metal and a halogen. Such compounds have a crystalline structure. Unlike some halides, such as potassium chloride, lithium fluoride does not occur as a decorative crystal. Its crystals, however, are important in the optics industry. Select Keyword **LiF** to learn more about its general properties, and Keyword **OPTICAL** to learn more about its optical properties.

Investigation 3
Copper and Sulfate, Plus a Look at Sulfur and Oxygen

Return to your favorite periodic table and look at copper. How does this transition metal differ from the alkali metals sodium and potassium? For more information, select Keyword **ELEMENTS**. Also look at sulfur and oxygen. How are these two elements related?

Investigation 4
Complex Ions

The sulfate, SO_4^{2-}, in copper sulfate is a polyatomic ion. Select the Keyword **COMPLEX** to learn about complex ions. Choose Cu and follow the directions to see some virtual experiments. The common form of copper sulfate is the pentahydrate, which has five bound water molecules. Heating this compound to drive off the water also causes the compound to lose its color. Select Keyword **CHALCANTHITE** for more information about this compound, which often occurs as a beautiful solid. Such *minerals* are often collected as well as studied.

COMMUNICATE YOUR RESULTS

Exercise 1
Atoms to Ions

Look at the cartoons in the sodium and potassium sections of *Web Investigation 1*. Use one of them to explain ions and ionic compounds in layperson's terms.

Exercise 2
Elements versus Compounds

Using the information available on the websites, prepare a table comparing the characteristics of lithium and fluorine with those of lithium fluoride, or of copper, oxygen, and sulfur with those of copper sulfate. Include atomic number and atomic mass, atomic and ionic radii, and electronic configuration, as well as some physical properties. Try to structure it so that a reader can easily see some of the differences between the elements and the compound. (You might also want to add a column for use or significance.)

Exercise 3
From Atom to Ion

Write an article for a home decoration magazine, advising readers about the use of copper or copper sulfate as a decorative item. Should climate play a role in the choice?

6 Chemical Accounting

Mass and Volume Relationships

Since molecules and atoms are impossible to count,
We use moles of substance as a measure of amount.

In this reaction two colorless liquids combine to produce a yellow solid. An aqueous solution of potassium chromate is poured into aqueous lead nitrate, giving brilliant yellow solid lead chromate as the product. As we will learn in this chapter, we can write an equation for the reaction and calculate an expected amount of product from known amounts of reactants.

M any areas of chemistry can be discussed without much use of mathematics, but the quantitative aspects of the subject are quite interesting, too. In this chapter we will consider some of the basic calculations used in chemistry and related fields such as biology and medicine. Chemists use all kinds of mathematics, from simple arithmetic to sophisticated calculus and complicated computer algorithms, but our calculations here will require no more than a knowledge of simple algebra.

6.1 Chemical Sentences: Equations

Chemistry is a study of matter and the changes it undergoes, and of the energy that brings about these changes or is released when these changes occur. In Chapter 5, we discussed the symbols and formulas used to represent elements and compounds. Now that we have learned the letters (symbols) and words (formulas) of our chemical language, we are ready to write sentences (chemical equations). A **chemical equation** is a shorthand way of describing chemical change using symbols and formulas to represent the elements and compounds involved in the change.

We can describe a chemical reaction in words. For example:

Carbon reacts with oxygen to form carbon dioxide.

We can also describe the same reaction in chemical shorthand.

$$C + O_2 \longrightarrow CO_2$$

The plus sign (+) indicates the addition of carbon to oxygen (or vice versa) or a combining of the two in some manner. The arrow ($\longrightarrow$) is often read "yields." Substances on the left of the arrow are **reactants** or *starting materials*. Those on the right are the **products** of the reaction.

At the submicroscopic (atomic or molecular) level, the chemical equation

$$C + O_2 \longrightarrow CO_2$$

means that one atom of carbon (C) reacts with one molecule of oxygen (O_2) to produce one molecule of carbon dioxide (CO_2).

Sometimes we indicate the physical states of the reactants and products by writing the initial letter of the state immediately following the formula. Thus, (g) indicates a gaseous substance, (l) a liquid, and (s) a solid. The label (aq) indicates an aqueous solution—that is, a water solution. Using these labels, our equation becomes

$$C(s) + O_2(g) \longrightarrow CO_2(g)$$

> Reactants and products need not be written in any particular order in a chemical equation, except that all reactants must be to the left of the arrow and all products to the right. In other words, we could also write the above equation as
>
> $$O_2 + C \longrightarrow CO_2$$

Balancing Chemical Equations

We can represent the reaction of carbon and oxygen to form carbon dioxide quite simply, but many other chemical reactions require more thought. For example, hydrogen reacts with oxygen to form water. Using formulas, we can write this reaction as

$$H_2 + O_2 \longrightarrow H_2O \qquad (not\ balanced)$$

However, this equation shows that two oxygen atoms react (as O_2), but only one is shown in the product (in H_2O). Recall from Chapter 2 that matter is neither created nor destroyed in a chemical reaction (the law of conservation of matter). To represent the chemical reaction correctly, the equation must be balanced. To balance the oxygen atoms, we need only place the coefficient 2 in front of the formula for water.

$$H_2 + O_2 \longrightarrow 2\,H_2O \qquad (not\ balanced)$$

This coefficient means that two molecules of water are produced. As is the case with subscripts, a coefficient of 1 is understood when no other number appears. A coefficient

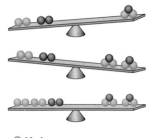

○ Hydrogen
● Oxygen

Figure 6.1 To balance the equation for the reaction in which hydrogen and oxygen react to form water, the same number of each kind of atom must appear on each side (atoms are conserved). When the equation is balanced, there are four hydrogen atoms and two oxygen atoms on each side.

preceding a formula multiplies everything in the formula. In the equation above, the coefficient 2 not only increases the number of oxygen atoms to two but also increases the number of hydrogen atoms to four.

But the equation is still not balanced. As we took care of the oxygen, we unbalanced the hydrogen. To balance the hydrogen, we place a coefficient 2 in front of H_2.

$$2 H_2 + O_2 \longrightarrow 2 H_2O \quad (balanced)$$

Now there are four hydrogen atoms and two oxygen atoms on each side of the equation. Atoms are conserved: the equation is balanced (Figure 6.1) the law of conservation of mass is obeyed. Figure 6.2 illustrates two common pitfalls in the process of balancing equations, as well as the correct method.

Example 6.1

Balance the following equation, which represents the chemical reaction involved when iron rusts.

$$Fe + O_2 \longrightarrow Fe_2O_3$$

Solution

For this sort of problem, we will use the concept of the least common multiple. We can begin with either element, but let's balance the oxygen atoms first. There are two oxygen atoms on the left and three on the right. The least common multiple of 2 and 3 is 6. We need *three* O_2 and *two* Fe_2O_3

$$\boxed{3 \times 2 = 6 \text{ O atoms}} \quad \boxed{2 \times 3 = 6 \text{ O atoms}}$$
$$Fe + 3 O_2 \longrightarrow 2 Fe_2O_3 \quad (not\ balanced)$$
$$\boxed{1 \text{ Fe atom}} \quad \boxed{2 \times 2 = 4 \text{ Fe atoms}}$$

We now have *four* iron atoms on the right side. We can get four on the left by placing the coefficient 4 in front of Fe.

$$4 Fe + 3 O_2 \longrightarrow 2 Fe_2O_3 \quad (balanced)$$

Checking, we count four Fe atoms and six O atoms on each side. The equation is balanced.

Figure 6.2 Balancing the chemical equation for the reaction between hydrogen and oxygen to form water. (a) Incorrect. There is no atomic oxygen (O) as a product. Extraneous products cannot be introduced simply to balance an equation. (b) Incorrect. The product of the reaction is water (H_2O), not hydrogen peroxide (H_2O_2). A formula can't be changed simply to balance an equation. (c) Correct. An equation can be balanced only through the use of correct formulas and coefficients.

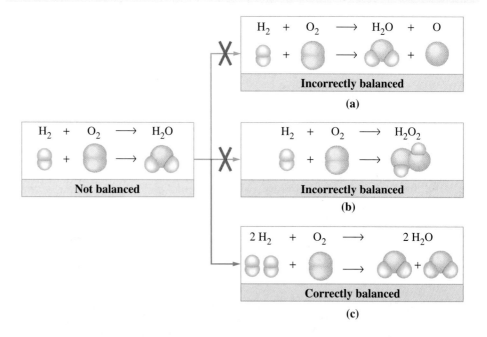

Exercise 6.1

A. The reaction between nitrogen and hydrogen to give ammonia, called the *Haber process*, is typically the first step in the industrial production of fertilizer. Balance the following equation for the Haber process.

$$N_2 + H_2 \longrightarrow NH_3 \quad (\textit{not balanced})$$

B. Balance the following equations.

    a. $P_4 + H_2 \longrightarrow PH_3$ **b.** $Mg + B_2O_3 \longrightarrow B + MgO$

Although simple equations can be balanced by trial and error, a couple of strategies often help. (1) If an element occurs in just one compound on each side of the equation, try balancing that element *first*. (2) Balance any reactants or products that exist as the free element *last*. Perhaps the most important step in any strategy is to check an equation to ensure that it is indeed balanced. Remember that for each element, the same number of atoms of the element must appear on each side of the equation; atoms are conserved in chemical reactions.

> An equation can be balanced only by writing coefficients (numbers) in front of terms in the equation wherever necessary.

Example 6.2

When a fuel such as methane is burned in sufficient air, the products are carbon dioxide and water. Balance the following equation for this combustion.

$$CH_4 + O_2 \longrightarrow CO_2 + H_2O \quad (\textit{not balanced})$$

Solution

In this equation, oxygen appears in two different products; we leave the oxygen for last and balance the other two elements first. Carbon is already balanced, with one atom on each side of the equation. For hydrogen, the least common multiple of 2 and 4 is 4, and so we place the coefficient 2 in front of H_2O to balance hydrogen. Now we have four hydrogen atoms on each side.

$$CH_4 + O_2 \longrightarrow CO_2 + 2\,H_2O \quad (\textit{not balanced})$$

Now for the oxygen. There are four oxygen atoms on the right. If we place a 2 in front of O_2 on the left, the oxygen atoms balance.

$$CH_4 + 2\,O_2 \longrightarrow CO_2 + 2\,H_2O \quad (\textit{balanced})$$

The equation is balanced.

Exercise 6.2

Balance the following equation.

$$C_3H_8 + O_2 \longrightarrow CO_2 + H_2O$$

In some reactions, such as those involving polyatomic ions, we can treat groups of atoms as a unit and balance them as a whole. Example 6.3 illustrates this kind of process.

Example 6.3

When a strong acid such as sulfuric acid is added to sodium cyanide, the highly poisonous gas hydrogen cyanide is produced. Balance the following equation for this reaction.

$$H_2SO_4 + NaCN \longrightarrow HCN + Na_2SO_4 \quad (\textit{not balanced})$$

Solution

This equation involves compounds with polyatomic ions. The SO_4 group (actually SO_4^{2-}) should be treated as a unit and balanced as a whole. The CN group (actually CN^-) can also be treated as a unit. As the equation is presently written, the SO_4 groups and the CN groups are balanced, but the hydrogen atoms and the sodium atoms are not. The least common multiple for sodium is 2, and so to get two sodium atoms on the left we place a 2 before NaCN. The least common multiple for hydrogen is 2, and so to get two hydrogen atoms on the right we place a 2 before HCN.

$$H_2SO_4 + 2\,NaCN \longrightarrow 2\,HCN + Na_2SO_4 \quad \textit{(balanced)}$$

The same coefficients balance the CN groups and the SO_4 groups, and the entire equation is balanced.

Exercise 6.3

Balance the following equation.

$$H_3PO_4 + Ca(OH)_2 \longrightarrow Ca_3(PO_4)_2 + H_2O$$

We have made the task of balancing equations deceptively easy by considering simple reactions. It is more important at this point for you to understand the principle than to be able to balance complicated equations. You should know what is meant by a balanced equation and be able to handle simple reactions.

6.2 Volume Relationships in Chemical Equations

Amedeo Avogadro (1776–1856) did not live to see his ideas accepted by the scientific community. Acceptance finally came in 1860 at a scientific conference at which Stanislao Cannizzaro (1826–1910) effectively communicated Avogadro's ideas from half a century earlier.

We have looked at chemical reactions in terms of individual atoms and molecules. In the real world, however, chemists work with quantities of matter that contain billions of atoms. John Dalton postulated that atoms of different elements had different masses. Therefore, equal masses of different elements would contain different numbers of atoms. Consider the analogous situation of golf balls and Ping-Pong balls. A kilogram of golf balls contains a smaller number of balls than a kilogram of Ping-Pong balls. One could determine the number of balls in each case simply by counting them. For atoms, however, such a straightforward method is not possible.

Experiments with gases provided the French scientist Joseph Louis Gay-Lussac (1778–1850) with an approach to quantifying atoms. In 1809 he announced the results of some chemical reactions that he had carried out with gases. He summarized these experiments as the **law of combining volumes**, which states that when all measurements are made at the same temperature and pressure, the volumes of gaseous reactants and products are in a small whole-number ratio. Two such experiments are illustrated in Figure 6.3. When hydrogen reacts with oxygen to form steam at 100°C, two volumes of hydrogen unite with one volume of oxygen to yield two volumes of steam. The small whole-number ratio is $2:1:2$. Similarly, when hydrogen reacts with nitrogen to form ammonia, the combining volumes are three of hydrogen with one of nitrogen to yield two of ammonia ($3:1:2$).

Gay-Lussac thought there must be some relationship between the numbers of molecules and the volumes of gaseous reactants and products. But it was Amedeo Avogadro who first explained the law of combining volumes in 1811. **Avogadro's hypothesis**, based on a shrewd interpretation of experimental facts, was that equal volumes of all gases, when measured at the same temperature and pressure, contain the same number of molecules (Figure 6.4).

The equation for the combination of hydrogen and oxygen to form water (steam) is

$$2\,H_2(g) + O_2(g) \longrightarrow 2\,H_2O(g)$$

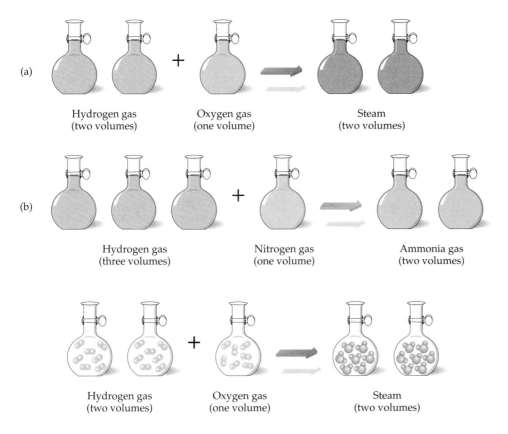

Figure 6.3 Gay-Lussac's law of combining volumes. (a) Two volumes of hydrogen gas react with one volume of oxygen gas to yield two volumes of steam. (b) Three volumes of hydrogen gas react with one volume of nitrogen gas to yield two volumes of ammonia gas.

Figure 6.4 Avogadro's explanation of Gay-Lussac's law of combining volumes. Equal volumes of each of the gases contain the same number of molecules.

The coefficients of the molecules are the same as the combining ratio of the gas volumes, 2:1:2 (Figure 6.3). Similarly, the formation of ammonia is described in the equation

$$3\ H_2(g)\ +\ N_2(g)\ \longrightarrow\ 2\ NH_3(g)$$

The coefficients are identical to the factors of the combining ratio. The equation says that a nitrogen molecule and three hydrogen molecules react to produce two ammonia molecules. If you had 1 million nitrogen molecules, you would need 3 million hydrogen molecules to produce 2 million ammonia molecules. The equation provides the combining ratios. According to the equation, one volume of nitrogen reacts with three volumes of hydrogen to produce two volumes of ammonia because identical volumes of gases contain identical numbers of molecules.

Example 6.4

What volume of oxygen is required to burn 0.556 L of propane if both gases are measured at the same temperature and pressure?

$$C_3H_8(g)\ +\ 5\ O_2(g)\ \longrightarrow\ 3\ CO_2(g)\ +\ 4\ H_2O(g)$$

Solution

The coefficients in the equation indicate that *five* volumes of $O_2(g)$ is required for every volume of $C_3H_8(g)$. Thus, we use 5 L $O_2(g)$/1 L $C_3H_8(g)$ as the ratio to find the volume of oxygen required.

$$?\ L\ O_2(g)\ =\ 0.556\ \cancel{L\ C_3H_8}(g)\ \times\ \frac{5\ L\ O_2(g)}{1\ \cancel{L\ C_3H_8}(g)}\ =\ 2.78\ L\ O_2(g)$$

Exercise 6.4

A. Using the equation in Example 6.4, calculate the volume of $CO_2(g)$ produced when 0.492 L of propane is burned if the two gases are compared at the same temperature and pressure.

B. Calculate the volume of $CO_2(g)$ produced when 5.42 L of ethane $[C_2H_6(g)]$ is burned if the two gases are compared at the same temperature and pressure. [*Hint*: First write a balanced chemical equation.]

6.3 Avogadro's Number: 6.02×10^{23}

Avogadro's hypothesis, which has been verified many times over the years, states that equal volumes of gas at the same temperature and pressure contain equal numbers of molecules. This means that if we weigh equal volumes of several gases, the ratio of their masses should be the same as the mass ratio of the molecules themselves.

Avogadro had no way of knowing how many molecules there were in a given volume of gas. Scientists since his time have determined the number of atoms in various weighed samples of substances. The numbers are extremely large, even for tiny samples. In defining atomic masses, the mass of a carbon-12 atom is defined as exactly 12 u. The number of carbon-12 atoms in a 12-g sample of carbon-12 is called **Avogadro's number** and has been determined experimentally to be 6.0221367×10^{23}. For our purposes, we round it off to three significant figures, 6.02×10^{23}.

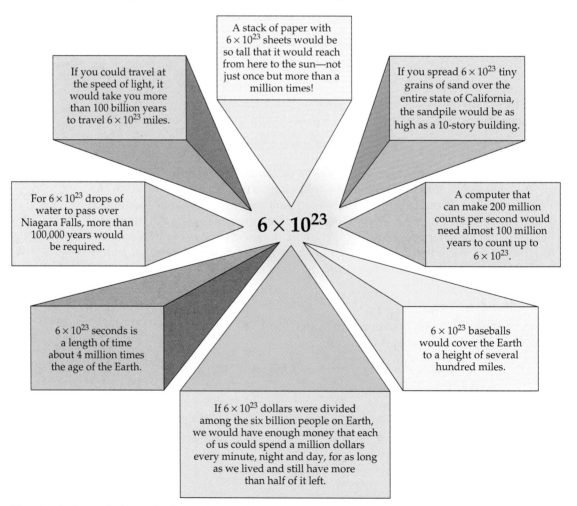

A stack of paper with 6×10^{23} sheets would be so tall that it would reach from here to the sun—not just once but more than a million times!

If you could travel at the speed of light, it would take you more than 100 billion years to travel 6×10^{23} miles.

If you spread 6×10^{23} tiny grains of sand over the entire state of California, the sandpile would be as high as a 10-story building.

For 6×10^{23} drops of water to pass over Niagara Falls, more than 100,000 years would be required.

6×10^{23}

A computer that can make 200 million counts per second would need almost 100 million years to count up to 6×10^{23}.

6×10^{23} seconds is a length of time about 4 million times the age of the Earth.

6×10^{23} baseballs would cover the Earth to a height of several hundred miles.

If 6×10^{23} dollars were divided among the six billion people on Earth, we would have enough money that each of us could spend a million dollars every minute, night and day, for as long as we lived and still have more than half of it left.

How big is Avogadro's number?

Whose Number Is It, Anyway?

Scientists knew that the number we now call Avogadro's number must be enormous, but they had no idea just how large it was for over half a century. In 1865 Josef Loschmidt (1821–1895) measured the size of air molecules and found them to be about a "millionth of a millimeter" in diameter. This rough measurement indicated a value of 4×10^{22}, not a bad estimate for a first attempt. Later measurements, using various approaches, have shown the actual diameter of air molecules to be a bit smaller than Loschmidt had determined, and the number to be closer to 6×10^{23}. In German-speaking countries, this value is usually called the *Loschmidt number*. In most places, however, it is called Avogadro's number in spite of the fact that it was not measured until after his death and Avogadro never knew how big the number was.

6.4 The Mole: "A Dozen Eggs and a Mole of Sugar, Please"

We buy socks by the pair (2 socks), eggs by the dozen (12 eggs), soda pop by the case (24 cans), pencils by the gross (144 pencils), and paper by the ream (500 sheets). A dozen is the same number whether we are counting a dozen melons or a dozen oranges. But a dozen oranges and a dozen melons do not weigh the same. If a melon weighs three times as much as an orange, a dozen melons will weigh three times as much as a dozen oranges.

Chemists count atoms and molecules by the *mole*. (A single carbon atom is much too small to see, but a mole of carbon atoms fills a tablespoon.) A mole of carbon and a mole of magnesium each contain the same number of atoms. But a magnesium atom has a mass twice that of a carbon atom, and so a mole of magnesium has a mass twice that of a mole of carbon.

According to the SI definition, a **mole (mol)** is an amount of substance that contains the same number of elementary units as there are atoms in exactly 12 g of carbon-12. That number is 6.02×10^{23}, Avogadro's number. The elementary units may be atoms (such as S or Ca), molecules (such as O_2 or CO_2), ions (such as K^+ or SO_4^{2-}), or any other kind of formula unit. A mole of NaCl, for example, contains 6.02×10^{23} NaCl formula units, which means that it contains 6.02×10^{23} Na^+ ions and 6.02×10^{23} Cl^- ions.

[Web Reference 1] The website of the National Mole Day Foundation, Inc. gives a light-hearted but informative tribute to the mole. It includes a surprising biographical sketch of Avogadro.

Molecular Masses and Formula Masses

Recall that each element has a characteristic atomic mass. Because chemical compounds are made up of two or more elements, the masses of compounds are combinations of atomic masses. For a molecular substance, the **molecular mass** is the average[1] mass of a molecule of a substance relative to that of a carbon-12 atom. More simply, it is the sum of the masses of the atoms represented in a molecular formula. For example, because the formula O_2 specifies two O atoms per molecule of oxygen, the molecular mass of oxygen (O_2) is twice the atomic mass of oxygen.

$$2 \times \text{atomic mass of O} = 2 \times 15.9994 \text{ u} = 31.9988 \text{ u}$$

The molecular mass of carbon dioxide (CO_2) is the sum of the atomic mass of carbon and twice the atomic mass of oxygen.

---

[1] We use the term "average" when speaking of the mass of an individual molecule because molecules of a compound may have different isotopes of one or more of their constituent elements.

$$
\begin{aligned}
1 \times \text{atomic mass of C} &= 1 \times 12.011 \text{ u} &&= 12.011 \text{ u} \\
2 \times \text{atomic mass of O} &= 2 \times 15.9994 \text{ u} &&= \underline{31.99988 \text{ u}} \\
\text{Molecular mass of CO}_2 &&&= 44.011 \text{ u}
\end{aligned}
$$

Example 6.5

Calculate the molecular mass of sulfur dioxide (SO_2), an irritating gas formed when sulfur is burned.

Solution

We think about the problem in the following way. Add the atomic mass of sulfur to twice the atomic mass of oxygen. However, if we use a calculator, we need only write down the final answer, 64.065 u. That is, we have no need to record the numbers 32.066 and 15.9994.

$$
\begin{aligned}
1 \times \text{atomic mass of S} &= 1 \times 32.066 \text{ u} &&= 32.066 \text{ u} \\
2 \times \text{atomic mass of O} &= 2 \times 15.9994 \text{ u} &&= \underline{31.9988 \text{ u}} \\
\text{Molecular mass of SO}_2 &&&= 64.065 \text{ u}
\end{aligned}
$$

Exercise 6.5

Calculate the molecular mass of (**a**) $C_6H_4Cl_2$, (**b**) $C_2H_4Cl_2$, and (**c**) H_3PO_4.

The term "molecular mass" is not appropriate for ionic compounds, such as NaCl and K_2O, in which individual molecules do not exist. For ionic compounds, we use the term *formula unit*. **Formula mass** is the average mass of a formula unit relative to that of a carbon-12 atom. In short, the formula mass is the sum of the masses of the atoms or ions represented by the formula.

Example 6.6

Calculate the formula mass of ammonium sulfate [$(NH_4)_2SO_4$] a fertilizer commonly used by home gardeners.

Solution

To determine a formula mass, we add the atomic masses of the constituent elements. In the summation below, remember that we must multiply everything within the parentheses by 2.

$$
\begin{aligned}
2 \times \text{atomic mass of N} &= 2 \times 14.0067 \text{ u} &&= 28.0134 \text{ u} \\
8 \times \text{atomic mass of H} &= 8 \times 1.00794 \text{ u} &&= 8.06352 \text{ u} \\
1 \times \text{atomic mass of S} &= 1 \times 32.066 \text{ u} &&= 32.066 \text{ u} \\
4 \times \text{atomic mass of O} &= 4 \times 15.9994 \text{ u} &&= \underline{63.9976 \text{ u}} \\
\text{Formula mass of } (NH_4)_2SO_4 &&&= 132.141 \text{ u}
\end{aligned}
$$

Exercise 6.6

Calculate the formula mass of (**a**) K_2CO_3, (**b**) $K_2Cr_2O_7$, and (**c**) $NaB(C_6H_5)_4$.

Molar Mass

The **molar mass** of a substance is the mass of 1 mol of that substance. The molar mass is numerically equal to the atomic mass, molecular mass, or formula mass, but it is expressed in the unit grams per mole (g/mol). The atomic mass of sodium is 22.99 u; its molar mass is 22.99 g/mol. The molecular mass of carbon dioxide is 44.01 u; its molar mass is 44.01 g/mol. The formula mass of magnesium chloride is 95.21 u; its molar mass is 95.21 g/mol. We can use these facts, together with the basic definition of the number of elementary units in a mole, to write the following relationships.

$$1 \text{ mol Na} = 22.99 \text{ g Na}$$
$$1 \text{ mol CO}_2 = 44.01 \text{ g CO}_2$$
$$1 \text{ mol MgCl}_2 = 95.21 \text{ g MgCl}_2$$

In turn, these relationships supply the conversion factors we need to make conversions between mass in grams and amount in moles, as illustrated in the following examples.

Example 6.7

How many grams of Na are there in 0.250 mol of Na?

Solution

Sodium has an atomic mass of 22.99 u and a molar mass of 22.99 g Na/mol Na. Therefore:

$$? \text{ g Na} = 0.250 \text{ mol Na} \times \frac{22.99 \text{ g Na}}{1 \text{ mol Na}} = 5.75 \text{ g Na}$$

Exercise 6.7

Calculate the mass, in grams, of (**a**) 55.5 mol H_2O, (**b**) 0.0102 mol $C_4H_{10}O$, and (**c**) 2.45 mol C_2H_6.

Example 6.8

Calculate the number of moles of CO_2 in a 225-g sample of the gas.

Solution

In this case we need the molar mass of CO_2. We determined the molecular mass of CO_2 to be 44.011 (page 157). Its molar mass is therefore 44.011 g CO_2/mol CO_2. To convert from a mass in grams to an amount in moles, we must use the *inverse* of the molar mass (1 mol CO_2/44.011 g CO_2) to get the proper cancelation of units.

$$? \text{ mol CO}_2 = 225 \text{ g CO}_2 \times \frac{1 \text{ mol CO}_2}{44.011 \text{ g CO}_2} = 5.11 \text{ mol CO}_2$$

Exercise 6.8

Calculate the amount, in moles, of (**a**) 3.71 g Fe, (**b**) 76.0 g phosphoric acid (H_3PO_4), and (**c**) 165 g C_4H_{10}.

Figure 6.5 is a photograph showing 1-mol samples of several different chemical substances. Each dish contains Avogadro's number of formula units of the substance.

Figure 6.5 One mole of each of several familiar substances: salt (left), sugar (top), copper (right), and carbon (center). Each dish contains Avogadro's number of formula units of the substance it contains. There are 6.02 × 10²³ formula units of NaCl, 6.02 × 10²³ molecules of sugar ($C_{12}H_{22}O_{11}$), 6.02 × 10²³ atoms of copper, and 6.02 × 10²³ carbon atoms in the respective samples.

[Web Reference 2] A more serious tutorial about the mole and Avogadro's constant.

There are just as many carbon atoms in the smallest dish as there are sugar molecules in the largest one.

What Is a Mole?

A mole is a particular amount
Of substance—just its formulary weight
Expressed in grams, with Avogadro's count
Of units making up the aggregate.
A mole is a specific quantity:
Its volume measures twenty-two point four
In liters, for a gas at STP.
A mole's a counting unit, nothing more.
A mole is but a single molecule
By Avogadro's number multiplied;
One entity, extremely minuscule,
A trillion trillion times intensified.
A mole is a convenient amount,
For molecules are just too small to count.

Molar Volume: 22.4 L at Standard Temperature and Pressure

A mole of gas contains Avogadro's number of molecules no matter what kind of gas it is. Furthermore, Avogadro's number of molecules of gas occupies the same volume (at a given temperature and pressure) regardless of how big, how small, or how heavy the individual molecules are. The volume occupied by 1 mol of gas is the **molar volume** of a gas.

Because the volume of a gas is altered by changes in temperature or pressure (Section 6.7), a particular set of conditions has been chosen for reference purposes as "standard." Standard pressure is 1 atmosphere (atm), which is the normal pressure of the air at sea level; and standard temperature is 0°C, which is the freezing point of water. A mole of any gas at **standard temperature and pressure** (**STP**) occupies a volume of about 22.4 L. This is known as the standard molar volume of a gas.

A cube that measures 28.2 cm along each edge has a volume of 22.4 L (Figure 6.6). It is just a bit smaller than a cubic foot. At a temperature of 0°C and 1 atm pressure,

Figure 6.6 The molar volume of a gas compared with a basketball, a football, and a soccer ball. At STP, the cube holds 16.0 g of CH_4, 28.0 g of N_2, 32.0 g of O_2, 39.9 g of Ar, and so on. It holds 1 mol of any gas at STP.

a 22.4-L container holds 28.0 g of N_2, 32.0 g of O_2, 44.0 g of CO_2, and so on; it holds 1 mol of any gas.

It is easy to calculate the density of a gas at STP. The mass of 1 mol of gas is its molecular mass, and the volume is 22.4 L. Just divide the molar mass by 22.4 L. Conversely, if you know the density of a gas at STP, it is a simple matter to calculate its molar mass. Merely multiply the mass per liter by 22.4 L.

Example 6.9

Calculate (**a**) the density of helium gas at STP and (**b**) the molar mass of oxygen, given the density of oxygen gas is 1.43 g/L at STP.

Solution

a. At STP, 1 mol (4.00 g) of He occupies 22.4 L.

$$\frac{4.00 \text{ g/mol}}{22.4 \text{ L/mol}} = 0.179 \text{ g/L}$$

b. Merely multiply the density (mass per liter) by 22.4 L.

$$\frac{1.43 \text{ g}}{1 \text{ L}} \times \frac{22.4 \text{ L}}{1 \text{ mol}} = 32.0 \text{ g/mol}$$

Exercise 6.9

A. Calculate the density of argon gas at STP.
B. The density of diethyl ether vapor at STP is 3.30 g/L. Calculate its molecular mass.

6.5 Mole and Mass Relationships in Chemical Equations

Chemists and other scientists and engineers are often confronted with questions such as: How many grams of phosphoric acid can I make from 660 g of phosphorus? or How many grams of hydrogen peroxide do I need to convert 5.82 g of lead sulfide to lead sulfate? It is not possible to calculate grams of one substance directly from grams of another. Because chemical reactions involve atoms and molecules, quantities in reactions are calculated in moles of atoms, ions, or molecules. To do such calculations, it is necessary to convert grams of a substance to moles, as we did in Example 6.8. We also must write balanced equations to determine molar ratios.

Chemical equations (Section 6.1) not only represent ratios of atoms and molecules but also give us mole ratios. For example, the equation

$$C + O_2 \longrightarrow CO_2$$

tells us that one atom of carbon reacts with one molecule (two atoms) of oxygen to form one molecule of carbon dioxide (one atom of carbon and two atoms of oxygen). The equation also indicates that 1 mol (6.02×10^{23} atoms) of carbon reacts with 1 mol (6.02×10^{23} molecules) of oxygen to yield 1 mol (6.02×10^{23} molecules) of carbon dioxide. Recall that the molar mass in grams of a substance is numerically equal to the formula mass of the substance in atomic mass units. The equation therefore also tells us (indirectly) that 12.0 g (1 mol) of carbon reacts with 32.0 g (1 mol) of oxygen to yield 44.0 g (1 mol) of CO_2 (Figure 6.7).

We need not use exactly 1 mol of each reactant. The important thing is to keep the ratio constant. For example, in the reaction above, the mass ratio of oxygen to carbon is 32.0 : 12.0 or 8.0 : 3.0. We could use 8.0 g of oxygen and 3.0 g of carbon to produce 11.0 g of CO_2. In fact, to calculate the amount of oxygen needed to react with a given amount of carbon, we need only multiply the amount of carbon by the factor 32.0/12.0. The following examples illustrate these relationships.

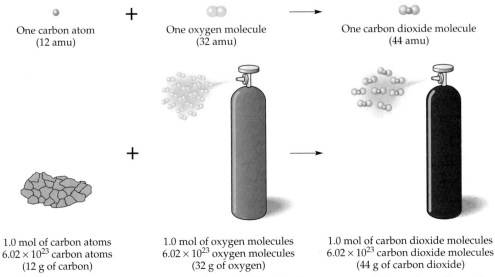

One carbon atom
(12 amu)

One oxygen molecule
(32 amu)

One carbon dioxide molecule
(44 amu)

1.0 mol of carbon atoms
6.02×10^{23} carbon atoms
(12 g of carbon)

1.0 mol of oxygen molecules
6.02×10^{23} oxygen molecules
(32 g of oxygen)

1.0 mol of carbon dioxide molecules
6.02×10^{23} carbon dioxide molecules
(44 g of carbon dioxide)

Figure 6.7   We cannot weigh single atoms or molecules, but we can weigh equal numbers of these fundamental particles.

Example 6.10

Nitrogen monoxide (nitric oxide), one of the air pollutants discharged by internal combustion engines, combines with oxygen to form nitrogen dioxide, a reddish-brown gas that irritates the respiratory system and eyes. The equation for this reaction is

$$2\,NO + O_2 \longrightarrow 2\,NO_2$$

State the molecular, molar, and mass relationships indicated by the equation.

Solution

Molecular: two molecules of NO react with one molecule of O_2 to form two molecules of NO_2.
Molar: 2 mol of NO react with 1 mol of O_2 to form 2 mol of NO_2.
Mass: 60.0 g of NO react with 32.0 g of O_2 to form 92.0 g of NO_2.

Exercise 6.10

Hydrogen sulfide, a gas that smells like rotten eggs, burns in air to produce sulfur dioxide and water according to the equation

$$2\,H_2S + 3\,O_2 \longrightarrow 2\,SO_2 + 2\,H_2O$$

State the molecular, molar, and mass relationships indicated by this equation.

Molar Relationships in Chemical Equations

Stoichiometry is the quantitative relationship between reactants and products in a chemical reaction. The ratio of moles of reactants and products is given by the coefficients in a balanced chemical equation. Consider the combustion of propane, the main component of bottled gas.

$$C_3H_8 + 5\,O_2 \longrightarrow 3\,CO_2 + 4\,H_2O$$

The coefficients in the balanced equation allow us to make statements such as

1 mol of C_3H_8 reacts with 5 mol of O_2.
3 mol of CO_2 is produced for every 1 mol of C_3H_8 that reacts.
4 mol of H_2O is produced for every 3 mol of CO_2 produced.

Moreover, we can turn these statements into conversion factors known as stoichio-metric factors. A **stoichiometric factor** relates the amounts of any two substances in-volved in a chemical reaction, on a *mole* basis. Note that we can set up conversion factors for any two compounds involved in a reaction. (Conversion factors are ex-plained in Appendix A.) Typical problems ask you to calculate how much of one compound is equivalent to a given amount of one of the other compounds. All you need to do to solve such a problem is put together a conversion factor relating the two compounds. In the examples that follow, stoichiometric factors are shown in color.

Example 6.11

When 0.105 mol of propane is burned in a plentiful supply of oxygen, how many moles of oxygen is consumed?

$$C_3H_8 + 5\,O_2 \longrightarrow 3\,CO_2 + 4\,H_2O$$

Solution

The equation tells us that 5 mol O_2 is required to burn 1 mol C_3H_8. We can write

$$1 \text{ mol } C_3H_8 \leftrightharpoons 5 \text{ mol } O_2$$

where we use the symbol $\leftrightharpoons$ to mean "is chemically equivalent to." From this re-lationship we can construct conversion factors to relate moles of oxygen to moles of propane. The possible conversion factors are

$$\frac{1 \text{ mol } C_3H_8}{5 \text{ mol } O_2} \quad \text{and} \quad \frac{5 \text{ mol } O_2}{1 \text{ mol } C_3H_8}$$

Multiply the given quantity (0.105 mol C_3H_8) by the factor on the right to get an answer with the asked-for units (moles of oxygen).

$$? \text{ mol } O_2 = 0.105 \text{ mol } C_3H_8 \times \frac{5 \text{ mol } O_2}{1 \text{ mol } C_3H_8} = 0.525 \text{ mol } O_2$$

Exercise 6.11

For the combustion of propane in Example 6.11: (**a**) How many moles of carbon dioxide is formed when 0.529 mol of C_3H_8 is burned? (**b**) How many moles of water is produced when 76.2 mol of C_3H_8 is burned? (**c**) How many moles of car-bon dioxide is produced when 1.010 mol of O_2 is consumed?

Mass Relationships in Chemical Equations

A chemical equation defines the stoichiometric relationship in terms of moles, but problems are seldom formulated in moles. Typically, you are given an amount of one substance in grams and asked to calculate how many grams of another sub-stance can be made from it. Such calculations involve several steps.

1. Write a balanced chemical equation for the reaction.
2. Determine the molar masses of the substances involved in the calculation.
3. Write down the given quantity and use the molar mass to convert this quanti-ty to moles.
4. Use the balanced chemical equation to convert moles of the given substance to moles of the desired substance.
5. Use the molar mass to convert moles of the desired substance to grams of the de-sired substance.

Figure 6.8 We can use a chemical equation to relate *moles* of any two substances represented in the equation. These substances may be a reactant and a product, two reactants, or two products. To obtain *mass* relationships, we must convert the mass of each substance to moles, relate moles of one substance to moles of the other through a stoichiometric factor, and then convert moles of the second substance to a mass.

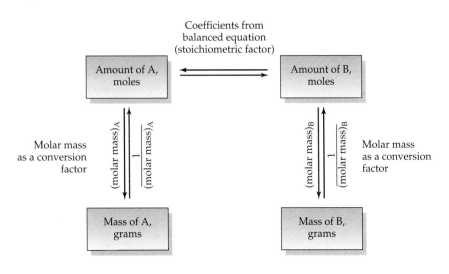

If we know the quantity of any substance in an equation, we can determine the quantities of all the other substances. The conversion process is diagrammed in Figure 6.8. It is best learned from examples and by working out exercises.

Example 6.12

Calculate the mass of oxygen needed to react with 10.0 g of carbon in the reaction that forms carbon dioxide.

Solution

Step 1. The balanced equation is

$$C + O_2 \longrightarrow CO_2$$

Step 2. The molar masses are $2 \times 15.9994 = 31.9988$ g/mol for O_2 and 12.011 g/mol for C.

Step 3. We convert the mass of the given substance, carbon, to an amount in moles.

$$? \text{ mol C} = 10.0 \text{ } \cancel{gC} \times \frac{1 \text{ mol C}}{12.011 \text{ } \cancel{gC}} = 0.833 \text{ mol C}$$

Step 4. We use coefficients from the balanced equation to establish the stoichiometric factor that relates the amount of oxygen to that of carbon.

$$0.833 \text{ } \cancel{\text{mol C}} \times \frac{1 \text{ mol O}_2}{1 \text{ } \cancel{\text{mol C}}} = 0.833 \text{ mol O}_2$$

Step 5. We convert from moles of oxygen to grams of oxygen.

$$0.833 \text{ } \cancel{\text{mol O}_2} \times \frac{31.9988 \text{ g O}_2}{1 \text{ } \cancel{\text{mol O}_2}} = 26.7 \text{ g O}_2$$

We can also combine the five steps into a single setup. Note that the units in the denominators of the conversion factors are chosen so that each cancels the unit in the numerator of the preceding term.

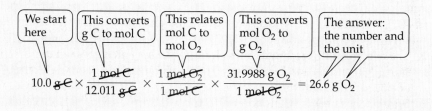

$$10.0 \ \cancel{g \, C} \times \frac{1 \ \cancel{mole \, C}}{12.011 \ \cancel{g \, C}} \times \frac{1 \ \cancel{mol \, O_2}}{1 \ \cancel{mole \, C}} \times \frac{31.9988 \ g \ O_2}{1 \ \cancel{mol \, O_2}} = 26.6 \ g \ O_2$$

(The slightly different answers are due to rounding in the intermediate steps.)

Exercise 6.12

A. Calculate the mass of oxygen (O_2) needed to react with 0.334 g of nitrogen (N_2) in the reaction that forms nitrogen dioxide.

B. Calculate the mass of carbon dioxide formed by burning 775 g of each of (**a**) methane (CH_4) and (**b**) butane (C_4H_{10}).

Example 6.13

Ammonia (NH_3) a common fertilizer, is made by causing hydrogen and nitrogen to react at a high temperature and pressure. What mass of ammonia, in grams, can be made from 60.0 g of hydrogen?

Solution

Step 1. We start by writing a chemical equation.

$$N_2 + H_2 \longrightarrow NH_3 \qquad \textit{(not balanced)}$$

Then we balance it.

$$N_2 + 3 \ H_2 \longrightarrow 2 \ NH_3 \qquad \textit{(balanced)}$$

Step 2. The molar masses are $2 \times 1.008 = 2.016$ g/mol for H_2 and $14.01 + (3 \times 1.008) = 17.03$ g/mol for NH_3.

Step 3. We convert the mass of the given substance, hydrogen, to an amount in moles.

$$60.0 \ \cancel{g \, H_2} \times \frac{1 \ mol \ H_2}{2.016 \ \cancel{g \, H_2}} = 29.8 \ mol \ H_2$$

Step 4. We use coefficients from the balanced equation to establish the stoichiometric factor that relates the amount of ammonia to that of hydrogen.

$$29.8 \ \cancel{mol \, H_2} \times \frac{2 \ mol \ NH_3}{3 \ \cancel{mol \, H_2}} = 19.9 \ mol \ NH_3$$

Step 5. We convert from moles of ammonia to grams of ammonia.

$$19.9 \ \cancel{mol \, NH_3} \times \frac{17.03 \ g \ NH_3}{1 \ \cancel{mol \, NH_3}} = 339 \ g \ NH_3$$

As is usually the case, all of the steps outlined above can be combined into a single setup.

$$60.0 \ \cancel{g \, H_2} \times \frac{1 \ \cancel{mol \, H_2}}{2.016 \ \cancel{g \, H_2}} \times \frac{2 \ \cancel{mol \, NH_3}}{3 \ \cancel{mol \, H_2}} \times \frac{17.03 \ g \ NH_3}{1 \ \cancel{mol \, NH_3}} = 338 \ g \ NH_3$$

Exercise 6.13

A. Ammonia reacts with phosphoric acid (H_3PO_4) to form ammonium phosphate [($NH_4)_3PO_4$]. How many grams of ammonia are needed to react completely with 74.8 g of phosphoric acid?

B. a. The decomposition of potassium chlorate ($KClO_3$) produces potassium chloride (KCl) and O_2 gas. How many grams of oxygen can be made from 2.47 g of potassium chlorate?

  b. Phosphorus reacts with oxygen to form tetraphosphorus decoxide. The equation is

$$P_4 + O_2 \longrightarrow P_4O_{10} \quad \text{(not balanced)}$$

How many grams of tetraphosphorus decoxide can be made from 3.50 g of phosphorus?

  c. What mass of magnesium metal is required to reduce 83.6 g of titanium(IV) chloride to titanium metal? The equation (not balanced) is

$$TiCl_4 + Mg \longrightarrow Ti + MgCl_2$$

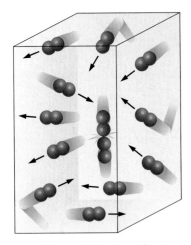

Figure 6.9 According to the kinetic–molecular theory, molecules of a gas are in constant, random motion. They move in straight lines and undergo collisions with each other and with the walls of the container.

[Web Reference 3] A simulation of ideal gas behavior.

6.6 The Gas Laws

Experiments with gases were instrumental in developing the concepts of Avogadro's number and of molar ratios in reactions. Let's now look at the behavior of gases more closely, using a model known as the **kinetic–molecular theory** (Figure 6.9). The basic postulates of this theory are the following.

1. All matter is composed of tiny, discrete particles called *molecules*.
2. Molecules of a gas are in rapid, constant motion and move in straight lines.
3. The molecules of a gas are tiny compared with the distances between them.
4. Because they are so far apart, there is very little attraction between molecules of a gas.
5. Molecules collide with one another. Energy is conserved in these collisions, although one molecule can gain energy at the expense of another.
6. Temperature is a measure of the average kinetic energy of the gas molecules.

We discussed (Section 5.15) how intermolecular forces hold molecules together in solids and liquids. In gases the particles are separated by relatively great distances and are moving about at random, and so they seldom interact. The behavior of gases can be described by the gas laws, which tell how the volume of a gas changes with temperature or pressure.

Boyle's Law: Pressure and Volume

A simple gas law, discovered by Robert Boyle in 1662, describes the relationship between the pressure and volume of a gas. **Boyle's law** states that *for a given amount of gas at a constant temperature, the volume of the gas varies inversely with its pressure.* That is, in a closed container of gas, when the pressure increases, the volume decreases; when the pressure decreases, the volume increases.

Think of gases as pictured in the kinetic–molecular theory. A gas exerts a particular pressure because its molecules bounce against the container walls with a certain frequency and speed (Figure 6.10). If the volume of the container is expanded while the amount of gas remains fixed, the number of molecules per unit volume of gas decreases. The frequency with which molecules strike a unit area of the container walls decreases, and the gas pressure decreases. Thus, as the volume of a gas is increased, its pressure decreases.

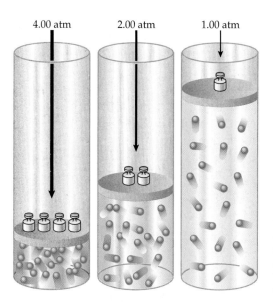

Figure 6.10 A kinetic theory view of Boyle's law. As the pressure is reduced from 4.00 atm to 2.00 atm and then to 1.00 atm, the volume of the gas doubles and then doubles again.

Mathematically, for a given amount of gas at a constant temperature, Boyle's law is written

$$V \propto \frac{1}{P}$$

where the symbol $\propto$ means "is proportional to." This relationship can be changed to an equation by inserting a proportionality constant, a.

$$V = \frac{a}{P}$$

Multiplying both sides of the equation by P, we get

$$PV = a \qquad \text{(with constant temperature and mass)}$$

Another way to state Boyle's law, then, is that for a given amount of gas at a constant temperature, the product of the pressure and volume is a constant. This is an elegant and precise, if somewhat abstract, way of summarizing a lot of experimental data. If the product $P \times V$ is to be constant, then if V increases P must decrease, and vice versa. This relationship is demonstrated in Figure 6.11 by a pressure–volume graph.

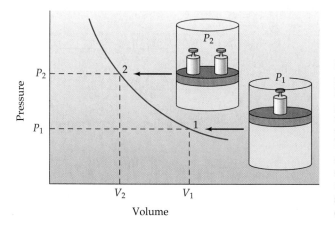

Figure 6.11 A graphic representation of Boyle's law. As the pressure of the gas is increased, its volume decreases. When the pressure is doubled ($P_2 = 2 \times P_1$), the volume of the gas decreases to one-half its original value ($V_2 = \frac{1}{2} \times V_1$). The pressure–volume product is a constant ($PV = a$).

Boyle's law has a number of practical applications perhaps best illustrated by some examples. In Example 6.14, we see how to estimate an answer. Sometimes this is all we need. Even when we want a quantitative answer, however, the estimate helps us to determine whether or not our answer is reasonable.

Example 6.14

A gas is enclosed in a cylinder fitted with a piston. The volume of the gas is 2.00 L at 0.524 atm. The piston is moved to increase the gas pressure to 5.15 atm. Which of the following is a reasonable value for the volume of the gas at the greater pressure?

0.20 L   0.40 L   1.00 L   16.0 L

Solution

The pressure increase is almost tenfold. The volume should drop to about one-tenth of the initial value. We estimate a volume of 0.20 L. (The calculated value is 0.203 L.)

Exercise 6.14

A gas is enclosed in a 10.2-L tank at 1208 mmHg. Which of the following is a reasonable value for the pressure when the gas is transferred to a 30.0-L tank?

25 lb/in.^{2} 300 mmHg   400 mmHg   3600 mmHg

Example 6.15 illustrates quantitative calculations using Boyle's law. Note that in these applications any units can be used for pressure and volume as long as the same units are used throughout a calculation. As long as we use the same sample of a confined gas at a constant temperature, the product of the initial volume (V_1) times the initial pressure (P_1) is equal to the product of the final volume (V_2) times the final pressure (P_2). Thus, the following useful equation representing Boyle's law can be written.

Gases are usually stored under high pressure, even though they are used at atmospheric pressure. This practice allows a large quantity of gas to be stored in a small volume.

$$V_1 P_1 = V_2 P_2$$

Example 6.15

A cylinder of oxygen has a volume of 2.25 L. The pressure of the gas is 1470 pounds per square inch (psi) at 20°C. What volume will the oxygen occupy at standard atmospheric pressure (14.7 psi) assuming no temperature change?

Solution

It is most helpful to first separate the initial from the final condition.

Hint: Regardless of which variable is to be determined, it is recommended that you solve the equation for the unknown *before* substituting values into the equation. Rearranging a few letters takes less time than rearranging and rewriting complex terms.

Initial	Final	Change
$P_1 = 1470$ psi	$P_2 = 14.7$ psi	⇓ The pressure goes down, therefore
$V_1 = 2.25$ L	$V_2 = ?$	⇑ the volume goes up.

Because the final pressure in Example 6.15 is *less than* the initial pressure, we expect the final volume to be *larger than* the original volume, and it is.

Then use the equation $V_1 P_1 = V_2 P_2$ and solve for the desired volume or pressure. In this case, we solve for V_2.

$$V_2 = \frac{V_1 P_1}{P_2}$$

$$V_2 = \frac{2.25 \text{ L} \times 1470 \text{ psi}}{14.7 \text{ psi}} = 225 \text{ L}$$

Exercise 6.15

A. A sample of air occupies 73.3 mL at 98.7 atm and 0°C. What volume will the air occupy at 4.02 atm and 0°C?

B. A sample of helium occupies 535 mL at 988 mmHg and 25°C. If the sample is transferred to a 1.05-L flask at 25°C, what will be the gas pressure in the flask?

Charles' Law: Temperature and Volume

In 1787 the French physicist Jacques Charles (1746–1823), a pioneer hot air balloonist, studied the relationship between the volume and temperature of gases. He found that when a fixed mass of gas is cooled at constant pressure, its volume decreases. When the gas is heated, its volume increases. Temperature and volume vary directly; that is, they rise or fall together. But this law requires a bit more thought. If a quantity of gas that occupies 1.00 L is heated from 100°C to 200°C at constant pressure, the volume does not double but only increases to about 1.27 L. The relationship between temperature and volume is not as tidy as it may seem at first.

Zero pressure or zero volume really means zero—no pressure or volume can be measured. Zero degrees Celsius (0°C) means only the freezing point of water. This zero point is arbitrarily set, much as mean sea level is set as the arbitrary zero for measuring altitudes on Earth. Temperatures below 0°C are often encountered, as are altitudes below sea level.

Charles noted that for each degree Celsius rise in temperature, the volume of a gas increases by $\frac{1}{273}$ of its volume at 0°C. If we plot volume against temperature, we get a straight line (Figure 6.12). We can extrapolate the line beyond the range of measured temperatures to the temperature at which the volume of the gas would become zero. This temperature is −273°C.[2] In 1848 William Thomson (Lord Kelvin) made this temperature the zero point on an absolute temperature scale now called the Kelvin scale. As noted in Chapter 1, the unit of temperature on this scale is the kelvin (K).

A modern statement of **Charles' law** is that *the volume of a fixed amount of a gas at a constant pressure is directly proportional to its absolute temperature.* Mathematically, this relationship is expressed as

$$V = bT \qquad \text{or} \qquad \frac{V}{T} = b$$

where b is a proportionality constant. To keep $\frac{V}{T}$ equal to a constant value, when the temperature increases, the volume must also increase. When the temperature decreases, the volume must decrease accordingly (Figure 6.13).

As long as we use the same sample of trapped gas at a constant pressure, the initial volume (V_1) divided by the initial absolute temperature (T_1) is equal to the final volume (V_2) divided by the final absolute temperature (T_2). We can use the following equation to solve problems involving Charles' law.

$$\frac{V_1}{T_1} = \frac{V_2}{T_2}$$

The kinetic–molecular model readily explains the relationship between gas volume and temperature. When we heat a gas, we supply the gas molecules with energy and they begin to move faster. These speedier molecules strike the walls of the container harder and more often. For the pressure to stay the same, the volume of the container must increase so that the increased molecular motion will be distributed over a greater space.

Any real gas liquefies—and the liquid freezes—before the gas ever reaches this zero-volume temperature. Extrapolation of the line to zero volume is therefore an exercise only for the imagination.

---

[2] The precise zero point on the Kelvin scale is equivalent to −273.15°C. For our purposes, we can round the value to −273°C.

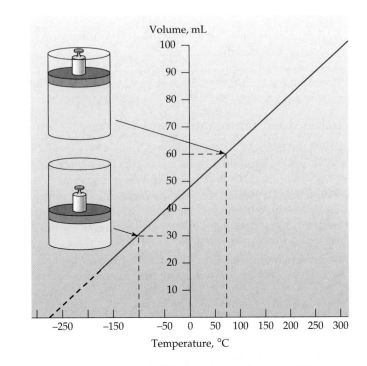

Figure 6.12 Charles' law relates gas volume to temperature at constant pressure. When the gas shown has been cooled to about 70°C, its volume is 60 mL. In the temperature interval from about 70°C to 100°C, the volume drops to 30.0 mL. The volume continues to fall as the temperature is lowered. The extrapolated line intersects the temperature axis (corresponding to a volume of zero) at about 270°C.

Figure 6.13 A dramatic illustration of Charles' law. (a) Liquid nitrogen (boiling point, -196°C) cools the balloon and its contents to a temperature far below room temperature. (b) As the balloon warms to room temperature, the volume of air increases proportionately (about fourfold).

(a)   (b)

Example 6.16

A balloon indoors, where the temperature is 27°C, has a volume of 2.00 L. What would its volume be (**a**) outdoors, where the temperature is −23°C, and (**b**) in a hot room where the temperature is 47°C? (Assume no change in pressure in either case.)

Solution

First, convert all temperatures to the Kelvin scale

$$T(K) = T(°C) + 273$$

The initial temperature is $(27 + 273) = 300$ K, and the final temperatures are (**a**) $(−23 + 273) = 250$ K and (**b**) $(47 + 273) = 320$ K.

a. We start by separating the initial from the final condition.

Initial	Final	Change
$t_1 = 27°C$	$t_2 = -23°C$	⇓
$T_1 = 300 \text{ K}$	$T_2 = 250 \text{ K}$	⇓
$V_1 = 2.00 \text{ L}$	$V_2 = ?$	⇓

Solving the equation

$$\frac{V_1}{T_1} = \frac{V_2}{T_2}$$

for V_2, we have

$$V_2 = \frac{V_1 T_2}{T_1}$$

$$V_2 = \frac{2.00 \text{ L} \times 250 \text{ K}}{300 \text{ K}} = 1.67 \text{ L}$$

As we expected, the volume decreased because the temperature decreased.

b. We have the same initial conditions as in **(a)**, but different final conditions.

Initial	Final	Change
$t_1 = 27°C$	$t_2 = 47°C$	⇑
$T_1 = 300 \text{ K}$	$T_2 = 320 \text{ K}$	⇑
$V_1 = 2.00 \text{ L}$	$V_2 = ?$	⇑

In this case, since the temperature increases, the volume must also increase.

$$V_2 = \frac{V_1 T_2}{T_1}$$

$$V_2 = \frac{2.00 \text{ L} \times 320 \text{ K}}{300 \text{ K}} = 2.13 \text{ L}$$

Exercise 6.16

A. a. A sample of oxygen gas occupies a volume of 2.10 L at 25°C. What volume will this sample occupy at 150°C? (Assume no change in pressure.)

 b. A sample of hydrogen occupies 692 L at 602°C. If the pressure is held constant, what volume will the gas occupy after being cooled to 23°C?

B. At what Celsius temperature will the initial volume of oxygen in Exercise 6.16 A occupy 0.750 L? (Assume no change in pressure.)

The Ideal Gas Law

Boyle's law and Charles' law are useful when the temperature or pressure can be held constant. Often, however, both the temperature and the pressure change at the same time, and the amount of gas may change as well. The **ideal gas law**, expressed in the equation

$$\frac{PV}{nT} = R \quad \text{or} \quad PV = nRT$$

involves four variables (R is a constant). The number of moles is given by n. The constant R (called the *universal gas constant*) can be calculated from the fact that 1 mol of gas occupies 22.4 L at 273 K (0°C) and 1 atm (Section 6.2). If P is in atmospheres, V in liters, and T in kelvins, then R has a value of

---

We read this R value as "0.0821 liter-atmosphere per mole-kelvin."

$$0.0821 \frac{\text{L} \cdot \text{atm}}{\text{mol} \cdot \text{K}}$$

The ideal gas equation can be used to calculate any of the four quantities—P, V, n, or T—if the other three are known.

Example 6.17

Use the ideal gas law to calculate (**a**) the volume occupied by 1.00 mol of nitrogen gas at 244 K and 1.00 atm pressure, and (**b**) the pressure exerted by 0.500 mol of oxygen in a 15.0-L container at 303 K.

Solution
a. We start by solving the ideal gas equation for V.

$$V = \frac{nRT}{P}$$

$$V = \frac{1.00 \text{ mol}}{1.00 \text{ atm}} \times \frac{0.0821 \text{ L} \cdot \text{atm}}{\text{mol} \cdot \text{K}} \times 244 \text{ K} = 20.0 \text{ L}$$

b. Here we solve the ideal gas equation for P.

$$P = \frac{nRT}{V}$$

$$P = \frac{0.500 \text{ mol}}{15.0 \text{ L}} \times \frac{0.0821 \text{ L} \cdot \text{atm}}{\text{mol} \cdot \text{K}} \times 303 \text{ K} = 0.83 \text{ atm}$$

Exercise 6.17
A. Determine (**a**) the pressure exerted by 0.0330 mol of oxygen in an 18.0-L container at 40°C, and (**b**) the volume occupied by 0.200 mol of nitrogen gas at 25°C and 0.980 atm.
B. At what temperature will 1.25 mol of helium gas exert a pressure of 5.00 atm in a 2.00-L container?

6.7 Solutions

We briefly introduced solutions in Chapter 5. Recall that a *solution* is a homogeneous mixture of two or more substances, the substance being dissolved is the *solute*, and the substance doing the dissolving is the *solvent*. The solute is usually the component present in the lesser quantity, and the solvent is usually present in the greater quantity. There are many solvents: Hexane dissolves grease. Ethanol dissolves many drugs. Isopentyl acetate, a component of banana oil, is a solvent for the glue used in making model airplanes. Water is no doubt the most familiar solvent, dissolving as it does many common substances such as sugar, salt, and ethanol. We focus our discussion here on **aqueous solutions**, those in which water is the solvent, and we take a more quantitative look at the relationship between solute and solvent.

Solution Concentrations

We say that some substances, such as sugar and salt, are *soluble* in water. Of course, there is a limit to the quantity of sugar or salt we can dissolve in a given volume of water, but we still find it convenient to say that they are soluble in water because an appreciable quantity dissolves. Other substances, such as an iron nail or sand (silicon dioxide), we consider to be *insoluble* because the limit of solubility is near zero. Such terms as "soluble" and "insoluble" are useful, but they are imprecise and must be used with care. Two other roughly estimated but sometimes useful terms are *dilute* and *concentrated*. A **dilute solution** is one that contains a little bit of solute in lots of solvent. For example, a pinch of sugar in a liter of water is a dilute, faintly sweet, sugar solution. A **concentrated solution** is one in which lots of solute is dissolved in a relatively small quantity of solvent. A sugar solution with lots of solute in a relatively small amount of water is a concentrated, very sweet solution. The dilute solution is quite "thin"—little changed in appearance from that of pure water. The concentrated solution is thick and rather syrupy.

Scientific work generally requires more precise measurement of quantities than "a pinch of sugar in a liter of water." Further, quantitative work often requires the unit amount of substance (moles) because substances enter into chemical reactions according to *molar* ratios.

Molarity

The concentration unit that chemists use most is *molarity*. For reactions involving solutions, the amount of solute is usually measured in moles and the quantity of solution in liters or milliliters. The **molarity (M)** is the amount of solute, in moles, per liter of solution.

$$\text{Molarity (M)} = \frac{\text{moles of solute}}{\text{liters of solution}}$$

Example 6.18

Calculate the molarity of a solution made by dissolving 3.50 mol of NaCl in enough water to produce 2.00 L of solution.

Solution

$$\text{Molarity (M)} = \frac{\text{moles of solute}}{\text{liters of solution}} = \frac{3.50 \text{ mol NaCl}}{2.00 \text{ L solution}} = 1.75 \text{ M NaCl}$$

We read 1.75 M NaCl as 1.75 molar NaCl.

Exercise 6.18

A. Calculate the molarity of a solution that has 0.0500 mol of NH_3 in 5.75 L of solution.
B. Calculate the molarity of a solution made by dissolving 0.750 mol of H_3PO_4 in enough water to produce 775 mL of solution.

We cannot determine moles of a substance directly; we usually work with a given mass and divide by the molar mass of the substance, as illustrated in Example 6.19.

Example 6.19

What is the molarity of a solution in which 333 g of potassium hydrogen carbonate is dissolved in enough water to make 10.0 L of solution?

Solution
First, prepare a setup to convert from mass of $KHCO_3$ to moles of $KHCO_3$.

$$333 \text{ g } \cancel{KHCO_3} \times \frac{1 \text{ mol } KHCO_3}{100.1 \text{ g } \cancel{KHCO_3}} = 3.33 \text{ mol } KHCO_3$$

Now use this value as the numerator in the defining equation for molarity. The solution volume, 10.0 L, is the denominator.

$$\text{Molarity} = \frac{3.33 \text{ mol KHCO}_3}{10.0 \text{ L solution}} = 0.333 \text{ M KHCO}_3$$

Exercise 6.19
Calculate the molarity of each of the following solutions.
a. 18.0 mol of H_2SO_4 in 2.00 L of solution
b. 3.00 mol of KI in 2.39 L of solution
c. 0.206 mol of HF in 752 mL of solution
d. 0.522 g of HCl in 0.592 L of solution
e. 4.98 g of $C_6H_{12}O_6$ in 224 mL of solution
f. 10.5 g of C_2H_5OH in 24.7 mL of solution

Frequently we need to know the *mass* of solute required to prepare a given volume of solution of a given molarity. In such calculations we can use molarity as a conversion factor between moles of solute and liters of solution. Thus, in Example 6.20, the expression 6.67 M NaOH means 6.67 mol of NaOH per liter of solution, expressed as the conversion factor

$$\frac{6.67 \text{ mol NaOH}}{1 \text{ L } solution}$$

Example 6.20
How many grams of NaOH is required to prepare 0.500 L of 6.67 M NaOH?

Solution
First we calculate the moles of NaOH.

$$0.500 \text{ L solution} \times \frac{6.67 \text{ mol NaOH}}{1 \text{ L solution}} = 3.34 \text{ mol NaOH}$$

Then we use the molar mass to calculate the grams of NaOH.

$$3.34 \text{ mol NaOH} \times \frac{40.01 \text{ g NaOH}}{1 \text{ mol NaOH}} = 133 \text{ g NaOH}$$

Exercise 6.20
How many grams of potassium hydroxide is required to prepare each of the following solutions?
a. 2.00 L of 6.00 M KOH   b. 100.0 mL of 1.00 M KOH
c. 10.0 mL of 0.100 M KOH   d. 33.0 mL of 2.50 M KOH

Quite often, solutions of known molarity are available. How would you calculate the *volume* needed to get a certain number of moles of solute? We can again rearrange the definition of molarity to obtain

$$\text{liters of solution} = \frac{\text{moles of solute}}{\text{molarity}}$$

Example 6.21

How many liters of 12.0 M HCl solution would one need to get 0.425 mol of HCl?

Solution

$$\text{Liters of HCl solution} = \frac{\text{moles of solute}}{\text{molarity}} = \frac{0.425 \text{ mol HCl}}{12.0 \text{ M HCl}}$$

$$= \frac{0.425 \text{ mol HCl}}{12.0 \text{ mol HCl}/\text{L}} = 0.0354 \text{ L}$$

We would need 0.0354 L (35.4 mL) of the solution to have 0.425 mol.

Exercise 6.21

A. How many liters of 15.0 M aqueous ammonia (NH_3) solution do you need to get 0.445 mol of NH_3?

B. How many milliliters of 15.0 M HNO_3 solution do you need to get 0.245 mol of HNO_3?

Remember that molarity is moles per liter of *solution*, not per liter of solvent. To make a liter of a 1 M solution, we weigh out 1 mol of solute and place it in a volumetric flask, which is a standard piece of laboratory glassware designed to contain a precisely specified volume of liquid (Figure 6.14). Enough water is added to dissolve the solute, and then more water is added to bring the volume up to the mark indicating 1 L of *solution*. (Simply adding 1 mol of solute to 1 L of water would, in most cases, give more or less than 1 L of solution.)

Figure 6.14 The 1.000 M NaOH solution in this flask was made by dissolving 1.000 mol of NaOH (40.01 g of NaOH) in water and then carefully diluting to a final volume of 1.000 L.

Percent Concentrations

For many practical applications, we often express solution concentrations in percentage composition. Then, if we require a precise quantity of solution, we simply measure out a mass or volume. If both the solute and solvent are liquids, **percent by volume** is often used because liquid volumes are so easily measured.

$$\text{Percent by volume} = \frac{\text{volume of solute}}{\text{volume of solution}} \times 100\%$$

Ethanol used for medicinal purposes is generally of a grade referred to as USP (an abbreviation of *United States Pharmacopoeia*, the official publication of standards for pharmaceutical products). USP ethanol is 95% CH_3CH_2OH, by volume.

Example 6.22

What is the percent by volume of a solution made by dissolving 235 mL of ethanol in enough water to make exactly 500 mL of solution?

Solution

$$\text{Percent by volume} = \frac{235 \text{ mL ethanol}}{500 \text{ mL solution}} \times 100\% = 47.0\%$$

Exercise 6.22

What is the percent by volume of a solution made by dissolving 11.7 mL of ethanol in enough water to make 25.0 mL of solution?

Example 6.23

Describe how to make 775 mL of a 40.0% by volume solution of acetic acid.

Solution

Let's begin by rearranging the equation for percent by volume to solve for volume of solute.

$$\text{Volume of solute} = \frac{\text{percent by volume} \times \text{volume of solution}}{100\%}$$

Substituting, we have

$$= \frac{40\% \times 775 \text{ mL}}{100\%} = 310 \text{ mL}$$

Take 310 mL of acetic acid and add enough water to make 775 mL of solution.

Exercise 6.23
Describe how to make 67.5 mL of a 33.0% by volume solution of acetic acid.

Many commercial solutions are labeled with the concentration in **percent by mass**. For example, sulfuric acid is sold as a solution that is 35.7% H_2SO_4 for use in storage batteries, 77.7% H_2SO_4 for the manufacture of phosphate fertilizers, and 93.2% H_2SO_4 for pickling steel. Each of these figures is a percent by mass: 35.7 g of H_2SO_4 per 100 g of sulfuric acid solution, and so on.

$$\text{Percent by mass} = \frac{\text{mass of solute}}{\text{mass of solution}} \times 100\%$$

Example 6.24
What is the percent by mass of a solution of 25.0 g of NaCl dissolved in 475 g (475 mL) of water?

Solution

$$\text{Percent by mass} = \frac{\text{mass of NaCl}}{\text{mass of solution}} \times 100\%$$

$$\text{Percent by mass} = \frac{25.0 \text{ g NaCl}}{500 \text{ g solution}} \times 100\% = 5.00\% \text{ NaCl}$$

Exercise 6.24
What is the percent by mass of a solution of 9.40 g of H_2O_2 dissolved in 335 g (335 mL) of water?

Example 6.25
Describe how to prepare 750 g of an aqueous solution that is 2.50% NaOH by mass.

Solution
Let's begin by rearranging the equation for percent by mass to solve for mass of solute.

$$\text{Mass of solute} = \frac{\text{percent by mass} \times \text{mass of solution}}{100\%}$$

Substituting, we have

$$\text{Mass of NaOH} = \frac{2.50\% \times 750 \text{ g}}{100\%} = 18.8 \text{ g NaOH}$$

The required mass of water is 750 g solution minus 18.8 g NaOH.

$$\text{Mass } H_2O = 750 \text{ g solution} - 18.8 \text{ g NaOH} = 731 \text{ g } H_2O$$

To make 750 g of solution, weigh out 18.8 g of NaOH and add it to 731 g of water.

Exercise 6.25
Describe how to prepare 275 g of an aqueous solution that is 5.50% glucose by mass.

Note that for percent concentrations, the mass of the solute needed doesn't depend on what the solute is. A 10% by mass solution of NaOH contains 10 g of NaOH per 100 g of total solution. Similarly, 10% HCl and 10% $(NH_4)_2SO_4$ and 10% $C_{110}H_{190}N_3O_2Br$ each contain 10 g of the specified solute per 100 g of solution. For *molar* solutions, however, the mass of solute in a solution of specified molarity is different for different solutes. A liter of a 0.10 M solution requires 4.0 g (0.10 mol) of NaOH, 3.7 g (0.10 mol) of HCl, 13.2 g (0.10 mol) of $(NH_4)_2SO_4$, or 166 g (0.10 mol) of $C_{110}H_{190}N_3O_2Br$.

Critical Thinking Exercises

Apply knowledge that you have gained in this chapter and one or more of the FLaReS principles (Chapter 1) to evaluate the following statements or claims.

6.1 Suppose that someone has published a paper claiming a new value for Avogadro's number. The author says that he has made some very careful laboratory measurements and his calculations indicate that the true value for the Avogadro constant is 3.01875×10^{23}. Is this claim credible in your opinion?

6.2 A chemistry teacher asked his students: "What is the mass, in grams, of a mole of bromine?" One student said "80"; another said "160"; and several others gave answers of 79, 81, 158, and 162. The teacher stated that all of these answers were correct. Do you believe his statement?

Summary

1. A chemical equation is a statement of chemical reactions written in a chemical shorthand of symbols and formulas. An equation is balanced by making the number of atoms of each element equal on both sides of the arrow.
2. Gay-Lussac's law of combining volumes states that gases at the same pressure and temperature combine in small whole-number ratios by volume.
3. A mole is an amount of substance that contains 6.02×10^{23} molecules (or formula units) of that substance. (The number of formula units in a mole is known as Avogadro's number.)
4. A mole is equal to the formula mass of the substance expressed in grams. This is the molar mass of the substance.
5. The molar volume of a substance that is a gas at standard temperature and pressure (STP) is 22.4 L, regardless of what gas it is.
6. In calculating how much product should be formed from a given amount of reactant, it is much more convenient to use moles than molecules.
7. One mole of a substance contains the same number of molecules (or formula units) as 1 mol of any other substance.
8. Stoichiometry is the mass relationship between reactants and products in chemical reactions.
9. Boyle's law states that for a given amount of gas at constant temperature, the volume varies inversely with its pressure.
10. Charles' law states that at constant pressure, the volume of a given amount of gas is directly proportional to its absolute temperature.
11. A solution is a homogeneous mixture of two or more substances.
12. In a solution, the solvent is usually present in a larger quantity, and one or more solutes in lesser quantities.
13. An aqueous solution is one in which the solvent is water.
14. A substance is soluble in a solvent if some appreciable quantity of the substance dissolves in the solvent. Any substance that does not dissolve significantly in a solvent is insoluble in that solvent.
15. A concentration expressed in moles of solute per liter of solution is called the molarity of the solution.

Key Terms

aqueous solution 6.7
Avogadro's hypothesis 6.2
Avogadro's number 6.3
Boyle's law 6.6
Charles' law 6.6
chemical equation 6.1
concentrated solution 6.7
dilute solution 6.7
formula mass 6.4
ideal gas law 6.6
kinetic–molecular theory 6.6
law of combining volumes 6.2
molarity 6.7
molar mass 6.4
molar volume 6.4
mole (mol) 6.4
molecular mass 6.4
percent by mass 6.7
percent by volume 6.7
products 6.1
reactants 6.1
standard temperature and pressure (STP) 6.4
stoichiometric factor 6.5
stoichiometry 6.5

Review Questions

1. Define or illustrate each of the following.
   a. binary compound   b. polyatomic ion
   c. mole   d. Avogadro's number
   e. molar mass   f. molar volume

2. Explain the difference between the atomic mass of oxygen and the formula mass of oxygen (gas).

3. What is Avogadro's hypothesis? How does it explain Gay-Lussac's law of combining volumes?

4. How do the law of combining volumes and Avogadro's hypothesis indicate that hydrogen gas is composed of diatomic molecules rather than individual atoms?

5. Consider the law of conservation of mass and explain why we must work with balanced chemical equations.

6. What is the molar volume of each of the following gases at STP?
   a. He   b. H_2 c. C_2H_6

7. What is the mass of a molar volume of each of the gases in Question 6?

8. State Boyle's law in words and as a mathematical equation.

9. Use the kinetic–molecular theory to explain Boyle's law.

10. State Charles' law in words and as a mathematical equation.

11. Use the kinetic–molecular theory to explain Charles' law.

12. Why must an absolute temperature scale rather than the Celsius scale be used for calculations involving Charles' law?

13. State the ideal gas law in words and in the form of a mathematical equation.

14. Define or explain—and, where possible, illustrate—the following terms.
   a. solution   b. solvent
   c. solute   d. aqueous solution

15. Define or explain—and, where possible, illustrate—the following terms.
   a. concentrated solution   b. dilute solution
   c. soluble   d. insoluble

16. Explain how to calculate each of the following concentrations.
   a. percent by mass   b. percent by volume
   c. molarity

17. How many oxygen molecules are there in 1.00 mol of O_2? How many oxygen atoms are there in 1.00 mol of O_2?

18. How many calcium ions and how many chloride ions are there in 1.00 mol of $CaCl_2$?

Problems

Formula Units

19. How many hydrogen atoms are indicated in one formula unit of each of the following?
   a. NH_4NO_3 b. CH_3OH
   c. $CH_3CH_2CH_3$ d. C_6H_5COOH

20. How many oxygen atoms are indicated in one formula unit of each of the following?
   a. $Al_2(C_2O_4)_3$ b. $Ca_3(PO_4)_2$
   c. $Zn_3(PO_4)_2$ d. $Al_2(SO_4)_3$

21. How many atoms of each kind (Al, C, H, and O) does the notation $2 Al(C_2H_3O_2)_3$ indicate?

22. How many atoms of each kind (N, P, H, and O) does the notation $6 (NH_4)_2HPO_4$ indicate?

Balancing Chemical Equations

23. Indicate whether the following equations are balanced. (You need not balance the equation; just determine whether it is balanced as written.)
   a. $Ca + 2 H_2O \longrightarrow Ca(OH)_2 + H_2$
   b. $2 LiOH + CO_2 \longrightarrow Li_2CO_3 + H_2O$
   c. $4 LiH + AlCl_3 \longrightarrow 2 LiAlH_4 + 2 LiCl$
   d. $2 Sn + 2 H_2SO_4 \longrightarrow 2 SnSO_4 + SO_2 + 2 H_2O$

24. Indicate whether the following equations are balanced as written.
   a. $2 KNO_3 + 10 K \longrightarrow 6 K_2O + N_2$
   b. $2 NH_3 + O_2 \longrightarrow N_2 + 3 H_2O$
   c. $SF_4 + 3 H_2O \longrightarrow H_2SO_3 + 4 HF$
   d. $4 BF_3 + 3 H_2O \longrightarrow H_3BO_3 + 3 HBF_4$
   e. $3 Cl_2 + 6 NaOH \longrightarrow 5 NaCl + NaClO_3 + 3 H_2O$

25. Balance the following equations.
   a. $Cl_2O_5 + H_2O \longrightarrow HClO_3$
   b. $V_2O_5 + H_2 \longrightarrow V_2O_3 + H_2O$
   c. $Al + O_2 \longrightarrow Al_2O_3$
   d. $Sn + NaOH \longrightarrow Na_2SnO_2 + H_2$
   e. $PCl_5 + H_2O \longrightarrow H_3PO_4 + HCl$
   f. $Na_3P + H_2O \longrightarrow NaOH + PH_3$
   g. $Cl_2O + H_2O \longrightarrow HClO$
   h. $CH_3OH + O_2 \longrightarrow CO_2 + H_2O$
   i. $Zn(OH)_2 + H_3PO_4 \longrightarrow Zn_3(PO_4)_2 + H_2O$
   j. $C_3H_8 + O_2 \longrightarrow CO_2 + H_2O$

26. Balance the following equations.
   a. $TiCl_4 + H_2O \longrightarrow TiO_2 + HCl$
   b. $C_4H_{10} + O_2 \longrightarrow CO_2 + H_2O$
   c. $WO_3 + H_2 \longrightarrow W + H_2O$
   d. $Al_4C_3 + H_2O \longrightarrow Al(OH)_3 + CH_4$
   e. $Al_2(SO_4)_3 + NaOH \longrightarrow Al(OH)_3 + Na_2SO_4$
   f. $Ca_3P_2 + H_2O \longrightarrow Ca(OH)_2 + PH_3$
   g. $Cl_2O_7 + H_2O \longrightarrow HClO_4$
   h. $MnO_2 + HCl \longrightarrow MnCl_2 + Cl_2 + H_2O$

i. $Fe + O_2 \longrightarrow Fe_3O_4$

j. $C_5H_{12} + O_2 \longrightarrow CO_2 + H_2O$

Formula Masses and Molecular Masses

You may round all atomic masses to one decimal place.

27. Calculate the molecular mass or formula mass of each of the following compounds.

    a. CH_4 **b.** AlF_3 **c.** UF_6

28. Calculate the molecular mass or formula mass of each of the following compounds.

    a. SO_3 **b.** $KBrO_3$ **c.** $CaSO_4$

29. Calculate the molecular mass or formula mass of each of the following compounds.

    a. C_6H_5Br **b.** $H_4P_2O_7$

    c. $K_2Cr_2O_7$ **d.** $Al_2(SO_4)_3$

30. Calculate the molecular mass or formula mass of each of the following compounds.

    a. $(NH_4)_3PO_4$ **b.** $Fe(NO_3)_3$

    c. $C_2H_5NO_2$ **d.** MgS_2O_3

Volume Relationships in Chemical Equations

31. Calculate the volume of methane that must decompose to produce 10.0 L of hydrogen in the following reaction, if the two gases are compared at the same temperature and pressure.

$$CH_4(g) \longrightarrow C(s) + 2 H_2(g)$$

32. Calculate the volume of $O_2(g)$ that must react to form 10.0 L of steam in the following reaction, if the two gases are compared at the same temperature and pressure.

$$CH_4(g) + O_2(g) \longrightarrow C(s) + 2 H_2O(g)$$

33. Consider the following equation.

$$2 C_4H_{10}(g) + 13 O_2(g) \longrightarrow 8 CO_2(g) + 10 H_2O(g)$$

    a. How many liters of $H_2O(g)$ is formed when 0.529 L of $C_4H_{10}(g)$ is burned? Assume both gases are measured under the same conditions.

    b. How many liters of $O_2(g)$ is required to burn 16.1 L of $C_4H_{10}(g)$? Assume both gases are measured under the same conditions.

34. Consider the following equation.

$$C_2H_4(g) + 3 O_2(g) \longrightarrow 2 CO_2(g) + 2 H_2O(g)$$

    a. How many liters of $CO_2(g)$ is formed when 2.93 L of $C_2H_4(g)$ is burned? Assume both gases are measured under the same conditions.

    b. How many liters of $O_2(g)$ is required to form 0.370 L of $CO_2(g)$? Assume both gases are measured under the same conditions.

Molar Volume, Gas Densities, and Molecular Masses

35. Calculate the density of argon (Ar) gas, in grams per liter, at STP.

36. Calculate the density of nitrogen (N_2) gas, in grams per liter, at STP.

37. Calculate the molecular mass of a gas that has a density of 2.35 g/L at STP.

38. Calculate the molecular mass of a gas that has a density of 1.98 g/L at STP.

39. Calculate the molecular mass of a liquid the vapor of which has a density of 3.02 g/L at STP.

40. Calculate the molecular mass of a liquid the vapor of which has a density of 2.81 g/L at STP.

Molar Masses

41. Calculate the molar mass of each of the following.

    a. C_6H_5Br **b.** H_3PO_4 **c.** $K_2Cr_2O_7$

42. Calculate the molar mass of each of the following.

    a. $C_2H_5NO_2$ **b.** $Na_2S_2O_3$ **c.** $(NH_4)_3PO_4$

43. Calculate the mass, in grams, of each of the following.

    a. 0.00500 mol MnO_2 **b.** 1.12 mol CaH_2

    c. 0.250 mol $C_6H_{12}O_6$

44. Calculate the mass, in grams, of each of the following.

    a. 4.61 mol $AlCl_3$ **b.** 0.615 mol Cr_2O_3

    c. 0.158 mol IF_5

45. Calculate the amount, in moles, of each of the following.

    a. 98.6 g HNO_3 **b.** 9.45 g CBr_4

    c. 9.11 g $FeSO_4$ **d.** 11.8 g $Pb(NO_3)_2$

46. Calculate the amount, in moles, of each of the following.

    a. 16.3 g SF_6 **b.** 25.4 g $Pb(C_2H_3O_2)_2$

    c. 35.6 g $FeCl_3$ **d.** 75.3 g $Co(ClO_3)_2$

Mole and Mass Relationships in Chemical Equations

47. Consider the reaction for the combustion of octane.

$$2 C_8H_{18} + 25 O_2 \longrightarrow 16 CO_2 + 18 H_2O$$

    a. How many moles of CO_2 is produced when 2.09 mol of octane is burned?

    b. How many moles of oxygen is required to burn 4.47 mol of octane?

48. Consider the reaction for the combustion of octane.

$$2 C_8H_{18} + 25 O_2 \longrightarrow 16 CO_2 + 18 H_2O$$

    a. How many moles of H_2O is produced when 2.81 mol of octane is burned?

    b. How many moles of CO_2 is produced when 4.06 mol of oxygen is consumed?

49. What mass of (**a**) ammonia, in grams, can be made from 440 g of H_2, and (**b**) hydrogen, in grams, is needed to react completely with 892 g of N_2?

$$N_2 + H_2 \longrightarrow NH_3 \quad (not\ balanced)$$

50. Toluene and nitric acid are used in the production of trinitrotoluene (TNT), an explosive.

$$C_7H_8 + HNO_3 \longrightarrow C_7H_5N_3O_6 + H_2O \quad (not\ balanced)$$
Toluene   TNT

    a. What mass of nitric acid, in grams, is required to react with 454 g of C_7H_8?

    b. What mass of TNT can be made from 829 g of C_7H_8?

51. What mass of quicklime (calcium oxide), in kilograms, can be made when 4.72×10^9 g of limestone (calcium carbonate) is decomposed by heating?

$$CaCO_3(s) \longrightarrow CaO(s) + CO_2(g)$$

52. What mass of nitric acid, in grams, can be made from 971 g of ammonia?

$$NH_3 + O_2 \longrightarrow HNO_3 + H_2O \quad \text{(not balanced)}$$

Boyle's Law

53. A sample of helium occupies 521 mL at 1572 mmHg. Assume that the temperature is held constant and determine (a) the volume of the helium at 752 mmHg and (b) the pressure, in mmHg, if the volume is changed to 315 mL.

54. A decompression chamber used by deep-sea divers has a volume of 10.3 m^3 and operates at an internal pressure of 4.50 atm. What volume, in cubic meters, would the air in the chamber occupy if it were at 1.00 atm pressure, assuming no temperature change?

55. Oxygen used in respiratory therapy is stored at room temperature under a pressure of 150 atm in gas cylinders with a volume of 60.0 L.
    a. What volume would the gas occupy at a pressure of 750.0 mmHg? Assume no temperature change.
    b. If the oxygen flow to the patient is adjusted to 8.00 L/min, at room temperature and 750.0 mmHg, how long will the tank of gas last?

56. The pressure within a 2.25-L balloon is 1.10 atm. If the volume of the balloon increases to 7.05 L, what will be the final pressure within the balloon if the temperature does not change?

Charles's Law

57. A gas at a temperature of 100°C occupies a volume of 154 mL. What will the volume be at a temperature of 10°C, assuming no change in pressure?

58. A balloon is filled with helium. Its volume is 5.90 L at 26°C. What will its volume be at 78°C, assuming no pressure change?

59. A 567-mL sample of a gas at 305°C and 1.20 atm is cooled at constant pressure until its volume becomes 425 mL. What is the new gas temperature?

60. A sample of gas at STP is to be heated at constant pressure until its volume triples. What is the new gas temperature?

The Ideal Gas Law

61. What effect will the following changes have on the volume of a fixed amount of gas?
    a. an increase in pressure at constant temperature
    b. a decrease in temperature at constant pressure
    c. a decrease in pressure coupled with an increase in temperature

62. What effect will the following changes have on the pressure of a fixed amount of a gas?
    a. an increase in temperature at constant volume

    b. a decrease in volume at constant temperature
    c. an increase in temperature coupled with a decrease in volume

63. According to the kinetic–molecular theory,
    a. what change in temperature is occurring if the molecules of a gas begin to move more slowly, on average?
    b. what change in pressure occurs when molecules of the gas strike the walls of the container less often?

64. Container A has twice the volume but holds twice as many gas molecules as Container B at the same temperature. Use the kinetic–molecular theory to compare the pressures in the two containers.

65. Calculate (a) the volume, in liters, of 1.12 mol $H_2S(g)$ at 62°C and 1.38 atm, and (b) the pressure, in atmospheres, of 4.64 mol CO(g) in a 3.96-L tank at 29°C.

66. Calculate (a) the volume, in liters, of 0.00600 mol of a gas at 31°C and 0.870 atm, and (b) the pressure, in atmospheres, of 0.0108 mol $CH_4(g)$ in a 0.265-L flask at 37°C.

67. How many moles of Kr(g) are there in 2.22 L of the gas at 0.918 atm and 45°C?

68. How many grams of CO(g) are there in 745 mL of the gas at 1.03 atm and 36°C?

Molarity of Solutions

69. Calculate the molarity of each of the following solutions.
    a. 6.00 mol of HCl in 2.50 L of solution
    b. 0.00700 mol of Li_2CO_3 in 10.0 mL of solution

70. Calculate the molarity of each of the following solutions.
    a. 2.50 mol of H_2SO_4 in 5.00 L of solution
    b. 0.200 mol of C_2H_5OH in 18.4 mL of solution

71. Calculate the molarity of each of the following solutions.
    a. 8.90 g of H_2SO_4 in 100.0 mL of solution
    b. 439 g of $C_6H_{12}O_6$ in 1.25 L of solution

72. Calculate the molarity of each of the following solutions.
    a. 44.3 g of KOH in 125 mL of solution
    b. 2.46 g of $H_2C_2O_4$ in 750.0 mL of solution

73. How many grams of solute are needed to prepare each of the following solutions?
    a. 2.00 L of 1.00 M NaOH
    b. 10.0 mL of 4.25 M $C_6H_{12}O_6$

74. How many grams of solute are needed to prepare each of the following solutions?
    a. 250 mL of 2.50 M $K_2Cr_2O_7$
    b. 20.0 mL of 0.0100 M $KMnO_4$

75. What volume of 6.00 M NaOH is required to contain 1.25 mol of NaOH?

76. What volume of 2.50 M NaOH is required to contain 1.05 mol of NaOH?

77. What volume of 0.0250 M $KMnO_4$ is needed to get 8.10 g of $KMnO_4$?

78. What volume of 4.25 M $C_6H_{12}O_6$ is needed to get 205 g of $C_6H_{12}O_6$?

Percent Concentrations of Solutions

79. What is the volume percent concentration of each of the following solutions?
    a. 35.0 mL of water in 725 mL of an ethanol–water solution
    b. 78.9 mL of acetone in 1.55 L of an acetone–water solution

80. What is the volume percent concentration of each of the following solutions?
    a. 58.0 mL of water in 625 mL of an ethanol–water solution
    b. 79.1 mL of methanol in 755 mL of a methanol–water solution

81. What is the mass percent concentration of each of the following solutions?
    a. 4.12 g of NaOH in 100.0 g of water
    b. 5.00 mL of ethanol (d = 0.789 g/mL) in 50.0 g of water

82. What is the mass percent concentration of each of the following solutions?
    a. 175 mg of NaCl per gram of solution
    b. 275 mL of methanol (d = 0.791 g/mL) per kilogram of water

83. Describe how you would prepare 775 g of an aqueous solution that is 10.0% NaCl by mass.

84. Describe how you would prepare 125 g of an aqueous solution that is 5.50% KOH by mass.

85. Describe how you would prepare exactly 2.00 L of an aqueous solution that is 2.00% acetic acid by volume.

86. Describe how you would prepare exactly 500 mL of an aqueous solution that is 30.0% isopropyl alcohol by volume.

Additional Problems

87. What is the formula mass of CO_2? What is the molar mass of CO_2? State in words how each is determined from the formula.

88. The following equation appeared in an article on the incineration of toxic gases.

$$(CH_3)_3As + 15\,O_2 \longrightarrow As_4O_6 + 12\,CO_2 + 6\,H_2O$$

Count the number of atoms of each kind on each side of the equation. Is the equation balanced?

89. Joseph Priestley discovered oxygen in 1774 by heating "red calx of mercury," mercury(II) oxide. The calx decomposed to its elements. The equation is

$$HgO \longrightarrow Hg + O_2 \quad \textit{(not balanced)}$$

How much oxygen is produced by the decomposition of 10.8 g of HgO?

90. How much iron can be converted to the magnetic oxide of iron (Fe_3O_4) by 8.80 g of pure oxygen? The equation is

$$Fe + O_2 \longrightarrow Fe_3O_4 \quad \textit{(not balanced)}$$

91. Laughing gas (dinitrogen monoxide, N_2O, also called nitrous oxide) can be made by heating ammonium nitrate with great care. The equation is

$$NH_4NO_3 \longrightarrow N_2O + H_2O \quad \textit{(not balanced)}$$

How much N_2O can be made from 4.00 g of ammonium nitrate?

92. Small amounts of hydrogen gas are often made by the reaction of calcium metal with water. The equation is

$$Ca + H_2O \longrightarrow Ca(OH)_2 + H_2 \quad \textit{(not balanced)}$$

How many grams of calcium is needed to make 0.413 g of hydrogen? What volume, in milliliters, will this hydrogen occupy at STP?

93. For many years the noble gases were called "inert gases," and it was thought that they formed no chemical compounds. Neil Bartlett made the first noble gas compound in 1962. Xenon hexafluoride is prepared according to the equation

$$Xe + F_2 \longrightarrow XeF_6 \quad \textit{(not balanced)}$$

How many grams of fluorine is required to make 0.112 g of XeF_6?

94. Phosphine gas (used as a fumigant to protect stored grain) is generated by the action of water on magnesium phosphide. The equation is

$$Mg_3P_2 + H_2O \longrightarrow PH_3 + Mg(OH)_2 \quad \textit{(not balanced)}$$

How much magnesium phosphide is needed to produce 134 g of PH_3?

95. In an oxyacetylene welding torch, acetylene (C_2H_2) burns in pure oxygen with a very hot flame.

$$C_2H_2 + O_2 \longrightarrow CO_2 + H_2O \quad \textit{(not balanced)}$$

What mass of oxygen, in grams, is required to react with 52.0 g of C_2H_2?

Project

96. Write a brief essay about one of the following.
    a. Joseph Louis Gay-Lussac
    b. Amedeo Avogadro
    c. Robert Boyle
    d. Jacques Charles

Online Project

97. In the view of many scientists, hydrogen holds great promise as a source of clean energy. Doing your own search, or beginning with the U.S. Department of Energy website at http://www.eren.doe.gov/hydrogen/in-fonet.htm, explore the proposals for using hydrogen as a fuel. Write a brief summary of what you find and use the appropriate balanced equation to explain why this is considered a clean, renewable source.

References and Readings

1. Brown, Theodore L., H. Eugene LeMay, Jr., and Bruce E. Bursten. *Chemistry: The Central Science*, 8th edition. Upper Saddle River, NJ: Prentice Hall, 2000. Chapters 3, "Stoichiometry: Calculations with Chemical Formulas and Equations" and Chapter 4, "Aqueous Reactions and Solution Stoichiometry."

2. Burns, Ralph A. *Fundamentals of Chemistry*, 3d edition. Upper Saddle River, NJ: Prentice Hall, 1999. Chapter 11, "Stoichiometry", Chapter 12, "Gases", and Chapter 14, "Solutions."

3. Gizara, Jeanne M. "Bridging the Stoichiometry Gap." *The Science Teacher*, April 1981, pp. 36–37. Presents a road map method for solving problems based on equations.

4. Hill, John W., and Ralph H. Petrucci. *General Chemistry*, 2nd edition. Upper Saddle River, NJ: Prentice Hall, 1999. Chapter 3, "Stoichiometry" and Chapter 5, "Gases."

5. Kolb, Doris. "The Mole." *Journal of Chemical Education*, November 1978, pp. 728–732.

Acids and Bases
Please Pass the Protons

Acids taste sour and turn litmus red, too.
Bases taste bitter and turn litmus blue.

The chemical substances we call acids and bases are all around us, even in the food we eat.

Have you ever tasted a lemon or a grapefruit? Or felt a burning sensation on your arm after using spray-on oven cleaner? These are but two examples of the presence of acids (citric acid in lemons and grapefruit) and bases (lye or sodium hydroxide in oven cleaner) in our daily lives. Other familiar acids are vinegar (acetic acid), vitamin C (ascorbic acid), and battery acid (sulfuric acid). Some familiar bases are drain cleaner (sodium hydroxide), baking soda (sodium bicarbonate), and ammonia.

From "acid indigestion" to "acid rain," the word "acid" occurs frequently in the news and in advertisements. Air and water pollution often involve acids and bases. Acid rain, for example, is a serious environmental problem, and alkaline (basic) water is sometimes undrinkable.

Did you know that the four tastes are related to acid–base chemistry? Acids taste sour, bases taste bitter, and the compounds formed when acids react with bases (salts) taste salty. The sweet taste is more complicated. To taste sweet, a compound must have both an acidic-type group and a basic-type group, plus just the right geometry to fit the sweet-taste receptor.

In this chapter we discuss some of the chemistry of acids and bases. You use them every day, and you will probably be hearing and reading about them as long as you live. We hope that what you learn here will help you gain a better understanding of these important classes of compounds.

7.1 Acids and Bases: Experimental Definitions

Acids and bases are chemical opposites, and so their properties are quite different—often opposite. Let us begin by listing a few of their properties.

An **acid** is a compound that

1. causes litmus indicator dye to turn red.
2. tastes sour.
3. dissolves active metals (such as zinc or iron), producing hydrogen gas.
4. reacts with bases to form water and ionic compounds called salts.

A **base** is a compound that

1. causes litmus indicator dye to turn blue.
2. tastes bitter.
3. feels slippery on the skin.
4. reacts with acids to form water and salts.

A sour taste identifies foods that are acidic. Vinegar and lemon juice are good examples. Vinegar is a solution of acetic acid (about 5%) in water. Lemons, limes, and other citrus fruits contain citric acid. Lactic acid gives yogurt its tart taste, and phosphoric acid is often added to carbonated drinks to impart tartness. The bitter taste of tonic water, on the other hand, is attributable to the presence of quinine, which is a base. Some common acids and bases are pictured in Figure 7.1.

Perhaps the most common way to identify a substance as an acid or a base is a litmus test. If you dip a strip of neutral (violet-colored) litmus paper into an unknown solution and it turns pink, the solution is acidic. If it turns blue, the solution is basic. If the strip does not turn pink or blue, the solution is neither acidic nor basic.

Litmus (Figure 7.2) is an **acid–base indicator**, one of several hundred such compounds. Many natural food colors, such as those in grape juice, red cabbage, and blueberries, are acid–base indicators. So are the colors in most flower petals.

SAFETY ALERT: Although all acids taste sour and all bases are bitter, a taste test is hardly the best general-purpose test for determining whether a substance is an acid or a base. Some acids and bases are highly poisonous, and many of them are quite corrosive unless they have been greatly diluted. As a general rule, you should NEVER TASTE LABORATORY CHEMICALS. Although many of them might be safe, there are too many that are toxic and there is always the risk of contamination.

 [Web Reference 1] An easy-to-do experiment with red cabbage as an acid–base indicator.

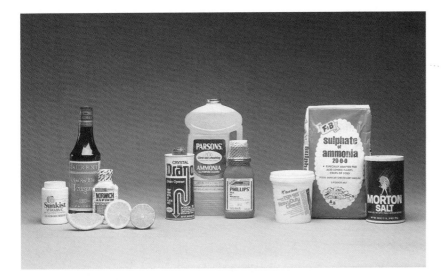

Figure 7.1 Some common acids (left), bases (center), and salts (right). Acids, bases, and salts are components of many familiar consumer products.

Figure 7.2 Strips of paper impregnated with litmus dye are often used to distinguish between acids and bases. The soil sample on the left turns red litmus blue and is therefore basic. The soil sample on the right turns blue litmus red and is acidic.

7.2 Acids, Bases, and Salts

We know that acids and bases have certain characteristic properties. But why do they have these properties? Svante Arrhenius developed the first successful theory of acids and bases in 1887. He proposed that an **acid** is a molecular substance that breaks up in aqueous solution into hydrogen ions (H^+) and anions. (Because hydrogen ions are hydrogen atoms from which the electrons have been removed, H^+ ions are also called *protons*.) The acid is said to *ionize*. Arrhenius viewed a **base** as a substance that releases hydroxide ions (OH^-) in aqueous solution. A base may either contain OH^-, as do ionic hydroxides such as $NaOH$, or it may ionize to produce OH^-. He proposed that the essential reaction between an acid and a base, *neutralization*, is the combination of H^+ and OH^- to form water. The cation originally associated with the OH^- and the anion associated with the H^+ give rise to an ionic compound, a **salt**.

Table 7.1 lists some common acids. Notice that each formula contains one or more hydrogen atoms. Chemists often indicate an acid by writing the formula with the H atoms first. HCl, H_2SO_4, and HNO_3 are acids; NH_3 and CH_4 are not. The formula $HC_2H_3O_2$ (acetic acid) indicates that one H atom ionizes and three do not.

Svante Arrhenius (1859–1927), a Swedish scientist, was the first to recognize that acids, bases, and salts, when dissolved in water, are dissociated into ions. Arrhenius was also the first to relate carbon dioxide in the atmosphere to the greenhouse effect (Chapter 12).

Ionizable (acidic) H atoms   Nonionizable (nonacidic) H atoms

H_2SO_4 HNO_3 $HC_2H_3O_2$ NH_3 CH_4

Table 7.1 ▌ Some Familiar Acids

Name	Formula	Acid Strength	Common Uses/Notes
Sulfuric acid	H_2SO_4 —	Strong	Battery acid; extremely corrosive
Nitric acid	HNO_3	Strong	Manufacture of fertilizers, explosives
Hydrochloric acid	HCl —	Strong	Cleaning of metals, bricks; removing scale from boilers
Phosphoric acid	H_3PO_4	Moderate	Manufacture of fertilizers; acidulant for foods
Hydrogen sulfate ion	HSO_4^-	Moderate	Toilet bowl cleaners
Lactic acid	$CH_3CHOHCOOH$	Weak	Acidulant for soda pop, foods
Acetic acid	CH_3COOH	Weak	Vinegar; acidulant
Carbonic acid	H_2CO_3 —	Weak	Unstable; formed in aqueous CO_2
Boric acid	H_3BO_3	Very weak	Antiseptic eye wash
Hydrocyanic acid	HCN	Very weak	None; extremely toxic

Example 7.1

Write an equation to show the ionization of HNO_3 (nitric acid) in water.

Solution

HNO_3 ionizes to form a hydrogen ion and a nitrate ion. Because this reaction occurs in water, we can use (aq) to indicate that these substances are in aqueous solution.

$$HNO_3(aq) \longrightarrow H^+(aq) + NO_3^-(aq)$$

Exercise 7.1

Write an equation to show the ionization of HBr (hydrobromic acid) in water.

In water the H^+ ion is probably associated with several H_2O molecules—for example, four H_2O molecules in the ion $H(H_2O)_4^+$ or $H_9O_4^+$. For most purposes, however, we simply use H^+ and ignore the associated water molecules. We can do this as long as we understand that we are using a simplification of the real situation. Thus, when we mention protons, we are really referring to their sources (hydronium ions).

The Arrhenius theory has its limitations. We now know, for example, that a simple free proton does not exist in water solution because H^+ has such a high positive charge density that it immediately seeks out a negative charge. It finds a lone pair of electrons on the O atoms of an H_2O molecule and attaches itself to form a *hydronium ion*, H_3O^+.

$$H\!:\!\overset{..}{\underset{..}{O}}\!: \; + \; H^+ \; \longrightarrow \; \left[H\!:\!\overset{..}{\underset{..}{O}}\!:\!H \right]^+$$
$$\overset{|}{H} \qquad\qquad\qquad \overset{|}{H}$$
Water Hydronium ion

In water, then, the properties of acids are those of the H^+ ion. It is the hydrogen ion that turns litmus red, tastes sour, and reacts with active metals and bases.

Experimental evidence indicates that the properties of bases in water are due to OH^-, the hydroxide ion. Table 7.2 lists some common bases. Most of these are ionic

Table 7.2 ▌ Common Bases

Name	Formula	Classification	Common Uses/Notes
Sodium hydroxide	$NaOH$	Strong	Acid neutralization; soap making; dehorning calves
Potassium hydroxide	KOH	Strong	Making liquid soaps; absorbing CO_2
Lithium hydroxide	$LiOH$	Strong	Alkaline storage batteries
Calcium hydroxide	$Ca(OH)_2$	Strong*	Mortar, plaster, cement; water purification
Magnesium hydroxide	$Mg(OH)_2$	Strong*	Antacid, laxative
Ammonia	NH_3	Weak	Fertilizer, household cleansers

*Although these bases are classified as strong, they are not very soluble. Calcium hydroxide is only slightly soluble in water, and magnesium hydroxide is practically insoluble.

compounds containing positive metal ions, such as Na^+ or Ca^{2+}, and negative hydroxide ions (OH^-). When the compounds dissolve in water, they all provide OH^- ions, and thus they are all bases. The properties of bases are those of hydroxide ions, just as the properties of acids are those of hydrogen ions.

Ammonia seems out of place in Table 7.2 because it contains no hydroxide ions. The Arrhenius theory doesn't really account for the basicity of ammonia and related compounds. It is also limited in that it applies only to reactions in aqueous solution. Like many scientific theories, a better one based on newer data has supplanted it.

The Brønsted–Lowry Acid–Base Theory

The shortcomings of the Arrhenius theory were largely overcome by a theory proposed independently, in 1923, by J. N. Brønsted in Denmark and T. M. Lowry in Great Britain. In their theory, an **acid** is a *proton donor*, and a **base** is a *proton acceptor*.

The Brønsted–Lowry theory describes the ionization of hydrogen chloride in this way:

$$HCl(aq) + H_2O \longrightarrow H_3O^+(aq) + Cl^-(aq)$$

The acid molecules donate hydrogen ions, or protons, to the water molecules, and so the acids act as proton donors. In the case of HCl, hydrogen chloride gas dissolves in water to form hydrochloric acid.

The "muriatic acid" sold in hardware stores (for cleaning bricks and concrete) is a solution of hydrochloric acid. Muriatic acid is an old name for hydrochloric acid.

Acid
(proton donor)

Hydrochloric acid

Notice that the HCl molecule donates a proton to the water molecule, producing a hydronium ion. Other acids react in a similar way, donating hydrogen ions to water to produce hydronium ions. Even when the solvent is something other than water, the acid acts as a proton donor, transferring H^+ ions to the solvent molecules.

Example 7.2

Write an equation to show the reaction of HNO_3 as a Brønsted–Lowry acid with water. What is the role of water in the reaction?

Solution

As a Brønsted–Lowry acid, HNO_3 donates a proton to water, forming a hydronium ion and a nitrate ion.

$$HNO_3(aq) + H_2O \longrightarrow H_3O^+(aq) + NO_3^-(aq)$$

The water molecule accepts a proton from HNO_3; water is a Brønsted-Lowry base in this reaction.

Exercise 7.2

Write an equation to show the reaction of HBr as a Brønsted–Lowry acid with water.

Where does the OH^- come from in the ionization of bases such as ammonia (NH_3)? The Arrhenius theory is inadequate in answering this question, but the Brønsted–Lowry theory explains how ammonia acts as a base in water. Ammonia is a gas

Figure 7.3 An acid is a proton donor. A base is a proton acceptor.

at room temperature. When it is dissolved in water, the following reaction occurs.

$$NH_3(aq) + H_2O \longrightarrow NH_4^+(aq) + OH^-(aq)$$

In its reaction with HCl, water acts as a base (proton acceptor). In its reaction with NH_3, water acts as an acid (proton donor). A substance, such as water, that can either donate a proton or accept a proton is said to be *amphiprotic*.

An ammonia molecule accepts a proton from a water molecule; it acts as a Brønsted–Lowry base. (Recall that the N atom of ammonia has a lone pair of electrons which can be used to attach a proton.) The water molecule acts as a proton donor— an acid. The ammonia molecule becomes an ammonium ion. When a proton leaves a water molecule, it leaves behind the electron pair that joined it to the O atom. The water molecule becomes a negatively charged hydroxide ion.

In general, then, a base is a proton acceptor (Figure 7.3). This definition includes not only hydroxide ions but also neutral molecules such as ammonia. It also includes other negative ions such as oxide (O^{2-}), carbonate (CO_3^{2-}), and bicarbonate (HCO_3^-). The idea of an acid as a proton donor and a base as a proton acceptor greatly expands our concept of acids and bases.

7.3 Acidic and Basic Anhydrides

In the Brønsted–Lowry view, many metal oxides act directly as bases because the oxide ion can accept a proton. These metal oxides also react with water to form metal hydroxides, compounds that are bases in the Arrhenius sense. Similarly, many non-metal oxides react with water to form acids.

Nonmetal Oxides: Acidic Anhydrides

Many acids are made by the reaction of nonmetal oxides with water. For example, sulfur trioxide reacts with water to form sulfuric acid.

$$SO_3 + H_2O \longrightarrow H_2SO_4$$

Similarly, carbon dioxide reacts with water to form carbonic acid.

$$CO_2 + H_2O \longrightarrow H_2CO_3$$

In general, nonmetal oxides react with water to form acids.

$$\text{Nonmetal oxide} + H_2O \longrightarrow \text{acid}$$

Nonmetal oxides that act in this way are called **acidic anhydrides**. Anhydride means "without water." These reactions explain why rainwater is acidic (Section 7.6).

Example 7.3

Give the formula for the acid formed when sulfur dioxide reacts with water.

Solution

Simply write the equation for the reaction, following the pattern in the preceding examples.

$$SO_2 + H_2O \longrightarrow H_2SO_3$$

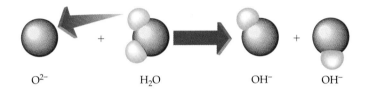

O^{2-} H_2O OH^- OH^-

Figure 7.4 Metal oxides are basic because the oxide ion reacts with water to form two hydroxide ions.

Exercise 7.3

Give the formula for the acid formed when dinitrogen pentoxide (N_2O_5) reacts with water. (*Hint:* Two molecules of acid are formed.)

Metal Oxides: Basic Anhydrides

Just as acids can be made from nonmetal oxides, many common hydroxide bases can be made from metal oxides. For example, calcium oxide (lime) reacts with water to form calcium hydroxide (slaked lime).

$$CaO + H_2O \longrightarrow Ca(OH)_2$$

Another example is the reaction of lithium oxide with water to form lithium hydroxide.

$$Li_2O + H_2O \longrightarrow 2\,LiOH$$

In general, metal oxides react with water to form bases (Figure 7.4). These metal oxides are called **basic anhydrides**.

$$\text{Metal oxide} + H_2O \longrightarrow \text{base}$$

Example 7.4

What base is formed by the addition of water to barium oxide (BaO)?

Solution
Simply write the equation for the reaction.

$$BaO + H_2O \longrightarrow Ba(OH)_2$$

Exercise 7.4

What base is formed by the addition of water to potassium oxide (K_2O)? (*Hint:* Two moles of base are formed for each mole of potassium oxide.)

7.4 Strong and Weak Acids and Bases

When gaseous hydrogen chloride (HCl) reacts with water, it reacts completely to form hydronium ions and chloride ions. Essentially no HCl molecules remain.

$$HCl + H_2O \longrightarrow H_3O^+ + Cl^-$$

For many purposes, we simply write the reaction as the ionization of HCl and use (aq) to indicate the involvement of water.

$$HCl(aq) \longrightarrow H^+(aq) + Cl^-(aq)$$

The poisonous gas hydrogen cyanide (HCN) also ionizes in water to produce hydrogen ions and cyanide ions.

$$HCN(aq) \longrightarrow H^+(aq) + CN^-(aq)$$

But HCN reacts only to a slight extent. In a solution that has 1 mol of HCN in 1 L of

A strong acid or base is one that is almost completely ionized in solution. The word "strong" does not refer to the amount of acid or base in the solution. A solution that contains a relatively large amount of acid or base in a given volume of solution is called a *concentrated* solution. (The acid or base may be strong or weak.) A solution with only a little solute in that same volume of solution is said to be a *dilute* solution.

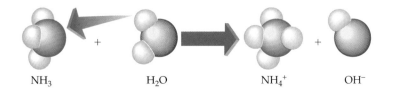

$$NH_3 \qquad H_2O \qquad NH_4^+ \qquad OH^-$$

Figure 7.5 Ammonia is a base because it accepts a proton from water. A solution of ammonia in water contains ammonium ions and hydroxide ions. Only a small fraction of the ammonia molecules react, however; most remain unchanged. Ammonia is therefore a weak base.

 [Web Reference 2] Two good sites with further information about acids and bases—good on terminology, especially as it applies to the acid–base balance in the body and good on concepts, with fairly extensive history.

water, only 0.0025% of the HCN molecules react to produce hydrogen ions. Most of them remain intact as HCN molecules. An acid such as HCl that reacts completely with water is called a **strong acid**. One that reacts only slightly with water is a **weak acid**. There are not many strong acids. The first three acids listed in Table 7.1 (sulfuric, nitric, and hydrochloric) are the only common ones. Most acids are weak acids.

Bases are also classified as strong or weak. A **strong base** is completely ionized in water, and a **weak base is** only slightly ionized. Perhaps the most familiar strong base is sodium hydroxide (NaOH). It exists as sodium ions and hydroxide ions even in the solid state. Other strong bases include potassium hydroxide (KOH) and the hydroxides of all the other Group 1A metals. Except for $Be(OH)_2$, Group 2A hydroxides are also strong bases. However, $Ca(OH)_2$ is only slightly soluble in water, and $Mg(OH)_2$ is nearly insoluble. The concentration of hydroxide ions in water from either is therefore not very high. The most familiar weak base is ammonia (NH_3). It reacts with water to a slight extent to produce ammonium ions (NH_4^+) and hydroxide ions (Figure 7.5).

$$NH_3 + H_2O \longrightarrow NH_4^+ + OH^-$$

Neutralization

When an acid reacts with a base, the products are water and a salt. If a solution containing hydrogen ions (an acid) is mixed with another solution containing exactly the same amount of hydroxide ions (a base), the resulting solution no longer affects litmus, and it no longer tastes sour or bitter (it tastes salty). It is no longer either acidic

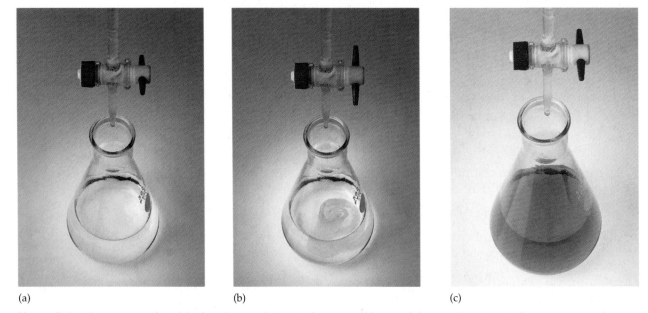

(a)   (b)   (c)

Figure 7.6 The amount of acid (or base) in a solution is determined by careful neutralization. In the experiment shown here, a 5.00-mL sample of vinegar, a small amount of water, and a few drops of phenolphthalein (an acid–base indicator) are added to a flask (a). A solution of 0.1000 M NaOH is added slowly from a buret (a device for precise measurement of volumes of solutions) (b). As long as the acid is in excess, the solution is colorless. When the acid has been neutralized and a tiny excess of base is present, the phenolphthalein indicator turns pink (c).

or basic; it is neutral. The reaction of an acid with a base is called **neutralization** (Figure 7.6). In water it is simply the reaction of hydrogen ions with hydroxide ions to form water molecules.

$$H^+ + OH^- \longrightarrow H_2O$$

If sodium hydroxide is neutralized by hydrochloric acid, the products are water and sodium chloride (ordinary table salt).

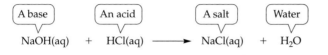

$$\underset{\text{A base}}{NaOH(aq)} + \underset{\text{An acid}}{HCl(aq)} \longrightarrow \underset{\text{A salt}}{NaCl(aq)} + \underset{\text{Water}}{H_2O}$$

Example 7.5

Write the equation for the neutralization reaction between potassium hydroxide and nitric acid.

Solution

The OH^- of the base and the H^+ of the acid combine to form water. The cation of the base (K^+) and the anion of the acid (NO_3^-) form a solution of the salt potassium nitrate (KNO_3).

$$KOH(aq) + HNO_3(aq) \longrightarrow KNO_3(aq) + H_2O$$

Exercise 7.5

Write the equation for the neutralization reaction between calcium hydroxide and hydrochloric acid.

7.5 The pH Scale

Recall (Chapter 6) that atoms, molecules, and ions are usually counted in moles. In solutions, the concentrations of ions are measured in moles per liter. Because hydrogen chloride is completely ionized in water, a solution of 1 molar hydrochloric acid (1 M HCl), for example, contains 1 mol of H^+ ions per liter of solution. In other words, 1 L of 1 M HCl contains 6×10^{23} hydrogen ions, and 0.5 L of 0.001 M HCl contains 3×10^{20} hydrogen ions.

We might describe the acidity of a particular solution in moles per liter: The hydrogen ion concentration of this solution is 1×10^{-3} mol/L. But such a statement is rare. Instead, the acidity of this solution is simply identified as pH 3.

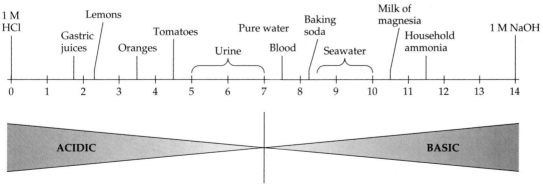

Figure 7.7 The pH scale. A change in pH of one unit means a tenfold change in the hydronium ion concentration.

The pH of a solution is defined as the negative logarithm of the hydrogen ion concentration.

$$pH = -\log[H^+]$$

The brackets indicate molar concentration. Perhaps the relationship is easier to see when the equation is written in the following form.

$$[H^+] = 10^{-pH}$$

[Web Reference 3]
An extensive tutorial about pH. This site also has some quizzes and animated reactions.

We usually use the **pH** scale, first proposed in 1909 by the Danish biochemist S. P. L. Sorensen, to describe the degree of acidity or basicity. The pH scale lies mainly in the range 0 to 14. The neutral point on the scale is 7, with values below 7 indicating increasing acidity and those above 7 increasing basicity. Thus, pH 6 is slightly acidic, whereas pH 12 is strongly basic (Figure 7.7).

The numbers on the pH scale are directly related to the hydrogen ion concentration. We might expect that pure water would be completely in the form of H_2O molecules, but it turns out that about 1 out of every 500 million molecules is split into H^+ and OH^- ions. This gives a concentration of hydrogen ions and of hydroxide ions in pure water of 0.0000001 mol/L, or 1×10^{-7} M. Can you see why 7 is the pH of pure water? It is simply the power of 10 for the molar concentration of H^+, with the negative sign removed. (The H in pH stands for "hydrogen," and the p represents "power.") Thus, we define pH as the negative logarithm of the molar concentration of hydrogen ions (Table 7.3).

Although pH is an acidity scale, note that its value goes down when acidity goes up. Not only is the relationship an inverse one, but it is also logarithmic. A decrease of 1 pH unit represents a tenfold increase in acidity, and when pH goes down by 2 units, acidity increases by a factor of 100. This relationship may seem strange at first, but once you understand the pH scale, you appreciate its convenience.

Table 7.3 summarizes the relationship between hydrogen ion concentration and pH. A pH of 4 means a hydrogen ion concentration of 1×10^{-4} mol/L, or 0.0001 M. If the concentration of hydrogen ions is 0.01 M, or 1×10^{-2} M, the pH is 2. The pH values for various common solutions are listed in Table 7.4.

Table 7.3 ▌
Relationship Between pH and Concentration of Hydronium Ions

Concentration of H_3O^+ (mol/L)	pH
1×10^0	0
1×10^{-1}	1
1×10^{-2}	2
1×10^{-3}	3
1×10^{-4}	4
1×10^{-5}	5
1×10^{-6}	6
1×10^{-7}	7
1×10^{-8}	8
1×10^{-9}	9
1×10^{-10}	10
1×10^{-11}	11
1×10^{-12}	12
1×10^{-13}	13
1×10^{-14}	14

Table 7.4 ▌ The Approximate pH Values of Some Common Solutions

Solution	pH
Hydrochloric acid (4%)	0
Gastric juice	1.6–1.8
Lemon juice	2.1
Vinegar (4%)	2.5
Soda pop	2.0–4.0
Rainwater (thunderstorm)	3.5–4.2
Milk	6.3–6.6
Urine	5.5–7.0
Rainwater*	5.6
Saliva	6.2–7.4
Pure water	7.0
Blood	7.4
Fresh egg white	7.6–8.0
Bile	7.8–8.6
Milk of magnesia	10.5
Washing soda	12.0
Sodium hydroxide (4%)	13.0

*Rainwater saturated with carbon dioxide from the atmosphere but unpolluted.

Example 7.6

What is the pH of a solution that has a hydrogen ion concentration of 1.0×10^{-5} M?

Solution

The exponent is -5; the pH is therefore 5.

Exercise 7.6

What is the pH of a solution that has a hydrogen ion concentration of 1.0×10^{-11} M?

Example 7.7

What is the hydrogen ion concentration of a solution that has a pH of 4?

Solution

The pH value is the negative exponent of 10, and so the hydrogen ion concentration is 1.0×10^{-4} M.

Exercise 7.7

What is the hydrogen ion concentration of a solution that has a pH of 2?

$$pH = -\log[H^+]$$

For coffee it's 5; for tomatoes it's 4;
While household ammonia's 11 or more.
It's 7 for water, if in a pure state,
But rainwater's 6 and seawater's 8.
It's basic at 10, quite acidic at 2,
And well above 7 when litmus turns blue.
Some find it a puzzlement. Doubtless their fog
Has something to do with that negative log!

A term akin to pH, called pOH, is related to $[OH^-]$ just as pH is related to $[H^+]$. In 0.01 M NaOH solution, $[OH^-] = 0.01$ M $= 1 \times 10^{-2}$ M and therefore pOH $= 2$. To convert pOH to pH, simply subtract pOH from 14.

$$pH + pOH = 14$$
$$pH = 14 - pOH$$

If pOH $= 2$, then pH $= 14 - 2 = 12$. We usually use pH rather than pOH regardless of whether the solution is acidic or basic.

 **[Web Reference 4]** Although publicity about acid rain has decreased in the last few years, it remains a problem.

7.6 Acid Rain

Rainwater saturated with carbon dioxide has a pH of 5.6. This slightly acidic pH is a result of the presence of carbon dioxide in the air, making the rain a dilute solution of a weak acid, carbonic acid. In many areas of the world, particularly those downwind from industrial centers, rainwater is much more acidic, with a pH as low as 3 or less. Rain with a pH below 5.6 is called *acid rain*.

Acid rain is due to acidic pollutants in the air. As we shall see in Chapter 12, several air pollutants are acid anhydrides. These include sulfur dioxide (SO_2) from burning high-sulfur coal in power plants and metal smelters, and nitrogen dioxide (NO_2) and nitric oxide (NO) from automobile exhaust fumes.

Some acid rain is due to natural pollutants, such as those resulting from volcano eruptions and lightning. Volcanoes give off sulfur oxides and sulfuric acid, and lightning produces nitrogen oxides and nitric acid.

Acid rain is an important environmental problem that involves both air pollution (Chapter 12) and water pollution (Chapter 13). It can have serious effects on plant and animal life.

You can make your own aspirin-free "Alka-Seltzer." Simply place half a teaspoon of baking soda in a glass of orange juice. (What is the acid and what is the base in this reaction?)

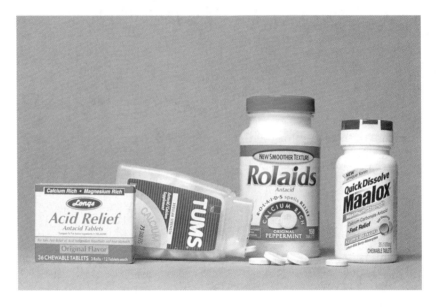

Figure 7.8 A great variety of antacids are available to consumers. All antacids are basic compounds.

Table 7.5 ▮ Some Common Antacids

Commercial Product	Antacid Ingredient(s)
Alka-Seltzer*	$NaHCO_3$, citric acid, aspirin
Amphojel	$Al(OH)_3$
Baking soda*	$NaHCO_3$
DiGel	$CaCO_3$
Maalox	$Al(OH)_3$, $Mg(OH)_2$
Milk of magnesia	$Mg(OH)_2$
Mylanta	$CaCO_3$, $Mg(OH)_2$
Rolaids	$CaCO_3$, $Mg(OH)_2$
Tums	$CaCO_3$

*Sodium-containing antacids are not recommended for people with hypertension (high blood pressure).

7.7 Antacids: A Basic Remedy

All antacids are basic compounds. They act by neutralizing hydrogen ions in stomach acid. Claims of "fast action" are almost meaningless. All acid–base reactions are almost instantaneous. Some tablets may dissolve a little slower than others. You can speed their action by chewing them. Drugs such as Zantac, Pepcid AC, and Tagamet are not antacids. These drugs inhibit the release of hydrochloric acid by the stomach.

The stomach secretes hydrochloric acid (HCl) to aid in the digestion of food. Sometimes overindulgence or emotional stress leads to a condition of *hyperacidity* (too much acid). Hundreds of brands of antacids (Figure 7.8) are sold in the United States to treat this condition. Despite the many brand names, there are only a few different antacid ingredients, primarily sodium bicarbonate, calcium carbonate, aluminum hydroxide, and magnesium hydroxide (Table 7.5).

Sodium bicarbonate ($NaHCO_3$), commonly called *baking soda*, is probably safe and effective for most people, but overuse can make the blood too alkaline, a condition called **alkalosis**.

Calcium carbonate ($CaCO_3$) is safe in small amounts, but regular use can cause constipation. It also appears that calcium carbonate can actually result in increased acid secretion after a few hours.

Aluminum hydroxide [$Al(OH)_3$], like calcium carbonate, can cause constipation. There is also some concern that antacids containing aluminum ions can deplete the body of essential phosphate ions.

A suspension of magnesium hydroxide [$Mg(OH)_2$] in water is sold as "milk of magnesia." Magnesium carbonate ($MgCO_3$) is also used as an antacid. In small doses, magnesium compounds act as antacids, but in large doses they act as laxatives.

Although antacids are generally safe for occasional use, they can interact with other medications; and anyone who has severe or repeated attacks of indigestion should consult a physician. Self-medication can sometimes be dangerous.

7.8 Acids and Bases in Industry and in Us

Acids and bases play an important role in industry, both as products for use in many areas and as by-products that can damage the environment. Their use requires caution, and their misuse can be dangerous to human health. Acids and bases are also important participants in the biochemistry of every living thing.

Acids and Bases in Industry and at Home

Sulfuric acid (H_2SO_4) is by far the leading chemical product in the United States. About 40 billion kg is produced each year, most of it for making fertilizers and other industrial chemicals. Around the home we use sulfuric acid in automobile batteries and in some special kinds of drain cleaners.

Hydrochloric acid (also called muriatic acid) is used in industry to remove rust from metal, in construction to remove excess mortar from bricks, and in the home to remove lime deposits from toilet bowls. Concentrated solutions (about 38% HCl) cause severe burns, but dilute solutions can be used safely in the home if handled carefully.

Lime (CaO) is the cheapest and most widely used commercial base. It is made by heating limestone ($CaCO_3$) to drive off CO_2.

$$CaCO_3(s) + heat \longrightarrow CaO(s) + CO_2(g)$$

It is the fifth most-produced industrial chemical in the United States, with an annual production of about 17 billion kg. It is used to make mortar and cement and also to "sweeten" acidic soil.

Sodium hydroxide (commonly known as *lye*) is the strong base most often used in the home. It is employed as an oven cleaner in products such as Easy Off, and it is used to open clogged drains in products such as Drano.

Acids and Bases in Health and Disease

When they are misused, acids and bases can be damaging to human health. Concentrated strong acids and bases are corrosive poisons that can cause serious chemical burns. Once the chemical agents are removed, the injuries are similar to burns caused by heat, and they are often treated in the same way. Besides being a strong acid, sulfuric acid is also a powerful dehydrating agent that can react with water in the cells.

Strong acids and bases, even in dilute solutions, break down the protein molecules in living cells. Generally, the fragments are not able to carry out the functions of the original proteins. In cases of severe exposure, this fragmentation continues until the tissue has been completely destroyed.

Acids and bases affect human health in more subtle—and ultimately more important—ways. A delicate balance must be maintained between acids and bases in the blood, body fluids, and cells. If the acidity of the blood changes too much, the blood loses its capacity to carry oxygen. In living cells, proteins function properly only at an optimum pH. If the pH changes too much in either direction, the proteins can't carry out their usual functions. Fortunately, the body has a complex but efficient mechanism for maintaining a proper acid–base balance. (Consult Reference 3 for an explanation of this mechanism.)

Treating the soil with lime makes it "sweeter" (less acidic).

SAFETY ALERT: Concentrated acids and bases can cause severe burns. They must be handled with great care. In working with them, always follow directions carefully. Wear safety goggles to protect the eyes and protective clothing to protect your skin and clothes.

Why Doesn't "Stomach Acid" Dissolve the Stomach?

We know that strong acids are corrosive to skin. The gastric juice in your stomach is a solution containing about 0.5% hydrochloric acid. Why doesn't the acid in your stomach destroy your stomach lining? The cells that line the stomach are protected by a layer of mucus, a viscous solution of a sugar–protein complex called *mucin*, and other substances in water. The mucus serves as a physical barrier, but its role is not simply passive. Rather, the mucin acts like a sponge that soaks up bicarbonate ions from the cellular side and hydrochloric acid from within the stomach. The bicarbonate ions neutralize the acid within the mucus. When aspirin, alcohol, bacteria, or other agents damage the mucus, the exposed cells are damaged and an ulcer can form.

Critical Thinking Exercises

Apply knowledge that you have gained in this chapter and one or more of the FLaReS principles (Chapter 1) to evaluate the following statements or claims.

7.1 A television advertisement claims that the antacid Maalox neutralizes stomach acid faster and therefore relieves heartburn faster than Pepcid AC, a drug that inhibits the release of stomach acid. To illustrate this claim, two flasks of acid are shown. In one, Maalox rapidly neutralizes the acid. In the other, Pepcid AC does not neutralize the acid. Does this visual demonstration validate the claim made in the advertisement?

7.2 Canadians in the province of Ontario claim that industrial plants in the United States are polluting the air in Canada and causing damage to their buildings, trees, fish, and other wildlife. Is their claim reasonable?

Summary

1. Acids taste sour, turn litmus red, react with active metals to form hydrogen, and react with bases to form salts and water.

2. Bases taste bitter, turn litmus blue, feel slippery to the skin, and react with acids to form salts and water.

3. According to Arrhenius's definition, acids produce hydrogen ions in water and bases produce hydroxide ions.

4. According to the Brønsted–Lowry theory, acids are proton donors and bases are proton acceptors.

5. When an acid dissolves in water, water molecules pick up hydrogen ions (H^+), or protons, from the acid molecules to produce hydronium ions (H_3O^+).

6. When a base dissolves in water, hydroxide ions (OH^-) are formed.

7. Nonmetal oxides (acidic anhydrides) react with water to form acids; metal oxides (basic anhydrides) react with water to form bases.

8. A strong acid or strong base, when dissolved in water, is almost completely in the form of ions.

9. A weak acid or weak base reacts only slightly with water to produce ions.

10. Sulfuric, hydrochloric, and nitric acids are strong acids. Acetic and most other acids are weak acids.

11. Group 1A hydroxides are strong bases and so are most Group 2A hydroxides. Ammonia is a weak base.

12. When an acid reacts with a base, the products are a salt and water, and the process is called neutralization.

13. The pH scale is an acidity scale. A pH of 7 is neutral, pH values lower than 7 are increasingly acidic, and pH values greater than 7 are increasingly basic.

14. Acid rain results from oxides of sulfur and nitrogen in the air.

15. Antacids are products used to neutralize excess stomach acid; but taking too much antacid can produce a condition of alkalosis.

Key Terms

acid 7.2
acid–base indicator 7.1
acidic anhydride 7.3
alkalosis 7.7

base 7.2
basic anhydride 7.3
neutralization 7.4

pH 7.5
salt 7.2
strong acid 7.4

strong base 7.4
weak acid 7.4
weak base 7.4

Review Questions

1. Define and illustrate the following terms.
   a. acid **b.** base **c.** salt
2. List four general properties of acidic solutions.
3. List four general properties of basic solutions.
4. What ion is responsible for the properties of acidic solutions (in water)?
5. What ion is responsible for the properties of basic solutions (in water)?
6. Can a substance be a Brønsted–Lowry acid if it does not contain H atoms? Are there any characteristic atoms that must be present in a Brønsted–Lowry base?
7. Give the formulas and the names of two strong acids and two weak acids.
8. Give the formulas and the names of two strong bases and one weak base.
9. Strong acids and weak acids both have properties characteristic of hydrogen ions. How do strong acids and weak acids differ?
10. Give the Brønsted–Lowry definition of an acid. Write an equation that illustrates the definition.
11. Give the Brønsted–Lowry definition of a base. Write an equation that illustrates the definition.

12. What is meant by the proton as used in acid–base chemistry? How does it differ from the proton of nuclear chemistry (Chapter 4)?
13. What is an acidic anhydride? A basic anhydride?
14. Describe the neutralization of an acid or base.
15. Describe the taste and the effect on litmus of a solution that has been neutralized.
16. Magnesium hydroxide is completely ionic, even in the solid state, yet it can be taken internally as an antacid. Explain why it does not cause injury as sodium hydroxide would.
17. What is the medical use of antacids?
18. Name some of the active ingredients in antacids.
19. What is alkalosis?
20. What is the leading chemical product of U.S. industry?
21. What are the effects of strong acids and strong bases on the skin?
22. According to the Arrhenius theory, all acids have one element in common. What is that element? Are all compounds containing that element acids? Explain.

Problems

Acids and Bases

23. Use the definitions of "acid" and "base" to identify the first compound in each equation as an acid or a base. (*Hint:* What is produced by the reaction?)
   a. $C_5H_5N + H_2O \longrightarrow C_5H_5NH^+ + OH^-$
   b. $C_6H_5OH + H_2O \longrightarrow C_6H_5O^- + H_3O^+$
   c. $CH_3COCOOH + H_2O \longrightarrow CH_3COCOO^- + H_3O^+$
24. Use the definitions of "acid" and "base" to identify the first compound in each equation as an acid or a base.
   a. $C_6H_5SH + H_2O \longrightarrow C_6H_5S^- + H_3O^+$
   b. $CH_3NH_2 + H_2O \longrightarrow CH_3NH_3^+ + OH^-$
   c. $C_6H_5SO_2NH_2 + H_2O \longrightarrow C_6H_5SO_2NH^- + H_3O^+$
25. Give formulas for the following acids.
   a. hydrochloric acid **b.** sulfuric acid
   c. carbonic acid **d.** hydrocyanic acid
26. Give formulas for the following acids.
   a. nitric acid **b.** sulfurous acid
   c. phosphoric acid **d.** hydrosulfuric acid

27. Give formulas for the following bases.
   a. lithium hydroxide **b.** magnesium hydroxide
   c. sodium hydroxide
28. Give formulas for the following bases.
   a. calcium hydroxide **b.** potassium hydroxide
   c. ammonia
29. Write the equation that shows hydrogen chloride gas reacts as a Brønsted–Lowry acid in water. What is the name of the acid formed?
30. Write the equation that shows how ammonia acts as a Brønsted–Lowry base in water.

Acidic and Basic Anhydrides

31. Give the formula for the compound formed when sulfur trioxide reacts with water. Is the product an acid or a base?
32. Give the formula for the compound formed when magnesium oxide reacts with water. Is the product an acid or a base?

33. Give the formula for the compound formed when potassium oxide reacts with water. Is the product an acid or a base?

34. Give the formula for the compound formed when carbon dioxide reacts with water. Is the product an acid or a base?

Strong and Weak Acids and Bases

35. Thallium hydroxide (TlOH) is ionic in the solid state and is quite soluble in water. Classify the compound as a strong acid, weak acid, weak base, or strong base.

36. Hydrogen iodide (HI) gas reacts completely with water to form hydronium ions and iodide ions. Classify the compound as a strong acid, weak acid, weak base, or strong base.

37. Hydrogen sulfide (H_2S) gas reacts slightly with water to form relatively few hydronium ions and hydrogen sulfide ions (HS^-). Classify the compound as a strong acid, weak acid, weak base, or strong base.

38. Methylamine (CH_3NH_2) gas reacts slightly with water to form relatively few hydroxide ions and methylammonium ions ($CH_3NH_3^+$). Classify the compound as a strong acid, weak acid, weak base, or strong base.

Ionization of Acids and Bases

39. Write equations showing the ionization of the following as Arrhenius acids.
    a. HI(aq)   b. HNO_2(aq)   c. $HClO_2$(aq)

40. Write equations showing the ionization of the following as Brønsted–Lowry acids.
    a. $HClO_2$(aq)   b. HNO_2(aq)   c. HCN(aq)

Neutralization

41. Write the equation for the reaction of sodium hydroxide with hydrochloric acid.

42. Write the equation for the reaction of lithium hydroxide with nitric acid.

43. Write the equation for the reaction of 1 mol calcium hydroxide with 2 mol hydrochloric acid.

44. Write the equation for the reaction of 1 mol sulfuric acid with 2 mol potassium hydroxide.

45. Write the equation for the reaction of 1 mol phosphoric acid with 3 mol sodium hydroxide.

46. Write the equation for the reaction of 1 mol sulfuric acid with 1 mol calcium hydroxide.

The pH Scale

47. Indicate whether each of the following pH values represents an acidic, basic, or neutral solution.
    a. 4   b. 7   c. 3.5   d. 9

48. Lime juice has a pH of about 2. Is it acidic or basic?

49. What is the pH of a solution that has a hydrogen ion concentration of 1.0×10^{-3} M?

50. What is the pH of a solution that has a hydrogen ion concentration of 1.0×10^{-10} M?

51. What is the hydrogen ion concentration of a solution that has a pH of 5?

52. What is the hydrogen ion concentration of a solution that has a pH of 11?

Additional Problems

53. Suggest some ways in which you might determine whether a particular water solution contains an acid or a base.

54. According to the Arrhenius theory, are all compounds containing OH groups bases? Explain.

55. Which, if any, of the properties listed in Questions 2 and 3 remain after an acid and a base neutralize each other?

56. What is the pH of a solution of 0.001 M KOH that has a pOH of 3?

57. What is the pH of a solution of 0.01 M NaOH with a hydroxide ion concentration of 1×10^{-2} M?

58. Like water, the hydrogen phosphate ion (HPO_4^{2-}) is amphiprotic. That is, it can act either as a Brønsted–Lowry acid or as a Brønsted–Lowry base. Write equations that illustrate this.

59. Look at the warning label on a can of oven cleaner. Using an equation as well as words, explain the antidote on the basis of the chemistry you learned in this chapter.

60. How is the term "pH" used in everyday life, including advertising? Is it used correctly?

61. A good site for advanced information about acids and bases is gasnet, an anesthesiology website. Speculate about why knowledge about acids and bases is so important for anesthesiologists.

Projects

62. Examine the labels of at least five antacid preparations. Make a list of the ingredients in each. Look up the properties (medical use, side effects, toxicity, and so on) of each ingredient in a reference book such as *The Merck Index*.

63. Examine the labels of at least five toilet bowl cleaners and five drain cleaners. Make a list of the ingredients in each. Look up the formulas and properties of each ingredient in a reference book such as *The Merck Index*. Which ingredients are acids? Which are bases?

Online Projects

64. Using your favorite search engine, do a search for "antacid" on the world wide web. How many sites can you find? How many appear to be sponsored by drug companies? Which seem to be unbiased?

65. Deterioration of paper in books and journals is a problem in libraries and in personal book collections. This is due to action of acid on the cellulose fibers comprising the paper. Using either the internet (preferred) or traditional library research, find out why this is happening and what is being done to counteract it.

References and Readings

1. "Antacids: Which Beat Heartburn Best?" *Consumer Reports*, July 1994, pp. 443–447.

2. Budvari, Susan (Ed.) *The Merck Index*, 12th edition. Rahway, NJ: Merck and Co., 1996.

3. Hill, John W., Stuart J. Baum, and Rhonda J. Scott-Ennis. *Chemistry and Life*, 6th edition. Upper Saddle River, NJ: Prentice Hall, 2000. Chapters 9 and 10.

4. "How to Choose an Antacid." *Consumer Reports*, August 1983, pp. 412–418.

5. Jensen, William B. "Acids and Bases: Ancient Concepts in Modern Science." *ChemMatters*, April 1983, pp. 14–15. Part of a special issue on acid–base chemistry.

6. Kolb, Doris. "Acids and Bases." *Journal of Chemical Education*, July 1978, pp. 459–464.

7. Kolb, Doris. "The pH Concept." *Journal of Chemical Education*, January 1979, pp. 49–53.

8. Lowenstein, Jerome. *Acids and Bases*. New York: Oxford University Press, 1993. A discussion of acid–base physiology.

MediaLab

Acid Rain: A Global Problem

Many environmental problems are not constrained by municipal, territorial, or national borders. Smokestack effluents rise far into the atmosphere, there to be carried into distant regions. Waste dumped into a river travels downstream, affecting aquatic life along the way, until it accumulates in a bay or lagoon, where it may kill the vegetation and fish. Plastic can holders thrown from a cruise ship may be carried by currents and end up around the neck of an Arctic seal.

Acid rain is one of the most obvious examples of the fact that we are indeed all one world. Recognition by one nation that its trees and fish are dying because of industrial activity in another country can lead to international disputes. In some cases the accusations and controversies have led to the signing of treaties. In this *Web Investigation*, you will look at the composition and sources of acid rain. You will determine its effects in several different countries and try to trace its origin. You will explore the ways in which it affects both natural and man-made treasures. You will also look at what is being done to correct the problem—and by whom. In *Communicate Your Results*, you will compare attempts by different nations to deal with acid rain. You will also have the opportunity to draft an international treaty on acid rain.

WEB INVESTIGATION

Investigation 1
The Chemistry of Acid Rain

Select Keyword **Chemistry** for the chemistry of acid rain.

Investigation 2
From Ohio's Smokestacks to Ontario's Lakes, and Beyond

Select Keyword **Overview** for overviews of the acid rain problem from several different nations. You need not look at every page, but looking at more than one will give you an idea of the similarity of the problem in various countries.

Investigation 3
Our Trees Are Dying!

Acid rain damages forests as well as lakes. This is seen to a distressing degree in Europe. Select Keyword **SOIL** for an overview of the effects of acid rain on soil and plants. Norway, a country with

lush forests, is suffering damage to both lakes and vegetation, most of which is due to transboundary pollution. Select Keyword **NORWAY** for a comprehensive look at acid rain conditions in Norway and the way pollution in one country affects the ecology of another.

Investigation 4
International Agreements

Recognition of the global effects of environmental pollution has led to several international agreements. Often these are changed over the years as the signatory nations learn more about the effects of pollutants and develop improved methods of measurement. A recent example is the Protocol to Abate Acidification, Eutrophication, and Ground-Level Ozone adopted and signed by 27 countries, including the United States, in December 1999. This is the eighth protocol adopted during the 20 years the Convention on Long-Range Transboundary Air Pollution has been in effect. Select Keyword **PROTOCOL** to read more about this protocol.

COMMUNICATE YOUR RESULTS

Exercise 1
Quantifying the Sources of Acid Rain

Integrate what you have learned about transboundary acidification by evaluating in writing the percentages of local versus transboundary acid rain and acid deposition in at least two countries of your choice. (Try to use tables, graphs, and/or maps.)

Exercise 2
Convincing the Unconvinced

What scientific evidence is needed to prove the necessity for taking steps against acid rain? How do scientists obtain this evidence? What are some of the measures needed to lessen the amount and/or impact of acid rain? Write an article showing what can be done about acid rain and why taking action is important. Assume that your reader is skeptical about all things environmental and requires good economic reasons for any change.

Exercise 3
Advocating Against Acid Rain

Take the viewpoint of a Canadian who sees the effects of acid rain on Canadian lakes and forests. Compose letters to appropriate officials or corporate executives in the United States explaining what you think should be done by the United States to lessen the impact of acid rain in Canada. Give good arguments for the statement that "Canada cannot win the fight against acid rain on its own." (If you prefer, use another pair of nations, such as Norway and Germany, or, doing some research on your own, a pair of states within the United States.)

Exercise 4 (Collaborative)
Drafting an International Agreement

Group yourselves into teams of at least three students. Each team will represent a specific nation. Using information from the *Web Investigations* and any additional data you find, draft some points of importance to "your" country for a hypothetical upcoming international meeting on acid rain. Select a team member to present these points before the "international assembly" of your classmates.

8

Oxidation and Reduction

Burn and Unburn

Reduction is electron gain; their loss is oxidation.
Each needs the other; neither can occur in isolation.
The term redox includes them both—a useful combination!

When coal burns, releasing its chemical energy as heat, the carbon in the coal is oxidized to carbon dioxide as oxygen from the air is reduced to water.

From the most advanced electric battery to a simple campfire, from the trees in a rain forest making oxygen to the people hurrying along a city street, we depend on an important group of reactions called *reduction–oxidation* (or *redox*) reactions. These reactions are extremely diverse: charcoal burns, iron rusts, bleach removes stains, the food we eat is converted to energy for our brain and muscles, and film is developed.

We sometimes consider oxidation and reduction as two processes, but in reality they always occur together. They are in fact opposite aspects of a single process, the redox reaction. You can't have one without the other. When one substance is oxidized, another is reduced. For convenience, however, we may choose to talk about only a part of the process—the oxidation part or the reduction part.

Our cells obtain energy to maintain themselves by oxidizing foods. Green plants, using energy from sunlight, produce food by the reduction of carbon dioxide. We extract metals from their ores by reduction and then lose them again to corrosion as they are oxidized. Today's technology depends on oxidizing fossil fuels (coal, natural gas, and petroleum) to obtain the chemical energy stored in these materials eons ago by green plants.

Reduced forms of matter—food, coal, and gasoline—are high in energy. Oxidized forms—carbon dioxide and water—are low in energy. Let's examine the processes of oxidation and reduction in some detail. By doing so, we can better understand the chemical reactions that keep us alive and enable us to maintain our civilization.

8.1 Oxidation and Reduction: Three Views

The term "oxidation" stems from the early recognition of oxygen's involvement in oxide formation; "reduction" then meant removal of oxygen from an oxide. When oxygen combines with other elements or compounds, the process is called **oxidation**.

Oxidation and reduction always occur together. Pictured here on the left is ammonium dichromate. In the reaction (center), the ammonium ion (NH_4^+) is oxidized and the dichromate ion ($Cr_2O_7^{2-}$) is reduced. Considerable heat and light are evolved. The equation for the reaction is

$$(NH_4)_2Cr_2O_7 \longrightarrow Cr_2O_3 + N_2 + 4\,H_2O$$

The water is driven off as vapor, and the nitrogen gas escapes, leaving pure Cr_2O_3 as the visible product (right).

(a)   (b)

(c)

Some reduced forms of matter. The energy in foods (a) and fossil fuels (b) and (c) is released when these materials are oxidized.

The substances that combine with oxygen are said to have been *oxidized*. Originally the term "oxidation" was limited to reactions involving combination with oxygen. Then as chemists came to realize that combination with chlorine (or bromine or other active nonmetals) was not all that different from reaction with oxygen, they broadened the definition of oxidation.

Reduction is the opposite of oxidation. When hydrogen burns, it combines with oxygen to form water.

$$2\,H_2 + O_2 \longrightarrow 2\,H_2O$$

The hydrogen is oxidized in this reaction, but at the same time the oxygen is reduced. Whenever oxidation occurs, reduction must occur also. Oxidation and reduction always happen at the same time and in exactly equivalent amounts.

Because oxidation and reduction are chemical opposites and constant companions, it is convenient to link their definitions together. We can view oxidation and reduction in at least three different ways (Figure 8.1).

1. **Oxidation** is a gain of oxygen atoms.
   Reduction is a loss of oxygen atoms.

At high temperatures (such as those in automobile engines), nitrogen, which is normally quite unreactive, combines with oxygen to form nitric oxide.

$$N_2 + O_2 \longrightarrow 2\,NO$$

Because nitrogen gains oxygen atoms, it is oxidized. Consider what happens when methane is burned to form carbon dioxide and water.

Figure 8.1 Three different views of oxidation and reduction.

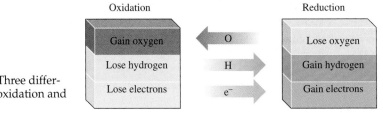

Oxidation		Reduction
Gain oxygen	O	Lose oxygen
Lose hydrogen	H	Gain hydrogen
Lose electrons	e^-	Gain electrons

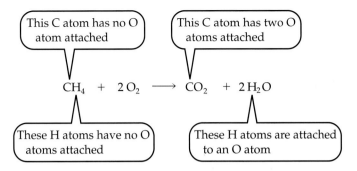

Both carbon and hydrogen gain oxygen atoms, and so both elements are oxidized. When lead dioxide is heated at high temperatures, it decomposes as follows.

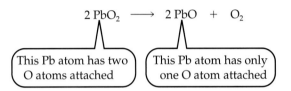

The lead dioxide loses oxygen, and so it is reduced.

Example 8.1

In each of the following reactions, is the reactant undergoing oxidation or reduction? (These are not complete chemical equations.)

a. $Pb \longrightarrow PbO_2$ b. $SnO_2 \longrightarrow SnO$
c. $KClO_3 \longrightarrow KCl$ d. $Cu_2O \longrightarrow 2\,CuO$

Solution

a. Lead gains oxygen atoms (it has none on the left and two on the right); it is oxidized.
b. Tin loses an oxygen atom (it has two on the left and only one on the right); it is reduced.
c. There are three oxygen atoms on the left and none on the right. The compound loses oxygen; it is reduced.
d. The two copper atoms on the left share a single oxygen atom; they have half an oxygen atom each. On the right, each copper atom has an oxygen atom all its own. Cu has gained oxygen; it is oxidized.

Exercise 8.1

In each of the following reactions, is the reactant undergoing oxidation or reduction? (These are not complete chemical equations.)

a. $3\,Fe \longrightarrow Fe_3O_4$ b. $NO \longrightarrow NO_2$
c. $Cr_2O_3 \longrightarrow CrO_3$ d. $C_3H_6O \longrightarrow C_3H_6O_2$

---

2. **Oxidation** is a loss of hydrogen atoms.
   Reduction is a gain of hydrogen atoms.

Look once more at the burning of methane:

$$CH_4 + 2\,O_2 \longrightarrow CO_2 + 2\,H_2O$$

While the carbon and hydrogen gain oxygen atoms, the oxygen gains hydrogen to form water. The oxygen is reduced.

Methyl alcohol (CH_3OH), when passed over hot copper gauze, forms formaldehyde and hydrogen gas.

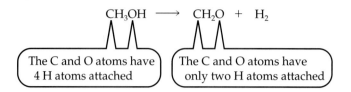

Because the methyl alcohol loses hydrogen, it is oxidized in this reaction.

Methyl alcohol can be made by reaction of carbon monoxide with hydrogen.

$$CO + 2\,H_2 \longrightarrow CH_3OH$$

Because the carbon monoxide gains hydrogen atoms, it is reduced.

Example 8.2

In each of the following reactions, is the reactant undergoing oxidation or reduction? (These are not complete chemical equations.)

a. $C_2H_6O \longrightarrow C_2H_4O$ **b.** $C_2H_2 \longrightarrow C_2H_6$

Solution

a. There are six hydrogen atoms in the compound on the left and only four in the one on the right. The compound loses hydrogen atoms; it is oxidized.

b. There are two hydrogen atoms in the compound on the left and six in the one on the right. The compound gains hydrogen atoms; it is reduced.

Exercise 8.2

In each of the following reactions, is the reactant undergoing oxidation or reduction? (These are not complete chemical equations.)

a. $C_6H_6 \longrightarrow C_6H_{12}$ **b.** $C_3H_6O \longrightarrow C_3H_4O$

---

3. Oxidation is a loss of electrons.
  Reduction is a gain of electrons.

When magnesium metal reacts with chlorine, magnesium ions and chloride ions are formed.

$$Mg + Cl_2 \longrightarrow Mg^{2+} + 2\,Cl^-$$

Because the magnesium atom loses electrons, it is oxidized; and because the chlorine atoms gain electrons, they are reduced.

It is easy to see that when magnesium atoms become Mg^{2+} ions they lose electrons, and that when chlorine atoms become Cl^- ions they must gain electrons. However, many students become confused as to which is oxidation and which is reduction. Perhaps Figure 8.2 can help. The charge on an ion is often referred to as its *oxidation number*. An increase in oxidation number (in positive charge) is oxidation; a decrease in oxidation number is reduction. For magnesium the charge is zero, and so conversion to Mg^{2+} is an increase in oxidation number, which is oxidation. For chlorine the charge is also zero, and so a change to Cl^- is a decrease in oxidation number, which is reduction.

Example 8.3

In each of the following reactions, is the reactant undergoing oxidation or reduction? (These are not complete chemical equations.)

a. $Zn \longrightarrow Zn^{2+}$ **b.** $Fe^{3+} \longrightarrow Fe^{2+}$

c. $S^{2-} \longrightarrow S$ **d.** $AgNO_3 \longrightarrow Ag$

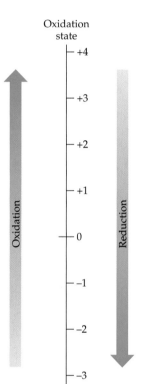

Figure 8.2 An increase in oxidation number means a loss of electrons and is therefore oxidation. A decrease in oxidation number means a gain of electrons and is therefore reduction.

Just remember Leo the lion.

LEO says GER

Loss of
  Electrons is
    Oxidation.

Gain of
  Electrons is
    Reduction.

Solution

a. To form a 2+ ion, zinc loses two electrons. Its oxidation number increases. Zinc is oxidized.
b. To go from a 3+ ion to a 2+ ion, iron gains an electron. Its oxidation number decreases. Iron is reduced.
c. To go from a 2− ion to an atom with no charge, sulfur loses two electrons. Its oxidation number increases. Sulfur is oxidized.
d. To answer this question, you must recognize that $AgNO_3$ is an ionic compound with Ag^+ and NO_3^- ions. In going from Ag^+ to Ag, silver gains an electron. Its oxidation number decreases. Silver is reduced.

Exercise 8.3

In each of the following reactions, is the reactant undergoing oxidation or reduction? (These are not complete chemical equations.)

a. $Cu^{2+} \longrightarrow Cu$
b. $MnO_4^{2-} \longrightarrow MnO_4^-$
c. $Sn^{2+} \longrightarrow Sn^{4+}$
d. $Cu \longrightarrow CuSO_4$

Why do we have so many different ways to look at oxidation and reduction? Oxidation as a gain of oxygen is historical and specific; but the definition in terms of electrons applies more broadly. Which one should we use? Whichever is clearest or most convenient. For the combustion of carbon

$$C + O_2 \longrightarrow CO_2$$

it is most convenient to see that carbon is oxidized by gaining oxygen atoms. Similarly, for the reaction

$$CH_2O + H_2 \longrightarrow CH_4O$$

it is easy to see that the reactant gains hydrogen atoms and is thereby reduced. Finally, in the case of the reaction

$$3\,Sn^{2+} + 2\,Bi^{3+} \longrightarrow 3\,Sn^{4+} + 2\,Bi$$

it is clear that tin is oxidized because it loses electrons, increasing in oxidation number from +2 to +4. Similarly, we see that bismuth is reduced because it gains electrons, going down in oxidation number from +3 to 0.

8.2 Oxidizing and Reducing Agents

Oxidation and reduction go hand in hand. You can't have one without the other. When one substance is oxidized, another is reduced. For example, in the reaction

$$CuO + H_2 \longrightarrow Cu + H_2O$$

copper oxide is reduced and hydrogen is oxidized. Further, if one substance is oxidized, the other must cause it to be oxidized. In the example above, CuO causes H_2 to be oxidized. Therefore, CuO is called the **oxidizing agent**. Conversely, H_2 causes CuO to be reduced, and so H_2 is the **reducing agent**. Each oxidation–reduction reaction has an oxidizing agent and a reducing agent among the reactants. The reducing agent is the substance being oxidized; the oxidizing agent is the substance being reduced.

Another mnemonic is

OIL RIG

Oxidation
    Is
        Loss of electrons.
Reduction
    Is
        Gain of electrons.

[Web Reference 1]
Good sites to expand your knowledge about oxidation–reduction reactions.

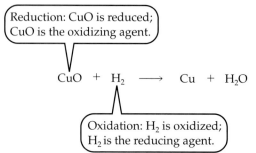

Reduction: CuO is reduced;
CuO is the oxidizing agent.

$$CuO + H_2 \longrightarrow Cu + H_2O$$

Oxidation: H_2 is oxidized;
H_2 is the reducing agent.

OXIDATION is electron DRAIN,
While REDUCTION is electron GAIN;
Forever linked, they must have one another,
For one cannot occur without the other.

OXIDATION
e^- e^-
REDUCTION

Example 8.4

Identify the oxidizing agents and reducing agents in the following reactions.

a. $2 C + O_2 \longrightarrow 2 CO$
b. $N_2 + 3 H_2 \longrightarrow 2 NH_3$
c. $SnO + H_2 \longrightarrow Sn + H_2O$
d. $Mg + Cl_2 \longrightarrow Mg^{2+} + 2 Cl^-$

Solution

We can determine the answers by one of the preceding methods.

a. C gains oxygen and is oxidized, and so it must be the reducing agent. O_2 is therefore the oxidizing agent.
b. N_2 gains hydrogen and is reduced, and so it is the oxidizing agent. H_2 therefore is the reducing agent.
c. SnO loses oxygen and is reduced, and so it is the oxidizing agent. H_2 is therefore the reducing agent.
d. Mg loses electrons and is oxidized, and so it is the reducing agent. Cl_2 is therefore the oxidizing agent.

Exercise 8.4

Identify the oxidizing agents and reducing agents in the following reactions.

a. $Se + O_2 \longrightarrow SeO_2$
b. $CH_3CN + 2 H_2 \longrightarrow CH_3CH_2NH_2$
c. $V_2O_5 + 2 H_2 \longrightarrow V_2O_3 + 2 H_2O$
d. $2 K + Br_2 \longrightarrow 2 K^+ + 2 Br^-$

Figure 8.3 The top photograph shows a blue solution of Cu^{2+} ions and a sample of zinc metal. When the zinc is added to the Cu^{2+} solution, the more active zinc displaces the less active copper from solution. The products of the displacement reaction (bottom) are a red–brown precipitate of copper metal and a colorless solution of Zn^{2+} ions. The equation for the reaction is
$Cu^{2+}(aq) + Zn(s)$
$\longrightarrow Cu(s) + Zn^{2+}(aq)$.

8.3 Electrochemistry: Cells and Batteries

An electric current in a wire is simply a flow of electrons. Oxidation–reduction reactions in which electrons are transferred from one substance to another can be used to produce electricity. This is what happens in dry cell and storage batteries.

When a strip of zinc metal is placed in a solution of copper(II) sulfate, the zinc atoms give up their outer electrons to the copper ions. (We can omit the sulfate ions from the equation because they do not change.) The zinc metal dissolves, going into solution as zinc ions, and the copper ions come out of solution as copper metal (Figure 8.3). The zinc is oxidized; the copper ions are reduced. The zinc atoms give up their outer electrons directly to the copper ions.

If we separate the copper ions from the zinc, placing them in two separate compartments but connecting them with a wire, the electrons must flow through

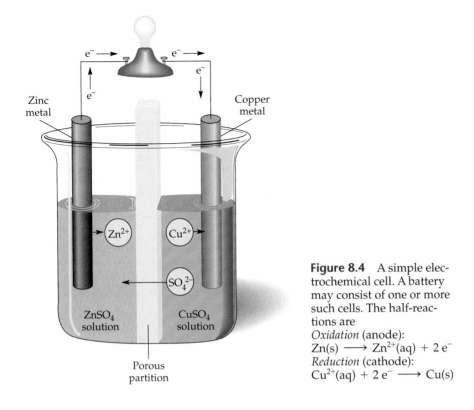

Figure 8.4 A simple electrochemical cell. A battery may consist of one or more such cells. The half-reactions are
Oxidation (anode):
$$Zn(s) \longrightarrow Zn^{2+}(aq) + 2\,e^-$$
Reduction (cathode):
$$Cu^{2+}(aq) + 2\,e^- \longrightarrow Cu(s)$$

the wire in order to get from the zinc to the copper ions. This flow of electrons through the wire constitutes an electric current, and it can be used to run a motor or light a lamp.

In the **electrochemical cell** pictured in Figure 8.4, there are two separate compartments. One contains zinc metal in a colorless solution of zinc sulfate, and the other contains copper metal in a blue solution of copper(II) sulfate. Zinc atoms give up electrons much more readily than copper atoms, and so electrons flow away from the zinc and toward the copper. The zinc metal slowly dissolves as zinc atoms give up electrons to form zinc ions. The electrons flow through the wire to the copper, where copper ions pick them up to become copper atoms. As time goes by, the zinc bar slowly disappears and the copper bar gets bigger. The blue solution gradually loses its color as the Cu^{2+} ions are converted to Cu atoms. Meanwhile, sulfate ions move from the copper sulfate solution to the zinc sulfate solution. More and more positively charged zinc ions are being added to the compartment at the left, whereas fewer and fewer copper ions are left in the compartment at the right. To keep the two solutions electrically neutral, some of the negative sulfate ions must move from the right to the left. Notice that the porous partition in the cell allows the sulfate ions to move through it. (If the sulfate ions were unable to move through the barrier, the cell would not work.) Each time a zinc atom gives up two electrons, a copper ion picks up two electrons, and one sulfate ion moves from the right compartment to the left compartment.

The two pieces of metal act as electrodes. The electrode where oxidation occurs is called the **anode**. The one where reduction occurs is the **cathode**. Because zinc gives up electrons, it is oxidized, and the zinc strip is therefore the anode. Copper ions gain electrons, and so they are reduced, and the copper strip is the cathode.

Electrochemical reactions are often represented as two *half-reactions*. The following representation shows the two half-reactions for the zinc–copper cell and how they are added to give the overall cell reaction.

We saw in Chapter 3 that electricity can produce chemical change, a process called *electrolysis*. For example, molten sodium chloride can be changed to sodium metal and chlorine gas by passing an electric current through it. Here we see the reverse process; chemical change can be used to produce electricity.

Note that *a*node and *o*xidation both begin with vowels, whereas *c*athode and *r*eduction begin with consonants.

Note that in an electro-chemical cell, anions (negative ions, SO_4^{2-} in this case) in solution move toward the anode, and cations (positive ions, Cu^{2+} in this case) move toward the cathode. To understand the cell, it is important to distinguish between what moves through the wire (electrons) and what moves through the solution (ions).

[Web Reference 2] An electrochemistry tutorial.

Oxidation: $Zn(s) \longrightarrow Zn^{2+}(aq) + 2e^-$

Reduction: $Cu^{2+}(aq) + 2e^- \longrightarrow Cu(s)$

Overall reaction: $Zn(s) + Cu^{2+}(aq) \longrightarrow Zn^{2+}(aq) + Cu(s)$

Note that the electrons cancel when the two half-reactions are added.

Example 8.5

Represent the following reaction as two half-reactions and label them as an oxidation half-reaction and a reduction half-reaction.

$$Mg + Cl_2 \longrightarrow Mg^{2+} + 2\,Cl^-$$

Solution
Magnesium is oxidized from Mg to Mg^{2+}, a process that involves loss of two electrons from the Mg atom. The oxidation half-reaction is therefore

$$Mg \longrightarrow Mg^{2+} + 2\,e^-$$

The reduction half-reaction must therefore involve chlorine. Two Cl atoms in the Cl_2 molecule must each gain an electron to form a Cl^- ion. The reduction half-reaction is therefore

$$Cl_2 + 2\,e^- \longrightarrow 2\,Cl^-$$

Example 8.6

Balance the following half-reactions and combine them to give a balanced overall reaction.

$$Sn^{2+} \longrightarrow Sn^{4+}$$

$$Bi^{3+} \longrightarrow Bi$$

Solution
Both are balanced as far as atoms are concerned, but they must also be balanced in electric charge. To do this requires two electrons on the right side in the first half-reaction.

$$Sn^{2+} \longrightarrow Sn^{4+} + 2\,e^-$$

We must add three electrons on the left side in the second half-reaction.

$$Bi^{3+} + 3\,e^- \longrightarrow Bi$$

Before we can combine the two, however, we must set electron loss equal to electron gain. (Electrons lost by the substance being oxidized must be gained by the substance being reduced.) To do this, we multiply the first half-reaction by 3 and the second by 2.

$$3 \times (Sn^{2+} \longrightarrow Sn^{4+} + 2\,e^-) = 3\,Sn^{2+} \longrightarrow 3\,Sn^{4+} + 6e^-$$
$$2 \times (Bi^{3+} + 3\,e^- \longrightarrow Bi) = 2\,Bi^{3+} + 6e^- \longrightarrow 2\,Bi$$

$$\overline{\quad 3\,Sn^{2+} + 2\,Bi^{3+} \longrightarrow 3\,Sn^{4+} + 2\,Bi \quad}$$

Note that both atoms and charges balance in the overall reaction.

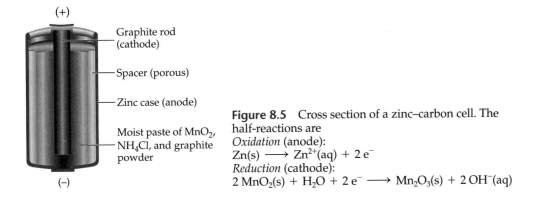

(+)

Graphite rod (cathode)

Spacer (porous)

Zinc case (anode)

Moist paste of MnO_2, NH_4Cl, and graphite powder

(−)

Figure 8.5   Cross section of a zinc–carbon cell. The half-reactions are
Oxidation (anode):
$$Zn(s) \longrightarrow Zn^{2+}(aq) + 2\ e^-$$
Reduction (cathode):
$$2\ MnO_2(s) + H_2O + 2\ e^- \longrightarrow Mn_2O_3(s) + 2\ OH^-(aq)$$

Dry Cells

The familiar dry cell (Figure 8.5) is used in flashlights and many other small portable devices. It has a zinc anode—in this case, the container itself. A carbon rod in the center of the cell is the cathode. The space between the cathode and the anode contains a moist paste of graphite powder (carbon), manganese dioxide (MnO_2), and ammonium chloride (NH_4Cl). The anode reaction is the oxidation of the zinc cylinder to zinc ions. The cathode reaction involves reduction of manganese dioxide. A simplified version of the overall reaction is

$$Zn + 2\ MnO_2 + H_2O \longrightarrow Zn^{2+} + Mn_2O_3 + 2\ OH^-$$

Alkaline cells are similar, but the zinc case is porous and the paste around the carbon cathode is moist manganese dioxide and potassium hydroxide. They are more expensive than ordinary zinc–carbon cells, but they maintain a high voltage longer.

Lead Storage Batteries

Although we often refer to dry cells as batteries, a **battery** is actually a series of electrochemical cells. The 12-volt (V) storage battery used in automobiles, for example, is a series of six 2-V cells. Each cell contains a pair of electrodes, one lead and the other lead dioxide, in a chamber filled with sulfuric acid. A distinctive feature of the lead storage battery is its capacity for being recharged. It discharges as it supplies electricity when you turn on the ignition to start a car or when the motor is off and the lights are on. But it is recharged when the car is moving and an electric current is supplied to the battery by the mechanical action of the car. The net reaction during discharge is

$$Pb + PbO_2 + 2\ H_2SO_4 \longrightarrow 2\ PbSO_4 + 2\ H_2O$$

The reaction during recharge is just the reverse.

$$2\ PbSO_4 + 2\ H_2O \longrightarrow Pb + PbO_2 + 2\ H_2SO_4$$

Lead storage batteries are durable, but they are heavy and contain corrosive sulfuric acid.

Other Batteries

Much of current battery technology involves the use of lithium, which has an extraordinarily low density as well as a fairly high voltage. Lithium–SO_2 cells are used in submarines and rocket vehicles (such as the Jupiter probe); a lithium–iodine cell is used in pacemakers; and lithium–FeS_2 batteries (such as the Energizer battery) are used in cameras, radios, and compact disc players.

One cell of a lead–acid battery has two anode plates and two cathode plates. Each cell produces about two volts, and six such cells make up the common 12-V car battery. Lead–acid batteries have been known for 150 years. They are used in power tools and certain appliances as well as in cars. Three hundred million of them are made per year. The half-reactions are

Oxidation (anode):
$$Pb(s) + SO_4^{2-}(aq) \longrightarrow PbSO_4(s) + 2\ e^-$$
Reduction (cathode):
$$PbO_2(s) + 4\ H^+(aq) + SO_4^{2-}(aq) + 2\ e^- \longrightarrow PbSO_4(s) + 2\ H_2O$$

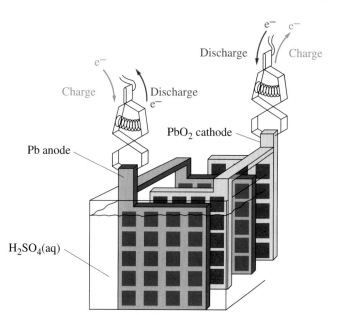

 [Web Reference 3] A historical time-line of developments in electrochemistry.

The rechargeable Ni–Cad cell (Cd anode, NiO cathode) has long been popular for portable radios and cordless appliances. Its biggest competitor is the nickel–metal hydride cell, which replaces cadmium with a hydrogen-absorbing nickel alloy such as $ZrNi_2$ or $LaNi_5$.

The small "button" cells used in hearing aids and hand calculators formerly contained mercury (zinc anode, HgO cathode), but they are gradually being phased out and replaced with zinc–air cells. Tiny silver oxide cells (zinc anode, Ag_2O cathode) are used mainly in watches and cameras.

Fuel Cells

An interesting kind of battery is the fuel cell. When fossil fuels, our major energy source, are burned to generate electricity, only about 35–40% of their energy of combustion is actually harnessed. In a **fuel cell**, the fuel is oxidized at the anode and oxygen is reduced at the cathode with 70–75% efficiency. Most fuel cells use hydrogen as fuel with platinum, nickel, or rhodium electrodes, in a solution of potassium hydroxide. Some systems produce hydrogen from methane or propane gas. Fuel cells have long been used in spacecraft, but they are now being used in automobiles and other applications. For example, a fuel cell supplies electricity to the police station in Central Park in New York City. Active research continues to develop fuel cells for many other energy needs.

Somewhat like a fuel cell is the metal–air battery. Typically, zinc or aluminum is used as the anode, which acts as a solid fuel, reacting with oxygen and water to form the metal hydroxide.

$$2\ Zn + O_2 + 2\ H_2O \longrightarrow 2\ Zn(OH)_2$$

Recharging the battery requires the addition of more metal and water. Batteries of this type have been used to power experimental electric cars.

A hydrogen–oxygen fuel cell has porous, inert electrodes to allow the reactant gases to reach the electrolyte. The electrodes also catalyze the reaction. The half-reactions are

Oxidation (anode):
$$H_2(g) + 2\ OH^-(aq) \longrightarrow 2\ H_2O + 2\ e^-$$
Reduction (cathode):
$$O_2(g) + 2\ H_2O + 4\ e^- \longrightarrow 4\ OH^-(aq)$$

8.4 Corrosion

An oxidation reaction of particular economic importance is the corrosion of metals. It is estimated that in the United States alone corrosion costs $100 billion a year. Perhaps 20% of all the iron and steel production in the United States each year goes to replace corroded items. Let's look first at the corrosion of iron.

The Rusting of Iron

In moist air, iron is oxidized, particularly at a nick or scratch.

$$Fe \longrightarrow Fe^{2+} + 2\,e^-$$

As iron is oxidized, oxygen is reduced.

$$O_2 + 2\,H_2O + 4\,e^- \longrightarrow 4\,OH^-$$

The net result, initially, is the formation of insoluble iron(II) hydroxide.

$$2\,Fe + O_2 + 2\,H_2O \longrightarrow 2\,Fe(OH)_2$$

This product is usually further oxidized to iron(III) hydroxide.

$$4\,Fe(OH)_2 + O_2 + 2\,H_2O \longrightarrow 4\,Fe(OH)_3$$

Iron(III) hydroxide [$Fe(OH)_3$], sometimes written as $Fe_2O_3 \cdot 3\,H_2O$, is the familiar iron rust.

Oxidation and reduction often occur at separate points on the metal's surface. Electrons are transferred through the iron metal. The circuit is completed by an electrolyte in aqueous solution. In the Snow Belt, this solution is often the slush from road salt and melting snow. The metal is pitted in an anodic area, where iron is oxidized to Fe^{2+}. These ions migrate to the cathodic area, where they react with hydroxide ions formed by the reduction of oxygen.

$$Fe^{2+} + 2\,OH^- \longrightarrow Fe(OH)_2$$

As indicated above, this iron(II) hydroxide is then oxidized to $Fe(OH)_3$, or rust. This process is diagrammed in Figure 8.6. Notice that the anodic area is protected from oxygen by a water film, whereas the cathodic area is exposed to air.

Protection of Aluminum

Aluminum is more reactive than iron, and yet corrosion is not a serious problem with aluminum. We use aluminum foil, we cook with aluminum pots, and we buy beverages in aluminum cans. Even after many years, they have not corroded. How is this possible?

The aluminum surface reacts with oxygen in the air to form a thin layer of oxide. However, instead of being porous and flaky like iron oxide, aluminum oxide is hard and tough, protecting the metal from further oxidation.

We should add, however, that corrosion can sometimes be a problem with aluminum. Certain substances, such as salt, can interfere with the protective oxide coating on aluminum, allowing the metal to oxidize. This problem has caused mag wheels on automobiles to crack and wheels to shear off planes with aluminum landing gear.

Silver Tarnish

The tarnish on silver results from oxidation of the silver surface by hydrogen sulfide (H_2S) in the air. It produces a film of black silver sulfide (Ag_2S) on the metal surface. You can use silver polish to remove the tarnish, but in doing so, you also lose part

Allis-Chalmers built experimental tractors powered by fuel cells as early as the 1950s. Fuel cells are discussed in more detail in Chapter 14.

[Web Reference 4] www An excellent review of fuel cells, with many links.

The formula $Fe_2O_3 \cdot 3\,H_2O$ is the same as $2\,Fe(OH)_3$.

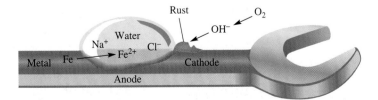

Figure 8.6 The corrosion of iron requires water, oxygen, and an electrolyte.

The Murrah Building in Oklahoma City following the bombing in 1995.

of the silver. An alternative method involves the use of aluminum metal to reduce the silver ions back to silver metal.

$$3\,Ag^+ + Al \longrightarrow 3\,Ag + Al^{3+}$$

This reaction also requires an electrolyte, and sodium bicarbonate ($NaHCO_3$) is usually used. The tarnished silver is placed in contact with aluminum foil and covered with a solution of sodium bicarbonate. A precious metal is conserved at the expense of a cheaper one.

8.5 Explosive Reactions

Chemical explosions (that is, those not based on nuclear fission or fusion) are usually the result of oxidation–reduction reactions. Explosions happen when a chemical reaction occurs rapidly and with a considerable increase in volume. They often involve compounds of nitrogen, such as nitroglycerine (the active ingredient in dynamite), ammonium nitrate (a common fertilizer), or trinitrotoluene (TNT). These compounds can decompose readily, yielding nitrogen gas as one of the products. Explosive mixtures are used for earthmoving projects and the demolition of buildings, as well as in making bombs. In 1995, for example, a truck bomb destroyed the Alfred P. Murrah Building in Oklahoma City, killing 168 people. The truck was loaded with two tons of ammonium nitrate (NH_4NO_3) mixed with fuel oil. This mixture, sometimes called ANFO (ammonium nitrate–fuel oil), is an explosive with enormous destructive power. The ammonium nitrate is both oxidized and reduced. Ammonium ion is the reducing agent, whereas nitrate ion is the oxidizing agent. The fuel oil provides additional oxidizable material. Using $C_{17}H_{36}$ as the formula for a typical molecule in fuel oil, we can write the equation for the explosive reaction as follows.

$$52\,NH_4NO_3(s) + C_{17}H_{36}(l) \longrightarrow 52\,N_2(g) + 17\,CO_2(g) + 122\,H_2O(g)$$

Notice that this reaction includes 52 mol of starting material, most of it solid and the rest in the liquid state, and 191 mol of products, all gases. A reaction of solid and/or liquid reactants that generates gaseous products involves a huge volume increase and a possibility of explosion. In this case, the reaction is rapid and the volume increase is enormous.

Ammonium nitrate is a widely used fertilizer, and fuel oil is used in many home furnaces. The fact that both these materials are so accessible to potential terrorists is cause for serious concern.

Cooking, burning of fuel, and breathing all involve oxidation.

8.6 Oxygen: An Abundant and Essential Oxidizing Agent

Oxygen itself is the most common oxidizing agent. Making up one-fifth of the air, it oxidizes the wood in our campfires and the gasoline in our automobiles. It even "burns" the food we eat to give us the energy to move and to think.

Surely one of the most important elements on Earth, oxygen is one of about two dozen elements essential to life. It is found in most compounds that are important to living organisms. Foodstuffs—carbohydrates, fats, and proteins—all contain oxygen. The human body is approximately 65% water (by mass). Because water is 89% oxygen by mass and many other compounds in your body also contain oxygen, almost two-thirds of your body mass is oxygen.

Oxygen comprises about half of the accessible portion of Earth by mass. It occurs in all three subdivisions of the outermost structure of Earth: Oxygen occurs as O_2 molecules in the atmosphere (the gaseous mass surrounding Earth). In the hydrosphere (the oceans, seas, rivers, and lakes), oxygen is combined with hydrogen in water. In the lithosphere (the outer solid portion), oxygen is combined with silicon (sand is SiO_2) and other elements.

The atmosphere is about 21% elemental oxygen (O_2) by volume. [The rest is mainly nitrogen (N_2), which is rather unreactive.] The free, uncombined oxygen in the air is taken into our lungs, passes into our bloodstreams, is carried to our body tissues, and reacts with the food we eat. This is the process that provides us with all our energy. Fuels such as natural gas, gasoline, and coal also need oxygen to burn and release their stored energy. Oxidation, or combustion, of these fossil fuels currently supplies about 86% of the energy that turns the wheels of civilization.

Pure oxygen is obtained by liquefying air and then letting the nitrogen and argon boil off. Nitrogen boils at −196°C, argon at −186°C, and oxygen at −183°C. More than 20 billion kg of oxygen is produced annually in the United States, but most of it is used directly by industry, much of it by steel plants. About 1% is compressed into tanks for use in welding, hospital respirators, and other purposes.

Not everything that oxygen does is desirable. In addition to causing corrosion (Section 8.4), it promotes food spoilage and wood decay.

Fuels burn more rapidly in pure oxygen than in air. Huge quantities of liquid oxygen (tank at the lower left) are used to burn the fuels that blast rockets into orbit.

The structure and chemistry of Earth are discussed in Chapter 11, the atmosphere in Chapter 12, and water in Chapter 13.

Physical activity under conditions in which plenty of oxygen is available is called *aerobic* exercise. When the available oxygen is insufficient, the exercise is called *anaerobic*, and it often leads to weakness and pain resulting from "oxygen debt." (See Chapter 18.)

Reactions with Other Elements

Many oxidation reactions involving atmospheric oxygen are quite complicated, but we can gain some understanding by looking at some of the simpler chemical reactions of oxygen. For example, as we have seen, oxygen combines with many metals to form metal oxides, and with nonmetals to form nonmetal oxides.

Example 8.7

Magnesium combines readily with oxygen when ignited in air. Write the equation for this reaction.

Solution

Magnesium is a Group 2A metal. Oxygen occurs as diatomic molecules (O_2). The two react to form MgO (Section 6.2). The reaction is

$$Mg(s) + O_2(g) \longrightarrow MgO(s) \ (\textit{not balanced})$$

To balance the equation, we need 2 MgO and 2 Mg.

$$2\,Mg(s) + O_2(g) \longrightarrow 2\,MgO(s)$$

Exercise 8.7

A. Zinc burns in air to form zinc oxide (ZnO). Write the equation for the reaction.
B. Selenium (Se) burns in air to form selenium dioxide. Write the equation for the reaction.

Reactions with Compounds

Oxygen reacts with many compounds, oxidizing one or more of the elements in the compound. As we have seen, combustion of a fuel is oxidation and produces oxides. There are many other examples. Hydrogen sulfide, a gaseous compound with a rotten-egg odor, burns, producing oxides of hydrogen (water) and of sulfur (sulfur dioxide).

$$2\,H_2S + 3\,O_2 \longrightarrow 2\,H_2O + 2\,SO_2$$

Example 8.8

Carbon disulfide is highly flammable; it combines readily with oxygen, burning with a blue flame. What products are formed? Write the equation.

Solution

Carbon disulfide is CS_2. The products are oxides of carbon (carbon dioxide) and of sulfur (sulfur dioxide). The balanced equation is

$$CS_2 + 3\,O_2 \longrightarrow CO_2 + 2\,SO_2$$

Exercise 8.8

When heated in air, lead sulfide (PbS) combines with oxygen to form lead(II) oxide (PbO) and sulfur dioxide. Write a balanced equation for the reaction.

Ozone

In addition to diatomic O_2, the normal form of oxygen, the element also has a triatomic form, O_3, called *ozone*. Ozone is a powerful oxidizing agent and a harmful air pollutant. It can be extremely irritating to both plants and animals and is especially destructive to rubber. On the other hand, a layer of ozone in the upper stratosphere serves as a shield that protects life on Earth against ultraviolet radiation from the sun (Chapter 12). Ozone is an excellent example of the fact that the same substance can be extremely beneficial or quite harmful, depending on where it happens to be.

8.7 Other Common Oxidizing Agents

Many oxidizing agents are important in the laboratory, in industry, and in the home. They are used as antiseptics, disinfectants, and bleaches and play a role in many chemical syntheses.

Hydrogen peroxide (H_2O_2) is a common oxidizing agent that has the advantage of being converted to water in most reactions. Pure hydrogen peroxide is a syrupy liquid. It is available (in the laboratory) as a dangerous 30% solution that has powerful oxidizing power, or as a 3% solution sold in stores for various uses around the home.

An oxidizing agent often used in the laboratory is potassium dichromate ($K_2Cr_2O_7$). It is an orange material that turns green when it is reduced to chromium(III) compounds. One of the compounds that potassium dichromate oxidizes is ethyl alcohol. The Breathalyzer test for intoxication makes use of a dichromate solution. The exhaled breath of the person being tested mixes with the acidic dichromate, and the degree of color change indicates the level of alcohol.

Many antiseptics are mild oxidizing agents. (Antiseptics are compounds applied to living tissue to kill microorganisms or prevent their growth.) For example, a 3% solution of hydrogen peroxide is often used to treat minor cuts, and tincture of iodine has long been a household antiseptic. Ointments for treating acne often contain 5–10% benzoyl peroxide, a powerful antiseptic and also a skin irritant. It causes old skin to slough off and be replaced by new, fresher-looking skin. When used on areas exposed to sunlight, however, benzoyl peroxide may promote skin cancer.

A *tincture* is a solution made up in alcohol.

Oxidizing agents are also used as disinfectants. A good example is chlorine, which is used to kill disease-causing microorganisms in drinking water. Swimming pools are often chlorinated with calcium hypochlorite [Ca(OCl)$_2$]. Because calcium hypochlorite is alkaline, it also raises the pH of the water. (When a pool becomes too alkaline, the pH is lowered by adding hydrochloric acid. Swimming pools are usually maintained at pH 7.2–7.8.)

Bleaches are oxidizing agents, too. A **bleach** removes unwanted color from fabrics or other material. Nearly any oxidizing agent could do the job, but some might be unsafe, or harmful to fabrics, or perhaps too expensive. Laundry bleaches are usually sodium hypochlorite (NaOCl) as an aqueous solution (in products such as Purex and Clorox) or calcium hypochlorite [Ca(OCl)$_2$], known as bleaching powder. The powder is usually preferred for large industrial operations, such as the whitening of paper or fabrics. Nonchlorine bleaches often contain sodium perborate (a combination of NaBO$_2$ and H$_2$O$_2$).

For lightening hair color, bleaches are usually 6 or 12% solutions of hydrogen peroxide, which oxidizes the dark pigment (melanin) in the hair to colorless products. Hydrogen peroxide can also be used to lighten certain paints by oxidizing sulfides (S^{2-}) to sulfates (SO$_4^{2-}$). When lead-based paints are exposed to air containing hydrogen sulfide (H$_2$S), they turn black because of the formation of lead sulfide (PbS). Hydrogen peroxide oxidizes the black sulfide to white sulfate.

$$PbS + 4 H_2O_2 \longrightarrow PbSO_4 + 4 H_2O$$
$$\text{Black} \qquad\qquad\qquad \text{White}$$

Stain removal is more complicated than bleaching. A few stain removers are oxidizing agents, but some are reducing agents, some are solvents or detergents, and some have quite different action. Some stains require rather specific stain removers.

8.8 Some Reducing Agents of Interest

In every reaction involving oxidation, the oxidizing agent is reduced, and the substance undergoing oxidation acts as a reducing agent. Let's now consider reactions in which the purpose of the reaction is reduction.

Most metals occur in nature as compounds. In order to prepare the free metals, the compounds must be reduced. Metals are often freed from their ores with coal or coke (elemental carbon obtained by heating coal to drive off volatile matter). Tin oxide is one of the many ores that can be reduced with coal or coke.

$$SnO_2 + C \longrightarrow Sn + CO_2$$

Sometimes a metal can be obtained by heating its ore with a more active metal. Chromium oxide, for example, can be reduced by heating it with aluminum.

$$Cr_2O_3 + 2 Al \longrightarrow Al_2O_3 + 2 Cr$$

Reduction in Photography

Perhaps a more familiar reducing agent, by use if not by name, is the developer used in black-and-white photography. Photographic film is coated with a silver salt (Ag$^+$Br$^-$). Silver ions that have been exposed to light react with the developer, a reducing agent (such as the organic compound hydroquinone), to form metallic silver.

$$C_6H_4(OH)_2 + 2 Ag^+ \longrightarrow C_6H_4O_2 + 2 Ag + 2 H^+$$
$$\text{Hydroquinone} \qquad\qquad\qquad \text{Silver}$$
$$\text{metal}$$

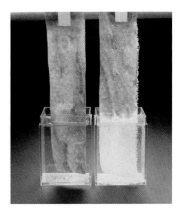

Pure water (left) has little effect on a dried tomato sauce stain. Sodium hypochlorite bleach [NaOCl(aq)] (right) effectively removes the stain by oxidizing the colored tomato pigments to colorless products.

Solutions of potassium permanganate (KMnO$_4$), a common laboratory oxidizing agent, are good disinfectants and bleaches. These solutions are a deep purplish pink. Even highly diluted solutions are a pale pink. (Used to disinfect foods in developing countries, these dilute solutions are called "pinky water" in India.) The use of KMnO$_4$ solutions is limited by the fact that the permanganate ion is reduced to brown manganese dioxide, leaving a brown stain on the skin or other surface being treated.

Hydrogen peroxide can be used to restore the once-white areas of old paintings that have darkened from the reaction of white lead compounds (paint pigments) with sulfur compounds. The darkened pigments (black PbS) are converted to white PbSO$_4$ by the hydrogen peroxide.

Figure 8.7   A photographic negative (left) and a positive print.

The developer does not reduce silver ions not exposed to light. The film is then treated with "hypo," a solution of sodium thiosulfate ($Na_2S_2O_3$), which washes out unexposed silver bromide to form the negative. This leaves the negative dark where the metallic silver was deposited (where it was originally exposed to light) and transparent where light did not strike it. Light is then shone through the negative onto light-sensitive paper to make the positive print. Figure 8.7 shows positive and negative prints.

Antioxidants

In food chemistry, certain reducing agents are called **antioxidants**. Ascorbic acid (vitamin C) can prevent the browning of fruit (such as sliced apples or pears) by inhibiting air oxidation. Whereas vitamin C is water-soluble, tocopherol (vitamin E) and beta-carotene (vitamin A precursor) are fat-soluble antioxidants. All these vitamins are believed to retard various oxidation reactions that are potentially damaging to vital components of living cells (Chapter 16).

Some antioxidants seem to act as anticarcinogens as well.

Hydrogen as a Reducing Agent

Hydrogen is an excellent reducing agent that can free many metals from their ores, but it is generally used to produce more expensive metals, such as tungsten (W).

$$WO_3 + 3\,H_2 \longrightarrow W + 3\,H_2O$$

Hydrogen can be used to reduce many kinds of chemical compounds. Ethylene, for example, can be reduced to ethane.

$$C_2H_4 + H_2 \longrightarrow C_2H_6$$

(Nickel is used as a *catalyst* in this reaction.) Hydrogen also reduces nitrogen, from the air, in the industrial production of ammonia.

$$N_2 + 3\,H_2 \longrightarrow 2\,NH_3$$

(An iron catalyst is used in this case.)

A **catalyst** is a substance that increases the rate of a chemical reaction without itself being used up. For example, unsaturated fats ordinarily react very slowly with hydrogen, but in the presence of nickel metal, the reaction proceeds readily. The nickel increases the rate at which the reaction takes place, but it does not increase the amount of product formed.

A stream of pure hydrogen burns quietly in air with an almost colorless flame; but when a mixture of hydrogen and oxygen is ignited by a spark or a flame, an explosion results. The product in both cases is water.

$$2\,H_2 + O_2 \longrightarrow 2\,H_2O$$

Hydrogen has such a strong attraction for oxygen that it can remove oxygen atoms from many metal oxides to yield the free metal. For example, when hydrogen is passed over heated copper oxide, metallic copper and water are formed.

$$CuO + H_2 \longrightarrow Cu + H_2O$$

Certain metals, such as platinum and palladium, have an unusual affinity for hydrogen, being able to absorb large volumes of the gas. Palladium can absorb up to 900 times its own volume of hydrogen. It is interesting to note that hydrogen and oxygen can be mixed at room temperature with no perceptible reaction. But if a piece of platinum gauze is added, the gases react violently at room temperature. The platinum is acting as a catalyst. It lowers the **activation energy** for the reaction. The heat from the initial reaction heats up the platinum, making it glow; it then ignites the hydrogen–oxygen mixture, causing an explosion.

Nickel, platinum, and palladium are often used as catalysts for reactions involving hydrogen. These metals have the greatest catalytic activity when they are finely divided and have lots of active surface area. Hydrogen adsorbed on the surface of these metals is more reactive than ordinary hydrogen gas.

> The **activation energy** for a chemical reaction is the minimum energy needed to get the reaction started. A catalyst acts by lowering the energy required to start a reaction.

8.9 A Closer Look at Hydrogen

Hydrogen is an especially vital element because it and oxygen are the components of water. By mass, hydrogen makes up only about 0.9% of Earth's crust. However, because of its low atomic mass, it ranks rather high in abundance by number of atoms. If we had a random sample of 10,000 atoms from the outer, accessible portion of Earth, more than half (5330) would be oxygen, 1590 would be silicon, and 1510 would be hydrogen. Unlike oxygen, hydrogen is seldom found as a free, uncombined element on Earth. Most of it is combined with oxygen in water. Some is combined with carbon in petroleum and natural gas, which are mixtures of hydrocarbons. Nearly all compounds derived from plants and animals contain combined hydrogen.

Because hydrogen has the lowest atomic number (1) and the smallest atoms, it is the first element in the periodic table. However, it is difficult to know exactly where to place it. Hydrogen is usually shown as the first member of Group 1A because it has the same valence electronic structure (ns^1) as alkali metals. Like the atoms of elements in Group 1A, a hydrogen atom can lose an electron to form a 1+ ion, but unlike the other Group 1A elements, hydrogen is not a metal.[1] Like the atoms of Group 7A elements, a hydrogen atom can pick up an electron to become a 1– ion, but unlike the other Group 7A elements, hydrogen is not a halogen. We place hydrogen in Group 1A in the periodic table on the inside front cover, but you should keep in mind that it is a unique element, in a class by itself.

Small amounts of elemental hydrogen can be made for laboratory use by reacting zinc with hydrochloric acid.

$$Zn(s) + 2\,HCl(aq) \longrightarrow ZnCl_2(aq) + H_2(g)$$

Because hydrogen does not dissolve in water, it can be collected by water displacement (Figure 8.8). Commercial quantities of hydrogen are obtained as by-products of petroleum refining or by the reaction of natural gas with steam. About 200 million kg of hydrogen is produced each year in the United States, at least two-thirds of it being used to make ammonia.

Hydrogen is a colorless, odorless gas and the lightest of all substances. Its density is only one-fourteenth that of air, and for this reason it was once used to fill lighter-than-air craft (Figure 8.9). Unfortunately, hydrogen can be ignited by a spark, which is what occurred in 1937 when the German airship *Hindenburg* was destroyed in a disastrous fire and explosion as it was landing in Lakehurst, New Jersey. The use of hydrogen in airships was discontinued after that, and the dirigible industry never recovered. Today the few airships that are still in service are filled with nonflammable helium, but they are used mainly in advertising.

> Hydrogen ranks low in abundance by mass on Earth. If we look beyond our home planet, however, hydrogen becomes much more significant. The sun, for example, is made up largely of hydrogen. The planet Jupiter is also mainly hydrogen. In fact, hydrogen is by far the most abundant element in the universe.

> Hot air balloons, as their name implies, are buoyed by hot air, which is less dense than the ambient air.

[1]When hydrogen is subjected to extreme pressures (greater than 1 million atm), it becomes an electric conductor, suggesting that it can become metallic.

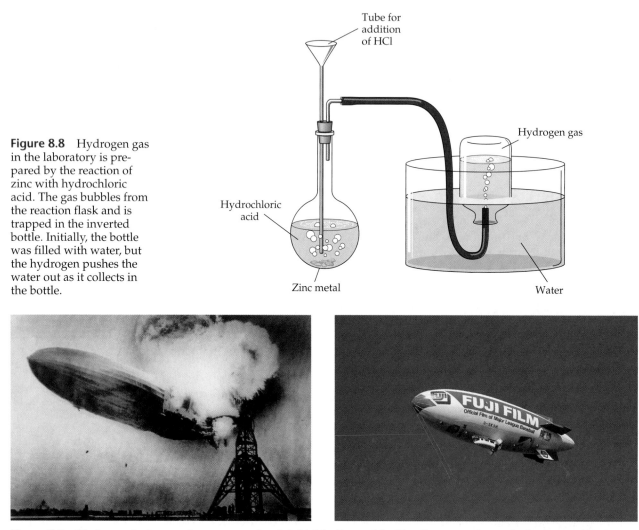

Figure 8.8 Hydrogen gas in the laboratory is prepared by the reaction of zinc with hydrochloric acid. The gas bubbles from the reaction flask and is trapped in the inverted bottle. Initially, the bottle was filled with water, but the hydrogen pushes the water out as it collects in the bottle.

Tube for addition of HCl

Hydrogen gas

Hydrochloric acid

Zinc metal

Water

Figure 8.9 Hydrogen is the most buoyant gas, but it is highly flammable. The disastrous fire in the hydrogen-filled German zeppelin *Hindenburg* led to the replacement of hydrogen by nonflammable helium, which buoys the Fuji blimp.

8.10 Oxidation, Reduction, and Living Things

Perhaps the most important oxidation–reduction processes are the ones that maintain life on this planet. We obtain energy for all our physical and mental activities by metabolizing food through respiration. The process has many steps, but eventually the food we eat is converted mainly into carbon dioxide, water, and energy.

Bread and many of the other foods we eat are largely made up of carbohydrates (Chapter 15). If we represent carbohydrates with the simple example glucose ($C_6H_{12}O_6$), we can write the overall equation for their metabolism as follows.

$$C_6H_{12}O_6 + 6\,O_2 \longrightarrow 6\,CO_2 + 6\,H_2O + \text{energy}$$

This process constantly occurs in animals, including humans. The carbohydrate is oxidized in the process.

Meanwhile, plants need carbon dioxide and water, from which they produce carbohydrates. The energy needed comes from the sun, and the process is called **photosynthesis**. The chemical equation is

Photosynthesis occurs in green plants. The chlorophyll pigments that catalyze the photosynthesis process give the green color to much of the land area of Earth.

$$6\,CO_2 + 6\,H_2O + energy \longrightarrow C_6H_{12}O_6 + 6\,O_2$$

Notice that this process in plant cells is exactly the reverse of the process going on inside animals. In food metabolism in animals we focus on an oxidation process. In photosynthesis we focus on a reduction process.

The carbohydrates produced by photosynthesis are the ultimate source of all our food because fish, fowl, and other animals either eat plants or eat other animals that eat plants. Note that the photosynthesis process not only makes carbohydrates but also yields free elementary oxygen (O_2). In other words, photosynthesis does not just provide all the food we eat, it also provides all the oxygen we breathe.

There are many oxidation reactions that occur in nature (with oxygen being reduced in the process). The net photosynthesis reaction is unique in that it is a natural reduction of carbon dioxide (with oxygen being oxidized). Many reactions in nature use oxygen. Photosynthesis is the only natural process that produces it.

Critical Thinking Exercises

Apply knowledge that you have gained in this chapter and one or more of the FLaReS principles (Chapter 1) to evaluate the following statements or claims.

8.1 For years some people have claimed that someone has invented an automobile engine that burns water instead of gasoline. They say that we have not heard about this wonderful invention because the oil companies have bought the rights to the engine from the inventor so that they can keep it from the public and people will continue to burn gasoline in their cars. Do you find this claim believable?

8.2 A friend tells you that if you take a sugar cube and hold it in a flame, it will melt but will not burn; but he claims that if you dip the sugar cube into cigarette ashes before you place it in the flame, it will burn. He explains that cigarette ashes contain something that is a catalyst for the burning of sugar. Do you think your friend's claim is valid?

Summary

1. When a substance gains oxygen atoms or loses hydrogen atoms, it is oxidized.
2. When a substance gains hydrogen atoms or loses oxygen atoms, it is reduced.
3. Oxidation is a loss of electrons; reduction is a gain of electrons.
4. An increase in oxidation number is oxidation; a decrease in oxidation number is reduction.
5. Whenever oxidation occurs, an equivalent amount of reduction must occur simultaneously.
6. The substance oxidized is the reducing agent; the substance reduced is the oxidizing agent.
7. In an electrochemical cell, oxidation occurs at the anode and reduction at the cathode.
8. An electrochemical cell generates an electric current by having oxidation and reduction occur in two different locations, so that electrons must flow through a wire to get from one to the other.
9. Metal corrosion is an oxidation process, the most familiar example being the rusting of iron.
10. Laundry bleaches and household antiseptics are common oxidizing agents; photographic developers and antioxidants in food are common reducing agents.
11. The reduction of carbon dioxide in photosynthesis is the most important reduction process on Earth. We could not exist without it.

Key Terms

activation energy 8.8
anode 8.3
antioxidant 8.8
battery 8.3
bleach 8.7
catalyst 8.8
cathode 8.3
electrochemical cell 8.3
fuel cell 8.3
oxidation 8.1
oxidizing agent 8.2
photosynthesis 8.10
reducing agent 8.2
reduction 8.1

Review Questions

1. Explain how we view oxidation and reduction in terms of the following.
   a. oxygen atoms gained or lost
   b. hydrogen atoms gained or lost
   c. electrons gained or lost
2. What is an electrochemical cell? An electrochemical battery?
3. What is the purpose of a porous plate between the two electrode compartments in an electrochemical cell?
4. What is a half-reaction? How are half-reactions combined to give an overall redox reaction?
5. How does an alkaline cell differ from a regular carbon–zinc dry cell?
6. From what material is the case of a carbon–zinc dry cell made? What purpose does this material serve? What happens to it as the cell discharges?
7. Describe what happens when a lead storage battery discharges.
8. What happens when a lead storage battery is charged?
9. Describe what happens when iron corrodes. How does road salt speed this process?
10. Why does aluminum corrode more slowly than iron, even though aluminum is more reactive than iron?
11. How does silver tarnish? How can tarnish be removed without the loss of silver?
12. List four common oxidizing agents.
13. List three common reducing agents.
14. Name some oxidizing agents used as antiseptics and disinfectants.
15. Relate the chemistry of photosynthesis to the chemistry that provides energy for your heartbeat.
16. Describe how a bleaching agent, such as hypochlorite (ClO^-), works.

Problems

Recognizing Oxidation and Reduction

17. Each of the following "equations" shows only part of a chemical reaction. Indicate whether the reactant shown is being oxidized or reduced. Explain.
   a. $C_2H_4O \longrightarrow C_2H_4O_2$
   b. $H_2O_2 \longrightarrow H_2O$

18. In which of the following partial reactions is the reactant undergoing oxidation? Explain.
   a. $C_2H_4O \longrightarrow C_2H_6O$ b. $WO_3 \longrightarrow W$

19. In which of the following partial reactions is the reactant undergoing oxidation? Explain.
   a. $Cl_2 \longrightarrow 2\,Cl^-$ b. $Fe^{3+} \longrightarrow Fe^{2+}$

20. In which of the following partial reactions is the reactant undergoing oxidation? Explain.
    a. $2 H^+ \longrightarrow H_2$ b. $CO \longrightarrow CO_2$

Oxidizing Agents and Reducing Agents

21. Identify the oxidizing agent and the reducing agent in each reaction.
    a. $4 Al + 3 O_2 \longrightarrow 2 Al_2O_3$
    b. $2 SO_2 + O_2 \longrightarrow 2 SO_3$

22. Identify the oxidizing agent and the reducing agent in each reaction.
    a. $Cl_2 + 2 KBr \longrightarrow 2 KCl + Br_2$
    b. $C_2H_4 + H_2 \longrightarrow C_2H_6$

23. Identify the oxidizing agent and the reducing agent in each reaction.
    a. $Fe + 2 HCl \longrightarrow FeCl_2 + H_2$
    b. $CS_2 + 3 O_2 \longrightarrow CO_2 + 2 SO_2$

24. Identify the oxidizing agent and the reducing agent in each reaction.
    a. $2 AgNO_3 + Cu \longrightarrow Cu(NO_3)_2 + 2 Ag$
    b. $CuCl_2 + Zn \longrightarrow ZnCl_2 + Cu$

Half-Reactions

25. Separate the following redox reactions into half-reactions and label each half as oxidation or reduction.
    a. $2 Ag^+(aq) + Cu(s) \longrightarrow Cu^{2+} + 2 Ag(s)$
    b. $Fe(s) + 2 H^+(aq) \longrightarrow Fe^{2+}(aq) + H_2(g)$

26. Separate the following redox reactions into half-reactions and label each half as oxidation or reduction.
    a. $3 Ag^+(aq) + Al(s) \longrightarrow Al^{3+} + 3 Ag(s)$
    b. $2 Al(s) + 6 H^+(aq) \longrightarrow 2 Al^{3+}(aq) + 3 H_2(g)$

27. Label each of the following half-reactions as oxidation or reduction and then combine them to obtain a balanced overall redox reaction.
    a. $2 I^- \longrightarrow I_2 + 2 e^-$ and $Cl_2 + 2 e^- \longrightarrow 2 Cl^-$
    b. $HNO_3 + H^+ + e^- \longrightarrow NO_2 + H_2O$ and
       $SO_2 + 2 H_2O \longrightarrow H_2SO_4 + 2 H^+ + 2 e^-$

28. Label each of the following half-reactions as oxidation or reduction and then combine them to obtain a balanced overall redox reaction.
    a. $2 H_2O_2 \longrightarrow 2 O_2 + 4 H^+ + 4 e^-$ and
       $Fe^{3+} + e^- \longrightarrow Fe^{2+}$
    b. $WO_3 + 6 H^+ + 6 e^- \longrightarrow W + 3 H_2O$ and
       $C_2H_6O \longrightarrow C_2H_4O + 2 H^+ + 2 e^-$

Oxidation and Reduction: Chemical Reactions

29. In the following reactions, which element is oxidized and which is reduced?
    a. $2 HNO_3 + SO_2 \longrightarrow H_2SO_4 + 2 NO_2$
    b. $2 CrO_3 + 6 HI \longrightarrow Cr_2O_3 + 3 I_2 + 3 H_2O$

30. In the following reactions, which substance is oxidized? Which is the oxidizing agent?
    a. $H_2CO + H_2O_2 \longrightarrow H_2CO_2 + H_2O$
    b. $5 C_2H_6O + 4 MnO_4^- + 12 H^+$
       $\longrightarrow 5 C_2H_4O_2 + 4 Mn^{2+} + 11 H_2O$

31. Acetylene (C_2H_2) reacts with hydrogen to form ethane (C_2H_6). Is the acetylene oxidized or reduced? Explain your answer.

32. Unsaturated vegetable oils react with hydrogen to form saturated fats. A typical reaction is

$$C_{57}H_{104}O_6 + 3 H_2 \longrightarrow C_{57}H_{110}O_6$$

Is the unsaturated oil oxidized or reduced? Explain.

33. To test for iodide ions (for example, in iodized salt), a solution is treated with chlorine to liberate iodine.

$$2 I^- + Cl_2 \longrightarrow I_2 + 2 Cl^-$$

Which substance is oxidized? Which is reduced?

34. Molybdenum metal, used in special kinds of steel, can be manufactured by the reaction of its oxide with hydrogen.

$$MoO_3 + 3 H_2 \longrightarrow Mo + 3 H_2O$$

Which substance is reduced? Which is the reducing agent?

35. Unripe grapes are exceptionally sour because of a high concentration of tartaric acid. As the grapes ripen, this compound is converted to glucose.

$$\begin{array}{ccc} C_4H_6O_6 & \longrightarrow & C_6H_{12}O_6 \\ \text{Tartaric acid} & & \text{Glucose} \end{array}$$

Is the tartaric acid oxidized or reduced?

36. When the water pump failed in the nuclear reactor at Three Mile Island in 1979, zirconium metal reacted with very hot water to produce hydrogen gas.

$$Zr + 2 H_2O \longrightarrow ZrO_2 + 2 H_2$$

What substance was oxidized in the reaction? What was the oxidizing agent?

37. Vitamin C (ascorbic acid) is thought to protect our stomachs from the carcinogenic effect of nitrite ions by converting the ions to NO gas.

$$NO_2^- \longrightarrow NO$$

Is the nitrite ion oxidized or reduced? Is ascorbic acid an oxidizing agent or a reducing agent?

38. In the preceding reaction (Problem 37), ascorbic acid is converted to dehydroascorbic acid.

$$C_6H_8O_6 \longrightarrow C_6H_6O_6$$

Is ascorbic acid oxidized or reduced in this reaction?

Combination with Oxygen

39. Give formulas for the products formed in the following reactions.
    a. $S + O_2 \longrightarrow$ b. $H_2 + O_2 \longrightarrow$
    c. $CH_4 + O_2 \longrightarrow$ d. $C_3H_8 + O_2 \longrightarrow$

40. Give formulas for the products formed in the following reactions.
    a. $C + O_2 \longrightarrow$ b. $N_2 + O_2 \longrightarrow$
    c. $CS_2 + O_2 \longrightarrow$ d. $C_6H_{12}O_6 + O_2 \longrightarrow$

Additional Problems

41. The dye indigo (used to color blue jeans) is formed by exposure of indoxyl to air.

$$2 \, C_8H_7ON + O_2 \longrightarrow C_{16}H_{10}N_2O_2 + 2 \, H_2O$$
Indoxyl Indigo

What substance is oxidized? What is the oxidizing agent?

42. Why do signs in hospitals warn against smoking? State a reason other than the long-term health risks.

43. Why do some mechanics lightly coat their tools with grease or oil before storing them?

44. When a strip of copper is placed in a solution of silver nitrate ($AgNO_3$), the copper is plated with silver. Write an equation for the reaction that occurs. (Nitrate ion is not involved in the reaction.)

45. Look at the caption of the drawing of a hydrogen–oxygen fuel cell on page 214. Show how the two half-reactions can be combined to give the overall cell reaction.

46. To oxidize 1.0 kg of fat, our bodies require about 2000 L of oxygen. A good diet contains about 80 g of fat per day. What volume of oxygen (at STP) is required to oxidize that fat?

47. The oxidizing agent we use to obtain energy from food is oxygen (from the air). If you breathe 15 times a minute (at rest), taking in and exhaling 0.5 L of air with each breath, what volume of air do you breathe each day? Air is 21% oxygen by volume. What volume of oxygen do you breathe each day?

48. The overall photosynthesis reaction is

$$6 \, CO_2 + 6 \, H_2O \longrightarrow C_6H_{12}O_6 + 6 \, O_2$$

Which substance is oxidized? Which is the oxidizing agent? Which substance is reduced? Which is the reducing agent?

Projects

49. Look up the process of galvanization. What does it do, and how is it related to oxidation and reduction?

50. Look up information about the restoration of the Statue of Liberty for its centennial in 1986. What role did oxidation–reduction play in the need for restoration?

Online Projects

51. Go to one or more of the websites about fuel cells. Write out the reactions that occur in the different types in development. Write an essay focusing on one of the following: (a) A cost–benefit analysis for their use in automobiles versus the use of gasoline, electricity via batteries, or natural gas. (b) Safety considerations of using a hydrogen–oxygen fuel cell. (c) Political-economic factors that may be hindering or promoting fuel cell research.

References and Readings

1. "Auto Batteries." *Consumer Reports*, October 1988, pp. 44–46.

2. DeLorenzo, Ronald. "Electrochemical Errors." *Journal of Chemical Education*, May 1985, p. 424.

3. "Dry Cell Batteries." *Consumer Reports*, December 1997, pp. 25–28

4. Ennis, John L. "Photography at Its Genesis." *Chemical and Engineering News*, 18 December 1989, pp. 26–42.

5. "Everyday Examples of Oxidation-Reduction Processes." *Journal of Chemical Education*, May 1978, pp. 332–333.

6. "Fade Creams." *Consumer Reports*, January 1985, p. 12. They contain oxidizing agents.

7. Hill, John W., and Ralph H. Petrucci. *General Chemistry*, 2d edition. Upper Saddle River, NJ: Prentice Hall, 1999. Chapter 4, "Chemical Reactions in Aqueous Solutions," and Chapter 18, "Electrochemistry."

8. Hoogers, Gregor, and Dave Thomsett. "Releasing the Potential of Clean Power." *Chemistry and Industry*, 18 October 1999, pp. 796–799. Discusses the latest developments in fuel cells.

9. Kolb, Doris. "The Chemical Equation: Part II—Oxidation–Reduction Reactions." *Journal of Chemical Education*, May 1978, pp. 326–331.

10. Worley, John D. "Hydrogen Peroxide in Cleansing Antiseptics." *Journal of Chemical Education*, August 1983, p. 678.

11. Zander, Ann. "What Do You Say After You've Said 'Oops'—Fabric Care and Stain Removal." Colorado State University Cooperative Extension, Boulder County, 27 December 1996.

MediaLab

Oxidation, Antioxidants, and Your Health

Oxidative processes run our batteries, produce warmth from our fuels, and turn the food we eat into energy. We could not live without oxidation within our bodies, and modern technology could not exist without oxidation of oil and gas.

We have long known that oxidation also results in metal corrosion and food decay. Now medical researchers are finding that reactive oxygen radicals within our cells may contribute to many of the diseases that confront us as we age. These *degenerative diseases* range from heart disease to cancer to Alzheimer's disease. This injurious reaction of oxygen with biological molecules is some-

times called *oxidative stress*, and research indicates that it can be prevented or minimized by antioxidants. Antioxidants have a greater tendency to react with oxygen than do biological molecules, so that the body's chemicals are protected. In the *Web Investigation* of this MediaLab, you will look at the possible role of oxidative stress in disease and at the evidence that antioxidants might help prevent some disorders. You will also look at some websites that exploit these findings to try to sell products. In *Communicate Your Results*, you will weigh the evidence about antioxidants and health and write or speak about your findings.

WEB INVESTIGATION

Investigation 1
Free Radicals

The forms of oxygen that are thought to lead to degenerative diseases are highly reactive species. These include superoxide radical, O_2^-; hydroxyl free radical, $\cdot OH$, and hydrogen peroxide, H_2O_2. Select Keyword **RADICAL** for an overview of what free radicals are and do. A more biological overview can be found at Keyword **ANTIOXIDANTS**.

Investigation 2
Antioxidants and Heart Disease

A great deal of research is aimed at determining the role that dietary antioxidants might play in preventing cancer or heart disease. Unfortunately, different studies often give apparently contradictory results. Select Keyword **CORO-**

NARY to look at some summaries of studies on heart disease and antioxidants. (*Hint*: Because this is an area of active research, always note the date when an article was written or posted.)

Investigation 3
Antioxidants: A "Fountain of Youth"?

Whenever the media announce new research showing possible benefits from a nutritional factor, commercial interests rush in to exploit the public's wish for a solution in a pill. Find some of these interests by selecting Keyword **SEARCH** and search for "oxidative stress," specifying com (commercial) sites. Also use subject limits Health & Medicine and Science & Mathematics. Look at a few of the top sites. How many are interested in giving solid health information, and how many are only selling something? Then repeat the search, specifying education and government sites. Compare the results.

Investigation 4
Food or Dietary Supplements?

Antioxidants, of course, are found in many foods. Select Keyword **FOOD** for an overview of food sources. You may get a sense of the vast amount of research being done on antioxidants in foods by selecting Keyword **SYMPOSIUM**.

Investigation 5
More About Supplements

If you have a strong interest in this topic and want to explore further, select Keyword **NIH** to download the proceedings of an NIH workshop on "The Role of Dietary Supplements for Physically Active People." For a briefer overview, select Keyword **DIET**.

Exercise 1
Free Radicals and You

Write a brief (about one or two pages) explanation of what free radicals do in the body, and why oxidative processes can be harmful. Aim at readers who know little or nothing about chemistry.

Exercise 2
Tabulating an Overview

Try to tabulate the information you've gathered, especially in *Web Investigation 2*, to give an overview of research findings about antioxidants and disease.

Exercise 3
Weighing the Facts

Select a commercial website from *Web Investigation 3* and evaluate its claims.

(Collaborative approach: Several of you might agree to evaluate different sites and present a combined analysis.)

Exercise 4
To Supplement or Not?

As a collaborative exercise, form two teams and debate one of the following:

- It is clear that oxidative processes contribute to cancer and heart disease.
- A healthy person who eats a balanced diet does not need to take dietary supplements.
- Even though the evidence is equivocal, a person should take supplements because "it can't hurt."

9

Organic Chemistry
The Infinite Variety of Carbon Compounds

Nowhere are there molecules with more variety
Than those we shall encounter in organic chemistry!

Two chemicals, vincristine and vinblastine, obtained from vinca plants (*Vinca rosea* Linn.), are effective against certain forms of cancer. Many organic chemicals are obtained from plants and animals, but many of the millions of organic compounds known today are synthetic.

The definition of organic chemistry has changed over the years. The changes serve as a good example of the dynamic character of science and of how scientific concepts change in response to experimental evidence. Until 1828 chemists believed that *organic* chemicals originated only in tissues of living *organisms* and required a "vital force" for their production. All chemicals not manufactured by living tissue were regarded as *inorganic*. Some chemists even believed that organic and inorganic chemicals followed different laws. This all changed in 1828, when Friedrich Wöhler synthesized the organic compound urea, which is found in urine, from ammonium cyanate, an inorganic compound. This important event led other chemists to attempt synthesis of organic chemicals from inorganic ones and changed the very definition of organic chemicals.

Organic chemistry is now defined as the chemistry of carbon-containing compounds. Most of them do come from living things or from things that were once living, but this is not necessarily the case. Perhaps the most remarkable thing about organic compounds is that there are so many of them. Of the more than 20 million known chemical compounds, over 95% are compounds of carbon.

9.1 The Unique Carbon Atom

Carbon atoms are unique in their ability to bond to each other so strongly that they can form long chains. Silicon and a few other elements can form chains, but only short ones; carbon chains often contain thousands of carbon atoms.

Carbon chains can also have branches or form rings of various sizes. Add to this the fact that carbon atoms also bond strongly to other elements, such as hydrogen, oxygen, and nitrogen, and that these atoms can be arranged in many different ways, and it soon becomes obvious why there are so many carbon compounds.

In addition to the millions of carbon compounds already known, new ones are being discovered every day. Carbon can form an almost infinite number of molecules of various shapes, sizes, and compositions.

We use thousands of carbon compounds every day without even realizing it because they are silently carrying out important chemical reactions within our bodies. Many of these carbon compounds are so vital that we literally could not live without them.

Hydrocarbons

The simplest organic compounds are hydrocarbons. A **hydrocarbon** contains only hydrogen and carbon. There are several kinds of hydrocarbons, classified according to the type of bonding between carbon atoms. We will consider several of these in turn in the following sections.

9.2 Alkanes

Each carbon atom forms four bonds, and each hydrogen atom forms only one bond, and so the simplest hydrocarbon molecule that is possible is CH_4. Called methane, CH_4 is the main component of natural gas. It has the structure

$$H - \underset{\underset{\displaystyle H}{|}}{\overset{\overset{\displaystyle H}{|}}{C}} - H$$

Methane is the first member of a group of related compounds called **alkanes**, hydrocarbons that contain only single bonds. The next member of the series is ethane (C_2H_6).

The word "organic" has several different meanings. Organic fertilizer is organic in the original sense; it is derived from living organisms. Organic foods are those grown without synthetic pesticides or fertilizers. Organic chemistry is simply the chemistry of carbon compounds.

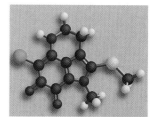

Organic compounds always contain carbon, and almost all also contain hydrogen. Many organic compounds contain oxygen and nitrogen, and quite a few contain sulfur and halogens. Nearly all the common elements are found in at least a few organic compounds. The compound batzelline, a molecular model of which is shown here, contains carbon (black), hydrogen (white), oxygen (red), nitrogen (blue), sulfur (yellow), and chlorine (green). Batzelline, now being evaluated as an anticancer drug, is obtained from a marine sponge; it is organic in both the modern and the original senses.

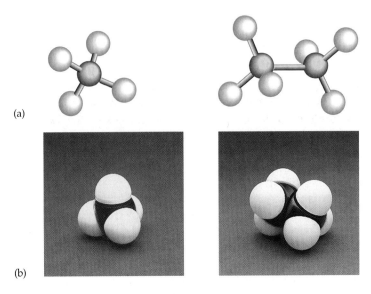

(a)

(b)

Figure 9.1 Ball-and-stick (a) and space-filling (b) models of methane (left) and ethane (right).

We saw in Chapter 5 that the methane molecule is tetrahedral, in accord with VSEPR theory. In fact, the tetrahedral shape results whenever a carbon atom is connected to four other atoms, be they hydrogen, carbon, or other elements. Figure 9.1 shows models of methane and ethane. The ball-and-stick models show the bond angles best, but the space-filling models more accurately reflect the shapes of the molecules. Ordinarily we use simple structural formulas such as the one shown above for ethane because they are much easier to draw. A **structural formula** shows which atoms are bonded to each other, but it does not attempt to show the actual shape of the molecule.

The three-carbon alkane is propane. Models of propane are shown in Figure 9.2, and its structural formula is

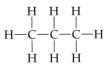

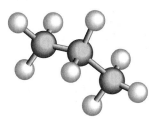

(a)

(b)

Figure 9.2 Ball-and-stick (a) and space-filling (b) models of propane.

Condensed Structural Formulas

Structural formulas show all the carbon and hydrogen atoms and how they are attached to one another. But these formulas take up a lot of space, and they are quite a bit of trouble to draw or to type. For these reasons, chemists usually prefer to use **condensed structural formulas**. Condensed structures show how many hydrogen atoms are attached to each carbon atom without showing the bonds to each hydrogen atom. For example, ethane and propane are written CH_3—CH_3 and CH_3—CH_2—CH_3. These formulas can be simplified even further by omitting some (or all) of the bond lines, resulting in CH_3CH_3 and $CH_3CH_2CH_3$.

Homologous Series

Note that there is a pattern with methane, ethane, and propane. We can build alkanes of any length simply by tacking carbon atoms together in long chains and adding sufficient hydrogen atoms to give each of the carbon atoms a total of four bonds.

Even the naming of these compounds follows a pattern (Table 9.1), with a stem name indicating the number of carbon atoms. For compounds of five carbon atoms or more, each stem is derived from the Greek or Latin name for the number. The compound names end in *-ane*, signifying that the compounds are *alkanes*. (An **alkane** is a hydrocarbon that has only single bonds; other hydrocarbons have double or triple bonds between carbon atoms.) Table 9.2 gives condensed structural formulas and names for continuous-chain (unbranched) alkanes up to ten carbon atoms in length.

Notice that the molecular formula for each alkane in Table 9.2 differs from the one preceding it by precisely one carbon atom and two hydrogen atoms—that is, by a CH_2 unit. Such a series of compounds has properties that vary in a regular and predictable manner. This principle, called *homology*, gives order to organic chemistry in much the same way that the periodic table gives organization to the chemistry of the elements. Instead of studying the chemistry of a bewildering array of individual carbon compounds, organic chemists study a few members of a **homologous series** from which they can deduce the properties of other compounds in the series.

We need not have stopped at 10 carbon atoms as we did in Table 9.2. One could hook together 100 or 1000 or 1 million carbon atoms. We can make an infinite number of alkanes simply by lengthening the chain. But lengthening the chain is not the only option. With four carbon atoms, chain branching is also possible.

Isomerism

When we extend the carbon chain to four atoms and add enough hydrogen atoms to give each carbon atom four bonds, we get $CH_3CH_2CH_2CH_3$. This formula represents butane, a compound that boils at about 0°C. A second compound, which has a boiling point of −12°C, has the same molecular formula, C_4H_{10}, as butane. The structural formula of the second compound, however, is not the same as that of butane. Instead of having four carbon atoms connected in a continuous chain, this new compound has a continuous chain of only three carbon atoms. The fourth carbon is branched off the middle carbon of the three-carbon chain. To show the two structures more clearly, let's use ordinary structural formulas showing the attachment of all the hydrogen atoms.

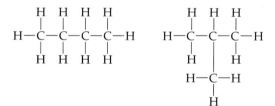

Table 9.1 ▮ Word Stems Indicating the Number of Carbon Atoms in Organic Molecules

Stem	Number
Meth-	One
Eth-	Two
Prop-	Three
But-	Four
Pent-	Five
Hex-	Six
Hept-	Seven
Oct-	Eight
Non-	Nine
Dec-	Ten

Alkanes often are called **saturated hydrocarbons**. They are hydrocarbons in which there are only single bonds. They can be represented by a general formula C_nH_{2n+2} in which n is the number of carbon atoms.

Table 9.2 ▮ The First Ten Continuous-Chain Alkanes

Name	Molecular Formula	Condensed Structural Formula	Number of Possible Isomers
Methane	CH_4	CH_4	—
Ethane	C_2H_6	CH_3CH_3	—
Propane	C_3H_8	$CH_3CH_2CH_3$	—
Butane	C_4H_{10}	$CH_3CH_2CH_2CH_3$	2
Pentane	C_5H_{12}	$CH_3CH_2CH_2CH_2CH_3$	3
Hexane	C_6H_{14}	$CH_3CH_2CH_2CH_2CH_2CH_3$	5
Heptane	C_7H_{16}	$CH_3CH_2CH_2CH_2CH_2CH_2CH_3$	9
Octane	C_8H_{18}	$CH_3CH_2CH_2CH_2CH_2CH_2CH_2CH_3$	18
Nonane	C_9H_{20}	$CH_3CH_2CH_2CH_2CH_2CH_2CH_2CH_2CH_3$	35
Decane	$C_{10}H_{22}$	$CH_3CH_2CH_2CH_2CH_2CH_2CH_2CH_2CH_2CH_3$	75

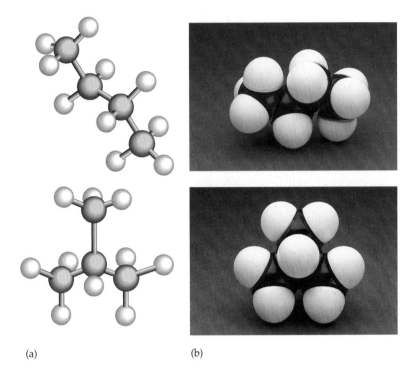

Figure 9.3　Ball-and-stick (a) and space-filling (b) models of butane (top) and isobutane (bottom).

(a)　　　　　　(b)

Compounds that have the same molecular formula but different structural formulas are called **isomers**. Because it is an isomer of butane, the branched four-carbon alkane is called isobutane. (As we shall see in Section 9.7, isobutane is also named 2-methylpropane because it has a methyl group attached to the second carbon of propane). Condensed structural formulas for the two isomeric butanes are written as follows.

$$CH_3CH_2CH_2CH_3 \qquad\qquad CH_3CHCH_3$$
$$| \qquad$$
$$CH_3$$

Butane　　　　　　　Isobutane

Figure 9.3 shows ball-and-stick and space-filling models of the two compounds.

The number of isomers increases rapidly with the number of carbon atoms (Table 9.2). There are three pentanes, five hexanes, nine heptanes, and so on. Isomerism is common in carbon compounds and provides another reason for the existence of millions of organic compounds.

Propane and the butanes are familiar fuels. Although they are gases at ordinary temperatures and under normal atmospheric pressure, they are liquefied under pressure and are usually supplied in tanks as liquefied petroleum (LP) gas. Gasoline is a mixture of hydrocarbons, mostly alkanes, with 5–12 carbon atoms.

We should point out that not all the possible isomers of the larger molecules have been isolated. Indeed, the task rapidly becomes more and more prohibitive as you proceed up the series. There are, for example, over 4 billion possible isomers with the molecular formula $C_{30}H_{62}$.

A propane torch. Propane burns in air with a hot flame.

Alkanes all have two more than twice as many hydrogen atoms as carbon atoms. That is, the general formula for an alkane is C_nH_{2n+2}.

Example 9.1

Without referring to Table 9.2, give the molecular formula, the structural formula, and the condensed structural formula for heptane.

Solution

The stem hept- means seven carbon atoms, and the ending -*ane* indicates an alkane. For the structural formula, we can write out a string of seven carbon atoms.

$$C—C—C—C—C—C—C$$

Then we attach enough hydrogen atoms to the carbons to give each carbon atom four bonds. This requires three hydrogen atoms on each end carbon and two each on the others.

For the condensed form, simply write each carbon atom's set of hydrogen atoms next to the carbon.

$$CH_3CH_2CH_2CH_2CH_2CH_2CH_3$$

For the molecular formula, we can simply count the carbon and hydrogen atoms to arrive at C_7H_{16}. Alternatively, we could use the general formula C_nH_{2n+2} with $n = 7$ to get C_7H_{16}.

Exercise 9.1

Give the molecular, structural, and condensed structural formulas for octane.

Properties of Alkanes

Note in Table 9.3 that after the first few members of the alkane series, the rest show about a 20–30°C increase in boiling point with each added CH_2 group. Note also that at room temperature alkanes with from 1 to 4 carbon atoms per molecule are gases, those with 5 to about 16 carbon atoms per molecule are liquids, and those with more than 16 carbon atoms per molecule are solids. The densities of liquid and solid alkanes are less than that of water. Alkanes are nonpolar molecules and are essentially insoluble in water; hence they float on top of water. They dissolve many organic substances of low polarity—such as fats, oils, and waxes. Alkanes undergo few chemical reactions. Their most important chemical property is that they burn, producing a lot of heat. They are used mainly as fuels.

Recall that water has a density of 1.00 g/mL at room temperature. All the alkanes listed in Table 9.3 have densities less than that.

Table 9.3 ▌ Physical Properties of Selected Alkanes

Name	Molecular Formula	Melting Point (°C)	Boiling Point (°C)	Density at 20°C (g/mL)
Methane	CH_4	−183	−162	(Gas)
Ethane	C_2H_6	−172	−89	(Gas)
Propane	C_3H_8	−188	−42	(Gas)
Butane	C_4H_{10}	−138	0	(Gas)
Pentane	C_5H_{12}	−130	36	0.626
Hexane	C_6H_{14}	−95	69	0.659
Heptane	C_7H_{16}	−91	98	0.684
Octane	C_8H_{18}	−57	126	0.703
Decane	$C_{10}H_{22}$	−30	174	0.730
Dodecane	$C_{12}H_{26}$	−10	216	0.749
Tetradecane	$C_{14}H_{30}$	6	254	0.763
Hexadecane	$C_{16}H_{34}$	18	280	0.775
Octadecane	$C_{18}H_{38}$	28	316	(Solid)
Eicosane	$C_{20}H_{42}$	37	343	(Solid)

The physiologic properties of alkanes vary. Methane appears to be inert physiologically. We probably could breathe a mixture of 80% methane and 20% oxygen without ill effect. This mixture would be flammable, however, and no fire or spark of any kind could be permitted in such an atmosphere. Breathing an atmosphere of pure methane (the "gas" of a gas-operated stove) can lead to death—not because of the presence of methane but because of the absence of oxygen (asphyxia). Light liquid alkanes, such as those in gasoline, dissolve and wash away body oils when spilled on the skin. Repeated contact may cause dermatitis. If swallowed, alkanes do little harm in the stomach; however, in the lungs, they cause chemical pneumonia by dissolving fatlike molecules from the cell membranes in the alveoli, allowing the lungs to fill with fluid. Heavier liquid alkanes, when applied to the skin, act as emollients (skin softeners). Petroleum jelly (Vaseline is one brand) is a semisolid mixture of hydrocarbons that can be applied as an emollient or simply as a protective film.

Skin lotions and creams are discussed in Chapter 17.

9.3 Cyclic Hydrocarbons: Rings and Things

The hydrocarbons we have encountered so far (alkanes) have been composed of open-ended chains of carbon atoms. Carbon atoms can also connect to form closed rings. The simplest possible ring-containing hydrocarbon, or **cyclic hydrocarbon**, has the molecular formula C_3H_6.

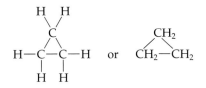

This compound is called cyclopropane (Figure 9.4).

Names of cycloalkanes (cyclic compounds containing only single bonds) are formed by adding the prefix *cyclo-* to the name of the open-chain compound with the same number of carbon atoms as are in the ring.

Chemists often use geometric figures to represent cyclic compounds (Figure 9.5). For example, a triangle is used to represent the cyclopropane ring, and a square to represent cyclobutane.

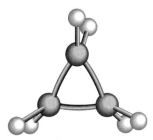

Figure 9.4 Ball-and-stick model of cyclopropane. Cyclopropane is a potent, quick-acting anesthetic with few undesirable side effects. It is no longer used in surgery, however, because it forms an explosive mixture with air at nearly all concentrations.

Example 9.2

Give the structural formula for cyclobutane.

Solution
Cyclobutane has four carbon atoms arranged in cyclic fashion.

$$\begin{array}{c} C-C \\ |\quad| \\ C-C \end{array}$$

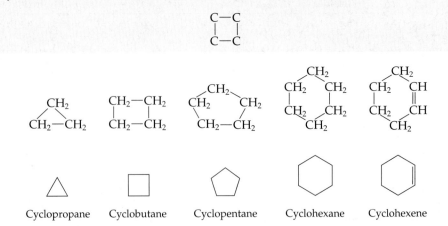

Figure 9.5 Structural formulas and symbolic representations of some cyclic hydrocarbons.

Each carbon atom needs two hydrogen atoms to complete its set of four bonds.

$$\begin{array}{c} CH_2-CH_2 \\ | \quad\quad | \\ CH_2-CH_2 \end{array}$$

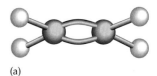

(a)

(b)

Figure 9.6 Ball-and-stick (a) and space-filling (b) models of ethylene.

Exercise 9.2

A. Give the structure for cyclopentane.
B. What is the general formula for cycloalkanes? (*Hint:* See Figure 9.5.)

9.4 Unsaturated Hydrocarbons: Alkenes and Alkynes

Two carbon atoms can share more than one pair of electrons. In ethylene (C_2H_4), the two carbon atoms share two pairs of electrons and are therefore joined by a double bond (Figure 9.6).

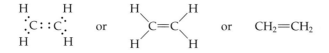

or $CH_2{=}CH_2$

Ethylene (also called ethene) is the simplest member of the alkene family. An **alkene** is a hydrocarbon that contains one or more carbon-to-carbon double bonds.

Ethylene is the most important commercial organic chemical. Annual U.S. production is over 20 billion kg. More than half goes into the manufacture of polyethylene, one of the most familiar plastics. Another 15% or so is converted to ethylene glycol, the major component of many formulations of antifreeze used in automobile radiators.

In acetylene (C_2H_2), the two carbon atoms share three pairs of electrons; the carbon atoms are joined by a triple bond (Figure 9.7).

$$H{:}C{:::}C{:}H \quad\quad \text{or} \quad\quad H-C{\equiv}C-H$$

(a)

(b)

Figure 9.7 Ball-and-stick (a) and space-filling (b) models of acetylene.

Acetylene (also called ethyne) is the simplest member of the alkyne family. An **alkyne** is a hydrocarbon that contains one or more carbon-to-carbon triple bonds. Acetylene is used in oxyacetylene torches for cutting and welding metals. Such torches can produce very high temperatures. Acetylene is also converted to a variety of other chemical products.

Alkenes occur widely in nature. Ripening fruits and vegetables give off ethylene, which triggers further ripening. Food processors artificially introduce ethylene to hasten the normal ripening process: 1 kg of tomatoes can be ripened by exposure to as little as 0.1 mg of ethylene for 24 hr. Unfortunately, the tomatoes don't taste much like those that ripen on the vine. The red color of tomatoes is due to lycopene, a hydrocarbon with many double bonds (a *polyene*).

Collectively, alkenes and alkynes are called *unsaturated hydrocarbons*. An **unsaturated hydrocarbon** is so called because it can add more hydrogen atoms. A **saturated hydrocarbon** (alkane) has the maximum number of hydrogen atoms attached to each carbon atom; it has no double or triple bonds.

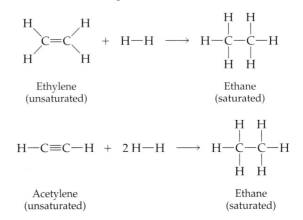

Ethylene
(unsaturated)

Ethane
(saturated)

Acetylene
(unsaturated)

Ethane
(saturated)

Alkenes with one double bond can be represented by a general formula C_nH_{2n}, and alkynes with one triple bond by C_nH_{2n-2}. In both formulas, n is the number of carbon atoms.

Double bonds are common in various biochemicals important to life. Unsaturated fats are a well-known example.

Properties of Alkenes and Alkynes

The physical properties of alkenes and alkynes are quite similar to those of the corresponding alkanes. Those with 2–4 carbon atoms per molecule are gases at room temperature, those with 5–18 carbon atoms are liquids, and those with more than 18 carbon atoms are solids. Like alkanes, alkenes and alkynes are insoluble in water and they float on water.

Alkenes and alkynes typically undergo addition reactions in which all the atoms of the reactants are incorporated into a single product. In the reactions here, ethylene (C_2H_4) adds hydrogen (H_2) to form ethane (C_2H_6), and acetylene (C_2H_2) adds two H_2 to form ethane (C_2H_6).

Like alkanes, both alkenes and alkynes burn. However, these unsaturated hydrocarbons undergo many more chemical reactions than alkanes. An alkene or alkyne can undergo an **addition reaction** across the double or triple bond. The addition of hydrogen to double and triple bonds is shown above. Chlorine, bromine, water, and many other kinds of molecules also add to double and triple bonds.

One of the most unusual features of alkene (and alkyne) molecules is that they can add to each other to form large molecules called *polymers*. These interesting molecules are discussed in Chapter 10.

9.5 Aromatic Hydrocarbons: Benzene and Relatives

Still another type of hydrocarbon is represented by benzene. Discovered by Michael Faraday in 1825, it has a molecular formula of C_6H_6. Many different structures can be drawn for this formula. Some examples are

All the possible structures for C_6H_6 have several double or triple bonds. But the remarkable thing about benzene is that it does not react as if it contains any double or triple bonds. It does not readily undergo addition reactions the way unsaturated compounds usually do.

For 40 years the structure of benzene remained a quandary. Then, in 1865, August Kekulé proposed a structure with a ring of six carbon atoms, each attached to one hydrogen atom.

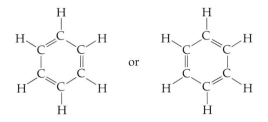

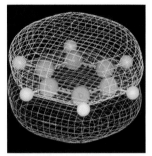

Figure 9.8 A computer-generated model of the benzene molecule. The six unassigned electrons occupy the yellow and blue areas below and above the plane of the carbon and hydrogen atoms.

The two structures shown above appear to contain double bonds, but in fact they do not. Both structures represent the same molecule, and the actual structure of this molecule is a hybrid of these two structures. The benzene molecule actually has six identical carbon-to-carbon bonds that are neither single bonds nor double bonds but something in between. In other words, the three pairs of electrons that would form the three double bonds are not tied down in one location but are spread around the ring (Figure 9.8). A popular modern way to represent the benzene ring is with a circle inside a hexagon.

The hexagon represents the ring of six carbon atoms, and the inscribed circle the ring of six unassigned electrons. Because its ring of electrons resists being disrupted, the benzene molecule is an exceptionally stable structure.

Benzene and similar compounds are called *aromatic hydrocarbons*. This is because quite a few of the first benzenelike substances to be discovered had strong aromas. Even though many benzene derivatives have turned out to be odorless, the name has stuck. Today an **aromatic compound** is any compound that contains a benzene ring or has certain properties similar to those of benzene.

Properties of Aromatic Hydrocarbons

Structures of some common aromatic hydrocarbons are shown in Figure 9.9. Benzene, toluene, and xylenes are all liquids that float on water. They are used mainly as solvents and fuels, but they are also used to make other benzene derivatives. Their vapors can act as narcotics when inhaled. Because benzene may cause leukemia after

A circle in a cyclic structure simply indicates an aromatic compound. It does not always mean six electrons. Naphthalene (Figure 9.9), for example, has a circle in each ring to indicate that it is an aromatic hydrocarbon. The total of unassigned electrons, however, is only ten. Some chemists still prefer to represent naphthalene (and other aromatic compounds) with alternate double and single bonds.

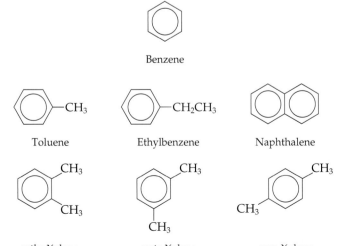

Figure 9.9 Some aromatic hydrocarbons. Benzene, toluene, and the three xylenes are components of gasoline and also serve as solvents. All these compounds are intermediates in the synthesis of polymers (Chapter 10) and other organic chemicals.

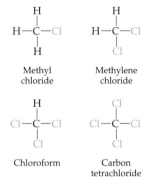

Methyl chloride

Methylene chloride

Chloroform

Carbon tetrachloride

Figure 9.10 Formulas for the four chlorine compounds derived from methane. Methyl chloride is a gas at room temperature; it is important as a reagent for introducing methyl groups into other molecules. The other three compounds are liquids that are frequently used as solvents.

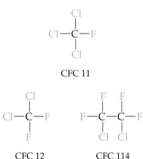

CFC 11

CFC 12

CFC 114

Figure 9.11 Three chlorofluorocarbons.

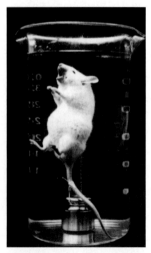

Figure 9.12 This mouse survives by breathing a liquid perfluoro compound saturated with oxygen.

long exposure, its use has been restricted. Naphthalene is a volatile, white, crystalline solid used as an insecticide, especially as a moth repellant.

9.6 Chlorinated Hydrocarbons: Many Uses, Some Hazards

Now that you know what hydrocarbons are, let's look at some chlorinated hydrocarbons. When chlorine gas (Cl_2) is mixed with methane (CH_4) in the presence of ultraviolet (UV) light, a reaction takes place at a very rapid (even explosive) rate. The result is a mixture of products (Figure 9.10), some of which may be familiar to you.

Methyl chloride (chloromethane, CH_3Cl) is used mainly in making silicone polymers (Chapter 10). Methylene chloride (dichloromethane, CH_2Cl_2) is a solvent used, for example, as a paint remover. Chloroform (trichloromethane, $CHCl_3$), also a solvent, was used as an anesthetic in earlier times, but such use is now considered dangerous. The dosage required for effective anesthesia is too close to a lethal dose. Carbon tetrachloride (tetrachloromethane, CCl_4), has been used as a dry-cleaning solvent and in fire extinguishers, but it is no longer recommended for either use. Exposure to carbon tetrachloride (or most other chlorinated hydrocarbons) can cause severe damage to the liver. The use of a carbon tetrachloride fire extinguisher in conjunction with water to put out a fire can be deadly. Carbon tetrachloride reacts with water to form phosgene ($COCl_2$), an extremely poisonous gas that was used during World War I.

A variety of more complicated chlorinated hydrocarbons are of considerable interest. DDT (dichlorodiphenyltrichloroethane) and other chlorinated hydrocarbons used as insecticides are discussed in Chapter 16. In Chapter 10, we study polychlorinated biphenyls (PCBs). For now, let's just say that all chlorinated hydrocarbons have similar properties. Most are only slightly polar, and they do not dissolve in water, which is highly polar. Instead, they dissolve (and dissolve in) fats, oils, greases, and other substances of low polarity. This is why certain chlorinated hydrocarbons make good dry-cleaning solvents; they remove grease and oily stains from fabrics. This is also why DDT and PCBs cause problems for fish and birds and perhaps for people; the toxic substances are concentrated in fatty animal tissues.

Chlorofluorocarbons

Carbon compounds containing fluorine as well as chlorine are called *chlorofluorocarbons (CFCs)*. CFCs have been used as the dispersing gases in aerosol cans, for making foamed plastics, and as refrigerants. Structures of three CFCs are illustrated in Figure 9.11. At room temperature, CFCs are gases or liquids with low boiling temperatures. They are essentially insoluble in water and inert toward most other substances. These properties make them ideal propellants for use in aerosol cans for deodorant, hair spray, and food products. Unfortunately, the inertness of these compounds allows them to persist in the environment. They diffuse into the stratosphere, where they undergo chemical reactions that lead to depletion of the ozone layer that protects Earth from harmful UV radiation. We look at this problem and some attempts to solve it in Chapter 12.

Perfluorocarbons

Fluorinated compounds have found some interesting uses. Some have been used as blood extenders. Oxygen is quite soluble in *perfluorocarbons*. (The prefix *per-* means that all hydrogen atoms have been replaced, in this case by fluorine atoms.) These compounds can therefore serve as temporary substitutes for hemoglobin, the oxygen-carrying protein in blood (Figure 9.12). Fluosol-SA, a mixture of perfluoro compounds, has been employed in Japan. In the United States, it has been used to treat premature babies, whose lungs are often underdeveloped.

Teflon (Chapter 10) is a perfluorinated polymer. It has many interesting applications because of its resistance to water and to high temperatures. It is especially noted for its unusual nonstick properties.

Oxygen-Containing Organic Compounds

Many organic compounds contain oxygen as well as carbon and hydrogen. In Sections 9.8 through 9.13, we introduce several important families of oxygen-containing organic compounds. First, however, we introduce another fundamental concept of organic chemistry in Section 9.7.

9.7 The Functional Group

Double and triple carbon–carbon bonds and halogen substituents are examples of **functional groups**, groups of atoms that give a family of organic compounds its characteristic chemical and physical properties. Table 9.4 lists a number of common functional groups found in organic molecules. Each of these groups is the basis for its own homologous series.

In many simple molecules, a functional group is attached to a hydrocarbon stem called an *alkyl group*. An **alkyl group** is just an alkane from which a hydrogen atom has been removed. The methyl group (CH_3—), for example, is derived from methane (CH_4), and the ethyl group (CH_3CH_2—) from ethane (CH_3CH_3). Propane yields two different alkyl groups, depending on whether the hydrogen atom is missing from the end or from the middle carbon atom.

$$CH_3\text{—OH} \qquad CH_3CH_2\text{—OH} \qquad CH_3CH_2CH_2\text{—OH}$$

Methanol — Ethanol — 1–Propanol

Table 9.5 lists the four alkyl groups mentioned here plus four others derived from the two butanes.

Often the letter R is used to stand for alkyl groups in general. Thus, ROH is a general formula for alcohols, and RCl represents any alkyl chloride.

9.8 The Alcohol Family

When a hydroxyl group (OH) is substituted for any hydrogen atom in an alkane, the molecule becomes an **alcohol** (ROH). Like many organic compounds, alcohols can be called by their common names or by the systematic names of IUPAC. The IUPAC names for alcohols are based on those of alkanes, with the ending changed from *-e* to *-ol*. If more than one location is possible, a number is used to indicate the location of the OH group. Methanol, ethanol, and 1-propanol are the first three members of the homologous series of alcohols. (The 1 in 1-propanol indicates that OH is attached to an end carbon atom of the three-carbon chain. An isomeric compound, 2-propanol, has an OH group on the second carbon atom of the chain.)

$$CH_3CH_2CH_2\text{—} \qquad CH_3\overset{|}{C}HCH_3$$

Propyl group — Isopropyl group

Common names for these alcohols, established by everyday usage, are methyl alcohol, ethyl alcohol, and propyl alcohol. The IUPAC names are the ones used in most scientific literature.

Table 9.4 ▌ Selected Organic Functional Groups

Name of Class	Functional Group	General Formula of Class
Alkane	None	$R-H$
Alkene	$-C=C-$	$R-\underset{\underset{R'}{\vert}}{C}=\underset{\underset{R''}{\vert}}{C}-R'''$
Alkyne	$-C\equiv C-$	$R-C\equiv C-R'$
Alcohol	$-C-O-H$	$R-O-H$
Ether	$-C-O-C-$	$R-O-R'$
Aldehyde	$-\overset{\overset{O}{\|\|}}{C}-H$	$R-\overset{\overset{O}{\|\|}}{C}-H$
Ketone	$-\overset{\overset{O}{\|\|}}{C}-$	$R-\overset{\overset{O}{\|\|}}{C}-R'$
Amine	$-C-N-$	$R-\overset{\overset{H}{\vert}}{N}-H$
		$R-\overset{\overset{H}{\vert}}{N}-R'$
		$R-\overset{\overset{R'}{\vert}}{N}-R''$
Carboxylic acid	$-\overset{\overset{O}{\|\|}}{C}-O-H$	$R-\overset{\overset{O}{\|\|}}{C}-O-H$
Ester	$-\overset{\overset{O}{\|\|}}{C}-O-C-$	$R-\overset{\overset{O}{\|\|}}{C}-O-R'$
Amide	$-\overset{\overset{O}{\|\|}}{C}-N-$	$R-\overset{\overset{O}{\|\|}}{C}-\underset{\underset{H}{\vert}}{N}-H$
		$R-\overset{\overset{O}{\|\|}}{C}-\underset{\underset{H}{\vert}}{N}-R'$
		$R-\overset{\overset{O}{\|\|}}{C}-\underset{\underset{R''}{\vert}}{N}-R'$

Table 9.5 ▮ Common Alkyl Groups

Name	Structural Formula	Condensed Structural Formula
Derived from Methane		
Methyl	$H-\overset{\overset{\displaystyle H}{|}}{\underset{\underset{\displaystyle H}{|}}{C}}-$	CH_3-
Derived from Ethane		
Ethyl	$H-\overset{\overset{\displaystyle H}{|}}{\underset{\underset{\displaystyle H}{|}}{C}}-\overset{\overset{\displaystyle H}{|}}{\underset{\underset{\displaystyle H}{|}}{C}}-$	CH_3CH_2-
Derived from Propane		
Propyl	$H-\overset{\overset{\displaystyle H}{|}}{\underset{\underset{\displaystyle H}{|}}{C}}-\overset{\overset{\displaystyle H}{|}}{\underset{\underset{\displaystyle H}{|}}{C}}-\overset{\overset{\displaystyle H}{|}}{\underset{\underset{\displaystyle H}{|}}{C}}-$	$CH_3CH_2CH_2-$
Isopropyl	$H-\overset{\overset{\displaystyle H}{|}}{\underset{\underset{\displaystyle H}{|}}{C}}-\overset{\overset{\displaystyle H}{|}}{C}-\overset{\overset{\displaystyle H}{|}}{\underset{\underset{\displaystyle H}{|}}{C}}-H$	$CH_3\underset{\underset{\displaystyle}{|}}{C}HCH_3$
Derived from Butane		
Butyl	$H-\overset{\overset{\displaystyle H}{|}}{\underset{\underset{\displaystyle H}{|}}{C}}-\overset{\overset{\displaystyle H}{|}}{\underset{\underset{\displaystyle H}{|}}{C}}-\overset{\overset{\displaystyle H}{|}}{\underset{\underset{\displaystyle H}{|}}{C}}-\overset{\overset{\displaystyle H}{|}}{\underset{\underset{\displaystyle H}{|}}{C}}-$	$CH_3CH_2CH_2CH_2-$
Secondary butyl (*sec*-butyl)	$H-\overset{\overset{\displaystyle H}{|}}{\underset{\underset{\displaystyle H}{|}}{C}}-\overset{\overset{\displaystyle H}{|}}{C}-\overset{\overset{\displaystyle H}{|}}{\underset{\underset{\displaystyle H}{|}}{C}}-\overset{\overset{\displaystyle H}{|}}{\underset{\underset{\displaystyle H}{|}}{C}}-H$	$CH_3CHCH_2CH_3$
Derived from Isobutane		
Isobutyl	$H-\overset{\overset{\displaystyle H}{|}}{C}-H$ above $H-\overset{\overset{\displaystyle H}{|}}{\underset{\underset{\displaystyle H}{|}}{C}}-\overset{\overset{\displaystyle H}{|}}{\underset{\underset{\displaystyle H}{|}}{C}}-\overset{\overset{\displaystyle H}{|}}{\underset{\underset{\displaystyle H}{|}}{C}}-$	$\overset{\overset{\displaystyle CH_3}{|}}{CH_3CHCH_2}-$
Tertiary butyl (*tert*-butyl)	$H-\overset{\overset{\displaystyle H}{|}}{C}-H$ above $H-\overset{\overset{\displaystyle H}{|}}{\underset{\underset{\displaystyle H}{|}}{C}}-\overset{\overset{\displaystyle H}{|}}{C}-\overset{\overset{\displaystyle H}{|}}{\underset{\underset{\displaystyle H}{|}}{C}}-H$	$CH_3-\overset{\overset{\displaystyle CH_3}{|}}{\underset{\underset{\displaystyle}{|}}{C}}-CH_3$

Methyl Alcohol (Methanol)

The simplest alcohol is methanol or methyl alcohol (CH_3OH) (Figure 9.13). Methanol is sometimes called wood alcohol because it can be made by destructive distillation of wood (Figure 9.14). The modern industrial process, however, makes methanol from two gases, carbon monoxide and hydrogen, at high temperature and pressure in the presence of a catalyst.

$$CO + 2\,H_2 \longrightarrow CH_3OH$$

Methanol is important as a solvent and as a chemical intermediate. It is also a possible replacement for gasoline in automobiles.

Ethyl Alcohol (Ethanol)

The next member of the homologous series of alcohols is ethyl alcohol (CH_3CH_2OH), also called ethanol or grain alcohol (Figure 9.15). Most ethanol is made by fermentation of grain (or other starchy or sugary materials). If the sugar is glucose, the reaction is

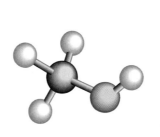

Figure 9.13   Ball-and-stick (a) and space-filling (b) models of the methanol molecule.

(a)   (b)

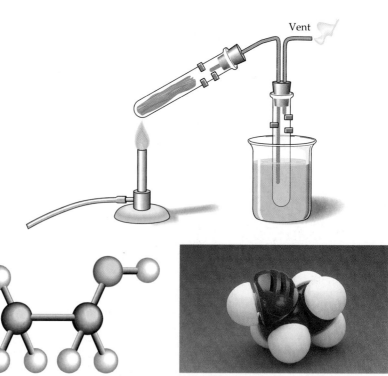

Figure 9.14   An apparatus for the destructive distillation of wood. The wood is heated in an enclosed tube, and the methanol is condensed in the second tube by the cold water in the beaker. Gases formed in the process are burned as they exit the vent tube.

Vent

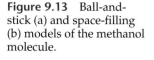

Figure 9.15   Ball-and-stick (a) and space-filling (b) models of the ethanol molecule.

(a)   (b)

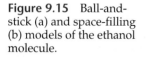

$$C_6H_{12}O_6 \xrightarrow{\text{yeast}} 2\ CH_3CH_2OH + 2\ CO_2$$

All ethyl alcohol for beverages and for automotive fuel is made in this way. A smaller amount of ethanol, for industrial use only, is made by reacting ethylene with water.

$$CH_2{=}CH_2 + H_2O \xrightarrow{\text{acid}} CH_3CH_2OH$$

This industrial alcohol is identical to that made by fermentation and is generally cheaper, but by law it cannot be used in alcoholic beverages. Since it carries no excise tax, the law requires that noxious substances be added to the alcohol to prevent people from drinking it. The resulting *denatured alcohol* is not fit to drink. This is the kind of alcohol commonly found on the shelves in chemical laboratories.

Gasoline in many parts of the United States contains up to 10% ethyl alcohol from fermentation. Because the United States has a large corn surplus, there is a generous government subsidy for gasoline producers who make their ethanol from corn. As a result, many factories now make ethanol by fermentation of corn.

Toxicity of Alcohols

Although ethyl alcohol is an ingredient in wine, beer, and other alcoholic beverages, alcohols are a rather toxic family. Methyl alcohol, for example, is oxidized in the body to formaldehyde (HCHO). Drinking as little as 1 oz (about 30 mL or 2 tablespoonfuls) can cause blindness and even death. Many poisonings each year result from mistaking methanol for its less toxic relative ethanol.

Ethyl alcohol is not as poisonous as methanol, but it is still toxic. One pint (about 500 mL) of pure ethyl alcohol, rapidly ingested, will kill most people. Of course, even the strongest alcoholic beverages seldom contain more than 45% ethanol (90 proof).

Generally, ethanol acts as a mild depressant; it slows down both physical and mental activity. Table 9.6 lists the effects of various doses. Although ethanol generally is a depressant, in small amounts it seems to act as a stimulant, perhaps by relaxing tensions and relieving inhibitions.

Excessive ingestion of ethanol over a long period of time can alter brain cell function, cause nerve damage, and shorten life span by contributing to diseases of the liver, cardiovascular system, and virtually every other organ of the body. In addition, about half of fatal automobile accidents involve at least one drinking driver. Babies born to alcoholic mothers often are small, deformed, and mentally retarded. Some investigators believe that this *fetal alcohol syndrome* can occur even if mothers drink only moderately. Ethanol, by far the most abused drug in the United States, is discussed more fully in Chapter 19.

Is it poison? seems to be a simple question, but it is a difficult one to answer. Toxicity depends on the nature of the substance, the amount, and the route by which it is taken into the body. Toxicity is discussed in detail in Chapter 20.

The *proof* of an alcoholic beverage is merely twice the percentage of alcohol by volume. The term has its origin in an old seventeenth-century English method for testing whiskey. Dealers were often tempted to increase profits by adding water to the booze. A qualitative method for testing the whiskey was to pour some of it on gunpowder and ignite it. If the gunpowder ignited after the alcohol had burned away, this was considered "proof" that the whiskey did not contain too much water.

Table 9.6 ▌ Approximate Relationship Among Drinks Consumed, Blood-Alcohol Level, and Behavior*

Number of Drinks[†]	Blood-Alcohol Level (percent by volume)	Behavior[‡]
2	0.05	Mild sedation; tranquility
4	0.10	Lack of coordination
6	0.15	Obvious intoxication
10	0.30	Unconsciousness
20	0.50	Possible death

[†] Rapidly consumed 30-mL (1-oz) "shots" of 90-proof whiskey, 360-mL (12-oz) bottles of beer, or 150-mL (5-oz) glasses of wine.

[‡] An inexperienced drinker would be affected more strongly, or more quickly, than one who is ordinarily a moderate drinker. Conversely, an experienced heavy drinker would be affected less.

*Data are for a 70-kg (154-lb) moderate drinker.

Warning labels on alcoholic beverages alert consumers to the hazards of excessive consumption.

Rubbing alcohol is a 70% solution (equivalent to 140 proof) of isopropyl alcohol (2-propanol). Because isopropyl alcohol is also more toxic than ethyl alcohol, it is not surprising that people become ill, and sometimes die, after drinking rubbing alcohol.

Multifunctional Alcohols

Several alcohols have more than one hydroxyl group. Examples are ethylene glycol, propylene glycol, and glycerol.

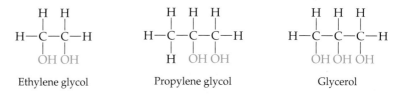

Ethylene glycol   Propylene glycol   Glycerol

Ethylene glycol is the main ingredient in most permanent antifreeze mixtures. Its high boiling point keeps it from boiling away in an automobile radiator. Ethylene glycol is a syrupy liquid with a sweet taste, but it is quite toxic. It is oxidized in the liver to oxalic acid.

$$\underset{\text{Ethylene glycol}}{\overset{\displaystyle \overset{\text{OH}}{\underset{\displaystyle CH_2}{|}}\quad \overset{\text{OH}}{\underset{\displaystyle CH_2}{|}}}{}} \xrightarrow{\text{liver enzymes}} \underset{\text{Oxalic acid}}{HO-\overset{O}{\overset{\|}{C}}-\overset{O}{\overset{\|}{C}}-OH}$$

Oxalic acid forms crystals of its calcium salt, calcium oxalate (CaC_2O_4), which can damage the kidneys, leading to kidney failure and death. Some manufacturers now market propylene glycol as a safer permanent antifreeze. It is a high-boiling antifreeze, and it is not poisonous.

Glycerol (or glycerin) is a sweet, syrupy liquid made as a by-product from fats during soap manufacture (Chapter 17). It is used in lotions to keep the skin soft and as a food additive to keep cakes moist. Its reaction with nitric acid makes nitroglycerin, the explosive material in dynamite. Incidentally, nitroglycerin is also important as a vasodilator, a medication taken by heart patients to relieve angina pain.

9.9 Phenols

When a hydroxyl group is attached to a benzene ring, the compound is called a **phenol**. Although it may appear to be an alcohol, it actually is not. The benzene ring greatly alters the properties of the hydroxyl group. In fact, phenol is a weak acid (sometimes called carbolic acid), and it is highly poisonous.

Phenol

Phenol was the first antiseptic used in an operating room—by Joseph Lister in 1867. Up until that time, surgery was not antiseptic, and many patients died from infections following surgical operations. Although phenol has a strong germicidal action, it is not an ideal antiseptic because it causes severe skin burns and it kills healthy cells along with harmful microorganisms.

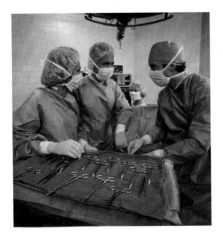

Figure 9.16 Phenolic compounds such as hexyl-resorcinol help to ensure antiseptic conditions in hospital operating rooms.

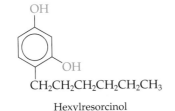

Hexylresorcinol   Hydroquinone

Phenol is still sometimes employed as a disinfectant for floors and furniture, but other phenolic compounds are now used as antiseptics. Hexylresorcinol, for example, is a more powerful germicide than phenol, and it is less damaging to the skin and has fewer other side effects (Figure 9.16).

A benzene ring with hydroxyl groups at both ends of the ring is called hydroquinone. It is commonly used as a photographic developer.

9.10 Ethers

Compounds with two alkyl groups attached to the same oxygen atom are called **ethers**. The general formula is ROR (or ROR' because the alkyl groups need not be alike). Best known is diethyl ether ($CH_3CH_2 - O - CH_2CH_3$), often called simply *ether*.

Diethyl ether is used mainly as a solvent because it dissolves many organic substances that are insoluble in water. It boils at 36°C, and so it evaporates readily, making it easy to recover dissolved materials. Although diethyl ether has little chemical reactivity, it is highly flammable, and great care must be used to avoid sparks or flames when it is in use.

Another problem with ethers is that over time they can react slowly with oxygen to form unstable peroxides, which may decompose explosively. Beware of previously opened containers of ether, especially old ones.

Diethyl ether was introduced in 1842 as a general anesthetic for surgery and was once the most widely used anesthetic. It is rarely used for humans today because it has undesirable side effects, such as postanesthetic nausea and vomiting. Many other general anesthetics are available these days, including several ethers. Methyl propyl ether ($CH_3OCH_2CH_2CH_3$) and methoxyflurane ($CHCl_2CF_2OCH_3$), for example, are anesthetics known as Neothyl and Penthrane, respectively.

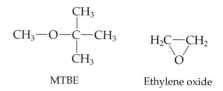

MTBE   Ethylene oxide

An ether once produced in large volume is methyl *tert*-butyl ether [MTBE, $CH_3OC(CH_3)_3$]. It was added to gasoline as an antiknock agent and to reduce the emission of carbon monoxide (CO) in automotive exhaust gas (Chapter 14).

The ether produced in the largest amount commercially is a compound called ethylene oxide. It is a cyclic ether with two carbon atoms and an oxygen atom forming a three-membered ring. Ethylene oxide is a toxic gas, more than 60% of which is used to make ethylene glycol.

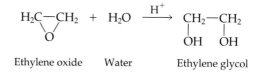

Ethylene oxide   Water   Ethylene glycol

Ethylene oxide is also used to sterilize medical instruments and as an intermediate in the synthesis of nonionic surfactants (Chapter 17). Most of the ethylene glycol produced is used in making polyester fibers (Chapter 10) and antifreeze.

Example 9.3

Classify each of the following as an alcohol, an ether, or a phenol.

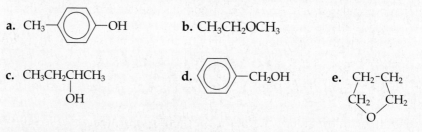

Solution

a. The functional group, an OH, is attached directly to the benzene ring; the compound is a phenol.
b. The O atom is between two alkyl groups; the compound is an ether.
c. The OH group is attached to an alkyl group; the compound is an alcohol.
d. An alcohol; the OH is not attached directly to the benzene ring.
e. An ether; the O atom is between two C atoms.

Exercise 9.3

Classify the following as alcohols, ethers, or phenols.

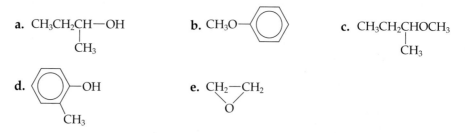

Example 9.4

Give the formula for isopropyl methyl ether.

Solution

Isopropyl methyl ether has an oxygen atom joined to an isopropyl group (three carbons joined to oxygen by the middle carbon) and a methyl group. The formula is

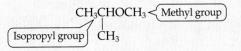

Exercise 9.4

A. Give the formula for methyl propyl ether.
B. Give the formula for ethyl *tert*-butyl ether.

9.11 Aldehydes and Ketones

Two families of organic compounds that share the same functional group are **aldehydes** and **ketones**. Both families contain the **carbonyl group** (C=O), but aldehydes have a hydrogen atom attached to the carbonyl carbon, whereas ketones have the carbonyl carbon attached to two other carbon atoms.

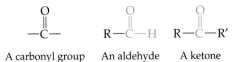

To simplify typing, these structures are often written on one line with the C=O understood:

$$-CO- \qquad R-CHO \qquad R-CO-R'$$

A carbonyl group   An aldehyde   A ketone

Models of three familiar carbonyl compounds are shown in Figure 9.17.

Some Common Aldehydes

The simplest aldehyde is formaldehyde (HCHO). It is a gas at room temperature but is readily soluble in water. As a 40% solution called formalin, it is used as a preservative for biological specimens and in embalming fluid.

   Formaldehyde is employed in making certain plastics (Chapter 10). It also is used to disinfect homes, ships, and warehouses. Commercially, formaldehyde is made by the oxidation of methanol. This is the same net reaction that occurs in the human body when methanol is ingested, and that accounts for the high toxicity of methyl alcohol.

$$CH_3OH \xrightarrow{\text{oxidation}} \underset{\text{Formaldehyde}}{H-\overset{\displaystyle O}{\overset{\|}{C}}-H}$$

Methanol

By the IUPAC system, formaldehyde is named methanal. The systematic names for aldehydes are based on those of alkanes, with the ending changed from *-e* to *-al*. These names are seldom used for simple aldehydes because of possible confusion with the corresponding alcohols. For example, methanal is easily confused with methanol, either orally or (especially) when handwritten.

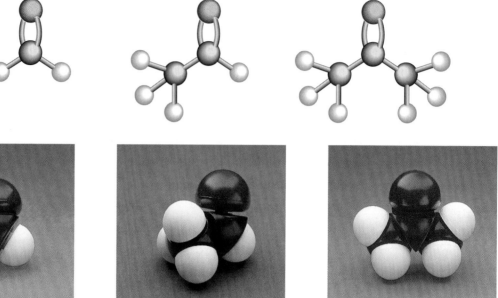

(a)

(b)

Figure 9.17 Ball-and-stick (a) and space-filling (b) models of formaldehyde (left), acetaldehyde (center), and acetone (right).

Organic chemists often write equations that show only the organic reactants and products. Inorganic substances are omitted for the sake of simplicity.

The next member of the homologous series of aldehydes is acetaldehyde (ethanal), formed by the oxidation of ethanol.

$$CH_3CH_2OH \xrightarrow{\text{oxidation}} CH_3-\overset{\overset{\displaystyle O}{\|}}{C}-H$$

Ethanol   Acetaldehyde

The next two members of the aldehyde series are propionaldehyde (propanal) and butyraldehyde (butanal). Both have strong, unpleasant odors. Benzaldehyde has an aldehyde group attached to a benzene ring. Also called (synthetic) oil of almond, benzaldehyde is used in perfumery and flavoring. It is the flavor ingredient in maraschino cherries.

$$CH_3CH_2-\overset{\overset{\displaystyle O}{\|}}{C}-H$$

Propionaldehyde

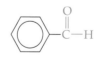

$$CH_3CH_2CH_2-\overset{\overset{\displaystyle O}{\|}}{C}-H$$

Butyraldehyde   Benzaldehyde

Some Common Ketones

The simplest ketone is acetone, made by the oxidation of isopropyl alcohol.

$$CH_3-\overset{\overset{\displaystyle OH}{|}}{CH}-CH_3 \xrightarrow{\text{oxidation}} CH_3-\overset{\overset{\displaystyle O}{\|}}{C}-CH_3$$

Isopropyl alcohol   Acetone

Acetone is a common solvent for such organic materials as fats, rubbers, plastics, and varnishes. It also finds use in paint and varnish removers. It is a common ingredient in some fingernail polish removers.

Two other familiar ketones are ethyl methyl ketone and isobutyl methyl ketone, which, like acetone, are frequently used as solvents.

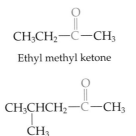

Ethyl methyl ketone

Isobutyl methyl ketone

By the IUPAC system, acetone is named propanone. The systematic names for ketones are based on those of alkanes, with the ending changed from -e to -one. When necessary, a number is used to indicate the location of the carbonyl group. For example, $CH_3CH_2COCH_2CH_2CH_3$ is 3-hexanone.

Example 9.5

Identify each of the following compounds as an aldehyde or a ketone.

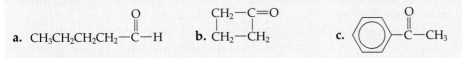

Solution

a. A hydrogen atom is attached to the carbonyl carbon atom; the compound is an aldehyde.
b. The carbonyl group is between two other carbon atoms; the compound is a ketone.
c. A ketone. (Remember that the corner of the hexagon stands for a carbon atom.)

Exercise 9.5

Identify each of the following compounds as an aldehyde or a ketone.

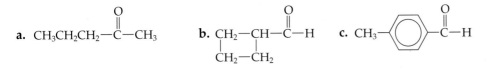

A ketone is a carbon chain
  With carbonyl inside.
If carbonyl is on the end,
  Then it's an aldehyde.

9.12 Carboxylic Acids

The functional group of organic acids is called the **carboxyl group**, and the acids are called **carboxylic acids**.

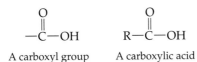

A carboxyl group   A carboxylic acid

Like aldehydes and ketones, these structures are often written on one line with the C=O understood:

−COOH   R−COOH

A carboxyl group   A carboxylic acid

The simplest carboxylic acid is formic acid (HCOOH) also called methanoic acid. It was first obtained by the destructive distillation of ants (the Latin word *formica* means "ant"). The sting of an ant smarts because the ant injects formic acid as it stings. The stings of wasps and bees also contain formic acid (as well as other poisonous compounds).

The systematic (IUPAC) names for carboxylic acids are based on those of alkanes, with the ending changed from -*e* to –*oic acid*. For example, $CH_3CH_2CH_2CH_2COOH$ is pentanoic acid.

$$H-\overset{\overset{\displaystyle O}{\|}}{C}-OH$$

Formic acid

$$\underset{\text{Benzoic acid}}{\overset{\overset{\displaystyle O}{\|}}{\bigcirc\!\!\!\!-C-OH}}$$

Benzoic acid

Acetic acid (ethanoic acid, CH_3COOH) can be made by the aerobic fermentation of a mixture of cider and honey. This produces a solution (vinegar) containing about 4–10% acetic acid plus a number of other compounds that give vinegar its flavor. Acetic acid is probably the most familiar weak acid used in academic and industrial chemistry laboratories.

The third member of the homologous series of acids, propionic acid (propanoic acid, CH_3CH_2COOH), is seldom encountered in everyday life. The fourth member is more familiar, at least by its odor. If you've ever smelled rancid butter, you know the odor of butyric acid (butanoic acid, $CH_3CH_2CH_2COOH$). It is one of the most foul-smelling substances imaginable. Butyric acid can be isolated from butterfat or synthesized in the laboratory. It is one of the ingredients of body odor, and extremely small quantities of this and other chemicals enable bloodhounds to track fugitives.

The acid with a carboxyl group attached directly to a benzene ring is called benzoic acid.

Carboxylic acid salts—calcium propionate, sodium benzoate, and others—are widely used as food additives to prevent mold (Chapter 16).

Acetic acid is a familiar weak acid in chemistry laboratories. It is also the principal active ingredient in vinegar.

Example 9.6

Give the structural formula for each of the following compounds.
a. propionaldehyde   **b.** butyl alcohol
c. acetic acid   **d.** ethyl methyl ketone

Solution
a. Propionaldehyde has three carbon atoms with an aldehyde function.

$$C-C-\overset{\overset{\displaystyle O}{\|}}{C}-H$$

Adding the proper number of hydrogen atoms to the other two carbon atoms gives the structure

or CH_3CH_2CHO

b. Butyl alcohol has four carbon atoms with an alcohol function at one end.

$$\underset{\substack{H \\ | \\ H}}{H-C}-\underset{\substack{H \\ | \\ H}}{C}-\underset{\substack{H \\ | \\ H}}{C}-\underset{\substack{H \\ | \\ H}}{C}-OH \quad or \quad CH_3CH_2CH_2CH_2OH$$

c. Acetic acid has two carbon atoms with a carboxylic acid function.

$$\underset{\substack{H \\ | \\ H}}{H-C}-\overset{O}{\underset{}{\underset{||}{C}}}-OH \quad or \quad CH_3COOH$$

d. Ethyl methyl ketone has a ketone function between an ethyl and a methyl group.

$$\underset{\substack{H \\ | \\ H}}{H-C}-\underset{\substack{H \\ | \\ H}}{C}-\overset{O}{\underset{}{\underset{||}{C}}}-\underset{\substack{H \\ | \\ H}}{C}-H \quad or \quad CH_3CH_2COCH_3$$

Exercise 9.6
A. Give the structural formula for each of the following compounds.
   a. butyric acid   **b.** propyl alcohol
   c. acetaldehyde   **d.** diethyl ketone
B. Give the structural formula for each of the following compounds.
   a. hexanoic acid   **b.** 1-pentanol
   c. 3-octanone   **d.** heptanal

9.13 Esters: The Sweet Smell of RCOOR'

Esters are derived from carboxylic acids and alcohols or phenols. The general reaction involves splitting out a molecule of water.

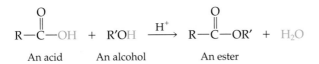

An acid   An alcohol   An ester

The name of an ester ends in *-ate* and is formed by naming the part from the alcohol first and the part from the carboxylic acid last. For example, the ester derived from butyric acid and methyl alcohol is methyl butyrate.

Although carboxylic acids often have powerfully unpleasant odors, the esters derived from them are usually quite fragrant, especially when dilute. Many esters have fruity odors and tastes. Some examples are given in Table 9.7. Esters are widely used as flavorings in cakes, candies, and other foods and as ingredients in perfumes.

Table 9.7 ▌ Ester Flavors and Fragrances

Ester	Formula	Flavor/Fragrance
Methyl butyrate	$CH_3CH_2CH_2COOCH_3$	Apple
Ethyl butyrate	$CH_3CH_2CH_2COOCH_2CH_3$	Pineapple
Propyl acetate	$CH_3COOCH_2CH_2CH_3$	Pear
Pentyl acetate	$CH_3COOCH_2CH_2CH_2CH_2CH_3$	Banana
Pentyl butyrate	$CH_3CH_2CH_2COOCH_2CH_2CH_2CH_2CH_3$	Apricot
Octyl acetate	$CH_3COOCH_2(CH_2)_6CH_3$	Orange
Methyl benzoate	$C_6H_5COOCH_3$	Ripe kiwi
Ethyl formate	$HCOOCH_2CH_3$	Rum
Methyl salicylate	$o\text{-}HOC_6H_4COOCH_3$	Wintergreen
Benzyl acetate	$CH_3COOCH_2C_6H_5$	Jasmine

Acids and Esters

Unless the hydrocarbon chain is reasonably long,
A carboxylic acid's apt to smell both foul and strong.
An ester, on the other hand, is so extremely sweet
It can be used in perfumes and delicious things to eat.

Salicylates: Pain Relievers Based on Salicylic Acid

Salicylic acid is both a carboxylic acid and a phenol. We can use it to illustrate some of the reactions of these two families of compounds.

Salicylic acid

Salicylic acid was first isolated from willow bark in 1860. Since ancient times, various peoples around the world had used willow bark to treat fevers. Edward Stone, an English clergyman, reported to the Royal Society in 1763 that an extract of willow bark was useful in reducing fever. Soon after its isolation, salicylic acid was used in medicine as an *antipyretic* (fever reducer) and as an *analgesic* (pain reliever). However, it is sour and irritating when taken orally. Chemists sought to use chemical reactions to modify its structure to reduce these undesirable properties while retaining or even enhancing its desirable properties. These reactions are summarized in Figure 9.18. The first such modification was simply to neutralize the acid (Reaction 1 in Figure 9.18). The resulting salt, sodium salicylate, was first used in 1875. It was less unpleasant to swallow than the acid but was still highly irritating to the stomach.

By 1886 chemists had produced another derivative (Reaction 2), the phenyl ester of salicylic acid, called phenyl salicylate or salol. Salol was less unpleasant to swallow, and it passed largely unchanged through the stomach. In the small in-

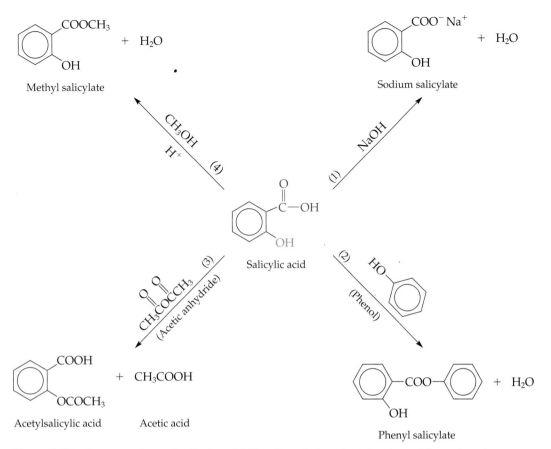

Figure 9.18 Some reactions of salicylic acid. Reactions 1, 2, and 4 take place at the carboxyl group; salicylic acid is reacting as a carboxylic acid. Reaction 3 takes place at the hydroxyl group; salicylic acid is reacting as a phenol.

testine, salol was hydrolyzed to the desired salicylic acid, but phenol was formed as a by-product. A large dose could produce phenol poisoning.

Acetylsalicylic acid, an ester of the phenol group of salicylic acid with acetic acid, was first produced in 1853. It is usually made by reacting salicylic acid with acetic anhydride, the acid anhydride of acetic acid (Reaction 3). The German Baeyer Company introduced acetylsalicylic acid as a medicine in 1899 under the trade name Aspirin. It soon became the best-selling drug in the world.

Another derivative, methyl salicylate, is made by reacting the carboxyl group of salicylic acid with methanol, producing a different ester (Reaction 4). Methyl salicylate, called oil of wintergreen, is used as a flavoring agent. It also finds use in rub-on analgesics. When applied to the skin, it causes a mild burning sensation, providing some relief for sore muscles.

Aspirin and its use as a drug are discussed in more detail in Chapter 19.

Nitrogen-Containing Organic Compounds

Many organic substances of interest to us in the chapters that follow contain nitrogen. It is the fourth most common element in organic compounds after carbon, hydrogen, and oxygen. We will look at a few simple examples in the following sections.

H
|
H—N—H

Ammonia

H
|
R—N—H

R'
|
R—N—H

R'
|
R—N—R''

Amines

Figure 9.19 Ammonia and three kinds of amines derived from it by replacing one, two, or all three of the hydrogen atoms by alkyl groups.

9.14 Amines and Amides

The two families we introduce in this section, amines and the amides, provide a vital background for the material ahead.

Amines

Amines contain the elements carbon, hydrogen, and nitrogen. An **amine** is derived from ammonia by replacing one, two, or three of the hydrogen atoms by one, two, or three alkyl or aromatic groups (Figure 9.19).

The simplest amine is methylamine (CH_3NH_2). The next higher homolog is ethylamine ($CH_3CH_2NH_2$). With two carbon atoms, however, we can also have dimethylamine (CH_3NHCH_3). Note that ethylamine and dimethylamine are isomers; both have the molecular formula C_2H_7N. With three carbon atoms, there are several possibilities, including trimethylamine [$(CH_3)_3N$].

Example 9.7

Give structures and names for the other three-carbon amines.

Solution

Three-carbon atoms can be in one alkyl group, and there are two such propyl groups.

$$CH_3CH_2CH_2NH_2 \qquad CH_3CHCH_3$$
$$\qquad\qquad\qquad\qquad\quad |$$
$$\qquad\qquad\qquad\qquad\quad NH_2$$

Propylamine   Isopropylamine

Three-carbon atoms can also be split into one methyl group and one ethyl group.

$$CH_3CH_2NHCH_3$$

Ethylmethylamine

Exercise 9.7

Give structures for the following amines.
a. butylamine   b. diethylamine
c. methylpropylamine   d. isopropylmethylamine

The amine with an NH_2 group attached directly to a benzene ring has the special name aniline.

$$\langle\!\bigcirc\!\rangle\!-\!NH_2$$

Aniline

Like many other aromatic amines, aniline is used in making dyes. Aromatic amines tend to be toxic, and some are strongly carcinogenic.

Simple amines are similar to ammonia in odor, basicity, and other properties. It is the higher amines that are the most interesting, though. Figure 9.20 shows a variety of these. Notice that each structure contains an **amino** (NH_2) **group**.

Among the most important kinds of organic molecules are amino acids. As the name implies, these compounds have carboxylic acid and amine functional groups. Amino acids are the building blocks from which proteins are constructed. The simplest amino acid, H_2NCH_2COOH, is glycine. Amino acids and proteins are considered in detail in Chapter 15.

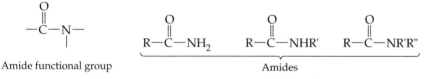

Figure 9.20 Some amines of interest. Amphetamine is a stimulant drug (Chapter 19). Cadaverine has the odor of decaying flesh. 1,6-Hexanediamine is used in the synthesis of nylon (Chapter 10). Pyridoxamine is a B vitamin (Chapter 18).

Amides

Another important group of nitrogen-containing compounds is *amides*, which contain oxygen bonded to the same carbon as nitrogen. Thus, an **amide** has the nitrogen atom attached directly to a carbonyl group.

Like those of other compounds that have C=O groups, the formulas of amides are often written on one line as $R—CONH_2$, $R—CONHR'$, and $R—CONR'R''$, respectively.

Note that urea (H_2NCONH_2), the compound that helped change the understanding of organic chemistry, is an amide.

Most simple amides are of little interest to us here, but complex amides are of tremendous importance. Your body contains many kinds of proteins, all held together by amide linkages (Chapter 15). Nylon, silk, and wool molecules also contain hundreds of amide functional groups.

[Web Reference 3] The structure and uses of urea are discussed.

Example 9.8

Which of the following are amides and which are amines? Identify the functional groups.

a. $CH_3CH_2CH_2NH_2$ b. CH_3CONH_2
c. $CH_3CH_2NHCH_3$ d. $CH_3COCH_2CH_2NH_2$

Solution

a. amine; the NH_2 is the amine function.
b. amide; the $CONH_2$ is the amide function.
c. amine; the NH is the amine function.
d. amine; the NH_2 is an amine function; there is a carbonyl group, but the NH_2 is not attached to it.

Exercise 9.8

Which of the following are amides and which are amines? Identify the functional groups.

a. CH_3NHCH_3 b. $CH_3CONHCH_3$
c. $CH_3CH_2N(CH_3)_2$ d. $CH_3NHCH_2CONH_2$

9.15 Heterocyclic Compounds: Alkaloids and Others

Cyclic hydrocarbons feature rings of carbon atoms. Now let's look at some ring compounds that have atoms other than carbon within the ring. These **heterocyclic compounds** usually have one or more nitrogen, oxygen, or sulfur atoms.

Example 9.9

Which of the following structures represent heterocyclic compounds?

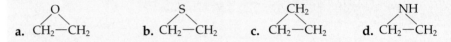

Solution

Compounds **a**, **b**, and **d** have oxygen, sulfur, and nitrogen atoms, respectively, in a ring structure; these represent heterocyclic compounds.

Exercise 9.9

Which of the following structures represent heterocyclic compounds?

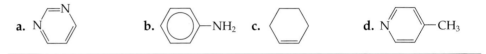

Many amines, particularly heterocyclic ones, occur naturally in plants. Like other amines, these compounds are basic. They are called **alkaloids**, which means "like alkalis." Among the familiar alkaloids are morphine, caffeine, nicotine, and cocaine. The actions of these compounds as drugs are considered in Chapter 19. Of more immediate interest are pyrimidine, which has two nitrogen atoms in a six-membered ring, and purine, which has four nitrogen atoms in two rings that share a common side. Compounds related to pyrimidine and purine are constituents of nucleic acids (Chapter 15).

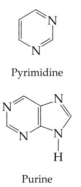

Pyrimidine

Purine

Critical Thinking Exercises

Apply knowledge that you have gained in this chapter and one or more of the FLaReS principles (Chapter 1) to evaluate the following statements or claims.

9.1 A television advertisement claims that gasoline with added ethanol burns cleaner than gasoline without added ethanol.

9.2 A news feature states that scientists have found that the odor of a certain ester improves workplace performance.

9.3 Alcohols, including ethanol, are toxic. An environmental activist states that all toxic chemicals should be banned from the home.

Summary

1. Organic chemistry is the study of carbon compounds—hydrocarbons and their derivatives. More than 95% of all known compounds contain carbon.

2. Hydrocarbons contain only carbon and hydrogen. They include alkanes, alkenes, alkynes, and aromatic hydrocarbons.

3. Alkanes, hydrocarbons that contain only single bonds, have names ending in -ane. They can have straight-chain, branched-chain, or cyclic structures.

4. Alkenes contain at least one double bond, and their names end in -ene. Alkynes contain at least one triple bond, and their names end in -yne. Both are unsaturated hydrocarbons.

5. The members of a homologous series of molecules differ one from another only by one or more CH_2 groups.

6. Isomers are molecules that have the same molecular formula but different structures. Because there is so much isomerism in organic chemistry, structural formulas are generally used rather than molecular formulas.

7. Aromatic hydrocarbons include benzene and its derivatives. The symbol for benzene is a hexagon with an inscribed circle representing a ring of six electrons.

8. Hydrocarbons form many kinds of derivatives, which behave according to the functional groups they contain.

9. Alcohols (ROH) are hydrocarbon derivatives containing OH groups.

10. An ether (ROR′) contains an oxygen atom between two alkyl groups.

11. Aldehydes (RCHO) and ketones (RCOR′) both contain the carbonyl group. In aldehydes, the carbonyl is on the end of the molecule.

12. Carboxylic acids (RCOOH) contain the carboxyl group, a carbonyl group with an OH group attached to the carbon atom.

13. An ester (RCOOR′) is a compound made by reacting a carboxylic acid with an alcohol. Most esters are fragrant.

14. An amine is a hydrocarbon derivative of ammonia.

15. An amide contains a carbonyl group, the carbon atom of which is attached to a nitrogen atom.

16. Alkaloids are amines, especially heterocyclic amines, that occur naturally in plants.

17. Amino acids, which have both carboxylic acid and amine functional groups, are the building blocks of proteins.

Key Terms

addition reaction 9.4	amino group	condensed structural	isomers 9.2
alcohol (ROH) 9.8	(—NH₂) 9.14	formula 9.2	ketone (R—CO—R′) 9.11
aldehyde (R—CHO) 9.11	aromatic	cyclic hydrocarbon 9.3	organic chemistry
alkaloid 9.15	compound 9.5	ester (RCOOR′) 9.13	(page 227)
alkane 9.2	carbonyl group	ether (ROR′) 9.10	phenol 9.9
alkene 9.4	(—CO—) 9.11	functional group 9.7	saturated hydrocarbon 9.2
alkyl group (R—) 9.7	carboxyl group	heterocyclic	structural formula 9.2
alkyne 9.4	(—COOH) 9.12	compound 9.15	unsaturated
amide 9.14	carboxylic acid	homologous series 9.2	hydrocarbon 9.4
amine 9.14	(RCOOH) 9.12	hydrocarbon 9.2	

Review Questions

1. What is organic chemistry?

2. List three characteristics of the carbon atom that make possible the existence of millions of organic compounds.

3. Define, illustrate, or give an example of each of the following terms.
   a. hydrocarbon   b. alkyne
   c. alkane   d. alkene

4. What is a homologous series? Give an example.

5. What are isomers? How can you tell whether or not two compounds are isomers?

6. What is an unsaturated hydrocarbon? Give examples of two types.

7. What is the meaning of the circle inside the hexagon in the modern representation of the structure of benzene?

8. What is an aromatic hydrocarbon? How can you recognize an aromatic compound from its structure?

9. Which alkanes are gases at room temperature? Which are liquids? Which are solids? State your answers in terms of the number of carbon atoms per molecule.

10. Compare the densities of liquid alkanes with that of water. When you add hexane to water in a beaker, what do you expect to observe?

11. What are the chemical names for the alcohols known by the following familiar names?
   a. grain alcohol   b. rubbing alcohol
   c. wood alcohol

12. What are some of the long-term effects of excessive ethanol consumption?

13. Give an important historical use for diethyl ether. What is its main use today?

14. Give an important use for phenols.

15. What structural feature distinguishes aldehydes from ketones?

16. How do carboxylic acids and esters differ in odor? In chemical structure?

17. For what family is RCOR′ the general formula?

18. What is an alkaloid?

Problems

Organic and Inorganic

19. Classify the following compounds as organic or inorganic.
   a. C_6H_{10} b. CH_3NH_2
   c. $CoCl_2$ d. $NaNH_2$

20. Classify the following compounds as organic or inorganic.
   a. CsCl   b. CH_3Cl
   c. $C_{12}H_{22}O_{11}$ d. $Cu(NH_3)_6Cl_2$

Names and Formulas: Hydrocarbons

21. How many carbon atoms are there in each of the following?
   a. ethane   b. cyclobutane
   c. heptane   d. pentane

22. How many carbon atoms are there in each of the following?
   a. propane   b. cyclopentane
   c. ethylene   d. nonane

23. Name the following hydrocarbons.

   a. $CH_3CH_2CH_3$ b. $HC{\equiv}CH$ c. $CH_2{=}CH_2$

24. Name the following hydrocarbons.

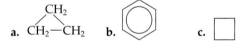

   a. $\overset{\displaystyle CH_2}{CH_2{-}CH_2}$ b.   c.

25. Give the structural formulas of four-carbon alkanes (C_4H_{10}). Identify butane and isobutane.

26. Give the molecular formulas and structures for the following hydrocarbons.
   a. pentane   b. octane

27. Give structures for the following alkyl groups.
   a. ethyl   b. isopropyl

28. Name the following alkyl groups.
   a. $CH_3CH_2CH_2{-}$ b. $CH_3{-}$

Names and Formulas: Oxygen-Containing Compounds

29. Name the following compounds.
   a. CH_3OH b. $CH_3CH_2CH_2OH$

30. Give the structural formula for each of the following alcohols.
   a. ethyl alcohol   b. butyl alcohol

31. Give the structure for phenol.

32. Give the structure for diethyl ether.

33. Give the structure of each of the following compounds.
   a. acetaldehyde   b. formaldehyde

34. Give the structure of each of the following compounds.
   a. butyraldehyde   b. propionaldehyde

35. Name these compounds.
   a. HCOOH   b. CH_3CH_2COOH

36. Name the following compounds by the IUPAC system.
   a. $CH_3CH_2CH_2CH_2COOH$
   b. $CH_3CH_2CH_2CH_2CH_2CH_2CH_2COOH$

37. Give the condensed structural formula for each of the following.
   a. heptanoic acid   b. decanoic acid

38. Name these compounds.

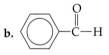

   a. $CH_3CH_2\overset{\displaystyle O}{\overset{\displaystyle \|}{C}}{-}H$ b.

39. Give the structural formula for each of the following.
   a. acetic acid   b. acetone

40. Give the structural formula for each of the following.
   a. butyric acid   b. ethyl methyl ketone

41. Give the structural formula for each of the following.
   a. ethyl acetate   b. methyl butyrate

42. Give the structural formula for each of the following.
   a. ethyl butyrate   b. methyl acetate

Names and Formulas: Nitrogen-Containing Compounds

43. Give the structural formula for each of the following.
   a. ethylamine   b. dimethylamine

44. Give the structural formula for each of the following.
   a. methylamine   b. isopropylamine

45. Name the following compounds.
   a. $CH_3CH_2CH_2NH_2$ b. $CH_3CH_2NHCH_2CH_3$

46. Name the following compounds.

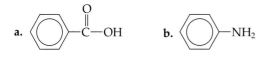

a.   b.

Isomers and Homologs

47. Indicate whether the structures in each set represent the same compound or isomers.

a. CH_3CH_3 $\underset{\underset{CH_3}{|}}{CH_3}$

b. $\underset{\underset{CH_3CH_2}{}}{\overset{\overset{CH_3}{|}}{}}$ $CH_3CH_2CH_3$

c. $\underset{CH_3CH_2\overset{\overset{CH_3}{|}}{C}HCH_2CH_3}{}$ $CH_3\overset{\overset{CH_3}{|}}{C}HCH_2CH_2CH_3$

48. Indicate whether the structures in each set represent the same compound or isomers.

a. $CH_3\overset{\overset{CH_3}{|}}{C}HCH_2CH_3$ $CH_3CH_2\underset{\underset{CH_3}{|}}{\overset{\overset{CH_3}{|}}{C}}H$

b. $CH_3CH_2\overset{\overset{CH_3}{|}}{C}H-\underset{\underset{CH_3}{|}}{C}H_2$ $CH_2CH_2\overset{\overset{CH_3}{|}}{C}HCH_3$ (with CH_3 top-left)

49. Classify the following pairs as homologs, identical, isomers, or none of these.

a. $CH_3CH_2CH_3$ and $CH_3CH_2CH_2CH_3$

b. 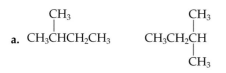 and $CH_3CH_2CH_3$

50. Classify the following pairs as homologs, identical, isomers, or none of these.

a. $CH_3\overset{\overset{CH_3}{|}}{C}HCH_3$ and CH_3CHCH_3 (with CH_3 below)

b. $CH_2{=}CHCH_3$ and (cyclopropane)

Classification of Hydrocarbons

51. Indicate whether each of the following compounds is saturated or unsaturated.

a. $CH_3C{=}CH_2$ (with CH_3 below)   b.

52. Indicate whether each of the following compounds is saturated or unsaturated.

a. $CH_3C{\equiv}CCH_3$ b. (cyclohexane)

53. Classify the compounds in Problem 51 as alkanes, alkenes, or alkynes.

54. Classify the compounds in Problem 52 as alkanes, alkenes, or alkynes.

Functional Groups

55. Give the structure of the carbonyl functional group.

56. Give the structure of the carboxyl group.

57. What is the name of the $-NH_2$ group?

58. Give the structure of the amide functional group.

59. Classify each of the following as a carboxylic acid, amine, amide, or ester. Identify the functional groups in each.

  a. $HCOOH$ b. CH_3CH_2COOH
  c. $CH_3CH_2NH_2$ d. $CH_3CH_2CONHCH_2CH_2CH_3$

60. Classify each of the following as a ketone, aldehyde, ester, carboxylic acid, or ether.

a. $CH_3CH_2\overset{\overset{O}{\|}}{C}OCH_3$ b. $CH_3CH_2\overset{\overset{O}{\|}}{C}H$

c. $CH_3CH_2\overset{\overset{O}{\|}}{C}CH_2CH_3$ d. $CH_3CH_2\overset{\overset{O}{\|}}{C}OH$

61. Classify each of the following as an alcohol, aldehyde, ester, ether, or carboxylic acid.
  a. $CH_3CH_2OCH_3$ b. $CH_3CH_2CH_2COOH$
  c. $CH_3CH_2CH_2CH_2OH$ d. $HCOOCH_2CH_2CH_3$

62. Which of the following represent heterocyclic compounds? Classify each also as an amine, ether, or (cyclo)alkane.

a. $\underset{CH_2-CH_2}{\overset{CH_2-NH}{|\quad\quad|}}$ b. $\underset{CH_2-CH_2}{\overset{CH_2-CH_2}{|\quad\quad|}}$

c. $\underset{CH_2-CH_2}{\overset{CH_2-O}{|\quad\quad|}}$ d. $\underset{CH_2-CH_2}{\overset{CH_2-CH-NH_2}{|\quad\quad|}}$

63. What is the percent by volume of alcohol in 80 proof vodka?

Additional Problems

64. Give the structure of the alcohol that can be oxidized to each of the following.

a. $CH_3\overset{O}{\underset{\|}{C}}CH_3$ **b.** $H-\overset{O}{\underset{\|}{C}}-H$

c. $CH_3\overset{O}{\underset{\|}{C}}-H$

65. Thymol is used as an antimold agent in preserving books.

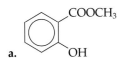

Thymol

To what family does thymol belong? Name the alkyl groups on the benzene ring.

66. Many organic molecules contain more than one functional group. For each of the following, identify and name the functional groups.

a.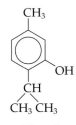

b.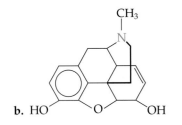

c.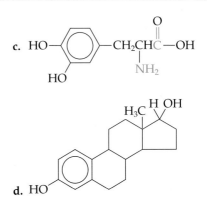

d. HO

67. Consider the following set of compounds. What principle does the series illustrate?

CH_3CH_2OH $CH_3CH_2CH_2OH$

$CH_3CH_2CH_2CH_2OH$ $CH_3CH_2CH_2CH_2CH_2OH$

68. Consider the following set of compounds. What principle does the series illustrate?

$CH_3CH_2CH_2OCH_3$

$CH_3\underset{\underset{CH_3}{|}}{C}HOCH_3$

$CH_3CH_2OCH_2CH_3$

$CH_3\overset{\overset{CH_3}{|}}{\underset{\underset{OH}{|}}{C}}-CH_3$

69. The complete combustion of benzene forms carbon dioxide and water.

$$C_6H_6 + O_2 \longrightarrow CO_2 + H_2O$$

Balance the equation. What mass of carbon dioxide is formed by the complete combustion of 39.0 g of benzene?

70. Water adds to ethylene to form ethyl alcohol.

$$CH_2{=}CH_2 + H_2O \longrightarrow CH_3CH_2OH$$

Balance the equation. What mass of ethyl alcohol is formed by the addition of water to 366 g of ethylene?

Online Projects

71. The theory that organic chemistry was the chemistry of living organisms, which was overturned by Wöhler's discovery in 1828, was known as vitalism. What was the effect of this philosophy on areas outside chemistry. You might begin your research at http://www.rit.edu/~flw-stv/biology.html, which has a long but interesting history of vitalism, biology, and the origin of life.

References and Readings

1. Asimov, Isaac. *A Short History of Chemistry*. Garden City, NY: Doubleday Anchor Books, 1965. Chapter 6 provides a brief history of organic chemistry.

2. Benfey, O. T. "August Kekulé and the Birth of the Structural Theory of Organic Chemistry in 1858." *Journal of Chemical Education*, January 1958, pp. 21–25.

3. Bruice, Paula Yurkanis. *Organic Chemistry, 2nd edition*. Upper Saddle River, NJ: Prentice Hall, 1998.

4. Hill, John W., Stuart J. Baum, and Rhonda J. Scott-Ennis. *Chemistry and Life*, 6th edition. Upper Saddle River, NJ: Prentice Hall, 2000. Chapters 13–17.

5. Julian, Maureen M. "What Compound Was Discovered as a Result of an Insurance Claim?" *Journal of Chemical Education*, October 1981, p. 793.

6. Wotiz, John H. (Ed.). *The Kekulé Riddle*. Vienna, IL: Cache River Press, 1993.

MediaLab

Fragrances: Stop and Smell the Roses

For millennia, fragrances have played a role in enhancing the lives of people. For centuries, women have worn flowers or rubbed petals on their skin to attract attention or hide body odor. Small perfume bottles are among the ancient Egyptian artifacts that have been recovered from the pyramids. Hunters have used animal scents to disguise their human odor. Cooks add spices and herbs to foods to preserve them and improve their flavor, but they also add exciting aromas to the food. And today, fragrances are a very big business!

The chemicals of fragrances and flavors are organic. (Can you think of an inorganic chemical with a fragrance? Does salt or copper have an odor?) These organic chemicals react with receptors in the nose or mouth of an animal to result in the sensation we know as odor or taste. Although anyone who has eaten with a head cold knows the dependence of taste on the sense of smell, the physiological mechanisms of the two senses are quite different. Many chemicals are used in both flavorings and fragrances. However, in this MediaLab, we shall concentrate on fragrance and the sense of smell.

In this *Web Investigation*, you will look at the chemical formulas of some familiar and unfamiliar scents and try to see similarities between those with similar fragrances. You will explore how these organic compounds interact with receptors in the body to produce what we actually smell. You will investigate the health effects—both beneficial and deleterious—of aromas. In *Communicate Your Results*, you will look more closely at the functional organic groups that result in fragrance. You will also explain, to the best of your understanding, why we react as we do to fragrances and flavors.

WEB INVESTIGATION

(You might find it fun!)

Investigation 1
Sweet-Smelling Molecules

Select Keyword **AROMA** for a partial list of fragrances available from one fragrance company. Then select **FRAGRANCE** for a choice of three databases of fragrant organic molecules arranged by chemical name. Look at the structures of a few chemicals. Scan the names for similar functional groups. (*Note*: The structures in these databases use greatly simplified bent-line structures, in which the lines are bonds between atoms, and all unmarked bends and ends indicate carbon atoms; all necessary hydrogen atoms are understood. Other atoms, such as oxygen, nitrogen, and sulfur are indicated by letter or color.) The manufacturer's ester databases have chemical structures that you can rotate and explore if you download the appropriate software.

Investigation 2
Is There an Odor If There's No One to Smell It?

An odor or flavor is perceived when the chemicals comprising it interact with appropriate receptors in the body. Select Keyword **SMELL** for an interesting discussion of the sense of smell. (This article is several web pages long; read at

264

least the first two screens.) Select Keyword **BIOLOGY** for a somewhat more technical treatment. If you are interested in recent research about olfactory receptors, select Keyword **RESEARCH**.

Investigation 3
Aromatherapy: Can Fragrance Heal?

The use of incense or fragrant herbs to promote healing or to heighten a religious experience extends back to the ancient Egyptians and probably beyond. Though still employed in some religions, scents are finding wider use today in the field of aromatherapy. Select Keyword **AROMATHERAPY** to learn more about this area. For a dissenting view of aromatherapy, select Keyword **QUACK**.

Investigation 4
Aromas as Allergens

A significant number of people are hypersensitive to certain perfumes, and fragrances are recognized as a trigger for some asthma and migraine attacks. However, we are seeing an increase in scented consumer products, from toilet tissue to laundry detergent. Select Keyword **ALLERGEN** for a look at the scope of this problem.

Olfactory Receptor

COMMUNICATE YOUR RESULTS

Exercise 1
The Importance of Structure

The odor of a particular chemical is determined by its structure. Using a database of *Web Investigation 1*, select two or three compounds with similar listed odors and copy their structures onto a single page. Try to determine which functional group or groups may contribute to the specified odor and circle them. Then select two or three *different* compounds with a similar functional group. (*Hint*: In the Bedoukian database, you can click on *Search* at the bottom of the page to search on a specific chemical name such as *acetate*.) Compare the odors described for these compounds.

Exercise 2
Is There an Odor If There's No One to Smell It?

Using what you have learned in *Web Investigation 2*, write a one-page answer to this question.

Exercise 3
Aromatherapy: Complementary Medicine or Worthless Placebo?

Write a one-to two-page paper about aromatherapy. The following are some possible questions you might want to consider: Does it qualify as a field of alternative medicine? Why or why not? For what medical or psychological conditions might it be valuable? What are the potential problems or even dangers of using aromatherapy to treat an illness? What type of studies might be used to determine the value or foolishness of this approach to healing?

Exercise 4 (Collaborative)
Too Many Scents?

Working by yourself or (preferably) in teams, look at some noncosmetic products (detergent, tissue, stationery) and determine how easy—or hard—it is to find varieties without added fragrance. Then do one of the following: (1) Write an advocacy letter to at least one of the following: a state or federal legislator, the Occupational Safety and Health Administration (OSHA), the American Lung Association, or a company with perfume-containing products. Explain why you think that household products should be fragrance-free or carry labels about possible allergens; (2) Discuss the pros and cons of adding fragrant chemicals to items other than perfumes and colognes. If you were a manufacturer of fine perfumes, would you approve of this trend?

10

Polymers
Giants Among Molecules

Plastics... textiles... gums... materials commonplace and rare...
Simply look around you... polymers are everywhere!

Polymers, in the form of films, fibers, and molded objects, pervade nearly every aspect of our lives. Foil-like sheets of conducting polymers can be made into light, flexible batteries. These batteries are much lighter than lead–acid batteries and can be crammed into almost any available space.

Look around you. Polymers are everywhere. Your clothes are made of polymers. In your car, the dashboard, seats, tires, steering wheel, floor mats, ceiling, and many parts that you cannot see are made of polymers. Around your home, carpets, curtains, upholstery, towels, sheets, floor tile, books, furniture, and most toys and containers (not to mention such things as telephones, toothbrushes, and piano keys) are made of polymers.

Polymers are an extremely important class of chemical compounds. Much of the food you eat contains polymers, and many important molecules in your body are polymers. You couldn't live without them. Some of the polymers that pervade our lives come from nature, but many are synthetic, made in chemical plants in an attempt to improve on nature in some way.

10.1 Polymerization: Making Big Ones Out of Little Ones

Polymers are *macromolecules* (from the Greek *makros*, meaning "large" or "long"). Macromolecules may not seem large to the human eye (in fact, many of these giant molecules are invisible), but when they are compared with other molecules, they are enormous.

Polymers (from the Greek *poly*, meaning "many," and *meros*, meaning "parts") are made from much smaller molecules called *monomers* (from the Greek *monos*, meaning "one"). Sometimes thousands of monomer units make up a single polymer molecule. **Monomers** are the building blocks of polymers. *Polymerization* is the process by which monomers are converted to polymers. A polymer is as different from its monomer as a long strand of spaghetti is from tiny specks of flour. For example, polyethylene, the familiar waxy material used to make plastic bags, is made from the monomer ethylene, which is a gas.

10.2 Natural Polymers

Polymers have served humanity for centuries in starches and proteins used for food; in wood used for shelter; and in wool, cotton, and silk used for clothing. Table 10.1 shows some natural polymers.

Starch is a polymer made up of glucose units. [(Glucose ($C_6H_{12}O_6$) is a simple sugar]. Cotton is made of cellulose, which is also a glucose polymer; wood is largely cellulose as well. Proteins are polymers made up of amino acid monomers. Wool and silk are two of the thousands of different kinds of proteins found in nature.

Living things could not exist without polymers. Each plant and animal requires many different specific types of polymers. Probably the most amazing natural polymers are nucleic acids, which carry the coded genetic information that makes each individual unique. The polymers found in nature are discussed in Chapter 15. In this chapter we focus mainly on macromolecules made in the laboratory.

10.3 Celluloid: Billiard Balls and Collars

The oldest attempts to improve on nature simply involved chemical modification of natural macromolecules. The synthetic material **celluloid**, as its name implies, was derived from natural cellulose (from cotton and wood, for example). When cellulose is treated with nitric acid, a derivative called cellulose nitrate is formed. In response to a contest to find a substitute for ivory for use in billiard balls, John Wesley Hyatt (1837–1920), an American inventor, found a way to soften cellulose nitrate by treating it with ethyl alcohol and camphor. The softened material could be molded into smooth, hard balls. Thus, Hyatt brought the game of billiards within the economic reach of more people—and possibly saved a few elephants.

[Web Reference 1]
The Macrogalleria, "a Cyberworld of polymer fun," has space-filling models and descriptions of many polymers, plus animations of some polymer syntheses.

A *plastic* material is composed mainly of large molecules that can be made to flow under heat and pressure. The material can then be shaped in a mold or in other ways (see page 275). Many different polymers are plastic.

A century ago, celluloid was widely used as a substitute for more expensive substances such as ivory, amber, and tortoiseshell. The movie industry was once known as the "celluloid industry," and even today a web search using the key word "celluloid" yields far more hits about movies than about the polymer. Disastrous fires in theaters, caused by ignition of the film, caused hundreds of deaths before celluloid film was banned.

[Web Reference 2]
More information about natural polymers can be found.

Table 10.1 ▌ Some Naturally Occurring Polymers*

Polymer Name	Unit Structure	Monomers	Functions
Protein	$—NHCH(R)CO—$	Amino Acids: $H_2NCH(R)COOH$ (about 20 different ones)	Components of all living organisms, from viruses to mammals: Structural proteins, enzymes, etc.
Nucleic Acid: Deoxyribonucleic Acid (DNA) or Ribonucleic Acid (RNA)	Base | $—sugar—phosphate—$	Nucleotides: with bases adenine, thymine (DNA) or uracil (RNA), guanine, cytosine	Contain and transmit information for all life processes
Starch (a polysaccharide)	$—C_6H_{10}O_5—$	α-Glucose$(C_6H_{12}O_6)$	Energy storage for plants; food for animals
Cellulose (a polysaccharide)	$—C_6H_{10}O_5—$	β-Glucose$(C_6H_{12}O_6)$	Structural material for plants

*These polymers are discussed in more detail in Chapter 15.

[Web Reference 3] An interesting biography of Hyatt, and information about his many inventions, can be found at the website of the National Plastics Center and Museum.

Celluloid was also used in movie film and for stiff collars (so they didn't require laundering and repeated starching). Because of its dangerous flammability (cellulose nitrate is also used as smokeless gunpowder), celluloid was removed from the market when safer substitutes became available. Today, movie film is made from cellulose acetate, another semisynthetic modification of cellulose. And high, stiff collars on men's shirts are out of fashion.

It didn't take long for the chemical industry to recognize the potential of synthetics. Scientists found ways to make macromolecules from small molecules rather than simply modifying large ones. The first such truly synthetic polymers were phenol–formaldehyde resins, first made in 1909. These complex polymers are discussed later in this chapter. Let's look at some simpler ones first.

10.4 Polyethylene: From the Battle of Britain to Bread Bags

The popular plastic polyethylene is the simplest and least expensive synthetic polymer. It is familiar today in the plastic bags used for packaging fruit and vegetables, in garment bags for dry-cleaned clothing, in garbage-can liners, and in many other items. Polyethylene is made from ethylene (CH_2=CH_2), an unsaturated hydrocarbon (Chapter 9) produced in large quantities from the cracking of petroleum.

With pressure and heat and in the presence of a catalyst, ethylene monomers join together in long chains.

The ellipses ($\cdots$) and tildes ($\sim$) are like et ceteras; they indicate that the number of monomers and the polymer structure are extended for many units in each direction.

$$\cdots + CH_2{=}CH_2 + CH_2{=}CH_2 + CH_2{=}CH_2 + CH_2{=}CH_2 + \cdots \longrightarrow$$
$$\sim CH_2CH_2{-}CH_2CH_2{-}CH_2CH_2{-}CH_2CH_2{-}CH_2CH_2\sim$$

Figure 10.1 shows models of a segment of a polyethylene molecule. The figure represents only a tiny part of a very long molecule, which can vary in number of carbon atoms from a few hundred to several thousand.

Polyethylene was invented shortly before the start of World War II. It proved to be tough and flexible, an excellent electric insulator, and able to withstand both high and low temperatures. Before long, it was used for insulating cables in radar, a top-secret invention that helped British pilots detect enemy aircraft before the aircraft

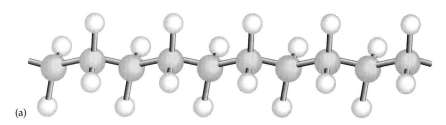

(a)

(b)

Figure 10.1 Ball-and-stick (a) and space-filling (b) models of a short segment of a polyethylene molecule.

could be spotted visually. Without polyethylene, the British could not have had effective radar, and without radar, the Battle of Britain might have been lost. The invention of this simple plastic helped to change the course of history.

Today, two principal kinds of polyethylene are produced by the use of different catalysts and different reaction conditions. *High-density polyethylenes* (*HDPEs*) have mostly linear molecules that pack closely together and can assume a fairly ordered, crystalline structure. HDPEs therefore are rather rigid and have good tensile strength. They are used for such items as threaded bottle caps, toys, bottles, and gallon milk jugs.

Low-density polyethylenes (*LDPEs*), on the other hand, have a lot of side chains branching off the polymer molecules. The branches prevent the molecules from packing closely together and assuming a crystalline structure. LDPEs are waxy, bendable plastics that are lower melting than high-density polyethylenes. Objects made of HDPE hold their shape in boiling water, whereas those made of LDPE are severely deformed (Figure 10.2). LDPEs are used to make plastic bags, plastic film, squeeze bottles, electric wire insulation, and many common household products.

A third type of polyethylene, linear low-density polyethylene (LLDPE), is actually a copolymer of ethylene with a higher alkene, such as 1-hexene. The structure of a segment of LLDPE would be something like

$$\sim CH_2CH_2 - CH_2CH - CH_2CH_2 \sim$$
$$| $$
$$CH_2$$
$$| $$
$$CH_2$$
$$| $$
$$CH_2$$
$$| $$
$$CH_3$$

where the horizontal portion shows an ethylene residue at each end and the middle portion (blue) depicts a segment from 1-hexene. More than 70% of the LLDPE produced is used to make film.

Figure 10.2 Both these bottles are made of polyethylene, and they were heated in the same oven for the same length of time. The one that melted has branched polyethylene molecules; the other is made from polyethylene with unbranched chains.

LDPE has densities ranging from 0.910 to 0.940 g/cm^3, and HDPE from 0.941 to 0.960 g/cm^3. LLDPE has densities ranging from 0.90 to 0.96 g/cm^3, depending on the higher alkene used with ethylene in the copolymerization.

Fullerenes: Buckyballs and Nanotubes

We saw in Chapter 9 that carbon atoms can form long chains, branched chains, and rings. Here we see in polymer molecules just how long some of these carbon chains can be. Carbon atoms, with their four bonds, can form still other structures. In 1985 a team of scientists discovered a variety of molecules formed exclusively of carbon atoms. A particularly prominent one had a molecular mass of 720 u, corresponding to the formula C_{60}. They proposed a structure for this molecule, which is a roughly spherical collection of hexagons and pentagons very much like a soccer ball. They named the molecule "buckminsterfullerene" in honor of architect R. Buckminster Fuller, who pioneered the geodesic-domed structures that C_{60} resembles. The general name *fullerenes* is now used for C_{60} and similar molecules with formulas such as C_{70}, C_{74}, and C_{82}. These substances are often colloquially called "buckyballs."

More recently, scientists have discovered tube-shaped carbon molecules called *nanotubes*. We can visualize a nanotube as a fullerene that has been stretched out into a hollow cylinder by the insertion of many, many more C atoms. We can also picture them as a two-dimensional array of hexagonal rings of carbon atoms, rather like ordinary "chicken wire." The "wire" is then rolled into a cylinder and capped at each end by half a C_{60} molecule. These nanotubes have unusual mechanical and electrical properties that are of great interest in current research.

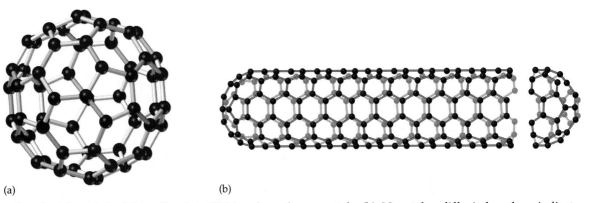

(a)   (b)

Ball-and-stick models of C_{60}, a "buckyball" (a) and a carbon nanotube (b). Nanotubes differ in length, as indicated here by the break in the structure.

Thermoplastic and Thermosetting Polymers

Polyethylene is one of several thermoplastic polymers listed in Table 10.2. A **thermoplastic polymer** can be softened by heat and pressure and then reshaped. It can be repeatedly melted down and remolded. Thermoplastics can be reshaped because their linear or branched polymers can slide past one another when heat and pressure are applied.

Not all plastics can be readily melted. Several thermosetting polymers are also listed in Table 10.2. *Thermosetting polymers* harden permanently when formed. They

Table 10.2 ▌ Production of Top Polymers in 1998*

Polymer	Billions of Kilograms
Thermoplastic resins	
Total polyethylene (PE)	12.60
LDPE	3.44
LLDPE	3.29
HDPE	5.87
Polyvinyl chloride (PVC)[†]	6.59
Polypropylene (PP)	6.28
Polystyrene (PS)[†]	4.29
Polyester (thermoplastic)	2.01
Polyamide (nylon type)	0.58
Thermosetting resins	
Phenolic resins	1.79
Urea resins	1.17
Polyester resins	0.78
Epoxy resins	0.29
Melamine	0.13

*Adapted from Reference 6.
[†]Includes copolymers.

cannot be softened by heat and remolded; instead, strong heating causes them to discolor and decompose. The permanence of thermosetting plastics is due to cross-linking of the polymer chains. We look at some thermosetting polymers later in this chapter.

10.5 Addition Polymerization: One + One + One + ⋯ Gives One!

There are two general types of polymerization reactions: addition polymerization and condensation polymerization. In **addition polymerization**, the building blocks (or monomers) add to one another in such a way that the polymeric product contains all the atoms of the starting monomers. The polymerization of ethylene to form polyethylene is an example. Notice that in the structure of polyethylene (Section 10.4), the two carbon atoms and the four hydrogen atoms of each monomer molecule are incorporated into the polymer structure. In *condensation polymerization* (Section 10.7), a portion of the monomer molecule is not incorporated in the final polymer but is split out as the polymer is formed.

Polypropylene

Most of the many familiar addition polymers are made from derivatives of ethylene in which one or more of the hydrogen atoms are replaced by another atom or group. Replacing one of the hydrogen atoms with a methyl group gives the monomer propylene (propene). Polypropylene looks like polyethylene, except that there are methyl groups (blue) attached to every other carbon atom.

$$\sim CH_2-CH-CH_2-CH-CH_2-CH-CH_2-CH\sim$$
$$\begin{array}{cccc} | & | & | & | \\ CH_3 & CH_3 & CH_3 & CH_3 \end{array}$$

Polypropylene is a tough plastic material that resists moisture, oils, and solvents. It is molded into hard-shell luggage, battery cases, and various kinds of appliance parts. It is also used to make packaging material, fibers for textiles such as upholstery fabrics and indoor–outdoor carpets, and ropes that float. Because of its high melting point (121°C), polypropylene objects can be sterilized with steam.

Polystyrene

Replacing one of the hydrogen atoms in ethylene with a benzene ring gives a monomer called styrene, with the formula $C_6H_5CH{=}CH_2$, where C_6H_5 represents the benzene ring. Polymerization of styrene produces polystyrene, a segment of which is illustrated below.

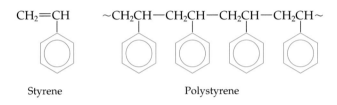

Styrene   Polystyrene

Polystyrene is the plastic material used to make transparent "throwaway" drinking cups, and, with color and filler added, it is used to make thousands of inexpensive toys and household items. When a gas is blown into polystyrene liquid, it foams and hardens into the familiar solid Styrofoam used to make disposable coffee cups. The foam is also used as packing material for shipping instruments and appliances, and it is widely used for home insulation.

Vinyl Polymers

Would you like a tough synthetic material that looks like leather yet costs only a fraction as much? Would you like a clear, rigid material from which unbreakable bottles could be made? Would you like an attractive, long-lasting floor covering? How about lightweight, rustproof plumbing? Polyvinyl chloride (PVC) has all these properties—and more.

Replacing one of the hydrogen atoms of ethylene with a chlorine atom gives vinyl chloride ($CH_2{=}CHCl$). This compound is a gas at room temperature. Polymerization of vinyl chloride yields the tough thermoplastic material PVC. A segment of the PVC molecule is illustrated below.

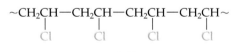

Polyvinyl chloride

PVC is readily formed into various shapes (Figure 10.3). The clear, transparent polymer is used in plastic wrap and clear plastic bottles. Adding color and other ingredients to vinyl plastics yields artificial leather. Most floor tile and shower curtains are also made from vinyl plastics, and they are widely used to simulate wood in home siding panels and window frames. About 40% of the PVC produced is molded into pipes.

The monomer from which vinyl plastics are made is a carcinogen. Several people who worked closely with vinyl chloride gas later developed a kind of cancer known as angiosarcoma. (Carcinogens are discussed in Chapter 20.)

Styrofoam insulation saves energy by reducing the transfer of heat from a warm house to the outside in winter or from the hot outside to the cooled inside in summer.

(a)   (b)

Figure 10.3 Polyvinyl chloride polymers can be formed into molded objects such as clear plastic shampoo bottles or made into films such as those used for shower curtains (a). They can even be made into fibers such as those used in the webbing for beach chairs (b).

Teflon: The Nonstick Coating

When all the hydrogen atoms in ethylene are replaced by fluorine, the formula is $CF_2{=}CF_2$ and the compound is called tetrafluoroethylene. Polymerization produces polytetrafluoroethylene (PTFE), or Teflon.

$$\sim CF_2{-}CF_2{-}CF_2{-}CF_2{-}CF_2{-}CF_2{-}CF_2{-}CF_2\sim$$

Teflon

Because its $C{-}F$ bonds are exceptionally strong and resistant to heat and chemicals, Teflon is a tough, unreactive, nonflammable material. It is used to make electric insulation, bearings, and gaskets. It is also widely used to coat surfaces on items such as skillets to give them nonsticking properties.

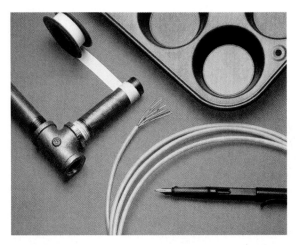

The coating on the muffin pan, the plumber's tape, and the insulation on the wire are all Teflon.

Polymerization Equations

The polymer structures and polymerization equations we have written are quite cumbersome. There are other ways of writing them that do not involve writing out long segments of the polymer chain. We can represent the polymerization of ethylene by the equation

$$n\, CH_2{=}CH_2 \longrightarrow {+}CH_2CH_2{+}_n$$

In the formula for the polymer product, the repeating polymer unit is placed within brackets with bonds extending to both sides. The subscript n indicates that this molecular fragment is repeated an unspecified number of times in the full polymer structure. It is certainly easier to write the bracketed unit than to draw the extended chain. The latter probably better conveys the concept of a polymer as a giant molecule, but even the extended chain represents only a small portion of the whole molecule.

The simplicity of the abbreviated formula facilitates certain comparisons between the monomer and the polymer. Note that the monomer ethylene contains a double bond and polyethylene does not. The double bond of the reactant contains two pairs of electrons. One of these pairs is used to connect one monomer unit to the next in the polymer (indicated by the lines sticking out to the sides in the repeating unit). This leaves only a single pair of electrons—in other words, a single bond—between the two carbon atoms of the polymer repeating unit. Note that each repeating unit in the polymer has the same composition (C_2H_4) as the monomer.

Example 10.1

What is the repeating unit in PVC? The polymer is

$$\sim CH_2CHCH_2CHCH_2CHCH_2CHCH_2CHCH_2CH \sim$$
$$\quad\ \ | \qquad\ | \qquad\ | \qquad\ | \qquad\ | \qquad\ |$$
$$\quad\ \ Cl \qquad Cl \qquad Cl \qquad Cl \qquad Cl \qquad Cl$$

Solution

The repeating unit is

$$CH_2CH$$
$$\quad\ |$$
$$\quad\ Cl$$

Exercise 10.1

What is the repeating unit in polyacrylonitrile? The polymer is

$$\sim CH_2CHCH_2CHCH_2CHCH_2CHCH_2CHCH_2CH \sim$$
$$\quad\ \ | \qquad\ \ | \qquad\ \ | \qquad\ \ | \qquad\ \ | \qquad\ \ |$$
$$\quad\ \ CN \qquad CN \qquad CN \qquad CN \qquad CN \qquad CN$$

Example 10.2

Give the structure of the polymer made from vinyl fluoride $(CH_2{=}CHF)$. Show at least four repeating units.

Solution

The carbon atoms become bonded in a chain with only single bonds between the carbon atoms. The fluorine atom is a substituent on the chain. (Two of the electrons in the double bond of the monomer are used to join the units together.) The polymer is

$$\sim CH_2CHCH_2CHCH_2CHCH_2CH \sim$$
$$\quad\ \ |\qquad\ \ |\qquad\ \ |\qquad\ \ |$$
$$\quad\ \ F\qquad\ F\qquad\ F\qquad\ F$$

Exercise 10.2

Give the structure of the polymer made from vinylidene chloride ($CH_2{=}CCl_2$). Show at least four repeating units.

Plastic products are usually made from powders. In *compression molding*, heat and pressure are applied directly to the polymer powder in the mold cavity. In *transfer molding*, the powder is softened by heating outside the mold and then poured into molds to harden. In *injection molding* the plastic is melted in a heating chamber and then forced by a plunger into cold molds to set. Another method, *extrusion molding*, the melted polymer is extruded through a die in continuous form to be cut into lengths or coiled.

Addition polymers have the highest production volume in the plastics industry, with polyethylenes leading the list (Table 10.2). We cannot mention all of the many kinds of addition polymers here, but Table 10.3 lists the more important ones, along with a few of their uses.

A giant bubble of tough, transparent plastic film emerges from a die of an extruding machine. The film is used in packaging, consumer products, and food services.

Conducting Polymers: Polyacetylene

Although acetylene has a triple bond instead of a double bond, it can still undergo addition polymerization, forming polyacetylene.

$$H{-}C{\equiv}C{-}H$$

$$\sim CH{=}CH{-}CH{=}CH{-}CH{=}CH{-}CH{=}CH{-}CH{=}CH{-}CH{=}CH\sim$$

Notice that, unlike the situation in polyethylene, which has a carbon chain containing only single bonds, every other bond in polyacetylene is a double bond (Figure 10.4).

The alternating double and single bonds make it easy for electrons to travel along the chain, and so this polymer is able to conduct electricity. (This is rather amazing, because plastics are normally electric insulators.) Polyacetylene can actually be used as a lightweight substitute for metal. In fact, the plastic even looks like metal, having a silvery luster. When formed into film, it looks like metal foil.

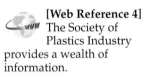

[Web Reference 4]
The Society of Plastics Industry provides a wealth of information.

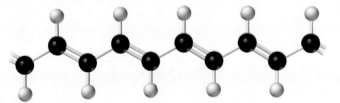

Figure 10.4 A ball-and-stick model of polyacetylene.

Polyacetylene, the first conducting polymer, was discovered in 1970. Since then, a number of other polymers with electric conductivity have been made.

Table 10.3 ▮ Some Addition Polymers

Monomer	Polymer	Polymer Name	Some Uses
$H_2C{=}CH_2$	$\begin{bmatrix} \overset{\displaystyle H}{\underset{\displaystyle H}{C}} - \overset{\displaystyle H}{\underset{\displaystyle H}{C}} \end{bmatrix}_n$	Polyethylene	Plastic bags, bottles toys electrical insulation
$H_2C{=}CH{-}CH_3$	$\begin{bmatrix} \overset{\displaystyle H}{\underset{\displaystyle H}{C}} - \overset{\displaystyle H}{\underset{\displaystyle CH_3}{C}} \end{bmatrix}_n$	Polypropylene	Indoor-outdoor carpeting, bottles, luggage
$H_2C{=}CH{-}\bigcirc$	$\begin{bmatrix} \overset{\displaystyle H}{\underset{\displaystyle H}{C}} - \overset{\displaystyle H}{\underset{\displaystyle \bigcirc}{C}} \end{bmatrix}_n$	Polystyrene	Simulated wood furniture, Styrofoam insulation, cups, toys packing materials
$H_2C{=}CH{-}Cl$	$\begin{bmatrix} \overset{\displaystyle H}{\underset{\displaystyle H}{C}} - \overset{\displaystyle H}{\underset{\displaystyle Cl}{C}} \end{bmatrix}_n$	Polyvinyl chloride (PVC)	Plastic wrap, simulated leather, plumbing, garden hoses, floor tile
$H_2C{=}CCl_2$	$\begin{bmatrix} \overset{\displaystyle H}{\underset{\displaystyle H}{C}} - \overset{\displaystyle Cl}{\underset{\displaystyle Cl}{C}} \end{bmatrix}_n$	Polyvinylidene chloride (Saran)	Food wrap, seatcovers
$F_2C{=}CF_2$	$\begin{bmatrix} \overset{\displaystyle F}{\underset{\displaystyle F}{C}} - \overset{\displaystyle F}{\underset{\displaystyle F}{C}} \end{bmatrix}_n$	Polytetrafluoroethylene (Teflon)	Nonstick coating for cooking utensils, electrical insulation
$H_2C{=}CH{-}CN$	$\begin{bmatrix} \overset{\displaystyle H}{\underset{\displaystyle H}{C}} - \overset{\displaystyle H}{\underset{\displaystyle CN}{C}} \end{bmatrix}_n$	Polyacrylonitrile (Orlon, Acrilan, Creslan, Dynel)	Yarns, wigs, paints
$H_2C{=}CH{-}O{-}\overset{O}{\overset{\|}{C}}{-}CH_3$	$\begin{bmatrix} \overset{\displaystyle H}{\underset{\displaystyle H}{C}} - \underset{\displaystyle O{-}\overset{}{C}{-}CH_3}{\overset{\displaystyle H}{C}} \end{bmatrix}_n$	Polyvinyl acetate	Adhesives, textile coatings, chewing gum resin, paints
$H_2C{=}\overset{CH_3}{\underset{O}{C}}{-}C{-}O{-}CH_3$	$\begin{bmatrix} \overset{\displaystyle H}{\underset{\displaystyle H}{C}} - \underset{\displaystyle C{-}O{-}CH_3}{\overset{\displaystyle CH_3}{C}} \end{bmatrix}_n$	Polymethyl methacrylate (Lucite, Plexiglas)	Glass substitute, bowling balls

10.6 Rubber

Although rubber is a polymer that comes from nature, it was the basis for much of the development of the synthetic polymer industry. During World War II, Japanese occupation of Malaysia and Indonesia cut off most of the Allies' supply of natural rubber. The search for synthetic substitutes resulted in much more than just a replacement for natural rubber. The plastics industry, to a large extent, developed out of the search for synthetic rubber.

Natural rubber can be broken down into a simple hydrocarbon called isoprene.

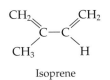

Isoprene

The macromolecules from which rubber is made are now known to have the structure

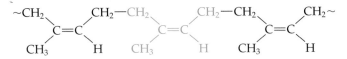

Isoprene is a volatile liquid, whereas rubber is a semisolid, elastic material. Chemists have learned to make polyisoprene, a substance identical in every way to natural rubber except that it comes from petroleum refineries rather than from plantations of rubber trees. They have also developed several synthetic rubbers and devised ways to modify these various polymers to change their properties.

Vulcanization: Cross-Linking

The long-chain molecules that make up rubber can be coiled and twisted and intertwined with one another. The stretching of rubber corresponds to the straightening of the coiled molecules. Natural rubber is soft and tacky when hot. It can be made harder by reaction with sulfur. This process, called **vulcanization**, cross-links the hydrocarbon chains with sulfur atoms (Figure 10.5). Charles Goodyear discovered vulcanization and was issued U.S. Patent 3633 in 1844.

Its three-dimensional cross-linked structure makes vulcanized rubber a harder, stronger substance that is suitable for automobile tires. Surprisingly, cross-linking also improves the elasticity of rubber. With just the right degree of cross-linking, the individual chains are still relatively free to uncoil and stretch. When a stretched piece

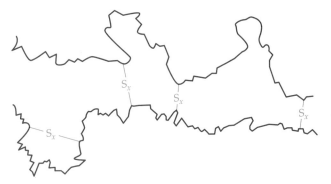

Figure 10.5 Vulcanized rubber has hydrocarbon chains (represented here by red lines) cross-linked by sulfur atoms. The subscript x indicates a small, indefinite number, usually not more than 4.

(a)   (b)   (c)

Figure 10.6 Vulcanization of rubber cross-links the molecular chains. (a) In unvulcanized rubber, the chains slip past one another when the rubber is stretched. (b) Vulcanization involves the addition of sulfur cross-linkages between the chains. (c) When vulcanized rubber is stretched, the sulfur cross-linkages prevent the chains from slipping past one another. Vulcanized rubber is stronger than unvulcanized rubber.

of this material is released, the cross-links pull the chains back to their original arrangement (Figure 10.6). Rubber bands owe their snap to this sort of molecular structure. Materials that act in this stretchable way are called **elastomers**.

Synthetic Rubber

Natural rubber is a polymer of isoprene, and some synthetic elastomers are closely related. For example, polybutadiene is made from the monomer butadiene ($CH_2=CH-CH=CH_2$), which differs from isoprene only in that it lacks a methyl group on the second carbon atom. Polybutadiene is made rather easily from the monomer.

$$n\ CH_2=CH-CH=CH_2 \longrightarrow \ +CH_2-CH=CH-CH_2+_n$$

However, it has only fair tensile strength and poor resistance to gasoline and oils. These properties limit its value for automobile tires, the main use of elastomers.

Another synthetic elastomer, polychloroprene (Neoprene), is made from a monomer similar to isoprene, but with a chlorine in place of the methyl group on isoprene.

$$n\ CH_2=\underset{Cl}{C}-CH=CH_2 \longrightarrow \ +CH_2-\underset{Cl}{C}=CH-CH_2+_n$$

Neoprene shows better oil and gasoline resistance than other elastomers. It is used to make gasoline hoses and similar items used at automobile service stations.

Another synthetic rubber illustrates the principle of *copolymerization*, in which the product (a **copolymer**) is formed from two or more different monomers. Styrene-butadiene rubber (SBR) is a copolymer of styrene (25%) and butadiene (75%). A segment of an SBR molecule might look something like this.

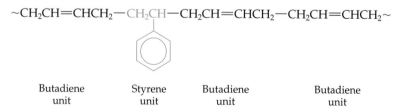

| Butadiene unit | Styrene unit | Butadiene unit | Butadiene unit |

SBR is more resistant to oxidation and abrasion than natural rubber, but its mechanical properties are less satisfactory.

Like those of natural rubber, SBR molecules contain double bonds and can be cross-linked by vulcanization. SBR accounts for more than half of the total production of elastomers and is used mainly for making tires.

One of the surprising uses for elastomers is in paints. The substance in a paint that hardens to form a continuous surface coating is called a *binder*. In traditional oil-base paints, the binder is usually linseed oil. Modern water-base latex paints contain a binder that is a synthetic polymer, usually one with rubberlike properties. Various kinds of polymers can be used as binders, depending on the specific qualities desired in the paint.

10.7 Condensation Polymers: Splitting Out Water

The polymers considered so far are all addition polymers. All the atoms of the monomer molecules are incorporated into the polymer molecules. In a condensation polymer, part of the monomer molecule is not incorporated in the final polymer. A small molecule, such as water, usually splits out during **condensation polymerization**.

Polyamides: Nylon

As an example, let's consider the formation of nylon. (There are several different nylons, each prepared from a different monomer or set of monomers, but all share certain common structural features.) The monomer in one type of nylon is a carboxylic acid with an amino group on the sixth carbon atom, 6-aminohexanoic acid. In the polymerization reaction, a carboxyl group of one monomer molecule forms an amide bond with the amine group of another.

Synthetic polymers serve as binders in paints. Pigments in paint make our world more colorful.

[Web Reference 5]
Massachusetts Institute of Technology has an interesting site, called The World of Materials, which includes information about polymers.

$$H_2NCH_2(CH_2)_3CH_2\overset{\overset{\displaystyle O}{\|}}{C}OH$$

6-Aminohexanoic acid

$$\cdots \; + \; HO-\overset{\overset{\displaystyle O}{\|}}{C}CH_2CH_2CH_2CH_2CH_2\overset{\overset{\displaystyle H}{|}}{N}-H \; + \; HO-\overset{\overset{\displaystyle O}{\|}}{C}CH_2CH_2CH_2CH_2CH_2\overset{\overset{\displaystyle H}{|}}{N}-H \; + \; \cdots \; \longrightarrow$$

$$\sim\overset{\overset{\displaystyle O}{\|}}{C}CH_2CH_2CH_2CH_2CH_2\overset{\overset{\displaystyle H}{|}}{N}-\overset{\overset{\displaystyle O}{\|}}{C}CH_2CH_2CH_2CH_2CH_2\overset{\overset{\displaystyle H}{|}}{N}\sim \; + \; n \, H_2O$$

Water molecules are formed as a by-product. It is this formation of a nonpolymeric by-product that distinguishes condensation polymerization from addition polymerization. Note that the formula of a repeating unit is not the same as that of the monomer.

The polymer formed in the reaction shown above is nylon 6. Because the linkages holding it together are amide bonds, nylon 6 is a **polyamide**. Another polyamide, nylon 66, is made by the condensation of two different monomers.

$$n \, H-\overset{\overset{\displaystyle H}{|}}{N}CH_2CH_2CH_2CH_2CH_2CH_2\overset{\overset{\displaystyle H}{|}}{N}-H \; + \; n \, HO-\overset{\overset{\displaystyle O}{\|}}{C}(CH_2)_4\overset{\overset{\displaystyle O}{\|}}{C}-OH \; \longrightarrow$$

1,6-Hexanediamine   Adipic acid

$$\{\!\!-\overset{\overset{\displaystyle H}{|}}{N}CH_2CH_2CH_2CH_2CH_2CH_2\overset{\overset{\displaystyle H}{|}}{N}-\overset{\overset{\displaystyle O}{\|}}{C}CH_2CH_2CH_2CH_2\overset{\overset{\displaystyle O}{\|}}{C}\!\!-\}_n \; + \; 2n \, H_2O$$

This was the original nylon polymer discovered in 1937 by DuPont chemist Wallace Carothers. Note that one monomer has two amino groups and the other has two carboxyl groups, but the product is still a polyamide, quite similar to nylon 6. Silk and wool, which are protein fibers, are natural polyamides.

Although nylon can be molded into various shapes, most nylon is made into fibers. Some is spun into fine thread to be woven into silklike fabrics, and some is made into yarn that is much like wool. Carpeting, which was once made primarily from wool, is now made largely from nylon. Less than 3% of today's carpets are made of wool, while 68% are nylon.

Polyesters: Polyethylene Terephthalate

Polyesters are also condensation polymers. The most common polyester is made by the condensation of ethylene glycol with terephthalic acid and is called polyethylene terephthalate (PET).

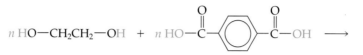

The hydroxyl groups in ethylene glycol react with the carboxylic acid groups in terephthalic acid to produce long chains held together by many ester linkages.

PET can be molded into bottles for beverages and other liquids. It can also be formed as a film. This film is used for the magnetically coated tapes used in audio- and videocassettes. Polyester molecules make excellent fibers that are widely used in wash-and-wear clothing. More than 50% of all synthetic fabrics contain polyester fibers.

Phenol–Formaldehyde and Related Resins

Let us now go back to Bakelite, the original synthetic polymer. Bakelite, a phenol–formaldehyde resin, was first synthesized by Leo Baekeland, who received U.S. Patent 942,699 for the process in 1909.

Phenol–formaldehyde resins are formed by splitting out water molecules, the hydrogen atoms coming from the benzene ring and the oxygen atoms from the aldehyde. The reaction proceeds stepwise, with formaldehyde adding first to the 2- or 4-position of the phenol molecule.

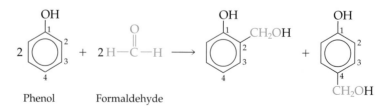

The substituted molecules then react by splitting out water. (Remember that there are hydrogen atoms at all the unsubstituted corners of a benzene ring.) The hookup of molecules continues until an extensive network is achieved.

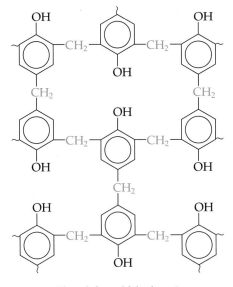

Phenol–formaldehyde resin

Water is driven off by heat as the polymer sets. The structure of the polymer is extremely complex, a three-dimensional network somewhat like the framework of a giant building. Note that the phenolic rings are joined together by CH_2 units from the formaldehyde. These polymers are **thermosetting resins**; they cannot be melted and remolded.

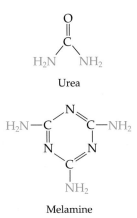

Urea

Melamine

Urea–formaldehyde and melamine–formaldehyde resins are similar to phenol–formaldehyde polymers. (The melamine molecule is formed by condensation of three molecules of urea.) Both these resins are complex three-dimensional networks formed by the splitting out of H_2O from formaldehyde ($H_2C{=}O$) molecules and amino ($-NH_2$) groups. Urea–formaldehyde resins are used to bind wood chips together in panels of particle board. Melamine–formaldehyde resins are used in plastic (Melmac) dinnerware and Formica countertops. Like Bakelite, both of these are thermosetting plastics.

Other Condensation Polymers

There are many other kinds of condensation polymers, but we will mention only a few.

Polycarbonates are tough, "clear as glass" polymers strong enough to be used as bulletproof windows. They are also used in protective helmets. One polycarbonate, commonly called Lexan, has the following repeating unit.

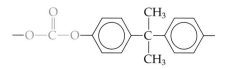

Note the carbonate-like structure on the left.

Polyurethanes are elastomers that are similar to nylon in structure. The repeating unit in a common polyurethane is

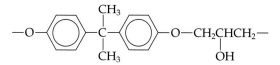

Polyurethanes are popular in foamed padding ("foam rubber") in cushions, mattresses, and padded furniture.

Epoxy resins make excellent surface coatings, and they are powerful adhesives. The following is a typical repeating unit.

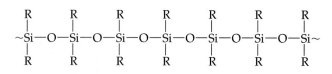

Epoxy adhesives have two components that are mixed just before they are used. The polymer chains become cross-linked, and the bonding is extremely strong.

Composites

Composite materials are made up of high-strength fibers (of glass, graphite, or ceramics) held together by a polymeric matrix, usually a thermosetting condensation polymer. The fiber reinforcement provides the support, and the surrounding plastic protects the fibers from breaking.

The most commonly used composite materials thus far have been polyester resins reinforced with glass fibers. These composites make up about 90% of all reinforced plastic materials. They are widely used in sports gear, such as tennis rackets, and in boat hulls, molded chairs, and automobile panels. Some composite materials have the strength and rigidity of steel, although they weigh only a fraction as much.

Silicones

Not all polymers are based on chains of carbon atoms. Silicones are a good example of a different type of polymer. A **silicone** is a polymer based on a series of alternating silicon and oxygen atoms.

Silicon is in the same group (4A) as carbon and, like carbon, is tetravalent and able to form chains. However, carbon can form chains of just carbon atoms (as in polyethylene), whereas silicone polymers have chains of alternating silicon and oxygen atoms.

$$\sim Si-O-Si-O-Si-O-Si-O-Si-O-Si-O-Si\sim$$

(R represents a hydrocarbon group, such as methyl, ethyl, or butyl.) Silicones can be linear, cyclic, or cross-linked networks. They are heat-stable and resistant to most chemicals, and they are excellent waterproofing materials. Depending on chain length

Various sports cars, including General Motors' Corvette and Fiero, have plastic composite bodies.

and amount of cross-linking, silicones can be oils or greases, rubbery compounds, or solid resins. Silicone oils are used as hydraulic fluids and lubricants, whereas other silicones are used in making such products as sealants, auto polish, shoe polish, and waterproof sheeting. Fabrics for raincoats and umbrellas are frequently treated with silicone.

An interesting silicone toy is Silly Putty. It can be molded like clay or rolled up and bounced like a ball. On standing, it flows like a liquid.

Perhaps the most remarkable silicones of all are the ones used for synthetic human body parts. Kinds of silicone replacements range from finger joints to eye sockets. Artificial ears and noses are also made from silicone polymers. They can even be specially colored to match the surrounding skin.

For many years, silicone gel was used for breast implants. In most cases, these implants have remained perfectly stable over the years. However, because of a manufacturing error, some batches of gel were not sufficiently cured. As a result, the gel implants later disintegrated, causing leakage into body tissues and leading to health problems in some cases. Silicone implants have become a topic of much controversy and litigation. Today only saline-filled implants are approved for use in the United States; they have a silicone–elastomer shell.

Fabrics for raincoats and umbrellas are often waterproofed by coating them with silicone.

10.8 Properties of Polymers

The many kinds of polymers vary greatly in structure and properties. We have already noted that some polymers are thermoplastic, able to be melted down and remolded, whereas other plastics are thermosetting (Section 10.4).

Crystalline and Amorphous Polymers

Some polymers are crystalline, and their molecules line up neatly to form long fibers of high tensile strength. Other polymers are amorphous, composed of randomly oriented molecules that get tangled up with one another (Figure 10.7). Crystalline polymers tend to make good synthetic fibers, whereas amorphous polymers make good elastomers.

Sometimes the same polymer is crystalline in one region and amorphous in another. For example, scientists have designed spandex fibers [used for stretch fabrics (Lycra) in ski pants, exercise clothing, and bathing suits] so as to combine the tensile strength of crystalline fibers with the elasticity of amorphous rubber. Two molecular structures are grafted onto one polymer chain, with blocks of crystalline character

Figure 10.7 Organization of polymer molecules. (a) Crystalline arrangement. (b) Amorphous arrangement.

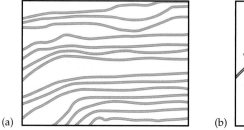

(a)

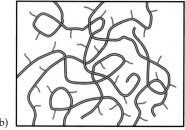

(b)

Spandex fiber is widely used in exercise clothing and sportswear.

Formation of fibers by extrusion through a spinneret. A melted polymer is forced through the tiny holes to make fibers that solidify as they cool.

We can make use of the T_g concept in everyday life. For example, we can remove chewing gum from clothing by applying a piece of ice to lower the temperature of the gum below the T_g of the polyvinyl acetate resin that makes up the bulk of the gum. The cold, brittle resin then crumbles readily and can be removed.

alternating with amorphous blocks. The resulting polymer exhibits both sets of properties—flexibility and rigidity.

The Glass Transition Temperature

An important parameter of most polymers is the glass transition temperature, T_g. Above this temperature, the polymer is rubbery and tough; below it, the polymer is like glass—hard, stiff, and brittle. Each polymer has a characteristic T_g. We want automobile tires to be tough and elastic, and so we use materials with low T_g values. On the other hand, we want plastic substitutes for glass to be glassy; thus, they have T_g values well above room temperature.

Fiber-Forming Properties

Not all synthetic polymers can be converted to useful fibers, but those that can often have properties superior to those of natural fibers. This is why about three-fourths of the fibers and fabrics used today in the United States are synthetic.

Silk fabrics are beautiful and have a luxurious "feel," but nylon fabrics are also beautiful and feel much like silk. Moreover, nylon fabrics wear longer, are easier to care for, and are less expensive than silk.

Polyesters such as PET can substitute for either cotton or silk, but they outperform the natural fibers in many ways. Polyesters are not subject to mildew as cotton is, and polyester fabrics do not need ironing. Fabric made by blending polyester fibers with 35–50% cotton combines the comfort of cotton with the no-iron easy care of polyester fabrics.

Acrylic fibers, made from polyacrylonitrile, can be spun into yarns that look like wool. Acrylic sweaters have the beauty and warmth of wool, but they do not shrink in hot water, are not attacked by moths, and do not cause the allergic skin reactions caused by wool in some people.

10.9 Disposal of Plastics

An advantage of plastics is that they are durable and resistant to many things in the environment. It may be that they are too good in this respect. Some of them last almost forever. Once they are dumped, they do not go away. You see them littering our parks, our sidewalks, and our highways; and if you should go out to the middle of the ocean, you would see them there, too. Small fish have been found dead with their digestive tracts clogged by bits of plastic foam ingested with their food.

Landfills

Plastics make up about 8% by mass of solid waste in the United States, but by volume they make up about 21%. This has created a problem because most solid waste goes into landfills, and we are running out of available landfill space.

Incineration

Another way to dispose of discarded plastics is to burn them. Most plastics have a high fuel value. For example, a pound of polyethylene has about the same energy content as a pound of fuel oil. Some communities actually generate electricity with the heat from garbage incinerators.

Some utility companies burn powdered coal mixed with a few percent of ground-up rubber tires. They not only obtain extra energy from the tires but also help solve the problem of tire disposal.

On the other hand, the burning of plastics and rubber can lead to some new problems. For example, PVC produces toxic hydrogen chloride gas when it burns, and burning automobile tires give off soot and a stinking smoke. Incinerators are corroded by acidic fumes and clogged by materials that are not readily burned.

Degradable Plastics

About half of our waste plastic is from packaging. One approach to the plastics disposal problem is to make plastic packages that are biodegradable or photodegradable (broken down in the presence of bacteria or light). Most biodegradable plastics are starch-based synthetic polymers. Photodegradable plastics usually contain a light-sensitive additive. Of course, it is important that the package remain intact and not start to decompose while it is still being used. So far many people seem reluctant to pay extra for garbage bags that are designed to fall apart.

Recycling

Recycling is perhaps the best way to handle waste plastics. The plastics must be collected, sorted, chopped, melted, and then remolded. Collection works well when there is strong community cooperation.

The separation step is simplified by code numbers (Table 10.4) stamped on plastic containers. Once the plastics have been separated, they can be chopped into flakes, melted, and remolded or spun into fiber.

At present, the only plastic items being recycled on a large scale are HDPE milk jugs, which are now remolded into detergent bottles, and PET soda bottles, which are being turned into fiber, mainly for carpets.

10.10 Plastics and Fire Hazards

The accidental ignition of fabrics, synthetic or otherwise, has caused untold human misery. The U.S. Department of Health and Human Services estimates that fires involving flammable fabrics annually injure as many as 150,000–200,000 people and kill several thousand.

A good deal of research has been done to develop flame-retardant fabrics. One common approach involves the incorporation of chlorine and bromine atoms within

Kevlar (see Problem 44), a fiber made by DuPont, is extremely strong. Among its many uses is in the making of bullet-resistant vests. The Kevlar Survivors' Club, jointly sponsored by the International Chiefs of Police and DuPont, has about 2000 members whose lives were saved by body armor based on Kevlar. DuPont chemist Stephanie Kwolek discovered a way to dissolve the polymer so that it could be made into fibers, a key step in the development of Kevlar.

Japanese scientists have developed a superior 50:50 blend of cotton and polyester by treating the fabric with formaldehyde (HCHO) in the vapor phase. After being washed and dried, the fabric looks as if it has been pressed. When given the formaldehyde treatment, even pure cotton, needs little or no ironing. The formaldehyde acts by cross-linking the strands of cellulose in the cotton.

Table 10.4 ▮ Code for Separating Plastics for Recycling		
Plastic	**Abbreviation**	**Code**
Polyethylene terephthalate	PET	1
High-density polyethylene	HDPE	2
Polyvinyl chloride	PVC	3
Low-density polyethylene	LDPE	4
Polypropylene	PP	5
Polystyrene	PS	6
All others	—	7

Piles of old discarded tires are not only unsightly but are ideal breeding grounds for disease-carrying mosquitoes and other insects and for rodents.

the giant molecules of the polymeric fiber. Federal regulations require that children's sleepwear, in particular, be made of such flame-retardant materials.

Another synthetic fabric, meta-aramid, or Nomex (from DuPont), has such heat resistance that it is used for protective clothing for firefighters and race car drivers. The fibers don't ignite or melt when exposed to flames or high heat. Nomex is also used in electric insulation and for machine parts exposed to high heat.

Burning plastics can also produce toxic gases. Laboratory tests show that hydrogen cyanide is formed in large quantities by the burning of polyacrylonitrile and other nitrogen-containing plastics. Lethal amounts of cyanide found in the bodies of plane crash victims have been traced to burned plastics. Firefighters often refuse to enter burning buildings without gas masks for fear of being overcome by fumes from burning plastics. Smoldering fires also produce lethal quantities of carbon monoxide.

10.11 Plasticizers and Pollution

Chemicals used in plastics manufacture can also present problems. Plasticizers, in particular polychlorinated biphenyls (PCBs), are an important example.

Some plastics, particularly vinyl polymers, are hard and brittle and thus difficult to process. **Plasticizers** can make them more flexible and less brittle by lowering the glass transition temperature, T_g. Thin sheets of pure PVC crack and break easily, but plasticizers make them soft and pliable. Plastic raincoats, garden hoses, and seat covers for automobiles can be made from modified PVC. Plasticizers are liquids of low volatility and are generally lost by diffusion and evaporation as a plastic article ages. The plastic becomes brittle and then cracks and breaks.

Once used widely as plasticizers, but now banned, PCBs are derived from a hydrocarbon called biphenyl ($C_{12}H_{10}$) by the replacement of hydrogen atoms with chlorine atoms. Some PCBs are shown in Figure 10.8. Note their structural similarity to the insecticide DDT.

Not all PCBs manufactured were used as plasticizers. Some became insulating materials in electric equipment (transformers, condensers, and other apparatus) because of their high electric resistance. The same properties that made PCBs so desirable as industrial chemicals cause them to be an environmental hazard. They degrade slowly in nature, and their solubility in nonpolar media—animal fat as well as vinyl plastics—leads to their concentration in the food chain. PCB residues have been found in fish, birds, water, and sediments. The physiologic effect of PCBs is similar to that of DDT. Monsanto Corporation, the only company in the United States that produced PCBs, discontinued production in 1977, but they still remain in the environment.

Thermoplastics can be recycled. After being separated, the polymers can be melted and reshaped.

Biphenyl

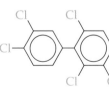

PCB₁

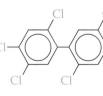

PCB₂

Figure 10.8 Biphenyl and some of the PCBs derived from it. These are but a few of the hundreds of possible PCBs. DDT is shown for comparison.

PCB₃

PCB₄

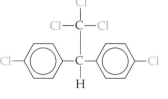

DDT

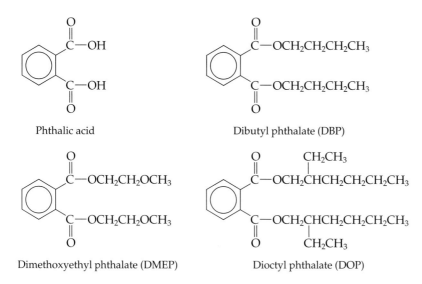

Figure 10.9 Phthalic acid and some esters derived from it. Dioctyl phthalate is also called di-2-ethyl-hexyl phthalate.

Today the most widely used plasticizers for flexible vinyl plastics are phthalate esters, a group of diesters derived from phthalic acid (Figure 10.9). Phthalate plasticizers seem to have low acute toxicity, although their long-term effects are unknown. Fortunately, they degrade fairly rapidly in the environment.

10.12 Plastics and the Future

Widely used today, synthetic polymers are the materials of the future. There is little doubt that there will be new kinds of polymers and even wider use in the future. We already have polymers that conduct electricity, amazing new adhesives, and synthetic materials that are stronger than steel but much lighter in weight. Although plastics present some problems, they have become such an important part of our daily lives that we would find it difficult to live without them.

In the field of medicine, body replacement parts made from polymers have become common, but in the future we will have artificial bones that can stimulate new bone growth so as to knit the new pieces more securely in place. We can also expect to have artificial lungs as well as artificial hearts.

Home construction today uses vinyl water pipes, siding, and window frames, plastic foam insulation, and polymeric surface coatings. Tomorrow's homes may also contain lumber and wall panels of artificial wood, perhaps made from recycled plastic.

Synthetic polymers are used extensively in airplane interiors, and lightweight composite materials comprise the bodies and wings of some planes. Several kinds of automobiles now have plastic bodies. It is likely that in the future many cars will have bodies made from plastic composites, and some may also have frames built from high-strength composite materials.

Electrically conducting polymers help make possible lighter-weight batteries for electric automobiles. The electronics industry will use increasing amounts of electrically conducting thermoplastics in their miniaturized circuits.

The new century will probably see the development of many more exciting polymers. Consider photosynthesis (Chapter 14), the process that uses energy from the sun to carry out a sequence of life-sustaining chemical reactions which provide the food we eat and all the oxygen we breathe. Many laboratories are trying to design polymers that will mimic nature in being able to convert sunlight directly to useful chemical energy.

But here is something to think about. Most synthetic polymers are made from petroleum or natural gas. Both these natural resources are nonrenewable, and our supplies are limited. We are likely to run out of petroleum during the twenty-first century. You might suppose that we would be actively conserving this valuable resource, but unfortunately, this is not the case. We are taking petroleum out of the ground at a rapid rate, converting most of it to gasoline and other fuels, and then simply burning it. There are other sources of energy, but is there anything that can replace petroleum as the raw material for making plastics?

Petroleum

That viscous, tarry liquid nature laid beneath the ground
A hundred million years before man came upon the scene
Can be transformed to marvelous new products we have found—
Like nylon, orlon, polyesters, polyethylene,
Synthetic rubber, plastics, films, adhesives, drugs, and dyes,
And other things, some that we now can only dream about.
Who knows what wondrous products man might some day synthesize
From oil! Except, alas, that our supplies are running out.
The time is near when Earth's prodigious flow of oil may stop
(We've taken so much from the ground, with no way to return it).
Meanwhile, we strive to find and draw out every precious drop,
And then…incredibly…we take the bulk of it and burn it.

Critical Thinking Exercises

Apply knowledge that you have gained in this chapter and one or more of the FLaReS principles (Chapter 1) to evaluate the following statements or claims.

10.1 An environmental activist states that all plastics should be banned.

10.2 A news report states that incinerators that burn plastics are a major source of chlorine-containing toxic compounds called dioxins (Chapter 16).

10.3 We know that vinyl chloride is a carcinogen. Parents fear that their children will get cancer from playing with toys made of PVC.

Summary

1. Polymers are giant molecules made up of many small units called monomers.

2. There are many polymers in nature, including starch, cellulose, nucleic acids, and proteins.

3. The first semisynthetic polymer was called celluloid. It was cellulose that had been modified by nitration.

4. The simplest, least expensive, and highest-volume synthetic polymer is polyethylene.

5. Addition polymers, such as polyethylene, polypropylene, polystyrene, and PVC, are made from monomers containing double bonds.

6. Condensation polymers are formed by splitting out small molecules from monomers containing at least two functional groups per molecule.

7. Nylon, polyethylene terephthalate, and Bakelite are all condensation polymers.

8. Phenol–formaldehyde resin (Bakelite), made in 1909, was the first truly synthetic polymer.

9. Thermoplastic polymers can be softened by heating and then remolded; thermosetting polymers cannot. Nylon and polyethylene terephthalate are both thermoplastic, but Bakelite is thermosetting.

10. Rubber materials are elastomers, cross-linked amorphous polymers that are free to coil and stretch.

11. Strong fibers are made from crystalline polymers that have molecules neatly aligned with one another.

12. Some plastics, especially vinyl polymers, need plasticizers in order to remain pliable. PCBs, once used as plasticizers, are now banned.

13. Disposal of waste plastics is a problem that can probably best be solved by recycling.

14. Most synthetic polymers come from petroleum, which is a limited and nonrenewable natural resource.

Key Terms

addition	copolymer 10.6	polyester 10.7	thermosetting
polymerization 10.5	elastomer 10.6	polymer 10.1	resins 10.4
celluloid 10.3	monomer 10.1	silicone 10.7	vulcanization 10.6
condensation	plasticizer 10.11	thermoplastic	
polymerization 10.7	polyamide 10.7	polymers 10.4	

Review Questions

1. Define the following terms.
   a. macromolecule   b. monomer
   c. polymer   d. elastomer
   e. copolymer   f. plasticizer
2. What does the word "plastic" mean?
3. What is celluloid? How is it made? Why is it no longer used to make movie film?
4. Describe the structure of high-density polyethylene. How does this structure explain the properties of this polymer?
5. Describe the structure of low-density polyethylene. How does this structure explain the properties of this polymer?
6. What is a thermoplastic polymer?
7. How does the structure of PVC differ from that of polyethylene? List several uses of PVC.
8. What is addition polymerization? What structural feature usually characterizes molecules used as monomers in addition polymerization?
9. What is Teflon? What unique property does it have?
10. What inexpensive plastic is used to make clear, brittle, disposable drinking cups? From what monomer are disposable foamed plastic coffee cups made?
11. What plastic is used to make (a) gallon milk jugs? (b) 2-L soda pop bottles?
12. Explain the elasticity of rubber.
13. Describe the process of vulcanization. How does vulcanization change the properties of rubber?
14. Name three synthetic elastomers.

15. How does polybutadiene differ from natural rubber in properties? In structure?
16. How does polychloroprene (Neoprene) differ from natural rubber in properties? In structure?
17. How is SBR made?
18. What substances other than elastomers are used in automobile tires?
19. What is condensation polymerization?
20. What is Bakelite? From what monomers is it made?
21. What is a thermosetting polymer?
22. What proportion of the fibers used in the United States are synthetic?
23. What is the glass transition temperature, T_g, of a polymer? For what uses do we want polymers with a low T_g? With a high T_g?
24. How do plasticizers make polymers less brittle?
25. What are silicones?
26. What are PCBs? Why are they no longer used as plasticizers?
27. What problems arise when plastics are (a) discarded into the environment? (b) disposed of in landfills? (c) disposed of by incineration?
28. What steps must be taken in order to recycle plastics?
29. Discuss plastics as fire hazards.
30. What elements are incorporated into polymer molecules to make flame-retardant fabrics?
31. How can using plastics save energy?
32. Where will plastics come from when Earth's supplies of coal and petroleum are exhausted?

Problems

Addition Polymerization

33. Give the structure of the monomer from which each of the following polymers is made.
    a. polyvinyl chloride   b. polystyrene
34. Give the structure of the monomer from which PTFE (Teflon) is made.
35. Give the structure of a segment of PVC that is at least five repeating units long.
36. Give the structure of a segment of polystyrene in which the carbon chain is at least ten carbon atoms long.

37. Give the structure of the polymer made from each of the following. Show at least four repeating units.
    a. acrylonitrile (CH_2=CHCN)
    b. vinyl acetate (CH_2=CHOCOCH$_3$)
    c. tetrafluoroethylene (CF_2=CF_2)
38. Give the structure of the polymer made from each of the following. Show at least four repeating units.
    a. 1-hexene (CH_2=CHCH$_2$CH$_2$CH$_2$CH$_3$)
    b. styrene

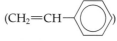

c. methyl methacrylate

$$CH_3 \quad O$$
$$(CH_2{=}C{-}COCH_3)$$

39. Give the structure of isoprene, the monomer of natural rubber.

40. Give the structure of the monomer from which polybutadiene is made.

Condensation Polymerization

41. Nylon 46 is made from the monomers $H_2NCH_2CH_2CH_2CH_2NH_2$ and $HOOCCH_2CH_2CH_2CH_2COOH$.
Give the structure of nylon 46 showing at least two units from each monomer.

42. Give the structure of a polymer made from glycolic acid (hydroxyacetic acid, $HOCH_2COOH$). Show at least four repeating units. (*Hint*: Compare with nylon 6 in Section 10.7.)

43. Kodel is a polyester fiber. The monomers are terephthalic acid and 1,4-cyclohexanedimethanol.

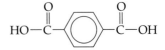

Terephthalic acid

1,4-Cyclohexanedimethanol

Write the structure of a segment of Kodel containing at least one of each monomer unit.

44. Kevlar, a polyamide used to make bulletproof vests, is made from terephthalic acid (Problem 43) and phenylenediamine. Write the structure for a segment of the Kevlar molecule containing at least one of each monomer unit.

Phenylenediamine

Additional Problems

45. In the following equation, identify the parts labeled a, b, and c as monomer, polymer, and repeating unit. What type of polymerization (addition or condensation) is represented?

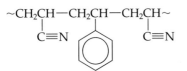

46. Is nylon 46 (Problem 41) an addition polymer or a condensation polymer? Explain.

47. From what monomers could the following copolymer be made?

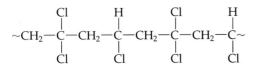

48. One type of Saran has the structure

$$\sim CH_2{-}\underset{\underset{Cl}{|}}{\overset{\overset{Cl}{|}}{C}}{-}CH_2{-}\underset{\underset{Cl}{|}}{\overset{\overset{H}{|}}{C}}{-}CH_2{-}\underset{\underset{Cl}{|}}{\overset{\overset{Cl}{|}}{C}}{-}CH_2{-}\underset{\underset{Cl}{|}}{\overset{\overset{H}{|}}{C}}\sim$$

Give the structures of the two monomers from which this Saran is made.

49. Polyethylene terephthalate has a high T_g. How can the T_g be lowered so that a manufacturer can permanently crease a pair of polyethylene terephthalate slacks?

50. Isobutylene [$CH_2{=}C(CH_3)_2$] polymerizes to form polyisobutylene, a sticky polymer used as an adhesive. Give the structure of polyisobutylene. Show at least four repeating units.

51. Copolymerized with isoprene, isobutylene (Problem 50) forms butyl rubber. Give the structure of butyl rubber. Show at least three isobutylene repeating units and one isoprene repeating unit.

52. The bacteria *Alcalgenes eutrophus* produce a polymer called polyhydroxybutyrate with the structure

$$\underset{n}{\overset{}{\left[O{-}\underset{\underset{CH_3}{|}}{\overset{\overset{}{CH}}{}}{-}CH_2{-}\overset{\overset{O}{\|}}{C}\right]}}$$

Give the structure of the hydroxy acid from which this polymer could be made.

53. The highly heat-resistant polyamide Nomex is a polymer of *meta*-phenylenediamine and *meta*-phthalic acid.

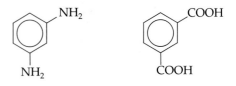

meta-phenylenediamine   *meta*-phthalic acid

Give the structure of a segment of the Nomex molecule containing at least one of each monomer unit.

Projects

54. Make a list of plastic objects (or parts of objects) that you encounter in your daily life. Try to identify a few of the kinds of polymers used in making these items. Compare your list with those of some of your classmates.

55. Do a risk-benefit analysis (Section 1.6) for the use of synthetic polymers as one or more of the following (as directed by your instructor).
  a. grocery bags   **b.** building materials
  c. clothing   **d.** carpets
  e. food packaging   **f.** picnic coolers
  g. automobile tires   **h.** artificial hip sockets

56. To what extent are plastics a litter problem in your community? Survey one city block (or other area as directed by your instructor) and inventory the litter found. (You might as well pick it up while you are at it.) What proportion of the trash is plastics? What proportion of it is fast-food containers?

57. To what extent are plastics recycled in your community? What factors limit further recycling?

Online Projects

58. Do some online research on Kevlar. Where is it used in addition to bullet-resistant vests? What makes it so strong? A possible starting point is http://www.lbl.gov/MicroWorlds/module_index.html (part of the Lawrence Berkeley Laboratory website).

59. Compare the treatment of environmental questions about plastics and other polymers at http://www.plasticsresource.com/ the website of the American Plastics Council, an industry group, and at an environmental site such as http://www.igc.org/pirg/masspirg/envi-ro/sw/pvc/index.htm (IGC stands for Institute for Global Communications) or http://www.greenpeace.org.

60. Environmental news can be found at http://www.planetark.org. Try a search on plastic.

61. Chitin is another natural polymer. Using either the internet or traditional sources, find out what kind of polymer it is and where it is found. You might also find it interesting to explore some of the research into modifying and using it. One interesting site is http://user.chollian.net/~chitin.

References and Readings

1. Bell, John. "Plastics: Waste Not, Want Not." *New Scientist*, 1 December 1990, pp. 44–47.

2. Bruice, Paula Yurkanis. *Organic Chemistry*, 2d edition, Upper Saddle River, NJ: Prentice Hall, 1999. Chapter 25, "Synthetic Polymers."

3. Carraher, Charles E., Jr. *Polymer Chemistry: An Introduction*, 4th edition. New York: Marcel Dekker, 1996.

4. Deanin, Rudolph D. "The Chemistry of Plastics." *Journal of Chemical Education*, January 1987, pp. 45–47.

5. Dinkel, John. "Where the Silica Meets the Road." *Discover*, April 1995, pp. 32–33. Tires are more than just rubber.

6. "Facts and Figures for the Chemical Industry." *Chemical and Engineering News*, 28 June 1999, pp. 32–73. Plastics are reported on p. 37. These figures are updated annually.

7. Friedel, Robert. "The Accidental Inventor." *Discover*, October 1996, pp. 58–69. Roy Plunkett and the invention of Teflon.

8. Hoffmann, Roald. "A Natural-Born Fiber." *American Scientist*, January–February 1997, pp. 21–22. Which is more natural, nylon or rayon?

9. Kauffman, George B., and Raymond B. Seymour. "Elastomers I: Natural Rubber." *Journal of Chemical Education*, May 1990, pp. 422–425.

10. Marvel, C. S. "The Development of Polymer Chemistry in America—The Early Days." *Journal of Chemical Education*, July 1981, pp. 535–539.

11. McCurdy, Patrick P. "Better Things for Better Living—How? Through Chemistry!" *Today's Chemist at Work*, July–August 1996, p. 84. Stephanie Kwolek and the Kevlar story.

12. "Polymers and People." Philadelphia: Beckman Center for the History of Chemistry, 1990.

13. Seymour, Raymond B., and George B. Kauffman. "Elastomers II: Synthetic Rubbers." *Journal of Chemical Education*, March 1991, pp. 217–220.

14. Stone, R. F., A. D. Sagar, and N. A. Ashford. "Recycling the Plastic Package." *Technology Review*, July 1992, pp. 48–56.

MediaLab

Synthetic Polymers: A Blessing or a Curse (or Both)?

Synthetic polymers, in particular plastics, have contributed much to modern civilization. They have enabled such diverse accomplishments as space missions, improved antiseptic conditions in medical treatment, affordable housing, and fat-free cooking. At the same time, plastics accumulate in landfills, may emit toxic fumes when burned, and are made from petroleum, a nonrenewable resource. Attitudes—and websites—about plastics range from enthusiasm to disapproval. (Although we focus on synthetic polymers here, do not forget the importance of natural polymers, ranging from those in our own bodies to such substances as rubber.)

In the following *Web Investigation*, you will explore some differing views about plastics in relation to the environment and human health. In *Communicate Your Results*, you will evaluate various viewpoints for yourself. (Remember the importance of knowing the source of your information!)

WEB INVESTIGATION

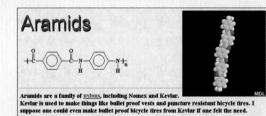

Page from the Macrogalleria,
http://www.psrc.usm.edu/macrog/aramid.htm

Investigation 1
The Benefits of Synthetic Polymers

Select Keyword **POLYMERS** to read an overview of the role of plastics in the modern world. From this site, choose one of the four chapters on "Giants" of polymers to go into more depth in one area.

Investigation 2
Recycling

One of the problems with most plastics is their longevity! Unlike most natural organic substances, they do not degrade readily; thus they contribute to growing landfills. One solution to the problem of increasing solid waste is recycling. Many communities now collect plastics for recycling. However, not all plastics are the same. Select Keyword **PLASTICS RECYCLING–Sorting** and look at the different categories of plastics. Follow some links to look at the subcategories used by the plastics recycling industry. How does this variety hinder community recycling programs? Find some of the products made from recycled plastics. Then select Keyword **PLASTICS RECYCLING—Challenges and Successes** and choose an article. What factors contribute to successful recycling programs?

Investigation 3
Biodegradable Synthetic Polymers?

The problem of solid waste has led to a search for biodegradable plastics. Much research focuses on plastics made wholly or partially from cornstarch or other plant material. Other research aims to develop genetically modified plants or bacteria that can produce polymers that in turn can be degraded by bacteria. Select Keyword **BIODEGRADABLE** for an overview of the various approaches or Keyword **PHA** (follow link to technology overview) for a closer look at the technology of polyhydroxyalkanoate production. Keyword **PROCESS** takes you to a flow diagram showing the use of lactic acid from carbohydrate fermentation. Select **COST** to read about arguments that the cost of plastics from plants may be more than that from petroleum. Follow links or do your own search to try to sort out these conflicting opinions. Keep in mind that environmental issues are seldom all black or all white.

Investigation 4
Can Plastics Poison You?

The potential human toxicity of various plastics is an area of continuing controversy and research. One example is the possibility that bisphenol-A (BPA), used in manufacture of polycarbonate, may leach into foods in polycarbonate containers. Several laboratories have shown that BPA mimics the female hormone estrogen in rats. Select Keyword **BPA** to read an overview of this controversy. Then use Keyword **INDUSTRY** for the plastics industry's viewpoint. Finally, use Keyword **PRECAUTION** for some insight into the scientific debate.

COMMUNICATE YOUR RESULTS

Exercise 1
Recycling: Costs and Benefits

We have seen that there are many types of plastics, which require different processes for recycling into other useful products. Does it make sense to collect *all* categories of plastic? Describe, in writing, what factors should be considered in planning or evaluating a recycling program in order to maximize the benefit–cost ratio. Do this from the perspective of a community, an industry, or an institution. Factor in environmental intangibles as well as actual steps in the recycling process.

Exercise 2 (Collaborative)
Recycling in Your Community

Working in teams, investigate the recycling programs in your community, both those run by the municipality and those run by private businesses. If possible, interview the responsible officials or business owners. Ask such questions as What is recycled? Where do the materials go after pickup? What are the costs or savings? Write an evaluation of the process for a community newspaper. (Depending on your interest and time, you may or may not wish to go beyond plastics.)

Exercise 3
Is Biodegradable Best?

Using what you have found about recycling of plastics and plant-based plastics, discuss the advantages and disadvantages of recycling versus plant-based plastics. Predict, giving your reasoning, which will be the main answer to the problem of plastic waste 10 years from now.

Exercise 4
Baby Bottles

Some publications have expressed concern about polycarbonate baby bottles. Why? Chemical toxicity can be difficult to evaluate. A chemical that kills a mouse may have no apparent effect in humans, and, conversely, something that seems safe for a rat may be carcinogenic for humans. Using information from *Investigation 4*, and what you have learned thus far in this course, list the types of data that might help clarify whether or not plastic baby bottles are safe.

If you wish to compare your ideas with some actual research, use Keyword **RESEARCH**.

Chemistry of Earth
Metals and Minerals

Hills and valleys all around,
Minerals beneath the ground,
Mountains, beaches, open space …
Earth is such a wondrous place!

Earth is a storehouse of chemicals—minerals, metals, and much more. Chemists modify many of these materials to make them more useful.

This wondrous world of ours is a fertile sphere blanketed in air, with about three-fourths of its surface covered by water. Although it is but a tiny, blue-green jewel in the vastness of space, our Spaceship Earth is about 40,000 km in circumference, with a surface area of 500 million km^2.

Human astronauts have walked on the barren surface of Earth's airless moon. Our space probes have explored the desolation of Mars and the crushing pressure and hellish heat of Venus, with its hurricane clouds of sulfuric acid. They have also given us close-up portraits of dry, pockmarked Mercury, and of Jupiter and Saturn with their horrendous lightning storms and their turbulent atmospheres of hydrogen and helium. Earth is but a small island in the inhospitable immensity of space, a tiny oasis uniquely suited to the life that inhabits it.

From earliest times, people have used Earth's resources, from surface stones to throw in hunting to the deepest ores employed in modern industry. Spaceship Earth carries 6 billion passengers, and their numbers are increasing rapidly. What kinds of materials do we have aboard this spaceship? Are they sufficient for this enormous load of passengers? Let us begin by looking at the composition of Earth.

Spaceship Earth is a blue jewel in the blackness of space. The desolate moon is in the foreground.

11.1 Spaceship Earth: The Materials Manifest

Earth is divided into three main regions: the core, the mantle, and the crust (Figure 11.1). The *core* is thought to consist largely of iron with some nickel. Because the core is not accessible and does not seem likely to become so, we won't consider it as a source of materials.

The *mantle* is mostly silicates—compounds of silicon and oxygen with a variety of metals. Although the mantle may be reached eventually, it probably has few useful materials that are not available in the relatively very thin but much more accessible crust.

The *crust* is the outer solid shell of Earth, often called the *lithosphere*. The watery part, made up of the oceans, seas, lakes, rivers, and so on, is the *hydrosphere*. The *atmosphere* is the air that surrounds the planet. In this chapter we focus mainly on the lithosphere. The lithosphere is about 35 km thick under the continents and about 10 km thick under the oceans. Through extensive sampling of the atmosphere, hydrosphere, and lithosphere, scientists have been able to estimate the elemental composition of the outer portion of Earth. Let's consider a random sample of 10,000 atoms. Of these, more than half (5330 atoms) are oxygen. This element occurs in the atmosphere as molecular oxygen (O_2), in the hydrosphere in combination with hydrogen as water, and in the lithosphere in combination with silicon (pure sand is largely SiO_2) and various other elements.

Silicon is the second most abundant element in the outer part of Earth, with 1590 atoms in our 10,000-atom sample. Hydrogen is third with 1510 atoms, most of them in combination with oxygen in water. Hydrogen is such a light element—the lightest of all—that it makes up only 0.9% of Earth's crust by mass. The nine most abundant elements and the numbers of their atoms in our sample of 10,000 are listed in Table 11.1. These nine elements account for 9630 of the atoms, leaving only 370 atoms of all the other elements. However, some very important elements, such as carbon, nitrogen, and phosphorus, are among the elements that are rather scarce on Earth.

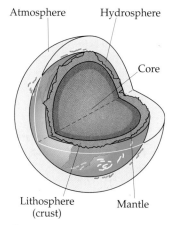

Figure 11.1 Diagram showing the structural regions of Earth. The drawing is not to scale.

Labels: Atmosphere, Hydrosphere, Core, Mantle, Lithosphere (crust)

11.2 The Lithosphere: Organic and Inorganic

The lithosphere is mainly rocks and minerals. Prominent among these are *silicate minerals* (compounds of metals with silicon and oxygen), *carbonate minerals* (metals combined with carbon and oxygen), *oxide minerals* (metals combined with oxygen), and *sulfide minerals* (metals combined with sulfur only). Thousands of these mineral

We discuss the atmosphere in Chapter 12, and the hydrosphere in Chapter 13.

Table 11.1 ▮ Elemental Composition of the Earth's Surface

Element	Symbol	Number of Atoms in a Sample of 10,000 Atoms	Percent by Mass
Oxygen	O	5,330	49.5
Silicon	Si	1,590	25.7
Hydrogen	H	1,510	0.9
Aluminum	Al	480	7.5
Sodium	Na	180	2.6
Iron	Fe	150	4.7
Calcium	Ca	150	3.4
Magnesium	Mg	140	1.9
Potassium	K	100	2.4
All others	—	370	1.4
Total		10,000	100.0

Organic chemistry was once simply the study of chemical compounds found in the life-based portion of the lithosphere. Today, *organic chemistry* (Chapter 9) is the chemistry of the carbon compounds, and *inorganic chemistry* is the chemistry of all the other elements and their compounds. (Simple carbon compounds such as carbon monoxide and carbon dioxide, and ionic compounds such as carbonates and cyanides, are often regarded as inorganic even though they contain carbon.)

compounds make up the *inorganic* portion of the solid crust. We discuss some silicate minerals in Section 11.4. Some typical nonsilicate minerals are listed in Table 11.2.

One of the more common rocks on Earth is limestone, which is calcium carbonate ($CaCO_3$). In nature, calcium carbonate is found in many different physical forms. Marble is calcium carbonate and so are seashells, eggshells, calcite, travertine, aragonite, coral, chalk, and the stalagmite and stalactite formations in caves (Figure 11.2).

Here is something to think about.

The Taj Mahal, resplendent in its solemn majesty;
Gibraltar's famous rock, a symbol of stability;
The snow white Cliffs of Dover in their bleak austerity;
Australia's mighty Barrier Reef that spans the Coral Sea;
The Roman Colosseum with its savage history;
The enigmatic pyramids enduring silently;
And Mammoth Cave's formations with their rich variety—
    The formula, in every case, is $CaCO_3$.

Although much, much smaller in quantity, the *organic* portion of Earth's outer layers includes all living creatures, their waste and decomposition products, and fossilized materials (such as coal, natural gas, petroleum, and oil shale) that once

Table 11.2 ▮ Some Nonsilicate Minerals of Economic Importance

Mineral Type	Name	Chemical Formula	Use
Oxide	Hematite	Fe_2O_3	Ore of iron; pigment
	Magnetite	Fe_3O_4	Ore of iron
	Corundum	Al_2O_3	Gemstone; abrasive
Sulfide	Galena	PbS	Ore of lead
	Chalcopyrite	$CuFeS_2$	Ore of copper
	Cinnabar	HgS	Ore of mercury
	Sphalerite	ZnS	Ore of zinc
Carbonate	Calcite	$CaCO_3$	Cement; lime

Figure 11.2 Limestone formations in a cave. Carbon dioxide in the air dissolves in rainwater to form a weakly acidic solution. As this acidic water seeps through limestone formations, insoluble calcium carbonate is converted to soluble calcium bicarbonate.

$$CaCO_3 + H_2O + CO_2$$
$$\rightleftharpoons Ca(HCO_3)_2$$

Over centuries, this action produces large caves. The reaction is reversible, however. As dripping water, saturated with calcium bicarbonate, evaporates, stalactites and stalagmites of insoluble calcium carbonate are formed. This photograph shows the Temple of the Sun, Carlsbad Cavern, New Mexico.

were living organisms. This organic material always contains the element carbon, nearly always has combined hydrogen, and often contains oxygen, nitrogen, and other elements.

11.3 Meeting Our Needs: From Sticks to Bricks

People long have been able to modify nature's materials to more nearly satisfy their needs and wants. For centuries, technological developments were based largely on trial and error. By developing an understanding of the structures of materials, modern science has greatly increased the human ability to modify natural materials. Thus, technological advances have been greatly accelerated by scientific knowledge.

The needs of early people were few. They obtained food by hunting and gathering. The skins of animals provided clothing when it was needed, and shelter was found in a convenient cave or constructed from available sticks and stones and mud.

After the agricultural revolution (about 10,000 years ago), people no longer were forced to search for food. Domesticated animals and plants supplied their needs for food and clothing. A reasonably assured food supply enabled them to live in villages and created a demand for more sophisticated building materials. People learned to convert natural materials into other products with superior properties. Adobe bricks, which serve well in arid areas, could be made simply by drying mud in the sun.

Along the way people discovered fire, and they learned quite early that cooking meat and grains improved their flavor and digestibility. Fire was one of the earliest agents of chemical change. With it people learned to heat their adobe bricks to make much stronger ceramic bricks. A similar process of firing clay at high temperatures produced ceramic pots, which made the cooking and storage of food much easier than using a spit for roasting it and gourds or wooden vessels for storing it.

With access to high temperatures, people later learned to make glass and to extract metals from ores. They were able to make tools from bronze, and later from iron.

11.4 Silicates and the Shapes of Things

There are thousands of different minerals in the lithosphere. We will consider only a few representative ones here. First, let's look at some silicates. The basic unit of silicate structure is an SiO_4 tetrahedron (Figure 11.3). As shown in Table 11.3, silicate tetrahedra can exist singly as silicate anions or can be joined in a variety of ways.

A prehistoric ceramic pot from Afghanistan.

We discuss the organic portion of the lithosphere in later chapters: petroleum under Energy (Chapter 14), living organisms under Biochemistry (Chapter 15), and edible products from nature under Food (Chapter 16). In this chapter we concentrate on the inorganic portion of the lithosphere.

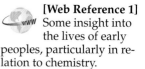

[Web Reference 1] Some insight into the lives of early peoples, particularly in relation to chemistry.

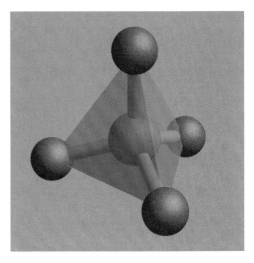

Figure 11.3 The silicate tetrahedron has an Si atom at the center and an O atom at each of the four corners. The shaded area (green) shows how the SiO_4 tetrahedron is typically represented in a mineral structure.

Table 11.3 ▮ SiO_4 Tetrahedra in Some Silicate Minerals

Mineral(s)	SiO_4 Arrangement	Formula	Uses
Zircon	Simple anion (SiO_4^{4-})	$ZrSiO_4$	Ceramics; gemstones
Spudomene	Long chains of tetrahedra	$LiAl(SiO_3)_2$	Source of lithium and its compounds
Chrysotile asbestos	Double chains of tetrahedra	$Mg_3(Si_2O_5)(OH)_4$	Fireproofing (now banned)
Muscovite mica	Sheets of tetrahedra	$KAl_2(AlSi_3O_{10})(OH)_4$	Insulation; fancy paints; packing (vermiculite)
Quartz	Three-dimensional array of tetrahedra	SiO_2	Making glass (sand); gemstones (amethyst, agate, citrine)

Quartz is pure silicon dioxide (SiO_2). The ratio of silicon to oxygen atoms is $1:2$. However, because each silicon atom is surrounded by *four* oxygen atoms, the basic unit of quartz is the SiO_4 tetrahedron. Crystals of pure quartz (rock crystal) are colorless, but various impurities produce amethyst (purple), citrine (yellow), rose quartz (pink), and smoky quartz (gray to black), which are some of the quartz crystals used as gems (Figure 11.4).

Agate, jasper, and onyx are variously colored forms of chalcedony, a kind of quartz with a waxy luster.

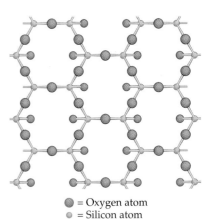

● = Oxygen atom
● = Silicon atom

Figure 11.4 (a) A variety of quartz crystals. Counterclockwise from upper right: citrine (yellowish quartz), colorless quartz, amethyst (purple quartz), and smoky quartz. (b) The chemical structure of quartz. Each silicon atom is bonded to four oxygen atoms, and each oxygen atom is bonded to two silicon atoms.

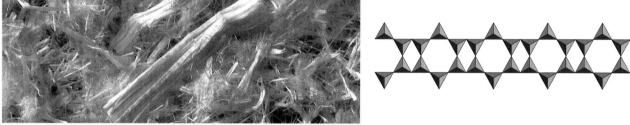

Figure 11.5 (a) A sample of mica, showing cleavage into thin, transparent sheets. (b) The chemical structure of mica shows sheets of SiO_4 tetrahedra. The sheets are bound together by cations, principally Al^{3+} (not shown).

Figure 11.6 (a) A sample of chrysotile asbestos. (b) The chemical structure of chrysotile shows double chains of SiO_4 tetrahedra. The two chains are joined to each other through oxygen atoms. The double chains in turn are bound to each other by cations, principally Mg^{2+} (not shown).

Micas are composed of SiO_4 tetrahedra arranged in two-dimensional sheetlike arrays. Micas are easily cleaved into thin, transparent sheets (Figure 11.5). Pieces of mica were once used as panels for lanterns and as windows in the doors of stoves.

Asbestos is a generic term for a variety of fibrous silicates. Perhaps the best known of these is chrysotile, a magnesium silicate (Figure 11.6). Note that chrysotile is a double chain of SiO_4 tetrahedra. The oxygen atoms that have only one covalent bond also bear a negative charge. Magnesium ions (Mg^{2+}) are associated with these negative charges.

Asbestos: Risks and Benefits

Asbestos is an excellent thermal insulator. It has been used widely to insulate furnaces, heating ducts, and steam pipes. It has been used to make protective clothing for firefighters and others who are exposed to flames and high temperatures. It was also used in brake linings for automobiles.

The health hazards to those who work with asbestos are well known. Inhalation of fibers 5–50 μm long over a period of 10–20 years causes asbestosis, and after 30–45 years, some asbestos workers contract lung cancer. Others get mesothelioma, a rare and incurable cancer of the linings of body cavities.

Long-term occupational exposure to asbestos increases the risk of lung cancer by a factor of 2. Cigarette smoking causes a tenfold increase in the risk of lung cancer. We would expect asbestos workers who smoke to have a risk 20 times that

Clothing made of asbestos fibers was once widely used to protect workers from high temperatures and hot materials. Asbestos for this purpose has been largely replaced by synthetic polymers such as Nomex (Chapter 10). A worker is shown here pouring molten gold in a refinery.

This ancient pottery jar is from San Jacinto, El Salvador.

 [Web Reference 2] An extensive website on ceramics, with illustrated articles on village pottery in the Andes.

Modern *bone china*, used to make fine dinnerware, is actually 50% bone (mainly calcium phosphate from animal bones), 25% kaolin (white clay), and 25% petuntse (decomposed granite). The mixture is blended with water, molded into shape, fired in a kiln, and then glazed.

---

of nonsmokers who do not work with asbestos, but instead their risk is increased by a factor of 90. Cigarette smoke and asbestos fibers act in such a way that each enhances the action of the other. Such a joint action is called a **synergistic effect**.

The harmful effects of asbestos are due mainly to a relatively rare form called crocidolite. Chrysotile, which makes up 95% of the asbestos used in the United States, appears not to be nearly as dangerous. Government regulations do not distinguish between the two types.

Public fears of developing cancer from asbestos have led to regulations that require its removal from schools and other public buildings. Costs of removal will eventually total $50–$150 billion. Benefits from this expenditure are doubtful (Reference 4). In fact, the risk to workers engaged in this removal may exceed the overall risk from leaving the asbestos in place.

11.5 Modified Silicates: Ceramics, Glass, and Cement

Manufacture of ceramic pots, glassmaking, and formulation of cement were among the earliest technologies developed. People learned to modify natural materials such as sand, clay, and limestone, mainly by mixing and heating, to make much more useful products.

Ceramics

Early potters used natural clays, which they merely hardened by heat. Clays are exceedingly complex and their compositions vary widely, but they are basically aluminum silicates. When clay is mixed with water, it can be molded into any shape. Firing leaves a hard, resistant (but porous) product. Bricks and tile are made in this manner. When porosity is not desirable—as in a cooking pot or a water jug—the pottery can be glazed by adding various salts to the surface. Heat then converts the entire surface to a glasslike matrix. Bricks and pottery are examples of **ceramics**, inorganic materials made by heating clay or other mineral matter to a high temperature at which the particles partially melt and fuse together.

Ceramic research has led to the development of some amazing new ceramic materials. Some have such high heat resistance that they can withstand the extreme temperatures of rocket reentry into the atmosphere. They are used to make rocket nose cones, surface tiles, and exhaust nozzles. Other ceramics have magnetic properties and can serve as memory elements in computers. Still others have such exceptional electric conductivity at the temperature of liquid nitrogen that they are known as "superconductors." Such superconducting materials are needed to build the powerful electromagnets used in particle accelerators and magnetically levitated ("maglev") trains.

Glass

Glass, a noncrystalline solid, is another technological development of ancient times. The first glass probably was made in ancient Egypt about 5000 years ago by heating a mixture of sand, sodium carbonate (Na_2CO_3), and limestone (calcium carbonate, $CaCO_3$). As the mixture melts, it becomes a homogeneous liquid. When the liquid cools, it becomes hard and transparent.

When crystalline materials are heated, they melt over a narrow temperature range. Glass is different; when heated, it gradually softens. While soft, it can be blown, rolled, pressed, or molded into almost any shape. The properties of glass result from an irregular arrangement, in three dimensions, of SiO_4 tetrahedra. The

Table 11.4 ▮ Compositions and Properties of Various Glasses

Type	Composition	Special Properties and Uses
Soda–lime glass	Sodium and calcium silicates (sand plus soda plus lime)	Ordinary glass (for windows, bottles, and so on)
Borosilicate glass (Pyrex, Kimax)	Boron oxide (instead of lime)	Heat-resistant (for laboratory ware and ovenware)
Aluminosilicate glass	Aluminum oxide (instead of soda)	More highly heat-resistant (for top-of-stove cookware and fiberglass)
Lead glass	Lead oxide (instead of lime)	Highly refractive (for optical glass, art glass, table crystal)
Alabaster glass	Sodium chloride (salt) added	White, opaque ("milk glass")
Colored glass	Selenium compounds added	Red color (ruby glass)
	Cobalt compounds added	Blue color (cobalt glass)
	Chromium compounds added	Green color
	Manganese compounds added	Violet color
	Cadmium sulfide added	Yellow color
	Carbon and iron oxide added	Brown color (amber glass)
Photochromic glass	Silver chloride or bromide added	Light-sensitive; darkens when exposed to light (for sunglasses, hospital windows)
Laser glass	Contains neodymium	Powerful lasers
Frosted glass	Etched with hydrofluoric acid (HF)	Satiny frosted surface

chemical bonds in this arrangement are not all equivalent. Thus, when glass is heated, the weaker bonds break first and the glass softens gradually.

The basic ingredients in glass can be used in different proportions. Oxides of various metals can be substituted in whole or in part for the lime, sodium carbonate ("soda"), or sand. Thus, many special types of glass can be made. Some examples are given in Table 11.4.

No vital raw materials are involved in the manufacture of ordinary glass, but the furnaces used to melt and shape glass require energy. Disposal presents a problem because glass is one of the most permanent materials known. It doesn't rot when discarded. Glass is easily recycled, however, as long as different kinds are separated. It can be melted and formed into new objects at considerable energy savings compared with the manufacture of new glass. In fact, it is standard practice to add "cullet" (broken glass) to each batch of sand, soda, and limestone to be melted. The cullet is usually scrap glass from previous melts. This practice not only gets rid of waste glass but also actually improves the melting process.

Heat-softened glass being shaped by blow-molding.

Optical Fibers

Modern glass research has resulted in many wonderful new products, but one of the most remarkable developments is a product that is practically invisible. It consists of pure threads of glass about as thick as a human hair. These optical fibers have replaced many old telephone lines. A bundle of glass fibers can carry several hundred times as many messages as a copper cable of the same size.

Optical fibers actually carry light. The messages they carry are intermittent bursts of light. When you speak into a telephone, the sound is translated into an

These optical fibers have become the preferred transmission medium for telecommunications. The glass fibers are generally protected with plastic coatings to preserve strength and to make them easier to handle.

 [Web Reference 3] Some of the varieties of glass can be seen at the website of the Corning Museum of Glass.

A long rotary kiln. Limestone is mixed with clay and sand to produce a complex mixture of calcium and aluminum silicates comprising a common kind of cement known as portland cement.

 [Web Reference 4] A discussion of the advantages and disadvantages of cement and concrete.

electric signal, which then pulses a laser to give off light messages that are carried by the glass fibers. At the end of the line, the light pulses are converted back to sound waves.

Optical fibers are also used in computer interfacing and are employed by medical personnel to view the inside of the body (blood vessels, intestines, and so on) without surgery. One can only imagine what other applications will be found for them in the future.

Cement and Concrete

Cement is still another ancient technological development. The Romans used cement to construct roads, aqueducts, and the famous Roman baths. The raw materials for the production of cement are limestone (calcium carbonate, $CaCO_3$) and clay (aluminum silicates). The materials are finely ground, mixed, and roasted at about 1500°C in a kiln heated by burning natural gas or powdered coal. The finished product is mixed with sand and gravel to form *concrete*.

Our understanding of the complex chemistry of cement is still imperfect. Nevertheless, cements with special properties are available, including a cement that sets quickly with high early strength, a white cement, a waterproof cement, and one that sets at high temperatures.

Concrete is used widely in the construction of buildings and roads because it is inexpensive, strong, chemically inert (and thus nonpolluting in itself), durable, and tolerant of a wide range of temperatures. However, its production involves extensive mining, with entire mountains being torn down for limestone rocks. The rotary kiln process consumes fossil fuels. Particulate matter from the crushing operations, and smoke and sulfur dioxide from the burning of fossil fuels, make air pollution from cement plants especially serious. In use, concrete covers what once were green acres. In cities, areas paved with asphalt and concrete are sufficiently extensive to change the climate (temperature and rainfall). Runoff of rainfall from paved areas is especially rapid and contributes to flash flooding. However, concrete can be broken up and used as rock fill; it is also sometimes used to encase hazardous waste for disposal.

11.6 Metals and Ores

Human progress through the ages is often described in terms of the materials used for making tools. We speak of the Stone Age, the Bronze Age, and the Iron Age. Ancient peoples knew about gold and silver because they are often found in the free state in nature, but these metals are too soft to use as tools. Metal tools were not possible until artisans learned to extract certain metals from ores by smelting.

Copper and Bronze

Copper was probably the first metal to be freed from its ore by early smelting techniques. We know from ancient Egyptian records that copper was known 5500 years ago. The Egyptians probably isolated copper by heating copper carbonate ($CuCO_3$) or copper sulfide (Cu_2S) ore.

$$Cu_2S(s) + O_2(g) \longrightarrow 2\,Cu(s) + SO_2(g)$$

Many copper ores are blue to green in color and easy to identify in the ground. The ancient Egyptians produced thousands of tons of copper metal.

Even more important than copper was **bronze**, a copper alloy containing about 10% tin. Bronze is harder than copper, and it could therefore be made into many use-

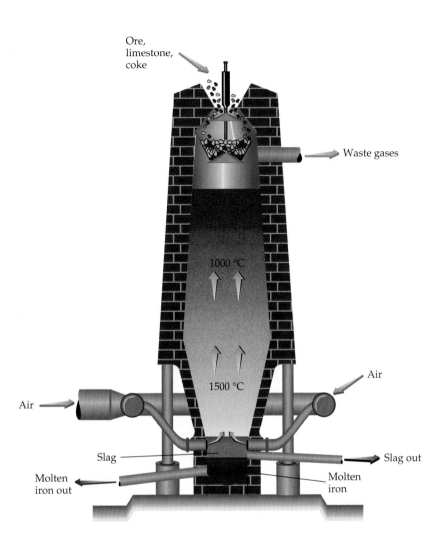

Ore, limestone, coke

Waste gases

1000 °C

1500 °C

Air

Air

Slag

Slag out

Molten iron out

Molten iron

Figure 11.7 A modern blast furnace. Iron ore, coke, and limestone are added at the top, and hot air is injected at the bottom.

This ancient bronze drum was made in Southeast Asia. The Bronze Age occurred at different times in different cultures. Some cultures seem to have skipped it altogether, going directly from the Stone Age to the Iron Age.

[Web Reference 5] An illustrated history of copper, part of the Copper Page, which covers many aspects of copper and the copper industry.

Today copper is important mainly because of its excellent electric conductivity. It is used primarily for electric wiring.

An **alloy** is a mixture of two or more elements, at least one of which is a metal. Alloys have metallic properties.

About half of all steel made today in the United States comes from "mini-mills," which make steel from scrap iron instead of ore. Almost all automobiles today are recycled.

Steel is an alloy of iron and carbon. Many varieties of steel also contain other elements.

ful tools. In some societies, bronze remained the most widely used metal for about 2000 years.

Iron and Steel

To produce iron from ore, carbon (coke) is employed as the reducing agent. First, the carbon is converted to carbon monoxide.

$$2\ C(s)\ +\ O_2(g)\ \longrightarrow\ 2\ CO(g)$$

Then the carbon monoxide reduces the iron oxide to iron metal.

$$Fe_2O_3(s)\ +\ 3\ CO(g)\ \longrightarrow\ 2\ Fe(s)\ +\ 3\ CO_2(g)$$

Iron can be made from ore in a huge chimneylike vessel called a *blast furnace* (Figure 11.7). The raw materials fed into the furnace are iron ore, coke, and limestone. The coke (which is mainly carbon) is made by heating coal in the absence of air to drive off coal oil and tar. The limestone is added to combine with silicate impurities to form a molten **slag** that floats on top of the iron and is drawn off. Molten iron is drawn off at the bottom of the furnace. The product of the blast furnace, called *pig iron*, has a fairly low melting point, and so it is easily cast into molds. Hence, it is also known as cast iron.

Cast iron is brittle and has many impurities, and so most iron from blast furnaces is converted to steel. In the steel furnace, pressurized oxygen reacts with

[Web Reference 6] For more information about metals from ancient times to the present.

A tin can—really steel with a thin coating of tin—rusts when you throw it away, eventually disintegrating. Scientists at Pennsylvania State University estimate that it would take 500 years for an aluminum can to degrade completely.

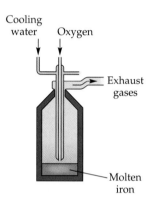

Cooling water · Oxygen · Exhaust gases · Molten iron

Figure 11.8 A basic oxygen furnace. The steel vessel is lined with dolomite, a mixed calcium and magnesium carbonate ($CaCO_3 \cdot MgCO_3$).

The method we still use to produce aluminum was discovered more than 100 years ago by a college student named Charles Martin Hall. Interestingly, the same process was discovered almost simultaneously in France by another college student named Paul Héroult. Both men were born in 1863, and they developed the process for manufacturing aluminum in 1886. But the coincidence doesn't end there. Both men died in 1914 at the age of 51.

impurities, such as phosphorus, silicon, and excess carbon. The properties of steel can be varied over a wide range by adjusting the amount of carbon in it. High-carbon steel is hard and strong, whereas low-carbon steel is ductile and malleable.

Steel is commonly alloyed with other metals to give it special properties. For example, manganese imparts hardness, tungsten gives high-temperature strength, and additions of chromium and nickel produce stainless steel. Because it is so abundant and can be made into so many different alloys, iron is the most useful of the metals.

Most steel producers now use the basic oxygen process (Figure 11.8). The furnace uses pig iron, an impure form of iron from a blast furnace. Powdered limestone is added just above the molten iron, and oxygen gas at about 10 atm pressure is injected there also. Metallic impurities in the pig iron are converted to oxides which react with SiO_2 to form a slag that floats above the liquid iron and is poured off. Then any desired alloying metals are added.

Some companies now make iron directly from iron ore in a single-step continuous process. They avoid the great energy requirements of the blast furnace by using temperatures below the melting points of any of the raw materials. The ore is reduced by H_2 and CO, which in turn are made from natural gas (Chapter 14). This *direct reduction* method is used mainly in developing countries that do not already have large iron and steel industries in place.

Aluminum: Abundant and Light

Aluminum, the most abundant metal in Earth's crust, has replaced iron for many purposes. However, aluminum ions are tightly bound in compounds in nature, and considerable energy is required to extract the metal from its ores. The principal ore of aluminum is bauxite, aluminum oxide. Al_2O_3 is extracted from impurities with a strong base, and then the melted oxide is reduced by passing electricity through it. It takes 2 metric tons (t) of aluminum oxide and 17,000 kilowatt hours (kWh) of electricity to produce 1 t of aluminum.

Aluminum is light and strong. A piece of aluminum weighs only one-third as much as a piece of steel the same size. Although it is considerably more active than iron, aluminum corrodes much more slowly. Freshly prepared aluminum metal reacts with oxygen to form a hard, transparent film of aluminum oxide (Al_2O_3) over its surface, and the thin film protects the metal from further oxidation. Iron, on the other hand, forms an oxide coating that is porous and flaky. Instead of protecting the metal, this coating flakes off, allowing further oxidation. Iron rusts; aluminum does not.

The Environmental Cost of Iron and Aluminum

Steel and aluminum play vital roles in the modern industrial world, and it would be hard to overemphasize their economic value. But shouldn't we include the environmental cost of producing, using, and discarding these materials?

Both steel mills and aluminum plants produce considerable waste. In days gone by, steel mills discharged lime, acids, grease, oil, and iron salts into waterways. They released carbon monoxide, nitrogen oxides, particulate matter, and other pollutants into the air. Aluminum plants discharged iron, aluminum, and other metal oxides into waterways, and particulate matter, fluorides, and other pollutants into the air. Modern treatment methods have reduced the quantity of pollutants released.

Scientists estimate that it takes 15 times as much fuel energy to produce aluminum as it does to produce a comparable weight of steel. Aluminum is less dense than steel, but making an aluminum can requires 6.3 times as much energy as making a steel can. On the other hand, when aluminum is recycled, the energy needed

Table 11.5 ▌ Some Other Important Metals

Metal	Symbol	Important Ore	Selected Properties	Typical Uses
Chromium	Cr	$FeCr_2O_4$	Shiny, resists corrosion	Chrome plating, stainless steel
Gold	Au	Au	Yellow metal, soft, dense	Coinage, jewelry, dentistry
Lead	Pb	PbS	Low-melting, dense, soft	Plumbing, batteries
Magnesium	Mg	$MgCl_2$	Light, strong	Auto wheels, luggage
Mercury	Hg	HgS	Dense liquid	Thermometers, barometers
Nickel	Ni	NiS	Resists corrosion	Coinage, alloy for stainless steel
Platinum	Pt	Pt	Inert, high-melting	Catalyst, instruments
Silver	Ag	Ag_2S	Excellent electric conductor	Electric contacts, mirrors, jewelry, coins
Sodium	Na	NaCl	Reactive, soft	Heat-transfer medium, reducing agent
Tin	Sn	SnO_2	Resists corrosion	Coating for steel cans
Tungsten	W	$CaWO_4$	Very high-melting	Light-bulb filaments
Uranium	U	U_3O_8	Fissionable	Energy source
Zinc	Zn	ZnS	Forms protective coating	Galvanizing coating

is only a fraction of that required to make the metal from ore. Further, a car made largely from aluminum is so much lighter than a similar car made entirely from steel that it takes a good deal less energy to operate it.

Other Important Metals

There are many other important metals. Some of them, with typical ores, properties, and uses, are listed in Table 11.5.

Metals and minerals are vital to a vigorous economy. The United States has depleted much of its high-grade ores and become increasingly dependent on other nations for these materials. No industrial nation is 100% self-sufficient in all the vital metals and minerals. The United States is more than 90% dependent on imports for chromium (Cr), molybdenum (Mo), and tungsten (W)—elements vital for steelmaking and other industries—and for platinum (Pt) and palladium (Pd)—elements used in making catalysts, including those needed to reduce pollution from automobile exhaust emissions. Metals and minerals are essential components of the modern global economy.

How can we ever run short of a metal? Aren't atoms conserved? Yes, there is as much iron on Earth as there was 100 years ago, but by using it we scatter the metal throughout the environment. Gathering it back to a factory to make new objects requires energy, just as obtaining metals from low-grade ores requires more energy than extracting them from high-grade ores.

11.7 Running Out of Everything: Earth's Dwindling Resources

Dramatic fuel shortages during the 1970s brought important changes in government policies and in industrial and individual practices. But now many of the conservation efforts begun at that time have largely been forgotten. We remain vulnerable to future fuel shortages that could be much worse than those of the 1970s.

We also face potential shortages of metals and minerals. The United States has exhausted its high-grade ores of metals such as copper and iron. When the Europeans first came to North America, the native Americans were mining nearly pure copper in Michigan. Now we mine ore with less than 0.5% copper. For much of the nineteenth century, we mined high-grade iron ore from the Mesabi range in Minnesota. Now we mine taconite, a hard rock that contains only small, dispersed

According to the Steel Recycling Institute, about 5740 kJ of energy is conserved for each pound of steel recycled. That is enough energy to light a 60-watt (W) bulb for more than 26 hr. Each ton of steel recycled saves 2500 lb of iron ore, 1000 lb of coal, and 40 lb of limestone.

In Chapter 16, we will discuss several metals (as metal ions) and minerals that are essential to life.

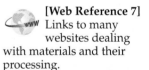 **[Web Reference 7]** Links to many websites dealing with materials and their processing.

These crystals of native copper are from Houghton, Michigan.

Waste cyanide solution from the leaching of gold from ore is stored in a large containment pond. This facility is part of a gold mining operation in the Mojave Desert of California. In January 2000 a dike holding cyanide-laced wastes gave way at a Romanian gold smelter. The wastes entered the Tisza River, forcing towns downstream to close their water intake systems. The cyanide also killed fish, birds, and other wildlife downriver in Romania and Hungary.

In 1974 the U.S. Central Intelligence Agency (CIA) used a putative search for manganese nodules as a cover for an attempt to raise a sunken Soviet submarine, the *K129*, from the depths of the Pacific Ocean. The recovery ship *Glomar Explorer* hoisted aboard a 100-ft section of the K129 containing two nuclear-tipped torpedoes and the bodies of six Russian sailors. Just what intelligence information the CIA obtained is still secret. The six sailors were buried at sea in a solemn ceremony filmed by the CIA.

Manganese-rich nodules on the ocean floor. They also contain iron and traces of cobalt, copper, and nickel.

amounts of iron oxide. Australia has vast deposits of high-grade iron ore, but the cost of importing it from there is high. Similarly, miners once found nuggets of pure gold and silver, but the "glory holes" of western North America are long gone. Now we mine ores with a fraction of an ounce of gold in a ton of gold ore (see Problem 52). Indeed, there are few high-grade deposits of any metal ores left that are readily accessible to the industrialized nations.

What's wrong with low-grade ores? Because more must be mined for the same final amount of metal, there is more environmental disruption, more energy is required to concentrate the ores, and both environmental cleanup and energy cost money. Consider an analogy. A bag of popcorn is useful. You can pop it and eat it. The same popcorn, scattered all over your room, would be less useful. Of course, you could gather it all up and then pop it, but that would take a lot of energy—perhaps more energy than you would care to expend and perhaps more energy than you would get back if you popped the corn and ate it.

We also must import quite a few metals because their ores are not found in the United States. For example, we get tin from Bolivia, chromium from Rhodesia, and platinum from South Africa and Russia. But the metal reserves in other countries are being depleted, too.

Where will we get metals in the future? The sea is one possible source. Nodules rich in manganese cover vast areas of the ocean floor. These nodules also contain copper, nickel, and cobalt. Questions of who owns them and how to mine them without major environmental disruption remain to be resolved.

11.8 Land Pollution: Solid Wastes

Productivity and creativity have enabled people in industrialized nations to have a seemingly endless variety of consumer goods in enormous quantities. These goods often are packaged in paper or plastic. Their wrappers litter the environment and fill our disposal sites. As the goods break, wear out, or merely become obsolete, they too add to our disposal problems. The approximate composition of municipal solid wastes in the United States is given in Figure 11.9.

In the past, most solid wastes were simply discarded in open dumps. This led to infestation by rats, flies, and other pests that often spread to nearby areas. Open burning led to offensive and unhealthful air pollution. These open dumps have now been largely phased out.

The primary method of disposal today is the *sanitary landfill*. Garbage and trash are piled into a trench, compacted, and covered over. This eliminates the problem of

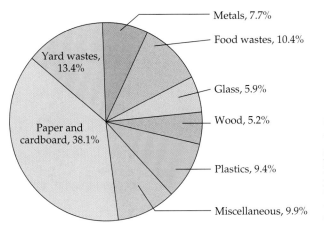

Figure 11.9 Municipal wastes in the United States are largely paper, cardboard, and yard wastes. Paper and cardboard can be recycled. Yard wastes and food wastes can be composted.

rats, flies, and odors. Landfills leak, however, leading to groundwater contamination. And materials in landfills decompose much more slowly than previously thought. Newspapers are still readable after being entombed for 20 years. Without water and oxygen, microorganisms are unable to carry out the normal decay processes.

We can also dispose of solid wastes by *incineration*. When this method is carried out in properly designed incinerators, air pollution is minimal. However, many incinerators are old, or poorly designed, and they generate considerable smoke and odor. Building new incinerators tends to be unpopular with local citizens.

Research in solid waste disposal has led to several new processes. The U.S. Bureau of Mines has a method for converting garbage to oil with an overall yield of 25%. Ground-up rubber tires can be mixed with powdered coal and used as fuel. Waste glass and rubber tires have both been added to road-paving mixtures for highway construction.

11.9 The Three R's of Garbage: Reduce, Reuse, Recycle

There are several ways to deal with our garbage problem. Not all are equally desirable. We should choose wisely among them in order to best save energy and protect the environment.

The best way to deal with our solid waste problem is to *reduce* the amount of throwaway materials produced. Decreasing the volume of materials produced and sold saves resources and energy and minimizes the disposal problem.

Next best is the *reuse* of materials. When possible, items should be made durable enough to withstand repeated use, rather than designed for a single use and disposal. Consider a glass beverage bottle. It takes about 6300 J (1500 kcal) of energy to make a nonreturnable half-liter bottle. It takes a little more energy, 8300 J (2000 kcal), to make a more durable, returnable bottle of the same capacity. The returnable bottle is used an average of 12.5 times before it is broken. That is about 660 J per use, only about 10% of the energy used to make the single-use nonreturnable bottle.

The third way to reduce the volume of wastes is to *recycle* them. Recycling requires energy, and some material is unavoidably lost, but recycling saves materials. Metals such as iron and steel, aluminum, copper, and lead are largely recycled. About 40% of paper and cardboard are recycled. Lawn trimmings and food wastes can be composted, converting them to soil. About 40% of yard trimmings are composted. In the future, as raw materials become more scarce, these proportions are bound to increase.

Recycling also saves energy. It takes only 5% as much energy to make new aluminum cans from old ones as it does to make them from aluminum ore. Using a ton

About 8 million automobiles are junked each year in North America, and 250 million tires are thrown away.

Some cities not only burn their combustible solid wastes but also use the heat from the incinerators to warm buildings, to generate power, or both. A pound of polyethylene contains about 20,000 British thermal units (Btu) of energy, about the same as a pound of no. 2 fuel oil.

[Web Reference 8]
Waste management information.

Every day in the United States, each person discards an average of about 2 kg of trash.
Every 3 months we throw away enough aluminum to rebuild our entire commercial air fleet.
The glass jars and bottles we discard each week would fill one of the towers at the New York Trade Center.
Every month we throw away a million tons of newspapers.

Waste materials separated for curbside collection. Once separated, the components are easily recycled.

Recycling a 1-m stack of newspapers saves the equivalent of a 10-m pine tree. Making paper from scrap instead of virgin wood pulp yields a 50% savings of water and energy.

of scrap to make new iron and steel saves 1.5 tons of iron ore and 0.3 ton of coal. It also results in a 74% savings in energy, an 86% reduction in air pollution, and a 76% reduction in water pollution.

Recycling of household wastes requires that materials be separated. Many communities have curbside pickups of separated paper, glass, aluminum, and other materials. This separation requires renewable human energy. The production of new materials requires nonrenewable energy from fossil fuels. Disposal of trash costs U.S. residents $4 billion each year. By following the three R's—reduce, reuse, and recycle—we can protect the environment at the same time that we save materials, energy, and money.

11.10 How Crowded Is Our Spaceship?

Each day, people die and other people are born, but the number born is about 215,000 more than the number that die. Every 4.7 days our world population goes up by a million people. In 1999 the world's population reached 6 billion.

Figure 11.10 is a graphic picture of population growth. For tens of thousands of years, human population was never more than a few hundred million. It did not reach the 1 billion mark until about 1800. Population growth peaked in the late 1960s

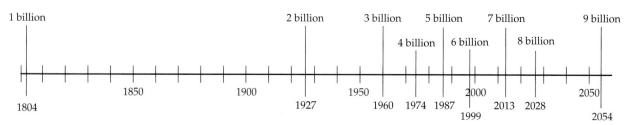

Figure 11.10   Two thousand years ago, Earth's population was about 300 million. It changed little over the next thousand years and is estimated to have been only 310 million at the end of the first millennium. It reached 1 billion about 1804. This graph shows the intervals in which an additional billion people were added to Earth's population, with projections to 2054. (*Source:* United Nations, *World Population Prospects: The 1998 Revision.* New York: December 1998.)

at about 2.0% a year. The present rate of growth is about 1.3% per year. The United Nations estimates a world population of 9 billion by 2050.

How many people can Spaceship Earth accommodate? Many believe that we have already gone beyond the optimum population for this planet. They predict increasing conflicts over resources and increasing pollution of the environment.

Why did Earth's population start increasing so rapidly after centuries of little change? Population growth depends on both the birth rate and the death rate. If they are equal, population growth is zero. They had been almost equal for thousands of years, but scientific progress in the 1800s led to a decline in the death rate in developed countries. Yet the birth rate changed little and the population grew rapidly. These changes reached the less developed countries in the 1950s, greatly lowering the death rate. However, the birth rate remained high and the population exploded. Population growth has slowed considerably in recent years, especially in developed countries. Our wonderful scientific progress led to a rapid decline in death rates, but a decline in birth rates has come more slowly.

Population and the food supply are discussed in Chapter 16. For a more detailed study of population issues, see the MediaLab on pages 312–313.

Critical Thinking Exercises

Apply knowledge that you have gained in this chapter and one or more of the FLaReS principles (Chapter 1) to evaluate the following statements or claims.

11.1 An economist has said that we need not worry about running out of copper because it can be made from other metals.

11.2 A citizen testifies against establishing a landfill

near his home, claiming that the landfill will leak substances into the groundwater, contaminating his water well (Chapter 16).

11.3 A citizen testifies against establishing an incinerator near her home, claiming that plastics burned in the incinerator will release hydrogen chloride into the air.

Summary

1. Earth is divided into three main regions: the core, the mantle, and the crust.

2. The core is thought to be made up largely of iron; the mantle is mostly silicate minerals.

3. The solid crust is the lithosphere; the liquid oceans, lakes, and rivers make up the hydrosphere; and gaseous air constitutes the atmosphere.

4. The lithosphere is made up largely of minerals such as silicates, carbonates, oxides, and sulfides.

5. Asbestos is composed of chains of SiO_4 tetrahedra, mica of sheets of tetrahedra, and quartz of a three-dimensional network of tetrahedra. Quartz is pure silicon dioxide, and sand is mainly quartz.

6. Ceramics, glass, and cement are modified silicates.

7. Most metals must be extracted from their ores before they can be made into tools. Ancient peoples used

bronze and, later, iron; modern societies use many metals, especially steel and aluminum.

8. Alloys are metallic mixtures such as bronze (copper and tin) and stainless steel (iron, carbon, chromium, and nickel).

9. Earth's resources are being used up at such a rate that we are bound to face shortages in the future unless we recycle materials more faithfully.

10. Our throwaway society generates mountains of solid waste each year, and we are running out of available landfill space.

11. Three ways to deal with our mounting garbage problems are to reduce our volume of disposables, to reuse items when we can, and to recycle our waste whenever possible.

Key Terms

alloy 11.6
asbestos 11.4
bronze 11.6

cement 11.5
ceramic 11.5
glass 11.5

mica 11.4
quartz 11.4
slag 11.6

steel 11.6
synergistic effect 11.4

Review Questions

1. What are the three main regions of Earth?
2. What are silicates?
3. What is a synergistic effect?
4. What environmental problems are associated with the manufacture of cement?
5. What is concrete?
6. What environmental problems are associated with steel mills?
7. What environmental problems are associated with aluminum production?
8. If matter is conserved, how can we ever run out of a metal?
9. List three methods of solid waste disposal. Give the advantages and disadvantages of each.
10. Explain why metals can be recycled fairly easily. What factors limit the recycling of metals?

Problems

Composition of Earth

11. Define lithosphere, hydrosphere, and atmosphere.
12. What element makes up most of the core of Earth?
13. What is the most abundant element in the outer portion of Earth?
14. Name four kinds of minerals found in Earth's crust and give an example of each.
15. What materials make up the organic portion of the lithosphere?
16. What is inorganic chemistry?

Silicate Minerals

17. What is the basic structural unit of silicate minerals?
18. What is the chemical composition of quartz? Describe its structure.
19. Why does mica occur in sheets?
20. Why does asbestos occur as fibers?

Modified Silicates

21. How does the structure of glass differ from that of crystalline silicates?
22. How do the properties of glass differ from those of crystalline substances?
23. What are the three principal raw materials used for making glass?
24. How is the basic recipe for glass modified to make glasses with special properties?
25. What is cement?
26. What are the two basic raw materials required for making cement?

Metals and Ores

27. What are the principal raw materials used in modern iron production? What is the purpose of each?
28. What is coke? What is slag?
29. By what kind of chemical process is a metal obtained from its ore?
30. By what chemical process does a metal corrode?
31. What is an alloy?

32. What is steel?
33. What is the most abundant metal in Earth's crust? Why is it not the most widely used?
34. Which metal is the most widely used? Why?
35. An aluminum plant produces 65 million kg of aluminum per year. How much aluminum oxide is required? How much bauxite is required? (It takes 2.1 kg of crude bauxite to produce 1.0 kg of aluminum oxide.)
36. It takes 17 kWh of electricity to produce 1 kg of aluminum. How much electricity does the plant in Problem 35 use for aluminum production in 1 year?

Oxidation and Reduction

For Problems 37–42, answer the following questions. (You may wish to review Chapter 8.) **(a)** What substance is reduced? **(b)** What is the reducing agent? **(c)** What substance is oxidized? **(d)** What is the oxidizing agent?

37. The Hunter method for the production of titanium uses the reaction

$$TiCl_4 + 4\,Na \longrightarrow Ti + 4\,NaCl$$

38. Pure chromium is prepared by reacting chromium(III) oxide with aluminum.

$$Cr_2O_3 + 2\,Al \longrightarrow 2\,Cr + Al_2O_3$$

39. Thorium metal is prepared by reacting thorium(IV) oxide with calcium.

$$ThO_2 + 2\,Ca \longrightarrow Th + 2\,CaO$$

40. Potassium metal is prepared by reacting molten potassium chloride with sodium metal.

$$KCl + Na \longrightarrow NaCl + K$$

41. Aluminum chloride reacts with manganese metal to form aluminum metal and manganese(II) chloride.

$$2\,AlCl_3 + 3\,Mn \longrightarrow 2\,Al + 3\,MnCl_2$$

42. The Toth Aluminum Company produces aluminum from kaolin. One step converts aluminum oxide to aluminum chloride by the following reaction.

$$Al_2O_3 + 3\,C + 3\,Cl_2 \longrightarrow 2\,AlCl_3 + 3\,CO$$

Mass Relationships

For Problems 43–46, you may wish to review Chapter 6.

43. Copper metal is prepared by blowing air through molten copper(I) sulfide.

$$Cu_2S + O_2 \longrightarrow 2\,Cu + SO_2$$

How much copper is obtained from 143 g of Cu_2S?

44. How much sodium is required to react with 1.73 g of KCl? Use the equation in Problem 40.

45. How much cesium chloride must be reduced to produce 0.529 g of cesium? The reaction is

$$2\,CsCl + Ca \longrightarrow 2\,Cs + CaCl_2$$

46. How much aluminum is required to reduce 41.7 g of calcium oxide? The reaction is

$$3\,CaO + 2\,Al \longrightarrow 3\,Ca + Al_2O_3$$

Additional Problems

47. An ore deposit at Crandon, Wisconsin, is estimated at 75 million tons and assays at 5.0% zinc, 0.4% lead, and 1.1% copper. What mass of each metal is contained in the deposit?

48. The Parc mine at Llanwrst, North Wales, United Kingdom, was closed in 1954, leaving behind a mound of about 250,000 tons of tailings. The tailings assayed at 3.22% Zn, 0.82% Pb, and 0.023% Cd. Erosion has washed away 13,000 tons of the tailings. How much of each of the three elements has been washed away?

49. Approximately 81,000 tons of gold has been mined throughout history. What size cube of gold would this make if it were all in one piece? The density of gold is 19.3 g/cm³.

50. The mass of gem-quality rough diamonds mined each year is about 10 million carats. What is the mass of these diamonds in kilograms? (1 carat = 200 mg.)

51. To obtain 1 carat of gem-quality diamonds, approximately 115 tons of earth has to be mined. How much earth must be mined to produce 10 million carats of diamonds?

52. Gold ore in the United States today assays at about 0.35 troy ounce of gold per ton of ore. How much earth is mined to produce the 0.7 million troy ounces of gold produced by mining each year? (Another 0.3 million troy ounces of gold is obtained as a by-product of mining other metal ores.)

Online Projects

53. Waste management has become big business. Find the websites of some corporations that specialize in waste management technology and evaluate their approaches to this problem. Possible starting places are http://www.environet.net and http://www.webdirectory.com/Pollution/Waste_Management. You might compare these with the approach of some ecological associations.

54. Using the internet to find opposing views, discuss one of the following controversies: (1) need to remove asbestos from all places where it is found; (2) whether Earth can support an ever-expanding population.

References and Readings

1. Baptiste, H. Prentice. "An African Chemistry Connection." *The Science Teacher*, February 1996, pp. 36–39. Africans extracted iron from its ore long before similar smelting was accomplished in Europe.

2. Hill, John W., and Ralph H. Petrucci. *General Chemistry*, 2d edition. Upper Saddle River, NJ: Prentice Hall, 1999. Chapter 24, "Chemistry of Materials," deals with some of the topics discussed in this chapter in more detail.

3. Kolb, Doris, and Kenneth E. Kolb. *Glass: Its Many Facets*. Hillside, NJ: Enslow, 1988.

4. Mossman, B. T., J. Bignon, M. Corn, A. Seaton, and J. B. L. Gee. "Asbestos: Scientific Developments and Implications for Public Policy." *Science*, 19 January 1990, pp. 294–301.

5. Ohashi, Nobuo. "Modern Steelmaking." *American Scientist*, November–December 1992, pp. 540–555.

6. Service, Robert F. "Plutonium Under Glass." *Science*, 15 December 1995, p. 1762. Glass as a storage medium for plutonium.

7. Studt, Tim. "New Advances Revive Interest in Cement-Based Materials." *R & D Magazine*, November 1992, pp. 74–78.

8. Tarbuck, Edward J. and Lutgens, Frederick K. *Earth Science*, 9th edition. Upper Saddle River, NJ: Prentice Hall, 2000.

MediaLab

Population Growth: Can Earth Support Us All?

Famine ... drought ... high oil prices .... From the catastrophic to the inconvenient, events throughout the world remind us that our resources are finite. Although we raise more food, produce more life-saving medicines, and rush to help after disasters in far-flung places, we must confront the issue of population growth. Can increases in production and improvements in technology keep pace with population growth indefinitely?

Most experts in many fields predict that we will one day exhaust Earth's carrying capacity—its ability to sustain increasing numbers of people. Others disagree, saying that technological advances can meet the planet's needs. Some even argue that we need more people to sustain economic growth.

In this *Web Investigation*, you will look at the rate of population increase in the United States and throughout the world. You will explore the relationship between population growth and the degradation of the environment. You will also consider some pro-growth arguments. In *Communicate Your Results*, you will write both a balanced report and an impassioned advocacy appeal about the topic.

WEB INVESTIGATION

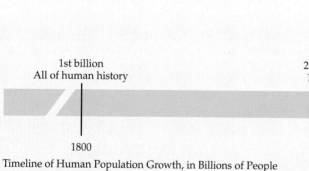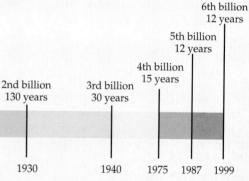

1st billion
All of human history

2nd billion
130 years

3rd billion
30 years

4th billion
15 years

5th billion
12 years

6th billion
12 years

1800   1930   1940   1975   1987   1999

Timeline of Human Population Growth, in Billions of People

Investigation 1
Growing ... Growing ... Growing

On October 12, 1999, the world population reached 6 billion, and it continues to increase. Select Keyword **POPULATION** to see what it is now. (Keep track of how many people are added during the time you spend working with this MediaLab.) For the current projected population of the United States, select Keyword **US**. Follow the *Documentation* link for a discussion of how population numbers are projected.

Investigation 2
Population and the Environment: Resources and Waste

Not surprisingly, population growth and the environment are intertwined. Population pressure leads to land clearing and habitat loss for other species, increased energy use with resultant harmful emissions, deterioration of water supply and quality, and other environmental problems. As our resources dwindle, our mountains of garbage grow! A large population, in particular a high density of people, can magnify small environmental problems to unmanageable proportions. For an overview of the problems resulting from a rising population in an interconnected world, select Keyword **ZPG**. For more details, with links for each problem, select Keyword **IMPACT**. All too often, however, environmental problems are considered separately from population growth. See, for example, the discussion at Keyword **ENVIRONMENT**.

Investigation 3
The Carrying Capacity of Earth: Will We Run Out of Food?

A food shortage or famine is perhaps the most visible and dramatic of the possible effects of overpopulation. In recent decades, improved farming technology has generally kept pace with the increasing numbers of people on our plan-
et. Fertilization, irrigation, and improved seeds and pesticides have enabled fewer farms to grow crops more efficiently. (Indeed, recent famines apparently have resulted from political disputes or sudden disasters, such as floods, rather than from overpopulation.) However, many experts foresee a time when an increasing population will outstrip our ability to provide food. Select Keyword **FOOD** for an overview of population and the food supply. (*Optional*: You may wish to go beyond the introduction to study the problems in depth.)

Investigation 4
A Different View: Population Growth Is Good

Some people argue that there is no population problem or that population growth is good. Select Keyword **PROGRESS** to read this point of view.

COMMUNICATE YOUR RESULTS

Exercise 1
Are There Too Many of Us?

Using figures and facts from *Web Investigations 1* and *2*, write a convincing report about overpopulation and its impact. Use a factual, journalistic style.

Exercise 2
Environmental Impact: Population Growth and Other Factors

Choose an environmental problem (such as acid rain, desertification, food shortages, or water pollution). Drawing on what you learned in *Web Investigation 2*, and any other sources you want to use, write a two- to three-page analysis of the problem. Emphasize the role of population growth, but look at other causes as well. Address such questions as: (1) Would this problem have arisen (or reached its current magnitude) without a large world population? (2) Can this problem be solved by technological advances without addressing the population question? (3) What other factors (such as unsustainable resource consumption, poor resource management, or inappropriate development policies) play a role? (4) Can population stabilization alone solve the problem? (5) Does this problem have a "point of no return"?

Exercise 3
Does Population Matter?

Write a one-page rebuttal (or, if you prefer, supporting arguments) to Simon's paper that you read in *Web Investigation 4*.

Exercise 4
What Can We Do?

Working by yourself or in teams, devise some approaches to solving the problem of overpopulation and environmental degradation. What can be done at the individual, community, state, or national level? Convey your ideas to a local newspaper, advocacy group, or governmental official. Try to be persuasive!

Air

The Breath of Life

Although you cannot see it,
It's around you everywhere.
You cannot live without it.
It's the breath of life. It's AIR!

This view from the space shuttle shows Earth's atmosphere as a thin, hazy, blue band. Florida is visible in the photograph.

W e could live about a month without food. We could even live for several days without water. But without air we cannot live more than a few minutes. Every breath we take is automatic; we seldom think about it. Under normal conditions, we cannot hold our breath for very long, even if we want to.

We may run short of food, and we may run short of fresh water, but we are not likely to run out of air. On the other hand, we might foul the air so badly in some places that it could become unfit to breathe. In some areas the air is so bad that people become sick from breathing it, and some even die because of it. Let's look more closely at the air we breathe.

12.1 Earth's Atmosphere: Divisions and Composition

As passengers on Spaceship Earth, we live under a thin blanket of air called the **atmosphere**. It is difficult to measure exactly how deep the atmosphere is. It does not end abruptly but gradually gets thinner as it fades into space. But we know that 99% of the atmosphere lies within 30 km of Earth's surface. This is a thin air layer indeed (a bit like the skin on a large apple, only much thinner, relatively speaking).

The atmosphere is divided into layers (Figure 12.1). The layer nearest Earth, the *troposphere*, harbors nearly all living things and nearly all human activity. (Although other planets in our solar system have atmospheres, Earth's atmosphere seems unique in its ability to support higher forms of life.) In the next region, the *stratosphere*, we find the **ozone layer** which shields living creatures from deadly ultraviolet radiation. Most of this chapter focuses on the troposphere, but we also examine some threats to the ozone layer in the stratosphere. We will not be concerned with higher regions of the atmosphere.

Air is a mixture of gases. Dry air is (by volume) about 78% nitrogen (N_2), 21% oxygen (O_2), and 1% argon (Ar). Damp air can contain up to about 4% water vapor. There are a number of minor constituents, the most important of which is carbon dioxide (CO_2). The concentration of carbon dioxide in the atmosphere increased from about 280 parts per million (ppm) in the preindustrial world to 296 ppm in 1900 to its present value of 363 ppm. It most likely will continue to rise as we burn more fossil fuels (coal, oil, and natural gas). The composition of the atmosphere is summarized in Table 12.1.

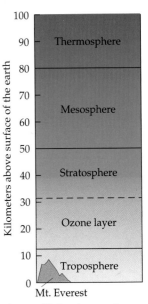

Figure 12.1 Approximate altitudes of the several layers of the atmosphere. These altitudes are only approximate. For example, the height of the troposphere varies from about 8 km at the poles to 16 km at the equator. The ozone layer is shown in color.

Component	Number of Molecules*
Nitrogen (N_2)	78,084
Oxygen (O_2)	20,946
Argon (Ar)	934
Carbon dioxide (CO_2)	36
Trace substances†	<1

Table 12.1 ▌ Composition of Clean Dry Air Near Sea Level*

*Values are number of molecules per 100,000 molecules of air.

†Total less than 1 molecule in 100,000: neon (Ne), helium (He), methane (CH_4), krypton (Kr), hydrogen (H_2), dinitrogen monoxide (N_2O), xenon (Xe), ozone (O_3), sulfur dioxide (SO_2), nitrogen dioxide (NO_2), ammonia (NH_3), carbon monoxide (CO), and iodine (I_2).

12.2 The Nitrogen Cycle

As mentioned in Chapter 9, nitrogen is a component of many important organic chemicals, including proteins and nucleic acids. Although nitrogen makes up 78% of the atmosphere, N_2 molecules can't be used directly by higher plants or by animals. They first have to be *fixed*—that is, combined with another element.

Certain types of bacteria convert N_2 to nitrates. Other bacteria convert the nitrogen in compounds back to N_2, establishing a nitrogen cycle (Figure 12.2). Lightning also serves to fix nitrogen by causing it to combine with oxygen. Nitric oxide (NO) and nitrogen dioxide (NO_2) are formed. The equations are

$$N_2 + O_2 + \text{energy (lightning)} \longrightarrow 2\,NO$$

$$2\,NO + O_2 \longrightarrow 2\,NO_2$$

Nitrogen dioxide reacts with water to form nitric acid (HNO_3).

$$3\,NO_2 + H_2O \longrightarrow 2\,HNO_3 + NO$$

The nitric acid falls in rainwater, adding to the supply of available nitrates in the oceans and the soil.

Humans have intervened in the nitrogen cycle by fixing nitrogen industrially in the manufacture of nitrogen fertilizers. This has greatly increased our food supply because the availability of fixed nitrogen is often the limiting factor in the production of food (Chapter 16). Not all the consequences of this intervention have been favorable, however; excessive runoff of nitrogen fertilizer has led to serious water pollution problems in some areas (Chapter 13).

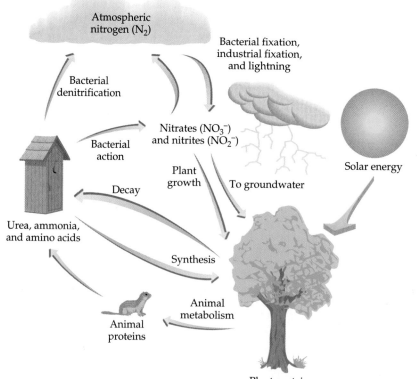

Figure 12.2 The nitrogen cycle.

12.3 The Oxygen Cycle

Oxygen makes up 21% of Earth's atmosphere. Green plants, including one-celled organisms (phytoplankton) in the sea, constantly replenish the oxygen supply. Both plants and animals use oxygen in the metabolism of foods. The decay and combustion of plant and animal materials consume oxygen, as do the rusting of metals and the weathering of rocks. A simplified oxygen cycle is illustrated in Figure 12.3.

In the stratosphere, oxygen is formed by the action of ultraviolet (UV) rays on water. Perhaps more importantly, some oxygen is converted to ozone (Section 12.11).

$$3\,O_2(g) + \text{energy (UV radiation)} \longrightarrow 2\,O_3(g)$$

The ozone, in turn, absorbs short-wavelength UV radiation that might otherwise make higher forms of life on Earth impossible.

We discuss the atmosphere in more detail in subsequent sections, paying particular attention to changes wrought by human activities.

12.4 Temperature Inversion

Normally, the air is warmest near the ground, and its temperature decreases with altitude. On clear nights, the ground cools rapidly, but the wind usually mixes the cooler air near the ground with the warmer air above it.

Once in a while the air is still, and the lower layer of cold air becomes trapped by the layer of warm air above it. This condition is known as an **atmospheric inversion** (or a thermal inversion). Pollutants in the cooler air are trapped near the ground, and the air can become quite seriously polluted if it takes several days for the wind to start to blow again (Figure 12.4).

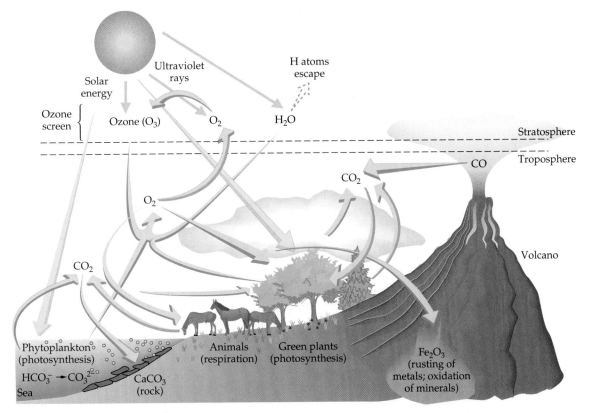

Figure 12.3 The oxygen cycle.

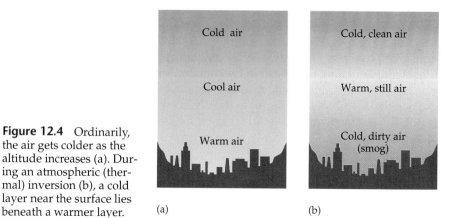

Figure 12.4 Ordinarily, the air gets colder as the altitude increases (a). During an atmospheric (thermal) inversion (b), a cold layer near the surface lies beneath a warmer layer.

(a)   (b)

A temperature inversion can also occur when a warm front collides with a cold front. The less dense warm air mass slides over the cold air mass, producing a condition of atmospheric stability. Because there is no vertical air movement, air stagnation results and air pollutants accumulate.

An Air Pollution Episode: Donora, Pennsylvania

The scene is a small industrial town at the bottom of a river valley. The time is late October. Dense smog settles over the valley. The air is pungent with sulfur dioxide. Dust settles over the land in a layer so thick that footprints and tire tracks are visible in it. A Halloween parade passes through the streets, the marchers holding handkerchiefs over their faces in an attempt to keep the smog out of their lungs. People become seriously ill. Before the smog lifts 5 days later, 17 people are dead. Four more, who became ill in October, die by Christmas Eve. Nearly half the population is made ill by smog. One in 10 is seriously ill.

What is this? A scene from a futuristic, nightmarish science fiction story? No— it is history. The scene actually took place in Donora, Pennsylvania, in 1948. A zinc production plant and a huge iron and steel mill contributed to the smog. So did trains, automobiles, and even barges on the Monongahela River. An atmospheric inversion—a still, warm upper layer of air over a cool lower layer—kept the smog socked in on the valley.

12.5 Natural Pollution

Even before there were people, there was air pollution. Wildfires, windblown dust, and volcanic eruptions added pollutants to the atmosphere, and they still do so. Volcanoes spew ash and poisonous gases into the atmosphere. In 1995 eruptions of the Soufriere Hills volcano on Montserrat covered much of the Caribbean island with ash. The ash and fear of further eruptions caused the evacuation of more than half the country. El Chichon in Mexico in 1982 ejected 3.3 million t of sulfur dioxide into the stratosphere where it was converted to sulfuric acid mist. Kilauea volcano in Hawaii emits 200–300 t of sulfur dioxide per day, leading to acid rain downwind which has created a barren region called the Kau Desert.

Dust storms, especially in arid regions, add massive amounts of particulate matter to the atmosphere. Dust from the Sahara Desert often reaches the Caribbean and South America. Swamps and marshes emit noxious gases. Nature isn't always benign.

The eruption of Mount St. Helens in 1980 ejected at least 1 km³ of material as ash that blanketed areas over several states. The volcano also spewed out 1000 t of sulfur dioxide per day, and it emitted 160 t/day for several years following the eruption.

12.6 The Air Our Ancestors Breathed

People have always altered their environment. The discovery of tools and the use of fire forever changed the balance between people and their environment. Their fires filled the air with smoke. They cleared land, making even larger dust storms possible. They built cities, and the soot from their hearths and the stench from their wastes filled the air. The Roman author Seneca wrote in 61 C.E. of the stink, soot, and "heavy air" of the imperial city. In 1257 the queen of England had to move away from the city of Nottingham because she could not endure the heavy smoke. The industrial revolution brought even worse air pollution. People burned coal to power factories and to heat homes. Soot, smoke, and sulfur dioxide filled the air. The good old days? Not in factory towns. But there were large rural areas that were relatively unaffected by air pollution.

Today air pollution is much more complex than the pollution our ancestors worried about. When society changes its activities, it often changes the nature of its waste materials, including the ones it dumps into the atmosphere. The world has gone through many changes. Change is not new, but the current rate of change is. The pace of change in today's world has reached a level that our environment may not be able to absorb.

12.7 Pollution Goes Global

The main difference between air pollution problems today and in the past is that it is now difficult to get away from them. Problems are worldwide. The air does purify itself, but we are pouring more pollutants into it than the atmosphere can handle readily. Contaminants from England and Germany pollute the snow in Norway. Air pollution originating in the United States contributes to acid rain in Canada and to strained relationships between the two countries. Orbiting astronauts visually traced eastward-drifting blobs of Los Angeles smog as far as Colorado. Other astronauts, over 100 km up, were able to see the plume of smoke from the Four Corners power plant near Farmington, New Mexico. From that distance this was the only evidence they could see that Earth was inhabited.

The U.S. Environmental Protection Agency (EPA) estimated in 1996 that 46 million Americans lived in counties that did not meet one or more of the EPA air quality standards.

With increasing population, the world is becoming more urban. Especially in developing countries, industrialization takes precedence over the environment. Cities in China, Iran, Mexico, Indonesia, and many other counties have experienced frightening episodes of air pollution.

Large metropolitan areas are most afflicted by air pollution, but rural areas are affected too. Neighborhoods around smoky factories have seen evidence of increased rates of spontaneous abortion and poor wool quality in sheep, decreased egg production and high mortality in chickens, and increased feed and care requirements for cattle. Plants are stunted, deformed, and even killed.

Air Pollution in China

As China strives to become an industrial power, its people are paying a heavy price in pollution. Coal burning supplies about three-fourths of China's commercial energy needs. The coal has a high sulfur content, and emission controls are often inadequate. As a result, levels of sulfur dioxide and particulate matter (ash, soot, and dust) are among the highest in the world. About half of China's largest cities have SO_2 levels above the World Health Organization (WHO) guidelines. Some cities, such as Lanzhou and Taiyuan have levels almost ten times the WHO

The polluted air of Shenyang, China, is conspicuous at sunrise.

Burning of coal in electric power plants and other industrial operations is the main source of industrial smog characterized by high levels of sulfur dioxide, particulate matter, and sometimes carbon monoxide. Industrial smog can be alleviated by the use of electrostatic precipitators, scrubbers, and other devices.

The other major type of smog, called *photochemical smog*, is discussed in Section 12.11.

standard. All but 2 of 87 cities monitored exceeded WHO guidelines for particulate matter.

China's growing fleet of motor vehicles also contributes to air pollution problems. Most vehicles are operated in large cities, and because few have effective emission controls, they contribute heavily to the smog in these cities.

China is beginning to attack the problem by closing heavily polluting factories in some larger metropolitan areas. The Chinese government has also invested in gas and in cleaner, more efficient briquettes as replacements for raw coal as a fuel for domestic cooking and heating.

What is a **pollutant**? It is merely a chemical in the wrong place. For example, ozone is a natural and important constituent of the stratosphere, where it shields Earth from life-destroying UV radiation. In the troposphere, however, ozone is a dangerous pollutant (Section 12.11).

12.8 Coal + Fire ⟶ Industrial Smog

There are two basic types of smog. Polluted air associated with industrial activities is often called **industrial smog**. It is characterized by the presence of smoke, fog, sulfur dioxide, and particulate matter such as ash and soot. The burning of coal, especially high-sulfur coal such as that found in the eastern United States, China, and eastern Europe, causes most industrial smog.

An Air Pollution Episode: London, England

The word "smog" is a contraction of the words "smoke" and "fog." This type of pollution was once common in London, and industrial smog was once called *London smog*. One of the more notorious episodes of this type of smog started in London on Thursday, 4 December 1952. A large, cold air mass moved into the valley of the Thames River. A temperature inversion placed a blanket of warm air over the cold air. With nightfall, a dense fog and below-freezing temperatures caused the people of London to heap coal into their small stoves. Millions of these fires burned through the night, pouring sulfur dioxide and smoke into the air. The next day (Friday), the people continued to burn coal because the temperature remained below freezing. Factories added their smoke and chemical fumes to the atmosphere.

Saturday was a day of darkness. For 20 mi around London, no light came through the smog. The air was cold and still. And the coal fires continued to burn throughout the weekend. On Monday, 8 December, more than 100 people died of heart attacks while desperately trying to breathe. People tried to sleep sitting up in chairs in order to breathe a little easier. The city's hospitals overflowed with patients with respiratory diseases.

By the time a breeze cleared the air on Tuesday, 9 December, more than 4000 deaths had been attributed to the smog. Other people afflicted at the time died later. The total death count has been estimated at 8000—more people than were ever killed in any single tornado, mine disaster, shipwreck, or airplane crash. Air pollution episodes may not be as dramatic as other disasters, but they can be just as deadly.

Chemistry of Industrial Smog

The chemistry of industrial smog is fairly simple. High-grade coal is mainly carbon, but it contains as much as 3% sulfur. Coal also contains varying amounts of mineral matter. When burned, the carbon in coal is oxidized to carbon dioxide.

$$C(s) + O_2(g) \longrightarrow CO_2(g) + heat$$

Heat is given off in this process. Not all the carbon is completely oxidized, and some of it winds up as carbon monoxide.

$$2\,C(s) + O_2(g) \longrightarrow 2\,CO(g)$$

Still other carbon, essentially unburned, ends up as soot.

Sulfur Oxides

The sulfur in coal also burns, forming sulfur dioxide, a choking, acrid gas.

$$S(s) + O_2(g) \longrightarrow SO_2(g)$$

Sulfur dioxide is readily absorbed in the respiratory system. It is a powerful irritant and is known to aggravate the symptoms of people who suffer from asthma, bronchitis, emphysema, and other lung diseases.

And things get worse. Some of the sulfur dioxide reacts further with oxygen in the air to form sulfur trioxide.

$$2\,SO_2(g) + O_2(g) \longrightarrow 2\,SO_3(g)$$

Sulfur trioxide then reacts with water to form sulfuric acid.

$$SO_3(g) + H_2O \longrightarrow H_2SO_4(l)$$

Fine droplets of this acid form an aerosol mist that is even more irritating to the respiratory tract than sulfur dioxide.

Particulate Matter

Usually, industrial smog is also characterized by high levels of **particulate matter (PM)**, solid and liquid particles of greater than molecular size. The largest particles are often visible in the air as dust and smoke. PM consists mainly of *soot* (unburned carbon) and the mineral matter that occurs in coal. These minerals do not burn, even

The incidence of asthma has increased dramatically in recent years. It is unlikely that air pollution is the direct cause of the increase. Rather, medical studies show that pollutants at moderate levels can exert a small but measurable effect on the lung function and level of symptoms of asthmatic individuals. It seems more likely that pollutants and allergens interact to enhance the severity of asthma attacks.

The two oxides of sulfur are often designated collectively as SO_x.

An *aerosol* is a suspension of particles of a liquid or solid 1 μm or less in diameter dispersed in a gas.

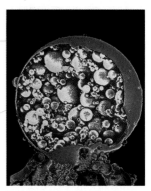

False-color scanning electron micrograph of PM from a coal-burning power plant.

According to WHO, air pollution kills 8000 people a day worldwide. About 90% of the deaths occur in developing countries.

Synergistic interactions of asbestos and cigarette smoke were discussed in Chapter 10, and synergistic interactions of drugs are discussed in Chapter 19.

The principal cause of emphysema is cigarette smoking (Chapter 18).

in the roaring fire of a huge factory or power plant boiler. Some of this solid mineral matter is left behind as *bottom ash*, but much of it is carried aloft in the tremendous draft created by the fire. This *fly ash* settles over the surrounding area, covering everything with dust. It is also inhaled, contributing to respiratory problems in animals and humans. Small particles—those less than 10 μm in diameter (PM10 particulates)—are believed to be especially harmful. They are thought to contribute to respiratory and heart disease. The EPA also monitors particles in the smaller part of this range by adding an indicator for PM2.5 (particles less than 2.5 μm in diameter). EPA studies indicate that as many as 40,000 premature deaths a year are linked to PM and that the majority of them are due to PM2.5. There are several uncertainties about these conclusions, but the combination of PM10 and PM2.5 standards will provide protection against a wide array of particles.

Health and Environmental Effects of Industrial Smog

Minute liquid droplets of sulfuric acid and smaller particulates (PM10s) are easily trapped in the lungs. Interaction may considerably magnify the harmful effects of pollutants. A certain level of sulfur dioxide, without the presence of PM, might be reasonably safe. A particular level of PM might be fairly harmless without sulfur dioxide around, but the effect of both together might well be deadly. *Synergistic effects* such as this are quite common whenever chemicals act together.

When the pollutants in industrial smog come into contact with the alveoli of the lungs, the cells are broken down. The alveoli lose their resilience, making it difficult for them to expel carbon dioxide. Such lung damage contributes to pulmonary emphysema, a condition characterized by increasing shortness of breath.

The oxides of sulfur and the aerosol mists of sulfuric acid also damage plants. Leaves become bleached and splotchy when exposed to sulfur oxides. The yield and quality of farm crops can be severely affected. These compounds are also major ingredients in the production of acid rain.

What to Do About Industrial Smog

Much effort goes into the prevention and alleviation of industrial smog. PM can be removed from smokestack gases in several ways, two of which are illustrated in Figure 12.5. An **electrostatic precipitator** induces electric charges on the particles, which are then attracted to oppositely charged plates and deposited. *Bag filtration* works much like the bag in a vacuum cleaner. Particle-laden gases pass through filters in a bag house. These filters can be cleaned by shaking and by periodically blowing air through them in the opposite direction. A third device, called a *cyclone separator*, is arranged so that the stack gases spiral upward with a circular motion. The particles hit the outer walls, settle out, and are collected at the bottom.

Still another device is a **wet scrubber**, which removes PM by passing stack gases through water. The water is usually sprayed in as a fine mist. The wastewater has to be treated to remove the particulates, which adds to the cost of this method.

The choice of device depends on the type of coal being burned, the size of the power plant, and other factors. All require energy—electrostatic precipitators use 10% of the plant's output—and the collected ash has to be put somewhere. Some of the ash is used. It can replace part of the clay in making cement, and some is melted and air-blown to make mineral wool for insulation. The rest has to be stored; most of it goes into ponds, and the rest into landfills.

It is harder to remove sulfur dioxide than PM. Sulfur can be removed by processing coal before burning, but both the flotation method and gasification or liquefaction processes (Chapter 14) are expensive. Another way to get rid of sulfur is to scrub sulfur dioxide out of stack gases after the coal has been burned. The most common scrubber uses the limestone–dolomite process. Limestone ($CaCO_3$) and dolomite

Fly ash particle

50,000 volts

Stack gas flow

Figure 12.5 Two devices for removing particulate matter from smokestack gases. **(a)** Cross section of a cylindrical electrostatic precipitator. Electrons from the negatively charged discharge electrode (in the center) become attached to the particles of fly ash, giving them a negative charge. The charged particles, are then attracted to and deposited on the positively charged collector plate.

(a mixed calcium–magnesium carbonate) are pulverized and heated. Heat drives off carbon dioxide to form calcium oxide (lime), a basic oxide that reacts with sulfur dioxide to form solid calcium sulfite ($CaSO_3$).

$$CaCO_3(s) + heat \longrightarrow CaO(s) + CO_2(g)$$

$$CaO(s) + SO_2(g) \longrightarrow CaSO_3(s)$$

This by-product presents a sizable disposal problem. Removal of 1 t of sulfur dioxide produces almost 2 t of solids. Calcium sulfate ($CaSO_4$) is a much more useful product than calcium sulfite because the sulfate is used to make commercial products such as plasterboard. Some modified scrubbers oxidize the calcium sulfite to calcium sulfate.

$$2\,CaSO_3(s) + O_2(g) \longrightarrow 2\,CaSO_4(s)$$

In nature, calcium sulfate occurs as gypsum ($CaSO_4 \cdot 2H_2O$), which has the same composition as hardened plaster.

12.9 Photochemical Smog: Making Haze While the Sun Shines

The other main type of smog is **photochemical smog**. Unlike industrial smog, which accompanies cold, damp air, photochemical smog usually occurs during dry, sunny weather. The principal culprits are unburned hydrocarbons and nitrogen oxides from automobiles. The warm, sunny climate that has drawn so many people to the Los Angeles area is also the perfect setting for photochemical smog.

The chemistry of photochemical smog is exceedingly complex. Let's look at the stuff that comes out of an automobile's exhaust pipe and examine the pollutants one at a time.

Carbon Monoxide: The Quiet Killer

When a hydrocarbon burns in sufficient oxygen, the products are carbon dioxide and water. Because both of these substances are normal constituents of air, they are not generally considered pollutants. Let's look at the combustion of octane, one of the hundreds of hydrocarbons that make up the mixture we call gasoline.

$$2\,C_8H_{18}(l) + 25\,O_2(g) \longrightarrow 18\,H_2O(g) + 16\,CO_2(g)$$

When insufficient oxygen is present, another oxide of carbon, carbon monoxide, is formed. Millions of metric tons of this invisible but deadly gas are poured into the atmosphere each year. In the United States, two-thirds of carbon monoxide emissions come from transportation sources, with the largest contribution coming from highway motor vehicles. In urban areas, the motor vehicle contribution to carbon monoxide pollution can exceed 90%. Cigarette smoke also contains a fairly high concentration of carbon monoxide.

The U.S. government has set danger levels of 9 ppm carbon monoxide (average) over 8 h and 35 ppm (average) over 1 h. On streets, danger levels are exceeded much of the time. Even in off-street urban areas, levels often average 7–8 ppm. Such levels do not cause immediate death, but over a long period exposure can cause physical and mental impairment.

Carbon monoxide is a nonirritating, invisible, odorless, tasteless gas. We can't tell that it is around except by using a CO detector or test reagents. Drowsiness is usually the only symptom, and drowsiness is not always unpleasant. How many auto accidents are caused by drowsiness or sleep induced by carbon monoxide? No one knows for sure.

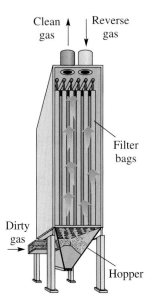

Figure 12.5 (b) A bag house. Vibration of the bags shakes loose the particles, which then collect in the dust hopper at the bottom. Air normally flows upward, but the flow can be reversed to remove more adhering particles.

Downtown Los Angeles on a smoggy day. Photochemical smog was first characterized in the Los Angeles area and was once known as Los Angeles smog.

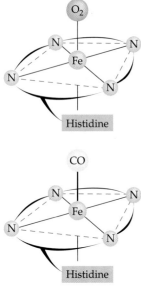

Figure 12.6 Schematic representations of a portion of the hemoglobin molecule. Histidine is an amino acid. Carbon monoxide bonds much more tightly than oxygen, as indicated by the heavier bond line.

In the United States, carbon monoxide makes up nearly half (by mass) of all air pollutants entering the atmosphere. About one-third of the U.S. population is exposed to an excessive quantity of CO from auto emissions. Americans drive as many miles as all the rest of the world combined.

Carbon monoxide exerts its insidious effect by tying up the hemoglobin in the blood. The normal function of hemoglobin is to transport oxygen (Figure 12.6). Carbon monoxide binds to hemoglobin so strongly that the hemoglobin is hindered in carrying oxygen. The symptoms of carbon monoxide poisoning are therefore those of oxygen deprivation. All except the most severe cases of acute carbon monoxide poisoning are reversible. The best antidote is the administration of pure oxygen. Artificial respiration may help if a tank of oxygen is not available.

In chronic carbon monoxide poisoning, the heart has to work harder to supply oxygen to the tissues. Chronic exposure to even low levels may put an added strain on the heart and lead to an increased chance of a heart attack. People already suffering from heart or lung disease may be severely affected. There are thousands of cases of carbon monoxide poisoning in the United States each year, and more than 100 of the victims die. Nonlethal levels of carbon monoxide have been shown to impair time-interval discrimination. In severe cases, performance on psychomotor tests is impaired.

Carbon monoxide is a local pollution problem. It is a severe threat in urban areas with heavy traffic, but it does not appear to be a global threat. In laboratory tests, carbon monoxide survives about 3 years in contact with air, but nature is somehow able to prevent its buildup.

Nitrogen Oxides: Some Chemistry of Amber Air

In addition to carbon dioxide, carbon monoxide, and unburned hydrocarbons, automobile exhaust contains oxides of nitrogen. Power plants that burn fossil fuels are another major source of nitrogen oxides. When nitrogen and oxygen combine during combustion, the main product is nitric oxide (NO).

$$N_2(g) + O_2(g) \longrightarrow 2\,NO(g)$$

The NO is then oxidized to nitrogen dioxide.

$$2\,NO(g) + O_2(g) \longrightarrow 2\,NO_2(g)$$

Nitrogen dioxide is an amber-colored gas that causes smarting eyes and a brownish haze. It plays a vital (villain's) role in photochemical smog. It absorbs a photon of sunlight and breaks down into nitric oxide and reactive oxygen atoms.

$$NO_2(g) + \text{energy (sunlight)} \longrightarrow NO(g) + O(g)$$

Photochemical smog results from the action of sunlight on oxides of nitrogen emitted from automobiles and other high-temperature combustion sources. An amber haze like that shown here over Houston, Texas, characterizes this type of smog.

The initiating step: NO_2 + sunlight $\longrightarrow$ NO + O

Nitrogen dioxide Nitric oxide Oxygen atom

Secondary reactions: O + O_2 $\longrightarrow$ O_3

Oxygen atom Oxygen molecule Ozone

O + RH $\longrightarrow$ R· + ·OH

Oxygen atom Hydrocarbon molecule Hydrocarbon radical Hydroxyl radical

Tertiary reactions: Hydrocarbons + O_2 + NO_2 $\longrightarrow$ $CH_3COO{-}ONO_2$

Peroxyacetyl nitrate (PAN)

Hydrocarbons + O_3 $\longrightarrow$ RCHO

(Aldehydes)

Figure 12.7 Some chemical processes in the formation of photochemical smog. The oxygen atoms formed in the initiating steps are highly reactive. Many other reactive intermediates have been omitted from this simplified scheme.

The oxygen atoms react with other components of automobile exhaust and the atmosphere to produce a variety of irritating and toxic chemicals (Figure 12.7).

At present concentrations, nitrogen oxides don't seem particularly dangerous in themselves. Nitric oxide at high concentrations reacts with hemoglobin; as in carbon monoxide poisoning, this leads to oxygen deprivation. Such high levels seldom, if ever, result from ordinary air pollution, but they might be reached in areas close to industrial sources. Nitrogen dioxide is an irritant to the eyes and the respiratory system. Tests with laboratory animals indicate that chronic exposure in the range of 10–25 ppm might lead to emphysema and other degenerative diseases of the lungs.

The most serious environmental effect of nitrogen oxides is that they lead to smog formation. These gases also contribute to the fading and discoloration of fabrics, and, by forming nitric acid, contribute to acid rain (Section 12.12). They also contribute to crop damage, although their specific effects are difficult to separate from those of sulfur dioxide and other pollutants.

> The two oxides of nitrogen are often designated collectively as NO_x.

> Although the actual reactions are more complicated than those we show here, the net results are the same.

Volatile Organic Compounds

A class of substances called **volatile organic compounds (VOCs)** are major contributors to smog formation. There are many sources of VOCs, including gasoline vapors, combustion products of fuels, and consumer products such as aerosol sprays. Many VOCs are hydrocarbons.

Hydrocarbons can even act as pollutants without burning. They are released from a variety of natural sources, such as decay in swamps. Only about 15% of all hydrocarbons found in the atmosphere are put there by people. In most urban areas, however, the processing and use of gasoline are the major sources of hydrocarbons in the environment. Gasoline can evaporate anywhere along the line, contributing substantially to the total amount of hydrocarbons in urban air. The automobile's internal combustion engine also contributes by exhausting unburned and partially burned hydrocarbons.

Certain hydrocarbons, particularly alkenes, combine with oxygen atoms or ozone molecules to form aldehydes. As a class, aldehydes have foul, irritating odors. Another series of reactions involving hydrocarbons, oxygen, and nitrogen dioxide leads

> Hydrocarbons from automobiles come largely from those with no pollution control devices or those with devices that are not operating properly. Significant amounts enter the atmosphere from spillage during automobile fueling.

to the formation of peroxyacetyl nitrate (PAN). Ozone, aldehydes, and PAN are responsible for much of the destruction wrought by smog. They make breathing difficult and cause the eyes to smart and itch. People who already have respiratory ailments may be severely affected, and the very young and the very old are particularly vulnerable.

$$CH_3\overset{\displaystyle O}{\overset{\|}{C}}\!-\!O\!-\!ONO_2$$

PAN

What to Do About Photochemical Smog

The principal culprits in photochemical smog are nitrogen oxides, hydrocarbons, and sunlight. Reducing the quantity of any of these would diminish the amount of smog. It isn't likely that we could—or would want to—reduce the amount of sunlight, and so let's focus on the other two.

Hydrocarbons have many important uses as solvents. If we could reduce the quantity of hydrocarbons entering the atmosphere, the amounts of aldehydes and PAN formed would also be reduced. Many industries that use hydrocarbons have substantially reduced emissions or switched to other solvents, and new storage and dispensing systems have decreased hydrocarbon emissions from gasoline stations. Similarly, modified gas tanks and crankcase ventilation systems have reduced evaporative emissions from automobiles. The main attack, however, has been through the use of **catalytic converters** to reduce hydrocarbon and carbon monoxide emissions in automotive exhausts. The catalyst in these converters is a precious metal such as platinum (Pt) and/or palladium (Pd). Hydrocarbons and carbon monoxide react rapidly with oxygen on the surfaces of the metal. With no catalyst, the unreacted pollutants are more likely to reach the atmosphere.

It is more difficult to reduce the quantity of nitrogen oxides. Whereas carbon monoxide and hydrocarbons are removed by oxidation, NO must be reduced to N_2. Lowering the operating temperature of an engine helps, but the engine then becomes less efficient. Running an engine on a richer (more fuel, less air) mixture lowers NO_x emissions but tends to raise carbon monoxide and hydrocarbon emissions. With a separate catalyst, such as rhodium (Rh) or palladium (Pd), some of the carbon monoxide can be used to reduce the quantity of nitric oxide.

$$2\,NO(g) + 2\,CO(g) \longrightarrow N_2(g) + 2\,CO_2(g)$$

The remaining carbon monoxide and the hydrocarbons are then oxidized over the platinum catalyst.

Several states, led by California, have mandated that increasing fractions of all cars sold in these states be zero-emission vehicles. In effect, this requires the sale of electric vehicles. These cars are expensive, with heavy batteries and a limited driving range between recharging. However, technology is advancing rapidly to develop less costly vehicles with a greater range. Hybrid vehicles, which use both a battery and a gasoline engine, may appeal to more consumers than cars run only on batteries. These cars use small engines that run at a steady rate, recharge during moderate-speed driving, and use little fuel during idling.

12.10 We Got the Lead Out

Our most successful attempt to reduce air pollution from vehicles involved lead, a heavy-metal poison that affects functioning of the blood, liver, kidneys, and brain (Chapter 20). For years, a lead compound, tetraethyllead, was used to improve the antiknock qualities of gasoline. As it comes from the distillation tower of a petrole-

A one-car/one-person transportation system is inherently inefficient. The best way to reduce emissions from automobiles is to drive less. Walk, ride a bicycle, carpool, or use public transportation whenever possible. And if you must drive, use a smaller car.

 [Web Reference 1] The U.S. Department of Energy provides a bibliography of resources about hybrid vehicles. The Natural Resources Defense Council has a page about "earth-smartcars" at its website. Another good site is a compilation of links about alternatives to the internal combustion engine.

um refinery, gasoline is called *straight-run gasoline* and has an octane rating of about 60. Modern high-compression engines cannot run on this fuel. Tetraethyllead is exceptionally effective at raising the octane level of gasoline, but its use created other problems. Automobiles spewed large amounts of lead into the environment, resulting in fairly high concentrations of lead along heavily traveled roads.

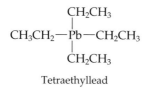

Tetraethyllead

In addition to its toxicity, lead fouls the catalysts in pollution control devices. We had two reasons to "get the lead out." The United States led the world in getting rid of leaded gasoline. Since 1974, all new cars in this country have used lead-free gasoline. Today leaded fuel is banned in the United States and most other developed countries. Many developing countries also have moved to eliminate leaded gasoline. The banning of lead from fuels has reduced lead as an air pollutant by 90% or more in the United States.

To achieve a high octane rating in lead-free gasoline, more branched-chain and aromatic hydrocarbons (Chapter 9) are used. Even without these extra aromatic hydrocarbons, the burning of gasoline produces 3,4-benzpyrene, a chemical that induces cancer in laboratory animals. Increasing the aromatic hydrocarbon content of gasoline probably increases the concentration of this carcinogen in the atmosphere.

3,4-Benzpyrene

The octane rating of gasoline is also increased by the use of oxygenates, compounds such as alcohols and ethers which contain oxygen as well as carbon and hydrogen.

We discuss gasoline as a fuel and gasoline octane boosters further in Chapter 14.

12.11 Ozone: Protector and Pollutant

The ordinary oxygen we breathe is made up of O_2 molecules. Ozone is a form of oxygen that consists of O_3 molecules. It is a natural component of the stratosphere, where it shields Earth from life-destroying UV radiation. Ozone is also a familiar constituent of photochemical smog. Inhaled, it is a toxic, dangerous chemical. Ozone is a good example of just what a pollutant is—a chemical substance out of place in the environment. In the stratosphere, it helps make life possible. In the lower troposphere—the part we breathe—it makes life difficult.

Ozone (O_3) and oxygen (O_2) are **allotropes**: two different forms of the same element.

In the mesosphere (Figure 12.1), some ordinary oxygen molecules are split into oxygen atoms by short-wavelength, high-energy UV radiation.

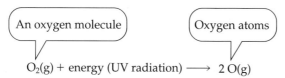

$$O_2(g) + \text{energy (UV radiation)} \longrightarrow 2\,O(g)$$

Some of these highly reactive atoms diffuse down to the stratosphere where they react with O_2 molecules to form ozone.

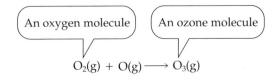

$$O_2(g) + O(g) \longrightarrow O_3(g)$$

The ozone in turn absorbs longer-wavelength, but still lethal, UV rays, thus shielding us from this harmful radiation. In absorbing the rays, ozone is converted back to oxygen molecules and oxygen atoms, reversing the previous reaction.

$$O_3(g) + \text{energy (UV radiation)} \longrightarrow O_2(g) + O(g)$$

Undisturbed, the concentration of ozone in the stratosphere is kept fairly constant by these cyclic processes. Human activity, however, threatens the existence of the protective shield. Before examining this problem, let's take a look at the effects of ozone down here where we live.

Ozone as an Air Pollutant

Ozone is a powerful oxidizing agent. It causes several health and environmental problems. Episodes of high ozone levels correlate with increased hospital admissions and emergency room visits for respiratory problems. Repeated exposures to ozone can make people more susceptible to respiratory infection, cause lung inflammation, aggravate asthma, decrease lung function, and increase chest pain and coughing. At low levels, it causes eye irritation. Inhalation of ozone is particularly dangerous during vigorous physical activity. School children in affected areas are not allowed to play outside when ozone reaches dangerous levels.

In addition to adversely affecting health, ozone causes economic damage. It causes rubber to harden and crack, shortening the life of automobile tires and other rubber items. Ozone also causes extensive damage to crops. Tobacco and tomatoes are particularly susceptible.

The Ozone Shield: Chlorofluorocarbons and Other Eradicators

Both excess ozone in the troposphere and too little ozone in the stratosphere threaten our health and our environment. The less ozone there is in the stratosphere, the more harmful UV radiation reaches Earth's surface. In humans, increased UV can lead to more cases of skin cancer, cataracts, and impaired immune systems. Increased UV also damages UV-sensitive crops, such as soybeans, leading to lower yields. A depletion of stratospheric ozone is thought to lead to a decreased population of phy-

UV light sometimes is used to sterilize tools in the hairdresser's shop and safety glasses in laboratories to prevent the transfer of infectious microorganisms from one person to another. We do not want the entire planet rendered sterile, however, for we are part of the life that would be harmed.

In 1999 the Texas gulf coast exceeded Los Angeles in number of days with ozone alerts. The Northeast, New York City, Denver, Salt Lake City, and the Chicago area also are afflicted. The American Lung Association links as many as 15,000 hospital admissions each year to ozone pollution.

Exposure of a plant to as little as 0.06 ppm of ozone can cut photosynthesis in half. Plant growth is reduced, and leaves often become spotted.

Ozone causes rubber to harden and crack. Tire manufacturers incorporate paraffin wax into the rubber to protect tires from ozone. The alkanes that make up the wax are among the few substances that are resistant to attack by ozone. As tires age, the paraffin wax migrates to the surface to replace that worn off.

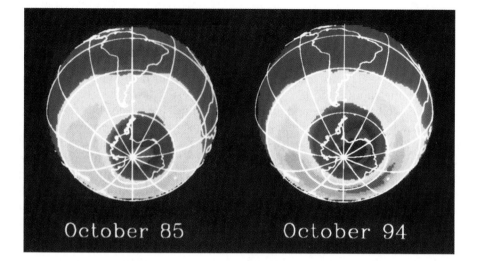

October 85   October 94

Figure 12.8 The ozone hole over Antarctica. The "holes" (the gray, pink, and purple areas at the center) are large areas of substantial ozone depletion. They occur mainly from late August through early October (the Antarctic spring) and fill in about mid-November. Ozone depletion also occurs over arctic regions.

In 1986 Susan Solomon, a chemist at the Oceanic and Atmospheric Administration in Boulder, Colorado, proposed a mechanism for ozone depletion over the poles by CFCs. Later experiments confirmed the theory and led to an international ban on the use of CFCs. Solomon was awarded a 1999 National Medal of Science for her work.

toplankton in the ocean, disrupting the web of life in the sea. Increased UV radiation can even lead to an increase in ground-level ozone.

Concern about ozone depletion escalated in 1986 with the discovery that ozone concentrations over Antarctica plummet each September. This thinning or "hole" in the **ozone layer** (Figure 12.8) has been getting worse since 1979. A similar but less dramatic thinning has been detected over the Arctic. Satellite data show that, overall, the ozone layer has been thinning at a rate of 0.5% per year since 1978. Evidence that this thinning is caused largely by chlorofluorocarbons (CFCs) (Chapter 9), is now compelling.

The United Nations has made addressing ozone depletion a priority through the Montreal Protocol, a series of international agreements on the reduction and eventual elimination of the production and use of ozone-depleting substances. This protocol became effective in 1989, and 160 countries currently participate. These efforts are expected to result in recovery of the ozone layer in about 50 years.

The U.S. National Research Council predicts a 2–5% increase in skin cancer for each 1% depletion of the ozone layer. Even a 1% increase in UV radiation can lead to a 5% increase in melanoma, a deadly form of skin cancer.

CFCs have been used as the dispersing gases in aerosol cans, as foaming agents for plastics, and as refrigerants. At room temperature, CFCs are gases or liquids with low boiling temperatures. They are essentially insoluble in water and inert toward most other substances. These properties make them ideal propellants for use in aerosol cans for deodorants, hair sprays, and food products. Unfortunately, the inertness of these compounds allows them to persist in the environment.

CFCs diffuse into the stratosphere where they are broken down by UV radiation. Chlorine atoms formed in this process break down the ozone that protects Earth from harmful UV radiation.

$$CF_2Cl_2 + \text{energy (UV light)} \longrightarrow CF_2Cl\cdot + Cl\cdot$$

$$Cl\cdot + O_3 \longrightarrow ClO\cdot + O_2$$

$$\cdot ClO + O \longrightarrow Cl\cdot + O_2$$

Note that the last step results in the formation of another chlorine atom that can break down another molecule of ozone. The second and third steps are repeated many times; the decomposition of one molecule of CFC can therefore result in the destruction of many molecules of ozone.

Why are there ozone holes at the poles? In the polar winter, a vortex of intensely cold air with ice crystals develops. Reactions occur on the surfaces of the ice crystals, where HCl and chlorine nitrate ($ClONO_2$) are adsorbed and react to form chlorine and nitric acid.

$$HCl + ClONO_2 \longrightarrow Cl_2 + HNO_3$$

The Cl_2 molecules are readily dissociated by light into chlorine atoms.

$$Cl_2 \longrightarrow 2\,Cl\cdot$$

These free radicals (Chapter 5) then enter into the ozone-destroying cycle.

[Web Reference 2]
The government climate prediction center has a website. Here you can find a chart of the area of the hole in the ozone layer over the Antarctic. This website is updated regularly so that you can see the current state of the hole.

Nitric oxide (NO) from agricultural activities and from the exhausts of airplanes that fly in the stratosphere also contribute to depletion of the ozone layer.

[Web Reference 3]
There are many websites about the hole in the ozone layer and related environmental problems: a website devoted to ozone, NASA's site (with the basic chemistry of ozone depletion), an animated model. Other sites offer a view from another country.

Normal rainwater, if saturated with dissolved carbon dioxide, has a pH of 5.6. During thunderstorms, the pH of rainwater can be much lower because of nitric acid formed by lightning. Recall (Chapter 7) that a solution with a pH value one unit lower than that of another is ten times as acidic. A decrease in pH from 5.6 to 4.6, for example, means an increase in acidity by a factor of 10. Two pH units lower means 100 times as acidic; and three pH units lower means 1000 times as acidic.

CFCs have been banned in the United States and many other countries. However, older refrigerators and air conditioners still use CFCs, which can escape into the atmosphere if these appliances are not serviced and discarded properly. The concentration of CFCs in the stratosphere has peaked, however, and will continue to decline slowly for decades. The ozone loss may also persist for many years, with normal levels returning about 2045.

CFCs and other chlorine- and bromine-containing compounds that might diffuse into the stratosphere have been replaced for the most part by more benign substances. Hydrofluorocarbons (HFCs) such as CH_2FCF_3, which have no chlorine or bromine to form ozone-destroying radicals, are one alternative. Others are hydrochlorofluorocarbons (HCFCs) such as $CHCl_2CF_3$. These compounds contain chlorine atoms, but they break down for the most part before reaching the stratosphere. All three classes of compounds—CFCs, HFCs, and HCFCs—are greenhouse gases (Section 12.15). HFCs and HCFCs are scheduled to be phased out between 2020 and 2040.

12.12 Acid Rain: Air Pollution ⟶ Water Pollution

We have seen (Section 12.8) how sulfur oxides are converted to sulfuric acid. Similarly, nitrogen oxides become nitric acid (Section 12.2). These acids fall on Earth as acid rain or acid snow or are deposited from acid fog or adsorbed on PM. **Acid rain** is defined as rain having a pH less than 5.6. Rain with a pH as low as 2.1 and fog with a pH of 1.8 have been reported (Figure 12.9). These values are lower than the pH of vinegar or lemon juice.

Acid rain comes mainly from sulfur oxides emitted from power plants and smelters and from nitrogen oxides discharged from power plants and automobiles. These acids are often carried far before falling as rain or snow (Figure 12.10).

Acids corrode metals and can even erode stone buildings and statues. Sulfuric acid eats away metal to form a soluble salt and hydrogen gas.

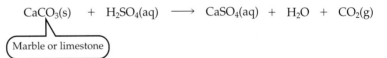

$$Fe(s) + H_2SO_4(aq) \longrightarrow FeSO_4(aq) + H_2(g)$$

(Iron (in steel))   (A soluble salt)

The reaction shown here is oversimplified. For example, in the presence of water and oxygen (air), the iron is converted to rust (Fe_2O_3). Marble buildings and statues are disintegrated by sulfuric acid in a similar reaction that forms calcium sulfate, a slightly soluble, crumbly compound.

$$CaCO_3(s) + H_2SO_4(aq) \longrightarrow CaSO_4(aq) + H_2O + CO_2(g)$$

(Marble or limestone)

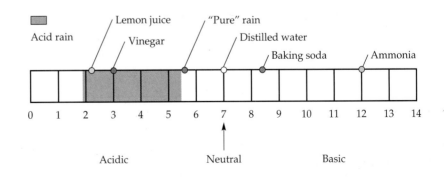

Figure 12.9 The pH of acid rain in relation to that of other familiar substances.

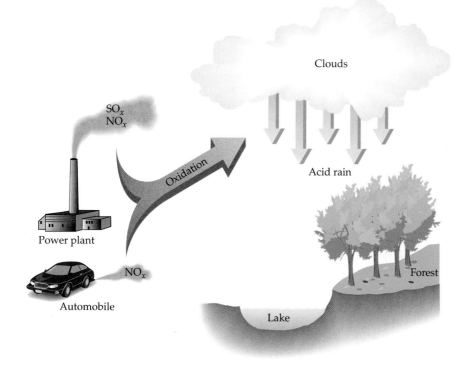

Figure 12.10 Acid rain. The two principal sources of acid rain are sulfur dioxide (SO_2) from power plants, and <u>nitric oxide</u> (NO) from <u>power plants</u> and automobiles. These oxides are converted to SO_3 and NO_2, which then react with water to form H_2SO_4 and HNO_3. The acids fall in rain, often hundreds of kilometers from their sources.

12.13 The Inside Story: Indoor Air Pollution

Indoor air pollution can pose a major health concern. It can be bad in the United States but is often much worse in developing countries where people burn wood, coal, or dung in their homes for cooking and/or heating.

The effects of acidic waters on plant and animal life are discussed in Chapter 13.

[Web Reference 4] The EPA has a website devoted to acid rain. Many ecology links about acid rain can be found.

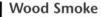

Wood Smoke

Many people like the smell of smoke from burning wood or leaves. However, wood smoke is far from benign. Each kilogram of wood burned yields 80–370 g of carbon monoxide, 7–27 g of VOCs, 0.6–5.4 g of aldehydes, 1.8–2.4 g of acetic acid, and 7–30 g of particulates. The smoke also contains hydrocarbons such as methane (14–25 g/kg wood), benzene (0.6–4.0 g), naphthalene (0.24–1.6 g), and a variety of alkylbenzenes (1–6 g) and other familiar pollutants such as nitrogen oxides (0.2–0.9 g) and sulfur oxides (0.16–0.24 g).

We have seen how aldehydes irritate the eyes and nasal passages, accounting in part for the effect of smoke on those gathered around a campfire. But wood smoke is more than an irritant; it contains a wide variety of carcinogenic polycyclic aromatic hydrocarbons (PAHs), chlorinated dioxins, and related compounds. The odor of wood smoke may evoke memories of autumns past or singing around a campfire, but it can also lead to asthma attacks and possible long-term harm to the lungs.

We can't necessarily escape air pollution by staying indoors. EPA studies on human exposure to air pollutants indicate that indoor air levels of many pollutants may be 2–5 times higher, and on occasion more than 100 times higher, than outdoor levels. Near high-traffic areas, the air indoors has the same carbon monoxide level

Marble statues are slowly eroded by the action of acid rain on the marble (calcium carbonate).

as the air outside. In some areas, office buildings, airport terminals, and apartments have indoor levels that exceed government safety standards.

The Home: No Haven from Air Pollution

We try to save energy by installing better insulation and by sealing air leaks around windows and doors, but in doing so we often make indoor pollution worse. We trap pollutants inside. A kitchen with a gas range often has levels of nitrogen oxides above U.S. government standards. Freestanding kerosene stoves produce levels of nitrogen oxides up to 20 times greater than those permitted by federal regulations for outdoor air. (These regulations do not apply to indoor air.)

Cigarettes and Secondhand Smoke

The carcinogens in tobacco smoke are similar to those in wood smoke. Tobacco smoke has PAHs, aromatic amines, nitrosamines, and radioactive polonium-210.

Another indoor polluter is tobacco. More than 40 carcinogens have been identified among the 4000 or so chemical compounds found in cigarette smoke.

The health effects of smoking on the smoker, widely publicized in a series of reports released by the Surgeon General of the United States, are well known (Chapter 18). The first report, published in 1964, led to a ban on television advertisements for cigarettes. Reports have also dealt with the effects of tobacco smoke as an air pollutant. Air quality in smoke-filled rooms is poor. Smoke from just one burning cigarette can raise the level of particulate matter above government standards. Even with good room ventilation, the level of carbon monoxide is often equal to, or greater than, the legal limits for ambient air.

There is a linear relationship between the number of cigarettes smoked per day and levels of damage to DNA found in smokers' lungs. This damaged DNA is most likely involved in the production of lung cancer.

The risk to nonsmokers from cigarette smoke is well established. Nonsmokers are exposed to significant levels of tar and nicotine. In fact, the average nonsmoker has 5% as much nicotine in the blood as a smoker. Workers at Cornell University Medical College found a glycoprotein (a protein molecule with sugar units attached) in tobacco that causes allergic reactions in many smokers and nonsmokers. The reaction is so severe that it may damage small blood vessels and lead to strokes and heart disease. Many states and municipalities have already banned or restricted smoking in many public places.

Radon and Her Dirty Daughters

A most enigmatic indoor air pollutant is radon. A noble gas, radon is colorless, odorless, tasteless, and unreactive chemically. Radon is, however, radioactive. The decay products of radon, called **daughter isotopes**, are the problem. Radon itself decays by alpha emission (Chapter 4) with a half-life of 3.8 days.

$$^{222}_{86}\text{Rn} \longrightarrow {}^{218}_{84}\text{Po} + {}^{4}_{2}\text{He}$$

When radon is inhaled, the polonium-218 and other daughters, including lead-214 and bismuth-214, are trapped in the lungs where their decay damages the tissues.

Radon is released naturally from soils and rocks, particularly granite and shale, and from minerals such as phosphate ores and pitchblende. The ultimate source is uranium atoms found in these materials. Radon is only one of several radioactive materials formed during the multistep decay of uranium. However, radon is unique in that it is a gas and escapes. The others are solids and remain in the soil and rock.

Outdoors, radon dissipates and presents no problems. Trapped inside a house, however, levels build up, sometimes reaching quantities several times the maximum safe level established by the EPA (0.15 disintegration per second per liter of air). At five times this level, the hazard is thought to equal that of smoking two packs of cigarettes a day. Scientists at the National Cancer Institute estimate that radon causes 15,000 lung cancer deaths a year; however, the exact threat isn't known. The "safe level" was set for people who work in uranium mines. These workers have high rates of lung cancer, but they also breathe radioactive dust and many smoke cigarettes. These risk factors are difficult to separate.

12.14 Who Pollutes? How Much?

The EPA lists six "criteria" pollutants (Table 12.2). Since the first Earth Day in 1970, we have improved air quality with regard to five of the six: carbon monoxide, lead, ground-level ozone, particulate matter, and SO_x. However, emissions of NO_x, which contribute to the formation of ozone, have increased 11%.

Improvements in air quality have come along with economic prosperity, demonstrating that economic growth and environmental protection can go hand in hand. Since 1970, national total emissions of the six criteria pollutants declined 32%, while the U.S. population increased 29%, the gross domestic product increased 104%, and vehicle miles traveled increased 121%.

Air pollution causes material damage by dirtying and destroying buildings, clothing, and other objects. It increases health hazards, especially for the very young, the old, and those already ill. It causes crop damage by stunting or killing green plants. Air pollution reduces visibility, thus increasing auto and air traffic accidents. With its ugly smoke plumes and unpleasant odors, it is even an aesthetic problem.

Who causes all this pollution? Table 12.3 summarizes the seven major air pollutants, their main sources, and their health and environmental effects. Note that motor vehicles are a prominent source. Indeed, they are the source of nearly one-half (by mass) of all air pollutants. Our transportation system accounts for about 80% of carbon monoxide emissions, 40% of hydrocarbon emissions, and 40% of nitrogen oxide emissions. Because most transportation in the United States is by private automobile, we can conclude that cars are a major source of pollution.

On the other hand, most PM comes from power plants (about 40%) and industrial processes (about 45%). Similarly, more than 80% of sulfur oxide emissions come from power plants, with an additional 15% coming from other industries. Power plants alone contribute about 55% of nitrogen oxide emissions. Who uses electricity from power plants? We all do. A 100-W bulb burning for 1 year uses the electricity generated by burning 275 kg of coal.

Table 12.2 ■ Ambient Air Quality Standards for Six "Criteria" Air Pollutants*

Pollutant	Limit	%Change (1970–1997)
Carbon monoxide		−35
8-h average	9 ppm	
1-h average	35 ppm	
Nitrogen dioxide		+11
Annual average	0.053 ppm	
Ozone		−15*
8-h average	0.08 ppm	
1-h average	0.12 ppm	
Particulate matter		−75
<10 μm, 24-h average	150 μg/m^3	
<2.5 μm, 24-h average	65 μg/m^3	
Sulfur dioxide		−35
Annual average	0.03 ppm	
24-h average	0.14 ppm	
3-h average	0.50 ppm	
Lead		−98
3-month average	1.5 μg/m^3	

*1987–1996.

Table 12.3 ▌ The Seven Major Air Pollutants

Pollutant	Formula or Symbol	Major Sources	Health Effects	Environmental Effects
Carbon monoxide	CO	Motor vehicles	Interferes with oxygen transport, causing dizziness, death; possibly contributes to heart disease	Slight
Hydrocarbons	C_nH_m	Motor vehicles, industry, solvents	Narcotic at high concentrations; some aromatics are carcinogens	Precursors of aldehydes, PAN
Sulfur oxides	SO_x	Power plants, smelters	Irritate respiratory system; aggravate lung and heart diseases	Reduce crop yields; precursors of acid rain, SO_4^{2-} particulates
Nitrogen oxides	NO_x	Power plants, motor vehicles	Irritate respiratory system	Reduce crop yields; precursors of ozone and acid rain; produce brown haze
Particulate matter	—	Industry, power plants, dust from farms and construction sites	Irritates respiratory system; synergistic with SO_2; contains adsorbed carcinogens, toxic metals	Impairs visibility
Ozone	O_3	Secondary pollutant from NO_2	Irritates respiratory system; aggravates lung and heart diseases	Reduces crop yields; kills trees (synergistic with SO_2); destroys rubber, paint
Lead	Pb	Motor vehicles, smelters	Toxic to nervous system and blood-forming system	Toxic to all living things

What is the worst pollutant? Carbon monoxide is produced in huge amounts and is quite toxic. Yet it is deadly only in concentrations approaching 4000 ppm. Its contribution to cardiovascular disease, by increasing stress on the heart, is difficult to measure. The WHO rates sulfur oxides as the worst pollutants. They are powerful irritants, and, according to WHO, people with respiratory illnesses are more likely to die from exposure to sulfur oxides than from exposure to any other kind of pollutant. Sulfur oxides and sulfuric acid formed from them have been linked to more than 50,000 deaths per year in the United States.

Who pollutes? We all do. In the words of Walt Kelley's comic-strip character Pogo, "We have met the enemy, and he is us." But we can all be part of the solution by conserving fuel and electricity. Many utility companies now offer suggestions for saving energy. Some even give free analyses of home energy use.

Despite the improvements, overall air quality still leaves much to be desired in many areas. The EPA estimates that one in three Americans—about 80 million—live in areas that do not meet federal health standards, and that 174 counties in 38 states have air that is unhealthy by these standards. Much remains to be done before we all have clean air to breathe.

12.15 Carbon Dioxide and Global Warming

No matter how clean an engine or a factory is, as long as it burns coal or petroleum products it produces carbon dioxide. The concentration of carbon dioxide in the atmosphere increased about 20% in the twentieth century. It continues to grow at an accelerating rate because of the increased burning of carbon fuels. We generally don't even consider carbon dioxide a pollutant because it is a natural component of the environment. Certainly its immediate effect on us is slight. But what about its long-term effects?

The Greenhouse Effect: Planet With a Fever

Carbon dioxide and other gases produce a **greenhouse effect**. They let the sun's rays (visible light) in to warm the surface of Earth. When Earth tries to radiate this heat (infrared energy) back out into space, however, molecules of these gases (Figure 12.11) trap the energy.

The greenhouse effect is natural. Without it, life on Earth would be much colder, with an average temperature of about −18°C. What many fear is an *enhanced* greenhouse effect caused by increased concentrations of carbon dioxide and other greenhouse gases in the atmosphere. This enhancement could lead to a rise in Earth's average temperature, an effect called **global warming**.

Human activities add 25 billion t of carbon dioxide to the atmosphere each year, of which 22 billion t comes from the burning of fossil fuels. About 15 billion t is removed by plants, the soil, and the oceans, leaving a net addition of 10 billion t/year. The concentration of carbon dioxide is therefore increasing at a rate of 1 ppm/year (Figure 12.12).

Methane, CFCs, and other trace gases also contribute to the greenhouse effect. The concentration of methane in the atmosphere has been increasing since 1977. Although present in much smaller amounts than carbon dioxide, these trace gases are much more efficient at trapping heat. Methane is 20–30 times, and CFCs 20,000 times, as effective as carbon dioxide at holding heat in Earth's atmosphere.

Water vapor also acts as a greenhouse gas. When released into the atmosphere, however, water soon falls back to Earth as rain. It therefore is unlikely to contribute to global warming.

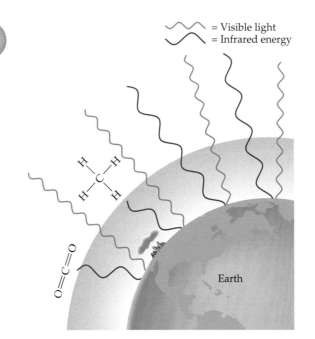

= Visible light
= Infrared energy

Earth

Figure 12.11 The greenhouse effect. Sunlight passing through the atmosphere is absorbed, warming Earth's surface. The warm surface emits infrared radiation. Some of this radiation is absorbed by CO_2, H_2O, CH_4, and other gases and retained in the atmosphere as heat energy.

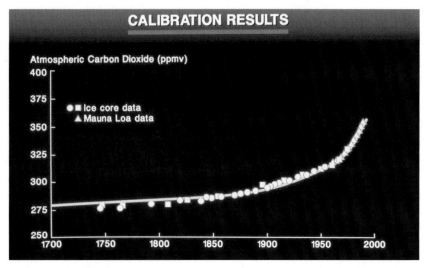

Figure 12.12 The increasing carbon dioxide content of the atmosphere. Shown here is a computer simulation (solid curve) of the atmospheric content of CO_2 (expressed in parts per million by volume). The curve closely matches concentrations measured in air extracted from Antarctic ice and at Mauna Loa, Hawaii. The increase in CO_2 content is most dramatic over the last 30 years or so.

Estimates of sea level rise by 2030 vary from 20 cm to 14 m, with a most probable rise of 28 cm. Differences arise from the use of various models of the atmosphere and from different estimates of the parameters used in the model. Most predictions are made by powerful computers, but scientists still are not sure they have input all the needed data.

 [Web Reference 5] An article from *Chemical and Engineering News* that evaluates the evidence about global warming. Also, the EPA's global warming site.

In 1995 a consensus of atmospheric scientists predicted a warming trend of 1–3°C by the year 2100. Some fear that global warming will melt the polar ice caps and flood coastal cities. The oceans are now rising 3 mm/year. This may not seem like much, but even slight rises increase tides and result in higher, more damaging storm surges. The 1990s was the warmest decade in more than a century, suggesting that global warming seems to be well underway.

12.16 The Ultimate Pollutant: Heat

Electric cars are sometimes advocated as a solution to air pollution. They could help to reduce air pollution in urban areas. But electric cars require electric power, and electric power requires power plants. Conventional power plants burn coal, gas, and oil. It might be argued that replacing cars that run on fossil fuel with ones that run on electricity would serve only to change the site of the pollution and perhaps spread it out a bit. Efficiency plays a role, however; if an electric vehicle, or a hybrid, is more efficient in its use of fuel, there may be a net benefit.

We could avoid considerable air pollution by using more nuclear power plants (Chapter 14). They do not spew soot, smoke, and poisonous chemicals into the atmosphere, but public fears of nuclear accidents and the problem of nuclear waste disposal render more nuclear power an unlikely choice.

There is one pollutant we cannot avoid: heat. According to the second law of thermodynamics, in any energy conversion some of the energy winds up as heat. All power plants dump residual heat into the environment as they produce electric energy, and this heat may be the ultimate pollutant. Even now human structures and activities often make urban areas "heat islands" that are 4–5°C warmer than the surrounding rural areas.

We use enormous amounts of energy—in cars, factories, homes, schools, and hospitals—and the release of energy heats up the environment. Eventually, this heat will contribute to global warming, change the climate of Earth, and affect the ecology of the planet. As René Dubos, world-renowned biologist, put it, "We will destroy our lives by producing more useless, destructive energy to make more and more needless things that do not increase the happiness of people."

12.17 Paying the Price

Air pollution costs us billions of dollars each year. It wrecks our health by causing or aggravating bronchitis, asthma, emphysema, and lung cancer. It destroys crops and sickens and kills livestock. It corrodes machines and blights buildings.

Elimination of air pollution will be neither cheap nor easy. It is especially difficult to remove the last fractions of pollutants. Cost curves are exponential; costs soar toward infinity as pollutants approach zero. For example, if it costs $200 per car to reduce emissions by 50%, it usually costs about $400 to reduce them by 75%, $800 to reduce them by 87.5%, and so on. It would be extremely expensive to reduce emissions by 99%. We can have cleaner air, but how clean depends on how much we are willing to pay.

What would we gain by getting rid of air pollution? How much is it worth to see the clear blue sky or to see the stars at night? How much is it worth to breathe clean, fresh air?

Critical Thinking Exercises

Apply knowledge that you have gained in this chapter and one or more of the FLaReS principles (Chapter 1) to evaluate the following statements or claims.

12.1 A citizen claims that we could solve air pollution problems in the Los Angeles area by allowing only zero-emission vehicles to be sold and licensed in the area.

12.2 Ozone in aerosol cans was once used as a room deodorizer. An inventor proposes that such preparations be reintroduced because any ozone entering the environment would help to restore the protective ozone layer in the stratosphere.

12.3 Although Earth's surface has grown warmer during the last few decades, satellite data collected over an 18-year period show a cooling trend in the upper atmosphere. On the basis of these data, some people claim that global warming as a result of an enhanced greenhouse effect is a myth.

12.4 A person claims that leaf burning in autumn should be allowed because smoke from burning leaves is natural and therefore not harmful.

12.5 Some experts believe that much of global warming comes from methane produced by such natural sources as plant decay in swamps and cattle flatulence. They say that industrial activities that produce carbon dioxide raise the standard of living and should be allowed to continue without interference since they aren't the only source of global warming.

Summary

1. The atmosphere is made up of layers: the troposphere (nearest Earth), the stratosphere (which includes the ozone layer), the mesosphere, and the thermosphere.

2. Clean, dry air contains about 78% nitrogen, 21% oxygen, and 1% argon (by volume), plus trace amounts of other gases.

3. Although the air is 78% nitrogen, animals and most plants cannot use this nitrogen unless it has been fixed (combined with other elements).

4. Nitrogen goes from the air into plants and animals and eventually back to the air via the nitrogen cycle.

5. Oxygen from the air is involved in oxidation (of plant and animal materials) and is eventually returned to the air via the oxygen cycle.

6. Oxygen is converted to ozone and then changed back to oxygen in the stratosphere.

7. Natural air pollution includes dust storms, noxious gases from swamps, and ash and sulfur dioxide from erupting volcanoes.

8. The primary air pollutants introduced by human activity are the smoke and gases produced by the burning of fuels and the fumes and particulates emitted by factories.

9. Atmospheric inversion is a condition in which cool, dirty air is held near the ground by a stable, warm, upper layer of air.

10. Industrial smog is the kind of smog produced in cold, damp air by excessive burning of fossil fuels. It consists of a combination of smoke and fog with sulfur dioxide, sulfuric acid, and soot.

11. Photochemical smog is produced in dry, sunny weather, mainly from automobile exhaust fumes.

12. Carbon monoxide (which is almost always produced when carbon-containing fuels are burned) is a poisonous gas that ties up hemoglobin in the blood so that it cannot transport oxygen to the cells.

13. Nitrogen oxides are present in automobile exhaust gases and emissions from power plants that burn fossil fuels. The brown color of nitrogen dioxide causes the brownish haze often seen over certain large cities.

14. Ozone, an allotrope of oxygen, is a harmful air pollutant in the troposphere but a protective layer in the stratosphere that shields Earth against UV radiation from the sun.

15. CFCs have been linked to the hole in the ozone layer and are now widely banned.

16. Acid rain is caused by sulfur oxides (mainly from power plants) and nitrogen oxides (mainly from power plants and automobiles).

17. Tetraethyllead, once added to gasoline as an antiknock agent, has been phased out because of lead pollution.

18. Because of the effects of secondhand smoke, most cities and states restrict tobacco smoking in public places.

19. Radon gas, an indoor pollutant in some places, is a problem because it is radioactive.

20. Gases such as carbon dioxide and methane can produce an enhanced greenhouse effect and lead to global warming.

Key Terms

acid rain 12.12	electrostatic	oxygen cycle 12.3	volatile organic
allotropes 12.11	precipitator 12.8	ozone layer 12.1	compounds (VOCs) 12.9
atmosphere 12.1	global warming 12.15	particulate matter	wet scrubber 12.8
atmospheric inversion 12.4	greenhouse effect 12.15	(PM) 12.8	
catalytic converters 12.9	industrial smog 12.8	photochemical smog 12.9	
daughter isotopes 12.13	nitrogen cycle 12.2	pollutant 12.7	

Review Questions

1. How does the ozone layer protect the inhabitants of Earth's surface?

2. What is an atmospheric (thermal) inversion? How is it related to air pollution?

3. How did an atmospheric inversion contribute to the 1948 air pollution episode in Donora, Pennsylvania?

4. Is all air pollution the result of human activity? Explain.

5. List two uses of CFCs.

6. What is tetraethyllead? Why was it used in gasoline?

7. Why has leaded gasoline been phased out in most countries?

8. Describe how **(a)** a bag filter and **(b)** a cyclone separator remove particulate matter from stack gases.

9. How does a wet scrubber remove particulate matter? What is a major disadvantage of wet scrubbers?

10. What is bottom ash? What is fly ash? Give two uses for fly ash.

11. What is smog?

12. What are the health effects of ozone in polluted air?

Problems

The Atmosphere: Composition and Cycles

13. Which layer of the atmosphere lies nearest Earth? Which contains the ozone layer?

14. List the three major components of dry air and give the approximate (nearest whole-number) percentage by volume of each.

15. What are the two most important variable components of the atmosphere? Give the approximate concentration range of each.

16. What is meant by nitrogen fixation? Why is it important?

17. How has industrial fixation of nitrogen to make fertilizers affected the nitrogen cycle?

18. Give equations for the reactions by which lightning fixes nitrogen.

19. How is the oxygen supply of Earth's atmosphere replenished?

20. What is a pollutant? Which of the following are pollutants? Explain.
    a. N_2 in the troposphere   **b.** O_3 in the troposphere
    c. NO_2 in the troposphere   **d.** O_3 in the stratosphere

Industrial Smog

21. What are the chemical components of industrial smog?

22. What weather conditions characterize industrial smog?

For Problems 23–26, give the equation for the indicated reaction.

23. Sulfur is oxidized to sulfur dioxide.

24. Sulfur dioxide is oxidized to sulfur trioxide.

25. Sulfur trioxide reacts with water to form sulfuric acid.

26. Sulfur dioxide reacts with hydrogen sulfide to form elemental sulfur.

27. Describe the synergistic action of sulfur dioxide and particulate matter.

28. What is an electrostatic precipitator? How does it work?

29. Describe how a limestone scrubber removes sulfur dioxide from stack gases. Give the chemical reaction(s) involved.

30. What is particulate matter? What is an aerosol?

Photochemical Smog

31. What are the chemical components of photochemical smog?

32. What weather conditions characterize photochemical smog?

33. Under what conditions do nitrogen and oxygen combine? Give the equation.

34. Give the equation for the reaction of nitric oxide with oxygen to form nitrogen dioxide.

35. What happens to nitrogen dioxide in sunlight? Give the equation.

36. What are VOCs?

37. What is the main source of hydrocarbons in polluted air?

38. What is PAN? From what is it formed? What are its health effects?

39. What are catalytic converters? How do they work?

40. Give two ways in which the level of nitrogen oxide emissions from an automobile can be reduced.

Carbon Monoxide

41. What is the main source of carbon monoxide in polluted air?

42. What are the health effects of carbon monoxide?

43. How might exposure to carbon monoxide contribute to heart disease?

44. Give the equation by which carbon monoxide reduces nitric oxide to nitrogen gas.

The Ozone Layer

For Problems 45–46, give the equation for the indicated reaction.

45. Oxygen is converted to ozone in the ozone layer.

46. CFCs destroy ozone.

47. What health effect might result from depletion of the ozone layer?

48. How might nuclear explosions in the stratosphere bring about an "ultraviolet summer"? (*Hint:* See the first margin note on page 330.)

Acid Rain

49. What is acid rain? How is it formed?

50. How might acid rain be alleviated?

51. What is the effect of acid rain on iron? Give an equation.

52. What is the effect of acid rain on marble? Give the equation.

Indoor Air Pollution

53. List several indoor air pollutants. What is the source of each?

54. List some risks associated with secondhand cigarette smoke.

55. What are the health effects of radon gas? How can a radon problem in a building be alleviated?

56. How does better insulation of buildings make indoor air pollution worse?

Global Warming

57. What is the greenhouse effect?

58. How can global warming be alleviated?

Additional Problems

59. Why is zero pollution not possible?

60. Is the electric car a solution to air pollution on a nation-wide basis? Within a given city?

61. The average person breathes about 20 m^3 of air a day. What weight of particulates would a person breathe in a day if the particulate level were 400 $\mu g/m^3$?

62. The atmosphere contains 5.2×10^{15} t of air. How much carbon dioxide (CO_2) is in the atmosphere if the concentration is 363 ppm?

63. The world's termite population is estimated to be 2.4×10^{17}. These termites produce an estimated 4.6×10^{16} g of carbon dioxide, 1.5×10^{14} g of methane, and 2×10^{14} g of hydrogen each year. How much of each gas is produced by each termite? What percentage increase in the total amount of carbon dioxide in the atmosphere (see Problem 62) would the termites cause in 1 year if none of it were removed?

64. Since 1978 the giant ice cap that floats in the Arctic Ocean has lost half its average thickness and has decreased in size by an area larger than the state of Texas. The ice cap atop the Antarctic continent also appears to be melting, dropping huge icebergs into the ocean. Which occurrence, if either, has caused a rise in sea level? (*Hint:* Consider whether the water level rises when an ice cube melts in a glass of water and when an ice cube is added to a glass of water.)

(The number of huge icebergs that break off the continental shelf in Antarctica may increase as a result of global warming.)

Projects

65. What are the average and peak concentrations of each of the following in your community or in a nearby large city? How can you find out?
    a. carbon monoxide   **b.** ozone
    c. nitrogen oxides   **d.** sulfur dioxide
    e. particulate matter

66. Does your community have an air pollution problem? If so, describe it. How could it be solved?

67. Should students be allowed to smoke in school buildings? On school grounds? Defend your position.

68. Using the internet or books, compare the amount of sulfur dioxide from Mt. Kilauea in Hawaii with that from all U.S. industry. (Remember to use the same time span for both.)

Online Projects

69. There is still controversy about whether government intervention is warranted to totally ban the use of CFCs. Find pro and con arguments on the internet. Discuss.

70. Substitutes for CFCs are more expensive than CFCs. What factors should you look at to do a cost–benefit analysis of this substitution? Should the government subsidize the switch?

71. Air quality standards are maximum allowable amounts of various pollutants. The federal government and some state governments issue these standards. Using the internet, find the current standards issued by the EPA. Does your state issue such standards? (*Hint:* Try the EPA site and your own state's environmental protection site.)

References and Readings

1. Beardsley, Tim. "It's Melting, It's Melting." *Scientific American*, July 1995, p. 28. Antarctic ice is melting.

2. Douglas, John. "Understanding the Global Carbon Cycle." *EPRI Journal*, July–August 1994, pp. 35–41.

3. Freemantle, Michael. "Total Life-Cycle Analysis Harnessed to Generate 'Greener' Automobiles." *Chemical and Engineering News*, 27 November 1995, pp. 25–29.

4. Hanson, David. "Conflict over Air Quality Rules." *Chemical and Engineering News*, 16 December 1996, pp. 27–28.

5. Herzog, Howard, Baldur Eliasson, and Olav Kaarstad. "Capturing Greenhouse Gases." *Scientific American*, February 2000, pp. 72–79.

6. Kiester, Edwin, Jr. "A Darkness in Donora." *Smithsonian*, November 1999, pp. 22–24.

7. Nemecek, Sasha. "Holes in Ozone Science." *Scientific American*, January 1995, pp. 26–27.

8. Sperling, Daniel. "The Case for Electric Vehicles." *Scientific American*, November 1996, pp. 54–59.

9. Spiro, Thomas G., and William M. Stigliani. *Chemistry of the Environment*. Upper Saddle River, NJ: Prentice Hall, 1996. Part II, the Atmosphere.

10. Tarbuck, Edward J., and Frederick K. Lutgens. *Earth Science*, 9th edition. Upper Saddle River, NJ: Prentice Hall, 2000. Part 3, The Atmosphere.

11. Trefil, James. "Modeling Earth's Future Climate Requires Both Science and Guesswork." *Smithsonian*, December 1990, pp. 28–37.

12. "Turning Up the Heat." *Consumer Reports*, September 1996, pp. 38–44. Global warming and what we can do about it.

13. Uppenbrink, Julia. "Arrhenius and Global Warming." *Science*, 24 May 1996, p. 1122. Svante Arrhenius was the first to link climate to increases in CO_2 concentration in the atmosphere.

14. Zurer, Pamela S. "NASA Cultivating Basic Technology for Supersonic Passenger Aircraft." *Chemical and Engineering News*, 24 April 1995, pp. 10–16. Considers the problem of NO_x in the stratosphere.

15. Zurer, Pamela S. "Air Pollution Link to Rise in Deaths Confirmed." *Chemical and Engineering News*, 7 August 1995, p. 9.

MediaLab

Global Warming: Is the World Heating Up?

From milder winters and melting glaciers, to northward movement of habitat boundaries, we see many signs that the temperature of our planet is increasing. Statistics indicate that the average temperature of Earth is about 0.5°C greater today than a century ago. Many scientists attribute this to the greenhouse effect caused by the accumulation of carbon dioxide and certain other gases in the atmosphere.

As with most ecological topics, there are profound disagreements among experts. These include whether the warming seen over the last few decades is a long-term trend or a short-term fluctuation, the extent to which increased carbon dioxide levels are responsible, and the role of human civilization in the rise in greenhouse gases.

In this *Web Investigation*, you will look at the data on global temperatures and greenhouse gas accumulations. You will examine the predicted effects of global warming and read about national and international efforts to slow the trend. In *Communicate Your Results*, you will evaluate the data and write letters or editorials promoting action.

WEB INVESTIGATION

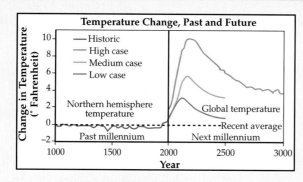

Investigation 1
Global Warming and the Greenhouse Effect

Select Keyword **OVERVIEW** for a concise description of global warming. For a more detailed description, with definitions and many links, select Keyword **DETAIL**. Or select Keyword **SCIENCE** for a review of the science and a table of greenhouse gases.

Investigation 2
More Than a Heat Wave

Global warming does not mean simply that we'll have to use our air conditioners more often. The potential consequences of this climate change are much broader. Select Keyword **CONSEQUENCE** to read about some of these problems, or Keyword **CHANGE** for several links to specific effects.

Investigation 3
The Global Warming Debate: Does the Science Mandate a Policy?

There is fairly wide agreement that Earth has been warmer in the 1990s, but much disagreement about the causes and remedies. Despite the mountain of data collected and analyzed about global warming, many uncertainties still exist. In addition to the scientific ambiguities, political and economic considerations enter into the debates about the existence of greenhouse warming, what it means for the future of the planet, and (especially!) what should be done. Select Keyword **SCIENCE** to read a long but valuable discussion about scientific uncertainty versus the desire for answers. (This is a chapter of a guidebook for reporters on the topic of climate change, but much is applicable to any scientific reporting.) The dilemma is heightened by the desire to act! Since the 1972 United Nations Conference on the Human Environment, nations have met to discuss ways to preserve the planet. Select Keyword **KYOTO** for varied pages about conferences on climate change, including the 1997 Kyoto conference. Some scientists—and many business-people—contest the view that human activities are responsible for any climate change and contend that it is premature to act. Select Keyword **DEBATE** for another view of the debate with some surprising conclusions.

Investigation 4
A Matter of Economics: Taking Action Against Global Warming

Although disagreement still exists about the role of human influence on the greenhouse effect, many individuals and organizations are taking action to reduce the energy use that pours carbon dioxide and other gases into the atmosphere. Often, the environmental approach turns out to be the economic one as well. Select Keyword **BOTTOM LINE** for a discussion of the economic benefits of cutting carbon emissions. Select Keyword **CREST** for case studies of efforts to reduce energy use. Choose one or more studies and note the details about emissions reduction and cost savings.

COMMUNICATE YOUR RESULTS

Exercise 1
Looking at the Data

Print out the graphs of temperature change and carbon dioxide concentration from *Web Investigation 1*. Superimpose them or place them side by side and compare the data. Climatologists claim that the twentieth century was the warmest in the millenium and that the 1990s was the warmest decade. Explain how the data were obtained. Discuss the correlation between increasing temperatures and increases in the atmospheric content of greenhouse gases. Why is the area of global warming so controversial?

Exercise 2
Communicating the Risks: More Than Heat and Humidity

As you have seen in *Web Investigation 2*, global warming means more than an increased number of hot days. Choose a possible effect of global warming and discuss some potential ripple effects. Use this information in a letter to an editor or government official advocating action about global warming.

Exercise 3
Acting with an Uncertain Scenario

The scientific uncertainties about global warming have exacerbated the dissension about possible courses of action.

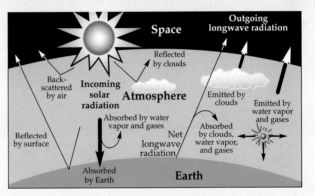

Many scientists and politicians advocate acting now to decrease carbon dioxide emissions. Others propose a "wait-and-study" course. A third approach is to find technological means to adapt to climate change. Draft a paper discussing these approaches and their potential consequences. (Are the action approach and the adaptive approach mutually exclusive?) Drawing on what you read in *Web Investigation 4*, discuss the probable cost–benefit ratio of each approach.

Exercise 4
To Slow Global Warming

Several of the sites in the *Web Investigations* have suggested ways to slow global warming. List some approaches for the individual, the community, and the government. What efforts are best undertaken by international treaty? Which are in the Kyoto protocols?

13

Water

Rivers of Life; Seas of Sorrows

You drink it. You pour it on flowers you've planted.
*   You wash your clothes in it. You take it for granted.*
"Earth has so much water!" you probably think;
*   But most of that water is not fit to drink!*

Water is the only substance on Earth that commonly exists in large amounts in all three physical states. Gaseous water vapor is invisible, but tiny droplets of liquid water are visible as clouds. Liquid water in the bay and solid water (ice and snow) on the mountains and in the glacier are also shown in this view of Paradise Bay in Antarctica.

E arth is a watery planet, with most of its surface covered with oceans and seas. People who live on this planet contain a lot of water, too. Did you know that about two-thirds of your body weight is water? The water in your blood is quite similar to water in the ocean. In fact, you might say that we are walking sacks of seawater.

The presence of large quantities of water makes our planet unique in the solar system, probably the only one capable of supporting higher forms of life. The nature of water makes it essential to life, and the nature of life as we know it makes it dependent on water. If we should someday discover life outside our solar system, it will likely be on a planet that has liquid water.

The properties of water that make it able to support life, however, also make it easy to pollute. Many substances are soluble in water and are therefore easily dispersed and eventually scattered to nearly infinite dilution in the ocean. Once they are there, it is not easy to remove dissolved substances from our water supplies. We start this chapter with a discussion of some of the unusual properties of water. Then we look at water pollution and water treatment. We should not underestimate the importance of clean drinking water. We could live for several weeks without food, but without water we would last only a few days.

Our search for life on Mars is based to a large extent on evidence that the planet once had vast quantities of water and that relics of that life might still exist below the dry surface.

13.1 Water: Some Unusual Properties

Although water is a familiar substance, it is a most unusual compound. It is the only common liquid on the surface of our planet. The solid form of water (ice) is less dense than the liquid because water expands when it freezes. The consequences of this peculiar characteristic are immense for life on this planet. Ice forms on the surfaces of lakes and insulates the lower layers of water. This enables fish and other aquatic organisms to survive winter in the temperate zones. If ice were denser than liquid water, it would sink to the bottom as it formed, and even the deeper lakes of the northern United States would freeze solid in winter.

The same property—ice being less dense than liquid water—has dangerous consequences for living cells. When living tissues freeze, ice crystals are formed, and the expansion ruptures and kills cells. The slower the cooling, the larger the crystals of ice and the more damage there is to the cell. Frozen-food manufacturers make "flash-frozen" products by freezing foods so rapidly that the ice crystals are kept very small and thus do minimum damage to the cellular structure of the food.

Water also has a higher density than most other familiar liquids. As a consequence, liquids less dense than water and insoluble in water float on its surface. The gigantic oil spills that occur when a tanker ruptures or when an offshore well gets out of control are a familiar problem. The oil, floating on the surface of the water, is often washed onto beaches where it does considerable ecological and aesthetic damage. If water were less dense than oil, the problem would certainly be different, although not necessarily less serious.

Another unusual property of water is its high heat capacity. Different substances have different capacities for storing energy absorbed as heat. The **heat capacity** of a substance is the quantity of heat required to raise the temperature of the substance 1°C. **Specific heat** is the heat capacity of a 1-g sample; that is, it is the quantity of heat required to change the temperature of 1 g of the substance 1°C. Table 13.1 gives the specific heats of several familiar substances. Note that it takes almost ten times as much heat to raise the temperature of 1 g of water 1°C as to raise the temperature of 1 g of iron the same amount. Conversely, much heat is given off by water for even a small drop in temperature. The vast amounts of water on the surface of Earth thus act as a giant heat reservoir to moderate daily temperature variations. You need only

Oil coats the water of Prince William Sound, Alaska, where the *Exxon Valdez* ran aground on Bligh Reef in 1989, spilling 42 million L of oil. One liter of oil can create a slick 2.5 hectares (6.2 acres) in size.

(a)

(b)

Figure 13.1 Hydrogen bonds in ice. (a) Oxygen atoms are arranged in layers of distorted hexagonal rings. Hydrogen atoms lie between pairs of oxygen atoms, closer to one (covalent bond) than to the other (hydrogen bond). The structure has large "holes." (b) At the macroscopic level, the hexagonal arrangement of water molecules in the crystal structure of ice is revealed in the hexagonal shapes of snowflakes.

Table 13.1 ∎ Specific Heats of Some Familiar Substances

Substance	Specific Heat (cal/g · °C)
Aluminum (Al)	0.216
Copper (Cu)	0.0920
Ethanol (CH_3CH_2OH)	0.588
Iron (Fe)	0.107
Lead (Pb)	0.0306
Silver (Ag)	0.0562
Sulfur (S)	0.169
Water	1.00*

*Note that in the indicated units the specific heat of water is 1.00. As the metric system was established, the properties of water were often taken as the standard.

consider the extreme temperature changes on the surface of the waterless moon to appreciate this important property of water.

Still another way in which water is unique is that it has a high **heat of vaporization**; that is, a large amount of heat is required to evaporate a small amount of water. This is of enormous importance to us because large amounts of body heat can be dissipated by the evaporation of small amounts of water (perspiration) from the skin. This effect also accounts for the climate-modifying property of lakes and oceans. A large portion of the heat that would otherwise heat up the land is used to vaporize water from the surface of lakes or seas. Thus, in summer it is cooler near a large body of water than in interior land areas.

All these fascinating properties of water depend on the unique structure of the highly polar water molecule (Chapter 5). In the liquid and solid states, water molecules are strongly associated through hydrogen bonds. In liquid water, the molecules are randomly associated but close together. When water freezes, its molecules take on a more ordered arrangement with large hexagonal holes (Figure 13.1). This three-dimensional structure extends out for billions and billions of molecules. The holes account for the fact that ice is less dense than liquid water. The hexagonal arrangement allows water to assume forms of exquisite beauty as snowflakes.

13.2 Water, Water, Everywhere

Three-fourths of the surface of Earth is covered with water. Nearly 98%, however, is salty seawater—unfit for drinking and not even suitable for most industrial purposes. Because of its polar nature, water tends to dissolve ionic substances. This solvent power of water accounts for the saltiness of the sea. Rainwater dissolves the ions of minerals, and these ions are carried by streams and rivers to the sea. There the heat of the sun evaporates part of the water, leaving the salts behind. The oceans grow saltier as the years go by, but the process is so slow that it is not noticeable in one human lifetime.

Rain falls on Earth in enormous amounts; however, most rainwater falls into the sea or on areas otherwise inaccessible. Less than 2% of Earth's water is frozen in the polar ice caps, leaving less than 1% available as fresh water. The average use of water per person (including that used for industrial, agricultural, and other purposes) in the United States is 8 million L/year, and this is increasing rapidly. Moreover, the available water is not always where the people are, and in many places freshwater supplies are polluted and unfit for human use.

13.3 The Water Cycle

Water is apportioned fairly constantly among the oceans, the ice caps, and freshwater rivers, lakes, and streams. However, it is dynamically cycled among these various repositories (Figure 13.2). Water constantly evaporates from both water and land surfaces. Water vapor condenses into clouds and returns to Earth as rain, sleet, and snow. This fresh water becomes part of the ice caps, runs off in streams and rivers, and fills lakes and underground aquifers.

Rainwater is not pure water, however. It carries dust particles from the atmosphere to the ground. It also dissolves some oxygen, nitrogen, and carbon dioxide as it falls through the atmosphere. During electric storms, lightning causes nitrogen, oxygen, and water vapor to combine to form nitric acid, which falls in rainwater as well.

As water moves along or beneath the surface of Earth, it dissolves minerals from rocks and soil. Recall that minerals (salts) are ionic and that ions have either positive or negative charges. The principal positive ions (cations) in natural water are ions of

At the present rate of accumulation, the seas will become saturated with salt (about 36%) in another 3.5 billion years. If our descendants are still around in that distant age, they will find the oceans to be much like today's Dead Sea.

About 10 million t of salt is used on roads each year during the winter months in the northern part of the United States. Much of this salt is washed into streams.

 [Web Reference 1] There are many websites dealing with the problem of water pollution. Among these are the Global Water Partnership and the WaterWeb consortium, created to promote sharing of information about water and other environmental aspects.

Earth is a watery world, but much of the water is salty or frozen in the polar ice caps. The fresh water from rainfall is not evenly distributed, nor does it always fall in predictable patterns. Often the land is too wet or too dry. Without an adequate supply of water, life becomes quite difficult. This scene is from drought-stricken Texas in 1996.

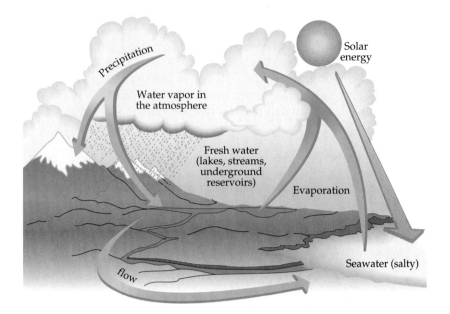

Figure 13.2 The hydrological (water) cycle.

Table 13.2 ▌ Some Substances Found in Natural Waters

Substance	Formula	Source
Carbon dioxide	CO_2	Atmosphere
Dust	—	Atmosphere
Nitrogen	N_2	Atmosphere
Oxygen	O_2	Atmosphere
Nitric acid (thunderstorms)	HNO_3	Atmosphere
Sand and soil particles	—	Soil and rocks
Sodium ions	Na^+	Soil and rocks
Potassium ions	K^+	Soil and rocks
Calcium ions	Ca^{2+}	Limestone rocks
Magnesium ions	Mg^{2+}	Dolomite rocks
Iron(II) ions	Fe^{2+}	Soil and rocks
Chloride ions	Cl^-	Soil and rocks
Sulfate ions	SO_4^{2-}	Soil and rocks
Bicarbonate ions	HCO_3^-	Soil and rocks

sodium (Na^+), potassium (K^+), calcium (Ca^{2+}), magnesium (Mg^{2+}), and sometimes iron (Fe^{2+} or Fe^{3+}). The negative ions (anions) are usually sulfate (SO_4^{2-}), bicarbonate (HCO_3^-), and chloride (Cl^-). Table 13.2 provides a summary of substances found in natural waters.

Water containing calcium, magnesium, or iron salts is called **hard water**. The positive ions react with the negative ions in soap to form a scum that clings to clothes and bathroom fixtures and leaves them dingy looking (Chapter 17).

The water cycle replenishes our supply of fresh water. When water evaporates from the sea, salts are left behind. When water moves through the ground, impurities are trapped in the rock, gravel, sand, and clay. This capacity to purify is not infinite, however.

Soft water may contain ions, such as Na^+ or K^+, but they do not form insoluble scum with soap.

Three-fourths of Earth's surface is covered with water, making the planet appear mostly blue when viewed from space. The world appears to be mainly water, but in fact only about 1/4000 of Earth's mass is water. If Earth were the size of a basketball and you could hold it in your hands, you would notice that it was wet on the surface, but the sphere would seem quite solid otherwise.

According to the Biblical story of Moses, getting a dependable supply of fresh water has long been a problem. When Moses led the Israelites out of Egypt into the wilderness, he encountered a desert area where potable water was scarce. While many people know the Biblical account of how Moses struck the rock to bring forth water, an incident at Marah, where the Israelites couldn't drink the water because it was bitter, is less well known. According to the account in Exodus, God commanded Moses to throw a tree into the water to purify it.

Recall from Chapter 7 that one property of basic solutions is that they are bitter. The water at Marah was probably basic, or *alkaline*. Such waters are common in desert areas. Attempts have been made to give a chemical explanation of Moses's purification of the brackish water. The tree was probably a dead one, bleached by the desert sun, with the alcohol groups in cellulose oxidized to carboxylic acid groups.

An adequate supply of water is essential to life. This photograph shows a community water tap in Asmara, Eritrea, in East Africa.

$$CH_2OH \qquad CH_2OH \qquad \longrightarrow \qquad COOH \qquad COOH$$

Cellulose $\qquad\qquad\qquad$ Oxidized cellulose

These acidic groups neutralize the alkali in the water. Moses didn't have to understand the science of water purification in order to apply the appropriate technology.

13.4 Biological Contamination: The Need for Clean Water

Rainwater dissolves matter from decaying plants and animals. In small quantities as part of the natural cycle in a forest, this enriches the soil, but in large quantities it contaminates soil and water. The pollutants in the waters of Marah were natural ones. Early people did little to pollute the water and the air, if only because their numbers were so few. The coming of the agricultural revolution and the rise of cities brought together enough *Homo sapiens* to pollute the environment seriously. Even then, pollution was mostly local and largely biological. Human wastes were dumped on the ground or into the nearest stream. Disease organisms were transmitted through food, water, and direct contact.

Waterborne Disease

Contamination of water supplies by microorganisms from human wastes was a severe problem throughout the world until about 100 years ago. During the 1830s, severe epidemics of cholera swept the Western world. Typhoid fever and dysentery

Contaminated beaches are closed to swimming and other recreational activities.

According to the WHO's Global Burden of Disease Study, unclean or inadequate water supplies, poor sanitation, and poor hygiene kill 2,668,000 people worldwide each year. These people die of bacterial, viral, and parasitic infections. Poor water supplies rank fourth in causes of death—after malnutrition (5,881,000 deaths), tobacco use (3,038,000 deaths), and hypertension (2,918,000 deaths).

were common. In 1900, for example, there were more than 35,000 deaths from typhoid in the United States. Today, as a result of chemical treatment, municipal water supplies in developed nations are generally safe. However, less than 10% of the people of the world have access to sufficient clean water. Waterborne diseases are still quite common in much of Asia, Africa, and Latin America, which still have epidemics of cholera, typhoid, and dysentery. Indeed, it is estimated that 80% of all the world's sickness is caused by contaminated water. People with waterborne diseases fill half the world's hospital beds and die at a rate of 25,000 people per day.

The threat of biological contamination has not been totally eliminated from developed nations. It is estimated that 30 million people in the United States are at risk because of bacterial contamination of drinking water. Hepatitis, a viral disease occasionally spread through drinking water, at times threatens to reach epidemic proportions, even in the most technologically advanced nations. Biological contamination also lessens the recreational value of water. Swimming is forbidden in many areas. Biological contamination may have reached a peak in the 1960s. Lake Erie, a notable example, was described as "America's Dead Sea" and "the world's largest cesspool."

Much improvement has been made in recent years. Fish thrive again in Lake Erie. Beaches on Lake Michigan near Chicago, closed in 1969 because of pollution, were reopened in 1975. The Mississippi River below Minneapolis–St. Paul has been reopened to recreation and swimming after being closed for many years. Shellfish are being harvested again in Maine's Belfast Bay. Shrimp and oysters are returning to Escambia Bay off Pensacola, Florida. Atlantic salmon have returned to the Connecticut River for the first time in 100 years. And, in England, fish once again inhabit the Thames. Despite all the good news, however, pollution from domestic sewage remains a threat to the water supply in the United States. We still have a long way to go.

Sewage and Dying Lakes

Pathogenic (disease-causing) *microorganisms* are not the only problem caused by the dumping of human sewage into our waterways. The breakdown of organic matter by bacteria depletes **dissolved oxygen** in the water and enriches the water with plant nutrients. A stream can handle a small amount of waste without difficulty, but when massive amounts of raw sewage are dumped into a waterway, undesirable changes occur.

Most organic material can be broken down (degraded) by microorganisms. This biodegradation can be either *aerobic* or *anaerobic*. **Aerobic oxidation** occurs in the

Common sewage from homes and businesses depletes the dissolved oxygen in water.

presence of dissolved oxygen. A measure of the amount of oxygen needed for this degradation is the **biochemical oxygen demand (BOD)**. The greater the quantity of degradable organic wastes, the higher the BOD. If the BOD is high enough, dissolved oxygen is depleted and no life (other than odor-producing anaerobic microorganisms) can survive in the lake or stream. Flowing streams can regenerate themselves. Streams with rapids soon come alive again as the moving water dissolves oxygen. Lakes with little or no flow can remain dead for years.

With adequate dissolved oxygen, aerobic bacteria (those that require oxygen) oxidize the organic matter to carbon dioxide, water, and a variety of inorganic ions (Table 13.3). The water is relatively clean, but the ions, particularly the nitrates and phosphates, may serve as nutrients for the growth of algae, which also cause problems. When the algae die, they become organic waste and increase the BOD. This process is called **eutrophication**. Runoff of fertilizers from farms and lawns stimulates algal bloom and die-off, leading to dead and dying streams and lakes that nature cannot purify nearly as quickly as we can pollute them.

Table 13.3 ▮ Some Substances Added to Water by the Breakdown of Organic Matter

Substance	Formula
Aerobic conditions	
Carbon dioxide	CO_2
Nitrate ions	NO_3^-
Phosphate ions	PO_4^{3-}
Sulfate ions	SO_4^{2-}
Bicarbonate ions	HCO_3^-
Anaerobic conditions	
Methane	CH_4
Ammonia	NH_3
Amines	RNH_2*
Hydrogen sulfide	H_2S
Methanethiol	CH_3SH

*See Chapter 9.

When too much organic matter depletes the dissolved oxygen in a body of water—whether from sewage, dying algae, or other sources—**anaerobic decay** processes take over. Instead of oxidizing the organic matter, anaerobic bacteria reduce it. Methane (CH_4) is formed. Sulfur is converted to hydrogen sulfide (H_2S) and other

The eutrophication of a lake is a natural process, but the action can be greatly accelerated by human wastes and the runoff of fertilizers from farms and lawns.

foul-smelling organic compounds. Nitrogen is reduced to ammonia and odorous amines (Chapter 9). The foul odors are a good indication that the water is overloaded with organic wastes. No life other than anaerobic microorganisms can survive in such water.

13.5 Ecological Cycles

Let's consider water pollution from the point of view of a fish. A simplified ecological cycle (Figure 13.3) could take place in a small lake. Fish in the water produce organic wastes. Bacteria break down these wastes into inorganic materials that serve as nutrients for the growth of algae. Fish eat the algae, balance is established, and the cycle is complete.

Now let's look at some of the ways people can disrupt the cycle. We can increase the organic wastes by dumping sewage into the water. In breaking down these wastes, bacteria consume all the dissolved oxygen and the fish die. Chemists can monitor the BOD of the water, thus identifying the problem and gauging its severity. Chemists can also contribute to the development of better methods of wastewater treatment.

13.6 Chemical Contamination: From Farm, Factory, and Home

Dumping our sewage into waterways isn't the only way we can foul up an ecological cycle. Fertilizer runoff and seepage from feedlots add inorganic nutrients to the cycle, and an algal bloom can lead to oxygen depletion and death for the fish. Perhaps the most enigmatic influence of all comes from the introduction of new substances into the ecological water cycle: pesticides, radioisotopes, detergents, toxic metals, and industrial chemicals.

The industrial revolution added a new dimension to our water pollution problems. Factories were often built on the banks of streams, and wastes were dumped into the water to be carried away. The rise of modern agriculture has led to increased

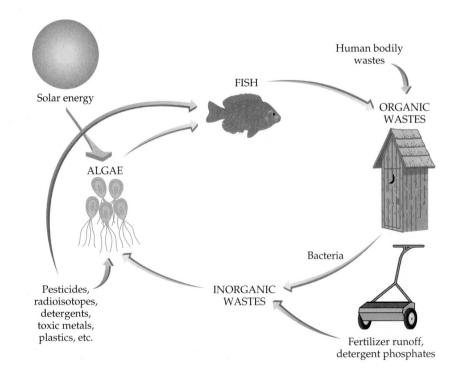

Figure 13.3 A simplified ecological cycle (green arrows) and some ways humans disrupt it (brown arrows).

contamination as fertilizers and pesticides have found their way into the water system. Transportation of petroleum results in oil spills in oceans, estuaries, and rivers. Acids enter waterways from mines and factories and from acid precipitation. Household chemicals also contribute to water pollution when detergents, solvents, and other chemicals are dumped down drains.

By creating new materials, chemists have brought on a variety of new ecological problems. But chemists and chemistry are also necessary to any understanding of our pollution problems. It is only through the development of sophisticated analytical methods and instruments that we have even become aware of some of our ecological problems. A stinking lake is obvious to everyone, but only a person trained in chemistry can determine the level of dangerous, yet invisible, materials in the water we drink (Table 13.4).

Table 13.4 ■ National Drinking Water Standards for Selected Substances (U.S. Environmental Protection Agency)*

Substance	Maximum Contaminant Level (mg/L)[†]
Primary standards: inorganic compounds	
Arsenic	0.05
Barium	2
Copper	1.3
Cyanide	0.2
Fluoride	4.0
Lead	0.015
Nitrate	10[‡]
Primary standards: organic compounds	
Atrazine	0.003
Benzene	0.005
p-Dichlorobenzene	0.075
Dichloromethane	0.005
Heptachlor	0.0004
Lindane	0.0002
Toluene	1
Trichloroethylene	0.005
Secondary standards (nonenforceable)	
Chloride	250
Iron	0.3
Manganese	0.05
Silver	0.10
Sulfate	250
Total dissolved solids	500
Zinc	5

*A more detailed list and a more detailed explanation of the rules can be found at http://www.epa.gov/OGWDW/wot/appa.html.

[†]1 mg/L is often called 1 part per million (1 ppm).

[‡]Measured as N.

13.7 Groundwater Contamination: Tainted Tap Water

About half the people in the United States drink surface water (from streams and lakes). The other half get their drinking water from groundwater via wells or springs. Groundwater is the main source of drinking water in 32 of the 50 states and 34 of our

Groundwater occurs in void spaces in the rocks below the water table. Because rocks have different porosity and permeability, water does not move around the same way in all rocks. When water-bearing rocks readily transmit water to wells and springs, they are called *aquifers*. Wells are drilled into an aquifer, and water is pumped out. Precipitation recharges the aquifer. For many aquifers, the rate of pumping water is much greater than the rate of recharge and the water table drops.

[Web Reference 2] More about groundwater can be found at the EPA's site. The worldwide scope of the problem can be seen by visiting the United Kingdom's Groundwater Forum.

Homes and a school were built on the site of an old chemical dump in the Love Canal section of Niagara Falls, New York. Now a contaminated area around the old dump is fenced off to keep people out.

Hydrocarbon and chlorinated hydrocarbons were discussed in Chapter 9.

largest cities. In rural areas, 97% of the population drink groundwater. The dropping water table in most parts of the country requires the digging of ever-deeper wells.

Well water can be contaminated. Toxic chemicals have been found in the groundwater in some areas. For example:

- Wells around the Rocky Mountain Arsenal, near Denver, are contaminated by wastes from the production of pesticides.
- Wells near Minneapolis, Minnesota, are contaminated with creosote, a chemical used as a wood preservative.
- Wells in parts of Wisconsin and on Long Island have been contaminated with aldicarb, a pesticide used on potato crops.
- Some community water supplies in New Jersey have been shut down because of contamination with industrial wastes.

Nitrates

In many agricultural areas, the concentration of nitrates (NO_3^-) in well water is above the maximum safe level (for infants) of 10 ppm. Even the maximum safe level of 45 ppm for adults is exceeded in some areas.

Excessive nitrates (which are reduced to nitrites in an infant's digestive tract) cause *methemoglobinemia*, the blue baby syndrome. In parts of Illinois, baby pigs turn blue and die after drinking the water. In California's Imperial Valley, some parents have to buy bottled water for their babies.

Nitrates in the groundwater come from fertilizers used on farms and lawns, from the decomposition of organic wastes in sewage treatment, and from runoff from animal feed lots. They are highly soluble and thus difficult to remove from water. Only an expensive treatment can remove nitrate compounds from water once they are there.

Volatile Organic Chemicals

We mentioned volatile organic compounds (VOCs) briefly as air pollutants (Chapter 12). VOCs are also water pollutants and add an undesirable odor to water. Many VOCs are suspected carcinogens. VOCs are used as solvents, cleaners, and fuels, and they are components of gasoline, spot removers, oil-based paints, thinners, and some drain cleaners. Common VOCs are hydrocarbon solvents, such as benzene and toluene, and chlorinated hydrocarbons, such as carbon tetrachloride (CCl_4), chloroform ($CHCl_3$), and methylene chloride (CH_2Cl_2). Especially common is trichloroethylene (CCl_2=$CHCl$), widely used as a dry-cleaning solvent and as a degreasing compound. When spilled or discarded, VOCs enter the soil and eventually get into the groundwater.

Chemicals buried in dumps—often years ago, before there was much awareness of environmental problems—have now infiltrated groundwater supplies. Often, as at the Love Canal site in Niagara Falls, New York, people built schools and houses on or near old dump sites. VOCs generally are only slightly soluble in water, with solubilities often in the parts-per-million or parts-per-billion range. These trace amounts are found in groundwater. Many VOCs are characterized by a lack of reactivity. They react so slowly that they are likely to be around for a long time.

Leaking Underground Storage Tanks

Another major source of groundwater contamination is underground storage tanks. Gasoline at service stations has traditionally been stored in buried steel tanks, which last an average of about 15 years before they rust through and begin to leak. Of perhaps 2.5 million such tanks in the United States, as many as 200,000 may now be leaking, many of them at stations that went out of business during the fuel shortage of the 1970s. Gasoline is now found in wells near these tanks in many areas of the country. Laws now require replacement of old gasoline tanks and proper cleanup of any contaminated ground.

Trace Toxics and Public Perception

Groundwater contamination is particularly alarming because, once contaminated, an underground aquifer may remain unusable for decades or longer. There is no easy way to remove the contaminants. Pumping out the water and purifying it could take years and cost billions of dollars.

The problem of groundwater pollution is serious, but it is often overdramatized by the news media. The amounts of contaminants are often minute. Chemists can now measure parts per million (ppm), parts per billion (ppb), and for some contaminants even parts per trillion (ppt). Without sophisticated chemical techniques, we wouldn't even know about some problems. For example, the EPA has set a safety limit of 10 ppb (10 mg in 1000 L of water) for aldicarb. You would have to drink 32,000 L of water to get as much aldicarb as there is aspirin in one tablet. Contamination of groundwater by toxic substances is serious and should not be taken lightly, but we must keep in mind that the amounts of toxic materials usually involved are infinitesimal. And biological contamination is much more widespread and often more deadly than chemical contamination.

> Many people drink bottled water to avoid real or perceived problems with public water supplies. Bottled water was largely unregulated until 1995, when the Food and Drug Administration set standards to ensure that its minimum quality was equal to that of public water supplies.

> Toxic substances are discussed further in Chapter 20.

Calculations of Parts per Million and Parts per Billion

We discussed solution concentrations in Chapter 6. For solutions that are extremely dilute, we often express concentrations in ppm, ppb, or even ppt. For example, in fluoridated drinking water, fluoride ion is maintained at about 1 ppm. A typical level of the contaminant chloroform ($CHCl_3$) in a municipal drinking water taken from the lower Mississippi River is 8 ppb. We can get a sense of the meanings of these terms by considering that one person in the city of San Diego, California, represents about 1 ppm, and one person in the People's Republic of China, about 1 ppb.

For aqueous solutions, ppm, ppb, and ppt are generally based on mass. Thus, 1 ppm of solute in a solution is the same as 1 g solute per 1×10^6 g solution. Calculations are much like those for percent by mass (Section 6.7). In Example 13.1, we introduce a useful relationship for aqueous solutions:

$$1 \text{ ppm} = 1 \text{ mg solute per liter of solution}$$

Example 13.1

The maximum allowable level of nitrates in drinking water set by the state of California is 45 mg NO_3^- per liter. What is this level expressed in ppm?

Solution

The density of water, even if it contains traces of dissolved substances, is essentially 1.00 g/mL. One liter of water has a mass of 1000 g. So, we can express the allowable nitrate level as the ratio

$$\frac{45 \text{ mg } NO_3^-}{1000 \text{ g water}}$$

To convert this nitrate level to ppm, we need to have the numerator and denominator in the same units. By using milligrams, we make the denominator 1 million mg. The numerator then expresses the ppm of solute, that is, ppm NO_3^-.

$$\frac{45 \text{ mg } NO_3^-}{1000 \text{ g water}} \times \frac{1 \text{ g water}}{1000 \text{ mg water}} = \frac{45 \text{ mg } NO_3^-}{1,000,000 \text{ mg water}} = 45 \text{ ppm } NO_3^-$$

Exercise 13.1

A. What is the concentration in **(a)** ppb and **(b)** ppt corresponding to a maximum allowable level in water of 0.1 µg/L of the pesticide chlordane?

B. What is the concentration of NaCl, in ppm, in 0.00152 M NaCl?

13.8 Acid Waters: Dead Lakes

Acids pour down on us from the sky as acid rain, fog, and snow (Chapter 12). These acids corrode metals, limestone, and marble, and even ruin the finishes on our automobiles. Acids also flow into streams from abandoned mines.

Acid water is detrimental to life in lakes and streams. More than 1000 bodies of water in the eastern United States are acidified, and 11,000 others have only a limited ability to neutralize the acids that enter them. In Ontario, Canada, alone, more than 100 lakes are devoid of any life, and another 48,000 lakes are threatened. The threat to these areas is presumed to be acid rain originating mainly in the Ohio River valley and Great Lakes regions. The sulfur oxides and nitrogen oxides from power plants, industries, and automobiles travel hundreds of kilometers downwind and fall as sulfuric and nitric acids. Acid rain has been linked to declining crop and forest yields.

The effects of acid waters on living organisms are hard to pin down precisely. Probably the greatest effect of acidity is that it causes the release of toxic ions from rocks and soil. For example, aluminum ions, which are tightly bound in clays and other minerals, are released by acid. Aluminum ions have low toxicity for humans, but they seem to be deadly to young fish. Many dying lakes have only old fish because none of the young survive. Ironically, lakes destroyed by excess acidity are often quite beautiful. The water is clear and sparkling—quite a contrast to those in which fish are killed by oxygen depletion following algal blooms.

Acids are no threat to lakes and streams in areas where the rock is limestone (calcium carbonate), which can neutralize excess acid.

$$CaCO_3(s) \ + \ 2\,H^+(aq) \ \longrightarrow \ Ca^{2+}(aq) \ + \ CO_2(g) \ + \ H_2O$$

Limestone   Acid

Where rock is principally granite, however, no such neutralization occurs.

Acidic waters can be treated with pulverized limestone or other basic substances. A few such attempts have been carried out, but the process is costly and the results last only a few years. An obvious way to alleviate the problem is to remove the sul-

Acidic water draining from an old mine.

fur from the coal before combustion or to scrub the sulfur oxides from the smokestack gases. We have made progress in this area, but these remedies are expensive and add to the cost of electricity.

13.9 Industrial Water Pollution

Industries in the United States have substantially reduced their contribution to water pollution. Most are in compliance with the Water Pollution Control Act, which requires that they use the best practicable technology. We examine here only a few examples of industrial pollution and some ways to alleviate it.

Automobiles and Water Pollution

It takes several hundred kilograms of steel to produce an automobile. To make a metric ton of steel requires about 100 t of water. About 4 t of water is lost through evaporation. The remainder is contaminated with acids, grease and oil, lime, and iron salts. This polluted water can be cleaned up, and most of it is recycled.

Chrome plating on bumpers, grills, and ornaments is also a source of pollution. Waste chromium, in the form of chromate ions (CrO_4^{2-}), and cyanide ions (CN^-) are products of this process. In the past, these toxic substances were dumped into waterways. Today, chemical treatment generally removes a large amount.

Cyanide is treated with chlorine and a base to form nitrogen gas, bicarbonate ions, and chloride ions.

$$10\,OH^- + 2\,CN^- + 5\,Cl_2 \longrightarrow N_2 + 2\,HCO_3^- + 10\,Cl^- + 4\,H_2O$$

The products are much less toxic than cyanide. The chromate is removed by reduction with sulfur dioxide to Cr^{3+} ion, and the sulfur dioxide is oxidized to sulfate.

$$2\,CrO_4^{2-} + 3\,SO_2 + 2\,H_2O \longrightarrow 2\,Cr^{3+} + 3\,SO_4^{2-} + 4\,OH^-$$

Sulfate is generally not a serious pollutant, and Cr^{3+} is relatively insoluble in alkaline solution but is soluble enough in acidic media to constitute a problem.

Now, add in the costs (environmental and economic) of elastomers for tires, fabrics for upholstery, glass for windows, electricity—the list goes on and on—and it is easy to see that the private automobile is an ecological problem even before it hits the road. And once on the road, it is a major contributor to air pollution (Chapter 12).

Most other industries also contribute to water pollution. Table 13.5 lists the water required (per metric ton) for the production of a variety of materials. Most of this water is cleaned up and recycled, but the need for clean water is still enormous.

Table 13.5 ▮ Water Required to Produce Various Materials

Material	Water Required*
Steel	100
Paper	20
Copper	400
Rayon	800
Aluminum	1280
Synthetic rubber	2400

*In cubic meters per metric ton. A cubic meter of water weighs 1000 kg, or 1 t.

Water Pollutants From Other Industries

Wastes from the textile industries include conditioners, dyes, bleaches, oils, dirt, and other organic debris. Most of these wastes can be removed by conventional sewage treatment. Wastes from meatpacking plants include blood and various animal parts.

The Soviet Union built so many dams and factories along the Volga River that it is no longer the mighty, vigorous waterway it once was. Much of the damage was hidden from the rest of the world until the USSR collapsed in 1991.

Other plants generate fruit and vegetable waste. Regular sewage plants also usually treat food industry wastes.

Oil refineries produce wastes that include dyes, oils, acids, brines, and sulfur compounds. Chemical plants produce a variety of waste materials. Most industries have made substantial reductions in the amount of wastes produced and are committed to further reductions.

Until the Soviet Union collapsed in 1991, their leaders covered up the environmental problems of eastern Europe. Let's consider the case of one of the most famous rivers in the world, the mighty Volga in Russia. So many dams have been built along the Volga that its flow has been slowed to a crawl. It used to take 50 days for water to travel the 2300 mi from the source to the mouth of the river, but now it takes 18 months. The sluggish flow plus the pollution from all the factories along the way have created an ecological catastrophe. Tons of industrial waste (cleaning fluids, fertilizers, pesticides, heavy metals, toxic chemicals, radioactive waste, and waste from pulp and paper mills) pour into the Caspian Sea, which appears to be dying. In some places you can see yellow, red, and black streams of water carrying sulfur, iron oxide, and various oils. The Volga once teemed with caviar-producing sturgeon, but pollution and poaching now threaten the sturgeon with extinction.

13.10 From Wastewater to Drinking Water

The water we drink often comes from reservoirs, lakes, and rivers. Many cities depend on water that has been used by other cities upstream. Such water may be polluted with chemicals and pathogenic microorganisms. In this section we will look first at the treatment of wastewater and then at water purification for municipal use.

Wastewater Treatment Plants

For many years, most communities simply held sewage in settling ponds for a while before discharging it into a stream, lake, or ocean. This type of facility (Figure 13.4) provides what now is called **primary sewage treatment**. Primary treatment removes some of the solids as *sludge*. The effluent contains a lot of dissolved and suspended organic matter and has a huge BOD. Often all the dissolved oxygen in the pond is used up, and anaerobic decomposition—with its resulting odors—takes over.

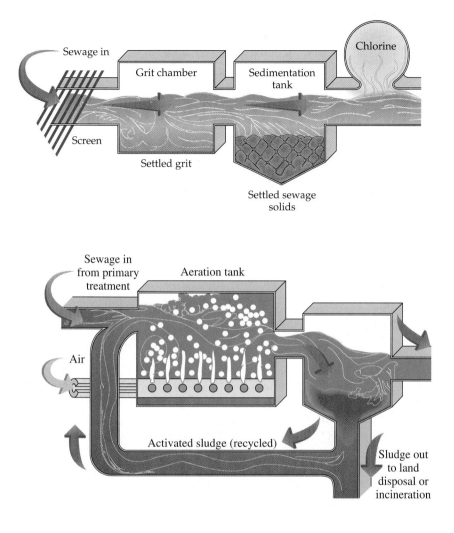

Figure 13.4 A diagram of a primary sewage treatment plant.

Figure 13.5 A diagram of a secondary sewage treatment plant that uses the activated sludge method of treatment.

A **secondary sewage treatment** plant passes effluent from the primary treatment facility through sand and gravel filters. There is some aeration in this step, and aerobic bacteria convert most of the organic matter to stable inorganic materials. In the **activated sludge method** (Figure 13.5), a combination of primary and secondary treatment methods, the sewage is placed in tanks and aerated with large blowers. This causes the formation of large, porous clumps called *flocs*, which filter and absorb contaminants. The aerobic bacteria further convert the organic material to sludge. Part of the sludge is recycled to keep the process going, but huge quantities must be removed for disposal. This sludge is stored on land (where it requires large areas), dumped at sea (where it pollutes the ocean), or burned in incinerators (where it requires energy—such as natural gas—and contributes to air pollution). Sometimes sludge is used as fertilizer.

In many areas, secondary treatment of wastewater is inadequate. To an increasing extent, federal mandates require **advanced treatment** (sometimes called *tertiary treatment*). Several advanced processes are in use, and most are quite costly.

- In **charcoal filtration**, charcoal adsorbs organic molecules that are difficult to remove by any other method.

- In **reverse osmosis**, pressure forces water through a semipermeable membrane, leaving impurities behind.

Charcoal filtration is especially effective for removing organic compounds from water. The organic molecules are *adsorbed* on the surface of the charcoal and thus removed from the water. After a period of time, the charcoal becomes saturated and is no longer effective. It can be regenerated by heating to drive off the adsorbed substances.

Table 13.6 ▮ Summary of Wastewater Treatment Methods

Method	Cost	Material Removed	Amount Removed (%)
Primary			
Sedimentation	Low	Dissolved organics	25–40
		Suspended solids	40–70
Secondary			
Trickling filters	Moderate	Dissolved organics	80–95
		Suspended solids	70–92
Activated sludge	Moderate	Dissolved organics	85–95
		Suspended solids	85–95
Advanced (tertiary)			
Carbon bed with regeneration	Moderate	Dissolved organics	90–98
Ion exchange	High	Nitrates and phosphates	80–92
Chemical precipitation	Moderate	Phosphates	88–95
Filtration	Low	Suspended solids	50–90
Reverse osmosis	Very high	Dissolved solids	65–95
Electrodialysis	Very high	Dissolved solids	10–40
Distillation	Extremely high	Dissolved solids	90–98

*See References and Readings for a discussion of methods not covered in the text.

Finding the money to finance adequate sewage treatment will be a major political problem for years to come. Table 13.6 summarizes wastewater treatment methods.

The effluent from sewage plants is usually treated with chlorine to kill any remaining pathogenic microorganisms before it is returned to a waterway. Chlorination has been quite effective in preventing the spread of waterborne infectious diseases such as typhoid fever. Further, some chlorine remains in the water, providing residual protection against pathogenic bacteria. However, chlorination is not effective against viruses such as those that cause hepatitis.

According to the WHO about 40,000 children around the world die each day for want of pure water.

A Drop to Drink

Most cities in developed nations treat their water supply again before it flows into homes (Figure 13.6). The water to be purified is usually placed in a settling basin where it is treated with slaked lime and a flocculent such as "alum." These materials react to form a gelatinous mass of aluminum hydroxide.

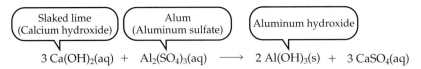

$$3\ Ca(OH)_2(aq)\ +\ Al_2(SO_4)_3(aq)\ \longrightarrow\ 2\ Al(OH)_3(s)\ +\ 3\ CaSO_4(aq)$$

The aluminum hydroxide carries down dirt particles and bacteria. The water is then filtered through sand and gravel.

Usually water is treated by *aeration*; it is sprayed into the air to remove odors and improve its taste (water without dissolved air tastes flat). Sometimes it is filtered through charcoal to remove colored and odorous compounds. In the final step, chlorine is added to kill any remaining bacteria. In some communities that use river water, a lot of chlorine is needed to kill all the bacteria, and you can taste the chlorine in the water. Some people question the use of chlorine because it converts dissolved organic compounds into chlorinated hydrocarbons. Analyses of the drinking water of several cities that take their water from rivers have found chlorinated hydrocarbons, including such known carcinogens as chloroform and carbon tetrachloride. The concentration is in the parts-per-billion range, probably posing only a small

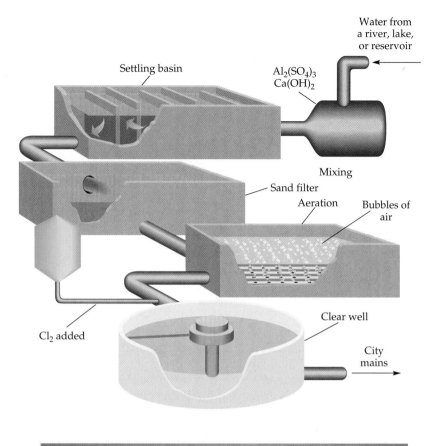

Figure 13.6 (a) A diagram of a municipal water purification plant. (b) In heavily populated areas, municipal sewage treatment and water purification plants take up large areas of land, often prime waterfront properties.

threat, but it is worrisome nonetheless. (It is not nearly so worrisome, however, as the waterborne diseases that prevail in much of the world where adequate water treatment is not available.)

Chlorination is not the only way to disinfect drinking water. Ozone (Chapter 12) is used by more than 1000 cities, mostly in Europe. Ozone is more expensive than chlorine, but less of it is needed. An added advantage is that ozone kills viruses on which chlorine has little, if any, effect. For example, ozone is 100 times as effective as chlorine in killing polioviruses.

Ozone (O_3) acts by transferring its "extra" oxygen to the contaminant. Oxidized contaminants are generally less toxic than chlorinated ones. In addition,

[Web Reference 3]
In the United States most water treatment plants are run by local governments. Many water utilities, both here and abroad, have their own websites.

ozone imparts no chemical taste to water. Unlike chlorine, however, ozone does not provide residual protection against microorganisms. If the problems with chlorine get worse, however, we may see a shift from chlorine to ozone in the treatment of drinking water and wastewater. Perhaps we could use ozone to disinfect water and then add just enough chlorine to provide residual protection.

Fluorides

Fluoride has been studied extensively as a possible carcinogen. The results are equivocal; we still don't know whether or not fluoride causes cancer. It does not seem likely, however, that levels found in fluoridated water can cause cancer in humans.

Dental caries (tooth decay) was once considered "the leading chronic disease of childhood." This is no longer true, thanks mainly to the addition of fluoride to municipal water supplies. Up to a point, the hardness of our tooth enamel can be correlated with the amount of fluoride present. Tooth enamel is a complex calcium phosphate called hydroxyapatite. Fluoride ions replace some of the hydroxide ions, forming a harder mineral called fluorapatite.

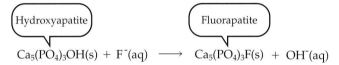

$$Ca_5(PO_4)_3OH(s) + F^-(aq) \longrightarrow Ca_5(PO_4)_3F(s) + OH^-(aq)$$

The fluoride concentration of the drinking water of many communities has been adjusted to 0.7–1.0 ppm (by mass) by adding fluoride, usually as H_2SiF_6 or Na_2SiF_6. Fluoridation at this level results in a reduction in the incidence of dental caries (cavities) by as much as two-thirds in some areas. While only 29% of the 9-year-olds in the United States were cavity-free in 1971, over 65% of that age group were cavity-free in 1987.

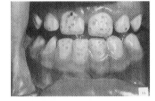

Excessive fluoride consumption in early childhood can cause mottling of tooth enamel. The severe case shown here was caused by continuous use during childhood of a water supply that had an excessive natural concentration of fluoride.

Some people object to water fluoridation. Fluoride salts are acute poisons in moderate to high concentrations. Indeed, sodium fluoride (NaF) is used as a poison for roaches and rats. Small amounts of fluoride ion, however, seem to contribute to our well-being.

There is some concern about the cumulative effects of consuming fluorides in drinking water, in the diet, in toothpaste, and from other sources. Excessive fluoride consumption during early childhood can cause mottling of tooth enamel. The enamel becomes brittle in certain areas and gradually discolors. Fluorides in high doses also interfere with calcium metabolism, with kidney action, with thyroid function, and with the actions of other glands and organs. Although there is little or no evidence that optimal fluoridation causes problems such as these, the fluoridation of public water supplies will most likely remain a subject of controversy.

[Web Reference 4]
The fluoride story is available at the National Institute of Dental and Craniofacial Research, National Institutes of Health's website.

A few communities treat their sewage by pumping it into a marshy area where the marsh plants use nutrients from the sewage as fertilizer. Some plants even remove toxic metals. Effluent from the marsh is often as clean as the water in municipal reservoirs.

13.11 Back to the Soil: An Alternative Solution

Why do we dump our wastes in waterways in the first place? Dumping fouls the water with wastes that contain nutrients that could fertilize the land. Why not return the nutrients to the soil? In many societies both human and animal wastes have always been used as fertilizer. It is only in "overdeveloped" countries that these vital resources are dumped into waterways on a large scale.

Many communities do return sludge to the soil. Sludge can be dried, sterilized, and transported to farmlands. Wet, suspended sludge also can be pumped directly to the fields. The water in the mixture irrigates the crops and the sludge is a source of nutrients and humus. One concern is that the pathogens in sludge will survive and spread disease. The biggest problem, however, is that sludge is often contaminated with toxic metals that could be taken up by plants and eventually end up in our food.

Other solutions are possible. Toilets that compost wastes and use no energy or water have been developed. The mild heat of composting drives off water from the wastes. The system is ventilated to keep the process aerobic, and no odors enter the house from a properly installed system. Dried waste is removed about once a year. The initial cost is much higher than that of a flushed toilet, however. For each person in the United States, we flush about 50,000 L of drinking-quality water annually. Perhaps we should consider an alternative to flushing our wastes into the water we drink.

13.12 We're the Solution to Water Pollution

We generally take our drinking water for granted. Perhaps we shouldn't. There are between 4000 and 40,000 cases of waterborne illnesses in the United States each year. Twenty million people have no running water at all, and many more obtain water from suspect sources. Another 30 million tap individual wells or springs which are often uncontrolled and of unknown quality. Much remains to be done before we can all be assured of safe drinking water.

How much water does one really need? Only about 1.5 L/day for drinking. In the United States each day, we use about 7 L per person for drinking and cooking, but we use directly a total of about 380 L. We use much more water *indirectly* in agriculture and industry to produce food and other materials (Table 13.7); it takes 800 L of water to produce 1 kg of vegetables, and 13,000 L of water to produce a steak. We also use water for recreation (for example, swimming, boating, and fishing). For most of these purposes, we need water that is free of bacteria, viruses, and parasitic organisms.

More than 400,000 people were made ill in 1993 when *Cryptosporidium*, a microorganism common in barnyard runoff, contaminated the water supply in the Milwaukee area. About 1000 people died.

Table 13.7 ▮ Average Daily per Person Use of Water in the United States

Use	Quantity (L)
Direct use	
Drinking and cooking	7
Flushing toilets	80
Swimming pools and lawn watering	85
Dishwashing	14
Bathing	70
Laundry	35
Miscellaneous	90
Total direct use	380
Indirect use	
Industrial	3800
Irrigation (agriculture)	2150
Municipal water (nonindustrial)	550
Total indirect use	6500
Total overall use	6900

The U.S. Congress has passed laws intended to drastically reduce water pollution. It can't be eliminated entirely, however. To use water is to pollute it. Each of us can do our share by conserving water and by minimizing our use of products that require vast amounts of water to make. As citizens we must be prepared to foot the bill for waste treatment. As our population grows, it will cost a lot just to maintain the present water quality. To clean our water up, and then keep it clean, will cost even more. Remember that the cost of unclean water is even higher—discomfort, loss of recreation, illness, and even death.

[Web Reference 5] Information about the Federal Clean Water Act of 1972 (renewed in 1987). Documents relating to it and other acts can be read at the Committee for the National Institute for the Environment's website.

Critical Thinking Exercises

Apply knowledge that you have gained in this chapter and one or more of the FLaReS principles (Chapter 1) to evaluate the following statements or claims.

13.1 An activist group claims that all chlorine compounds should be banned because they are toxic.

13.2 A mother discovers that tests have found 1.0 ppb of trichloroethylene in the well water the family uses for drinking, bathing, cooking, and so on. They have used the well for 2 years. She decides that the family members should be tested for cancer.

13.3 A group of citizens want to ban fluoridation of the community water supply because fluorides are toxic.

Summary

1. Water covers three-fourths of Earth's surface, but only about 1% of it is available as fresh water.

2. Water has a higher density than most other liquids.

3. Water has an unusually high heat capacity and heat of vaporization.

4. Unlike most liquids, water expands when it freezes.

5. The unique properties of water arise from its highly polar structure, which enables the molecules to associate by hydrogen bonding.

6. Seawater is very salty because water is a good solvent for ionic compounds.

7. Water evaporated from oceans and lakes condenses into clouds and then returns to Earth as rain or snow. This is the water cycle.

8. Fresh water can readily become contaminated by various kinds of chemicals and microorganisms, some from natural processes and others from human activities.

9. Hard water contains salts of calcium, magnesium, and/or iron.

10. Dumping sewage into water increases its biochemical oxygen demand (BOD).

11. Water pollutants that are nutrients for the growth of algae can lead to eutrophication of a lake or stream.

12. Acid rain can cause lakes and streams to become so acidic that they damage fish populations and other aquatic life.

13. Groundwater obtained from wells provides drinking water for many cities and most rural households.

14. Wastewater treatment usually includes sludge removal, sand and gravel filtration, aeration, and chlorination.

15. Advanced treatment methods include charcoal filtration, ion exchange, and reverse osmosis.

16. Although fluoride is poisonous in large concentrations, the addition of 1 ppm to municipal water supplies has greatly reduced the incidence of dental caries.

17. Maintaining water quality is expensive and will become more so; but not maintaining the quality of our water will cost much more in terms of our health and comfort.

Key Terms

activated sludge
  method 13.10
advanced treatment 13.10
aerobic oxidation 13.4
anaerobic decay 13.4

biochemical oxygen
  demand (BOD) 13.4
charcoal filtration 13.10
dissolved oxygen 13.4
eutrophication 13.4

hard water 13.3
heat capacity 13.1
heat of vaporization 13.1
primary sewage
  treatment 13.10

reverse osmosis 13.10
secondary sewage
  treatment 13.10
specific heat 13.1

Review Questions

1. What proportion of Earth's water is seawater?

2. How is our supply of fresh water replenished by natural processes?

3. List some waterborne diseases. Why are these diseases no longer common in developed countries?

4. What is BOD? Why is a high BOD undesirable?

5. What are pathogenic microorganisms?

6. What is eutrophication?

7. List some ways in which groundwater is contaminated.

8. What problems do leaking underground storage tanks cause?

9. List some common industrial contaminants of groundwater.

10. Why do chlorinated hydrocarbons remain in groundwater for such a long time?

Problems

Properties of Water

11. Ice is less dense than liquid water. What consequences does this property have for life in northern lakes?
12. Why should foods be flash-frozen rather than frozen slowly?
13. A barge filled with gasoline sinks and breaks open. Will the gasoline dissolve in the water? Will it float or sink?
14. Define heat capacity. Why is the high heat capacity of water important to planet Earth?
15. Why is the high heat of vaporization of water important to our bodies?
16. Why is it cooler near a lake than inland during the summer?
17. Why is ice less dense than liquid water?
18. Why are seas salty?

Natural Waters

19. What impurities are present in rainwater?
20. List four cations and three anions present in groundwater.
21. What is hard water? Why is it sometimes undesirable?
22. What are the products of the breakdown of organic matter by aerobic bacteria? List some of the products of anaerobic decay.

Acidic Waters

23. List two ways in which lakes and streams have become acidic.
24. Why is acidic water especially harmful to fish?
25. List several ways in which the acidity of rain can be reduced.
26. What kind of rocks tend to neutralize acidic waters?
27. How can we restore (at least temporarily) lakes that are too acidic?
28. List two toxic compounds found in wastes from the chrome plating process. How is each removed?

Wastewater Treatment

29. Describe a primary sewage treatment plant. What impurities does it remove?
30. Describe a secondary sewage treatment plant. What impurities does it remove?
31. Describe the activated sludge method of sewage treatment.

32. What substances remain in wastewater after effective secondary treatment?
33. Why is wastewater chlorinated before it is returned to a waterway?
34. What is meant by advanced treatment of wastewater? List two methods of advanced treatment.
35. What kinds of substances are removed from wastewater by charcoal filtration?
36. What are the advantages and disadvantages of spreading sewage sludge on farmlands?

Municipal Water Supplies

37. Why are municipal water supplies treated with aluminum sulfate and slaked lime?
38. Why are municipal water supplies aerated?
39. List the advantages and disadvantages of chlorination of drinking water.
40. List the advantages and disadvantages of ozone as a disinfectant of drinking water.
41. How does fluoride strengthen tooth enamel?
42. What are some health effects of too much fluoride in the diet?
43. What is the source of nitrate ions in well water?
44. What is methemoglobinemia? How is it caused?

Chemical Equations

45. Give the equation for the neutralization of acidic water (H^+) by limestone (calcium carbonate).
46. Give the equation for the reaction of slaked lime (calcium hydroxide) with alum (aluminum sulfate) to form aluminum hydroxide.

Parts per Million and Parts per Billion

47. Express the following aqueous concentrations in the unit indicated.
    a. 1 μg benzene per liter of water, as ppb benzene
    b. 0.0035% NaCl, by mass, as ppm NaCl
48. Express the following aqueous concentrations in the units indicated.
    a. 5 μg trichloroethylene/L water, as ppb trichloroethylene
    b. 0.0025 g KI/L water, as ppm of KI

Additional Problems

49. Describe an alternative to the flush toilet.
50. Express the following aqueous concentrations in the unit indicated.
    a. 2.4 ppm F^-, as molarity of fluoride ion, $[F^-]$
    b. 45 ppm NO_3^-, as molarity of nitrate ion, $[NO_3^-]$
51. Wastewater disinfected with chlorine must be dechlorinated before it is returned to sensitive bodies of water. The dechlorinating agent is often sulfur dioxide. The

reaction is

$$Cl_2 + SO_2 + 2\,H_2O \longrightarrow 2\,Cl^- + SO_4^{2-} + 4\,H^+$$

Is the chlorine oxidized or reduced? Identify the oxidizing agent and reducing agent in the reaction.
52. Radioactive aluminum-26 is used to study the mobilization of aluminum by acidic waters. When it decays, the isotope is transmuted into magnesium-26. What kind of particle is emitted by aluminum-26?

Projects

53. Consult the references or other sources and prepare a report on the desalination of seawater by one of the following methods.
    a. distillation
    b. freezing
    c. electrodialysis
    d. reverse osmosis
    e. ion exchange

54. What sort of wastewater treatment is used in your community? Is it adequate?

55. Where does your drinking water come from? What steps are used in purifying it?

56. Call your water utility office and obtain a chemical analysis of your drinking water. What substances are monitored? Are any of these substances considered problems? (If you use water from a private well, has the water been analyzed? If so, what were the results?)

Online Projects

57. The website of the EPA's Office of Water, http://www.epa.gov/owm/, has a "drop-down" menu of water topics. First, notice how many water-related topics there are. Then, choose one and write a few paragraphs about that aspect of water management.

58. Using the EPA's Office of Water site (http://www.epa.gov/owm/) or your state's environmental website, see what you can learn about the water quality where *you* live.

59. http://yosemite.epa.gov/water\surfnote.nsf is the EPA's directory of links to other environmental websites. Find a website related to water quality and write a few paragraphs about its sponsoring organization.

60. Using http://www.nrdc.org/ or other websites, investigate and write about pollution from urban stormwater runoff, which some say rivals sewage plants and factories as a source of water contamination.

References and Readings

1. Adler, Tina. "The Expiration of Respiration: Oxygen—the Missing Ingredient in Many Bodies of Water." *Science News*, 10 February 1996, pp. 88–89.

2. Boutacoff, David. "Working with the Watershed." *EPRI Journal*, January–February 1990, pp. 28–33. Liming the watershed keeps a lake neutral longer than liming the lake directly.

3. Conkling, Winifred. "Water: How Much Do You Need?" *American Health*, May 1995, pp. 62–64.

4. Ertle, James. "Using 'Bugs' to Treat Wastewater." *Environmental Technology*, January/February 2000, pp. 13–14.

5. Mitchell, John G. "Our Polluted Runoff." *National Geographic*, February 1996, pp. 106–125.

6. Ouellette, Robert P. "A Perspective on Water Pollution." *National Environmental Journal*, September–October 1991, pp. 20–24.

7. Pope, Victoria. "Poisoning Russia's River of Plenty." *U.S. News & World Report*, 13 April 1992, pp. 49–51.

8. Postel, Sandra. *Last Oasis: Facing Water Scarcity*. New York: W. W. Norton, 1992.

9. Spiro, Thomas G., and William M. Stigliani. *Chemistry of the Environment*. Upper Saddle River, NJ: Prentice Hall, 1996. Part III, Hydrosphere.

10. Tarbuck, Edward J., and Frederick K. Lutgens. *Earth Science*, 9th edition. Upper Saddle River, NJ: Prentice Hall, 2000. Chapter 4, "Running Water and Groundwater," and Chapter 12, "The Ocean Environment."

11. Thurman, Harold V. *Introduction to Oceanography*, 7th edition. Upper Saddle River, NJ: Prentice Hall, 1996.

12. Van der Leeden, F., F. L. Troise, and D. K. Todd. *The Water Encyclopedia*, 2nd edition. Chelsea, MI: Lewis Publishers, 1990.

MediaLab

Will We Have Wars Over Water?

Water has long been recognized as a precious substance. Humans have built their settlements on the banks of rivers or near flowing springs or wells. Prayers for rain are familiar in many religions. All living things depend on water. Without water, our crops wither and die. Although we can go without food for many days—or weeks—we can survive only a few days without water.

Our watery planet seems to have a great deal of this precious liquid, but only a small fraction is potable. As we humans use more and more water and pollute many of our freshwater supplies, we draw nearer to the day when we may fight over water rights, as the early settlers did in the American West. Disputes over water rights and use are occurring with increasing frequency between farmers and urban dwellers, between the upper and the lower classes, and between or among regions or nations.

In this *Web Investigation*, you will look at conflicts over water in several regions of the world. You will examine efforts to reconcile opposing sides and consider what can be done to preserve our waters. In *Communicate Your Results*, you will write about the common factors in water disputes and evaluate attempts to settle them.

WEB INVESTIGATION

Traditional Spray Irrigation

Investigation 1
The Water Wars: Present and Future Conflicts

Select Keyword **CYCLE** for a look at the hydrological cycle—the circulation and conservation of Earth's water. In particular, note that throughout this cycle, the total amount of Earth's water remains constant. Although technology can change the distribution or the salt content of water, the total supply cannot be increased. Select Keyword **CONFLICT** for a discussion of the potential for conflicts over water.

Investigation 2
The Middle East: Lack of Water Adds to the Tinderbox

In our exploration of current or potential conflicts around the world, we look first at the Middle East. Among the many barriers to lasting peace in this region is the question of water rights. From ancient times, such as the time when Abraham reproved Abimelech about a well (Genesis 21:25), people have argued over water. Over the centuries, the need for sources of fresh water in an arid land has contributed to many conflicts. For an overview of the problem of water in the Middle East, select Keyword **MIDEAST**. The peace treaty between the nations of Jordan and Israel attempted to allocate water rights in a fair manner, but problems continue, particularly in regard to use of the Jordan River. For a detailed history and analysis, select **JORDAN**.

Investigation 3
The Colorado River

Extending from Wyoming to Mexico, the Colorado River supplies water to seven states and to Mexico. It has been the subject of controversy for decades, with each state claiming a share of its water. Select Keyword **COLORADO** for a history and analysis of the discords and agreements.

Investigation 4
The Cauvery River Dispute: Conflict Over Water in India

Since 1974, two states in India, Karnataka and Tamil Nadu, have been fighting over the waters of the Cauvery River, an important watershed in southern India. Select Keyword **CAUVERY** for an analysis of this controversy, or Keyword **INDIA** for a more detailed treatment with history and some generalizations about any water rights conflict.

Investigation 5
Saving Water: What Is Being Done?

Steps to save water must occur at the local, regional, national, and international levels. In some cases, a national government must mediate regional conflicts; in others, international treaties are essential. Select Keyword **ACTION** to see a booklet called "Tackling Urgent Water Priorities." (Skim the headings and select a section to read.)

The Grand Canyon of the Colorado River

COMMUNICATE YOUR RESULTS

Exercise 1
Conflicts Over Water: Similarities and Differences

Choose two of the conflicts in the *Web Investigations* (or find another, if you wish) and compare and contrast them. How are these conflicts similar in terms of their natural, cultural, and political aspects? How do they differ? Which conflict is more amenable to resolution? How? Use a factual, journalistic style.

Exercise 2
Whose Water Is It?

Drawing on what you learned in the *Web Investigations*, discuss water ownership. Traditional law in the Western United States granted ownership to the person who first used the water. Do traditional rules of ownership work in today's world?

Exercise 3
The Web of Nature

If you explored the MediaLabs on acid rain, global warming, and population, you will probably realize that the environment and its problems are an interconnected web. In one to two pages, discuss this interrelatedness, showing how one form of environmental insult can cause or exacerbate another. In other words, pull together some of the things you have learned into a unified report.

Exercise 4
What Can We Do?

Using what you have learned in the *Web Investigations*, write an op-ed article describing the urgent need for action to preserve our water. Suggest some specific measures that readers can take.

14 Energy
A Fuels Paradise

Gas and oil and coal supply
Most of the energy
We need for transportation, heat,
And electricity.

Coal, petroleum, and natural gas are the principal sources of the energy that sustains modern civilization. Petrochemical plants supply the liquid and gaseous fuels that power our machines and heat our homes and businesses. They also provide the chemicals from which many modern materials are made.

W e were tempted to call this chapter Fire. Then the titles of Chapters 11, 12, 13, and 14 would have been Earth, Air, Water, and Fire—the four elements of the ancient world.

Actually, fire would not have been too inappropriate a title for this chapter. We obtain most of our energy by burning fuels, which certainly involves fire. However, many sources of energy do not involve burning of anything, and we will depend more heavily on these other sources as our reserves of fossil fuels are depleted.

Energy is the ability to do work. It takes energy to make an automobile move or an airplane fly, or to make electrons flow through a copper wire. The United States, with less than 5% of the world's population, uses almost one-fourth of all the energy currently generated. Abundant energy has enabled the United States to build its industrial base and provide its people with one of the highest standards of living in the world.

Industry uses about 28% of all energy produced in the United States. This energy is used to convert raw materials to the many products our society seems to demand. Transportation uses about 28%, to power automobiles, trucks, trains, airplanes, and buses. Private homes and commercial spaces use about 19%. Utilities use about 33% of the nation's energy production, primarily to generate electricity.

These figures add up to more than 100% because the electricity is used in industry and homes and commercial spaces and some of it is therefore counted twice.

Energy lights our homes, heats and cools our living spaces, and makes us the most mobile society in the history of the human race. It powers the factories that provide us with abundant material goods. Indeed, energy is the basis of modern civilization.

14.1 Heavenly Sunlight Flooding Earth with Energy

Did you know that you and all other living things on Earth depend on nuclear energy for survival? Most of the energy available to us on this planet comes from the giant nuclear reactor we call the sun. Although it is about 150 million km away, the sun has been supplying Earth with most of its energy for billions of years, and it will likely continue to do so for billions more (Table 14.1).

The SI unit of energy is the joule (J). A watt (W), the SI unit of power, is 1 joule per second (J/s).

$$1W = 1\frac{J}{s}$$

The sun is a nuclear fusion reactor that steadily converts hydrogen to helium (Chapter 4) with a power output of 4×10^{26} W. Earth receives about 1.73×10^{17} W [173,000 terawatts (TW); 1 TW = 10^{12} W] from the sun—an amount equivalent to the output of 115 million nuclear power plants. In 3 days, Earth receives energy from the sun equivalent to all our fossil fuel reserves. Yet this is only about 1 part in 50 billion of the sun's output!

The sun fuses more than 600 million t of hydrogen into helium each second. It has enough hydrogen to last another 5 billion years.

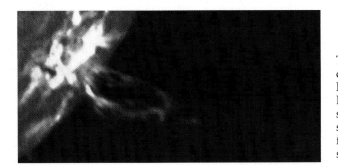

The sun is fueled by nuclear fusion reactions. Nearly all the energy on Earth originated in the sun. This false-color image shows an enormous solar flare leaping out from the surface of the sun.

Table 14.1 ▌ Earth's Energy Ledger (Rough Estimates)

Item	Energy (TW)	Approximate Percent
Energy in		
Solar radiation	173,000	99+
Internal heat	32	0.02
Tides	3	0.002
Energy out		
Direct reflection	52,000	30
Direct heating*	81,000	47
Water cycle*	40,000	23
Winds*	370	0.2
Photosynthesis*	40	0.02

Source: M. King Hubbert, "The Energy Resources of the Earth," *Scientific American,* September 1971.
*This energy is eventually returned to space by means of long-wave radiation (heat).

Example 14.1

How many joules of electric energy does a 75-W bulb burning for 1.0 hr (3600 s) consume?

Solution
From the definition of a watt, 1 W = 1 J/s, we know that

$$75 \text{ W} = 75 \frac{\text{J}}{\text{s}}$$

$$\frac{75 \text{ J}}{1 \text{ s}} \times 3600 \text{ s} = 270,000 \text{ J}$$

Exercise 14.1
A. How many joules of electric energy does a 650-W microwave oven consume in heating a cup of coffee for 75 s?
B. How many kilojoules of electric energy does a 1200-W space heater consume when it runs for 8.0 h?

Energy and the Life-Support System

Only a small fraction of the energy the biosphere receives is used to support life. In fact, about 30% of incident radiation is immediately reflected back into space as short-wave radiation (ultraviolet and visible light). Nearly half is converted to heat, making the third planet a warm and habitable place. About 23% of solar radiation powers the water cycle (Chapter 13), evaporating water from land and seas. The radiant energy of the sun is converted to the potential energy of water vapor, water droplets, and ice crystals in the atmosphere. This potential energy is converted to the kinetic energy of falling rain and snow and of flowing rivers.

The biosphere is the thin film of air, water, and soil in which all life exists. It is only about 15 km thick.

A tiny—but most important—fraction of solar energy is absorbed by green plants, which use it to power **photosynthesis**. In the presence of green chlorophyll pigments, this energy converts carbon dioxide and water to glucose, a simple sugar rich in energy.

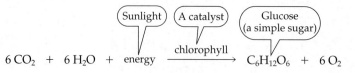

$$6 \text{ CO}_2 + 6 \text{ H}_2\text{O} + \text{energy} \xrightarrow{\text{chlorophyll}} \text{C}_6\text{H}_{12}\text{O}_6 + 6 \text{ O}_2$$

This reaction also serves to replenish oxygen in the atmosphere. Glucose can be stored, or it can be converted to more complex foods and structural materials. All animals depend on the stored energy of green plants for survival.

Energy: Some Scientific Laws

A study of chemistry is incomplete without a discussion of energy. In calculations based on chemical equations, we can include quantities of energy. In addition, we find two scientific laws to be of considerable help in understanding chemical processes. We examine some of these ideas in the following sections.

14.2 Energy and Chemical Reactions

How fast—or how slowly—chemical reactions take place depends on several factors. One factor is *temperature*. Reactions generally take place faster at higher temperatures. For example, coal (carbon) reacts so slowly with oxygen (from the air) at room temperature that the change is imperceptible. However, if coal is heated to several hundred degrees, it reacts rapidly. The heat evolved in the reaction keeps the coal burning smoothly.

We use the kinetic–molecular theory to explain the effect of temperature on the rates of chemical reactions. At high temperatures, molecules move more rapidly. Thus, they collide more frequently, increasing the chance for reaction. The increase in temperature also supplies more energy for the breaking of chemical bonds—a condition necessary for most reactions.

Another factor affecting the rate of a chemical reaction is the *concentration of reactants*. The more molecules there are in a given volume of space, the more likely they are to collide. With more collisions, there are more reactions. For example, if you light a wood splint and then blow out the flame, the splint continues to glow as the wood reacts slowly with oxygen in the air. When the glowing splint is placed in pure oxygen, the splint bursts into flame, indicating a much more rapid reaction. We use the concentration of oxygen to interpret the phenomenon: air is about 21% oxygen, and so the concentration of O_2 molecules in pure oxygen is almost five times as great as that in air.

Catalysts (Chapter 8) also affect the rates of chemical reactions. Catalysts are of great importance in the chemical industry. Appropriate catalysts increase the rate of reactions that otherwise would be so slow as to be impractical.

We make use of our knowledge of the effect of temperature on chemical reactions in our daily lives. For example, we freeze foods to retard chemical reactions that lead to spoilage. When we want to speed up the reactions involved in cooking food, we turn up the heat.

Catalysts are even more important in living organisms. Biological catalysts, called *enzymes*, mediate nearly all the chemical reactions that take place in living systems (Chapter 15).

(a)   (b)

The action of a catalyst is illustrated. Hydrogen peroxide decomposes slowly to water and oxygen (a). When platinum metal is inserted into a solution of hydrogen peroxide (b), the reaction proceeds rapidly. The heat released during the process produces steam, and the solution froths as oxygen gas is evolved.

Energy Changes and Chemical Reactions

The energy changes associated with chemical reactions are quantitatively related to the *amounts* of chemicals involved. For example, burning 1.00 mol (16.0 g) of methane to form carbon dioxide and water releases 192 kcal (803 kJ) of energy as heat.

$$CH_4(g) + 2\,O_2(g) \longrightarrow CO_2(g) + 2\,H_2O(g) + 192 \text{ kcal (803 kJ)}$$

Burning 2.00 mol (32.0 g) of methane produces twice as much heat, 384 kcal.

Example 14.2

Burning 1.00 mol of propane releases 526 kcal of energy.

$$C_3H_8(g) + 5\,O_2(g) \longrightarrow 3\,CO_2(g) + 4\,H_2O(g) + 526 \text{ kcal}$$

How much energy is released when 15.0 mol of propane is burned?

Solution

$$\frac{526\text{kcal}}{1 \text{ mol}} \times 15.0 \text{ mol} = 7890 \text{ kcal}$$

Exercise 14.2

The reaction of nitrogen and oxygen to form nitrogen monoxide (nitric oxide) requires an input of energy.

$$N_2(g) + O_2(g) + 4.32 \text{ kcal} \longrightarrow 2\,NO(g)$$

A. How much energy is absorbed when 5.05 mol of N_2 reacts with oxygen to form NO?
B. How much energy is involved when 4.39 mol of NO is decomposed into N_2 and O_2? Is the energy absorbed or released?

For an exothermic reaction, heat is often listed as a product in the chemical equation. Chemical energy is converted to heat. An endothermic reaction has energy as a reactant in the chemical equation. Heat is converted to chemical energy.

Chemical reactions that result in the release of heat to the surroundings are **exothermic reactions**. The burning of methane, gasoline, and coal (Figure 14.1) are all exothermic reactions. In each case, chemical energy is converted to heat energy.

In other reactions, such as the decomposition of water, energy must be supplied to the reactants from the surroundings.

$$2\,H_2O(g) + 137 \text{ kcal} \longrightarrow 2\,H_2(g) + O_2(g)$$

If energy is supplied as heat, such reactions are **endothermic reactions**. It takes 137 kcal of energy to decompose 36.0 g (2.00 mol) of water into hydrogen and oxygen. It should be noted that the same amount of energy is released when enough hydrogen is burned to form 36.0 g of water.

$$2\,H_2(g) + O_2(g) \longrightarrow 2\,H_2O(g) + 137 \text{ kcal}$$

Physical processes can also be either exothermic or endothermic. Table 14.2 lists several examples of physical and chemical processes and classifies them as exothermic or endothermic.

Figure 14.1 Coal burns in a highly exothermic reaction. The heat released can convert water to steam that can turn a turbine to generate electricity.

Example 14.3

How much energy is released when 225 g of propane (see Example 14.2) is burned?

Solution

First calculate the formula mass of propane.

$$(3 \times C) + (8 \times H) = (3 \times 12.0 \text{ u}) + (8 \times 1.0 \text{ u}) = 36.0 + 8.0 \text{ u} = 44.0 \text{ u}$$

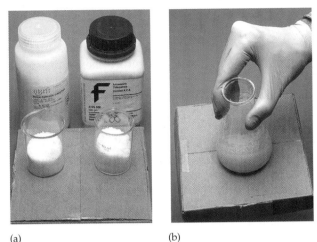

(a)   (b)

A striking endothermic reaction occurs when stoichiometric amounts of barium hydroxide octahydrate react with ammonium thiocyanate to produce barium thicyanate, ammonia gas, and water.

$$\text{Heat} + Ba(OH)_2 \cdot 8H_2O(s) + 2\,NH_4SCN(s)$$
$$\longrightarrow Ba(SCN)_2(s) + 2\,NH_3(g) + 10\,H_2O(l)$$

When the reaction is carried out in a beaker placed on a wet board, the temperature drops well below the freezing point of water, thus freezing the beaker to the board.

Table 14.2 ▌ Some Exothermic and Endothermic Processes

Exothermic Processes	Endothermic Processes
Freezing of water	Melting of ice
Condensation of water vapor	Evaporation of water
Burning of wood	Cooking of food
Metabolism in animals	Photosynthesis in plants
Making chemical bonds	Breaking chemical bonds
Discharging of a battery	Charging a battery
Oxidation of Mg to MgO	Decomposition of HgO
Mixing CaO (or H_2SO_4) with water	Mixing NH_4Cl with water
Explosion of dynamite	Evaporation of a chlorofluorocarbon
Nuclear fission and nuclear fusion	—

The molar mass of propane is therefore 44.0 g. Next, we use the molar mass to convert grams of propane to moles of propane.

$$225\ \text{g}\ C_3H_8 \times \frac{1\ \text{mol}\ C_3H_8}{44.0\ \text{g}\ C_3H_8} = 5.11\ \text{mol}\ C_3H_8$$

Now, proceeding as in Example 14.2:

$$5.11\ \text{mol}\ C_3H_8 \times \frac{526\ \text{kcal}}{1\ \text{mol}\ C_3H_8} = 2690\ \text{kcal}$$

Photosynthesis is an endothermic reaction. Energy in sunlight converts carbon dioxide and water to sugars in a reaction that requires 15,000 kJ (3600 kcal) of energy for each kilogram of glucose produced.

Exercise 14.3

A. How much energy, in kilocalories, is released when 4.42 g of methane is burned?

$$CH_4 + 2\,O_2 \longrightarrow CO_2 + 2\,H_2O + 192\ \text{kcal}$$

B. How much energy, in kilojoules, is absorbed when 0.528 g of N_2 is converted to NO (see Exercise 14.2)?

14.3 Energy and the First Law: Energy Is Conserved

Occasionally someone claims that we can somehow get around the law of conservation of energy. That seems highly unlikely. The law has been verified experimentally to 1 part per quadrillion (1 part in 10^{15}), and any exception to it would have to be less than this. How small is the possible deviation? A typist with such a rate of precision could type 100 words per minute for 30 million years and make only one mistake. For all practical purposes, the law is absolute.

We will take a look at our present energy sources and our future energy prospects shortly. To do so as scientifically as possible, let's first examine some natural laws. Recall that natural laws merely summarize the results of many experiments. We won't recount all those experiments here; we will merely state the laws and some of their consequences.

The **first law of thermodynamics** ("thermo" refers to heat; "dynamics", to motion) grew out of a variety of experiments conducted during the early 1800s. By 1840 it was clear that, although energy can be changed from one form to another, it is neither created nor destroyed. This law (also called the **law of conservation of energy**) has been restated in several ways, including "You can't get something for nothing" and "There is no such thing as a free lunch." Energy can't be made from nothing. Neither does it just disappear, although it may go someplace else.

From the first law of thermodynamics we can conclude that we can't "win." We can't come out ahead by making a machine that produces more energy than it takes in. From the first law alone, however, we might conclude that we can't possibly run out of energy because energy is conserved. This is true enough, but it doesn't mean that we don't have problems. There is another long-armed law from which we cannot escape.

14.4 Energy and the Second Law: Things Are Going to Get Worse

Despite innumerable attempts, no one has ever built a successful "perpetual-motion" machine. You can't make a machine that produces more energy than it consumes. Even if an engine isn't doing any work, it loses energy (as heat) because of the friction of its moving parts. In fact, in any real engine, you can't get as much useful energy out as you put in. You can't even break even.

If energy is neither created nor destroyed, why do we always need more? Won't the energy we have now last forever? The answer lies in the facts that (1) energy can be changed from one form to another, (2) not all forms are equal, and (3) high-grade forms of energy are constantly being degraded into low-grade forms. Energy flows downhill. Mechanical energy is eventually changed into heat energy. Hot objects cool off by transferring their heat to cooler objects. There is a tendency toward an even distribution of energy. Energy always flows from a hot object to a cooler one (Figure 14.2). The reverse never occurs spontaneously.

Observations of heat flow led to formulation of the **second law of thermodynamics**. In one form (of many), this law states that energy does not flow spontaneously from a cold object to a hot one. It is true that we can make energy flow from a cold

Figure 14.2 Energy always flows spontaneously from a hot object to a cold one, never the reverse. It flows from a hot fire to cold hands. (A spontaneous event is one that occurs without outside influence.)

region to a hot one—that's what refrigerators are all about—but we cannot do so without producing changes elsewhere. This sort of reversal of a natural process can be done only at a price. The price, in the case of refrigerators, is the consumption of electricity; that is, you must use energy to make energy flow from a cold space to a warmer one.

When we change energy from one form to another, we can't concentrate all the energy in a particular source to do the job we want it to do. For example, we use energy in a fuel to push a piston in a car engine or water rushing down from the top of a dam to run dynamos to generate electricity. In either case—indeed in all cases—some of the energy is wasted, mainly as unusable heat released to the environment. ("Wasted" isn't always the proper word. We use the energy from food for all our activities, but the waste heat helps us to maintain our body temperature at about 37°C.)

Another way to look at the second law is in terms of disorder. Scientists use the term **entropy** to indicate the randomness in position or energy of a system. The more mixed-up a system is, the higher its entropy. This concept is illustrated in Figure 14.3. Natural processes tend toward greater entropy—toward disorder.

The energy to run a car or any other kind of engine comes from the concentrated chemical energy inside molecules of oil or coal. In biochemistry, food molecules are the concentrated energy source. In either case, energy is spread out during the process. The spread-out energy in the product gases—CO_2 and H_2O molecules in each case—is less useful. We can't run a car very well on exhaust gases, nor can organisms obtain energy for life processes from respiratory products.

(a)

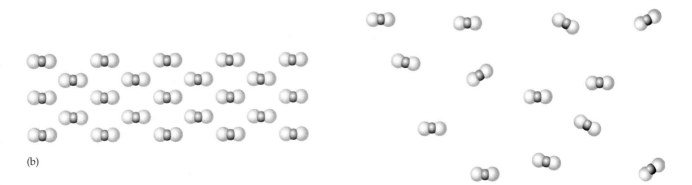

(b)

Figure 14.3 Any spontaneous change is from a more ordered state to a less ordered one. A sample (a) of dry ice [CO_2(s)] at room temperature spontaneously changes to CO_2(g). The CO_2 molecules in the solid are highly ordered (b); those in the gaseous state are disordered (c). We would never expect CO_2(g) at room temperature to reorganize itself into an ordered solid spontaneously. We can make CO_2(s) from CO_2(g), but only by compressing the gas and lowering its temperature.

Of course, we can often reverse the tendency toward randomness—but only through an input of energy.

Entropy

Entropy is best expressed quantitatively by way of abstract mathematical equations that are well beyond the scope of this text. Strictly speaking, the concept applies only to the behavior of internally energetic, mobile molecules and atoms. These particles indeed tend to be as random and disorderly as possible. Macro objects, such as books and stones stay exactly where they are unless some energy flow from outside forces them to move. Macro objects have no inherent tendency toward disorder. Rather the energy flow that moves them causes the disorder. There are simply many, many more "disorderly" arrangements than "orderly" ones.

This tendency toward disorder helps to explain why it is relatively easy to pollute air or water and so difficult to clean it up once it is polluted. For example, it would take little energy to dump a ton of a chlorofluorocarbon (CFC) into the air. The rapidly moving CFC molecules would quickly disperse among the nitrogen, oxygen, and other molecules in the air. Once the CFC molecules were scattered, it would take a lot of energy to concentrate them from the air. It costs a lot to clean up polluted water, soil, or air. Prevention of pollution is by far the better alternative.

14.5 People Power: Early Uses of Energy

Early peoples obtained their energy (food and fuel) by collecting wild plants and hunting wild animals. They expended the energy so obtained in hunting and gathering. Domestication of horses and oxen increased the availability of energy only slightly. The raw materials used by these work animals were natural—and replaceable—plant materials.

Plants were also the first fuels. Tree branches and other combustible materials from dead trees and bushes kept early fires burning. As late as 1760, wood was almost the only fuel being used. Even today, wood remains the primary fuel in some parts of the world.

Even in 1850, the burning of wood still accounted for more than 90% of the energy produced in the United States.

One of the first mechanical devices used to convert energy to useful work was the waterwheel. The Egyptians first used waterpower about 2000 years ago, primarily for grinding grain. Later on, waterpower was used for sawmills, textile mills, and other small factories. Windmills were introduced into western Europe during the Middle Ages, primarily for pumping water and grinding grain. More recently, wind power has been used to generate electricity.

Windmills and waterwheels are fairly simple devices for converting the kinetic energy of blowing wind and flowing water to mechanical energy. They were sufficient to power the early part of the industrial revolution, but development of the steam engine freed factories from locations along waterways. Since 1850, turbines turned by water, steam, and gas, the internal-combustion engine, and a variety of other energy-conversion devices have boosted the amount of energy available for use by an estimated factor of 10,000.

 [Web Reference 1] An extensive website about the physics of energy and the environment, with many links to other sites.

Let's turn our attention now to the fossils that fueled the industrial revolution and still serve as the basis for modern civilization.

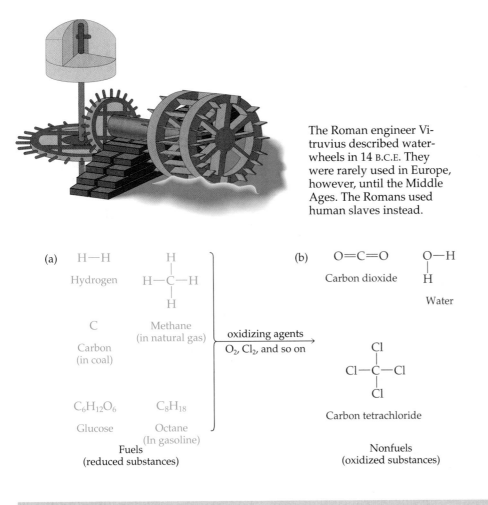

The Roman engineer Vitruvius described water-wheels in 14 B.C.E. They were rarely used in Europe, however, until the Middle Ages. The Romans used human slaves instead.

Figure 14.4 Some fuels (a). Fuels are reduced forms of matter that release relatively large quantities of heat when burned. Some nonfuels (b); these compounds are oxidized forms of matter.

Fossil Fuels

Fossils fuel modern industrial civilization. Almost 90% of the energy used to sustain our way of life comes from **fossil fuels**—coal, petroleum, and natural gas. In the following sections, we consider the origin and chemical nature of these fuels and how they are burned to release energy that plants of ages past captured from rays of sunlight.

Fuels are substances that burn readily with the release of significant amounts of energy. Fuels are *reduced* forms of matter, and the burning process is oxidation (Chapter 8). If an atom already has its maximum number of bonds to oxygen (or to other electronegative atoms such as chlorine or bromine), the atom cannot serve as a fuel. Indeed, such substances can be used to put out fires. Figure 14.4 shows some representative fuels and nonfuels.

14.6 Fossil Fuels: Reserves and Consumption Rates

Earth has only a limited supply of fossil fuels. Estimated U.S. and world reserves and annual U.S. and world consumption are given in Table 14.3. Estimates of reserves vary greatly, depending on the assumptions made. Even the most optimistic estimates, however, lead to the conclusion that nonrenewable energy resources are being depleted rapidly. Indeed, in just a century, we will have used up more than half the fossil fuels that were formed over the ages. In only a few hundred more years, we will have removed from Earth and burned virtually all the remaining recoverable fossil fuels. Of all that ever existed, about 90% will have been used in a period of 300

The troubled lands of the Middle East control more than half of the world's petroleum reserves.

Table 14.3 ▮ Estimated U.S. and World Reserves of Economically Recoverable Fuels and Annual Consumption of Fossil Fuels*

Fuel	Reserves		Consumption	
	United States	**World**	**United States**	**World**
Coal	240,558	1,031,610	936	4,697
Petroleum	11,390	475,230	3,287	7,374
Natural gas	8,720	265,300	1,132	4,160
Total	260,668	1,772,140	5,355	16,231

Source: Oil and Gas Journal, World Oil, and Energy Information Administration, *International Energy Outlook 1996,* Washington, DC: U.S. Department of Energy, 1996. See the last-mentioned for assumptions made in making these estimates.

*Expressed in millions of metric tons of coal equivalents for ready comparison of energy content.

By energy *production*, we mean conversion of some form of energy into a more useful form. For example, production of petroleum means pumping it from the ground and transporting it to a refinery. Energy *consumption* means using it in a way that changes it to a less useful form. In either case, the law of conservation of energy holds; we are neither making energy nor destroying it.

[Web Reference 2]
The U.S. Department of Energy has a fossil energy website.

years. Within your lifetime, it is likely that natural gas and petroleum will become so scarce and so expensive that they won't be used as fuels. At the current rate of production, U.S. reserves of petroleum and natural gas will run out sometime during this century. Coal reserves should last 300 years at the present rate of use. However, the rate of use of all fossil fuels is increasing, especially in developing nations.

Presumably, fossil fuels are still being formed in nature, a process that is perhaps most evident in peat bogs. The rate of formation is extremely slow, however. In fact, it has been estimated that we are using fossil fuels 50,000 times as fast as they are being formed.

14.7 Coal: The Carbon Rock of Ages

People probably have used small amounts of coal since prehistoric times. After the steam engine came into widespread use (by about 1850), the industrial revolution was powered largely by coal. By 1900 about 95% of the world's energy production came from the burning of coal.

Coal is a complex combination of organic materials that burn and inorganic materials that produce ash. Its main element is carbon, but it also contains small percentages of other elements. The quality of coal as an energy source is based on its carbon content. Complete combustion of the carbon produces carbon dioxide.

Carbon (from coal)   Oxygen (from air)

$$C(s) + O_2(g) \longrightarrow CO_2(g)$$

In limited quantities of air, however, carbon monoxide and soot are formed.

$$2\,C(s) + O_2(g) \longrightarrow 2\,CO(g)$$

Soot is mostly unburned carbon.

Coal is ranked by carbon content from low-grade peat and lignite to high-grade anthracite (Table 14.4). The energy obtained from coal is roughly proportional to its carbon content. Soft (bituminous) coal is much more plentiful than hard coal (anthracite). Lignite and peat have become increasingly important as the supplies of higher grades of coal have been depleted.

The supply of coal is limited. Deposits that exist today are less than 600 million years old. For part of the last 600 million years, Earth was much warmer than it is now, and plant life flourished. Most plants lived, died, and decayed—playing their

Burning pure carbon (graphite) to carbon dioxide and water yields 393.5 kJ/mol C (94.05 kcal/mol C). Coal yields less energy, on a mass basis, because it is not pure carbon.

Table 14.4 ▮ Approximate Composition (Percent by Mass) and Energy Content of Typical Grades of Coal (Dry Basis)

Grade of Coal	Carbon	Hydrogen	Oxygen	Nitrogen	Energy Content (MJ/kg)
Wood (for comparison)	50	6	43	1	—
Peat	60	6	33	2	14.7
Lignite (brown coal)	69	5	25	1	23
Bituminous (soft) coal	88–89	5	5–15	1	36
Anthracite (hard coal)	95	2–3	2–3	Trace	35.2

Source: Diessel, C. F. K. *Coal-Bearing Depositional Systems.* New York: Springer-Verlag, 1992.

Giant ferns, reeds, and grasses grew during the Pennsylvanian period 300 million years ago. These plants were buried and through the ages were converted to the coal we burn today.

normal role in the carbon cycle (Figure 14.5). But a portion of the plant material became buried under mud and water. There, in the absence of oxygen, it decayed only partially. The structural material of plants is largely cellulose—a compound of carbon, hydrogen, and oxygen. Under increasing pressure, as the material was buried more deeply, the cellulose molecules broke down. Small molecules rich in hydrogen and oxygen escaped, leaving behind a material increasingly rich in carbon. Thus, peat is a young coal, only partly converted, with plant stems and leaves clearly visible. Anthracite, on the other hand, has been almost completely carbonized.

Abundant but Inconvenient Fuel

Coal is our most plentiful fossil fuel. The United States is estimated to have a quarter of the world's reserves of coal, and there is probably enough easily recoverable coal to last us several hundred years. Electric utilities in the United States burn nearly 700 million t of coal each year, generating 2000 TW of electricity (56% of the total). Unfortunately, this coal use is associated with serious environmental problems.

Coal, a solid, is an inconvenient fuel. Dangerous deep mining or devastating strip mining must be used to remove it. Most coal is hauled in trains, barges, and trucks. It takes quite a bit of energy to extract coal from the ground and transport it to the power plant or factory where it is to be used.

Some coal is pulverized into a powder, mixed with water, and transported by pumping the slurry through a pipeline.

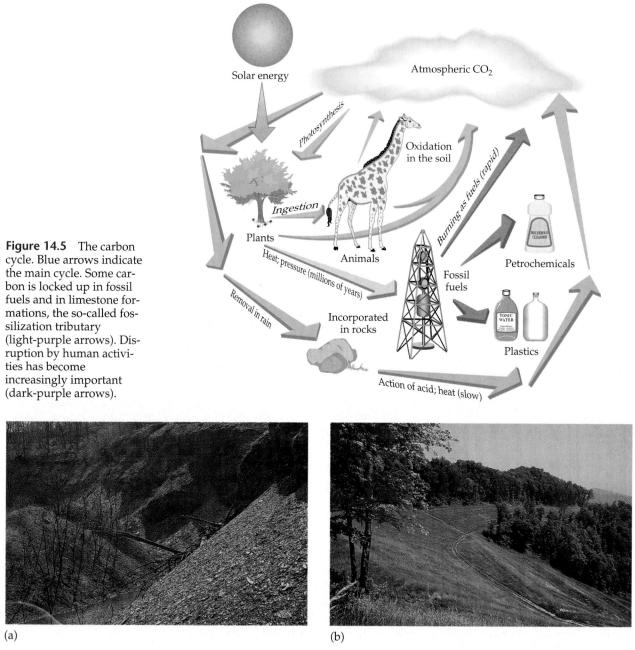

Figure 14.5 The carbon cycle. Blue arrows indicate the main cycle. Some carbon is locked up in fossil fuels and in limestone formations, the so-called fossilization tributary (light-purple arrows). Disruption by human activities has become increasingly important (dark-purple arrows).

(a)   (b)

Strip mining, the least expensive way to obtain coal, bares vast areas of all vegetation, and the exposed soil washes away, filling streams with mud and silt. Laws in the United States now require the (expensive) restoration of most stripped areas and ban stripping on steep slopes where restoration is not feasible. Unfortunately, a lot of damage was done before these laws were passed, and many people consider the present laws inadequate. These photographs show the site of a strip mine in Morgan County, Tennessee. (a) The abandoned mine in 1963, and (b) the same area after it was reclaimed in a demonstration project in 1971.

Source of Pollution

Coal, a solid, contains many impurities. Unlike natural gas and petroleum, it contains minerals which are left as ash when the coal is burned. Some minerals enter the air as particulate matter, constituting a major pollution problem (Chapter 12).

Perhaps even worse, most of our remaining coal is high in sulfur. When it burns, unless the stack gases are scrubbed, choking sulfur dioxide pours into the atmosphere. Reacting with oxygen and moisture in the air, sulfur dioxide is converted to sulfuric acid. This acid slowly damages steel and aluminum structures, marble buildings and statues, and human lungs.

Coal can be cleaned before it is burned. A flotation method makes use of the different densities of coal and its major impurities. Coal has a density of about 1.3 g/cm^3. Shale, a rock formed from hardened clay, has a density of about 2.5 g/cm^3. Pyrite (FeS$_2$), the major source of sulfur in coal, has a density of about 5.0 g/cm^3. A solution of the proper density, with detergents added to cause the coal to float, allows the coal to be floated off, leaving the heavier minerals behind. Coal also can be converted to more convenient gaseous and liquid fuels, again leaving the minerals behind. All these processes add to the cost of the coal.

> The flotation method removes most inorganic sulfur compounds, such as pyrite, from coal. It does not remove sulfur atoms covalently bonded to the carbon atoms in the coal. This bound sulfur still contributes to pollution.

Source of Chemicals

But coal is not just a fuel. When coal is heated in the absence of air, volatile material is driven off, leaving behind a mostly carbon product called *coke*, which is used in the production of iron and steel. The more volatile material is condensed to liquid coal oil and a gooey mixture called *coal tar*, both of which are good sources of organic chemicals.

14.8 Natural Gas: Mostly Methane

Natural gas is the cleanest of the fossil fuels. It is composed principally of methane, but its composition varies greatly. In North America, natural gas typically contains 60–80% methane, 5–9% ethane, 3–18% propane, and 2–14% butane and pentane. It burns with a relatively clean flame, and its products are mainly carbon dioxide and water.

Natural gas burns with a relatively clean flame.

$$CH_4(g) + 2\,O_2(g) \longrightarrow CO_2(g) + 2\,H_2O(g) + heat$$

As in any combustion in air, some nitrogen oxides are formed. When natural gas is burned in limited amounts of air, carbon monoxide and even elemental carbon (soot) can be major products.

$$2\,CH_4(g) + 3\,O_2(g) \longrightarrow 2\,CO(g) + 4\,H_2O(g)$$

$$CH_4(g) + O_2(g) \longrightarrow C(s) + 2\,H_2O(g)$$

The gas, as it comes from the ground, often contains sulfur and nitrogen compounds as impurities.

Natural gas was most likely formed—ages ago—by the actions of heat, pressure, and perhaps bacteria on buried organic matter. The gas is trapped in geological formations capped by impermeable rock. It is removed through wells drilled into the gas-bearing formations.

> The role of nitrogen oxides, CO, and soot in air pollution is discussed in Chapter 12.

Most natural gas is used as fuel, but it is also an important raw material. In North America, some higher alkanes are separated out of natural gas. Ethane and propane are cracked—decomposed by heat in the absence of air—to form ethylene and propylene. These alkenes are intermediates in the synthesis of plastics (Chapter 10) and many other useful commodities. Natural gas is also the raw material from which methanol and many other one-carbon compounds are made. Like other fossil fuels, its supply is not unlimited (Table 14.3).

14.9 Petroleum: Liquid Hydrocarbons

With the development of the internal combustion engine, petroleum became increasingly important and by 1950 had replaced coal as the principal fuel. **Petroleum** is an exceedingly complicated liquid mixture of organic compounds, mainly hydrocarbons. The hydrocarbons, chiefly alkanes but some cyclic compounds, burn readily. As with natural gas, complete combustion yields mainly carbon dioxide and water. A representative reaction is that of an octane.

$$2\ C_8H_{18}(l)\ +\ 25\ O_2(g)\ \longrightarrow\ 16\ CO_2(g)\ +\ 18\ H_2O(g)$$

Combustion in air also produces nitrogen oxides. Incomplete burning yields carbon monoxide and soot. Petroleum usually contains small amounts of sulfur compounds that produce sulfur dioxide when burned.

We have seen that coal is primarily of plant origin, but petroleum is thought to be of animal origin. Most likely it is formed primarily from the fats (Chapter 15) of ocean-dwelling, microscopic animals because it is nearly always found in rocks of oceanic origin. Fats are made up mainly of compounds of carbon and hydrogen, with a little oxygen (for example, $C_{57}H_{110}O_6$). Removal of the oxygen and slight rearrangement of the carbon and hydrogen atoms in these fats form typical petroleum hydrocarbon molecules.

Efficiently burned, petroleum products are rather clean fuels. Fuel oil, used for heating homes or to produce electricity, can be burned efficiently and thus contributes only moderately to air pollution. Gasoline, however, the major fraction of petroleum, is used to power automobiles; because the internal combustion engines in most automobiles are rather inefficient, the combustion of gasoline contributes heavily to air pollution (Chapter 12). Petroleum-derived fuels are generally dirtier than natural gas because they contain more impurities.

Air pollution isn't the only problem: burning up our petroleum reserves will leave us without a ready source of many familiar materials. Most industrial organic chemicals come from petroleum (smaller amounts are derived from coal and natural gas); and plastics, synthetic fibers, solvents, and many other consumer products are made from these chemicals.

Worldwide, petroleum still appears to be quite abundant (Table 14.3). However, U.S. petroleum reserves have been declining since 1972. Offshore deposits along the Atlantic coast of the United States would produce only a few years' supply, and offshore drilling requires a great deal more energy (and money) than drilling on land.

The developed countries of North America, western Europe, and Japan depend heavily on oil imports from developing nations, some of which are politically unstable. This dependence makes the industrial nations vulnerable economically and politically.

Obtaining and Refining Petroleum

Crude oil is liquid, a convenient form for its transportation. Petroleum is pumped easily through pipelines or hauled across the oceans in giant tankers. Pumping petroleum from the ground requires little energy. As domestic supplies diminish, however, we will have to expend an increasing fraction of the energy content of petroleum to transport it to where it is needed. Also, as petroleum is pumped out of a well, it becomes increasingly difficult to extract the remainder. The oil that is left behind is more scattered, and more energy is required to collect it.

As it comes from the ground, crude oil is of limited use. To make it better suit our needs, we separate it into fractions by boiling it in a distillation column (Figure 14.6). Petroleum deposits nearly always have associated natural gas, part or all of which is sent through a separation process. The lighter hydrocarbon molecules come off at the top of the column, and the heavier ones at the bottom (Table 14.5.) Gasoline is usually the fraction of petroleum most in demand, and fractions that boil at higher temperatures are often in excess supply. These higher-boiling fractions are often converted to gasoline by heating in the absence of air. The *cracking* process breaks down bigger molecules into smaller ones. The effect of this process, with $C_{14}H_{30}$ as an example, is illustrated in Figure 14.7. Cracking not only converts some molecules to those in the gasoline range (C_5H_{12} through $C_{12}H_{26}$), but it also produces a variety of useful by-products. The unsaturated hydrocarbons are starting materials for the manufacture of a host of petrochemicals.

Our dependence on imported oil leads to economic and political problems. Political events in the oil-rich Middle East have caused huge oil price increases three times. In 1973–1974, following the Yom Kippur War between Israel and the Arab states and the subsequent Arab oil embargo, prices quadrupled from about $3.00 a barrel to $12.00. Crises in Iran and Iraq led to another round of crude oil price increases. The Iranian revolution and the Iran–Iraq war resulted in crude oil prices more than doubling from $14.00 in 1978 to $35.00 per barrel in 1981. The invasion of Kuwait by Iraq in 1990 and the subsequent defeat of Iraq by the United States-led coalition in early 1991 caused prices to almost double.

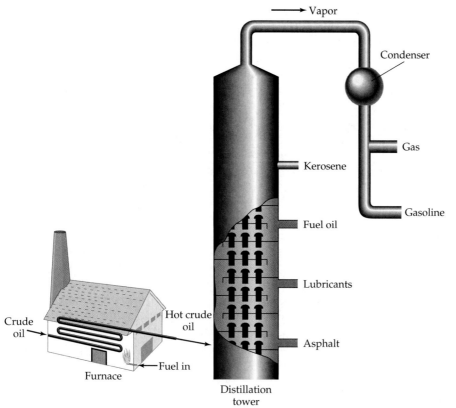

Figure 14.6 The fractional distillation of petroleum. Crude oil is vaporized, and the column separates the components according to their boiling points. The lower-boiling constituents reach the top of the column, and the higher-boiling components come off lower in the column. A nonvolatile residue collects at the bottom.

$$\text{CH}_3\text{CH}_2\text{CH}_2\text{CH}_2\text{CH}_2\text{CH}_2\text{CH}_2\text{CH}_2\text{CH}_2\text{CH}_2\text{CH}_2\text{CH}_2\text{CH}_2\text{CH}_3 \xrightarrow[\text{catalyst}]{\text{heat}}$$

$$\text{CH}_3\text{CH}_2\text{CH}_2\text{CH}_2\text{CH}_2\text{CH}_2\text{CH}_2\text{CH}_2\text{CH}_2\text{CH}_2\text{CH}_2\text{CH}_3 \;+\; \text{CH}_2\!=\!\text{CH}_2$$

and

$$\text{CH}_3\text{CH}_2\text{CH}_2\text{CH}_2\text{CH}_2\text{CH}_2\text{CH}_2\text{CH}_2\text{CH}_2\text{CH}_2\text{CH}_3 \;+\; \text{CH}_3\text{CH}\!=\!\text{CH}_2$$

and

$$\text{CH}_3\text{CH}_2\text{CH}_2\text{CH}_2\text{CH}_2\text{CH}_2\text{CH}_2\text{CH}_3 \;+\; \text{CH}_3\text{CH}_2\text{CH}_2\text{CH}_2\text{CH}\!=\!\text{CH}_2$$

and so on

Figure 14.7 Formulas of a few of the possible products formed when $C_{14}H_{30}$, a typical molecule in kerosene, is cracked. In practice, a wide variety of hydrocarbons (most of which have fewer than 14 carbon atoms), hydrogen gas, and char (mostly elemental carbon) are formed. Cyclic and branched-chain hydrocarbons are also produced.

Table 14.5 ∎ Typical Petroleum Fractions

Fraction	Typical Range of Hydrocarbons	Approximate Range of Boiling Points (°C)	Typical Uses
Gas	CH_4 to C_4H_{10}	Less than 40	Fuel, starting materials for plastics
Gasoline	C_5H_{12} to $C_{12}H_{26}$	40–200	Fuel, solvents
Kerosene	$C_{12}H_{26}$ to $C_{16}H_{34}$	175–275	Diesel fuel, jet fuel, home heating; cracking to gasoline
Heating oil	$C_{15}H_{32}$ to $C_{18}H_{38}$	250–400	Industrial heating, cracking to gasoline
Lubricating oil	$C_{17}H_{36}$ and up	Above 300	Lubricants
Residue	$C_{20}H_{42}$ and up	Above 350 (some decomposition)	Paraffin, asphalt

The cracking process is an example of how chemists modify nature's materials to meet human needs and desires. Starting with coal tar or petroleum, chemists can create a dazzling array of substances with a wide variety of properties. They can make plastics, pesticides, herbicides, perfumes, preservatives, painkillers, antibiotics, stimulants, depressants, and detergents.

The ability to modify hydrocarbon molecules enables the petroleum industry to produce increased amounts of whatever fraction they desire. They can, on demand, increase the proportion of gasoline or of fuel oil from a given supply of petroleum. They can even make gasoline from coal. They can't, however, increase the amount of fossil fuels aboard Spaceship Earth. Scientists are seeking new sources of energy that do not depend on petroleum. Perhaps we can soon stop this profligate waste of resources. Spaceship Earth has aboard it all the supplies it will ever have. We must use them wisely.

Gasoline

Gasoline, like the petroleum from which it is derived, is a mixture of hydrocarbons. Typical alkanes in gasoline range from C_5H_{12} to $C_{12}H_{26}$. Because there are many isomeric forms, particularly for the higher members of the group, gasoline is an exceedingly complex mixture of alkanes. There are also small amounts of other kinds of hydrocarbons present, and even some sulfur- and nitrogen-containing compounds. The gasoline fraction of petroleum as it comes from a distillation column is called *straight-run gasoline*. It doesn't burn very well in modern high-compression automobile engines, however, chemists have learned how to modify it in a variety of ways to make it burn more smoothly.

The Octane Ratings of Gasolines

Early in the development of the automobile engine, scientists learned that some types of hydrocarbons burned more evenly and were less likely to ignite prematurely than others. Ignition before the piston was in proper position led to a knocking in the engine. Scientists were soon able to correlate good performance with a branched-chain structure in hydrocarbon molecules. An arbitrary performance standard, called the **octane rating**, was established in 1927. Isooctane was assigned a value of 100 octane. An unbranched-chain compound, heptane, was given an octane rating of 0. A gasoline rated 90 octane was one that performed the same as a mixture that was 90% isooctane and 10% heptane.

$$\underset{\text{Isooctane}}{\overset{\begin{array}{cc}CH_3 & CH_3 \\ | & | \end{array}}{CH_3-C-CH_2-CH-CH_3}} \qquad \underset{\text{Heptane}}{CH_3CH_2CH_2CH_2CH_2CH_2CH_3}$$

During the 1930s, chemists discovered that the octane rating of gasoline could be improved by heating it in the presence of a catalyst such as sulfuric acid (H_2SO_4) or aluminum chloride ($AlCl_3$). This increase in octane rating was attributed to conversion (*isomerization*) of part of the unbranched structures to highly branched molecules. Heptane molecules can be isomerized to branched structures, for example.

$$CH_3CH_2CH_2CH_2CH_2CH_2CH_3 \xrightarrow[\text{heat}]{H_2SO_4} \underset{\overset{|}{CH_3}}{\overset{\overset{CH_3}{|}}{CH_3-CH_2-CH-CH-CH_3}}$$

Chemists are also able to combine small hydrocarbon molecules (below the gasoline range) into larger ones more suitable for use as fuel. This process is called *alkylation*. In a typical alkylation reaction, shown below, isobutylene is reacted with propane.

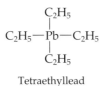

The product molecules are in the right size range for gasoline, and they are also highly branched; therefore they are high in octane number.

Other chemicals also were found to substantially improve the antiknock quality of gasoline. Tetraethyllead was especially effective. As little as 1 mL tetraethyllead per liter of gasoline (one part per thousand) increases the octane rating by 10 or more.

$$C_2H_5-\overset{\overset{\textstyle C_2H_5}{|}}{\underset{\underset{\textstyle C_2H_5}{|}}{Pb}}-C_2H_5$$

Tetraethyllead

Lead is toxic, however, and it fouls the catalytic converters used in modern automobiles. In the United States unleaded gasoline became available in 1974, and today almost all leaded gasoline has been phased out.

Scientists have developed other ways to get high octane ratings in unleaded fuels. For example, petroleum refineries use **catalytic reforming** to convert low-octane alkanes to high-octane aromatic compounds. Hexane (with an octane number of 25) is converted to benzene (with an octane number of 106).

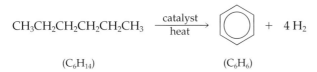

$$CH_3CH_2CH_2CH_2CH_2CH_3 \xrightarrow[\text{heat}]{\text{catalyst}} \quad + \quad 4\,H_2$$

$$(C_6H_{14}) \qquad\qquad\qquad (C_6H_6)$$

Catalytic reforming was an extremely important oil refining process for half a century. Now, however, we are trying to reduce the amount of aromatics, especially benzene, in gasoline because of health concerns.

Octane boosters to replace tetraethyllead include methyl *tert*-butyl ether (MTBE), methanol, ethanol, and *tert*-butyl alcohol (Chapter 9). None of these compounds is nearly as effective as tetraethyllead in boosting the octane rating. They must therefore be used in fairly large quantities. The amount that can be used in gasoline is limited by solubility problems. Methanol in excess of 5% and ethanol in excess of 10% tend to separate from the gasoline, especially if moisture gets into the fuel.

Unlike gasoline, which is made up of hydrocarbons, methyl *tert*-butyl ether and the various alcohols and their derivatives all contain oxygen. They are sometimes called "oxygenates." These oxygenates are added to gasoline not only to improve the octane rating but also, and perhaps more importantly, because of their ability to decrease the amount of carbon monoxide in auto exhaust gas.

14.10 Convenient Energy: Electricity

The convenience of a fossil fuel depends on its physical state: gases are most convenient, liquids next, and solids least convenient. Perhaps the most convenient form of energy of all is electricity (Table 14.6). With electricity we can have light and hot water, and we can run motors of all sorts. We can use it to heat and cool our homes and workplaces. When looking at future energy sources, then, we look mainly at ways of generating electricity.

Lead as a pollutant is discussed in Chapter 12. Lead is especially toxic to the brain. Even small amounts can lead to learning disabilities in children.

Ethylene dibromide was added to gasoline along with tetraethyllead in order to keep lead metal from fouling the spark plugs in car engines. The lead was converted to lead bromide, which left the engine in the exhaust gas. Instead of fouling the spark plugs, the lead just fouled up the atmosphere.

MTBE, like ethers in general (Chapter 9), is rather unreactive chemically, but it is soluble in water to the extent of 4.8 g per 100 g of water. When gasoline spills or leaks from storage tanks, MTBE enters the groundwater, leading to widespread contamination. MTBE is listed as a hazardous substance under the Federal Superfund law and is considered a potential human carcinogen by the EPA. The exact threat to human health is far from clear, but some areas have banned the use of MTBE in gasoline, and the EPA has recommended a nationwide ban.

Electricity is a secondary energy source. A primary source, such as coal, must be used to produce it.

Table 14.6 ▌ Convenience of Fuels in Various Physical States

Physical State	Extraction	Transportation to Cities	Distribution Within a City	In Use Convenience	Cleanliness
Solids (coal, wood)	Shovels, borers, blasting	Trucks, trains, barges (slurry with water in pipe)	Trucks, buckets	Least	Dirtiest
Liquids (gasoline, fuel oil)	Pumps	Pipelines, tankers, barges, trucks	Trucks		
Gases (natural gas)	Pumps	Pipeline	Pipes		
Electricity* (electron flow)	—	Wires	Wires	Most	Cleanest

*Produced by burning any of the primary fuels, and included for comparison.

Any fuel can be burned to boil water, and the steam produced (in great enough quantities) can turn a turbine to generate electricity. Figure 14.8 shows a coal-fired steam power plant. At present, about 51% of U.S. electric energy comes from coal-burning plants (Figure 14.9). Such facilities are at best only about 40% efficient; 60%

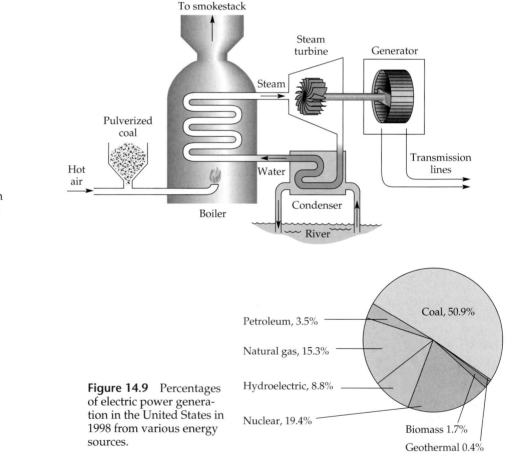

Figure 14.8 A diagram of a coal-burning power plant for generating electricity.

Figure 14.9 Percentages of electric power generation in the United States in 1998 from various energy sources.

of the energy of the fossil fuel is wasted as heat. Today, some power installations use the waste heat generated to warm buildings, a technique called *cogeneration*.

Nuclear Energy

In Chapter 4 we explored some nuclear reactions. As noted there, both nuclear fission and nuclear fusion can be used to power bombs. Nuclear fission can also be controlled to generate power. So far nuclear fusion reactions cannot be controlled; they can be employed only in bombs.

14.11 Nuclear Fission

Nuclear fission reactions can be controlled in **nuclear reactors**. The energy released during fission can be used to generate steam, which can turn a turbine to generate electricity (Figure 14.10).

At the dawn of the nuclear age, nuclear power was envisioned by some to be destined to fulfill the biblical prophecy of a fiery end to our world. While some predicted doom, others saw nuclear power as a source of unlimited energy. During the late 1940s, some claimed that electricity from nuclear plants would become so cheap that it eventually would not have to be metered. Nuclear power has not yet brought us either paradise or perdition, but it has become a most controversial issue. Our great demand for energy indicates that the controversy will continue for years to come.

At present, nearly 20% of U.S. electricity comes from nuclear power plants. The eastern seaboard and upper midwestern states, many of which have minimal fossil fuel reserves, are heavily dependent on nuclear power for electricity.

We could rely more on nuclear power as other nations do (Table 14.7), but the public is quite fearful of nuclear power plants. This apprehension was exacerbated by the accident at Chernobyl, Ukraine. The United States has 104 operating nuclear reactors, but no new plants have been ordered since 1978. It takes about 10 years to build a nuclear power plant. There will have to be a dramatic change in public attitude if nuclear power is to play a big role in our future.

[Web Reference 3] There are many websites about nuclear power plants, and it may be difficult to find a balanced view in a single site. It is important to note the sponsor and orientation of any site. With this in mind, some interesting sites are: (a) the International Nuclear Information System (INIS), operated by the International Atomic Energy Agency (IAEA), an autonomous organization within the United Nations system, (b) American Nuclear Society, (c) Nuclear Physics: Past, Present, and Future, an impressive site developed and maintained by high school students, (d) Institute for Energy and Environmental Research, and (e) a Japanese forum on nuclear issues.

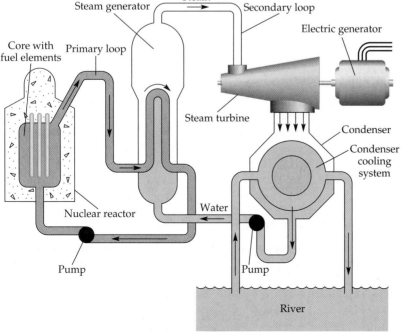

Figure 14.10 A diagram of a nuclear power plant for generating electricity.

Table 14.7 ∎ Electricity Generation Using Nuclear Power Plants (Selected Countries)

Country	Total Electricity From Nuclear Power (%)	Number of Operating Nuclear Power Plants	Number of Nuclear Power Plants Under Construction*
Belgium	55	7	0
Bulgaria	42	6	0
Czech Republic	21	4	2
France	76	58	1
Germany	28	20	0
Japan	36	53	0
Korean Republic	41	16	3
Lithuania	77	2	0
Slovak Republic	42	5	3
Spain	32	9	0
Sweden	46	12	0
Switzerland	38	5	0
Ukraine	45	16	4
United Kingdom	27	35	0
United States	19	104	0

Source: International Atomic Energy Agency, *PRIS Database,* Vienna: 1998.

*China has 6 nuclear power plants under construction; India and Russia, 4 each; Iran and Iraq, 2 each; and Argentina, Brazil, Pakistan, and Romania, 1 each.

Types of Nuclear Power Plants

There are several types of nuclear power plants, but we won't attempt to discuss all the possible types here. The one illustrated in Figure 14.10 is a pressurized water reactor. Earlier models were mainly boiling water reactors in which the steam from the reactor was used to power the turbine directly.

Nuclear power plants utilize the same fission reactions employed in nuclear bombs. A *moderator* is used to slow down the fission neutrons. The reaction is also controlled by the insertion of boron steel or cadmium *control rods*. Boron and cadmium absorb neutrons readily, and so the rods prevent the neutrons from participating in the chain reaction. These rods are installed when the reactor is built. Removing them part way starts the chain reaction; the reaction is stopped when the rods are pushed in all the way.

The Nuclear Advantage: Minimal Air Pollution

The United States has 104 operating nuclear power plants. If they were replaced by coal-burning plants, airborne pollutants would increase by 18,000 tons/day.

The main advantage of nuclear power plants over those that burn fossil fuels is in what they do not do. Unlike fossil-fuel-burning plants, nuclear power plants produce no carbon dioxide to add to the greenhouse effect, and they add no soot, fly ash, sulfur oxides, or nitrogen oxides to the atmosphere. They contribute almost nothing to global warming, air pollution, or acid rain.

If used in place of oil-burning plants, nuclear power plants could also reduce our dependence on foreign oil and lower our trade deficit.

Problems with Nuclear Power

Nuclear power plants have some disadvantages. Elaborate and expensive safety precautions must be taken to protect plant workers and the inhabitants of surrounding areas from radiation.

The reactor itself must be heavily shielded and housed inside a containment building of metal and reinforced concrete. Because loss of coolant water can result in a meltdown of the reactor core, backup emergency cooling systems have to be

The cloud above this power plant is just that—a cloud composed of condensed droplets of water. Nuclear power plants emit a minuscule quantity of radiation but none of the noxious pollutants that come from coal-burning power plants.

Nuclear reactors are housed in containment buildings made of steel and reinforced concrete. The structures are designed to withstand nuclear accidents without releasing radioactive substances into the environment.

constructed. Despite what proponents call utmost precautions, some opponents of nuclear power still fear a runaway nuclear reaction in which the containment building is breached and massive amounts of radioactivity escape into the environment. The chance of such an accident is probably exceedingly small, but if one did occur, thousands of people could be killed and large areas rendered uninhabitable for centuries. The benefit of nuclear power—abundant electric energy—is clear, but the small probability of an accident causes scientists and others to endlessly debate its desirability.

Another problem with nuclear power is that the fission products are highly radioactive and must be isolated from the environment for centuries. Again, scientists disagree about the feasibility of nuclear waste disposal. Proponents of nuclear power say that such wastes can be safely stored in old salt mines or other geologic formations. Opponents fear that the wastes may arise from their "graves" and eventually contaminate the groundwater. It is impossible to do a million-year experiment in a few years to determine who is right.

There are also wastes from the mining of uranium ore. Over 200 million tons of tailings (wastes left from uranium ore processing) now plague ten western states. These tailings are mildly radioactive, giving off radon gas and gamma radiation. Dust from these tailings carries problems to surrounding areas.

Still another problem, thermal pollution (Chapter 13), is unavoidable. As the energy from any material is converted to heat to generate electricity, some of the energy is released into the environment as waste heat. Nuclear power plants generate

The production of nuclear weapons also results in radioactive wastes. Even if we had never used nuclear reactions to produce electricity, we would still have nuclear wastes to store.

Mildly radioactive tailings are a by-product of the processing of uranium ores.

The United States has tentatively decided to store its nuclear wastes deep within Yucca Mountain in a remote area of Nevada. Scientific evaluation of the site, lawsuits, and other considerations will likely delay the construction and subsequent use of this storage site for years. In the meantime, wastes are being stored under water at the sites of the plants. (This on-site storage was supposed to be temporary, just long enough for short-lived radioisotopes to decay.) These temporary tanks are rapidly being filled.

more thermal pollution than plants that burn fossil fuels, but the difference is perhaps not as important, as the other problems we have mentioned.

Nuclear Accidents: Real and Imagined Risks

In 1979 a loss-of-coolant accident at the Three Mile Island nuclear power plant near Harrisburg, Pennsylvania, released a tiny amount of radioactivity into the environment. Although no one was killed or seriously injured, this accident whetted public fear of nuclear power. The accident at Chernobyl was much more frightening. There a reactor core meltdown killed several people outright. Others died from radiation sickness in the following weeks and months, and thousands were evacuated. A large area will remain contaminated for decades. Radioactive fallout spread across much of Europe. Thousands of people, particularly those close to the accident, have a greatly increased risk of cancer because of exposure to radiation. At Three Mile Island, a

An accident at this nuclear power plant at Chernobyl, Ukraine, increased public fears of nuclear power. It released considerable radioactivity because it had no reinforced containment building. The plant used graphite as a moderator (U.S. plants use water), and the burning carbon hindered efforts to tame the runaway reaction.

containment building kept most of the radioactive material inside. The Chernobyl plant had no such protective structure.

The fear that nuclear power plants might blow up like the bombs that devastated Hiroshima and Nagasaki is unfounded. The uranium used in these plants is enriched to only 3 or 4% uranium-235. To make a bomb, you need about 90% uranium-235.

There is considerable controversy over most aspects of nuclear power. Although scientists may be able to agree on the results of laboratory experiments, they don't always agree on what is best for society.

Breeder Reactors: Making More Fuel than They Burn

The supply of fissionable uranium-235 isotope is limited, making up less than 1% of naturally occurring uranium. Separation of uranium-235 leaves behind large quantities of uranium-238, which is not fissionable. However, uranium-238 can be converted to fissile plutonium-239 by bombardment with neutrons. Uranium-239 is formed initially, but it rapidly decays to neptunium-239, which quickly decays to plutonium. The reactions are shown in Figure 14.11a.

If a reactor is built with a core of fissionable plutonium surrounded by uranium-238, neutrons from the fission of plutonium convert the uranium-238 shield to more plutonium. In this way, the reactor breeds more fuel than it consumes. There is enough uranium-238 to last several centuries, and so one of the disadvantages of nuclear plants could be overcome by the use of **breeder reactors**.

Breeder reactors have some problems of their own, however. Plutonium is fairly low melting (640°C), and a plant is therefore limited to fairly cool—and inefficient—operation. Water is not adequate as a coolant, and it is necessary to use molten sodium metal in the primary loop. These reactors are often called *liquid-metal fast breeder reactors*. If an accident occurred in such a breeder, the sodium could react violently with both the water and the air.

Because plutonium is low melting, a failure of the cooling system could cause the reactor's core to melt. All reactors are required to have an emergency backup core-cooling system. Whether these systems work or not is a principal area of controversy.

Plutonium is highly toxic and has a half-life of about 25,000 years. It emits alpha particles, making it especially dangerous if ingested. It is estimated that 1 μg in the lungs of a human is enough to induce lung cancer. A further hazard is that a nation or a terrorist group could readily convert reactor-grade plutonium, unlike the uranium used in nuclear reactors, to a nuclear bomb. Plutonium is produced for nuclear weapons, some in ordinary (nonbreeder) reactors.

Another possible breeder reaction is that in which thorium-232 is converted to fissile uranium-233 (Figure 14.11b). It is thought to be rather difficult to make bombs from reactor-grade uranium-233. However, uranium-233 is like plutonium in that it emits damaging alpha particles.

No breeder reactors are operating in the United States, but they are used in France and other countries.

14.12 Nuclear Fusion: The Sun in a Magnetic Bottle

In Chapter 4, we discussed the thermonuclear reactions that power the sun and that have been adapted by scientists and engineers to make hydrogen bombs. If we could find a way to control these fusion reactions and use them to produce electricity, we would have a nearly unlimited source of power. To date, fusion reactions have been useful only for making bombs, although research on the control of nuclear fusion is progressing (Figure 14.12). Controlled fusion would have several advantages over nuclear fission reactors. The principal fuel, deuterium ($^2_1$H), is plentiful and is obtained

A 20-year study of 70,000 nuclear shipyard workers found that they had 24% *lower* mortality than non-nuclear workers in the general population. Those with the highest chronic radiation exposure had the *lowest* mortality from all causes—including cancer. Could it be that small doses of radiation are beneficial?

$$^{238}_{92}\text{U} + ^1_0\text{n} \longrightarrow ^{239}_{92}\text{U}$$

$$^{239}_{92}\text{U} \longrightarrow ^{239}_{93}\text{Np} + ^0_{-1}\text{e}$$

$$^{239}_{93}\text{Np} \longrightarrow ^{239}_{94}\text{Pu} + ^0_{-1}\text{e}$$

Plutonium-breeding reactions
(a)

$$^{232}_{90}\text{Th} + ^1_0\text{n} \longrightarrow ^{233}_{90}\text{Th}$$

$$^{233}_{90}\text{Th} \longrightarrow ^{233}_{91}\text{Pa} + ^0_{-1}\text{e}$$

$$^{233}_{91}\text{Pa} \longrightarrow ^{233}_{92}\text{U} + ^0_{-1}\text{e}$$

Uranium-233-breeding reactions
(b)

Figure 14.11 Reactions that breed fissile fuels from nonfissile isotopes. In each case, neutron bombardment first converts a nonfissile isotope to an unstable isotope. This isotope decays into another unstable isotope, which in turn decays into the fissile isotope.

Figure 14.12 The most promising fusion reaction is the deuterium–tritium reaction. A hydrogen-2 (deuterium) nucleus fuses with a hydrogen-3 (tritium) nucleus to form a helium-4 nucleus. A neutron is released along with a considerable quantity of energy.

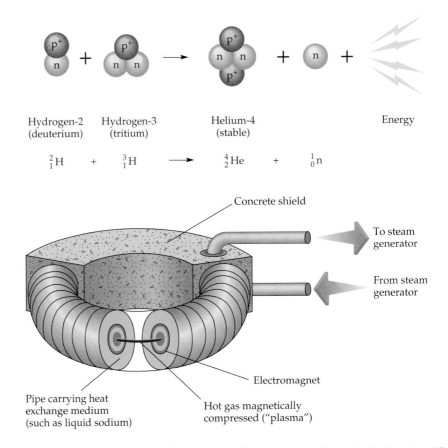

Hydrogen-2 (deuterium) Hydrogen-3 (tritium) Helium-4 (stable) Energy

$$^{2}_{1}\text{H} + {}^{3}_{1}\text{H} \longrightarrow {}^{4}_{2}\text{He} + {}^{1}_{0}\text{n}$$

Figure 14.13 A giant, donut-shaped electromagnet called a *tokamak* is designed to confine plasma at the extremely high temperatures and pressures required for nuclear fusion.

by the fractional electrolysis (splitting apart by means of electricity) of water. (Only 1 hydrogen atom in 6500 is a deuterium atom, but we have oceans of water to work with.) The problem of radioactive wastes would be minimized. The end product, helium, is stable and biologically inert. Escape of tritium ($^{3}_{1}\text{H}$) might be a problem because this hydrogen isotope would be readily incorporated into organisms. Tritium undergoes beta decay, with a half-life of 12.3 years. We would still be concerned with thermal pollution: the unavoidable loss of part of the energy as heat.

Great technical difficulties have to be overcome before a controlled fusion reaction can be used to produce energy. Temperatures of 50,000,000°C must be attained, and no molecule could hold together at that temperature. Even atoms are unstable. No material on Earth can withstand more than a few thousand degrees. All atoms would be stripped of their electrons, and the nuclei and free electrons would form a mixture called a **plasma**. Because plasma is made of charged particles (nuclei and electrons), it could be contained by a strong magnetic field (Figure 14.13), however. Scientists in several nations are getting closer to developing the environment necessary for controlled fusion.

Nuclear fusion may well be our best hope for relatively clean, abundant energy in the future; however, much work remains to be done. Even when controlled fusion is achieved in the laboratory, it will still be decades before it becomes a practical source of energy.

[Web Reference 4] A good site about nuclear fusion.

Renewable Energy Sources

Burning fossil fuels leads to air pollution and to the depletion of vital resources. Using nuclear power also presents problems, and nuclear fuel is not unlimited. Are there no renewable energy resources? There are indeed, and we shall devote the remainder of this chapter to them.

14.13 Harnessing the Sun: Solar Energy

At the beginning of this chapter, we saw that nearly all the energy on Earth comes from the sun. With all that energy from our celestial power plant, why do we need fossil fuels or nuclear power? The answer lies in the fact that this solar energy is thinly spread out and difficult to concentrate.

Solar Heating

Diffuse energy is not very useful. As it arrives on the surface of Earth, about half of solar energy is converted to heat. Another 30% is simply reflected back into space. We can increase the efficiency of this conversion rather easily. A black surface absorbs radiation better than a colored one. To make a simple solar collector, we need only cover a metal surface, painted black, with a glass plate. The glass is transparent to the incoming solar radiation, but it partially prevents the heat from escaping back into space. The hot surface is used to heat water or other liquids, and the hot liquids are usually stored in an insulated reservoir.

Water heated in this manner can be used directly for bathing, dish washing, and laundry. A building can be heated by passing air around the warm reservoir. The warmed air is then circulated through the building (Figure 14.14). Even in cold northern climates, solar collectors could meet about 50% of home heating requirements. These installations are expensive but could pay for themselves, through fuel savings, in a few years.

Solar Cells: Electricity from Sunlight

Sunlight also can be converted directly to electricity by devices called **photovoltaic cells** or *solar cells*. These devices can be made from a variety of substances, but most are made from elemental silicon. In a crystal of pure silicon, each silicon atom has four

Solar cells using amorphous silicon can be more than 10% efficient, and they are much easier to produce than those made from silicon crystals.

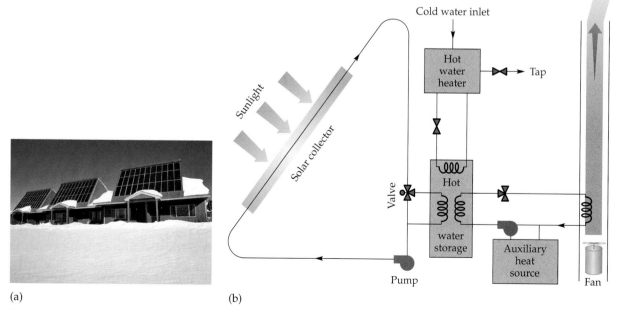

(a)   (b)

Figure 14.14 Energy in sunlight is absorbed by solar collectors and used to heat water. The hot water can be used directly, or it can be circulated to partially heat the building. (a) Solar collectors on a rooftop. (b) This diagram shows how a solar collector furnishes hot water and warm air for heating the building.

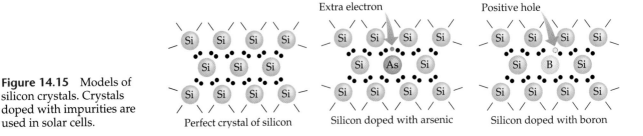

Extra electron

Positive hole

Perfect crystal of silicon   Silicon doped with arsenic   Silicon doped with boron

Figure 14.15 Models of silicon crystals. Crystals doped with impurities are used in solar cells.

valence electrons and is covalently bonded to four other silicon atoms (Figure 14.15). To make a solar cell, extremely pure silicon is doped with small amounts of specific impurities and formed into single crystals. One type of crystal has about 1 ppm of arsenic added. Arsenic atoms have five valence electrons, four of which are used to form bonds to silicon atoms. The fifth electron is relatively free to move around. This type of crystal, with extra electrons, is called a *donor crystal*. Adding about 1 ppm of boron makes another crystal. Boron atoms have three valence electrons. These three electrons are used to bond silicon atoms, but there is a shortage of one electron, leaving a *positive hole* in the crystal. This boron-doped crystal is called an *acceptor crystal*.

When the two types of crystals (acceptors and donors) are joined, electrons tend to flow from the donor to the acceptor. However, the holes near the junction are quickly filled by nearby mobile electrons, and the flow ceases. When sunlight strikes the cell, more electrons are dislodged, creating more mobile electrons and more positive holes. When an external circuit connects the two crystals, electrons flow from the donor to the acceptor (Figure 14.16).

An array of solar cells, combined to form a solar battery, can produce about 100 W/m² surface; it takes a square meter of cells to power one 100-W light bulb. Solar batteries have been used for years to power spaceships. They are also used to provide electricity for weather instruments in remote areas. Prices are decreasing, and solar batteries are now widely used to power small devices such as electronic calculators.

Solar cells are not very efficient (usually about 10%). Much of the sunlight striking them is reflected back into space. The generation of enough energy to meet a significant portion of our demands would require covering vast areas of desert land with solar cells. It would require 2000 hectares (about 5000 acres) of cloud-free desert land to produce as much energy as one nuclear power plant. Research has produced more efficient solar cells, and their cost is decreasing.

Utilizing solar energy requires storage for use at night and on cloudy days. One solution to this problem involves storage of energy as heat. While the sun shines, energy would be transferred to tanks of molten salts. Then, heat from these salts would, as needed, be used to produce steam to run a turbine, which would generate electricity.

The technology is now available for the use of solar energy for space heating and for providing hot water, however, widespread use of electricity from this source is probably still a few years away.

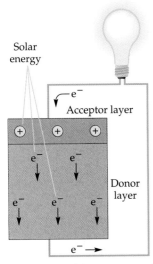

Solar energy

e⁻

Acceptor layer

Donor layer

Figure 14.16 Diagram of a solar cell. Electrons flow from the donor crystal to the acceptor crystal through the external circuit.

[Web Reference 5] Among the many websites dealing with solar energy are those maintained by (a) the Renewable Energy Network of the U.S. Department of Energy; (b) the American Solar Energy Society (ASES); (c) the Alliance to Save Energy, a coalition promoting energy efficiency, sustainable energy resources, and clean energy; (d) the Arizona Solar Center, dealing with solar and geothermal energy and other renewable energy technologies and resources; and (e) the National Solar Power Research Institute, Inc.

14.14 Biomass: Photosynthesis for Fuel

Why bother with solar collectors and photovoltaic cells to capture energy from the sun when green plants do it every day? Indeed, dry plant material burns quite well. It could be used to fuel a power plant for the generation of electricity. "Energy plantations" could grow plants for use as fuel. Plant **biomass** is a renewable resource whose production is powered by the sun (Figure 14.17).

Unfortunately, there are several disadvantages to this scheme, too. Most available land is needed for the production of food. Even where productive land is avail-

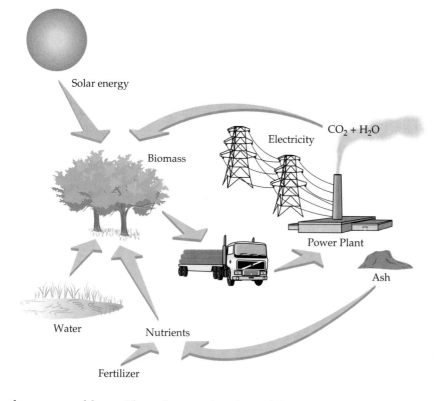

Figure 14.17 Energy in sunlight can be converted to electric energy by way of green plants. Ash from the burned plants should be returned to the soil to prevent nutrient depletion and to avoid having to fertilize the soil. In reality, most of the soil under consideration for biomass production is too poor to be productive without considerable application of fertilizers.

able, there are problems. Plants have to be planted, harvested, and transported to the power plant. Often, the land is far from where the energy is needed, and the overall efficiency is even less than that of solar cells—only about 3% at best.

Nevertheless, there is considerable research activity in the area of biomass. Scientists in upstate New York, for example, are actively investigating fast-growing trees as fuel for utility companies.

Plant material does not have to be burned directly. Starches and sugars from plants can be fermented to form ethanol. Wood can be distilled in the absence of air to produce methanol. Both alcohols are liquids and convenient to transport. Both are also excellent fuels that burn relatively cleanly. Bacterial breakdown of plant material produces methane. Under proper conditions, this process can be controlled to produce a clean-burning fuel similar to natural gas. Each of these conversions, however, must result in the loss of a portion of the useful energy. The laws of thermodynamics tell us that we would get the most energy by burning the biomass directly rather than converting it to a more convenient liquid or gaseous fuel.

The shortage of land probably means that we will never obtain a major portion of our energy from biomass. We could, however, supplement our other sources by burning agricultural wastes, fermenting some to ethanol, producing methanol from wood where wood is plentiful, and fermenting human and animal wastes to produce methane. The technology for all these processes is readily available. Each has been used in the past, and all are now being used on a limited scale.

14.15 Other Energy Sources

Modern civilization requires substantial energy. We are depleting our reserves of fossil fuels, and many people fear nuclear power. Using the sun's energy seems to be a good idea, but it would be difficult to meet all our needs with solar energy. In this section we look at other kinds of energy sources. Some are already in use. Indeed, some have been used for centuries.

Hemp, the fiber crop from which rope is made, is a good energy source. It has woody stalks that are 77% cellulose and can produce 10 t of biomass per acre in just 4 months. The oil from hemp seeds can also be used as diesel fuel. The problem is that it is illegal to raise hemp in the United States. (The leaves and flowers constitute the drug marijuana.)

Tumbleweed has been pressed into logs, peanut shells have been formed into briquettes, and mutton fat has been added to fuel oil.

The Grand Coulee Dam on the Columbia River in Washington State produces over 5000 MW of electric energy.

Wind Power and Waterpower

The sun, by heating Earth, causes winds to blow and water to evaporate and rise into the air, later to fall as rain. The blowing wind and the flowing water can be used as sources of energy. Indeed, they have been used that way for centuries. Waterwheels were used to lift water into the irrigation ditches of ancient Egypt. Windmills were used in tenth-century Persia to grind grain.

What is the connection between waterpower and wind power and chemistry? Chemists are involved in the production of new materials such as metal alloys, reinforced plastics, and lubricants vital to the construction and operation of dams, windmills, and turbines. Chemists also monitor water quality above and below hydroelectric dams, and they participate in other activities vital to protection of the environment around generating facilities.

Why not use wind power and waterpower to solve the energy crisis? We do. Waterpower provides nearly 9% of our current electricity production, most of it in the mountainous western United States. In a modern hydroelectric plant, water is held behind huge dams. Some of this stored water is released through penstocks against the blades of water turbines. The potential energy of the stored water is converted to the kinetic energy of flowing water. The moving water imparts mechanical energy to the turbine. The turbine drives a generator that converts mechanical energy to electric energy.

Hydroelectric plants are relatively clean, but most of the best dam sites in the United States have already been used. To obtain more hydroelectric energy, we would have to dam up scenic rivers and flood valuable cropland and recreational areas. Reservoirs silt up over the years, and sometimes dams break, causing catastrophic floods. Even hydroelectric power has its problems.

We could use more wind power. The kinetic energy of moving air is readily converted to mechanical energy to pump water and grind grain. Windmills have been so used for centuries. Wind also can be used to turn turbines and generate electricity. Giant windmills have been built, and smaller units are available for use on farms and for rural dwellings.

Wind turbine generators at Tracy, California. More than 50,000 wind turbines have been installed worldwide since 1974 to generate electricity, with 13,000 located in California.

Properly developed, wind power could supply about 10% of our energy needs. Wind is clean, free, and abundant. However, the wind does not always blow, and some means of energy storage or an alternative source of energy is needed. Land use might become a problem if wind power were used widely, but land under windmills could be used for farming or grazing.

The Tides: Moon Power

[Web Reference 6] The website of the American Wind Energy Association.

The power of the tides is another potential source of energy. One tide-powered generating plant is now in operation in France. Although tides produce great amounts of energy, there are few appropriate sites available in the United States. Some sites that could be used, such as Passamaquoddy Bay, are prized for their beauty; many people would oppose any attempt to construct power plants at such locations. Another drawback is that electricity can be generated only when the tide is coming in

The tide comes in on a rocky coast. Are we willing to give up such a view for a tide-powered electricity-producing plant?

or going out. There would be breaks in the production of electricity, and a system of energy storage would be required.

Geothermal Energy

The interior of Earth is heated by immense gravitational forces and by natural radioactivity. This heat comes to the surface in some areas through geysers and volcanoes. **Geothermal energy** has long been used in Iceland, New Zealand, and Italy. It has some potential in the United States—indeed, it is being used now in California (Figure 14.18)—but in the near future this potential could be realized only in areas where steam or hot water is at or near the surface. One drawback of geothermal energy is that the wastewater is quite salty; its disposal could be a problem.

The Geysers in northern California supply about 7% of the state's electric power. This geothermal power is low in cost (second only to hydropower).

(a)

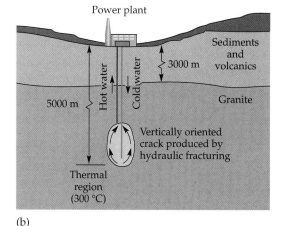

(b)

Figure 14.18 Geothermal energy contributes somewhat to our energy supply, and it has the potential to do so even more. (a) A geothermal plant at the Geysers in Sonoma County, California. (b) Energy is extracted from dry, hot rock 5 km below the surface by pumping water into the rock to be heated.

Oil shale is a rock rich in kerogen. Pyrolysis (heating in the absence of air) of the rock yields a liquid from which fuels can be made.

Tar sands are composed of bitumen, a tarry hydrocarbon mixture, and sand. Oil from tar sands makes up 19% of Canada's total crude oil production.

Brazil already makes considerable use of soybean oil as diesel fuel.

Oil Shale and Tar Sands

There are fossil carbon compounds other than coal, petroleum, and natural gas. For example, there are immense reserves of oil shale in Colorado, Utah, and Wyoming and of tar sands in Alberta, Canada. The problem with both these resources is that their extraction is difficult, requires a lot of energy, and can cause severe environmental disruption.

Actually, **oil shale** contains no oil. The organic matter is present as a complex material called **kerogen**, which has an approximate composition of $(C_6H_8O)_n$, where n is a large number. When heated in the absence of air, kerogen breaks down, forming hydrocarbon oils similar to petroleum. Most of the kerogen is thinly distributed throughout the rock. Although it does not seem to be covalently bonded to the rock, it can't be pumped. The energy content of the shale oil—by the time it is refined into gasoline and fuel oil—is not much greater than the amount of energy put into producing it, and there is a lot of waste rock left for disposal.

Tar sands are quite different from oil shale, but the problems associated with extracting fuel from them are similar. The organic matter of tar sands is present as **bitumen**, a hydrocarbon mixture. As in oil shale, the organic matter is thinly distributed in a lot of inorganic waste.

We may someday obtain a significant portion of our energy from oil shale and tar sands. It will be expensive energy, however. The thinly distributed, intimately mixed organic matter is high-entropy material. Converting it to a useful, low-entropy (purified) form will require a lot of energy.

Oil from Seeds

Many plants produce oil-bearing seeds. Cotton, peanuts, soybeans, and sunflowers are a few examples. Usually these oils are extracted and bottled as salad or cooking oil, but they also have value as fuels. Some farmers in North Dakota have used sunflower oil to power diesel tractors.

Because plants are a renewable resource, fast-growing plants with high oil yields might be developed for producing oil-seed fuels. If these plants could also be made to grow in areas and climates unsuitable for most food crops, perhaps they might help to solve part of our energy problem.

Coal Gasification and Liquefaction

Quite often, in discussions of future energy supplies, we hear about converting coal to gas or oil. When we run short of gas and petroleum, why not make them out of coal? The technology has been around for years. Gasification and liquefaction do have some advantages: gases and liquids are easy to transport, and the process of conversion leaves much of the sulfur and minerals behind, thus overcoming a serious disadvantage of coal as a fuel.

A variety of experimental projects are under way to convert coal to a synthetic gaseous fuel. The basic process is a reduction of carbon by hydrogen.

$$C(s) + 2 H_2(g) \longrightarrow CH_4(g)$$

Passing steam over hot charcoal can produce the hydrogen.

$$C(s) + H_2O(g) \longrightarrow CO(g) + H_2(g)$$

Coal also can be reacted with hydrogen to form petroleum-like liquids from which gasoline and fuel oil can be made.

Coal can also be converted to methanol, which can be used directly as fuel or converted, in turn, to gasoline-like hydrocarbons. In the first step, the coal reacts with extremely hot steam to form carbon monoxide and hydrogen.

$$C(s) + H_2O(g) \longrightarrow CO(g) + H_2(g)$$

Part of the carbon monoxide is then reacted with water to form more hydrogen.

$$CO(g) + H_2O(g) \longrightarrow CO_2(g) + H_2(g)$$

The remaining carbon monoxide is combined with hydrogen to form methanol.

$$CO(g) + 2\,H_2(g) \longrightarrow CH_3OH(l)$$

Finally, the methanol is converted to a hydrocarbon mixture that we represent as C_nH_m.

$$n\,CH_3OH(l) \longrightarrow C_nH_m(l) + x\,H_2O(l)$$

The Mobil process is actually being used in New Zealand. However, the fuel is very expensive.

Mobil Oil Corporation invented this process. Each step requires an appropriate catalyst.

Both gasification and liquefaction of coal require a lot of energy: up to one-third of the energy content of the coal is lost in the conversion. Liquid fuels from coal are high in unsaturated hydrocarbons and in sulfur, nitrogen, and arsenic compounds, and their combustion products are high in particulate matter. Coal conversions also require large amounts of water, yet the large coal deposits on which they would be based are in arid regions. Further, the conversions are messy. Without stringent safeguards, plants would seriously pollute both air and water.

Other processes that have been used to make liquid fuels from coal are the Bergius and Fischer–Tropsch methods. In the Bergius process coal is reacted with hydrogen. In the Fischer–Tropsch method coal is reacted with steam to make a mixture of carbon monoxide and hydrogen, which is then reacted over an iron catalyst to produce a mixture of hydrocarbons. Both processes were used in Germany during World War II. Since 1955, the SASOL process, which uses the Fischer–Tropsch method, has provided a significant portion of liquid fuels for South Africa.

Hydrogen as Fuel

Just as natural gas is sent through pipes to wherever it is needed, so could other fuel gases be sent through these same pipes. One such gas is hydrogen.

When hydrogen burns, it produces water and gives off energy.

$$2\,H_2(g) + O_2(g) \longrightarrow 2\,H_2O(g) + 572\ kJ$$

On a mass basis, hydrogen has more energy than any other chemical fuel. It is also a clean fuel, yielding only water as a chemical product. Because hydrogen can be made from seawater, the supply is virtually inexhaustible. However, hydrogen is strictly a secondary energy source because elemental hydrogen does not exist in significant amounts on Earth. It must be produced from a hydrogen compound (usually water), and this requires energy. Hydrogen, therefore, is not a basic energy source but just a convenient energy delivery system.

Several automobile companies have built experimental cars that operate on hydrogen fuel.

Solar cells or nuclear reactors might be used to produce hydrogen from water by electrolysis. The hydrogen gas could then be delivered though pipelines. It costs only about one-fourth as much to pipe hydrogen across a long distance as it does to transmit high-voltage electricity.

Some people are afraid of using hydrogen as fuel for home heating because of the possibility of explosion, but natural gas leaks can also lead to serious explosions. Hydrogen escapes more readily than natural gas or gasoline because of its smaller molecular size, but it also dissipates more readily, so that it never forms a flammable low-lying vapor as gasoline sometimes does.

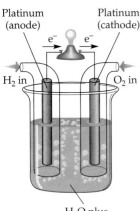

Platinum (anode) Platinum (cathode)

H_2 in O_2 in

H_2O plus an electrolyte

Figure 14.19 A hydrogen–oxygen fuel cell. The electrodes are porous to allow the gases to reach the electrolyte. The electrodes also catalyze the reaction.

[Web Reference 7] More information about fuel cells.

A bicycle not only has no fuel tank to fill but it also provides wonderful exercise.

According to a study by the World Energy Council, the global demand for energy in 2050 will be 58–175% higher than it was in 1990 because of a population increase and improved living conditions, especially in developing countries.

[Web Reference 8] Some websites deal with general energy efficiency and renewable resources, including: (a) the American Council for an Energy-Efficient Economy; (b) the U.S. Department of Energy's Renewable Energy Links; and (c) the CEERT, a collaboration of major environmental organizations, public interest groups, and clean technology companies.

Alcohol as Fuel

Ethanol is used as an octane booster in gasoline. Up to 10% of this alcohol can be added to gasoline without any need to adjust the automobile engines using it. When their engines are adjusted, however, cars can run on ethanol alone. Many of Brazil's cars run on alcohol made from sugarcane. (The alcohol they use is 95%, rather than the more expensive absolute alcohol used as a gasoline additive.) In the United States ethanol is usually made by the fermentation of corn.

Methanol is also used as automobile fuel. It is less popular than ethanol, however, because it is more toxic and promotes the rusting of engine parts.

Fuel Cells

A **fuel cell** is a device in which fuel is oxidized in an electrochemical cell (Chapter 8) so as to produce electricity directly. Fuel cells differ from the usual electrochemical cell in two ways. First, the fuel and oxygen are fed continuously. Second, the electrodes are made of an inert material such as platinum that does not react during the process. The electrodes serve to conduct the electrons to and from the solution.

Consider a fuel cell using hydrogen and oxygen (Figure 14.19). Hydrogen is oxidized at the anode; oxygen is reduced at the cathode. The overall reaction is similar to combustion, but not all the chemical energy is converted to heat.

$$2 H_2(g) + O_2(g) \longrightarrow 2 H_2O(g)$$

Rather, about 40–55% of the chemical energy is converted to electricity.

Fuel cells are used on spacecraft to produce electricity. The water produced can be used for drinking. Fuel cells overall have a weight advantage over storage batteries, an important consideration when launching a spaceship.

On Earth, research is underway to reduce cost and design long-lasting cells. Perhaps someday fuel cells will provide electricity to meet peak needs in large power plants. Unlike huge boilers and nuclear reactors, they can be started and stopped simply by turning the fuel on or off.

14.16 Energy: How Much Is Too Much?

Although there are many problems, science and technology probably will be able to provide us with a plentiful supply of energy for the foreseeable future. This energy cannot be pollution-free.

How do we choose the best method of energy production? The task is certainly difficult. Informed citizens who have examined the process from beginning to end should make the choice. We must know what is involved in the construction of power plants, the production of fuels, and the ultimate use of energy in our homes and factories. We must know that energy is wasted (as heat) at every step in the process.

Will our profligate consumption of energy affect Earth's climate? Our activities have already modified the climate in and around metropolitan areas. The worldwide effects of our expanding energy consumption are harder to estimate.

What can we do as individuals? We can conserve. We can walk more and use cars less. We can reduce our wasteful use of electricity. We can buy more efficient appliances, and we can avoid purchasing energy-intensive products.

Over the past quarter–century we have made significant progress in energy conservation. Appliances are more energy-efficient. New furnaces are 90–95% efficient, as compared with the 60% efficiency of 20-year-old furnaces. New refrigerators and air conditioners use about 35% less energy than earlier models. More people use fluorescent lamps, which are at least four times as energy-efficient as incandescent lamps. Overall, American industry has reduced its energy consumption per product by almost 30%.

We could greatly reduce energy consumption in the United States by greater use of public transportation. In Europe, where there is much wider use of public transport systems, per capita energy use is considerably less. So far, few people used to the convenience of personal cars seem eager to start using buses and trains. In Europe the average price for gasoline approaches $4.00 per gallon. If gasoline were that expensive in the United States, perhaps public transportation would seem more attractive.

The simple facts are that our population on Spaceship Earth is going up and our fuel resources are going down. People in developing countries want to raise their standard of living, and that will require more energy. We need to conserve energy and to look for new energy sources.

Critical Thinking Exercises

Apply knowledge that you have gained in this chapter and one or more of the FLaReS principles (Chapter 1) to evaluate the following statements or claims.

14.1 In 1989 chemists Stanley Pons and Martin Fleischman announced that they had achieved nuclear fusion in an ordinary laboratory at ordinary temperatures, a process known as "cold fusion." They had electrolyzed heavy water using palladium electrodes and claimed that much more energy was produced at the cathode than was used for the electrolysis. Many scientists have tried to repeat this work. Some seem to have had some success, but most have not. The system is unpredictable, and excess heat is not given off every time. In addition, Pons and Fleischman claim to have observed neutrons being emitted during the reaction. A few scientists have verified one or more of their claims, but most have failed in attempts to reproduce their experiments.

14.2 An inventor claims to have discovered a chemical that converts water to gasoline.

14.3 An inventor claims to have discovered a catalyst that speeds the conversion of linear alkanes, such as hexane, to branched-chain isomers.

Summary

1. The United States, with less than 5% of the world's population, uses almost one-quarter of all the energy being generated.
2. Most of the energy on Earth originated in the sun.
3. Photosynthesis is a process whereby energy is stored in plants. Animals depend on this stored energy for survival.
4. A chemical reaction is either exothermic (giving off energy) or endothermic (requiring energy).
5. The first law of thermodynamics deals with conservation of energy: energy can be neither created nor destroyed.
6. The second law of thermodynamics states that in natural processes energy is degraded from more useful forms to less useful forms.
7. Ancient societies used slave labor and animal power for doing work. Wood was the fuel burned.
8. Waterwheels have been used for about 2000 years, and windmills for about half that long.
9. Fossil fuels are natural fuels derived from once-living plants. They include coal, natural gas, and petroleum.
10. Coal is a black, burnable rock in which the main element is carbon. Coal is very slowly made from peat.
11. Natural gas is mainly methane. It is the cleanest of the fossil fuels.
12. Petroleum is a black liquid that is a complex mixture of organic compounds, mainly hydrocarbons.
13. Gasoline is a mixture of hydrocarbons, from about C_5 to C_{12}, obtained from petroleum.
14. Petroleum is fractionally distilled to separate gasoline from higher-boiling heating and lubricating oils. Then part of the higher-boiling oils are cracked to increase the yield of gasoline.
15. In order to raise the octane rating (antiknock quality) of gasoline, oil refiners use isomerization (to increase branching), catalytic reforming (to convert straight-chain hydrocarbons to aromatic rings), and alkylation (to convert gases to highly branched gasoline molecules).
16. Tetraethyllead was added to gasoline to improve its octane rating for more than 50 years (from the late 1920s until the mid-1980s), but its use has been discontinued.
17. Burning of fossil fuels produces mainly carbon dioxide and water, but it also releases other compounds that pollute the air.
18. Most electric power is generated by boiling water to produce steam, which turns the turbines that produce electricity. Most power plants are fired with coal.

19. Nuclear reactors produce energy by nuclear fission, which provides the heat that turns steam turbines in power plants. These reactors create very little air pollution, but they must be well shielded so as to contain all radioactivity.

20. Breeder reactors make more fissile fuel (plutonium-239) than they use.

21. Nuclear fusion generates energy by fusing small atoms to form larger ones. Thus far, attempts to control fusion for commercial purposes have not been successful.

22. With solar cells (photovoltaic cells), sunlight can be converted directly to electricity.

23. Burning biomass to generate heat might involve growing plants specifically for use as fuel, or just burning paper and other combustible wastes.

24. Energy can also be obtained as hydroelectric power from dams or as wind power from windmills.

25. Other natural energy sources include geothermal power (geysers and volcanoes), tides, and the fossil fuels in oil shale and tar sands.

26. Synthetic fuels can be made by liquefying or gasifying coal.

27. Fuel cells generate electricity by reacting a fuel such as hydrogen with oxygen in an electrochemical cell.

Key Terms

biomass   14.15
bitumen   14.16
breeder reactor   14.11
catalytic reforming   14.9
coal   14.7
endothermic
   reaction   14.2
entropy   14.4

exothermic reaction   14.2
first law of
   thermodynamics   14.3
fossil fuel   14.6
fuel   14.6
fuel cell   14.16
gasoline   14.9
geothermal energy   14.15

kerogen   14.16
law of conservation of
   energy   14.3
natural gas   14.8
nuclear reactor   14.12
octane rating   14.9
oil shale   14.16
petroleum   14.9

photosynthesis   14.1
photovoltaic (solar)
   cells   14.14
plasma   14.12
second law of
   thermodynamics   14.4
tar sands   14.16

Review Questions

1. What is a fuel?

2. What kind of reaction powers the sun?

3. What was the principal fuel used in the United States before 1800?

4. What was the principal fuel used in the United States from about 1850 to 1950? What has been our principal fuel since 1950?

5. List some advantages and disadvantages of hydrogen as a vehicle fuel.

6. What is the ultimate source of nearly all the energy on Earth?

Problems

Fuels and Combustion

7. Which of the following are fuels?
   a. C_3H_8 b. CCl_4
   c. C_8H_{18} d. CO_2

8. Which of the following are fuels?
   a. H_2 b. C_2H_2
   c. C_3F_8 d. H_2O

9. Give the equation for the complete combustion of coal. Assume that the coal is carbon (C).

10. Give the equation for the incomplete combustion of coal to form carbon monoxide. Assume that the coal is carbon (C).

11. Give the equation for the complete combustion of methane.

12. Give the equation for the incomplete combustion of methane to form carbon monoxide.

Energy and Chemical Reactions

13. How does temperature affect the rate of a chemical reaction?

14. Why does wood burn more rapidly in pure oxygen than in air?

15. What is an exothermic reaction? Give an example.

16. What is an endothermic reaction? Give an example.

17. Burning 1.00 mol of methane releases 192 kcal of energy. How much energy is released by burning 5.25 mol of methane?

$$CH_4(g) + 2\,O_2(g) \longrightarrow CO_2(g) + 2\,H_2O(g) + 192\ kcal$$

18. It takes 137 kcal of energy to decompose 2.00 mol of water. How much energy does it take to decompose 55.5 mol of water?

$$2 \, H_2O(g) + 137 \text{ kcal} \longrightarrow 2 \, H_2(g) + O_2(g)$$

Energy and the Laws of Thermodynamics

19. State the first law of thermodynamics.

20. State the second law of thermodynamics in terms of energy flow and in terms of degradation of energy.

21. What is entropy? Is entropy increased or decreased when a fossil fuel is burned?

22. Energy is conserved. How can we ever run out of energy?

Fossil Fuels

23. What are the advantages and disadvantages of coal as a fuel?

24. Which of the fossil fuels is most plentiful in the United States?

25. What is the physical state of each of the three fossil fuels?

26. Why did the United States shift from coal to petroleum and natural gas when we have much larger reserves of coal than of the two hydrocarbon fuels?

27. The estimated world oil reserves in 1999 were 975 billion barrels. The world rate of annual use was 26.1 billion barrels. How long will these reserves last if this rate of use continues? (The actual rate of use is increasing.)

28. The estimated U.S. oil reserves in 1999 were 22.5 billion barrels. How long will these reserves last if there are no imports or exports and if the U.S. annual rate of use of 6.9 billion barrels continues?

29. The estimated world natural gas reserves in 1999 were 5087 trillion ft^3. The world rate of annual use was 68.4 trillion ft^3. How long will these reserves last if this rate of use continues?

30. The U.S. proved natural gas reserves in 1997 were 167.2 trillion ft^3. How long will these reserves last if there are no imports or exports and if the U.S. annual rate of use of 21.7 trillion ft^3 continues?

Natural Gas

31. What is the main component of natural gas?

32. What are the advantages and disadvantages of natural gas as a fuel?

Petroleum

33. What is thought to be the origin of petroleum?

34. What are the advantages and disadvantages of petroleum as a source of fuels?

35. How is crude petroleum modified to better meet our needs and wants? What are the advantages and disadvantages of tetraethyllead as an octane booster?

36. What is meant by the octane rating of a gasoline?

Nuclear Power

37. What proportion of U.S. electricity is generated by nuclear power plants?

38. What proportion of electricity in France is generated by nuclear power plants?

39. Can a nuclear power plant explode like a nuclear bomb? Explain your answer.

40. Can nuclear bombs be made from reactor-grade uranium? Explain your answer.

41. How does a breeder reactor produce more fuel than it consumes?

42. List some advantages of nuclear power plants over coal-fired plants.

43. What are some of the disadvantages of breeder reactors?

44. Can a nuclear bomb be made from reactor-grade plutonium?

45. List some possible advantages of a nuclear fusion reactor over a fission reactor. What is plasma?

46. List some possible problems with nuclear fusion reactors.

47. Give nuclear equations showing how thorium-232 is converted to fissile uranium-233.

48. Give nuclear equations showing how uranium-238 is converted to fissile plutonium-239.

49. Give the nuclear equation that shows how deuterium and tritium fuse to form helium and a neutron.

50. Give the nuclear equation that shows how four protons fuse to form helium and two positrons.

Renewable Energy Sources

51. What are some problems associated with the use of solar energy?

52. What is a photovoltaic cell?

53. What is plant biomass?

54. List some advantages and disadvantages of the use of biomass as a source of energy.

55. List two ways that fuel cells differ from electrochemical cells.

56. List the advantages, disadvantages, and limitations of each of the following as an energy source.
    a. wind power   b. geothermal power
    c. power from tides   d. hydroelectric power

Synthetic and Converted Fuels

57. Give the chemical equation for the basic process by which coal is converted to methane.

58. Give the chemical equation for the conversion of coal (carbon) to carbon monoxide and hydrogen.

59. Give the chemical equation for the conversion of carbon monoxide and hydrogen to methanol.

60. Give the chemical equation for the reaction that occurs in a hydrogen–oxygen fuel cell.

Additional Problems

61. A woman uses 1125 kcal in running 10.0 km in 39 min 18 s. Calculate her average power output in watts. (1.000 kcal = 4184 J.)

62. The energy requirements of the human brain are about 20% of the total body metabolism. Calculate the power output (in watts) of your brain if you use 2175 kcal of energy per day.

63. What mass of carbon dioxide is formed by the combustion of 1250 kg of coal that is 51.2% carbon?

$$C + O_2 \longrightarrow CO_2$$

64. What mass of methanol can be made from 437 g of carbon monoxide if sufficient hydrogen is available?

65. Heats of combustion of several gaseous fuels, in kilojoules per mole, are given below. Which yields the most energy (a) per kilogram and (b) per liter when the volume is measured under the same conditions of temperature and pressure?

Hydrogen (H_2), 286.6 kJ

Isobutane [$CH_3CH(CH_3)_2$], 2868 kJ

Neopentane [$CH_3C(CH_3)_3$], 3515 kJ

66. Indicate a natural process or processes by which carbon atoms are (a) removed from the atmosphere; (b) returned to the atmosphere; (c) effectively withdrawn from the carbon cycle.

67. The United States leads the world in per capita emissions of $CO_2(g)$ with 19.8 t per person per year (1 t = 1000 kg). What mass, in metric tons, of each of the following fuels would yield this quantity of CO_2?
a. CH_4
b. C_8H_{18}
c. coal that is 94.1% C by mass

68. A large coal-fired electric plant burns 2500 tons of coal per day. The coal contains 0.65% S by mass. Assume that all the sulfur is converted to SO_2. What mass of SO_2 is formed? If a thermal inversion traps all this SO_2 in a parcel of air that is 45 km × 60 km × 0.40 km, will the level of SO_2 in the air exceed the primary national air quality standard of 365 $\mu g\ SO_2/m^3$ air?

69. How much the combustion of various fuels contributes to the buildup of CO_2 in the atmosphere can be assessed in different ways. One relates the mass of CO_2 formed to the mass of fuel burned; another relates the mass of CO_2 to the quantity of heat evolved in the combustion. Which of the three fuels C(graphite), $CH_4(g)$, or $C_4H_{10}(g)$ produces the smallest mass of CO_2 (a) per gram of fuel and (b) per kilojoule of heat evolved? The heat released *per mole* of the three substances is C(graphite), 393.5 kJ; $CH_4(g)$, 803 kJ; and $C_4H_{10}(g)$, 2877 kJ.

Projects

70. Which would you rather have in your neighborhood, a nuclear power plant or a coal-burning plant? Why? Explain your choice fully.

71. What characteristics should an ideal energy source have? Which of these characteristics do the different forms of energy have?

72. Social cost pricing takes into account all the costs of a product or activity that are external to market costs. For example, one of the factors in the cost of gasoline-based transportation is air pollution. Select three forms of energy and list the possible social costs that should be included.

73. Compare the quantity of electricity used for two appliances that are alike except for different efficiencies. Extend this comparison to CO_2 emissions from the appliances, assuming the electricity is generated by burning coal.

74. Compare qualitatively the efficiency of heating with natural gas versus electricity produced from burning natural gas at a power plant.

75. Compare qualitatively the efficiency of using solar panels on the roof versus photovoltaic cells in a "solar farm."

76. Which of the following is the best fuel for heating your home? What problems with supply, use, and waste products are involved in each case?
a. natural gas
b. electricity
c. coal
d. fuel oil

Online Projects

77. There was an accident at a uranium processing plant in Tokai, Japan, on September 30, 1999. Using your favorite search engine, find out what happened and compare this accident with the ones at Chernobyl and Three Mile Island. (*Hint*: Search on the name, followed by "nuclear accident.") How could these accidents have been avoided? Do these incidents prove that nuclear power plants should be phased out? Why or why not?

78. The use of appliances and heating equipment that employ alternative forms of energy (such as solar power) can help in individual energy conservation. Using internet sources, find such equipment. Do companies selling these products appear to be doing so for mostly idealistic—or mainly commercial—reasons? (A good place to begin is SolarAccess.com, which features solar industry directories.)

References and Readings

1. Asimov, Isaac. "In Dancing Flames a Greek Saw the Basis of the Universe." *Smithsonian*, November 1971, pp. 52–57. Discusses heat, from the Greek "element" fire to the first law of thermodynamics. Vintage Asimov.

2. Bailey, Maurice E. "The Chemistry of Coal and Its Constituents." *Journal of Chemical Education*, July 1974, pp. 446–448.

3. Bent, Henry A. "Haste Makes Waste: Pollution and Entropy." *Chemistry*, October 1971, pp. 6–15. A classic.

4. Bozak, Richard E., and Manuel Garcia, Jr. "Chemistry in the Oil Shales." *Journal of Chemical Education*, March 1976, pp. 154–155.

5. Brown, Lester R., Michael Renner, and Brian Halwell. *Vital Signs 1999*. New York: W. W. Norton, 1999, "Energy Trends," pp. 47–55.

6. Campbell, Colin J. *The Coming Oil Crisis*. Multi-Science Publishing and Petroconsultants, Brentwood, England, 1997.

7. Freemantle, Michael. "Ten Years After Chernobyl, Consequences Are Still Emerging." *Chemical and Engineering News*, 29 April 1996, pp. 18–28.

8. Hill, John W., and Ralph H. Petrucci. *General Chemistry*. 2d edition. Upper Saddle River, NJ: Prentice Hall, 1999. Chapter 6, "Thermochemistry", and Chapter 17, "Thermodynamics."

9. Ivanhoe, L. F. "Updated Hubbert Curves Analyze World Oil Supply." *World Oil*, November 1996, pp. 91–94.

10. Kolb, Doris, and Kenneth E. Kolb. "Chemical Principles Revisited: Petroleum Chemistry." *Journal of Chemical Education*, July 1979, pp. 465–469.

11. Siuru, Bill. "Power Supply for the Mid-21st Century: Lunar Solar Collectors." *Public Power*, May–June 1996, pp. 16–18.

12. Song, Chunshan, and Harold H. Schobert. "Synthesis of Chemicals and Materials from Coal." *Chemistry and Industry*, 1 April 1996, pp. 253–257.

13. Spiro, Thomas G., and William M. Stigliani. *Chemistry of the Environment*. Upper Saddle River, NJ: Prentice Hall, 1996. Part I, Energy.

14. Westbrook, Charles K. "The Chemistry Behind Engine Knock." *Chemistry and Industry*, 3 August 1992, pp. 562–566.

15. "What's Happening to Gasoline?" *Consumer Reports*, November 1996, pp. 54–57. Discusses the new "cleaner burning" fuels.

Biochemistry
A Molecular View of Life

Each cell in every form of life
Acts in a certain way,
According to the blueprints
Coded in its DNA.

Earth has millions of forms of life. Each has its own distinctive set of DNA "blueprints."

Y our body is an incredible chemical factory, more complex than any industrial plant. To stay in good condition and perform its varied tasks, it needs many specific chemical compounds, and it manufactures most of them in an exquisitely organized network of chemical production lines.

Every minute of every day, thousands of chemical reactions take place within each of the 100 trillion tiny cells in your body. The study of these reactions and the chemicals they produce is called *biochemistry*.

15.1 The Cell

Biochemistry is the chemistry of living things and life processes. The structural unit of all living things is the *cell*. Every cell is enclosed in a membrane, and plant cells also have walls made of cellulose. Each kind of tissue is made up of cells specific to the function of that particular tissue. Muscle cells are different from nerve cells, and red blood cells are different from skin cells.

Not many years ago, scientists thought that a cell was a fairly simple entity, with an outer membrane and a central nucleus. All the space in between was thought to be filled with a substance called *cytoplasm*. With the development of the electron microscope came the discovery of many more cell parts (Figure 15.1). We can consider only a few of them here.

Each animal cell gains nutrients and gets rid of wastes through the *cell membrane* that encloses it. Various interior structures serve a multiplicity of functions. The largest structure is usually the *cell nucleus*, which contains the material that controls heredity. Protein synthesis takes place in the *ribosomes*. The *mitochondria* are the cell "batteries" where energy is produced.

Plant cells (but not animal cells) also contain *chloroplasts* in which energy from the sun is converted to chemical energy, which is stored in the plant in the form of carbohydrates.

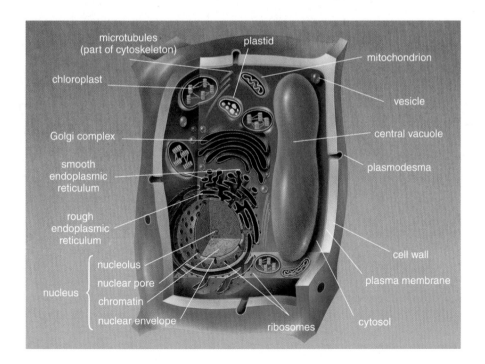

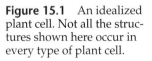

Figure 15.1 An idealized plant cell. Not all the structures shown here occur in every type of plant cell.

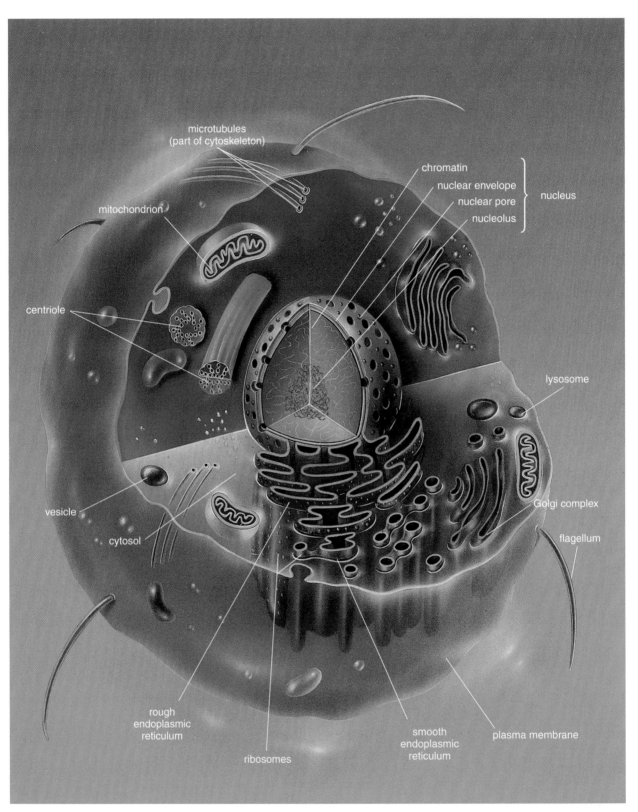

An animal cell. The entire range of structures shown here seldom occurs in a single cell, and we discuss only a few of them in this text. Each kind of plant or animal tissue has cells specific to the function of that tissue. Muscle cells differ from nerve cells, nerve cells differ from red blood cells, and so on.

15.2 Energy in Biological Systems

Life requires energy. Living cells are inherently unstable and avoid falling apart only because of a continued input of energy. Living organisms are restricted to using certain forms of energy. Supplying a plant with heat energy by holding it in a flame will do little to prolong its life. On the other hand, a green plant is uniquely able to tap sunlight, the richest source of energy on Earth. Chloroplasts in green plant cells capture the radiant energy of the sun and convert it to chemical energy, which is then stored in carbohydrate molecules. The photosynthesis of glucose is represented by the equation

$$6\,CO_2 + 6\,H_2O \longrightarrow C_6H_{12}O_6 + 6\,O_2$$

Plant cells can also convert these carbohydrate molecules to fat molecules and, with the proper inorganic nutrients, to protein molecules (Figure 15.2).

Animals cannot directly use the energy of sunlight. They must get their energy by eating plants or by eating other animals that eat plants. There are three major types of foods from which animals obtain energy: carbohydrates, fats, and proteins.

Once digested and transported to a cell, a food molecule can be used as a building block to make new cell parts or to repair old ones, or it can be "burned" for energy. The entire series of coordinated chemical reactions that keep cells alive is called **metabolism**. In general, metabolic reactions are divided into two classes. The degrading of molecules to provide energy is called **catabolism**, and the process of building up or synthesizing the molecules of living systems is called **anabolism**.

15.3 Carbohydrates: A Storehouse of Energy

Composed of the elements carbon, hydrogen, and oxygen, carbohydrates include sugars, starches, and cellulose. Usually, the atoms of these elements are present in a ratio expressed by the formula $C_x(H_2O)_y$. Glucose, a simple sugar that has the formula $C_6H_{12}O_6$, could be written $C_6(H_2O)_6$. It is from formulas such as these that the term

Carbohydrates, fats, and proteins as foods are discussed in Chapter 16. In this chapter, we deal mainly with the synthesis and structure of these vital materials.

 [Web Reference 1] The description of the important metabolic process of glycolysis gives much more than you need to learn in this course, but you might find it interesting.

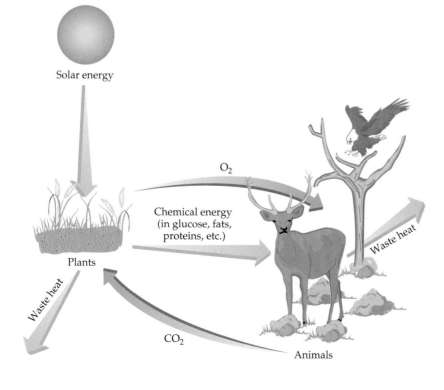

Figure 15.2 Some energy transformations in living systems.

[Web Reference 2]
You can get an idea of how many different monosaccharides there are.

It is difficult to give a simple formal definition of **carbohydrates**. Chemically, they are polyhydroxy aldehydes or ketones or compounds that can be hydrolyzed to form such compounds. ("Hydrolyzed" means "split by water.") Note the hydroxyl groups and aldehyde or ketone functions in the compounds in Figure 15.3.

"carbohydrate" is derived. We use it even though carbohydrates are not hydrates of carbon.

Some Simple Sugars

Sugars are sweet-tasting carbohydrates. The simplest of these are **monosaccharides**, carbohydrates that cannot be further hydrolyzed (split by water). Familiar monosaccharides include glucose, fructose, and galactose (Figure 15.3). Although represented in Figure 15.3 as open-chain compounds (to show the aldehyde or ketone functional groups), these sugars exist mainly as cyclic molecules (Figure 15.4).

Sucrose and lactose are examples of **disaccharides**, carbohydrates consisting of molecules that can be hydrolyzed to two monosaccharide units (Figure 15.5). Sucrose is split into glucose and fructose. Hydrolysis of lactose gives glucose and galactose.

$$\text{Sucrose} + \text{H}_2\text{O} \longrightarrow \text{Glucose} + \text{Fructose}$$

$$\text{Lactose} + \text{H}_2\text{O} \longrightarrow \text{Glucose} + \text{Galactose}$$

Polysaccharides: Starch and Cellulose

Polysaccharides are composed of molecules that yield many monosaccharide units on hydrolysis. Polysaccharides include starches, which comprise the main energy storage system of many plants, and cellulose, which is the structural material of

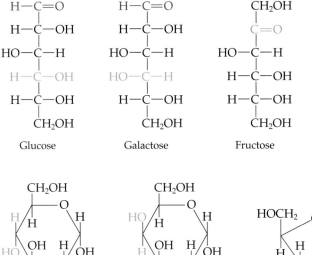

Figure 15.3 Three common monosaccharides. All have hydroxyl groups. Glucose and galactose have aldehyde functions (red), and fructose has a ketone group (blue). Glucose and galactose differ only in the arrangement of the H and OH on the fourth carbon (green) from the top. Glucose is also called dextrose, fructose is fruit sugar, and galactose is a component of milk sugar.

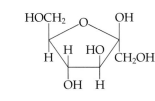

Figure 15.4 Cyclic structures for glucose, galactose, and fructose. A corner with no letter represents a carbon atom. Glucose and galactose are represented as six-membered rings. They differ only in the arrangement of the H and OH on the fourth carbon (green). Fructose is shown as a five-membered ring. Some sugars exist in more than one cyclic form. For simplicity, we show only one form of each here.

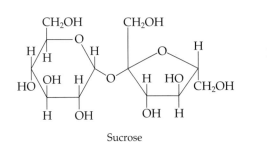
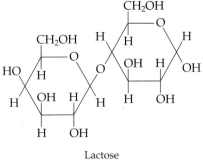

Figure 15.5 Sucrose and lactose are disaccharides. On hydrolysis, sucrose produces glucose and fructose, whereas lactose yields glucose and galactose. Sucrose is cane or beet sugar and lactose is milk sugar.

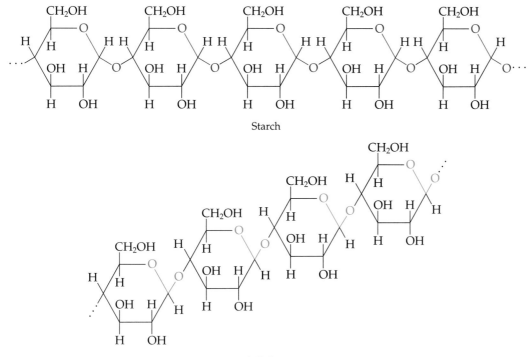

Starch

Cellulose

Figure 15.6 Both starch and cellulose are polymers of glucose. They differ in that the glucose units are joined by alpha linkages (red) in starch and by beta linkages (blue) in cellulose.

Figure 15.7 An electron micrograph of the cell wall of an alga. The wall is made up of successive layers of cellulose fibers in parallel arrangement.

plants. Figure 15.6 shows short segments of starch and cellulose molecules. Notice that both are polymers of glucose. Starch molecules generally have from 100 to about 6000 glucose units. Cellulose molecules are composed of 1800–3000 or more glucose units.

A crucial structural difference between starch and cellulose is in the way the glucose units are hooked together. Consider starch first. Using the CH_2OH as a reference, the oxygen atom joining the glucose units is pointed down. This arrangement is called an *alpha linkage*. In cellulose, again with the CH_2OH as a point of reference, the oxygen atom connecting the glucose segments is pointed up, an arrangement called a *beta linkage*. This difference in linkages may seem to be a minor one, but it determines whether or not the materials can be digested (Chapter 16). The different linkages also result in different three-dimensional forms for cellulose and starch. For example, cellulose in the cell walls of plants is arranged in fibrils, bundles of parallel chains. Fibrils in turn lie parallel to each other in each layer of the cell wall (Figure 15.7). In alternate layers, the fibrils are perpendicular, an arrangement that imparts great strength to the wall.

As foods (Chapter 16), polysaccharides are called *complex carbohydrates*. Starches serve as foods, and cellulose serves as dietary fiber. Simple sugars are rapidly absorbed into the body from the digestive tract. Starches must be digested before absorption. Humans cannot digest cellulose.

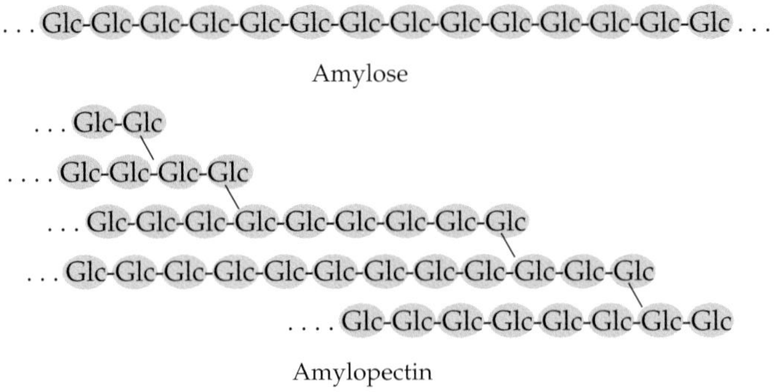

... Glc-Glc-Glc-Glc-Glc-Glc-Glc-Glc-Glc-Glc-Glc-Glc-Glc-Glc ...

Amylose

... Glc-Glc

... Glc-Glc-Glc-Glc

... Glc-Glc-Glc-Glc-Glc-Glc-Glc-Glc-Glc

... Glc-Glc-Glc-Glc-Glc-Glc-Glc-Glc-Glc-Glc-Glc-Glc

... Glc-Glc-Glc-Glc-Glc-Glc-Glc-Glc

Amylopectin

Figure 15.8 Schematic representations of amylose and amylopectin. Glc stands for a glucose unit. The structure of glycogen is similar to that of amylopectin.

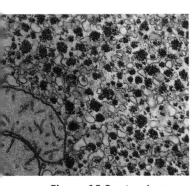

Figure 15.9 An electron micrograph of glycogen granules in a liver cell of a rat.

There are two kinds of plant starch. One, called *amylose*, has glucose units joined in a continuous chain like beads on a string. The other kind, *amylopectin*, has branched chains of glucose units. These starches are perhaps best represented schematically, as in Figure 15.8 where each glucose unit is represented by the abbreviation Glc.

Animal starch is called *glycogen*. Like amylopectin, it is composed of branched chains of glucose units. Glycogen in muscle and liver tissue is arranged in granules (Figure 15.9), clusters of small particles. Plant starch, on the other hand, forms large granules. Granules of plant starch rupture in boiling water to form a paste, and on cooling, the paste gels. Potatoes and cereal grains form this type of starchy broth. All forms of starch are hydrolyzed to glucose during digestion.

15.4 Fats and Other Lipids

Fats are the predominant forms of a class of compounds called *lipids*. These substances are not defined structurally; rather, they have common solubility properties. A **lipid** is a cellular constituent that is soluble in organic solvents of low polarity such as hexane, diethyl ether, or carbon tetrachloride. Lipids are insoluble in water. In addition to fats, the lipid family contains **fatty acids** (long-chain carboxylic acids), steroids such as cholesterol and sex hormones (Chapter 18), fat-soluble vitamins (Chapter 16), and other substances.

A **fat** is an ester of fatty acids and the trihydroxy alcohol glycerol (Figure 15.10). A fat may contain up to three fatty acid chains joined to glycerol through ester linkages. Fats are classified according to the number of fatty acid chains they contain. A *monoglyceride* has one fatty acid chain, a *diglyceride* has two, and a *triglyceride* (also called a triacylglycerol) has three.

Naturally occurring fatty acids nearly always have an even number of carbon atoms. Representative ones are listed in Table 15.1. Animal fats are generally rich in saturated fatty acids and have a smaller proportion of unsaturated fatty acids. At room temperature, most animal fats are solids. Liquid fats, called **oils**, are obtained principally from vegetable sources. Structurally, oils are identical to fats except that they incorporate a higher proportion of unsaturated fatty acid units.

Fats are often classified according to the degree of unsaturation of the fatty acids they incorporate. A *saturated fatty acid* contains no carbon-to-carbon double bonds, a *monounsaturated fatty acid* has one carbon-to-carbon double bond per molecule, and a *polyunsaturated fatty acid* molecule has two or more carbon-to-carbon double bonds. A *saturated fat* contains a high proportion of saturated fatty acids; these fat molecules have relatively few double bonds. A *polyunsaturated fat* (oil) incorporates mainly unsaturated fatty acids; these fat molecules have many double bonds.

The iodine number usually measures the degree of unsaturation of a fat or oil. The **iodine number** is the number of grams of iodine that are consumed by 100 g of fat or oil. Chlorine and bromine add readily to carbon-to-carbon double bonds.

Table 15.1 ▎ Some Fatty Acids in Natural Fats

Number of Carbon Atoms	Condensed Structure	Common Name	Source
4	$CH_3CH_2CH_2COOH$	Butyric acid	Butter
6	$CH_3(CH_2)_4COOH$	Caproic acid	Butter
8	$CH_3(CH_2)_6COOH$	Caprylic acid	Coconut oil
10	$CH_3(CH_2)_8COOH$	Capric acid	Coconut oil
12	$CH_3(CH_2)_{10}COOH$	Lauric acid	Palm kernel oil
14	$CH_3(CH_2)_{12}COOH$	Myristic acid	Oil of nutmeg
16	$CH_3(CH_2)_{14}COOH$	Palmitic acid	Palm oil
18	$CH_3(CH_2)_{16}COOH$	Stearic acid	Beef tallow
18	$CH_3(CH_2)_7CH=CH(CH_2)_7COOH$	Oleic acid	Olive oil
18	$CH_3(CH_2)_4CH=CHCH_2CH=CH(CH_2)_7COOH$	Linoleic acid	Soybean oil
18	$CH_3CH_2(CH=CHCH_2)_3(CH_2)_6COOH$	Linolenic acid	Fish oils
20	$CH_3(CH_2)_4(CH=CHCH_2)_4CH_2CH_2COOH$	Arachidonic acid	Liver

(a)

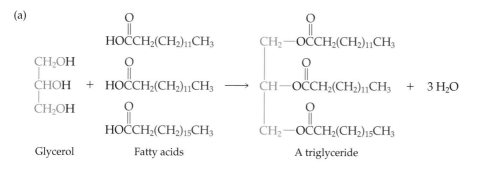

(b)

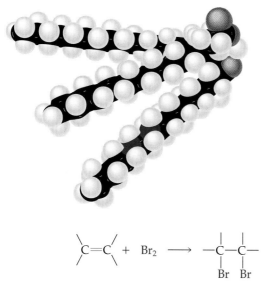

Figure 15.10 Triglycerides are esters in which the alcohol glycerol, with three OH groups, is esterified with three fatty acid groups. (a) The equation for the formation of a triglyceride. (b) Space-filling model of a triglyceride.

Iodine also adds, but less readily. The more double bonds a fat contains, the more iodine is required for the addition reaction; thus, a high iodine number means a high degree of unsaturation. Representative iodine numbers are listed in Table 15.2. Note the generally lower values for animal fats (butter, tallow, and lard) compared with

Table 15.2 ▪ Typical Iodine Numbers for Some Fats and Oils*

Fat or Oil	Iodine Number	Fat or Oil	Iodine Number
Coconut oil	8–10	Cottonseed oil	100–117
Butter	25–40	Corn oil	115–130
Beef tallow	30–45	Fish oils	120–180
Palm oil	37–54	Canola oil	125–135
Lard	45–70	Soybean oil	125–140
Olive oil	75–95	Safflower oil	130–140
Peanut oil	85–100	Sunflower oil	130–145

*Oils shown in blue are from plant sources. Three fats and one oil come from animals.

those for vegetable oils. Coconut oil, which is highly saturated, and fish oils, which are relatively unsaturated, are notable exceptions to the general rule.

Fats and oils feel greasy. They are less dense than water and float on it.

Proteins: The Stuff of Life

Proteins are vital components of all life. No living part of the human body—or of any other organism, for that matter—is completely without protein. There is protein in blood, muscles, brain, and even tooth enamel. The smallest cellular organisms—bacteria—contain protein. Viruses, so small that they make bacteria look like giants, are nothing but proteins and nucleic acids. This combination of nucleic acids and proteins is found in all cells and is the stuff of life itself.

Each type of cell makes its own kinds of proteins. Proteins serve as the structural material of animals, much as cellulose does for plants. Muscle tissue is largely protein; so are skin and hair. Silk, wool, nails, claws, feathers, horns, and hooves are proteins. All proteins contain the elements carbon, hydrogen, oxygen, and nitrogen, and most also contain sulfur. The structure of a short segment of a typical protein molecule is shown in Figure 15.11.

15.5 Proteins: Polymers of Amino Acids

Proteins are copolymers of about 20 different amino acids (Table 15.3). An **amino acid** has two functional groups: an amino group ($-NH_2$) and a carboxyl group ($-COOH$) attached to the same carbon atom (called the *alpha carbon*).

$$H_2N-\overset{\displaystyle |}{\underset{\displaystyle |}{C}}-\overset{\displaystyle O}{\overset{\displaystyle \|}{C}}-OH$$

This partial formula indicates the proper placement of these groups, but the structure is not really correct. Acids react with bases to form salts; the carboxyl group is acidic, and the amino group is basic. Therefore, these two functional groups interact, the acid transferring a proton to the base. The resulting product is an inner salt, or **zwitterion**, a compound in which the anion and the cation are parts of the same molecule. Each unique amino acid has a different group (symbolized by the R) attached to the alpha carbon atom.

$$H_3N^+-\overset{\displaystyle R}{\underset{\displaystyle H}{\overset{\displaystyle |}{\underset{\displaystyle |}{C}}}}-\overset{\displaystyle O}{\overset{\displaystyle \|}{C}}-O^-$$

A zwitterion

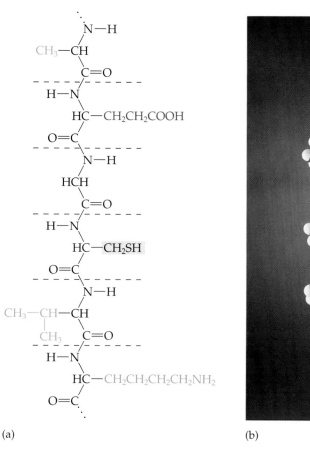

(a)

(b)

Figure 15.11 Structural formula (a) and space-filling model (b) of a short segment of a protein molecule. In the structural formula, hydrocarbon side chains (green), an acidic side chain (red), a basic side chain (blue), and a sulfur-containing side chain (amber) are highlighted. The broken lines in (a) indicate where two amino acid units join (see Section 15.6).

Plants are able to synthesize proteins from carbon dioxide, water, and minerals [such as nitrates (NO_3^-) and sulfates (SO_4^{2-})]. Animals require proteins as one of the three major classes of foods.

15.6 The Peptide Bond: Peptides and Proteins

The human body contains tens of thousands of different proteins. Each of us has a tailor-made set. Proteins are polyamides. The amide linkage ($-CONH-$) is called a **peptide bond** when it joins two amino acid units.

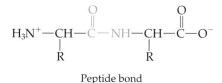

Peptide bond

Note that there is still a reactive amino group on the left and a carboxyl group on the right. Each of these can react further to join more amino acid units. This process can continue until thousands of units have joined to form a giant molecule—a polymer called a *protein*.

N-terminal   C-terminal

Table 15.3 ▋ The 20 Amino Acids Specified by the Genetic Code*

Glycine	Gly	No	CH_2-COO^- $^+NH_3$
Alanine	Ala	No	$CH_3-CH-COO^-$ $^+NH_3$
Phenylalanine	Phe	Yes	⬡$-CH_2-CH-COO^-$ $^+NH_3$
Valine	Val	Yes	$CH_3-CH-CH-COO^-$ $CH_3\ ^+NH_3$
Leucine	Leu	Yes	$CH_3CHCH_2-CH-COO^-$ $CH_3\quad ^+NH_3$
Isoleucine	Ile	Yes	$CH_3CH_2CH-CH-COO^-$ $CH_3\ ^+NH_3$
Proline	Pro	No	CH_2-CH_2 ... $C-COO^-$ CH_2 $^+NH_2$ H
Methionine	Met	Yes	$CH_3-S-CH_2CH_2-CH-COO^-$ $^+NH_3$
Serine	Ser	No	$HO-CH_2-CH-COO^-$ $^+NH_3$
Threonine	Thr	Yes	$CH_3CH-CH-COO^-$ $OH\ ^+NH_3$
Asparagine	Asn	No	$H_2N-\overset{O}{\overset{\|}{C}}-CH_2-CH-COO^-$ $^+NH_3$
Glutamine	Gln	No	$H_2N-\overset{O}{\overset{\|}{C}}-CH_2CH_2-CH-COO^-$ $^+NH_3$
Cysteine	Cys	No	$HS-CH_2-CH-COO^-$ $^+NH_3$

(continued)

*Amino acids essential to adult humans are shown in red.
†Essential to growing children but not to adult humans.

Table 15.3 ▍ (continued)

Tyrosine	Tyr	No	$HO-\langle\bigcirc\rangle-CH_2-CH-COO^-$ with $^+NH_3$
Tryptophan	Trp	Yes	indole ring $-CH_2-CH-COO^-$ with $^+NH_3$
Lysine	Lys	Yes	$H_3\overset{+}{N}CH_2CH_2CH_2CH_2-CH-COO^-$ with NH_2
Arginine	Arg	___*	$H_2N-C(=\overset{+}{N}H_2)-NHCH_2CH_2CH_2-CH-COO^-$ with NH_2
Histidine	His	___†	imidazole $-CH_2-CH-COO^-$ with $^+NH_3$
Aspartic acid	Asp	No	$HOOC-CH_2-CH-COO^-$ with $^+NH_3$
Glutamic acid	Glu	No	$HOOC-CH_2CH_2-CH-COO^-$ with $^+NH_3$

When only two amino acids are joined, the product is a *dipeptide*.

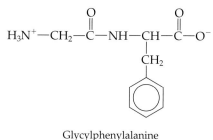

Glycylphenylalanine
(a dipeptide)

When three amino acids are combined, the substance is a *tripeptide*.

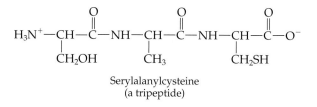

Serylalanylcysteine
(a tripeptide)

In describing peptides and proteins, scientists find it simpler to indicate the amino acids in a chain by using abbreviations (Table 15.3). Thus, glycylphenylalanine is written Gly-Phe, and serylalanylcysteine is written Ser-Ala-Cys. Example 15.1 further illustrates how these substances are named.

There is an alternate single-letter set of abbreviations for amino acids. Many, but not all, are simply the initial letters of the name. Using these abbreviations, the pentapeptide Met-Gly-Phe-Ala-Cys is simply MGFAC. Those who are interested may consult Reference 7, pages 570–571.

Example 15.1

Name the pentapeptide Met-Gly-Phe-Ala-Cys. You may use Table 15.3.

Solution

The endings of the names for all except the last amino acid are changed from -*ine* to -*yl*. The name is therefore methionylglycylphenylalanylalanylcysteine.

Exercise 15.1

A. Name the tetrapeptide His-Pro-Val-Ala.
B. Use abbreviations to indicate the amino acids in the peptide threonylglycyl-alanylalanylleucine.

Some protein molecules are enormous, with molecular weights in the tens of thousands. The molecular formula for hemoglobin, the oxygen-carrying protein in red blood cells, is $C_{3032}H_{4816}O_{780}N_{780}S_8Fe_4$, corresponding to a molar mass of 64,450 u. Although they are huge compared with ordinary molecules, a billion average-sized protein molecules could still fit on the head of a pin.

The end of a protein molecule with a free carboxyl group ($-COOH$) is called the C-terminal end. The end with a free amino group ($-NH_2$) is designated the N-terminal end.

The primary structure of angiotensin II, a peptide produced in the kidneys that causes powerful constriction of blood vessels, is

Asp-Arg-Val-Tyr-Ile-His-Pro-Phe

This sequence specifies an octapeptide with aspartic acid at the N-terminal, joined to arginine, valine, tyrosine, isoleucine, histidine, and proline and ending with phenylalanine at the C-terminal.

 [Web Reference 3] Using insulin as an example, this site gives a very good idea of the different levels of protein structure.

A molecule with more than ten amino acid units is often simply called a **polypeptide**. When the molecular weight of a polypeptide exceeds 10,000, it is called a **protein**. The distinction between polypeptides and proteins is an arbitrary one, and it is not always precisely applied.

The Sequence of Amino Acids

For peptides and proteins to function properly, it is not enough that they incorporate certain amounts of specific amino acids. The order or *sequence* in which the amino acids are connected is also of critical importance. Glycylalanine is different from alanylglycine, for example. Although the difference seems minor, the two substances behave differently in the body. The sequence for glycylalanine is written Gly-Ala, and that for alanylglycine is Ala-Gly. It is understood from this shorthand that the peptide is arranged with the free amino group (N-terminal) to the left and the free carboxyl group (C-terminal) to the right.

As the length of a peptide chain increases, the number of possible sequential variations becomes very large. Just as we can make millions of different words with our 26-letter English alphabet, we can make millions of different proteins with 20 amino acids. Just as one can write gibberish with the English alphabet, one can make nonfunctioning proteins by putting together the wrong sequence of amino acids.

Although the correct sequence is ordinarily of utmost importance, it is not always absolutely required. Just as you can sometimes make sense of incorrectly spelled English words, a protein with a small percentage of "incorrect" amino acids may continue to function. It may not function as well, however. And sometimes a seemingly minor change can have a disastrous effect. Some people have hemoglobin with 1 incorrect amino acid unit out of about 300. That "minor" error is responsible for sickle cell anemia, an inherited condition that ordinarily proves fatal.

15.7 Structure of Proteins

The structure of proteins has four organizational levels. The **primary structure** of a protein molecule is its amino acid sequence. To specify the primary structure, one merely writes out the sequence of amino acids in the long-chain molecule. The primary structure is held together by the peptide links between the amino acid units.

Molecules of proteins aren't just arranged at random as tangled threads. The chains are held together in unique configurations. The **secondary structure** of a protein refers to the arrangement of chains about an axis. Common arrangements are a *pleated sheet*, as in silk (Figure 15.12), and a *helix*, as in wool (Figure 15.13).

In the pleated sheet conformation, protein chains exist in an extended zigzag arrangement. The molecules are stacked in extended arrays, with hydrogen bonds

Interchain hydrogen bonds

Top view

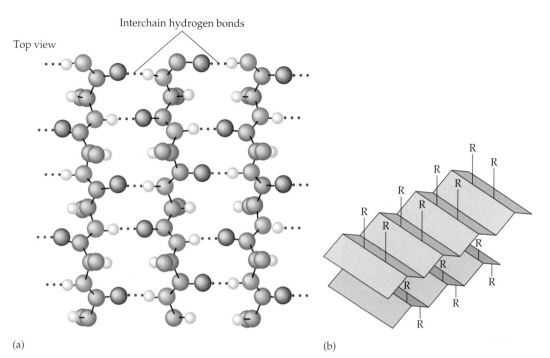

(a)

(b)

Figure 15.12 Pleated sheet conformation of protein chains. (a) Ball-and-stick model. (b) Schematic drawing emphasizing the pleats. The peptide bonds lie in the plane of the pleated sheet. The side chains extend above or below the sheet and alternate along the chain. The protein chains are held together by interchain hydrogen bonds.

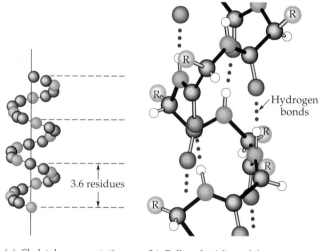

(a) Skeletal representation   (b) Ball-and-stick model

Figure 15.13 Two representations of the α-helical conformation of a protein chain. (a) The skeletal representation best shows the helix. (b) Intrachain hydrogen bonding between turns of the helix is shown in the ball-and-stick model.

holding adjacent chains together. The appearance gives this type of secondary structure its name, the **pleated sheet** arrangement. This structure, with its multitude of hydrogen bonds, makes silk strong and flexible.

The protein molecules in wool, hair, and muscle contain large segments arranged in the form of a right-handed helix, or **alpha helix**. Each turn of the helix requires 3.6 amino acid units. The N—H groups in one turn form hydrogen bonds to carbonyl (C=O) groups in the next turn. These helices wrap around one another in threes or sevens like strands of a rope. Unlike silk, wool can be stretched, much as we can stretch a spring by pulling the coils apart.

We can relate three levels of protein structure to a more familiar object, a telephone cord. Think of the coiled cord that connects the handset to the base of the instrument. The cord starts out as a long, straight wire (primary structure). The wire is coiled into a helical arrangement (secondary structure). When the handset is hung up, the coiled cord folds in a particular pattern (tertiary structure).

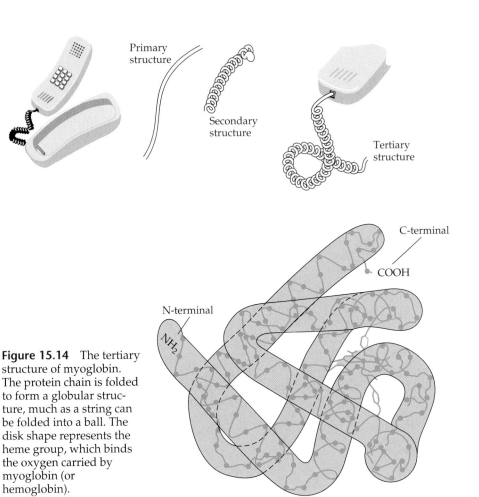

Figure 15.14 The tertiary structure of myoglobin. The protein chain is folded to form a globular structure, much as a string can be folded into a ball. The disk shape represents the heme group, which binds the oxygen carried by myoglobin (or hemoglobin).

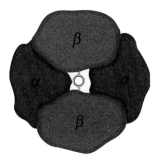

Figure 15.15 The quaternary structure of hemoglobin has four coiled chains, each analogous to myoglobin (Figure 15.14), stacked in a nearly tetrahedral arrangement.

[Web Reference 4] This site presents a comprehensive article on protein folding.

We introduced ionic and covalent bonds in Chapter 5, where we also discussed hydrogen bonds and dispersion forces as intermolecular forces. In proteins, all these forces can be intramolecular as well.

The **tertiary structure** of a protein refers to the spatial relationships of amino acid units that are relatively far apart in the protein chain. In describing tertiary structure, we frequently talk about how the molecule is folded. An example is the protein chain in **globular proteins**. Figure 15.14 shows the structure of myoglobin, which is folded into a compact, spherical shape.

Some proteins contain more than one polypeptide chain; that is, a protein molecule can be an aggregate of subunits. Hemoglobin is the most familiar example. A single hemoglobin molecule contains four polypeptide units, and each unit is roughly comparable to a myoglobin molecule. The four units are arranged in a specific pattern (Figure 15.15). When we describe the *quaternary structure* of hemoglobin, we describe the way the four units are stacked in the hemoglobin molecule.

Four Ways to Link Protein Chains

Peptide bonds fix the primary structure of proteins. What forces hold proteins in the structural arrangements we have referred to as secondary, tertiary, and quaternary? There are four kinds of forces—hydrogen bonds, ionic bonds, disulfide linkages, and dispersion forces. Dispersion forces are the only forces operating between nonpolar side chains. The various forces are illustrated in Figure 15.16.

Some amino acids have side chains that participate in *hydrogen bonding* (the hydroxyl group of serine is one example). Nonetheless, the hydrogen bonds of greatest importance are those that involve an interaction between the atoms of one peptide bond and those of another. Thus, the amide hydrogen (N—H) of one peptide link

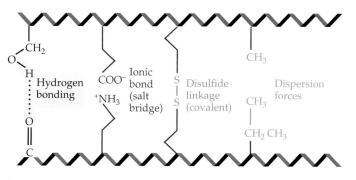

Figure 15.16 The tertiary structure of proteins is maintained by four different types of interactions.

can form a hydrogen bond to a carbonyl ($C=O$) oxygen located (a) some distance away on the same chain, an arrangement that occurs in the secondary structure of wool (alpha helix), or (b) on an entirely different chain, as in the secondary structure of silk (pleated sheet). In either case, there is usually a pattern of such interactions. Because peptide links are regularly spaced along the chain, the repetition of hydrogen bonds is also regular, both intramolecularly (in wool) and intermolecularly (in silk).

Ionic bonds, sometimes called *salt bridges,* occur when an amino acid with a basic side chain appears opposite one with an acidic side chain. Proton transfer results in opposite charges, which then attract one another. These interactions can occur between relatively distant groups that happen to come in contact because of folding or coiling of a single chain. They also occur between chains.

A **disulfide linkage** is formed when two cysteine units (whether on the same chain or on two different chains) are oxidized. A disulfide bond is a covalent bond, and the strength of this bond is much greater than that of a hydrogen bond. The much larger number of hydrogen bonds compensates somewhat for the individual weakness of these bonds.

Still weaker are the *dispersion forces* between nonpolar side chains. These interactions can be important, however, when other types of interactions are missing or are minimized. They are made stronger by the cohesiveness of the water molecules surrounding the protein. Nonpolar side chains minimize their exposure to water by clustering together on the inside folds of the protein in close contact with one another (Figure 15.17). Dispersion forces become fairly significant in structures such

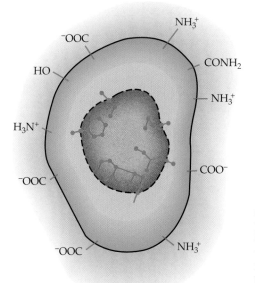

Figure 15.17 A protein chain often folds with nonpolar groups on the inside (red) where they are held together by dispersion forces. The outside has polar groups extending into the water (blue).

We have emphasized protein structure and function in this chapter. Proteins as foods are discussed in Chapter 16.

as that of silk, in which a high proportion of amino acids in the protein have nonpolar side chains.

15.8 Enzymes: Exquisite Precision Machines

A highly specialized class of proteins, **enzymes**, are biological catalysts produced by cells. They have enormous catalytic power and are ordinarily highly specific, catalyzing only one reaction or a closely related group of reactions. Nearly all known enzymes are proteins.

Enzymes enable reactions to occur at much more rapid rates and lower temperatures than they otherwise would. They do this by changing the reaction path. An enzyme reacts with a compound, called the **substrate**, and bonds to it. This newly formed complex then separates into products and the regenerated enzyme. Schematically, this can be written

$$\text{Enzyme } + \text{ Substrate} \longrightarrow \text{Enzyme–substrate complex} \rightleftharpoons \text{Enzyme } + \text{ Products}$$

The double arrows indicate that the steps are reversible; the enzyme catalyzes both the forward and reverse reactions.

The mechanism of enzyme action is a complex area of ongoing research, but it is often explained using the analogy of a lock and key. The substrate must fit a portion of the enzyme, called the **active site**, quite precisely, just as a key must fit a tumbler lock in order to open it (Figure 15.18). Not only must the enzyme and substrate fit precisely, but they are also likely to be held together by electrical attraction. This requires that certain charged groups on the enzyme complement certain charged or partially charged groups on the substrate. The formation of new bonds to the enzyme by the substrate weakens bonds within the substrate, and these weakened bonds can then be more easily broken to form products.

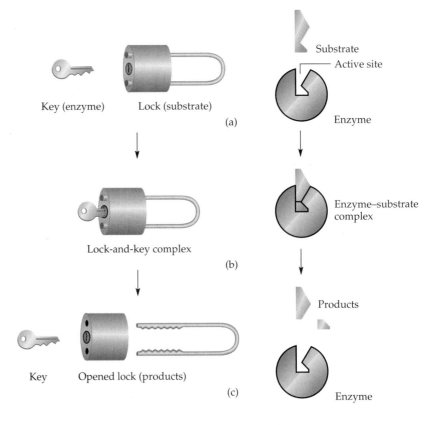

Figure 15.18 The lock-and-key model of enzyme action. (a) The substrate and the active site of the enzyme have complementary structures and bonding groups; they fit together as a key fits a lock. (b) The catalytic reaction occurs as the two are bonded together in the enzyme-substrate complex. (c) The products of the reaction leave the surface of the enzyme, freeing it to combine with another substrate molecule.

Portions of the enzyme molecule other than the active site may be involved in the catalytic process. Some interaction at a position remote from the active site can change the shape of the enzyme and thus change its effectiveness as a catalyst. In this way it is possible to slow or stop the catalytic action. Figure 15.19 offers a model for the inhibition of enzyme catalysis. An inhibitor molecule attaches to the enzyme at a position remote from the active site where the substrate is bound. The enzyme changes shape as it accommodates the inhibitor, and the substrate is no longer able to bind to the enzyme. This is one of the mechanisms by which cells "turn off" enzymes when their work is done.

Some enzymes consist entirely of protein chains. In others, another chemical component, called a **cofactor**, is necessary for proper function of the enzyme. This cofactor may be a metal ion such as zinc (Zn^{2+}), manganese (Mn^{2+}), magnesium (Mg^{2+}), iron(II) (Fe^{2+}), or copper(II) (Cu^{2+}). An organic cofactor is called a **coenzyme**. By definition, coenzymes are nonprotein. The pure protein part of an enzyme is called the **apoenzyme**. Both the coenzyme and the apoenzyme must be present for enzymatic activity to take place.

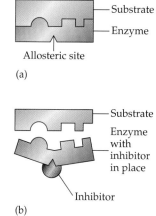

Figure 15.19 A model for the inhibition of an enzyme. (a) The enzyme and its substrate fit like a lock and key. (The allosteric site is where the inhibitor binds.) (b) With the inhibitor bound to the enzyme, the active site of the enzyme is distorted, and the enzyme cannot bind to its substrate.

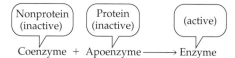

Many coenzymes are vitamins or are derived from vitamin molecules.

Clinical analysis for enzymes in body fluids or tissues is a common diagnostic technique in medicine. For example, blood levels of an enzyme involved in muscle metabolism rise during a heart attack in which the heart muscle has been damaged. However, many forms of strenuous (and healthful) physical activity also result in elevated blood levels of this enzyme. Despite problems of interpretation, analysis for specific enzymes is a valuable diagnostic tool.

Nucleic Acids: The Chemistry of Heredity

Nucleic acids, found in every living cell, serve as the information and control centers of the cell. There are actually two kinds of nucleic acids: *deoxyribonucleic acid (DNA)*, which serves as the blueprint for all the proteins of an organism, and *ribonucleic acid (RNA)*, which carries out protein assembly. DNA is found primarily in the cell nucleus, and RNA is found in all parts of the cell.

Some viruses have only RNA.

15.9 Nucleic Acids: Parts and Structure

Both DNA and RNA are long chains of repeating units called **nucleotides**. Each nucleotide in turn consists of three parts (Figure 15.20): a pentose sugar, a heterocyclic amine base, and a phosphate unit. The sugar is either ribose (in RNA) or deoxyribose (in DNA). Looking at the sugar in the nucleotide, note that the hydroxyl group on the first carbon atom is replaced by one of five bases. The bases with two fused rings, adenine and guanine, are classified as **purines**. Cytosine, thymine, and uracil have only one ring; and they are called **pyrimidines**.

The hydroxyl group on the fifth carbon of the sugar unit is converted to a phosphate ester group. Adenosine monophosphate (AMP) is a representative nucleotide. In AMP, the base (red) is adenine and the sugar (blue) is ribose.

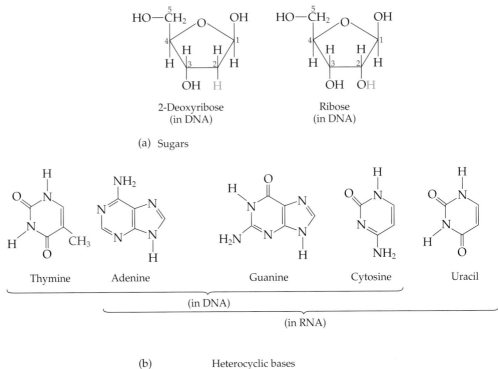

(a) Sugars

2-Deoxyribose
(in DNA)

Ribose
(in DNA)

Thymine　　Adenine　　　　　　Guanine　　　　　Cytosine　　　　Uracil

(in DNA)

(in RNA)

(b)　　　　　　Heterocyclic bases

(c)

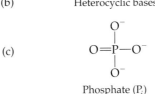

Phosphate (P_i)

Figure 15.20 The components of nucleic acids. (a) The sugars are 2-deoxyribose (found in DNA) and ribose (found in RNA). Note that deoxyribose differs from ribose in that it lacks an oxygen atom on the second carbon atom. "Deoxy" indicates that an oxygen atom is "missing." (b) Heterocyclic bases found in nucleic acids. Adenine, guanine, and cytosine are found in both DNA and RNA. Thymine occurs only in DNA, and uracil only in RNA. Note that thymine has a methyl group (red) that is lacking in uracil. (c) Phosphate units, often abbreviated as P_i, a compact designation of "inorganic phosphate."

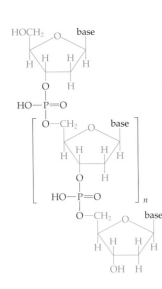

Figure 15.21 The backbone of a DNA molecule. The n indicates that the unit is repeated many times. Each repeating unit is composed of a base, phosphate unit, and sugar.

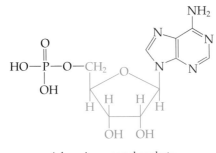

Adenosine monophosphate

Nucleotides are joined to one another through the phosphate group to form nucleic acid chains. The phosphate unit on one nucleotide forms an ester linkage to the hydroxyl group on the third carbon atom of the sugar unit in a second nucleotide. This unit is in turn joined to another nucleotide, and the process is repeated to build up a long nucleic acid chain (Figure 15.21). The backbone of the chain consists of alternating phosphate and sugar units, and the heterocyclic bases branch off this backbone.

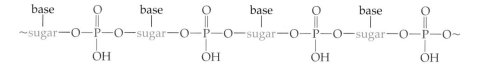

As we have seen, the sugar in DNA is deoxyribose, and that in RNA is ribose. The bases in DNA are adenine, guanine, cytosine, and thymine, and those in RNA are adenine, guanine, cytosine, and uracil.

Example 15.2

Identify the sugar and the base in the following nucleotide.

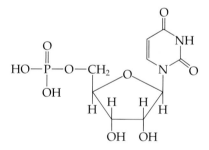

Solution

The sugar has two H atoms (no OH) on one C atom and therefore is deoxyribose. The base is a pyrimidine (one ring). The methyl group helps identify it as thymine.

Exercise 15.2

Identify the sugar and the base in the following nucleotide.

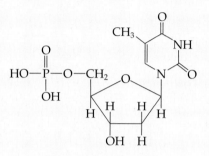

Base Sequence in Nucleic Acids

All the vast genetic information needed to build living organisms is stored in the sequence of the four bases along the nucleic acid strand. Not surprisingly, these molecules are huge, with molecular weights ranging into the billions for mammalian DNA. Along these great chains, the four bases can be arranged in essentially infinite variations. Before we examine this aspect of nucleic acid chemistry, let us consider one more important feature of nucleic acid structure.

The Double Helix

In an experiment designed to probe the structure of DNA, it was determined that the molar amount of adenine (A) in DNA corresponded to the molar amount of thymine (T). Similarly, the molar amount of guanine (G) is essentially the same as that of cytosine (C). To maintain this balance, the bases in DNA must be paired, A to T and

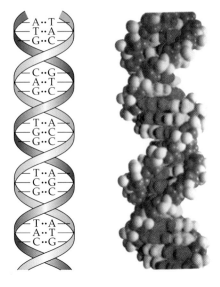

Figure 15.22 (a) A schematic representation of the double helix, showing the hydrogen bonds that connect the complementary base pairs. (b) A model of a portion of the DNA double helix.

(a)   (b)

James D. Watson (1928–) (right) and Francis H. C. Crick (1916–) (seated) proposed the double helix model of DNA in 1953. Their model was based on data obtained by British chemist Rosalind Franklin (1920–1958). Franklin had used a technique called X-ray diffraction to show DNA's helical structure. She also showed the location of phosphate sugars in DNA. In 1953, without her permission, her colleague Wilkins shared some of her work with Watson and Crick. These data, together with their own findings, allowed Watson and Crick to decipher DNA's structure. Four years after Franklin's death, Wilkins, Watson, and Crick were awarded the Nobel Prize for the discovery of DNA's structure. Franklin might have shared in this prize had she lived.

G to C. But how? At the midpoint of the twentieth century, it was quite clear that the answer to this question would bring with it a Nobel prize. Although many illustrious researchers worked on the problem, two relatively unknown scientists announced in 1953 that they had worked out the structure of DNA. Using data that involved quite sophisticated chemistry, physics, and mathematics, and working with models not unlike a child's construction set, James D. Watson and Francis H. C. Crick determined that DNA must be composed of two helices wound about one another. The phosphate and sugar backbone of the polymer chains form the outside of the structure, which is rather like a spiral staircase. The heterocyclic amines are paired on the inside—with guanine always opposite cytosine and adenine always opposite thymine. In our staircase analogy, these base pairs are the steps (Figure 15.22).

Why do the bases pair in this precise pattern, always A to T and T to A, and always G to C and C to G? The answer is hydrogen bonding and a truly elegant molecular design. Figure 15.23 shows the two sets of base pairs. You should notice two things. First, a pyrimidine is paired with a purine in each case, and the long dimensions of both pairs are identical (1.085 nm).

The second thing you should notice in Figure 15.23 is the hydrogen bonding between the bases in each pair. When guanine is paired with cytosine, three hydrogen bonds can be formed between the bases. No other pyrimidine–purine pairing permits such extensive interaction. Indeed, in the combination shown in the figure, both pairs of bases fit like locks and keys.

Other scientists around the world accepted the Watson–Crick structure almost immediately because it answers so many crucial questions. It explains how cells are able to divide and go on functioning, how genetic data are passed on to new generations, and even how proteins are built to required specifications. It all depends on the base pairing.

Structure of RNA

The molecules of RNA consist of single strands of the nucleic acid. Some internal (intramolecular) base pairing may occur in sections where the molecule folds back on itself. Portions of the molecule may exist in double-helical form (Figure 15.24).

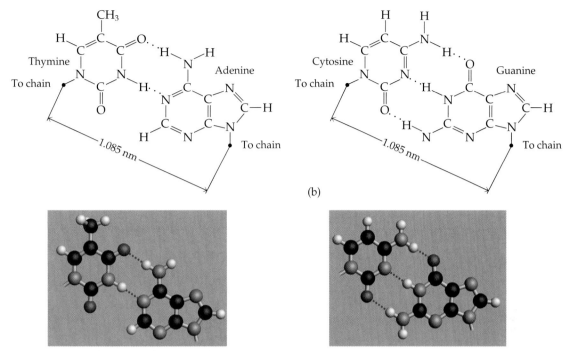

Figure 15.23   Pairing of the complementary bases thymine and adenine (a) and cytosine and guanine (b). The pairing involves hydrogen bonding, as in DNA.

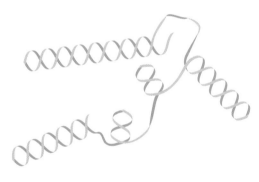

Figure 15.24   RNA occurs as single strands that can form double-helical portions by internal pairing of bases.

15.10 DNA: Self-Replication

Cats have kittens that grow up to be cats. Bears have cubs that grow up to be bears. How is it that each species reproduces its own kind? How does a fertilized egg "know" that it should develop into a kangaroo and not a koala?

The physical basis of heredity has been known for a long time. Most higher organisms reproduce sexually. A sperm cell from the male unites with an egg cell from the female. The fertilized egg so formed must carry all the information needed to make the various cells, tissues, and organs necessary for the functioning of a new individual. In addition, if the species is to survive, information must be passed along in germ cells—both sperms and eggs—for the production of new individuals.

Chromosomes and Genes

The hereditary material is found in the nuclei of all cells, concentrated in elongated, threadlike bodies called *chromosomes*. The number of chromosomes varies with the

Humans have 80,000 or more active genes. Scientists are now concluding an ambitious endeavor, called the *Human Genome Project*, to map the location of each of these genes on the chromosomes. (The *genome* of an organism is its complete set of genes.) They are also determining the sequence of nucleotides in all human DNA.

 [Web Reference 5] The website of the Human Genome Project.

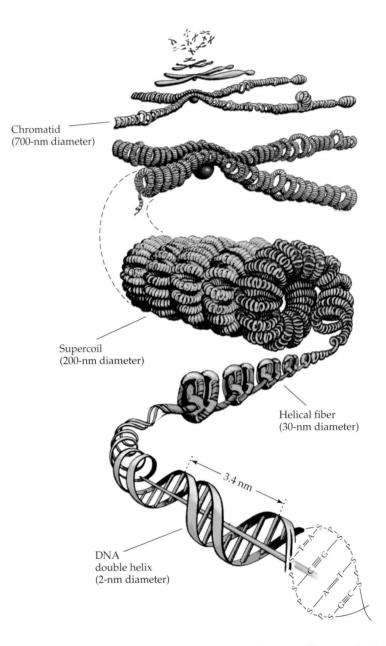

Chromatid
(700-nm diameter)

Supercoil
(200-nm diameter)

Helical fiber
(30-nm diameter)

3.4 nm

DNA
double helix
(2-nm diameter)

Figure 15.25 Chromosomes and DNA. Successive enlargements progress from chromosomes (top) to a portion of the DNA double helix (bottom). Each chromosome is made up of two chromatids united at the centromere. A chromatid is a protein-coated strand of multi-coiled DNA.

species. Human body cells have 46 chromosomes, and germ cells carry half that number. Thus, in sexual reproduction, the entire complement of chromosomes is achieved only when the egg and sperm combine; a new individual receives half its hereditary material from each parent.

Chromosomes are made of nucleic acids and proteins. The nucleic acid in chromosomes is DNA, the primary hereditary material (Figure 15.25). Arranged along the chromosomes are the basic units of heredity, genes. Structurally, a **gene** is a section of the DNA molecule (some viral genes contain only RNA). When cell division occurs, each chromosome produces an exact duplicate of itself. Transmission of genetic information therefore requires the **replication** (copying or duplication) of DNA molecules.

The Watson–Crick double helix provides a ready model for the process of replication. If the two chains of the double helix are pulled apart, each chain can direct the synthesis of a new DNA chain using the nucleotides available in the cellular fluid sur-

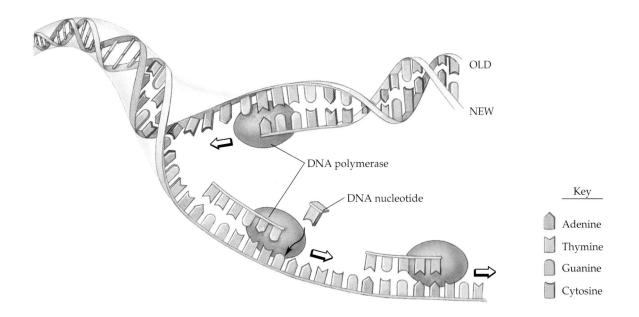

Figure 15.26 A schematic diagram of DNA replication. The original double helix "unzips" and the new nucleotides are brought into position by an enzyme (DNA polymerase). Phosphate bridges are formed, thus restoring the original double helix configuration. Each newly formed double helix consists of one old strand and one new strand. (This representation is simplified; the nucleotides are actually triphosphate derivatives.)

rounding the DNA. Each of the separating chains serves as a template, or pattern, for the formation of a new complementary chain. Synthesis begins with a base on a nucleotide pairing with its complementary base on the DNA strand—adenine with thymine and guanine with cytosine (Figure 15.26). Each base unit in the separated strand can pick up only a unit identical to the one it was paired with before.

Genes have functional regions called *exons* interspersed with inactive portions called *introns*. During protein synthesis, introns are snipped out and not translated.

As the nucleotides align, enzymes connect them to form the sugar–phosphate backbone of the new chain. In this way, each strand of the original DNA molecule forms a duplicate of its former partner. Whatever information was encoded in the original DNA double helix is now contained in each of the replicates. When the cell divides, each daughter cell gets one of the DNA molecules and all the information that was available to the parent cell.

How is information stored in DNA? The sequence of bases along the DNA chain encodes the directions for building all the proteins that comprise an organism and enable it to function. Just as "cat" means one thing in English and "act" means another, the sequence of bases CGT means one amino acid, and GCT means another. Although there are only four "letters"—the four bases—in the genetic codes of DNA, their sequence along the long strands can vary so widely that essentially unlimited information storage is available. Each cell carries in its DNA all the information it needs to determine all its hereditary characteristics.

15.11 RNA: Protein Synthesis and the Genetic Code

DNA carries a message that must somehow be relayed and acted on in a cell. Because DNA does not leave the cell nucleus, its information, or "blueprint," must be transported by something else, and it is. In the first step, called **transcription**, DNA transfers its information to a special RNA molecule called **messenger RNA (mRNA)**. The base sequence of DNA specifies the base sequence of mRNA. Thymine in DNA calls for adenine in mRNA, cytosine specifies guanine, guanine calls for cytosine, and adenine requires uracil (Table 15.4). Remember that in RNA molecules, uracil is used in place of DNA's thymine. Notice the similarity in the structure of these two bases (Figure 15.20).

Table 15.4 ▌ DNA Bases and Their Complementary RNA Bases

DNA Base	Complementary RNA Base
Adenine (A)	Uracil (U)
Thymine (T)	Adenine (A)
Cytosine (C)	Guanine (G)
Guanine (G)	Cytosine (C)

[Web Reference 6] This website shows the first complete images of a ribosome in action, first "caught in the act" of making proteins in 1999.

The next step in protein synthesis involves interpretation of the code that has been copied by mRNA and **translation** of this code into a protein structure. The mRNA travels from the nucleus to structures called *ribosomes* in the cytoplasm of the cell. Ribosomes also are constructed of nucleic acids and proteins. The mRNA becomes attached to a ribosome, and it is here that the genetic code is deciphered. Each mRNA molecule contains all the information necessary to make one particular protein.

There is another type of RNA molecule, called **transfer RNA (tRNA)**, located in the cytoplasm of cells. tRNA is responsible for translating the specific base sequence of mRNA into a specific amino acid sequence in a protein. (The function of proteins is dependent on the sequence of their amino acid building blocks.) The tRNA molecule has a looped structure (Figure 15.27) containing a critical set of three bases, called a **base triplet**, that pair with a set of three complementary bases on mRNA. This

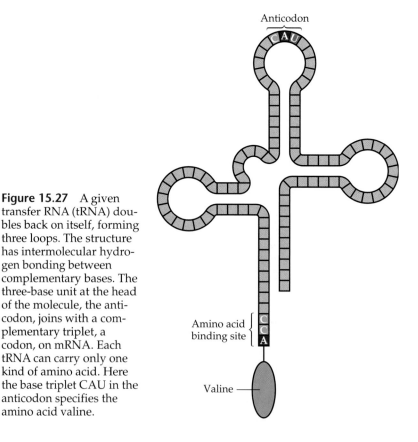

Figure 15.27 A given transfer RNA (tRNA) doubles back on itself, forming three loops. The structure has intermolecular hydrogen bonding between complementary bases. The three-base unit at the head of the molecule, the anticodon, joins with a complementary triplet, a codon, on mRNA. Each tRNA can carry only one kind of amino acid. Here the base triplet CAU in the anticodon specifies the amino acid valine.

Anticodon

CAU

Amino acid binding site

C
C
A

Valine

triplet determines which amino acid is carried at the tail of the tRNA. For example, the base triplet GUA on a segment of mRNA, called a **codon**, pairs with the base triplet CAU, called an **anticodon**, on a tRNA molecule. tRNA molecules with the anticodon CAU always carry the amino acid valine.

We can now describe in a little more detail how cells build proteins. Imagine that, after having been formed from a DNA molecule by transcription, an mRNA molecule migrates out from the nucleus into the cytoplasm and binds ribosomes along its chain. Suppose that a portion of the mRNA base sequence reads

$$\cdots C-G-C-G-G-A-G-G-C\cdots \quad \text{(mRNA strand)}$$

Floating around in the cytoplasm surrounding the mRNA are tRNA molecules, each carrying its own amino acid. The first codon of the mRNA segment, CGC, can pair, through hydrogen bonding, with a tRNA molecule having the anticodon GCG. This tRNA molecule carries the amino acid arginine. The triplets of bases on the tRNA and mRNA pair up.

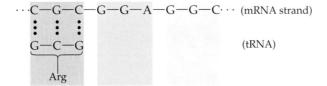

The next three bases of the mRNA, GGA, pair up with a tRNA molecule having the triplet CCU and carrying the amino acid glycine.

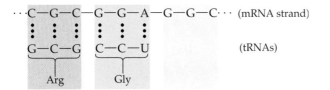

When the two tRNA molecules are appropriately lined up along the mRNA, a peptide (amide) bond forms between the two amino acids.

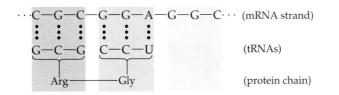

After the peptide bond is formed, the tRNA with the GCG triplet returns to the cytoplasm, and the next three bases on the mRNA (GGC) pair with a tRNA molecule having the anticodon CCG.

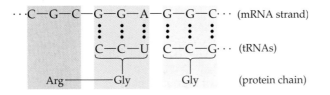

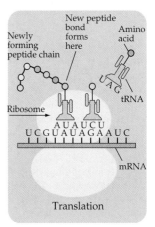

A ribosome (shaped like a figure 8) moves along on a mRNA molecule. tRNA molecules bind to the codon by base pairing. An amino acid borne by a tRNA molecule is linked to the growing protein chain.

Why is the genetic code based on a triplet of bases? The code must specify 20 different amino acids. Four "letters" can be arranged in $4 \times 4 \times 4 = 64$ possible three-letter "words." In a "doublet" code, four letters can be arranged in only $4 \times 4 = 16$ different words. In a "quadruplet" code, four letters can be arranged in $4 \times 4 \times 4 \times 4 = 256$ different ways. The triplet code can call for 20 different amino acids with some redundancy. A doublet code is inadequate, and a quadruplet code would provide for too much redundancy.

The third amino acid now bonds to the two previously linked.

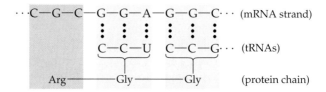

The protein chain gradually built up in this way is released from the tRNA and mRNA as it is formed.

A complete dictionary of the genetic code has now been compiled (Table 15.5). It shows which amino acids are specified by all the possible mRNA codons. There are 64 possible codons and only 20 amino acids, and so there is some redundancy in the code. Three amino acids (serine, arginine, and leucine) are each specified by six different codons. Two others (tryptophan and methionine) have only one codon each. Three codons, termed *stop signals*, call for termination of the protein chain. The codon AUG signals "start" as well as specifying methionine in the chain.

Table 15.5 ▌ The Genetic Code

		Second Base			
First Base		U	C	A	G	Third Base
U	U	UUU ⎱ Phe UUC ⎰ UUA ⎱ Leu UUG ⎰	UCU ⎱ UCC ⎰ Ser UCA UCG	UAU ⎱ Tyr UAC ⎰ UAA Termination UAG Termination	UGU ⎱ Cys UGC ⎰ UGA Termination UGG Trp	U C A G
	C	CUU ⎱ CUC ⎰ Leu CUA CUG	CCU ⎱ CCC ⎰ Pro CCA CCG	CAU ⎱ His CAC ⎰ CAA ⎱ Gln CAG ⎰	CGU ⎱ CGC ⎰ Arg CGA CGG	U C A G
	A	AUU ⎱ AUC ⎰ Ile AUA AUG Met	ACU ⎱ ACC ⎰ Thr ACA ACG	AAU ⎱ Asn AAC ⎰ AAA ⎱ Lys AAG ⎰	AGU ⎱ Ser AGC ⎰ AGA ⎱ Arg AGG ⎰	U C A G
	G	GUU ⎱ GUC ⎰ Val GUA GUG	GCU ⎱ GCC ⎰ Ala GCA GCG	GAU ⎱ Asp GAC ⎰ GAA ⎱ Glu GAG ⎰	GGU ⎱ GGC ⎰ Gly GGA GGG	U C A G

We can compare the genetic code to a book containing directions for putting together a model airplane or knitting a sweater. Knitting directions store information as words on paper. Letters of the alphabet are arranged in a certain way (for example, "knit one, purl two" or K1, P2), and these words direct the knitter to carry out a particular operation with needles and yarn. When all the directions are correctly followed, the ball of yarn becomes a sweater.

15.12 Genetic Engineering

Over 3000 human diseases have clear genetic components. With the completion of the Human Genome Project, researchers have located genes on the DNA molecule and have identified the function of many of them. The ability to use this information to diagnose and cure genetic diseases will revolutionize medicine.

Genetic Testing

The first chromosome to be sequenced was number 22, one of the smaller ones. It has 34,491,000 bases, 97% of which have been identified. These bases comprise 545 genes that code known proteins, and perhaps 325 more that code for proteins as yet unidentified. As many as 35 genetic diseases are thought to be associated with genes on chromosome 22. Isolating the one defective gene that causes a particular genetic disease is still not easy, but once all the genes are identified, it will be a less monumental task.

DNA is sequenced by using enzymes to cleave it into segments of about 500 nucleotides each. A technique called the **polymerase chain reaction (PCR)**[1] is now used to duplicate and amplify any specific DNA sequence in a test tube. PCR employs bacterial enzymes, called *DNA polymerases*, to multiply the small amount of DNA extracted from a sample into the large quantities needed for DNA sequencing. The technique is intricate and requires several chemical steps. In the final step, a "print" is represented as a series of horizontal bars resembling the bar codes imprinted on packaged goods sold in supermarkets.

Patterns of DNA fragments are characteristic of certain families and can be used to look for inherited diseases. If the DNA pattern of a relative resembles that of a person with a genetic disease closely enough, that relative will probably develop the disease. It is thus possible to identify and predict the occurrence of a genetic disease.

All 50 states and the District of Columbia screen newborn babies for genetic diseases. About 4 million infants are checked for metabolic and other disorders that could cause mental retardation, disability, or death. Many states screen for disorders such as phenylketonuria, galactosemia, tyrosinemia, homocystinuria, hypothyroidism, maple syrup urine disease, congenital adrenal hyperplasia, and sickle cell disease.

DNA Fingerprinting

In 1985 the British biologist Alec Jeffreys invented a technique for which he coined the term "DNA fingerprinting." Like fingerprints, each person's DNA is unique. Any cells—skin, blood, semen, saliva, and so on—can supply the necessary DNA sample.

DNA samples from evidence found at the crime scene are amplified by PCR and compared with the DNA obtained from a suspect. The DNA fingerprints are then compared to those of any suspects and of people known to have been at the scene.

DNA fingerprinting is a major advance in criminal investigation, and several hundred criminal cases have been solved with this technology. Also, because children inherit half their DNA from each parent, DNA fingerprinting has been used to establish the parentage of a child of contested origin. The odds in favor of being right in such cases are excellent—at least 100,000 : 1.

DNA fingerprinting has led to a number of criminal convictions, but perhaps more important, it can readily prove someone innocent. If the DNA does not match, the suspect could not have left the biological sample. About 70 people have been shown to be innocent and freed, some after spending years in prison. The technique is a major advance in the search for justice.

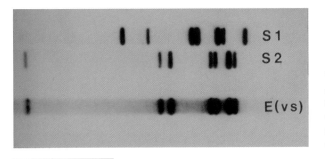

This set compares DNA segments of two suspects, S 1 and S 2, to DNA evidence (E) collected at the scene of a crime. Which one matches?

[1]For a concise description of the polymerase chain reaction, see Reference 7, page 644.

Recombinant DNA: Using Organisms as Chemical Factories

All living organisms (except some viruses) have DNA as their hereditary material. It should be possible, then, to place a gene from one organism into the genetic material of another. Recombinant DNA technology does just that.

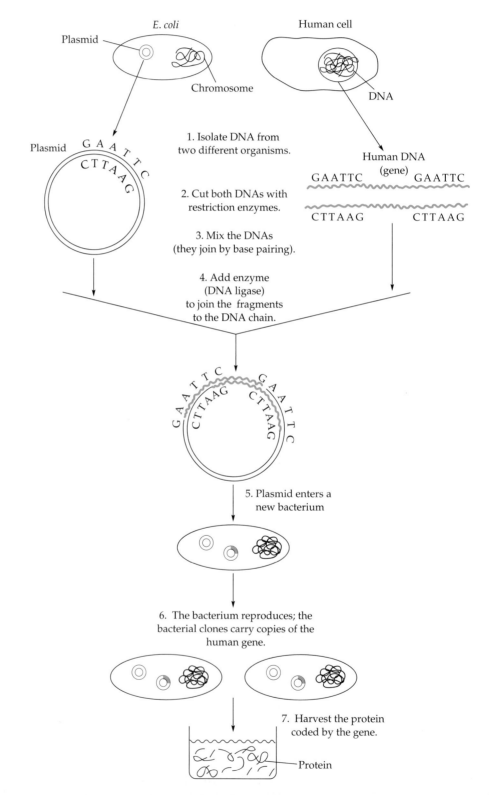

Figure 15.28 Recombinant DNA: cloning a human gene into a bacterial plasmid. The bacteria multiply, and in doing so, make multiple copies of the human gene. The gene can be one that codes for insulin, human growth hormone, or another valuable protein. Potentially, a gene from any organism—microbe, plant, or animal—can be incorporated into any other organism.

Scientists can work backward from the amino acid sequence of a particular protein to determine the base sequence of the gene that codes for that protein. After isolation and amplification by PCR, the gene is spliced into a special kind of bacterial DNA called a *plasmid*. The recombined plasmid is then inserted into the host organism (Figure 15.28). Once inside the host, the plasmid replicates, making multiple exact copies (clones) of itself. As the engineered bacteria multiply, they become effective factories for producing the desired protein.

Insulin, a protein coded by DNA, is required for the proper use of glucose by cells. People with diabetes, an insulin-deficiency disease, formerly had to use insulin from pigs or cattle. Now human insulin is made using recombinant DNA technology. Scientists take the human gene for insulin production and paste it into the DNA of *Escherichia coli*, a bacterium commonly found in the human digestive tract. The bacterial cells multiply rapidly, making billions of copies of themselves, and each new *E. coli* cell carries in its DNA a replica of the gene for human insulin. All newly diagnosed insulin-dependent diabetics in the United States are now treated with human insulin made by recombinant DNA technology. The hope for the future is that a functioning gene for insulin can be incorporated directly into the cells of insulin-dependent diabetics.

In addition to insulin, many other valuable materials difficult to obtain in any other way are now made using recombinant DNA technology. Human growth hormone, used to treat children who fail to grow properly, was formerly available only in tiny amounts obtained from cadavers. Now it can be readily obtained through recombinant DNA technology. This technology also yields interferon, a promising anticancer agent. The gene for epidermal growth factor, which stimulates the growth of skin cells, has also been cloned and is used to speed the healing of burns and other skin wounds.

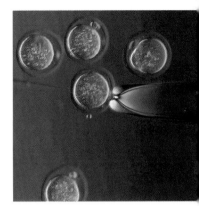

Under a powerful microscope, a tiny pipet is used to introduce foreign genetic material into animal cells.

Gene Therapy

Gene therapy involves introducing a functioning gene into a person's cells to correct the action of a defective gene. Viruses are commonly used to carry DNA into cells. The first human gene transplant, a demonstration project to prove that the technique worked, was carried out in 1989 but was not used to treat disease. The first human gene therapy was attempted in 1990. Gene therapy has been employed successfully to treat a form of hemophilia. Doctors injected patients with a gene that produced a blood-clotting protein. A preliminary but promising study used gene therapy to revive failing heart cells.

The stakes are huge. We all carry some defective genes. About one in ten people either has or will develop an inherited genetic disorder. Gene therapy could become an efficient way of curing disease. Instead of receiving debilitating chemotherapy or radiation for cancer, replacement of defective genes could permanently cure cancer in a relatively painless fashion.

To date, several thousand people have been treated with gene therapy. However, the death of a gene-therapy patient in 1999, associated with the virus used to transport the gene, introduced a sobering note of caution to researchers.

Controversy and Promise

Concern about the potential for disaster in this type of research has lessened somewhat in recent years. Initially, scientists worried about the possibility of producing a deadly "artificial" organism. What if a gene that causes cancer was spliced into the DNA of a bacterium that normally inhabits our intestine? To protect against such a development, strict guidelines for recombinant DNA research have been instituted.

Genes can now be transferred from nearly any organism to any other. Transgenic cows can produce proteins for the treatment of human ailments. New animals can be cloned from cells in adult animals. Genes for herbicide resistance are now incorporated into soybeans so that herbicide application will kill the weeds without killing the soybean plants. People worry that these genes will escape into wild plants (weeds), making them too resistant to herbicides. Genes for bacterial toxins are now incorporated into corn plants to kill caterpillars that would otherwise feed on the plants.

People worry that these toxins will hasten the time when insects will become resistant to the bacterial toxin and butterflies and other desirable insects will also be killed.

Molecular genetics has already resulted in some impressive achievements. Its possibilities are mind-boggling—elimination of genetic defects, a cure for cancer, and who knows what else? Knowledge gives power, but it does not necessarily give wisdom. Who will decide what sort of creature the human species should be? The greatest problem we are likely to face in our use of bioengineering is choosing who is to play God with the new "secret of life."

Critical Thinking Exercises

Apply knowledge that you have gained in this chapter and one or more of the FLaReS principles (Chapter 1) to evaluate the following statements or claims.

15.1 In a paternity case, a woman states that a certain man is the father of her child and that DNA fingerprinting will prove her case.

15.2 DNA is an essential component of every living cell. An advertisement claims that DNA is a useful diet supplement.

15.3 A man convicted of rape claims he is innocent because the DNA fingerprinting of a semen sample shows that it contains DNA from another man.

Summary

1. The human body is made up of about 100 trillion cells, each with its own nucleus, mitochondria, ribosomes, and other parts.

2. Cell metabolism includes many processes of both anabolism (building up of molecules) and catabolism (breaking down of molecules).

3. The three basic foodstuffs are carbohydrates, fats, and proteins.

4. Sugars are simple carbohydrates that have a sweet taste.

5. Starches are polymers of glucose held together by alpha linkages; cellulose is a glucose polymer held together by beta linkages.

6. Lipids include solid fats and liquid oils, which are both triglycerides, esters of fatty acids with glycerol.

7. Iodine number measures the degree of unsaturation (double bonds) in a fat or oil. Oils tend to have more, and fats fewer, double bonds.

8. Proteins are polymers of amino acids, with 20 amino acids forming thousands of different proteins.

9. The primary structure of a protein is its sequence of amino acids, which are held together by peptide bonds (amide linkages).

10. The secondary structure of a protein is its conformation (for example, alpha helix or pleated sheet), held in place by hydrogen bonds.

11. The tertiary structure of a protein is its folding pattern, determined by the formation of hydrogen bonds, ionic bonds, disulfide linkages, and dispersion forces.

12. An enzyme is a biological catalyst made up of an apoenzyme (protein) and a coenzyme (often a vitamin) and sometimes containing an inorganic cofactor (such as a metal ion).

13. Nucleic acids are nucleotide polymers.

14. A nucleotide is a compound made from phosphoric acid, a pentose sugar, and a nitrogen base (purine or pyrimidine).

15. The nitrogen bases in nucleic acids are adenine and guanine (purines) and thymine, cytosine, and uracil (pyrimidines).

16. DNA (deoxyribonucleic acid) is a nucleic acid in which the pentose sugar is deoxyribose and the nitrogen bases are adenine, thymine, guanine, and cytosine.

17. RNA (ribonucleic acid) is a nucleic acid in which the pentose sugar is ribose and the nitrogen bases are the same as those in DNA, except that uracil replaces thymine.

18. The sequence of nitrogen bases on the sugar–phosphate stem carries the message on the DNA molecule, with each triplet of bases acting as the code for an amino acid.

19. Nitrogen bases always pair up the same way, guanine with cytosine and adenine with thymine (in DNA) or with uracil (in RNA).

20. DNA has a double-helical structure that can unwind during replication or transcription to RNA so that a new strand of nucleic acid can be produced.

21. RNA carries out protein synthesis, with mRNA carrying the blueprint from DNA to the ribosomes and tRNA delivering amino acids to the growing protein chain.

22. Each base triplet on mRNA is a codon for a specific amino acid. The base triplets on tRNA are called anticodons.

23. Through genetic engineering, using recombinant DNA technology, it may be possible someday to eliminate hundreds of hereditary diseases.

Key Terms

active site   15.8
alpha helix   15.7
amino acid   15.5
anabolism   15.2
anticodon   15.11
apoenzyme   15.8
base triplet   15.11
biochemistry   15.1
carbohydrate   15.3
catabolism   15.2
codon   15.11
coenzyme   15.8

cofactor   15.8
disaccharide   15.3
disulfide linkage   15.7
enzyme   15.8
fat   15.4
fatty acid   15.4
gene   15.10
globular protein   15.7
iodine number   15.4
lipid   15.4
messenger RNA
   (mRNA)   15.11

metabolism   15.2
monosaccharide   15.3
nucleic acid   15.8
nucleotide   15.9
oils   15.4
peptide bond   15.6
pleated sheet   15.7
polymerase chain reaction
   (PCR)   15.12
polypeptide   15.6
polysaccharide   15.3
primary structure   15.7

protein   15.6
purine   15.9
pyrimidine   15.9
replication   15.10
secondary structure   15.7
substrate   15.8
tertiary structure   15.7
transcription   15.11
transfer RNA
   (tRNA)   15.11
translation   15.11
zwitterion   15.5

Review Questions

1. What is biochemistry?

2. Briefly identify and state a function of each of the following parts of a cell.
   a. cell membrane   b. cell nucleus
   c. chloroplasts   d. mitochondria
   e. ribosomes

3. How do green plants obtain food?

4. What are the three major types of foods from which animals obtain energy?

5. Define each of the following.
   a. anabolism   b. catabolism
   c. metabolism

6. What are carbohydrates?

7. What is glycogen? How does it differ from amylose? From amylopectin?

8. What is a lipid? Name at least three kinds of lipids.

9. What are fatty acids?

10. To what class of compounds do fats belong?

11. Define monoglyceride, diglyceride, and triglyceride.

12. List some of the properties of fats and oils. How do fats and oils differ in structure? In properties?

13. What does the iodine number of a fat mean? In the determination of the iodine number of a fat, what part of the fat molecule reacts with the reagent?

14. Define saturated fat. Define polyunsaturated fat (oil).

15. In what parts of the body are proteins found? What tissues are largely proteins?

16. What is the chemical nature of proteins? How many different amino acids are incorporated in proteins by the genetic code?

17. How does the elemental composition of proteins differ from those of carbohydrates and fats?

18. What functional groups are found on amino acid molecules? What is a zwitterion?

19. What is a peptide bond?

20. What is a dipeptide? A tripeptide? A polypeptide?

21. Of what importance is the sequence of amino acids in a protein molecule?

22. What is the difference between a polypeptide and a protein?

23. What is the primary structure of a protein? The secondary structure?

24. What is the tertiary structure of a protein? The quaternary structure?

25. Describe the pleated sheet arrangement in a protein.

26. Describe the alpha helix arrangement in a protein.

27. What is a globular protein?

28. List the four different ways protein chains are bonded to one another.

29. Describe the lock-and-key model of enzyme action.

30. Describe how an inhibitor deactivates an enzyme.

31. Name the two kinds of nucleic acids. Which is found primarily in the nucleus of the cell?

32. What is the sugar unit in RNA? What sugar is found in DNA?

33. List the heterocyclic bases found in DNA. List those found in RNA.

34. How do DNA and RNA differ in structure?

35. What kind of intermolecular force is involved in base pairing?

36. Describe the process of replication.

37. In replication, a parent DNA molecule produces two daughter molecules. What is the fate of each strand of the parent DNA double helix?

38. We say DNA controls protein synthesis, yet most DNA resides within the cell nucleus and protein synthesis occurs outside the nucleus. How does DNA exercise its control?

39. Explain the role of (**a**) mRNA, and (**b**) tRNA in protein synthesis.

40. Which nucleic acid or acids are involved in (**a**) the process referred to as transcription, and (**b**) the process referred to as translation?

41. Which nucleic acid contains the codon? Which contains the anticodon?

42. What is the relationship among the cell parts called chromosomes, the units of heredity called genes, and the nucleic acid DNA?

43. What is the polymerase chain reaction (PCR)?

44. How do DNA fragment patterns indicate whether a person will probably develop a genetic disease?

45. List the four steps in recombinant DNA technology.

46. What is a plasmid?

47. How can a virus be used to replace a defective gene in a human?

48. Discuss some applications of genetic engineering.

Problems

Carbohydrates

49. In what way are amylose and cellulose similar? What is the main structural difference between starch and cellulose?

50. What are the two kinds of plant starch? How do they differ?

51. Which of the following are monosaccharides?
    a. amylose
    b. cellulose
    c. fructose
    d. galactose

52. Which of the following are monosaccharides?
    a. glucose
    b. glycogen
    c. lactose
    d. sucrose

53. Which of the carbohydrates in Problems 51 and 52 are disaccharides?

54. Which of the carbohydrates in Problems 51 and 52 are polysaccharides?

55. What are the hydrolysis products of each of the following?
    a. amylose
    b. lactose
    c. glycogen

56. What are the hydrolysis products of each of the following?
    a. amylopectin
    b. sucrose
    c. cellulose

57. Give the formula for the open-chain form of glucose. What functional groups are present?

58. Give the formula for the open-chain form of fructose. What functional groups are present?

59. Give the formula for the open-chain form of galactose. What functional groups are present?

60. Mannose differs from glucose only in that the H and OH on the second carbon atom are reversed. Give the formula for the open-chain form of mannose. What functional groups are present?

Lipids: Fats and Oils

61. Which of the following fatty acids are saturated? Which are unsaturated?
    a. linolenic acid
    b. linoleic acid
    c. oleic acid
    d. palmitic acid
    e. stearic acid

62. Which of the fatty acids in Problem 61 are monounsaturated? Which are polyunsaturated?

63. How many carbon atoms are there in a molecule of each of the following fatty acids?
    a. linolenic acid
    b. palmitic acid
    c. stearic acid

64. How many carbon atoms are there in a molecule of each of the following fatty acids?
    a. linoleic acid
    b. oleic acid
    c. butyric acid

65. Which would you expect to have a higher iodine number—corn oil or beef tallow? Explain your reasoning.

66. Which would you expect to have a higher iodine number—lard or liquid margarine? Explain your reasoning.

Proteins

67. Is the dipeptide represented by Ser-Ala the same as the one represented by Ala-Ser? Explain.

68. A chemist is asked to make Phe-Ile-Leu. He makes Leu-Ile-Phe. Evaluate the chemist for possible job advancement.

69. Give structural formulas for the following amino acids.
    a. alanine
    b. serine

70. Give structural formulas for the following amino acids.
    a. glycine
    b. phenylalanine

71. Give structural formulas for the following dipeptides.
    a. glycylalanine
    b. alanylserine

72. Give structural formulas for the following dipeptides.
    a. alanylglycine
    b. phenylalanylserine

Nucleic Acids

73. Which of the following are nucleotides?

a.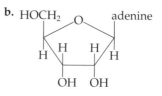

b.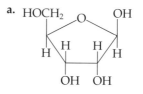

c.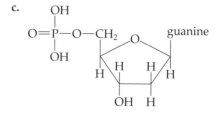

74. Which of the following are nucleotides?

a.

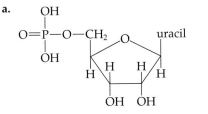

uracil

b.

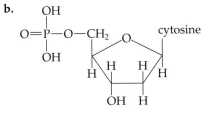

cytosine

c.

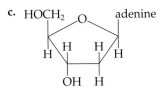

adenine

75. For each of the structures in Problem 73, identify the sugar unit as ribose or deoxyribose.

76. For each of the structures in Problem 74, identify the sugar unit as ribose or deoxyribose.

77. Identify the sugar and the base in the following nucleotide.

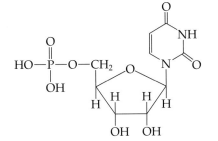

78. Identify the sugar and the base in the following nucleotide.

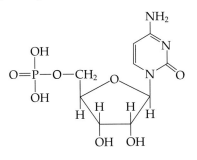

79. In DNA, which base would be paired with the base listed?
  a. cytosine   **b.** adenine
  c. guanine   **d.** thymine

80. In an RNA molecule, which base would pair with the base listed?
  a. adenine   **b.** guanine
  c. uracil   **d.** cytosine

81. The base sequence along one strand of DNA is AATTCG. What would be the sequence of the complementary strand of DNA?

82. What sequence of bases would appear in the mRNA molecule copied from the original DNA strand shown in Problem 81?

83. If the sequence of bases along a mRNA strand is UCC-GAU, what was the sequence along the DNA template?

84. What are the complementary triplets on tRNA for the following triplets on mRNA?
  a. UUU   **b.** CAU
  c. AGC   **d.** CCG

85. What are the complementary codons on mRNA for the following anticodons on tRNA molecules?
  a. UUG   **b.** GAA

86. What are the complementary codons on mRNA for the following anticodons on tRNA molecules?
  a. UCC   **b.** CAC

Additional Problems

87. Which of the following bases is purine and which is pyrimidine?

a.

b.

88. Answer the following questions for the molecule shown.

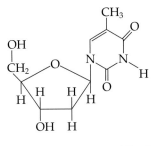

  a. Is the base a purine or a pyrimidine?
  b. Would the compound be incorporated in DNA or in RNA?

89. Answer the questions posed in Problem 88 for the following compound.

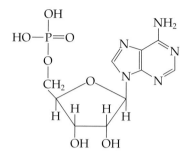

90. In the schematic below, Glc represents glucose. What substance is indicated?

. . . Glc-Glc
            \
. . . Glc-Glc-Glc-Glc
                     \
  . . . Glc-Glc-Glc-Glc-Glc-Glc-Glc-Glc-Glc
                                    \
     . . . Glc-Glc-Glc-Glc-Glc-Glc-Glc-Glc-Glc-Glc-Glc-Glc
                                                \
        . . . Glc-Glc-Glc-Glc-Glc-Glc-Glc

91. Which of the following is a lipid?

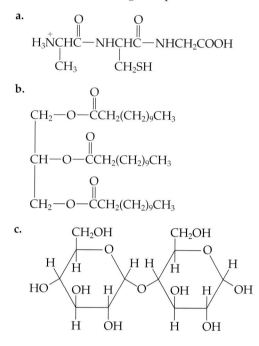

92. Write the abbreviated version of the following structural formula.

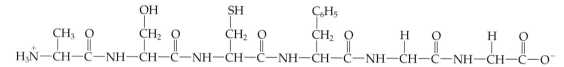

Online Project

93. Genetic testing carries both promise and peril and is an important area of bioethics today. After some online research, write a brief essay on one of the following—or on a similar topic assigned by your instructor. A good site to begin with is http://www.accessexcellence.org/AE/AEPC/NIH/index.html—Understanding Gene Testing, from the National Cancer Institute and the National Center for Human Genome Research.

a. In what cases might a person not want to be tested for a disease his or her parent had?

b. When is genetic testing most valuable? What privacy issues need to be addressed in this area?

c. Does testing negative for the breast cancer genes mean that a woman doesn't need mammograms?

References and Readings

1. Anderson, W. French. "Gene Therapy." *Scientific American*, September 1995, pp. 124–128.

2. Brennan, Mairen. "Deciphering the Human Code." *Chemical & Engineering News*, 6 December 1999, p. 9. Describes the sequencing of chromosome 22.

3. Cohen, Jack, and Ian Stewart. "Our Genes Aren't Us." *Discover*, April 1994, pp. 78–84. Discusses temperature, maternal hormones, and other factors that influence the expression of genes.

4. Copeland, Robert A. "Enzymes, the Catalysts of Life." *Today's Chemist at Work*, March 1994, pp. 51–53.

5. Diamond, Jared. "The Cruel Logic of Our Genes." *Discover*, November 1989, pp. 72–78. Discusses genetic diseases.

6. Gasser, Charles S., and Robert T. Fraley. "Transgenic Crops." *Scientific American*, June 1992, pp. 62–69.

7. Hill, John W., Stuart J. Baum, and Rhonda J. Scott-Ennis. *Chemistry and Life*, 6th edition. Upper Saddle River, NJ: Prentice Hall, 2000. Chapters 19 through 28 discuss biochemical topics.

8. Lee, Kenneth B., and Lily S. Hu. "Biotechnology: Past, Present, and Future." *Chemistry & Industry*, 6 May 1996, pp. 334–338.

9. Roberts, Leslie. "The Gene Hunters." *U.S. News & World Report*, 3–10 January 2000, pp. 34–44. Includes attractive graphics in a centerfold primer.

10. Trefil, James. "How the Body Defends Itself from the Risky Business of Living." *Smithsonian*, December 1995, pp. 42–49. Describes how special enzymes repair defective or damaged DNA.

11. Vile, Richard G. "Gene Therapy for Treating Cancer—Hope or Hype?" *Chemistry & Industry*, 15 April 1996, pp. 285–289. Part of a special issue on gene therapy.

12. Watson, James D. *The Double Helix*. New York: New American Library, 1968. Watson's personal account of the discovery of the structure of DNA.

MediaLab

Genetic Recombination: Promise or Peril?

Genetic recombination has become one of the controversies of the new century. In universities and industries, research leaps ahead—determining the sequence and meaning of the human genome, refining cloning techniques, and experimenting with genetic modification of plants and animals. Meanwhile, the public is growing more and more wary. The technique of DNA recombination has many applications, from the production of insulin and other vital polypeptides by bacteria with inserted human genes to attempts to replace defective genes in people. The production of medicines is probably the least controversial of the applications. Leading to greater, often heated, dispute are technologies involving genetically altered plants and animals and employing gene therapy.

In this *Web Investigation*, you will look at several applications of recombinant DNA techniques. You will examine the attitudes of the government, the public, and bioethicists toward several of these biotechnology areas. In *Communicate Your Results*, you will explain genetic engineering and write about its controversial aspects.

Investigation 1
Gene Therapy:
A Promise Not Yet Fulfilled

Many diseases are hereditary, and still others have a genetic component. Many genetic diseases, such as Huntington's disease, are truly heartbreaking—and inescapable for the person inheriting the gene. Thus, great excitement has accompanied each report of apparently successful gene therapy trials. However, though the concept of injecting replacement genes seems straightforward, many obstacles lie in the way, including the body's own defenses against intruders. Select Keyword **THERAPY** for a look at early successes and disappointments. This site also has a nice animated illustration of the technique. Select Keyword **NEWS** for current news and bioethical debates. (Read what seems interesting and relevant at this site.)

Investigation 2
A New Gene in My Food?

Some of the fiercest reactions from the public, particularly in Europe, have

arisen in response to genetic engineering of various foods. For centuries, people have improved plants and animals by selective breeding. Sometimes, however, breeding for one trait has diminished another desirable attribute. Genetic engineering, however, speeds up the process and enables the insertion of totally new genes, such as genes for selective pest resistance. Select Keyword **FARMING** to read about the benefits of biotechnology in farming and to explore the concerns some people have about it. For an interesting account of the use of gene transfer to protect plants against pests, select Keyword **PLANTS**.

Investigation 3
Cloning: Techniques and Ethics

Among the most controversial topics in genetic engineering is that of cloning. Since Dolly the sheep was created in 1997 by the nuclear replacement technique, interest has focused on the possibility of cloning human beings, and governments around the world have attempted to legislate limits. For a clear but lengthy description of the cloning technique, select Keyword **CLONING**. This British document discusses potential benefits and perils of cloning and related genetic research and technology. It also includes a glossary of genetic engineering terms and a brief overview of national laws concerning

Dolly and Her Surrogate Mother

human cloning. If you wish to see a list of U. S. federal bills, with links to transcripts, select Keyword **BILLS**.

COMMUNICATE YOUR RESULTS

Exercise 1
Risk–Benefit Ratio

Little controversy exists over the use of recombinant methods to produce human insulin, growth hormone, and other essential medications. Is this because people realize that in these cases the benefits of the drugs are worth any perceived risk from genetic engineering? List some benefits of current recombinant DNA techniques, primarily as used in human medicine, and discuss any accompanying risks in terms of these benefits.

Exercise 2
Cloning: Benefits and Pitfalls

Looking at *Web Investigation 3*, discuss the potential benefits and ethical pitfalls of genetic engineering in general and vertebrate cloning in particular. For this discussion, concentrate on techniques used in animals. What are some potential benefits to agriculture? Are these worth less risk than medical applications? Should the risks dictate a "go slow" approach or lead to an outright ban of some applications?

Exercise 3
Bioethics and the Law

In the not-so-distant past, scientists generally regulated themselves, enforcing intellectual integrity by the peer review process. Sometimes they seem to have given little thought to the impact of their findings on society. With an accelerated pace of research and development, including genetic recombination, is this sufficient? Drawing on the *Web Investigations* and any other resources you wish to use, discuss the role of law in regulating the course of biotechnical endeavors. (You might evaluate the degree to which scientific research has moved from academic laboratories into the realm of big business. Does this make federal controls more essential?)

Exercise 4
It's in the Genes

To what degree is the fear of genetic engineering due to the fact that it involves the *genes*—the "stuff of life"—and the concern that any changes will be passed on to future generations? How much concern is due to the fact that many technological advances in the past (DDT, for example) had hidden costs? Write an op-ed article on this topic.

Food
Those Incredible Edible Chemicals

The biosphere of Earth is rich, and much of it is edible.
Our wide variety of foods is really quite incredible!

The food we eat supplies all the building blocks from which our bodies are made and all the energy for our life activities. That energy comes ultimately from the sun through the remarkable process of photosynthesis.

C hemistry has contributed much to the quality, variety, and abundance of our food supply.

FOOD! From holiday feasts to midnight snacks, so many of life's joys seem to involve food. Yet the prime purpose of food is not to give pleasure. It is to sustain life. Food supplies all the building blocks from which our bodies are made and all the energy for our life activities. That energy comes ultimately from the sun through the remarkable process of photosynthesis.

There are some places in the world where people never have enough food. They are always hungry. Meanwhile, other places have a surplus. In the Western world we have such an overabundance of food that we often eat too much. And frequently we eat the wrong kinds of food. It is estimated that more than 30% of the adults in the United States are overweight. Obesity contributes to poor health, increased risk of diabetes and heart attacks, and increased susceptibility to other diseases. Our diet (too rich in fat, sugar, alcohol, and cholesterol) has been linked to five of the ten leading causes of death in the United States.

The human body is a conglomeration of chemicals. If we did a chemical analysis of the human body, we would find that it is largely—about two-thirds—water, along with a number of minerals and thousands of other chemicals. Our foods are also made up of chemicals. They enable children to grow, they provide people with energy, and they supply the chemicals needed for the repair and replacement of body tissues.

In this chapter we discuss food—what it's made of and where it comes from.

Worldwide, four children die every minute from eating food contaminated with microorganisms, and another two die from drinking contaminated water. One of every eight people on Earth suffers from malnutrition severe enough to stunt physical and mental growth and to shorten life.

The Food We Eat

The three main classes of foods are carbohydrates, fats, and proteins. These, too, are chemicals. For proper nutrition, our diet should include balanced proportions of these three foodstuffs, plus water, vitamins, minerals, and fiber.

16.1 Carbohydrates: The Preferred Fuels

Dietary **carbohydrates** include sugars and starches. Sugars are monosaccharides and disaccharides; starches are polysaccharides.

Some of the chemistry of carbohydrates is discussed in Section 15.3.

Sweet Chemicals: Sugars

Sugars have been used for ages to make food sweeter. Two monosaccharides, *glucose* (or *dextrose*) and *fructose* (fruit sugar), and the disaccharide *sucrose* are the most common dietary sugars (Figure 16.1). Common table sugar is sucrose, usually obtained from sugarcane or sugar beets. Glucose is the sugar used by the cells of our bodies for energy. Because it is the sugar that circulates in the bloodstream, it is often called **blood sugar**. Fructose is found in honey and in some fruits, but much of it is made from glucose. Corn syrup, made from starch, is mainly glucose. *High-fructose corn syrup* is made by treating corn syrup with enzymes to convert much of the glucose to fructose. Fructose is sweeter than sucrose or glucose. Foods sweetened to the same degree with high-fructose corn syrup have somewhat fewer calories than those sweetened with sucrose.

During the three decades from 1968 to 1998, per-capita consumption of sugars in the United States increased from 11 kg/year to 69 kg/year, and it now exceeds 70 kg/year. Most are consumed in soft drinks, presweetened cereals, candy, and other highly processed foods with little or no other nutritive value. The sugars in sweetened foods provide empty calories and contribute heavily to tooth decay and obesity.

Glucose and fructose are isomers; both have the molecular formula $C_6H_{12}O_6$. Sucrose has the molecular formula $C_{12}H_{22}O_{11}$. Structures of these and other sugars are given in Section 15.3.

Figure 16.1 The principal sugars in our diet are sucrose (cane or beet sugar), glucose (corn syrup), and fructose (fruit sugar, often in the form of high-fructose corn syrup).

Hydolysis means "splitting by water." When we say that sucrose is hydrolyzed, we mean that it reacts with water and is broken down into glucose and fructose.

Digestion and Metabolism of Carbohydrates

Glucose and fructose are absorbed directly into the bloodstream from the digestive tract. Sucrose is hydrolyzed during digestion to glucose and fructose.

$$\text{Sucrose} + H_2O \longrightarrow \text{Glucose} + \text{Fructose}$$

The disaccharide *lactose* occurs in milk. During digestion, it is hydrolyzed to two simpler sugars, glucose and galactose.

$$\text{Lactose} + H_2O \longrightarrow \text{Glucose} + \text{Galactose}$$

Nearly all human babies have the enzyme necessary to accomplish this breakdown, but many adults do not. People who lack the enzyme get digestive upsets from drinking milk. This condition is called *lactose intolerance*. When milk is cooked or fermented, the lactose is at least partially hydrolyzed. People with lactose intolerance may still be able to enjoy cheese, yogurt, or cooked foods containing milk with little or no discomfort.

All monosaccharides are converted to glucose during metabolism. Some babies are born with *galactosemia*, a deficiency of the enzyme that catalyzes the conversion of galactose to glucose. For proper nutrition, they require a synthetic formula (Figure 16.2) in place of milk.

Figure 16.2 Infants born with galactosemia can thrive on a milk-free substitute formula.

Lactose-free milk, made by treating milk with an enzyme that hydrolyzes lactose, is available in most grocery stores. The enzyme itself also can be bought in many stores.

Complex Carbohydrates: Starch and Cellulose

Starch and cellulose are both polymers of glucose, but the connecting links between the glucose units are different. Humans can digest starch, but not cellulose. **Starch** is an important part of any balanced diet (Figure 16.3). **Cellulose** is an important component of dietary fiber.

Starch is hydrolyzed to glucose when it is digested. The body then metabolizes the glucose, using it as a source of energy. Glucose is broken down through a complex set of more than 50 chemical reactions to produce carbon dioxide and water, with the release of energy.

$$C_6H_{12}O_6 + 6 O_2 \longrightarrow 6 CO_2 + 6 H_2O + \text{Energy}$$

These reactions are essentially the reverse of photosynthesis. In this way, animal organisms are able to make use of the energy from the sun that was captured by plants in the process of photosynthesis.

Figure 16.3 Bread, flour, cereals, and pasta are rich in starches.

Carbohydrates supply 4 kcal of energy per gram. When we eat more than we can use, small amounts of carbohydrates can be stored in our livers and muscle tissues as **glycogen** (animal starch). Large excesses, however, are converted to fat for storage. There is no set dietary recommendation for carbohydrates, but at least 65%, and perhaps as much as 80%, of our calories should come from them—because these are our bodies' preferred fuels. The majority of our carbohydrate intake should be the complex carbohydrates found in cereal grains, rather than the simple sugars found in many prepared foods.

Cellulose

Cellulose is the most abundant carbohydrate. It is present in all plants, forming their cell walls and other structural features. Wood is about 50% cellulose, and cotton is almost pure cellulose.

Like starch, cellulose is a polymer of glucose. But, unlike starch, it cannot be digested by humans or any other meat-eating animals. We can eat cellulose, but we get no nutrient value from it. It is just *fiber*. Whereas the glucose units in starch molecules are joined by alpha-type bonds, cellulose molecules have beta bonds holding them together, and most animals lack the enzymes needed to break this beta glycoside bonding (Figure 15.6). Certain bacteria have such enzymes, however, and they are present in grazing animals, such as cows, and in termites, so that these animals can also convert cellulose to glucose.

16.2 Fats: Energy Reserves, Cholesterol, and Cardiovascular Disease

Fats are high-energy foods (Figure 16.4). They yield about 9 kcal of energy per gram. Some fats are "burned" as fuel for our activities. Others are used to build and maintain important constituents of our cells, such as cell membranes. The fat in our diet

Figure 16.4 Cream, butter, margarine, cooking oils, and foods fried in fat are rich in fats.

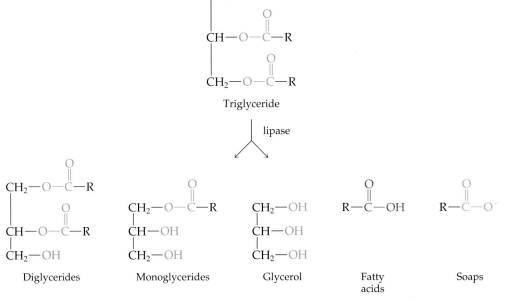

Figure 16.5 Products formed in the digestion of fats. A triglyceride (fat) is hydrolyzed to a variety of products in a reaction catalyzed by the enzyme lipase. The R stands for the hydrocarbon portion of fatty acid molecules.

The average American diet contains too much fat; about 34% of our calories come from fat, with about 13% of calories from saturated fats.

comes from butter, cream, margarine, cheese, vegetable oils and shortenings, meat products, and some seeds and nuts. The American Heart Association and most other health authorities recommend that no more than 30% of our calories come from fat, and no more than 10% from saturated fats, with 10% from monounsaturated and 10% from polyunsaturated fats.

Digestion and Metabolism of Fats

Dietary **fats** are mainly **triglycerides**, esters of glycerol and fatty acids (Chapter 15). Fats are digested, by enzymes called *lipases*, to fatty acids and glycerol, along with some monoglycerides and diglycerides (Figure 16.5). The products of fat digestion are reassembled into triglycerides, which are attached to proteins for transportation through the bloodstream.

Fats are stored throughout the body, principally in **adipose tissue**, located in places called **fat depots**. Considerable fat is stored around vital organs, such as the heart, kidneys, and spleen, where it cushions and helps to prevent injury to these organs. Fat is also stored under the skin, where it helps insulate against temperature changes.

When fat reserves are called on for energy, fat molecules are hydrolyzed once more to glycerol and fatty acids. The glycerol can be burned for energy or converted to glucose. The fatty acids enter a process called the *fatty acid spiral* that chops off carbon atoms two at a time. The two-carbon fragments can be used for energy or for the synthesis of new fatty acids.

Emulsions

When oil and water are violently shaken together, the oil is broken up into tiny, submicroscopic droplets and dispersed throughout the water. Such a mixture is called an *emulsion*. Unless a third substance has been added, the emulsion usually breaks down rapidly, the oil droplets recombining and floating to the surface of the water.

Emulsions can be stabilized by adding certain types of gum, a soap, or a protein that can form a protective coating around the oil droplets and prevent them from coming together. Compounds called *bile salts* keep tiny fat droplets suspended in aqueous media during human digestion.

Many foods are emulsions. Milk is an emulsion of butterfat in water. The stabilizing agent is a protein called *casein*. Mayonnaise is an emulsion of salad oil in water, stabilized by egg yolk.

Fats, Cholesterol, and Human Health

Dietary **saturated fats** have been implicated, along with cholesterol, in *arteriosclerosis* (hardening of the arteries). As this condition develops, deposits form on the inner walls of arteries. Eventually these deposits harden, and the vessels lose their elasticity (Figure 16.6). Blood clots tend to lodge in the narrowed arteries, leading to a heart attack (if the blocked artery is in heart muscle) or a stroke (if the blockage occurs in an artery that supplies the brain).

The plaque in clogged arteries is rich in cholesterol. High blood levels of cholesterol, like those of triglycerides, correlate closely with the risk of cardiovascular disease. Like fats, cholesterol is also insoluble in water. It is transported in blood by water-soluble proteins, which are usually classified according to their density (Table 16.1). Very-low-density lipoproteins (VLDLs) serve mainly to transport triglycerides,

Saturated fats are those made of a large proportion of saturated fatty acids, such as palmitic and stearic. **Polyunsaturated fats** (oils) have a high proportion of unsaturated fatty acids such as linoleic and linolenic. (The structures of these acids are given in Section 15.5).

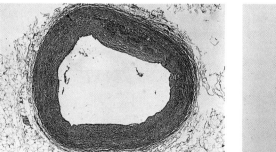 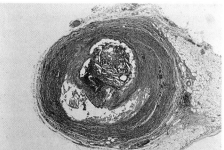

Figure 16.6 Photomicrographs of cross sections of a normal artery and a hardened artery, showing deposits of plaque that contain cholesterol.

Table 16.1 ▌ Lipoproteins in the Blood

Class	Abreviation	Protein (%)	Density (g/mL)	Main Function
Very low density	VLDL	5	1.006–1.019	Transport triglycerides
Low density	LDL	25	1.019–1.063	Transport cholesterol to the cells for use
High density	HDL	50	1.063–1.210	Transport cholesterol to the liver for processing for excretion

A **lipoprotein** is any of a group of proteins combined with a lipid (Section 15.5), such as cholesterol or a triglyceride.

whereas low-density lipoproteins (LDLs) are the main carriers of cholesterol. LDLs carry cholesterol to the cells for use, and it is these lipoproteins that are thought to deposit cholesterol in arteries, leading to cardiovascular disease. High-density lipoproteins (HDLs) also carry cholesterol, but they carry it to the liver for processing and excretion. Exercise is thought to increase the levels of HDL, the lipoprotein sometimes called "good" cholesterol. High levels of LDL, called "bad" cholesterol, put one at increased risk of heart attack and stroke.

There is also statistical evidence that fish oils can prevent heart disease. Researchers at the University of Leiden in the Netherlands have found that Greenlanders who eat a lot of fish have a low risk of heart disease despite a diet that is high in total fat and cholesterol. The probable effective agents are polyunsaturated fatty acids such as eicosapentaenoic acid.

$$CH_3CH_2CH{=}CHCH_2CH{=}CHCH_2CH{=}CHCH_2CH{=}CHCH_2CH{=}CHCH_2CH_2CH_2COOH$$

One large egg contains about 213 mg of cholesterol. Most health authorities recommend no more than 300 mg/day of cholesterol in the diet.

Other studies have shown that diets with added fish oil lead to lower cholesterol and triglyceride levels in the blood.

Example 16.1

A McDonald's Big Mac sandwich furnishes 541 kcal, of which 279 kcal comes from fat. (Recall that a kilocalorie is a food calorie.) What percentage of the total calories is from fat?

Solution

Simply divide the calories from fat by the total calories. Then multiply by 100 to get the percentage (parts per 100).

$$\% \text{ calories from fat} = \frac{279 \text{ kcal}}{541 \text{ kcal}} \times 100 = 51.6\%$$

Exercise 16.1

A Burger King Whopper sandwich furnishes 606 kcal, of which 288 kcal comes from fat. What percentage of the total calories is from fat?

Example 16.2

You can calculate your maximum fat intake in grams by multiplying your daily calorie intake by 0.30 (30%) and then by 1 g fat/9 kcal. What are the maximum amounts of fat and saturated fat that should be included in the diet of a person who needs 1800 kcal/day to maintain his proper weight?

Solution

$$1800 \text{ kcal} \times 0.30 \times \frac{1 \text{ g fat}}{9 \text{ kcal}} = 60 \text{ g fat}$$

Only one-third of the total fat should be saturated fat. The person should eat no more than 60 g of total fat, of which 20 g is saturated fat, each day.

Exercise 16.2

What are the maximum amounts of fat and saturated fat that should be included in the diet of a person who needs 2500 kcal/day to maintain his proper weight?

Example 16.3

The label on a macaroni and cheese dinner indicates that each $\frac{3}{4}$ cup serving (as pre-pared) furnishes 290 kcal, 9 g protein, 34 g carbohydrate, and 13 g fat. Calculate the percentage of calories.

Solution

First, calculate the calories from each kind of nutrient.

$$9 \text{ g} \times 4 \text{ kcal/g} = 36 \text{ kcal} \qquad (\textit{from protein})$$
$$34 \text{ g} \times 4 \text{ kcal/g} = 136 \text{ kcal} \qquad (\textit{from carbohydrate})$$
$$13 \text{ g} \times 9 \text{ kcal/g} = 117 \text{ kcal} \qquad (\textit{from fat})$$

Then calculate the percentage of calories from fat

$$\frac{117 \text{ kcal}}{290 \text{ kcal}} \times 100 = 40\%$$

Exercise 16.3

The label on a can of cream-style corn indicates that each $\frac{1}{2}$-cup serving furnishes 90 kcal, 2 g protein, 22 g carbohydrate, and 1 g fat. Calculate the percentage of calories from fat.

Fake Fats

Despite the lack of evidence that artificial sweeteners help dieters lose weight, there is considerable interest in fake fats as diet aids. Several fat substitutes are now available.

Part of our enjoyment of fats is the feel of the food in the mouth (think of the feel of chocolate as its cocoa butter melts in the mouth). Simplesse, made by G. D. Searle, producers of NutraSweet, is a low-calorie fat substitute made from egg white or milk proteins. The shape of the protein particles gives Simplesse a creamy feel in the mouth, but it has only 15% of the calories of real fat. It is useful only in cold foods such as frozen desserts because cooking makes it tough and leathery like overcooked egg white.

Olestra, made by Procter and Gamble, is a polyester of sucrose. Under the trade name OLEAN, it is used as a substitute for fats and oils in cooking potato chips and other snack foods. Because it is not digested, olestra contributes no calories to foods cooked in it. It provides the rich taste and sensory effect of fat-fried foods without the calories of foods fried in real fat. Because it is not absorbed,

Sucrose polyester, also known by the generic name olestra, is an ester of the alcohol groups on a sucrose molecule with six, seven, or eight fatty acid molecules. Unlike triglyceride molecules, which are readily broken down by digestive enzymes, olestra molecules are too large to fit the active sites of these enzymes. Olestra is therefore not digested or absorbed, and so it adds no fat or calories to the diet.

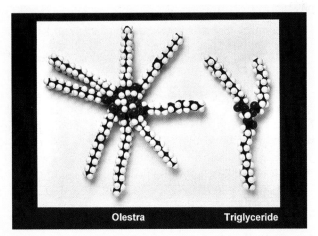

Olestra   Triglyceride

olestra in large quantities can cause gastrointestinal discomfort and diarrhea in some people. It also carries fat-soluble vitamins—A, D, E, and K—right through the digestive tract, lessening their absorption. Procter and Gamble adds enough of these vitamins to offset the lower absorption.

Another entry in the fake fat derby is Z-Trim, an ultrafine powder made from oat and soybean hulls. This powder can absorb up to 24 times its weight in water, and it forms a fatlike gel when cooled. Z-Trim can be used to replace some of the fat in baked goods such as cookies and cakes. It adds no calories to such foods; rather, it provides fiber that many of us need in our diets.

Other fat substitutes include emulsions of starches, which are used to make light versions of mayonnaise, salad dressings, and ice cream; and emulsions of gelatin, which are used to make light margarine. With an increasing proportion of the population suffering from obesity, we probably can expect still more fake fats to reach the marketplace.

Half of the total fat, three-fourths of the saturated fat, and all the cholesterol in a typical human diet come from animal products such as meat, milk, cheese, and eggs. Advertising that a vegetable oil (for example) contains no cholesterol is silly. No vegetable product contains cholesterol.

16.3 Proteins: Muscle and Much More

Proteins (Chapter 15) are polymers of amino acids. Each gene carries the blueprint for a specific protein, and each protein serves a particular purpose. We require protein in our diet in order to provide the amino acids needed to make muscles, hair, enzymes, and many other cellular components vital to life.

Protein Metabolism: Essential Amino Acids

Proteins are broken down in the digestive tract into their component amino acids. From these amino acids, our bodies synthesize proteins for growth and repair of tissues. When a diet contains more protein than is needed for the body's growth and repair, the leftover protein is used as a source of energy.

The adult human body can synthesize all but eight of the amino acids needed for making proteins. These eight are called **essential amino acids**—isoleucine, lysine, phenylalanine, tryptophan, leucine, methionine, threonine, and valine—(see Table 15.3 for structures) and must be included in our diet. Each of the essential amino acids is a **limiting reagent** in protein synthesis. When the body is deficient in one of them, it can't make proper proteins.

An *adequate* (or *complete*) *protein* supplies all the essential amino acids in the quantities needed for the growth and repair of body tissues. Most proteins from plant sources are deficient in one or more amino acids. For example, corn protein has insufficient lysine and tryptophan. People whose diets consist chiefly of corn may suffer from malnutrition even though the calories supplied by the food are adequate. Protein from rice is short of lysine and threonine. Wheat protein lacks enough lysine. Even soy protein, probably the best nonanimal protein, is deficient in the essential amino acid methionine.

Most proteins from animal sources contain all the essential amino acids in adequate amounts. Lean meat, milk, fish, eggs, and cheese supply adequate protein. In fact, gelatin is one of the few inadequate animal proteins. It contains almost no tryptophan and has only small amounts of threonine, methionine, and isoleucine.

Protein Deficiency: Kwashiorkor

Our requirement for protein is about 0.8 g per kilogram of body weight. Diets with inadequate protein are common in some parts of the world. A protein-deficiency disease called *kwashiorkor* (Figure 16.7) is common at times in parts of Africa where corn is the major food.

Nutrition is especially important in a child's early years. This is readily apparent from a consideration of Figure 16.8, which shows that the human brain reaches nearly full size by the age of 2 years. Protein deficiency leads to both physical and mental retardation.

Vegetarian Diets

Green plants such as grasses and grains trap a small fraction of the energy that reaches them from the sun. They use this energy to convert carbon dioxide, water, and mineral nutrients (including nitrates, phosphates, and sulfates) to proteins. Cattle eat plant protein, digest it, and convert a small portion of it to animal protein. People eat this animal protein, digest it, and reassemble some of the amino acids into human protein. Some of the energy originally transformed by green plants is lost as heat at every step of the food chain (Table 16.2). If people ate the plant protein directly, one highly inefficient step would be skipped. Vegetarianism is in harmony with energy conservation.

Figure 16.7   An extreme lack of proteins and vitamins causes a deficiency disease called kwashiorkor. The symptoms include retarded growth, discoloration of skin and hair, bloating, a swollen belly, and mental apathy.

Table 16.2 ▮ Efficiency of Protein Conversions

Food	Efficiency of Production (%)*
Beef or veal	4.7
Pork	12.1
Chicken or turkey	18.2
Milk	22.7
Eggs	23.3

*Calculated by dividing the weight of edible protein by the weight of the protein feed required to produce it, and then multiplying the result by 100.

Source: President's Science Advisory Committee. *The World Food Problem,* Vol. 2, 1967.

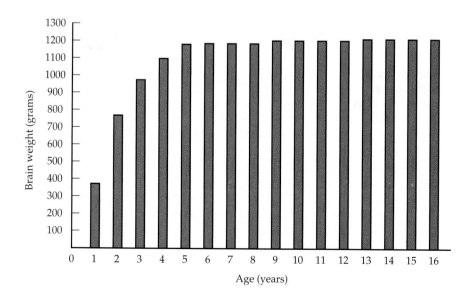

Figure 16.8   Growth of the human brain according to age. Most of the growth occurs in the first 2 years of life.

Table 16.3 ▌ Ethnic Foods that Combine a Cereal Grain with a Legume

Group	Food*
Mexicans	Corn tortillas and refried beans
Japanese	Rice and soybean curds (tofu)
English working classes	Baked beans on toasted bread
American Indians	Corn and beans (succotash)
Western Africans	Rice and peanuts (ground nuts)
Cajuns (Louisiana)	Red beans and rice

*Cereal grains are in red, and legumes are in blue.

[Web Reference 1] Two good websites about vegetarianism.

It is interesting to note that a variety of ethnic dishes supply relatively good protein by combining a cereal grain with a legume (beans, peas, peanuts, and so on). The grain is deficient in tryptophan and lysine, but it has sufficient methionine. The legume is deficient in methionine, but it has enough tryptophan and lysine. A few such combinations are listed in Table 16.3. Peanut butter sandwiches are a popular American example of a legume–cereal grain combination.

Although complete proteins can be obtained by eating a careful mixture of vegetable foods, extreme vegetarianism can be dangerous, especially for young children. Even when the diet includes a wide variety of plant materials, an all-vegetable diet lacks vitamin B_{12} because this nutrient is not found in plants. Other nutrients scarce in all-plant diets include calcium, iron, riboflavin, and (required by children not exposed to sunlight) vitamin D. A modified vegetarian (*ovo-lacto*) diet that includes milk, eggs, cheese, and fish can provide excellent nutrition, with red meat totally excluded.

16.4 Minerals: Important Inorganic Chemicals in Our Lives

Most of the chemicals in food are organic. But several inorganic chemicals, called **dietary minerals**, are also vital to sustaining life. It is estimated that minerals represent about 4% of the weight of the human body. Some of these minerals, such as chlorides (Cl^-), phosphates (PO_4^{3-}), bicarbonates (HCO_3^-), and sulfates (SO_4^{2-}), occur in the blood and other body fluids. Others, such as iron (as Fe^{2+}) in hemoglobin and phosphorus in nucleic acids (DNA and RNA), are constituents of complex organic compounds.

Table 16.4 lists the 30 elements known to be essential to one or more living organisms. The 11 that comprise the structural elements and the macrominerals make up more than 99% of all the atoms in the human body. The 19 trace elements include iron, zinc, copper, and 16 others called *ultratrace elements*.

Minerals serve a variety of functions. The function of iodine is quite dramatic. A small amount of iodine is necessary for the thyroid gland to function properly. A deficiency of iodine has dire effects, of which goiter is perhaps the best known (Figure 16.9). Iodine is available naturally in seafood, but, to guard against iodine deficiency, a small amount of potassium iodide (KI) is added to table salt. The use of iodized salt has greatly reduced the incidence of goiter.

Iron(II) ions (Fe^{2+}) are necessary for proper function of the oxygen-transporting compound hemoglobin. Without sufficient iron, not enough oxygen is supplied to body tissues, and anemia, a general weakening of the body, results. Foods especially rich in iron compounds include red meat and liver. It appears that most adult males need very little dietary iron because iron seems to be retained by the body and lost mainly through bleeding.

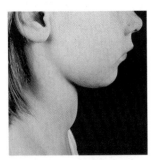

Figure 16.9 A person with goiter. The swollen thyroid gland in the neck results from a dietary deficiency of the trace element iodine.

Table 16.4 ▌ Elements Essential to Life

Element	Symbol	Form Used	Element	Symbol	Form Used
Bulk Structural Elements			**Ultratrace Elements**		
Hydrogen	H	Covalent	Manganese	Mn	Mn^{2+}
Carbon	C	Covalent	Molybdenum	Mo	Mo^{2+}
Oxygen	O	Covalent	Chromium	Cr	?
Nitrogen	N	Covalent	Cobalt	Co	Co^{2+}
Phosphorus*	P	Covalent	Vanadium	V	?
Sulfur*	S	Covalent	Nickel	Ni	Ni^{2+}
			Cadmium	Cd	Cd^{2+}
Macrominerals			Tin	Sn	Sn^{2+}
Sodium	Na	Na^+	Lead	Pb	Pb^{2+}
Potassium	K	K^+	Lithium	Li	Li^+
Calcium	Ca	Ca^{2+}	Fluorine	F	F^-
Magnesium	Mg	Mg^{2+}	Iodine	I	I^-
Chlorine	Cl	Cl^-	Selenium	Se	SeO_4^{2-}?
Phosphorus*	P	$H_2PO_4^-$	Silicon	Si	?
Sulfur*	S	SO_4^{2-}	Arsenic	As	?
			Boron	B	H_3BO_3
Trace Elements					
Iron	Fe	Fe^{2+}			
Copper	Cu	Cu^{2+}			
Zinc	Zn	Zn^{2+}			

*Note that phosphorus and sulfur each appear twice; they are structural elements and are also components of the macrominerals phosphate and sulfate.

Calcium and phosphorus are necessary for the proper development of bones and teeth. Growing children need about 1.5 g of each mineral per day. These elements are available from milk. The calcium and phosphorus needs of adults are less widely known but are very real just the same. For example, calcium ions are necessary for the coagulation of blood (to stop bleeding) and for maintenance of the rhythm of the heartbeat. Phosphorus, as phosphate, is the "energy currency" of the body and is necessary for the body to obtain, store, and use energy from foods. As mentioned in Section 15.9, phosphorus is an important part of the nucleic acid backbone.

Sodium chloride—in moderate amounts—is essential to life. It is important in the exchange of fluids between cells and plasma, for example. The presence of salt increases water retention, however, and a high volume of retained fluids can cause swelling and high blood pressure (hypertension). An estimated 28 million people in the United States suffer from hypertension. Most physicians agree that our diets generally contain too much salt.

Iron, copper, zinc, cobalt, manganese, molybdenum, calcium, and magnesium are essential to the proper functioning of metalloenzymes, which are life-sustaining. A great deal remains to be learned about the role of inorganic chemicals in our bodies. Bioinorganic chemistry is a flourishing area of research.

In the United States more than 120 million prescriptions are written each year for diuretics, drugs that increase urine production in order to reduce body fluids. Another 30 million prescriptions are written for potassium (K^+) supplements to replace the potassium ions washed out in the excess urine.

16.5 The Vitamins: Vital, but Not All Are Amines

Why are British sailors called "limeys"? And what does this term have to do with food? Sailors have been plagued since early times by *scurvy*. In 1747 Scottish navy surgeon James Lind showed that this disease could be prevented by including fresh fruit and vegetables in the diet. Fresh fruits conveniently carried on long voyages

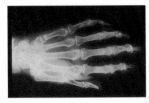

(a)

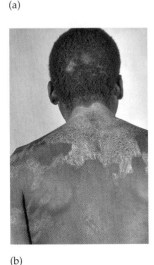

(b)

Figure 16.10 (a) This X-ray shows softened bones caused by a deficiency of vitamin D. (b) Inflammation and abnormal pigmentation characterize pellagra, caused by a deficiency of niacin.

(there was no refrigeration) were limes, lemons, and oranges. British ships put to sea with barrels of limes aboard, and sailors ate a lime or two every day. That is how they came to be known as "lime eaters," or simply limeys.

In 1897 the Dutch scientist Christiaan Eijkman showed that polished rice lacked something found in the hull. Lack of that something caused the disease *beriberi*, which was a serious problem in the Dutch East Indies at that time. A British scientist, F. G. Hopkins, found that rats fed a synthetic diet of carbohydrates, fats, proteins, and minerals were unable to sustain healthy growth. Again, something was missing.

In 1912 Casimir Funk, a Polish biochemist, coined the word vitamine (from the Latin word *vita*, meaning "life") for these missing factors. Funk thought all these factors contained the amine group. In the United States, the final *e* was dropped after it was found that not all the factors were amines. The generic term became *vitamin*. Eijkman and Hopkins shared the 1929 Nobel prize in physiology and medicine for their important discoveries relating to vitamins.

Vitamins are specific organic compounds that are required in the diet (in addition to the usual proteins, fats, carbohydrates, and minerals) to prevent specific diseases. Some vitamins, along with their structures, sources, and deficiency symptoms, are shown in Table 16.5. The role of vitamins in the prevention of deficiency diseases (Figure 16.10) has been well established.

As you can see from Table 16.5, vitamins do not share a common chemical structure. They can, however, be divided into two broad categories: *fat-soluble vitamins*—including A, D, E, and K—and *water-soluble vitamins*—made up of B complex vitamins and vitamin C. All fat-soluble vitamins incorporate a high proportion of hydrocarbon elements. They contain one or two oxygen atoms but are nonpolar as a whole. In contrast, a water-soluble vitamin contains a high proportion of the electronegative atoms oxygen and nitrogen, which can form hydrogen bonds to water; therefore, the molecule as a whole is soluble in water.

Fat-soluble vitamins dissolve in the fatty tissue of the body, where reserves can be stored for future use. For example, an adult can store several years' supply of vitamin A. If the diet becomes deficient in vitamin A, these reserves are mobilized, and the adult remains free of the deficiency disease for quite a while. On the other hand, a small child who has not had the opportunity to build up a store of the vitamin soon exhibits symptoms of the deficiency. Many children in developing countries have been permanently blinded as a result of vitamin A deficiency.

Because fat-soluble vitamins are efficiently stored in the body, overdoses can result in adverse effects. Large excesses of vitamin A cause irritability, dry skin, and a feeling of pressure inside the head. Massive doses given to pregnant rats result in malformed offspring. Vitamin D, like vitamin A, is fat-soluble. Too much vitamin D can cause pain in the bones, hard deposits in the joints, nausea, diarrhea, and weight loss. Even though vitamins E and K are also fat-soluble, they are metabolized and excreted. They are not stored to the extent that vitamins A and D are, and excesses seldom cause problems.

The body has a limited capacity to store water-soluble vitamins. It excretes anything over the amount that can be used immediately. Water-soluble vitamins must be taken at frequent intervals, whereas a single dose of a fat-soluble vitamin can be used by the body over several weeks. Some foods lose their vitamin content when they are cooked in water and then drained. The water-soluble vitamins go down the drain with the water.

Table 16.5 ∎ Some of the Vitamins (Polar groups are in red.)

Vitamin	Structure and Name	Sources	Deficiency Symptoms
Fat-Soluble Vitamins			
A	Retinol	Fish, liver, eggs, butter, cheese; also a vitamin precursor in carrots and other vegetables	Night blindness
D	Vitamin D_2 (calciferol)	Cod liver oil, irradiated ergosterol (milk supplement)	Rickets
E	α-Tocopherol	Wheat germ oil, green vegetables, egg yolks, meat	Sterility, muscular dystrophy
K	Vitamin K_1 (phylloquinone)	Spinach, other green leafy vegetables	Hemorrhage
Water-Soluble Vitamins			
The B Complex	B_1 (thiamine)	Germ of cereal grains, legumes, nuts, milk, and brewer's yeast	Beriberi—polyneuritis resulting in muscle paralysis enlargement of heart, and ultimately heart failure
	B_2 (riboflavin)	Milk, red meat, liver, egg white, green vegetables, whole wheat flour (or fortified white flour), and fish	Dermatitis, glossitis (tongue inflammation)

(continued)

Table 16.5 ▌ (continued)

Vitamin	Structure and Name	Sources	Deficiency Symptoms

Water-Soluble Vitamins (*cont.*)

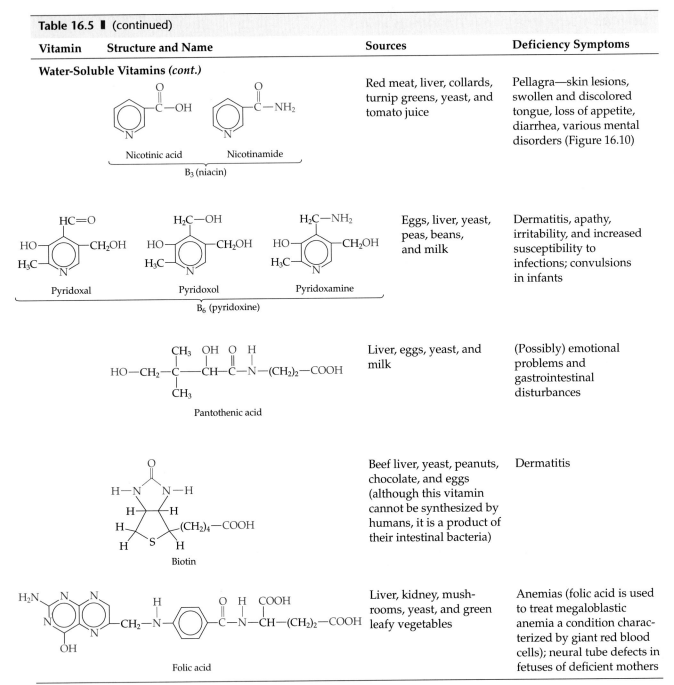

Nicotinic acid   Nicotinamide

B₃ (niacin)

Red meat, liver, collards, turnip greens, yeast, and tomato juice

Pellagra—skin lesions, swollen and discolored tongue, loss of appetite, diarrhea, various mental disorders (Figure 16.10)

Pyridoxal   Pyridoxol   Pyridoxamine

B₆ (pyridoxine)

Eggs, liver, yeast, peas, beans, and milk

Dermatitis, apathy, irritability, and increased susceptibility to infections; convulsions in infants

Pantothenic acid

Liver, eggs, yeast, and milk

(Possibly) emotional problems and gastrointestinal disturbances

Biotin

Beef liver, yeast, peanuts, chocolate, and eggs (although this vitamin cannot be synthesized by humans, it is a product of their intestinal bacteria)

Dermatitis

Folic acid

Liver, kidney, mushrooms, yeast, and green leafy vegetables

Anemias (folic acid is used to treat megaloblastic anemia a condition characterized by giant red blood cells); neural tube defects in fetuses of deficient mothers

(continued)

Table 16.5 ▌ (continued)

Vitamin	Structure and Name	Sources	Deficiency Symptoms
Water-Soluble Vitamins (*cont.*)			

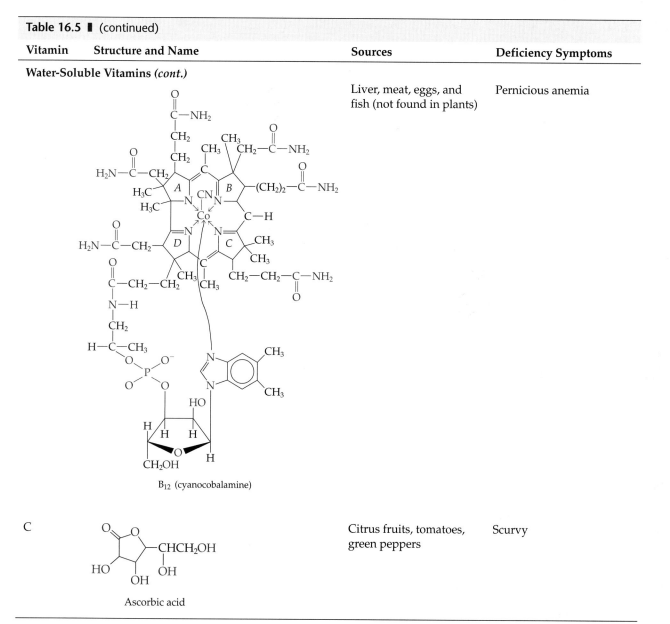

B₁₂ (cyanocobalamine)

Liver, meat, eggs, and fish (not found in plants) — Pernicious anemia

C — Ascorbic acid — Citrus fruits, tomatoes, green peppers — Scurvy

16.6 Other Essentials

In addition to carbohydrates, proteins, fats, minerals, and vitamins, there are other things in our food that are important. One of them is *fiber*; another is *water*.

Dietary Fiber

Dietary fiber may be soluble or insoluble. The insoluble fiber is usually cellulose, while the soluble fibers are generally hemicelluloses, polymers of pentoses (five-carbon sugars).

In the mid-1970s some scientists claimed to have found a link between a lack of fiber in the diet and the incidence of colon cancer. They noted that people in developed countries, who eat diets rich in highly processed, low-fiber foods, are much more likely to get cancer of the colon than are people in underdeveloped nations

What we now call dietary fiber was once called *roughage* or referred to as *bulk in the diet.*

Wheat bran is about 90% cellulose. It does not dissolve in water or digestive juices but passes unchanged into the colon. (It is *insoluble fiber*.) The fibers in oat bran, rice bran, and fruits, on the other hand, are mainly hemicelluloses, which are more soluble than cellulose. (They are *soluble fibers*.) Large amounts of soluble fibers in the diet are sometimes helpful in lowering blood cholesterol levels.

 [Web Reference 2] The National Cancer Institute's program to increase consumption of fruits and vegetables.

where diets are usually high in fiber. A plausible mechanism was formulated to explain the correlation between low-fiber diets and colon cancer.

In the year 2000, however, two studies involving thousands of people failed to confirm the theory that eating low-fat, high-fiber foods reduces the risk of colon cancer. (It should be noted that all of the participants in both of these studies had already had at least one polyp removed from the colon. Perhaps it was already too late in the case of these people for a low-fat, high-fiber diet to make a difference.)

Water

Much of the food we eat is mainly water. It should come as no surprise that tomatoes are 90% water or that melons, oranges, and grapes are largely water. But water is one of the main ingredients in practically all foods, from roast beef and seafood to potatoes and onions.

In addition to the water we get in our food, it is recommended that we drink about 1.0–1.5 L of water each day. We could satisfy this need by drinking plain water, but often we choose other beverages. We drink coffee, tea, milk, fruit juice, and soft drinks. In 1985 carbonated soft drinks replaced water as the kind of beverage consumed most in the United States.

Many people also drink beverages that contain ethyl alcohol. These are made by fermenting grains or fruit juices. Beer is usually made from malted barley, and wine from grape juice. Even these alcoholic beverages are mainly water.

16.7 Starvation and Fasting

When the human body is totally deprived of food, whether voluntarily or involuntarily, the condition is known as **starvation**. Involuntary starvation is a serious problem in much of the world. Although starvation is seldom the sole cause of death, those weakened by malnutrition succumb readily to disease. Even a seemingly minor disease, such as measles, can become life-threatening.

Metabolic changes similar to those occurring in starvation take place during fasting. During total fasting, the body's glycogen stores are depleted rapidly and the body calls on its fat reserves. Fat is first taken from around the kidneys and the heart. Then it is removed from other parts of the body, even the bone marrow.

Increased dependence on stored fats as an energy source leads to *ketosis*, a condition characterized by the appearance of compounds called *ketone bodies* in the blood and urine (Figure 16.11). Note that two of the ketone bodies are acids and that β-hydroxybutyric acid is not a ketone. Ketosis rapidly develops into *acidosis*; the blood pH drops, and oxygen transport is hindered. Oxygen deprivation leads to depression and lethargy.

In the early stages of a *total* fast, body protein is metabolized at a relatively rapid rate. After several weeks, the rate of protein breakdown slows considerably as the brain adjusts to using the breakdown products of fatty acid metabolism for its energy source. When fat reserves are substantially depleted, the body must again draw heavily on its structural proteins for its energy requirements. The emaciated appearance of a starving individual is due to the depletion of muscle proteins.

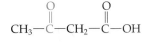

$$CH_3-\overset{\displaystyle O}{\overset{\displaystyle \|}{C}}-CH_3$$

Acetone

$$CH_3-\overset{\displaystyle O}{\overset{\displaystyle \|}{C}}-CH_2-\overset{\displaystyle O}{\overset{\displaystyle \|}{C}}-OH$$

Acetoacetic acid

$$CH_3-\overset{\displaystyle OH}{\overset{\displaystyle |}{CH}}-CH_2-\overset{\displaystyle O}{\overset{\displaystyle \|}{C}}-OH$$

β-Hydroxybutyric acid

Figure 16.11 The three ketone bodies. Only two have ketone functional groups (blue); the third has an alcohol function. Two of the compounds have carboxylic acid functional groups (red).

Fad Diets

Weight loss diets are often lacking in balanced nutrition and can be harmful to one's health. In the more extreme low-carbohydrate diets, ketosis is deliberately induced. Understandably, two of the possible side effects of these diets are depression and lethargy. In the early stages of a diet deficient in carbohydrates, the body converts amino acids to glucose. The brain must have glucose; it can't obtain sufficient energy from fats. If there are enough adequate proteins in the diet, tissue proteins are spared. (Liquid protein diets, from which several people died, were based on the inadequate proteins obtained from beef hides and connective tissues.)

Even low-carbohydrate diets high in adequate proteins are hard on the body, which must rid itself of the nitrogen compounds—ammonia and urea—formed by the breakdown of proteins. This puts an increased stress on the liver, where the waste products are formed.

Contrary to a popular notion, fasting does not cleanse the body. Indeed, quite the reverse occurs. A shift to fat metabolism produces ketone bodies, and protein breakdown produces ammonia, urea, and other wastes. You can lose weight by fasting, but the process should be carefully monitored by a physician.

Weight loss through diet and exercise is discussed in Chapter 18.

Processed Food: Less Nutrition

Malnutrition need not be due to starvation or dieting. It can be the result of eating too much highly processed food.

Whole wheat is an excellent source of vitamin B_1 and other vitamins. To make white flour, the wheat germ and bran are removed from the grain. The remaining material has few minerals and no vitamins. We eat the starch and use the germ and bran for animal food. Our cattle and hogs do get good nutrition. Similarly, polished rice has had most of its protein and minerals removed, and it has virtually no vitamins. The disease beriberi became prevalent when polished rice was introduced into Southeast Asia.

When many fruits and vegetables are peeled, they lose most of their vitamins, minerals, and fibers. The peels are often dumped (directly or through a garbage disposal) into a sewage system where they contribute to water pollution (Chapter 13). The heat used to cook food also destroys some vitamins. If water is used in cooking, some of the water-soluble vitamins (C and B complex) and some of the minerals are often drained off and discarded with the water.

It is estimated that well over half of the diet of an average person in the United States consists of processed foods. A diet of hamburgers, potato chips, and colas is lacking in many essential nutrients. Highly processed convenience foods threaten to leave the people of developed nations poorly nourished despite their abundance of food.

[Web Reference 3] Several good sites on the subject of general nutrition.

Food Additives

The label reads "egg whites, vegetable oils, nonfat dry milk, lecithin, mono- and diglycerides, propylene glycol monostearate, xanthan gums, sodium citrate, aluminum sulfate, artificial flavor, iron phosphate, niacin, riboflavin, and irradiated ergosterol." You recognize only a few of the ingredients. Most of the substances listed are **food additives**, substances other than basic foodstuffs that are present in food as a result of some aspect of production, processing, packaging, or storage.

16.8 Additives to Enhance Our Food

Because food processing results in the removal or destruction of certain essential food substances, it is sometimes necessary to include additives in prepared food to increase its nutritional value. Other chemicals are added to enhance color and flavor, to retard spoilage, to provide texture, to sanitize, to bleach, to ripen (or to prevent

Ellen Swallow Richards (1842–1911) was the first woman graduate of Massachusetts Institute of Technology (1873). She was one of the first to apply the methods of science to nutrition. She was also a pioneer in the scientific analysis of waste water and thus helped lay the foundation for environmental chemistry.

ripening), to control moisture or dryness, or to prevent (or enhance) foaming. There are several thousand different additives. Sugar, salt, and corn syrup are used in the greatest amounts. These three, plus citric acid, baking soda, vegetable colors, mustard, and pepper, make up more than 98% (by weight) of all additives.

Food additives are not a recent development. Salt has been used to preserve meat and fish since ancient times. Spices have been employed since earliest recorded history to flavor and preserve foods. Other additives have been used throughout the centuries. Movement of the population from farms to cities has increased the necessity for using additives to preserve foods. The increased consumption of convenience foods has also led to greater use of additives.

In the United States, food additives are regulated by the Food and Drug Administration (FDA). The original Food, Drug, and Cosmetic Act was passed by Congress in 1938. Under this act, the FDA had to prove that an additive was unsafe before its use could be prevented. The Food Additives Amendment of 1958 shifted the burden of proof to the food industry. Now, a company that wishes to use a food additive must first furnish proof to the FDA that the additive is safe for the intended use. The FDA can also regulate the amount of chemicals that can be added.

Some people express concern about the "chemicals in our food." Food itself consists of chemicals, however. Table 16.6 shows the chemical composition of a typical breakfast. Many of the chemicals in this breakfast might be harmful in large amounts but are harmless in the trace amounts that occur naturally in foods. Indeed, some make important contributions to delightful flavors and aromas.

Our bodies are also collections of chemicals. If broken down into its elements (Table 16.7), your body would be worth only a few dollars. It is the unique combination and arrangement of the elements in every human body that make you different from everyone else and make each individual's worth beyond measure. Since food is chemical and we are chemical, we shouldn't have to worry about chemicals in our food, except perhaps for some specific ones.

Additives That Improve Nutrition

The first nutrient supplement approved by the Bureau of Chemistry of the U.S. Department of Agriculture, potassium iodide (KI), was added to table salt in 1924 to reduce the incidence of goiter (Figure 16.9). (The Bureau of Chemistry later became the FDA.)

A number of other chemicals have been added to foods specifically to prevent deficiency diseases. The addition of vitamin B_1 (thiamine) to polished rice is essential in the Far East, where beriberi still is a problem. The replacement of the B-complex vitamins thiamine, riboflavin, folic acid, and niacin (which are removed in processing) and the addition of iron (usually ferrous carbonate, $FeCO_3$) to flour is called **enrichment**. Enriched bread or pasta made from this flour still isn't as good a food as bread made from whole wheat. It lacks vitamin B_6, pantothenic acid, zinc, magnesium, and fiber, nutrients usually provided by whole-grain flour. Despite these shortcomings, the enrichment of bread, corn meal, and cereals has virtually eliminated pellagra, a disease that once plagued the southern United States.

Vitamin C (ascorbic acid) is frequently added to fruit juices, flavored drinks, and beverages. Although our diets generally contain enough ascorbic acid to prevent scurvy, a number of scientists recommend a much larger intake than minimum daily requirements. Vitamin D is added to milk in developed countries, and this use of fortified milk has led to the virtual elimination of rickets. Similarly, vitamin A is added to margarine. (This vitamin occurs naturally in butter. It is added to margarine so that the substitute more nearly matches butter in nutritional quality.)

If we ate a balanced diet of fresh foods, we probably wouldn't need nutritional supplements. For people in large cities, however, such a diet might be impossible, and people everywhere find that convenience foods are indeed convenient. With our

In 1996 the FDA ordered the addition of folic acid to enriched foods after it was shown that folic acid deficiency can lead to spina bifida, a condition in which the spinal tube fails to close properly during fetal development.

Table 16.6 ▌ Your Breakfast—As Seen by a Chemist*

Chilled Melon

Starches	Anisyl propionate
Sugars	Amyl acetate
Cellulose	Ascorbic acid
Pectin	Vitamin A
Malic acid	Riboflavin
Citric acid	Thiamine
Succinic acid	

Scrambled Eggs

Ovalbumin	Lecithin
Conalbumin	Lipids (fats)
Ovomucoid	Fatty acids
Mucin	Butyric acid
Globulins	Acetic acid
Amino acids	Sodium chloride
Lipovitellin	Lutein
Livetin	Zeazanthine
Cholesterol	Vitamin A

Sugar-Cured Ham

Actomyosin	Adenosine triphosphate
Myogen	(ATP)
Nucleoproteins	Glucose
Peptides	Collagen
Amino acids	Elastin
Myoglobin	Creatine
Lipids (fats)	Pyroligneous acid
Linoleic acid	Sodium chloride
Oleic acid	Sodium nitrate
Lecithin	Sodium nitrite
Cholesterol	Sodium phosphate
Sucrose	

Coffee

Caffeine	Acetone
Essential oils	Methyl acetate
Methanol	Furan
Acetaldehyde	Diacetyl
Methyl formate	Butanol
Ethanol	Methylfuran
Dimethyl sulfide	Isoprene
Propionaldehyde	Methylbutanol

Cinnamon Apple Chips

Pectin	Propanol
Cellulose	Butanol
Starches	Pentanol
Sucrose	Hexanol
Glucose	Acetaldehyde
Fructose	Propionaldehyde
Malic acid	Acetone
Lactic acid	Methyl formate
Citric acid	Ethyl formate
Succinic acid	Ethyl acetate
Ascorbic acid	Butyl acetate
Cinnamyl alcohol	Butyl propionate
Cinnamic aldehyde	Amyl acetate
Ethanol	

Toast and Coffee Cake

Gluten	Mono- and diglycerides
Amylose	Methyl ethyl ketone
Amino acids	Niacin
Starches	Pantothenic acid
Dextrins	Vitamin D
Sucrose	Acetic acid
Pentosans	Propionic acid
Hexosans	Butyric acid
Triglycerides	Valeric acid
Sodium chloride	Caproic acid
Phosphates	Acetone
Calcium	Diacetyl
Iron	Maltol
Thiamine	Ethyl acetate
Riboflavin	Ethyl lactate

Tea

Caffeine	Phenyl ethyl alcohol
Tannin	Benzyl alcohol
Essential oils	Geraniol
Butyl alcohol	Hexyl alcohol
Isoamyl alcohol	

*The chemicals listed are those found normally in the foods. The chemical listings are not necessarily complete.
Source: Manufacturing Chemists Association, Washington, DC.

usual diets rich in highly processed foods, we need the nutrients provided by vitamin and mineral food additives.

Additives That Taste Good

Spice cake, soda pop, gingerbread, sausage, and many other foods depend on spices and other additives for most of their flavor. Cloves, ginger, cinnamon, and nutmeg are examples of natural spices, and basil, marjoram, thyme, and rosemary are some widely used herbs. Natural flavors are also extracted from fruits and other plant materials.

Table 16.7 ▌ Approximate Elemental Analysis of the Human Body	
Element	**Percent by Weight in Human Body**
Oxygen	65
Carbon	18
Hydrogen	10
Nitrogen	3
Calcium	1.5
Phosphorus	1
Potassium	0.35
Sulfur	0.25
Chlorine	0.15
Sodium	0.15
Magnesium	0.05
Iron	0.004
Trace elements to make 100%	

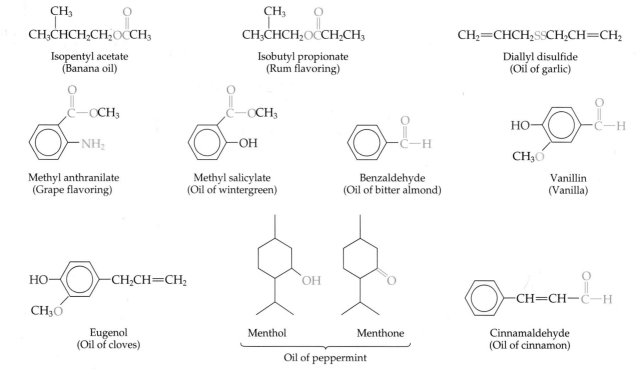

Figure 16.12   Some molecular flavorings.

Chemists sometimes analyze natural flavors and then synthesize the components to make mixtures that resemble the natural products. Major components of natural and artificial flavors are often identical. For example, both vanilla extract and imitation vanilla owe their flavor mainly to vanillin (Figure 16.12). The natural flavor is often more complex because it contains a wider variety of chemicals than the imitation. Flavor additives, whether natural or synthetic, probably present little hazard when used in moderation, and they contribute considerably to our enjoyment of food.

Artificial Sweeteners

Obesity is a major problem in most developed countries. Presumably, we could reduce the intake of calories by replacing sugars with noncaloric sweeteners. Although there is little evidence that artificial sweeteners are of any value in controlling obesity, they have become a significant part of our culture.

For many years, the major artificial sweeteners were saccharin and cyclamates. Cyclamates were banned in the United States in 1970 after studies showed they caused cancer in laboratory animals. (Subsequent studies have failed to confirm these findings, but the FDA has not lifted the ban.) In 1977 saccharin was shown to cause bladder cancer in laboratory animals. The move by the FDA to ban it, however, was blocked by Congress because at that time saccharin was the only approved artificial sweetener. Its ban would have meant the end of diet soft drinks and low-calorie products.

In 1981 the FDA approved aspartame (the methyl ester of the dipeptide aspartylphenylalanine) as an artificial sweetener. Aspartame is about 160 times sweeter than sucrose and has largely replaced saccharin as the artificial sweetener of choice. There are anecdotal reports of problems with aspartame, but repeated studies have shown it to be generally safe, except for people with phenylketonuria, an inherited condition in which phenylalanine cannot be metabolized properly. Acesulfame K has also been approved for use in the United States. It can survive the high temperatures of cooking processes, whereas aspartame is broken down by heat. The structures of four artificial sweeteners are given in Figure 16.13.

Table 16.8 compares the sweetness of a variety of substances. What makes a compound sweet? There is little structural similarity among the compounds. Saccharin, cyclamates, and compound P-4000 bear little resemblance to sugars. Even more baffling are the tastes of two compounds that closely resemble P-4000: compound I, which has the NO_2 and NH_2 groups reversed, is tasteless; compound II, which has

> Even though the use of artificial sweeteners has more than doubled during the past decade, the per-capita use of sugar, in the United States has increased more than 30%.

> The popular commercial products Nutrasweet and Equal are both aspartame.

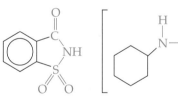 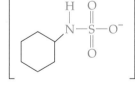 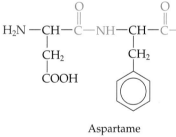 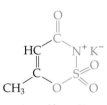

| Saccharin | Calcium cyclamate | Aspartame | Acesulfame K |

Figure 16.13 Four artificial sweeteners.

Table 16.8 ▌ Sweetness of Some Compounds	
Compound	**Relative Sweetness***
Lactose	16
Maltose	33
Glucose	74
Sucrose	100
Fructose	173
Aspartame	16,000
Acesulfame K	20,000
Saccharin	50,000
Sucralose	60,000
P-4000	400,000

*Sweetness is relative to sucrose at a value of 100.

> Sucralose (Splenda) is a chlorinated derivative of sucrose. It is the only artificial sweetener actually made from sucrose.
>
>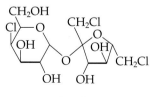
>
> Sucralose

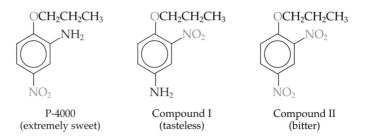

Figure 16.14 Taste responses to these three compounds are dramatically different.

two NO_2 groups, is extremely bitter (Figure 16.14). P-4000 is one of the sweetest substances known, but in 1974 its use was banned in the United States because of possible toxic effects.

Recall from Chapter 15 that sugars are polyhydroxy compounds. Indeed, many compounds with hydroxyl groups on adjacent carbon atoms are sweet. Even ethylene glycol ($HOCH_2CH_2OH$) is sweet, although it is quite toxic. Glycerol, obtained from the hydrolysis of fats (Section 16.2), is also sweet. It is used as a food additive, principally because of its properties as a **humectant** (moistening agent), but not as a sweetener.

Other polyhydroxy alcohols used as sweeteners are *sorbitol*, made by the reduction of glucose, and *xylitol*, which has five carbon atoms with a hydroxyl group on each. These compounds have an advantage over sugars in that they are not broken down in the mouth and thus do not contribute to tooth decay. This makes them useful in sugar-free chewing gums.

CH₂OH
|
CHOH   CH₂OH
| |
CHOH   CHOH
| |
CHOH   CHOH
| |
CHOH   CHOH
| |
CH₂OH   CH₂OH

Sorbitol   Xylitol

Flavor Enhancers

Some chemical substances, although not particularly flavorful themselves, are used to enhance other flavors. Common table salt (sodium chloride) is a familiar example. In addition to being a necessary nutrient, salt seems to increase sweetness and helps to mask bitterness and sourness.

Another popular flavor enhancer is monosodium glutamate (MSG). MSG is the sodium salt of glutamic acid, one of about 20 amino acids that occur naturally in proteins. It is used in many convenience foods and is also found in foods for which the FDA has established *standards of identity*. Such foods must contain certain ingredients in established proportions. They can, however, contain certain other substances as well. (Mayonnaise, for example, is defined by a standard of identity. It can contain MSG without listing it as an ingredient.)

Because MSG is used heavily in some Chinese food, it was once thought to be responsible for a set of symptoms (headaches, weakness) known as Chinese restaurant syndrome. After carefully reviewing several hundred studies, experts have declared MSG "not guilty."

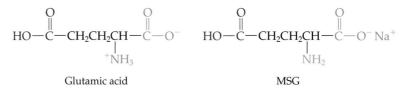

Glutamic acid   MSG

Although glutamates are found naturally in proteins, there is evidence that huge excesses can be harmful. MSG can numb portions of the brains of laboratory animals. It may also be teratogenic, causing birth defects when eaten in large amounts by women who are pregnant.

Spoilage Inhibitors

Food spoilage can result from the growth of molds (fungi) or bacteria. Propionic acid and its sodium and calcium salts are added to bread and cheese to act as mold and yeast inhibitors, and sorbic acid, benzoic acid, and their salts are also used (Figure 16.15).

Some inorganic compounds are also added as spoilage inhibitors. Sodium nitrite (NaNO$_2$) is used in meat curing and to maintain the pink color of smoked hams, frankfurters, and bologna. It also contributes to the tangy flavor of processed meat products. Nitrites are particularly effective as inhibitors of *Clostridium botulinum*, the bacterium that produces botulism poisoning. However, only about 10% of the amount used to keep meat pink is needed to prevent botulism. Nitrites have been investigated as possible causes of cancer of the stomach. In the presence of the hydrochloric acid (HCl) in the stomach, nitrites are converted to nitrous acid,

$$NaNO_2 + HCl \longrightarrow HNO_2 + NaCl$$

which may then react with secondary amines (amines with two alkyl groups on nitrogen) to form nitroso compounds.

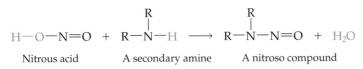

H—O—N=O	+	R—N—H (with R on top)	⟶ R—N—N=O (with R on top) + H$_2$O
Nitrous acid		A secondary amine	A nitroso compound

The R groups can be alkyl groups such as methyl (CH$_3$—) or ethyl (CH$_3$CH$_2$—), or they can be more complex. In any case, these nitroso compounds are among the most potent carcinogens known. It is a fact that the rate of stomach cancer is higher in countries that use prepared meats than in developing nations where people eat little or no cured meat. The incidence of stomach cancer is *decreasing* in the United States, however, and we aren't quite sure why. We do know that the reaction between nitrous acid and amines to form nitrosamines is inhibited by ascorbic acid (vitamin C). Perhaps a similar inhibition occurs in our stomachs. Nevertheless, problems with nitrites have led the FDA to approve sodium hypophosphite (NaH$_2$PO$_2$) as a substitute.

Sulfur dioxide is another inorganic food additive. A gas at room temperature, sulfur dioxide serves as a disinfectant and preservative, particularly for dried fruits such as peaches, apricots, and raisins. It is also used as a bleach to prevent the browning of wines, corn syrup, jelly, dehydrated potatoes, and other foods. Sulfur dioxide seems safe for most people when ingested with food, but it is a powerful irritant when inhaled and is a damaging ingredient of polluted air in some areas (Chapter 12). Sulfite salts were once used in restaurants to keep salad vegetables appearing crisp and fresh, until they were shown to cause severe allergic reactions in some people. The FDA banned this use of sulfites in 1986, and their use in other foods must be indicated on the label.

Antioxidants: BHA and BHT

Antioxidants are added to foods (or their packaging) to prevent fats and oils from forming rancid products that make the food unpalatable. Packaged foods that contain fats or oils (bread, potato chips, sausage, breakfast cereal) often have antioxidants added. Two compounds commonly used are butylated hydroxytoluene (BHT) and butylated hydroxyanisole (BHA) (Figure 16.16).

Fats turn rancid, in part as a result of oxidation, a process that occurs through the formation of molecular fragments called **free radicals**, which have an unpaired electron as a distinguishing feature. (Recall from Chapter 5 that covalent bonds are shared *pairs* of electrons.) We need not concern ourselves with the details of the structures of radicals, but we can summarize the process. First, a fat molecule reacts with oxygen to form a free radical.

$$Fat + O_2 \longrightarrow Free\ radical$$

Then, the radical reacts with another fat molecule to form a new free radical that can repeat the process. A reaction such as this, in which intermediates are formed that

O ‖
CH$_3$CH$_2$C—OH
Propionic acid

O ‖
CH$_3$CH$_2$C—O$^-$ Na$^+$
Sodium propionate

O ‖
CH$_3$CH=CHCH=CHC—OH
Sorbic acid

O ‖
CH$_3$CH=CHCH=CHC—O$^-$ K$^+$
Potassium sorbate

O ‖
⬡—C—OH
Benzoic acid

O ‖
⬡—C—O$^-$ Na$^+$
Sodium benzoate

Figure 16.15 Most spoilage inhibitors are carboxylic acids or salts of carboxylic acids.

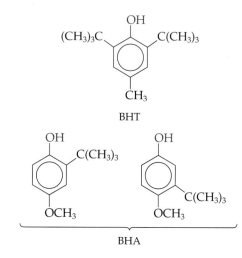

Figure 16.16 Two common synthetic antioxidants. BHA is a mixture of two isomers.

keep the reaction going, is called a **chain reaction**. One molecule of oxygen can lead to the decomposition of many fat molecules.

To preserve foods containing fats, processors package the products to exclude air, but it cannot be excluded completely, and so chemical antioxidants are used to stop the chain reaction by reacting with the free radicals.

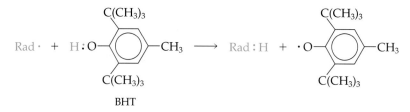

The new radical formed from BHT is rather stable. The unpaired electron doesn't have to stay on the oxygen atom but can move around in the electron cloud of the benzene ring. The BHT radical doesn't react with fat molecules, and the chain is broken.

Why are the butyl groups important? Without them, the phenols would simply couple when exposed to an oxidizing agent.

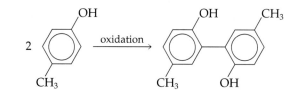

In the laboratory, BHT has shown antitumor activity. Some scientists have speculated that the increased use of antioxidants as food additives has contributed to a decline in the incidence of stomach cancer in the United States.

Antioxidants have been promoted as anticarcinogens (Chapter 20). There is strong evidence that eating lots of fruits and vegetables protects against cancer, but there is little compelling evidence that taking antioxidant vitamin supplements does so.

With the bulky butyl groups aboard, the rings can't get close enough together for coupling. They are free, then, to trap free radicals formed by the oxidation of fats.

Many food additives have been criticized as being harmful, and BHA and BHT are no exceptions. They have been reported to cause allergic reactions in some people. In one study, pregnant mice fed diets containing 0.5% BHA or BHT gave birth to offspring with brain abnormalities. On the other hand, when relatively large amounts of BHT were fed to rats daily, their life spans were increased by a human equivalent of 20 years. One theory about aging is that it is caused in part by the formation of free radicals. BHT retards this chemical breakdown in cells in the same way that it retards spoilage in foods. Although BHA and BHT are synthetic chemicals, there are also natural antioxidants, such as vitamin E (Section 16.5). Vitamin E (like BHT) is a phenol, with several substituents on the benzene ring. Presumably, its action as an antioxidant is quite similar to that of BHT.

Food Colors

Some foods are naturally colored. For example, the yellow compound β-carotene (read "beta-carotene") occurs in carrots. β-Carotene is used as a color additive in other foods such as butter and margarine. (Our bodies convert β-carotene to vitamin A; thus, it is a vitamin additive as well as a color additive.) Other natural food colors include beet juice, grape-hull extract, and saffron (from autumn-flowering crocus flowers).

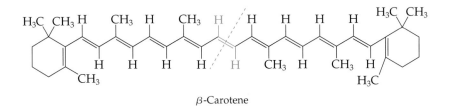

β-Carotene

We expect many foods to have characteristic colors. To increase the attractiveness and acceptability of its products, the food industry has used synthetic food colors for decades. Since the Food and Drug Act of 1906, the FDA has regulated the use of these coloring chemicals and set limits on their concentrations. But the FDA is not infallible. Some colors once on the approved list were later shown to be harmful and were removed from the list. In 1950 a candy company used large amounts of Food, Drug, and Cosmetic (FD & C) Orange No. 1 in an attempt to give Halloween candy a bright orange color like that of pumpkins. Although safe in the smaller amounts previously used, the dye caused gastrointestinal upsets in several children and was then banned by the FDA.

In following years, other dyes were banned. FD & C Yellow Nos. 3 and 4 were found to contain small amounts of β-naphthylamine, a carcinogen that causes bladder cancer in laboratory animals. Furthermore, they reacted with stomach acids to produce more β-naphthylamine. FD & C Red No. 2 was also shown to be a weak carcinogen in laboratory animals. (The structures of these dyes are given in Figure 16.17.) Under the 1958 **Delaney Amendment** to the Food and Drug Act of 1906, substances are automatically banned as food additives if they have been shown to cause cancer in laboratory animals.

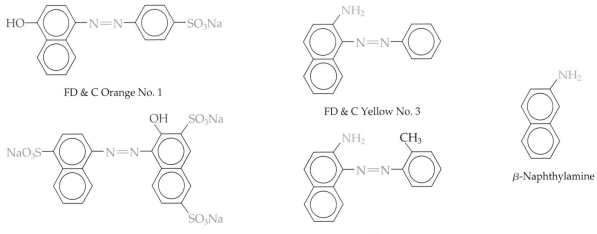

Figure 16.17   Four synthetic food colors that have been banned by the U.S. FDA. Note that the two yellow dyes are related to β-naphthylamine, a carcinogen.

Food colors, even those that have been banned, present little risk. They have been used for years with apparent safety and are normally used in very small amounts. Although the risk is low, however, the benefit is largely aesthetic. Any foods that contain artificial colors must say so on the label, and so you can avoid them if you want to.

The GRAS List

Some food additives have been used for many years without apparent harmful effects. In 1958 the U.S. Congress established a list of additives "generally recognized as safe." This compilation became known as the **GRAS list**. However, some of the additives that were once on the list have since been removed. There were some deficiencies in the original testing procedures. New research findings—and greater consumer awareness—have led the FDA to reevaluate many of the substances on the list.

Cyclamate sweeteners and certain food colors are among the additives that were once on the GRAS list but are now banned. Improved instruments and better experimental designs have revealed possible harm, however slight, where none was thought to exist formerly. Most of the newer experiments involve feeding massive doses of additives to small laboratory animals, and these studies have been criticized for that reason.

16.9 Poisons in Your Food

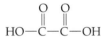

Oxalic acid

People have been trying to deal with poisons in their food for millennia. Early foragers learned—by the painful process of trial and error—that some plants and animals were poisonous. Rhubarb leaves contain toxic oxalic acid. Celery produces psoralens in response to injury. These compounds are powerful mutagens and carcinogens. The Japanese relish a variety of puffer fish that contains a deadly poison in its ovaries and liver. More than 100 Japanese die each year from improperly prepared puffers.

The most toxic substance known is the toxin produced by the bacterium *C. botulinum*. This organism grows in improperly canned food by a perfectly natural process. If the food isn't properly sterilized before it is sealed in jars or cans, the microorganism flourishes under anaerobic (without air) conditions. The poison it produces is so toxic that 1 g of it could kill more than a million people. The point is that a food is not inherently good simply because it is natural. Neither is it necessarily bad because a synthetic chemical substance has been added to it.

Carcinogens

There is little chance that we will suffer acute poisoning from approved food additives. But what about cancer? Could all these chemicals in our food increase our risk of cancer? The possibility exists, even though the risk is low.

We should recognize, however, that carcinogens occur in food naturally. A charcoal-broiled steak contains 3,4-benzpyrene, a carcinogen also found in cigarette smoke and automobile exhaust fumes. Cinnamon and nutmeg contain safrole, a carcinogen that has been banned as a flavoring in root beer.

Among the most potent carcinogens are **aflatoxins**, compounds produced by molds growing on stored peanuts and grains (Figure 16.18). Aflatoxin B_1 is estimat-

3,4-Benzpyrene

Safrole

Aflatoxin B$_1$

Figure 16.18 Aflatoxins, toxic and carcinogenic compounds, are produced by molds that grow on peanuts and stored grains.

ed to be 10 million times as potent a carcinogen as saccharin, and there is no way to completely keep it out of our food. The FDA sets a tolerance of 20 ppb for aflatoxins.

It is estimated that our consumption of natural carcinogens is 10,000 times as great as that of synthetic carcinogens. Should we ban steaks, spices, peanuts, and gains because they contain naturally occurring carcinogens? Probably not. The risk is slight, and life is filled with more serious risks. Should we ban additives that have been shown to be carcinogenic? Certainly we should carefully weigh the risks against any benefits the additives might provide.

Incidental Additives

There are two major categories of food additives: *intentional additives* are put in a product on purpose to perform a specific function; *incidental additives* get in accidentally during production, processing, packaging, or storage. Pesticide residues, insect parts, and antibiotics added to animal feeds are examples of incidental additives that sometimes get into our food. There are about 2800 intentional additives, and perhaps 10,000 incidental additives. The incidental additives often receive wide publicity.

In 1989 the discovery of residues of daminozide (Alar) on apples caused great public concern. Daminozide is a plant growth regulator that causes apples to ripen at the same time and have a better appearance than untreated apples.

$$HO-\overset{\overset{O}{\|}}{C}CH_2CH_2\overset{\overset{O}{\|}}{C}-NHN(CH_3)_2$$

Daminozide

The compound itself appears to be harmless, but it breaks down to yield dimethyl-hydrazine, a suspected carcinogen.

$$H_2NN(CH_3)_2$$

Dimethylhydrazine

To get a dose equivalent to that fed to laboratory animals, a person would have to eat 13,000 kg of apples a day for 70 years. To most scientists, the risk seemed vanishingly small, but many consumers were unwilling to assume the risk, however small. Falling sales of apples led to a withdrawal of the chemical from use by apple growers and removal of the compound from the market by the manufacturer.

Alar was not the first incidental additive to cause concern. In 1959 the sale of cranberries was forbidden after some shipments were found to be contaminated by the herbicide aminotriazole, a compound shown to be carcinogenic in tests on laboratory animals. In 1969 coho salmon taken from Lake Michigan were shown to contain DDT above the tolerance level, and the sale of these fish was banned also.

Polychlorinated biphenyls (PCBs; Chapter 10) have been found in poultry and eggs. These products, too, have been seized and destroyed. Related compounds, polybrominated biphenyls (PBBs), meant to be used as fire retardants, were accidentally mixed with animal feed in western Michigan. Many farm animals were destroyed, and yet PBBs still got into the food supply.

Aminotriazole

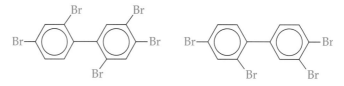

Typical PBB compounds

Chemical substances in animal feed often show up in the meat we eat. Antibiotics are added to animal feed to promote weight gain. In fact, nearly 50% of the 9 million kg of antibiotics produced annually in the United States goes into the low-level dosages in animal feeds. Residues of these antibiotics have been found in meat. These residues may result in the sensitization of individuals who eat the meat, thus hastening the development of allergies. The use of antibiotics in animal feeds may also hasten the process by which bacteria become drug-resistant.

Diethylstilbestrol (DES), a synthetic female hormone, was once added to animal feeds to promote weight gain. It was banned after evidence showed that DES caused vaginal cancer in the daughters of women who had taken it during pregnancy. DES has also been shown to cause testicular cancer in the sons of women who took it while pregnant.

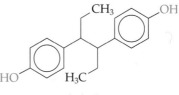

Diethylstilbestrol

It should be noted that the effect of DES did not show up for 15 years and that even then it appeared in the *offspring* of women who took the drug. This points out some of the problems involved in evaluating a chemical for its possible harmful effects.

16.10 A World Without Food Additives

Could we get along without food additives? Some of us could. But food spoilage might drastically reduce the food supply in an already hungry world, and diseases due to vitamin and mineral deficiencies might flourish again. Foods could cost more and be less nutritious. Food additives seem to be a necessary part of modern society. It is true that there are potential hazards associated with the use of some food additives, but the major problem with our food supply is still contamination by rodents, insects, and harmful microorganisms. Indeed, there are about 9000 deaths and 6 million illnesses annually in the United States resulting from food poisoning caused by bacterial toxins. Few, if any, deaths associated with the use of intentional food additives have been documented.

What should we do about food additives? We should be sure that the FDA is staffed with qualified personnel to ensure the adequate testing of proposed food additives. People trained in chemistry are necessary for the control and monitoring of food additives and the detection of contaminants. Research on the analytical techniques necessary for the detection of trace quantities is vital to adequate consumer protection. We should also work for laws adequate to prevent the unnecessary and excessive use of pesticides and other agricultural chemicals that might contaminate our food. Above all, we should be alert and informed about these problems that are so vital to our health and well-being.

> In 1993 improperly cooked hamburgers contaminated with a virulent strain of *Escherichia coli* bacteria made several hundred people ill in Washington, Idaho, and Nevada, including two children who died.

16.11 Plants: Sun-Powered Food-Making Machines

All food comes ultimately from green plants. Although all organisms can transform one type of food into another, only green plants can use sunlight to convert carbon dioxide and water to the sugars that directly or indirectly fuel all living things.

$$6\,CO_2 + 6\,H_2O \xrightarrow{\text{photosynthesis}} C_6H_{12}O_6 + 6\,O_2$$

This reaction also replenishes the oxygen in the atmosphere and removes carbon dioxide (Chapter 12). With other nutrients, particularly compounds of nitrogen and phosphorus, plants can convert the sugars from photosynthesis to proteins, fats, and other chemicals that we use as food.

The structural elements of plants—carbon, hydrogen, and oxygen—are derived from air and water. Other plant nutrients are taken from the soil, and energy is supplied by the sun. In early agricultural societies, people grew plants for food and obtained energy from the food. Nearly all this energy was reinvested in the production of food, although a portion went into making clothing and building shelter. Figure 16.19 shows a simplified diagram of the energy flow in such a society. Obviously, a real society would be much more complicated than indicated in the diagram. Some plants, for example, might be fed to animals, and human energy would in turn be obtained from animal flesh or animal products (such as milk and eggs).

In early societies, nearly all the energy came from renewable resources. One unit of human work energy, supplemented liberally by energy from the sun, might produce ten units of food energy. The surplus energy could be used to make clothing or to provide shelter. It also might be used in games or cultural activities.

The flow of nutrients is also rather simple. Unused portions of plants and human and animal wastes are returned to the soil. These are broken down by bacteria to provide nutrients for the growth of new plants (Figure 16.20). Thus, the humus content of the soil is maintained. Properly practiced, this kind of agriculture could be continued for centuries without seriously depleting the soil. The only problem is that farming at this level supports relatively few people.

> Soil is rock broken down by weathering, plus various types of decaying plant and animal matter. It takes many years to build even a few millimeters of soil from rocks. The United States loses 5 billion t of soil each year to erosion. Developing countries often suffer far more devastating losses as more land is cleared for farming in an attempt to feed an increasing population.

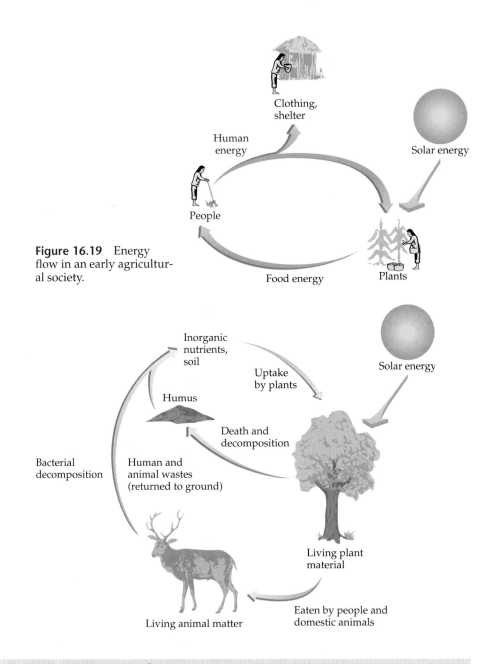

Figure 16.19 Energy flow in an early agricultural society.

Figure 16.20 Flow of nutrients in a simple system.

Growing Our Food

16.12 Farming with Chemicals: Fertilizers

To replace nutrients lost from the soil and to increase crop production, modern farmers use a variety of chemical fertilizers. The three **primary plant nutrients** are nitrogen, phosphorus, and potassium. Let's consider nitrogen first.

Nitrogen Fertilizers

Nitrogen, although present in air in the elemental form (N_2), is not generally available to plants. There are some forms of bacteria that are able to fix nitrogen—that is, to convert it to a combined, soluble form. Colonies of bacteria that can perform this vital function grow in nodules on the roots of legumes (plants such as clovers and peas) (Figure 16.21). Thus, farmers are able to restore fertility to the soil by crop ro-

Figure 16.21 Bacteria in nodules on the roots of legumes fix atmospheric nitrogen by converting it to soluble compounds that plants can use as nutrients.

tation. A nitrogen-fixing crop (such as clover) is alternated with a nitrogen-consuming crop (such as corn).

Legumes are still used to some extent to supply nitrogen to the soil. But the modern methods of high-yield production demand chemical fertilizers. Why get a corn crop every other year when you can have one every year?

Plants usually take up nitrogen in the form of nitrate ions (NO_3^-) or ammonium ions (NH_4^+). These ions are combined with carbon compounds from photosynthesis to form amino acids, the building blocks of proteins (the polymeric compounds essential to all life processes). For years, farmers were dependent on manure as a source of nitrates. The discovery of deposits of sodium nitrate (called Chile saltpeter) in the deserts of northern Chile led to use of this substance as a source of nitrogen.

A rapid rise in population growth during the late nineteenth and early twentieth centuries led to increasing pressure on the available food supply, and this pressure led to an increasing demand for nitrogen fertilizers. The atmosphere offered a seemingly inexhaustible supply of nitrogen—if only it could be converted to a form useful to humans. Every flash of lightning forms some nitric acid in the air (Figure 16.22). Lightning is now known to contribute about a billion tons of fixed nitrogen each year—more than is provided by nitrogen-fixing bacteria.

The first real breakthrough in nitrogen fixation came in Germany on the eve of World War I. A process developed by Fritz Haber made possible the combination of nitrogen and hydrogen to make ammonia.

$$3\,H_2 + N_2 \longrightarrow 2\,NH_3$$

By 1913 one nitrogen-fixation plant was in production and several more were under construction. The Germans were able to make ammonium nitrate (NH_4NO_3), an explosive, by oxidizing part of the ammonia to nitric acid.

$$NH_3 + 2\,O_2 \longrightarrow HNO_3 + H_2O$$

The nitric acid was then reacted with ammonia to produce ammonium nitrate.

$$HNO_3 + NH_3 \longrightarrow NH_4NO_3$$

The Germans were interested in ammonium nitrate mainly as an explosive, but it turned out to be a valuable nitrogen fertilizer as well.

A gas at room temperature, ammonia is easily compressed into a liquid that can be stored and transported in tanks. In this form, called anhydrous ammonia, it is applied directly to the soil as fertilizer (Figure 16.23). Some ammonia is reacted with

Figure 16.22 The intense heat in a lightning flash fixes nitrogen that then falls to Earth as nitric acid, enriching the soil.

"Anhydrous" means "without water."

Figure 16.23 A farmer applies anhydrous ammonia, a source of the plant nutrient nitrogen, to his fields.

carbon dioxide to form urea, an organic compound that releases nitrogen into the soil slowly (rather than all at once as inorganic forms do).

$$2\,NH_3 \ + \ CO_2 \ \longrightarrow \ NH_2-\overset{\displaystyle O}{\overset{\|}{C}}-NH_2 \ + \ H_2O$$
Urea

Some ammonia, as we mentioned, is converted to ammonium nitrate, and some to ammonium sulfate [$(NH_4)_2SO_4$] by reaction with sulfuric acid. Both of these ammonium salts are crystalline solids. These ammonia products (Table 16.9) can be applied separately or combined with other plant nutrients to make a more complete fertilizer.

The atmosphere has vast amounts of nitrogen, but current industrial methods of fixing it require a lot of energy and use up nonrenewable resources. (The hydrogen needed for the synthesis of ammonia is made from natural gas.) Genes for nitrogen fixation have been transferred from bacteria into higher (nonlegume) plants, and perhaps some day soon corn and cotton will be able to produce their own nitrogen fertilizer the way clovers and peas do.

Phosphorus Fertilizers

The availability of phosphorus is often the limiting factor in plant growth. In plants phosphates are incorporated into DNA and RNA (Chapter 15) and other compounds essential to plant growth. Phosphates have probably been used as fertilizers since ancient times—in the form of bone, guano (bird droppings), or fish meal. However, phosphates as such were not recognized as plant nutrients until 1800. Following this discovery, the great battlefields of Europe were dug up and bones were shipped to chemical plants for processing into fertilizer.

Table 16.9 ▌ Various Forms of Nitrogen Fertilizers Made from Ammonia

Reagent	Product Fertilizer	Formula
None	Anhydrous ammonia	NH_3
Carbon dioxide	Urea	NH_2CONH_2
Sulfuric acid	Ammonium sulfate	$(NH_4)_2SO_4$
Nitric acid	Ammonium nitrate	NH_4NO_3
Phosphoric acid	Ammonium monohydrogen phosphate	$(NH_4)_2HPO_4$

Fritz Haber was awarded the Nobel prize in chemistry in 1918. There are a number of ironies in this. Alfred Nobel, the Swedish inventor and chemist who died in 1896, endowed the Nobel prize (including the peace prize) with a fortune derived from his own work with explosives. During his lifetime, he was bitterly disappointed that the explosives he had developed for excavation and mining were put to such destructive uses in war. And Haber, a man who helped his country during World War I to the extent of going to the front to supervise the release of chlorine in the first poison-gas attack, was exiled from his native land in 1933. He was Jewish, and Nazi racial laws forced him out of his position as director of the Kaiser Wilhelm Institute of Physical Chemistry. He accepted a post at Cambridge University in England but died of a stroke less than a year later.

Fritz Haber (1868–1934), a German chemist, invented a process for manufacturing ammonia from N_2 and H_2.

Animal bones are rich in phosphorus, but the phosphorus is tightly bound and not readily available to plants. In 1831 the Austrian chemist Heinrich Wilhelm Kohler treated animal bones with sulfuric acid to convert them to a more soluble form called superphosphate. In 1843 John Lawes applied the same treatment to phosphate rock. The essential reaction for forming superphosphate is

$$Ca_3(PO_4)_2 + 2\,H_2SO_4 \longrightarrow \underbrace{Ca(H_2PO_4)_2 + 2\,CaSO_4}$$

Phosphate rock or bone (insoluble)   Superphosphate (more soluble)

Modern phosphate fertilizers are often produced by treating phosphate rock with phosphoric acid to make water-soluble calcium dihydrogen phosphate.

$$Ca_3(PO_4)_2 + 4\,H_3PO_4 \longrightarrow 3\,Ca(H_2PO_4)_2$$

Even more common today is the use of ammonium monohydrogen phosphate [$(NH_4)_2HPO_4$], which supplies both nitrogen and phosphorus.

Phosphate is quite common in the soil, but available forms are often at concentrations too low for adequate support of plant growth. Fortunately, there are deposits that are more concentrated. The presence of bones and teeth from early fish and other animals in these ores indicates that the deposits are largely the skeletal remains of sea creatures of ages past (Figure 16.24). Unfortunately, fluorides are often associated with phosphate ores, and they create a serious pollution problem involving the production of phosphate fertilizers.

About 90% of all phosphates produced are used in agriculture. The United States is the leading producer and user, but the rich phosphate deposits in the United States may soon be depleted. Large offshore deposits have been discovered in the Atlantic Ocean off North Carolina. Worldwide, Morocco has two-thirds of the reserves. All the high-grade phosphate reserves may be exhausted in 30 or 40 years. Our use of phosphates scatters them irretrievably throughout the environment.

Potassium Fertilizers

The third major element necessary for plant growth is potassium. Plants use it in the form of the simple ion K^+. Generally, potassium is abundant, and there are no problems with solubility. Potassium ions, along with Na^+ ions, are essential to the fluid balance of cells. They also seem to be involved in the formation and transport of

Figure 16.24 Phosphate fertilizers are obtained from an ore called rock phosphate which is largely the skeletal remains of ancient sea creatures.

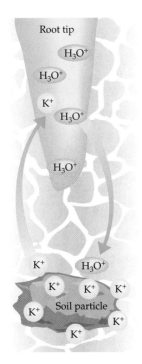

Figure 16.25 When a root tip takes up K^+ from the soil, H_3O^+ ions are transferred to the soil. The uptake of potassium ions therefore tends to make the soil acidic.

In 1912 American farmers produced an average of 26 bushels of corn per acre. Today the yield per acre is about 100 bushels. The fourfold increase is mainly due to the increased use of fertilizers.

carbohydrates. Also, K^+ may be necessary for the assembly of proteins from amino acids. Uptake of potassium ions from the soil leaves the soil acidic; each time one potassium ion enters the root tip, a hydronium ion must leave in order for the plant to maintain electrical neutrality (Figure 16.25).

The usual chemical form of potassium in commercial fertilizers is potassium chloride (KCl). Vast deposits of this salt occur in Stassfurt, Germany, and for years this source supplied nearly all the world's potassium fertilizer. With the coming of World War I, the United States sought supplies within its own borders. Deposits at Searles Lake, California, and Carlsbad, New Mexico, now supply much of the needs of the United States. Canada has vast deposits in Saskatchewan and Alberta. Beds of potassium chloride up to 200 m thick lie about 1.5 km below the Canadian prairies (Figure 16.26). Although reserves are large, potassium salts are a nonrenewable resource and we should use them wisely.

Other Essential Elements

In addition to the three major nutrients (nitrogen, phosphorus, and potassium), a variety of other elements are necessary for proper plant growth. Three **secondary plant nutrients**—magnesium, calcium, and sulfur—are needed in moderate amounts. Calcium, in the form of lime (calcium oxide), is used to neutralize acidic soils.

$$CaO + 2 H_3O^+ \longrightarrow Ca^{2+} + 3 H_2O$$

Calcium ions are also necessary plant nutrients. Magnesium ions (Mg^{2+}) are incorporated into chlorophyll molecules and therefore are necessary for photosynthesis. Sulfur is a constituent of several amino acids, and it is necessary for protein synthesis.

Eight other elements, called **micronutrients**, are needed in small amounts. These elements are summarized in Table 16.10. Many soils contain these trace elements in sufficient quantities, but some are deficient in one or more, and their productivity can be markedly increased by adding small amounts of the needed elements.

Plants may also require other elements, including sodium, silicon, vanadium, chromium, selenium, cobalt, fluorine, and arsenic. Most of these elements are present in soil, but it is not known whether they are necessary for plant growth.

Fertilizers: A Mixed Bag

Most farmers buy *complete fertilizers*, which, despite the name, usually contain only the three main nutrients. There are usually three numbers on fertilizer bags

Figure 16.26 Mining potassium chloride about 1.5 km below the prairies of Saskatchewan, Canada.

Table 16.10 ▌ Eight Micronutrients Necessary for Proper Plant Growth

Element	Form Used by Plants	Function	Deficiency Symptoms
Boron	H_3BO_3	Required for protein synthesis; essential for reproduction and for carbohydrate metabolism	Death of growing points of stems, poor growth of roots, poor flower and seed production
Copper	Cu^{2+}	Constituent of enzymes; essential for reproduction and for chlorophyll production	Twig dieback, yellowing of newer leaves
Iron	Fe^{2+}	Constituent of enzymes; essential for chlorophyll production	Yellowing of leaves, particulariy between veins
Manganese	Mn^{2+}	Essential for redox reactions and for the transformation of carbohydrates	Yellowing of leaves, brown streaks of dead tissue
Molybdenum	$MoO_4{}^{2-}$	Essential in nitrogen fixation by legumes and reduction of nitrates for protein synthesis	Stunting, pale-green or yellow leaves
Nickel	Ni^{2+}	Required for iron absorption; constituent of enzymes	Failure of seeds to germinate
Zinc	Zn^{2+}	Essential for early plant growth and maturing	Stunting, reduced seed and grain yields
Chlorine	Cl^-	Increases water content of plant tissue; involved in carbohydrate metabolism	Shriveling

(Figure 16.27). The first number represents the percent of nitrogen (N); the second, the percent of phosphorus (calculated as P_2O_5); and the third, the percent of potassium (calculated as K_2O). So 5-10-5 means that a fertilizer contains 5% N, 10% P_2O_5, and 5% K_2O; the rest is inert material.

Fertilizers must be water-soluble to be used by plants. When it rains, nutrients from fertilizers can be washed into streams and lakes, where they can stimulate blooms of algae. These chemicals, particularly nitrates, also enter the groundwater.

16.13 The War Against Pests

Since the earliest days of recorded human history (and surely even before that), people have been plagued by insect pests. Three of the ten plagues of Egypt (described in the Book of Exodus) were insect plagues—lice, flies, and locusts. The decline of Roman civilization has been attributed in part to malaria, a disease carried by mosquitoes, which destroys vigor and vitality when it does not kill. Bubonic plague carried by rats (and by fleas from rats to humans) swept through the Western world repeatedly during the Middle Ages. One such plague (during the 1660s) is estimated to have killed 25 million people—25% of the population of Europe at that time. The first attempt to dig a canal across Panama (made by the French during the 1880s) was defeated by outbreaks of yellow fever and malaria.

The use of modern chemical **pesticides** may be the only thing that stands between us and some of these insect-borne plagues. Pesticides also prevent the consumption of a major portion of our food supply by insects and other pests. Crop losses to insects in the United States are estimated to be $4 billion a year. When losses to other pests—rodents, fungi, nematodes, and weeds—are included, losses total $15 billion a year.

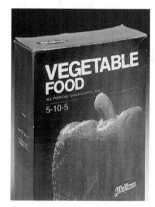

Figure 16.27 The numbers on this bag of fertilizer indicate that the fertilizer is 5% nitrogen (N), 10% P_2O_5, and 5% K_2O. There really is no K_2O or P_2O_5 in fertilizer. These formulas are used merely as a basis for calculation. The actual form of potassium is nearly always KCl, although any potassium salt would furnish the needed K^+ ion. Phosphorus is supplied as one of several salts.

In earlier days, people tried to control insect pests by draining swamps, pouring oil on ponds (to kill mosquito larvae), and using various chemicals. Most of these chemicals were compounds of arsenic. Lead arsenate [$Pb_3(AsO_4)_2$] is a particularly effective poison because both the lead and the arsenic in it are toxic. A few pesticides, such as pyrethrum (used in mosquito control) and nicotine sulfate (Black Leaf 40), are obtained from plant matter.

Only a few insect species are harmful. Many are beneficial, and others play important roles in ecological systems and are indirectly beneficial. Most poisons are indiscriminate. They kill all insects, not just those we consider pests. Many are also toxic to humans and other animals. Some say we should call such poisons *biocides* (because they kill living things) rather than insecticides. Table 16.11 lists the toxicities of some insecticides.

Table 16.11 ▌ Toxicity of Insecticidal Preparations Administered Orally to Rats

Pesticide	LD_{50}*		
Pyrethrins[1]	1200		Least toxic
Malathion	1000	(1375)	
Lead arsenate	825		
Diazinon	285	(250)	
Carbaryl	250		
Nicotine[2]	230		
DDT[3]	118	(113)	
Lindane	91	(88)	
Methyl parathion	14	(24)	
Parathion	3.6	(13)	
Carbofuran[4]	2		
Aldicarb	1		Most toxic

*Dose in milligrams per kilogram of body weight that will kill 50% of test population (see Section 20.13). Values in parentheses are for male rats.

Notes:
(1) Active ingredients of pyrethrum.
(2) In mice. Nicotine is much more toxic by injection.
(3) Estimated LD_{50} for humans is 500 mg/kg.
(4) In mice.
Source: Susan Budavari (ED.), *The Merck Index*, 12th edition. Rahway, NJ: Merck and Co., 1996

Example 16.4

What amount of the pesticide lindane could lead to the death of a 15-kg (33-lb) child if the lethal dose is the same as it is for male rats (Table 16.11)?

Solution
The lethal dose is given as 88 mg of lindane per kilogram of body weight.

$$15 \text{ kg} \times \frac{88 \text{ mg}}{1 \text{ kg}} = 1320 \text{ mg} \quad \text{or} \quad 1.32 \text{ g}$$

Exercise 16.4
How much parathion would it take to kill a 75-kg (165-lb) farm worker if the lethal dose were the same as it is for rats (Table 16.11)?

DDT: The Dream Insecticide

Shortly before World War II, Swiss scientist Paul Müller (1899–1965) found that DDT (dichlorodiphenyltrichloroethane), a chlorinated hydrocarbon, is a potent insecticide. DDT was soon used effectively against grapevine pests and against a particularly severe potato beetle infestation.

When the war came, supplies of pyrethrum, a major insecticide of the time, were cut off by the Japanese occupation of Southeast Asia and the Dutch East Indies (now Indonesia). Lead, arsenic, and copper that went into insecticides were needed for armaments and other military purposes.

The Allies desperately needed an insecticide to protect soldiers from disease-bearing lice, ticks, and mosquitoes. They obtained a small quantity of DDT and quickly tested it. Combined with talcum, DDT was an effective delousing powder. Clothing was impregnated with DDT, and it seemed to have no harmful effects even on those exposed to large doses. Allied soldiers were nearly free of lice, but German troops were heavily infested and many were sick with typhus. In wars before World War II, more soldiers probably died from typhus than from bullets.

DDT is easily synthesized from cheap, readily available chemicals. Chlorobenzene and chloral hydrate are warmed in the presence of sulfuric acid. When the reaction is complete, the mixture is poured into water, and the DDT separates out (like other chlorinated hydrocarbons, DDT is essentially insoluble in water).

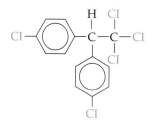

DDT
(Dichlorodiphenyltrichloroethane)

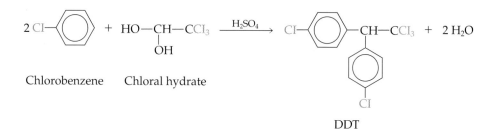

Chlorobenzene   Chloral hydrate

DDT

A cheap **insecticide** effective against a variety of insect pests, DDT came into widespread use after the war (Figure 16.28). Other chlorinated hydrocarbons were synthesized, tested, and used in the war against insects. Although invaluable to farmers in the production of food and fiber, chlorinated hydrocarbons won their most dramatic victories in the field of public health. According to the World Health Organization, approximately 25 million lives have been saved and hundreds of millions of illnesses prevented by the use of DDT and related pesticides.

Figure 16.28 Insecticides are often applied by aerial spraying. Winds can spread the pesticide over a wide area.

DDT seemed to be a dream come true. The world would be free at last from insect plagues and insect-borne diseases. Crops would be protected from the ravages of insects, and food production would be increased. In recognition of his discovery, Paul Müller was awarded the Nobel prize in physiology and medicine in 1948.

The Decline and Fall of DDT

Before Müller had even received his prize, however, there were warnings that all was not well. Houseflies resistant to DDT were reported as early as 1946, and DDT's toxicity to fish had been reported by 1947. In subsequent years, 11 additional resistant species of insects were discovered. Such early warnings were largely ignored, and it was also assumed that the toxicity would disappear soon after the chemical was discharged into the environment. DDT was used extensively to protect crops and control mosquitoes and to try to prevent Dutch elm disease. By 1962, the year Rachel Carson's book *Silent Spring* appeared, U.S. production of DDT had reached 76 million kg/year.

Today, in developed countries, DDT is known mostly for its harmful environmental effects. Birds are threatened because DDT causes their eggs to have thin shells that are poorly formed and easily broken. Even a few parts per billion of DDT interferes with the growth of plankton and the reproduction of crustaceans such as shrimp. As bad as DDT sounds, however, it has probably saved the lives of more people than any other chemical substance.

Chlorinated hydrocarbons generally are unreactive. This lack of reactivity was a major advantage of DDT. Sprayed on a crop, DDT stayed there and killed insects for weeks. This *pesticide persistence* was also a major disadvantage: the substance did not break down readily in the environment. Although not very toxic to humans and other warm-blooded creatures, DDT is much more toxic to cold-blooded organisms, which include insects, of course, and fish. DDT's lack of reactivity toward oxygen, water, and components of the soil led to its buildup in the environment where it threatened fish, birds, and other wildlife.

Biological Magnification: Concentration in Fatty Tissues

Chlorinated hydrocarbons are good solvents for fats, and fats are good solvents for chlorinated hydrocarbons such as DDT. When these compounds are ingested as contaminants in food or water, they are concentrated in fatty tissues. Their fat-soluble nature causes chlorinated hydrocarbons to be concentrated up the food chain. This *biological magnification* was graphically demonstrated in California in 1957. Clear

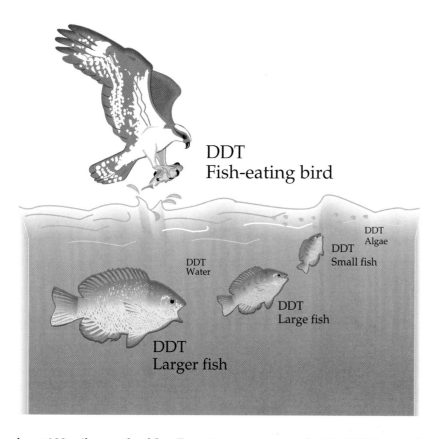

DDT
Fish-eating bird

DDT
Algae

DDT
Small fish

DDT
Water

DDT
Large fish

DDT
Larger fish

Figure 16.29 Concentration of DDT up the food chain. Animals at the top of the food chain have the highest concentrations of the pesticide.

Lake, about 100 miles north of San Francisco, was sprayed with DDT in an effort to control gnats. After spraying, the water contained only 0.02 ppm of DDT, but the microscopic plant and animal life contained 5 ppm—250 times as much. Fish feeding on these microorganisms contained up to 2000 ppm. Grebes, the diving birds that ate the fish, died by the hundreds (Figure 16.29).

DDT and other chlorinated hydrocarbons such as PCBs (Chapter 10) are nerve poisons. They are concentrated in the fatlike compounds that make up nerve sheaths (Figure 16.30). DDT also interferes with calcium metabolism essential to the formation of healthy bones and teeth. Although no harm to humans has ever been conclusively demonstrated, the disruption of calcium metabolism in birds has been disastrous for some species. The shells of their eggs, composed mainly of calcium compounds, are thin and poorly formed. The bald eagle, peregrine falcon, and other birds became endangered species. All have made dramatic recoveries since DDT was banned in the United States in 1972.

The ban was not total, however, DDT is still produced and exported, particularly to developing countries where it still sometimes plays a role in malaria control. But the problem of obtaining a worldwide ban on DDT may solve itself as more and more insect species develop resistance. Its use may be discontinued simply because it is no longer effective. Most chlorinated hydrocarbon insecticides have been banned or severely restricted in developed countries.

Figure 16.30 DDT dissolves in the fatty membranes of cells. The wedge-shaped DDT molecules open channels in the membranes, causing them to leak.

Organic Phosphorus Compounds

Bans and restrictions on chlorinated hydrocarbon insecticides have led to increased use of organic phosphorus compounds such as malathion, diazinon, and parathion (Figure 16.31). More than two dozen of these insecticides are available commercially. They have been extensively studied and evaluated for effectiveness against insects and for toxicity to people, laboratory animals, and farm animals.

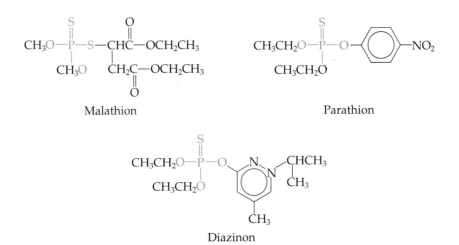

Figure 16.31 Three organic phosphorus compounds used as insecticides.

Organic phosphorus compounds generally are more toxic to mammals than chlorinated hydrocarbons (Table 16.11).(Malathion, used to control fruit fly infestations in California, is a notable exception; it is less toxic than DDT.) Like chlorinated hydrocarbons, organic phosphorus pesticides concentrate in fatty tissues. Phosphorus compounds are less persistent in the environment; they break down in days or weeks, whereas chlorinated hydrocarbons often persist for years. Residues of organic phosphorus compounds are seldom found in food.

Carbamates

To protect themselves from being eaten, some plants produce their own pesticides, often making up 5–10% of the dry weight of the plants. Nicotine protects tobacco plants. Pyrethrins are also plant-produced pesticides. Physostigmine is a naturally occurring carbamate. Our consumption of natural pesticides is probably 10,000 times as great as that of synthetic ones.

Carbamates and organic phosphorus compounds are nerve poisons. They are discussed in Section 20.11.

Another group of compounds that have gained prominence since the ban on DDT are carbamates. Examples are carbaryl (Sevin), carbofuran (Furadan), and aldicarb (Temik) (Figure 16.32). Carbaryl has a low toxicity to mammals, but carbofuran and aldicarb have toxicities similar to that of parathion. Most carbamates are *narrow-spectrum insecticides* directed specifically at one, or a few, insect pests. Chlorinated hydrocarbons and organic phosphorus compounds that kill many kinds of insects are called *broad-spectrum insecticides*. Carbamates break down rapidly in the environment. Unfortunately, carbaryl, the most widely used carbamate, is particularly toxic to honeybees. Millions of these valuable insects have been wiped out by spraying crops with carbaryl.

The essential feature of the carbamate molecule is the group

Many variations can be obtained by changing the groups attached to the oxygen and the nitrogen. Generally, they break down rapidly in the environment and do not accumulate in fatty tissue, which is an advantage of these insecticides over chlorinated hydrocarbons and organic phosphorus compounds.

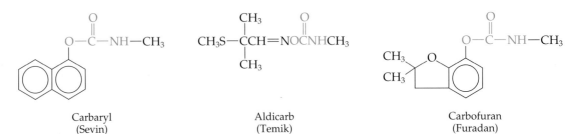

Carbaryl (Sevin) Aldicarb (Temik) Carbofuran (Furadan)

Figure 16.32 Three carbamate insecticides.

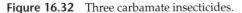

16.14 Biological Insect Controls

A rapidly developing technique is the use of natural enemies to control pests. Praying mantises and ladybugs are sold commercially and are used to destroy garden pests. Biological controls also include bacteria and viruses, often specific for a given species of pest. *Bacillus thuringiensis*, sold under trade names such as Dipel, is widely used by home gardeners against cabbage loopers, hornworms, and other moth caterpillars. The gene for the toxin produced by this bacterium has also been inserted into cotton, corn, and potato plants. These genetically engineered cotton plants can then protect themselves against the cotton bollworm. To date, this technique has shown only limited success, and insects are already developing a resistance to the toxin.

Viruses that attack insect pests can be grown in culture and applied to crops. These viral agents are generally specific to their insect hosts and therefore harmless to people and other animals. They are also completely biodegradable. The biggest problem is that the production of viral pesticides is expensive.

A highly successful biological approach is the breeding of insect- and fungus-resistant plants. Yields of corn (maize), wheat, rice, and other grains have been increased substantially in this manner. Genetic engineering should lead to even more rapid development of disease-resistant plants. Not all such research is successful, however. Plant breeders have produced a potato that is insect-resistant, but it had to be taken off the market because it also is toxic to people.

Sterilization

Sterilization is effective in a few insect species. Large numbers of males are sterilized by radiation, chemicals, or cross-breeding. Then, enough are released so that they far outnumber the local fertile males. If a female mates with a sterile male, no offspring are produced.

Radiation sterilization has virtually eliminated the screwworm fly, once a serious pest that affected cattle in the southern United States. Tropical fruit flies also have been eradicated by this method in some areas. Sterilization involves raising vast numbers of insects, separating the males from the females, and then sterilizing and releasing the males. The great expense and limited applicability of this process probably mean that it will not become a major method of insect control.

Pheromones: The Sex Trap

One intriguing area of research on insect control is the investigation of **pheromones**. These chemicals, secreted externally by insects, mark a trail, send an alarm, or attract a mate. Perhaps the most interesting pheromones are insect **sex attractants**, usually secreted by females to attract males (Figure 16.33). Chemical research can identify sex attractants and determine their structures. These compounds can then be synthesized and used to lure males into traps. Alternatively, the attractant can be used in quantities sufficient to confuse and disorient males, who detect a female in every direction but can't find one to mate with.

Some sex attractants are amazingly simple. Others are complex. The sex attractant for the common housefly is a simple, unsaturated hydrocarbon.

$$CH_3(CH_2)_7CH{=}CH(CH_2)_{12}CH_3$$

Although this molecule contains 23 carbon atoms (with a double bond between the ninth and tenth), it is fairly easy to synthesize. Quantities of it are now available for testing and development.

Most research on sex attractants is not easy. Some attractants have complicated structures, and all are secreted in extremely tiny amounts. For example, a team of U.S. Department of Agriculture researchers had to use the tips of 87,000 female gypsy moths to isolate a minute amount of a powerful sex attractant. Pheromones are effective at extremely low concentrations. A male silkworm moth can detect as few as

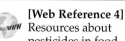

Carbamates were being made at a chemical plant in Bhopal, India, in December 1984 when an explosion released large amounts of the toxic gas methyl isocyanate.

 [Web Reference 4] Resources about pesticides in food and other aspects of food safety.

[Web Reference 5] Website of the nonprofit Environmental Working Group known for its "computer-powered research."

Figure 16.33 A male gypsy moth uses its large antennae to detect pheromones from a female.

40 molecules per second. If a female releases as little as 0.01 mg, she can attract every male within 1 km. Gypsy moth larvae have defoliated great forests, mainly in the northeastern United States, and have now spread over much of the country. The gypsy moth pheromone has been used mainly in traps to monitor insect populations.

Much more research has to be done before pheromones can play a major role in insect control. The method is expensive, and research is painstaking and time-consuming. Workers must be careful not to get the attractants on their clothes. Who wants to be attacked on a warm summer night by a million sex-crazed gypsy moths?

Juvenile Hormones

Another approach to insect control is the use of **juvenile hormones**. Hormones are the chemical messengers that control many life functions in plants and animals, and minute quantities produce profound physiological changes. In the insect world, juvenile hormones control the rate of development of the young. Normally, production of the hormone is shut off at the appropriate time to allow proper maturation to the adult stage.

Chemists have been able to isolate insect juvenile hormones and determine their structures. With knowledge of the structure, they can synthesize the hormone or an analog. The application of juvenile hormones to ponds where mosquitoes breed keeps mosquitoes in the harmless preadult stage. Because only adult insects can reproduce, juvenile hormones appear to be a nearly perfect method of mosquito control.

A growing number of "biopesticides" act by triggering the natural defenses of plants. In some cases, the products can turn on the plants' own immune systems.

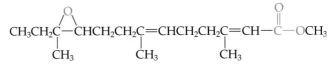

A natural juvenile hormone

Methoprene, a juvenile hormone analog, is approved by the EPA for use against mosquitoes and fleas.

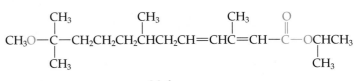

Methoprene

The synthesis of juvenile hormones is difficult and expensive. These hormones are for use against insects that are pests at the adult stage. Little would be gained by keeping a moth or a butterfly in the caterpillar stage for a longer period of time; caterpillars have voracious appetites and do a lot of damage to crops.

16.15 Herbicides and Defoliants

The United States produces about 700 million kg of pesticides annually, including nearly 400 million kg of **herbicides**.

Herbicides are chemicals used to kill weeds. Closely related are **defoliants**, which cause leaves to fall off plants.

2,4-D and 2,4,5-T

Crops produce more abundant harvests when they have no competition from weeds. Removing weeds by hand is tedious, backbreaking work, and so chemical herbicides have been employed for a number of years to kill unwanted plants. Solutions of copper salts, sulfuric acid, and sodium chlorate ($NaClO_3$) have been used, but it wasn't until the introduction of 2,4-D (2,4-dichlorophenoxyacetic acid or one of its derivatives in 1945 that the use of herbicides became common). These chemicals are growth regulator herbicides and are especially effective against newly emerged, rapidly growing broad-leaved plants.

A relative of 2,4-D, called 2,4,5-T (2,4,5-trichlorophenoxyacetic acid), is especially effective against woody plants; it works by causing leaves to fall off plants (defoliation). These two chemicals, combined in a formulation called **Agent Orange**, were used extensively in Vietnam to remove enemy cover and to destroy crops that maintained enemy armies. In addition to causing vast ecological damage, 2,4-D and 2,4,5-T were suspected of causing birth defects in children born to both American soldiers and Vietnamese exposed to the herbicides. Laboratory studies show that these compounds, when pure, do not cause abnormalities in fetuses of laboratory animals. Extensive birth defects are caused, however, by contaminants called **dioxins**, once frequently found in the herbicides. These dioxins are also chlorinated compounds. Continuing concern about dioxin contamination led the EPA to ban 2,4,5-T in 1985.

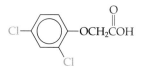

2,4-Dichlorophenoxyacetic acid
(2,4-D)

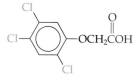

2,4,5-Trichlorophenoxyacetic
acid
(2,4,5-T)

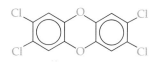

2,3,7,8-Tetrachlorodibenzopara-dioxin
(a "dioxin")

Pesticides—Risks and Benefits

In the United States about 45,000 pesticide poisonings of humans are reported each year, and about 200 of them are fatal. Pesticides are usually poisonous because their purpose is to kill—harmful insects perhaps, or troublesome weeds. There are about 50,000 different pesticide products on the market, and they contain about 1400 active ingredients, most of them quite toxic. Some are used in homes or industrial buildings, but 77% of all pesticides are used in agriculture.

Do we really need to use pesticides? Insect pests cause billions of dollars of damage each year, and weeds can greatly reduce the yields of agricultural products, but can we really justify adding so much toxic material to our environment in order to get rid of them? Consider the case of DDT, which was banned because it had done so much environmental damage. It was used to kill mosquitoes, but it also killed birds.

It is true that DDT killed mosquitoes that spread malaria, but is malaria really a health problem today? If you consider the whole world, malaria is still a major health problem. Worldwide at least 100 million people have the disease, and each year it causes more than a million deaths, mostly among young children.

What about herbicides? Except for problems with dioxin contamination in 2,4,5-T, herbicides seem to have a better safety record than insecticides. Roadsides, vacant lots, and industrial areas have been kept free of weeds, and the value of agricultural crops has been increased by billions of dollars. People who suffer from allergies to ragweed, poison ivy, or other noxious weeds have been helped greatly by the use of herbicides. But we still don't know the ultimate effect on the environment of their long-term use.

One of the earliest defoliant chemicals to be widely used was calcium cyanamide (CaCNCN), which causes cotton plants to lose their leaves when the cotton bolls become mature. This defoliant makes it possible to use mechanical cotton pickers to harvest cotton. If leaves were left on the plants, they would be crushed by the machinery and would stain the cotton.

Atrazine and Glyphosate

Another widely used herbicide is atrazine, which binds to a protein in chloroplasts in plant cells, shutting off the electron transfer reactions of photosynthesis. Atrazine is often used on corn crops. Corn plants deactivate atrazine by removing the chlorine atom. Weeds cannot deactivate the compound and are killed.

Glyphosate, a derivative of the amino acid glycine, is used to control perennial grasses.

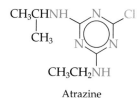

Atrazine

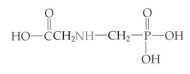

Glyphosate

It is not selective, however, and kills all vegetation. Its isopropylamine salt is sold by Monsanto under the trade name Round-Up.

Paraquat: A Preemergent Herbicide

Another type of herbicide is the **preemergent herbicide** paraquat. This ionic compound is toxic to most plants, but it is rapidly broken down in the soil. Therefore, paraquat can be used to kill weed plants before crop seedlings emerge. Paraquat inhibits photosynthesis by accepting the electrons that otherwise would reduce carbon dioxide.

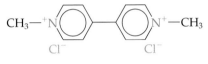

Paraquat

16.16 Alternative Agriculture

One-third of the cost of food in the United States goes to pay for transportation. We spend $6 million and use 3.6 million L of fuel each year just to transport broccoli from California to New York.

Problems with pesticides have led to calls for alternatives to conventional farming, which uses pesticides and fertilizers from chemical plants. This call was heightened in 1989 when the Board of Agriculture of the National Research Council urged farmers to consider other ways to deal with pests and provide plant nutrients. Among the Board's suggestions were crop diversification, integrated pest management (using a mixture of biological controls and synthetic chemicals), disease prevention by careful crop management, and genetic improvement of crops. Much of this advice is similar to that given to organic farmers and gardeners for generations.

Modern agriculture is also energy-intensive. Agriculture uses more petroleum than any other basic industry, accounting for about 13% of all our energy use. This nonrenewable energy is required for the production of fertilizers, pesticides, and farm machinery. Energy is also required to run the machinery needed to till, harvest, dry, and transport the crops and to process and package the food (Figure 16.34).

Organic farmers use manure from farm animals for fertilizer, and they rotate other crops with legumes to restore nitrogen to the soil. They control insects by planting a variety of crops, alternating the use of fields. (A corn pest has a hard time surviving during the year that its home field is planted in soy beans.) Organic farming also is less energy-intensive. According to a study by the Center for the Biology of Natural Systems at Washington University, comparable conventional farms used 2.3 times as much energy as organic farms. Production on organic farms was 10% lower, but costs were lower by a comparable percent. Organic farms require 12% more labor than conventional ones, but human labor is a renewable resource, whereas petroleum energy is not.

Conventional agriculture also often results in severe soil erosion and is the source of considerable water pollution. No doubt we should practice organic farming to the limit of our ability to do so. But we should not delude ourselves. Banning synthetic fertilizers and pesticides would likely lead to a drastic drop in food production.

As far as human energy is concerned, U.S. agriculture is enormously efficient. Each farm worker produces enough food for about 80 people. But this productivity is based on fossil fuels; about ten units of petroleum energy are required to produce

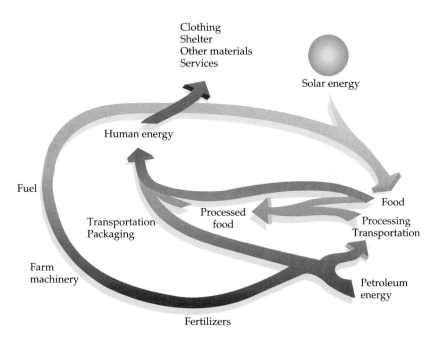

Figure 16.34 Energy flow in modern agriculture. Petroleum energy has largely supplanted human energy.

one unit of food energy. If we consider production per hectare, modern farming is marvelously efficient. If we consider the energy used in relation to the energy produced, it is remarkably inefficient. It should be noted, however, that in an energy-efficient early agricultural society, nearly all human energy went into food production. In modern societies, only about 10% of human energy is devoted to producing food. The other 90% is used to provide the materials and services that are so much a part of our civilization. We should try to make our food production more energy-efficient, but it is unlikely that we will want to return to an outmoded way of life.

[Web Reference 6] A good website about sustainable agriculture.

Organic farming is carried out without the use of synthetic fertilizers or pesticides.

16.17 Some Malthusian Mathematics

In 1830 Thomas Robert Malthus, an English clergyman and political economist, made the statement that population increases faster than the food supply. Unless the birthrate was controlled, he said, poverty and war would have to serve as restrictions on the increase.

Malthus's predictions were based on simple mathematics. Population, he said, grows geometrically, while the food supply increases arithmetically. In **arithmetic growth**, a constant amount is added during each growth period. As an example, consider a cookie jar savings account. The first week, little Lavinia puts in the 25 ¢ she received for her birthday. Each week thereafter, she adds 25 ¢. The growth of the savings is arithmetic; it increases by a constant amount (25 ¢) each week. At the end of the first week, she will have 25 ¢; at the end of the second, 50 ¢; the third, 75 ¢; the fourth, $1.00; the fifth, $1.25; the sixth, $1.50; and so on.

Now let's consider an example of **geometric growth,** in which the increment increases in size for each growth period. Again, let's use little Lavinia's bank as an example. The first week she puts in 25 ¢. The second week she puts in another 25 ¢ to double the amount (to 50 ¢). The third week she puts in 50 ¢ to double this amount again. Each week, she puts in an amount equal to what is already there; and doubles the amount in the bank. At the end of the first week, she has 25 ¢; the second, 50 ¢; the third, $1.00; the fourth, $2.00; the fifth, $4.00; the sixth, $8.00; and so on. Before long, little Lavinia will have to start robbing banks to keep up her geometrically growing deposits.

Table 16.12 ▌ Arithmetic and Geometric Growth Through Ten Periods (Starting with One Unit)

				Growth Period						
	0	1	2	3	4	5	6	7	8	9	10
Arithmetic growth	1	2	3	4	5	6	7	8	9	10	11
Geometric growth	1	2	4	8	16	32	64	128	256	512	1024

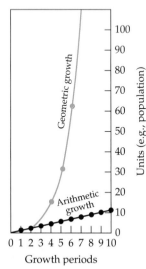

Figure 16.35 Arithmetic growth and geometric growth through ten periods, each starting with one unit.

Table 16.12 compares arithmetic and geometric growth through ten growth periods. These data are shown graphically in Figure 16.35. Note that arithmetic growth is slow and steady; geometric growth starts slowly and then shoots up like a rocket.

For a population growing geometrically, we can calculate the **doubling time** from the **rule of 72**. Simply divide the percent of annual growth into 72. For example, Earth's population is growing 1.5% per year. If it continues to grow at this rate, it will double in 48 years.

Example 16.5

The population of the United States was 250 million in 1990 and was growing at a rate of 0.8% per year. If growth continues at the same rate, when will the population have doubled to 500 million?

Solution

Divide the annual growth rate into 72.

$$\frac{72}{0.8} = 90$$

The population will have doubled to 500 million within 90 years, or by the year 2080.

Exercise 16.5

The population of Malaysia was 21 million in 1999 and was growing at a rate of 2.3% per year. If growth continues at the same rate, when will the population have doubled to 42 million?

Figure 16.36 In developed countries, science has thwarted the Malthusian prediction of hunger. But starvation is still a fact of life for many people in developing countries, particularly for those in areas wracked by war.

Earth's population has grown enormously since the time of Malthus, reaching an estimated 6 billion in 1999. Famine has brought death to millions in war-ravaged areas around the world (Figure 16.36), but modern farming has piled up surpluses in developed countries. Even developing nations have made great progress in food production, many of them becoming self-sufficient. Despite surpluses in some parts of the world, an estimated 800 million people are seriously malnourished. Half are children under 5 years old, who will carry the physical and mental scars of this deprivation for the rest of their lives.

Scientific developments, such as those outlined in this chapter, have brought abundance to many, but millions still go hungry.

16.18 Can We Feed a Hungry World?

If Earth's population continues at its present growth rate, it is expected to reach 9 billion by midcentury and double again to 18 billion before the end of this century. Can we feed all those people?

An alliance between science and agriculture has brought an increase in food supplies beyond the imagination of people of a few generations ago. Through the use of irrigation, synthetic fertilizers, pesticides, and improved genetic varieties of plants and animals, most of the people of the world have abundant food.

We may be able to increase food production even more through genetic engineering (Chapter 15). Scientists can design plants that produce more food, that are resistant to disease and insect pests, and that grow well in hostile environments. They can design animals that grow larger and produce more meat and milk. They may some day provide us with food in unimaginable abundance. But the hungry are with us now, and 100 million more people come to dinner each year.

Even by quadrupling present food production, we could meet our needs only until 2050. Virtually all the world's available arable land is now under cultivation, and we lose farm land every day to housing, roads, erosion, encroaching deserts, and the increasing salt content of irrigated soils. It seems unlikely that we will be able to keep food production ahead of the high rate of population growth, particularly in developing countries.

We seem to be hooked on a high-energy form of agriculture that uses synthetic fertilizers, pesticides, and herbicides and depends on the power of machinery that burns fossil fuels. Ultimately, the only solution to the problem is population stabilization. Obviously, that will come someday. The only questions are when and how. Population control could come through decreasing the birthrate. This is happening in almost all the developed world and in many developing nations. Other nations still have populations growing at explosive rates that far outstrip their food supplies. Birth control isn't the only way to limit populations. Another possibility is an increase in the death rate—through catastrophic war, famine, pestilence, or the poisoning of the environment by the wastes of an ever-expanding population. The ghost of Thomas Malthus haunts us yet.

Critical Thinking Exercises

Apply knowledge that you have gained in this chapter and one or more of the FLaReS principles (Chapter 1) to evaluate the following statements or claims.

16.1 The author of a new diet book claims that he and many other people have been able to lose a substantial amount of weight by following a diet very low in carbohydrates (such as bread, cereals, pasta) but high in proteins (such as beef and lamb).

16.2 An advertisement claims that a certain organically grown cereal provides a "life energy" that does not exist in ordinary processed cereals.

16.3 A physician noted that the blood cholesterol of one of his patients had gone down dramatically. On the advice of a friend, the patient had been taking daily supplements of niacin, one of the B vitamins. Although he was skeptical, the doctor found six other patients who were willing to try the niacin treatment without any further medication. A year later he was surprised to find that five of these patients had lowered their cholesterol levels significantly. The doctor says he believes that niacin can be used to lower blood cholesterol.

Summary

1. The three main classes of foods are carbohydrates, fats, and proteins.
2. We also need to include vitamins, minerals, fiber, and water in our diets.
3. Carbohydrates include sugars, starches, and cellulose.
4. Dietary fats and oils are mainly triglycerides, esters of fatty acids and glycerol.
5. Solid fats (mostly from animals) are largely saturated; liquid oils (mostly from plants) are usually unsaturated or polyunsaturated.
6. Proteins are amino acid polymers. Dietary proteins supply all the amino acids needed for making body proteins and other nitrogen compounds.
7. There are 30 elements essential to life: some structural elements, some macrominerals, and some trace elements.
8. Vitamins are compounds that our bodies need but cannot make. Most are needed as coenzymes.
9. Vitamins B and C are water-soluble, whereas A, D, E, and K are all fat-soluble.
10. Dietary fiber may be soluble (hemicelluloses) or insoluble (cellulose).
11. Drinking enough water is very important, but it need not be plain water. Most drinks are mainly water.
12. Food additives are used to add nutritional enrichment, to inhibit spoilage, to add color or flavor, to add sweetness, to bleach, or to provide texture.

13. Toxic substances found in food are sometimes natural ingredients, but sometimes they are added inadvertently to our food in the form of pesticide residues or animal feed additives.

14. All food originates in plants. Photosynthesis occurring in green plants is the reaction that stores energy from the sun, providing fuel for all living things.

15. Fertilizers add the essential elements nitrogen, phosphorus, and potassium to the soil (in the forms of ammonium, nitrate, phosphate, and potassium salts).

16. Plants also need calcium, magnesium, and sulfur, as well as a number of micronutrients.

17. Various toxic substances are added to farm crops in order to control insects, rodents, and weeds.

18. Although DDT saved millions of lives during and after World War II, its harmful effect on the environment has led to a government ban.

19. Certain organic phosphorus compounds and carbamates are nerve poisons currently used as insecticides.

20. Pheromones (sex attractants), juvenile hormones, viruses, radiation, and other kinds of insects (enemies) have been used to help control insect populations.

21. Herbicides are used to kill weeds, and defoliants are used to remove leaves.

22. According to Thomas Malthus, because population increases geometrically and food production arithmetically, food production cannot keep up with population growth.

Key Terms

adipose tissue   16.2
aflatoxins   16.9
Agent Orange   16.15
antioxidants   16.8
arithmetic growth   16.17
blood sugar   16.1
carbohydrates   16.1
cellulose   16.1
chain reaction   16.8
defoliants   16.15
Delaney
    Amendment   16.8
dietary minerals   16.4

dioxins   16.15
doubling time   16.17
enrichment   16.8
essential amino
    acids   16.3
fat depots   16.2
fats   16.2
food additives   16.8
free radicals   16.8
geometric growth   16.17
glycogen   16.1
GRAS list   16.8
herbicides   16.15

humectant   16.8
insecticide   16.13
juvenile hormones   16.14
limiting reagent   16.3
lipoprotein   16.2
micronutrients   16.12
organic farming   16.16
pesticides   16.13
pheromones   16.14
polyunsaturated
    fats   16.2
preemergent
    herbicides   16.15

primary plant
    nutrients   16.12
rule of 72   16.17
saturated fats   16.2
secondary plant
    nutrients   16.12
sex attractants   16.14
starch   16.1
starvation   16.7
triglycerides   16.2
vitamins   16.5

Review Questions

1. List the three major types of food.

2. What is the role of carbohydrates in the diet?

3. What are the chemical names of the following sugars?
   a. blood sugar   b. table sugar
   c. fruit sugar

4. What is the principal sugar in corn syrup?

5. How is high-fructose corn syrup made?

6. What is the dietary difference between starch and cellulose?

7. What sugar is formed when starch is digested?

8. What are the ultimate products formed when cells metabolize glucose?

9. How much energy is supplied by 1 g of carbohydrates?

10. What proportion of our diet should be carbohydrates?

11. What functions do fats serve?

12. What are polyunsaturated fats?

13. How do animal fats differ from vegetable oils?

14. How much energy is supplied by 1 g of fat?

15. Where is fat stored in the body?

16. What is the maximum percentage of calories in our diet that should come from fats?

17. What is the maximum percentage of calories in our diet that should come from saturated fats?

18. What are essential amino acids?

19. What is a limiting reagent?

20. What is an adequate protein?

21. List some foods that contain adequate proteins.

22. Which essential amino acids are likely to be lacking in corn? In beans?

23. A new type of bread is made by adding pea flour to wheat flour. Will the bread provide adequate protein? Why or why not?

24. In general, what problems are associated with a strict vegetarian diet?

25. A diet high in meat products makes less efficient use of the energy originally captured by plants through photosynthesis than a vegetarian diet does. Explain why.

26. Vitamins and minerals are discussed in this chapter. Which are organic and which are inorganic?

27. Indicate a biological function for each of the following.
   a. iodine   **b.** iron
   c. calcium   **d.** phosphorus

28. Which of the following minerals would you expect to find in relatively large amounts in the human body? Explain your reasoning.
   a. Ca   **b.** Cl   **c.** Co   **d.** Mo
   e. Na   **f.** P   **g.** Zn

29. Match the compound with its designation as a vitamin.

Compound	Designation
Ascorbic acid	Vitamin A
Calciferol	Vitamin B_{12}
Cyanocobalamine	Vitamin C
Retinol	Vitamin D
Tocopherol	Vitamin E

30. Which of the following are B vitamins?
   a. folic acid   **b.** riboflavin
   c. β-hydroxybutyric acid   **d.** thiamine
   e. niacin

31. Identify the vitamin deficiency associated with each of the following diseases.
   a. scurvy   **b.** rickets
   c. night blindness

32. In each case, identify the deficiency disease associated with a diet lacking in the indicated vitamin.
   a. vitamin B_1 (thiamine)   **b.** niacin
   c. vitamin B_{12} (cyanocobalamine)

33. Identify each of the following vitamins as water-soluble or fat-soluble.
   a. vitamin A   **b.** vitamin B_6
   c. vitamin B_{12} **d.** vitamin C
   e. vitamin K

34. Identify each of the following vitamins as water-soluble or fat-soluble.
   a. calciferol   **b.** niacin
   c. riboflavin   **d.** tocopherol

35. Is an excess of a water-soluble vitamin or an excess of a fat-soluble vitamin more likely to be dangerous? Why?

36. What is starvation?

37. In fasting, which stores are depleted first—fats or glycogen?

38. What is a food additive?

39. What are the two major categories of food additives? Give an example of each.

40. What is the function of each of the following food additives?
   a. potassium iodide   **b.** vanilla extract
   c. MSG   **d.** sodium nitrite

41. What is the function of each of the following food additives?
   a. $FeCO_3$ **b.** SO_2
   c. potassium sorbate

42. What is the purpose of each of the following food additives?
   a. BHA   **b.** FD & C Yellow No. 5
   c. saccharin

43. What is the purpose of each of the following food additives?
   a. aspartame   **b.** vitamin D
   c. sodium hypophosphite

44. What is enriched bread? Is it equal in nutritional value to bread made from whole grain?

45. What is MSG?

46. What is botulism?

47. What are antioxidants?

48. What is a chain reaction?

49. What vitamin serves as a fat-soluble antioxidant?

50. Name three artificial sweeteners. Which are approved for current use in the United States?

51. What is the chemical nature of aspartame?

52. What is the GRAS list?

53. What are aflatoxins?

54. List some incidental additives that have been found in foods.

55. What is the Delaney Amendment to the Food and Drug Act of 1906?

56. What United States government agency regulates the use of food additives?

57. What must be done before a food company can use a new food additive?

58. List five functions of food additives.

59. Where does most of the matter of a growing plant come from?

60. List the three structural elements of a plant.

61. In what form is nitrogen used by plants?

62. How is ammonia made? What raw materials are required for ammonia synthesis?

63. List several ways that nitrogen can be fixed.

64. What is urea? From what components is synthetic urea made?

65. What is anhydrous ammonia? How is it used?

66. What is the source of phosphate fertilizers?

67. What is the role of nitrogen in plant nutrition?

68. What is the role of phosphorus in plant nutrition?

69. Why does soil become acidic when potassium ions are absorbed by plants?

70. List the advantages and disadvantages of DDT as an insecticide.

71. Why is DDT especially harmful to birds?

72. Describe how chlorinated hydrocarbons become concentrated in a food chain.

73. What is a narrow-spectrum insecticide? Name one.

74. What is a broad-spectrum insecticide? Name two.

75. What are pheromones?

76. What are juvenile hormones? How are they used against insect pests?

77. Describe the technique of sterilization as a method for controlling insects. Why is it not more widely used?

78. What are herbicides?

79. What is a defoliant? Why are defoliants used on cotton crops?

80. What is a preemergent herbicide?

81. List the four main features of alternative agriculture.

82. Compare organic farming with conventional farming. Consider energy requirements, labor, profitability, and crop yields.

83. What did Thomas Malthus predict in his famous 1830 statement?

84. Why has Malthus's prediction not come true?

85. What is arithmetic growth? Give an example.

86. What is geometric growth? Give an example.

Problems

Chemical Equations

87. Give the chemical equation for the photosynthesis reaction.

88. Give the chemical equation for the synthesis of ammonia.

Calorie Calculations

89. The label on a can of soup indicates that each half-cup portion supplies 60 kcal and has 3 g of protein, 8 g of carbohydrate, and 1 g of fat. Calculate the percent of calories from fat.

90. The label on a can of milk substitute indicates that each half-cup serving supplies 150 kcal and has 8 g of protein, 12 g of carbohydrate, and 8 g of fat. Calculate the percent of calories from fat.

Nutrient Calculations

91. Our requirement for protein is about 0.8 g per kilogram of body weight. What is the protein requirement of a 50-kg (110-lb) person?

92. Graham crackers are 8% protein. Assume that your Recommended Daily Allowance (RDA) is 60 g of protein. What percent of your RDA for protein would you receive if you ate 150 g of graham crackers?

93. Assume that 1 cup of skim milk contains 225 g of milk. Skim milk is 3.6% protein. How many grams of protein are there in 4 cups of skim milk?

94. Campbell's Chunky Vegetable Beef Soup is available in regular and low-sodium versions. The regular has 935 mg of Na^+, and the low-sodium has 90 mg of Na^+ per 10.75-oz portion. What percent of the recommended daily maximum of 3300 mg of Na^+ would you get from one portion of each?

95. One cup of whole milk supplies 34 mg of cholesterol. The American Heart Association recommends a maximum of 300 mg of cholesterol per day. What percentage of the maximum recommendation is met by 1 cup of whole milk?

Toxicity

96. How much DDT would it take to kill a person weighing 60 kg if the lethal dose is 0.5 g per kilogram of body weight?

97. How much parathion would it take to kill the person in Problem 96 if the lethal dose is 5 mg per kilogram of body weight?

Growth

98. You start a stamp collection, planning to buy two stamps each week. How many weeks will it take to acquire 100 stamps? Does the collection grow arithmetically or geometrically?

99. You are raising rabbits. Starting with two, you have four at the end of the first month, eight at the end of the second month, and so on; the rabbit population doubles each month. How many rabbits will you have at the end of 12 months? Is the rabbit population growing arithmetically or geometrically?

100. The populations of some Latin American, African, and Asian nations are growing at 3.5% a year. At this rate, how many years of growth will it take for these populations to double?

101. The population of China, now 1.25 billion, is growing at an annual rate of 1.4%. At this rate, how many years of growth will it take for the population to double?

Additional Problems

102. Structural formulas for two vitamins are shown. Identify each of them as water-soluble or fat-soluble.

a.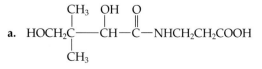

b.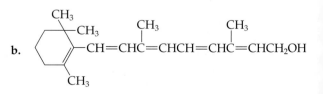

103. Identify the functional group in the sex attractant for the common housefly (Section 16.14).

104. Identify the functional groups in the juvenile hormone molecule (Section 16.14).

105. Identify the functional groups in the methoprene molecule (Section 16.14).

Projects

106. Consult a recent issue of *FDA Consumer* and make a list of the incidental food additives reported in that issue. What are the most common causes of contamination?

107. Examine the label on a sample of each of the following.

a. a can of soft drink **b.** a can of beer

c. a dried soup mix **d.** a can of soup

e. a can of fruit drink **f.** a cake mix

Make a list of the food additives in each. Try to determine the function of each additive.

108. Maraschino cherries are bleached (with sulfur dioxide) and then dyed with an organic food coloring. Should the dye be banned? Justify your conclusion.

109. When Russian refugee Alexander Jourjine escaped from the former USSR by hiking for 23 days across Finland to Sweden, he carried 9 lb of lard and cheese, several loaves of dried black bread, tea, sugar, and 12 chocolate bars. His diet was supplemented with berries picked along the way. Defend or criticize his choice of each food item.

110. Examine the label on a package of yard or garden pesticide. List its trade name and active ingredients.

111. Examine the label on a bag of fertilizer. What is its composition?

112. If you grew a garden, would you use chemical fertilizers? Pesticides? Why or why not?

Online Projects

113. Search the web for information about the following pesticides.

a. aldrin **b.** heptachlor

c. methoxychlor

114. The Haber process is one of the most important industrial developments of the twentieth century. What can you find out about Haber and his famous process on the web beyond what you have learned from this book?

115. Look for several websites with different opinions about the use of fertilizers and pesticides.

References and Readings

1. Ainsworth, Susan J. "Changing Technologies Help Herbicide Producers Compete in a Mature Market." *Chemical & Engineering News*, 29 April 1996, pp. 36–42.

2. Barer-Stein, Thelma. *Eat What You Are: People, Culture, and Food Traditions*. Buffalo, NY: Firefly, 1999.

3. Belitz, H.D., and W. Grosch. *Food Chemistry*, 2nd edition. New York: Springer, 1999. Comprehensive reference book with hundreds of tables and figures.

4. Brody, Jane. *Jane Brody's Nutrition Book*. New York: Bantam Books, 1981. A good general source.

5. Brown, Lester R. *Tough Choices: Facing the Challenge of Food Scarcity*. Washington: Worldwatch Institute, 1996.

6. Cohen, Joel E. "Ten Myths of Population." *Discover*, April 1996, pp. 42–47.

7. "The Facts About Fats." *Consumer Reports*, June 1995, pp. 390–393.

8. Gorman, Christine. "The Joy of Soy." *Time*, June 7, 1999, pp. 68–69.

9. Hannon, Kerry. "Pure and Unadulterated." *U.S. News & World Report*, 15 May 1995, pp. 86–94. Discusses organic foods.

10. Jones, Julie Miller. "Fat Substitutes: Nutritional Promise or Potential Disaster?" *Chemistry & Industry*, 1 July 1996, pp. 494–498.

11. Kourik, Robert. "Noxious Naturals." *Garbage*, September–October 1990, pp. 54–57. Describes how plants make their own pesticides.

12. McGee, Harold. *On Food and Cooking: The Science and Lore of the Kitchen*. New York: Scribner's, 1984.

13. McGee, Harold. *The Curious Cook: More Kitchen Science and Lore*. New York: Collier, 1990.

14. McKone, Harold T. "Copper in the Candy, Chromium in the Custard." *Today's Chemist*, October 1990, pp. 22–25. A history of food colors before synthetic organic dyes.

15. Mela, David J. "Implications of Fat Replacement for Food Choice and Energy Balance." *Chemistry & Industry*, 6 May 1996, pp. 329–332.

16. Rouhi, A. Maureen. "Biotechnology Steps up Pace of Rice Research." *Chemical & Engineering News*, 7 October 1996, pp. 10–14. About 2.4 billion people eat rice every day, and biotechnology is the only hope to feed the additional 80 million to 100 million more people added each year.

17. Sheldon, Richard P. "Phosphate Rock." *Scientific American*, June 1982, pp. 45–51.

18. Vaughan, John. *The New Oxford Book of Food Plants*. Oxford: Oxford University Press, 1999.

17 Household Chemicals
Helps and Hazards

Solvents, paint, detergents, wax, cosmetics, bleach, shampoo...
There are chemicals for almost everything we do.

The household chemicals sold in greatest volume are cleaning products—soaps, detergents, and various special-purpose and multipurpose cleaning mixtures.

I f someone offered you a job in a place where poisonous chemicals were used every day, where toxic vapors and harmful dusts were common, where corrosive acids and alkalis were often used, and where highly flammable liquids and vapors posed a fire hazard, would you take the job? You probably have, many times. This is a description of a typical American home.

People tend to be careful in a chemistry laboratory because they know the place can be dangerous. They are much less cautious in their own homes, even though some of the chemicals on the shelves can be just as toxic and hazardous as those in a laboratory. There are perhaps half a million chemical products available for use in the American home. They include waxes, wax removers, paints, paint removers, bleaches, insecticides, rodenticides, spot removers, solvents, disinfectants, detergents, toothpaste, shampoo, perfumes, lotions, shaving cream, deodorants, hair spray, and many other products (Figure 17.1). Some of these products are quite harmless, but others contain corrosive or toxic chemicals or present fire hazards. Some cause environmental problems when they are used or discarded.

Extensive use of chemicals in the home has led to many accidents. Studies have shown that chemicals are often used in the home without regard to the directions or precautions given on their labels. Frequently the labels aren't read at all, and *misuse* of household chemicals can sometimes end in tragedy.

We have already discussed the chemical composition of food and the chemicals that are used in producing it (Chapter 16). Some of these agricultural chemicals are also used around the home, especially in the yard and garden. We have discussed polymers present in our clothing and home furnishings (Chapter 10) and fuels used in our furnaces and automobiles (Chapter 14). We have even discussed the chemistry of our own bodies (Chapter 15). But there are many other chemicals around the house—cleaning materials, personal care products, and an assortment of other things. In this chapter we look at some household chemicals. Let's begin with the ones that make up the largest volume—cleaning compounds.

Figure 17.1 A modern home is stocked with a variety of chemical products.

A typical U.S. supermarket offers about 5000 different consumer products.

17.1 A History of Cleaning

In developing societies, even today in some places, clothes are cleaned by beating them with rocks in the nearest stream. Sometimes plants, such as the soapworts of Europe or the soapberries of tropical America, are used as cleansing agents. The leaves of soapworts and soapberries contain **saponins**, chemical compounds that produce a soapy lather. These saponins were probably the first detergents.

Ashes of plants contain potassium carbonate (K_2CO_3) and sodium carbonate (Na_2CO_3). The carbonate ion, present in both these compounds, reacts with water to form an alkaline solution. The basic solution has detergent properties. These alkaline plant ashes were used as cleansing agents by the Babylonians at least 4000 years ago. Europeans used plant ashes to wash their clothes as recently as 100 years ago. Sodium carbonate is still sold today as washing soda.

Although soap has been known for hundreds of years, it was first used mainly as a medicine. The discovery of disease-causing microorganisms, and subsequent public health practices, increased interest in cleanliness during the eighteenth century. By the middle of the nineteenth century soap was in common use.

Personal Cleanliness

The history of cleanliness of body and clothing is rather spotty. The Romans, with their great public baths, probably did not use any sort of soap. They covered their bodies with oil, worked up a sweat in a steam bath, and then had the oil wiped off by a slave. They finished by taking a dip in a pool of fresh water. And the slaves? They probably didn't bathe at all.

During the Middle Ages body cleanliness was prized if not always attained. Twelfth-century Paris, with a population of about 100,000, had many public bathhouses. The Renaissance, which began in the fourteenth century and extended to the seventeenth, was noted for a revival of learning and art. However, it was not noted for cleanliness. Queen Elizabeth I of England (1558–1603) bathed once a month, a habit that caused many to think her overly fastidious. Queen Isabella of Castille (1474–1504), who supported the 1492 voyage of Columbus to the New World, is reported to have bathed only twice in her life. People sometimes did use perfume to mask body odors.

17.2 Fat + Lye ⟶ Soap

The first written record of soap is found in the writings of Pliny the Elder, the Roman who described the Phoenicians' synthesis of soap using goat tallow and ashes. By the second century C.E., sodium carbonate (produced by the evaporation of alkaline water) was heated with lime (from limestone or seashells) to produce sodium hydroxide (lye).

$$Na_2CO_3 + Ca(OH)_2 \longrightarrow 2\,NaOH + CaCO_3$$

The sodium hydroxide was heated with animal fats or vegetable oils to produce soap (Figure 17.2). [Note that a **soap** is a salt of a long-chain carboxylic acid

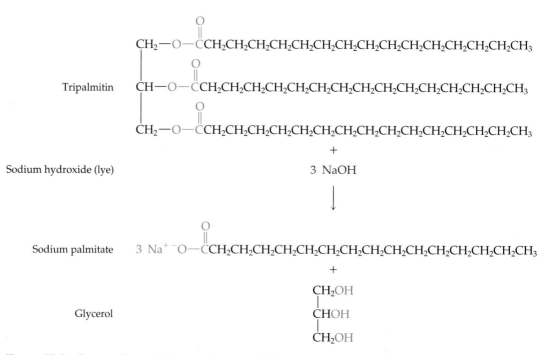

Figure 17.2 Soap can be made by reacting animal fat or vegetable oil with sodium hydroxide. Vegetable oils, with unsaturated carbon chains, generally produce softer soaps. Coconut oils, with shorter carbon chains, yield soaps that are more soluble in water.

(Chapter 9).] American pioneers made soap in much the same manner. Lye was added to animal fat in a huge iron kettle, and the mixture was cooked over a wood fire for several hours. The soap rose to the surface and, on cooling, solidified. The glycerol remained as a liquid on the bottom of the pot. Both the glycerol and the soap often contained unreacted alkali, which ate away the skin. Grandma's lye soap is not just a myth.

In modern commercial soapmaking, fats and oils are often hydrolyzed with superheated steam. The fatty acids are then neutralized to make soap. Toilet soaps usually contain such additives as dyes, perfumes, creams, and oils. Scouring soaps contain abrasives such as silica and pumice. Many soaps claim to have deodorant action, but few have any active deodorant other than the soap itself. Some soaps have air blown in before they solidify to lower their density so that they float. Some bath bars contain synthetic detergents. Their action is similar to that of soap.

Potassium soaps are softer than sodium soaps, and they produce a finer lather. They are used alone, or in combination with sodium soaps, in liquid soaps and shaving creams. Soaps are also made by reacting fatty acids with triethanolamine. These substances are used in shampoos and other cosmetics.

[Web Reference 1] Some people still make their own soap! Some fascinating recipes and photos.

How Soap Works

Dirt and grime usually adhere to skin, clothing, and other surfaces because they are combined with greases and oils—body oils, cooking fats, lubricating greases, and a variety of similar substances—that act a little like sticky glues. Since oils are not miscible with water, washing with water alone does little good.

A soap molecule has a dual nature. One end is ionic and therefore polar and **hydrophilic** (water-attracting). The rest of the molecule is hydrocarbon-like and therefore nonpolar and **hydrophobic** (water-repelling). The hydrophilic "head" dissolves in water, while the hydrophobic "tail" dissolves in nonpolar substances, such as oils (Figure 17.3). The cleansing action of soap is illustrated in Figure 17.4. A spherical collection of molecules like this is called a **micelle**. The hydrocarbon tails stick in the oil, with the ionic heads remaining in the aqueous phase. In this manner, the oil is broken into tiny droplets and dispersed throughout the solution. The droplets don't coalesce because of the repulsion of the charged groups (the carboxyl anions) on their surfaces. The oil and water form an emulsion, with soap acting as the emulsifying agent. With the oil no longer "gluing" it to the surface, the dirt can be removed easily. Any agent, including soap, that stabilizes the suspension of nonpolar substances—such as oil and grease—in water is called a **surface-active agent** or a **surfactant**. If not mixed into the water, soap tends to act at the liquid surface, the ionic heads interacting with the water, while the hydrocarbon tails stick up out of the water.

$$CH_3CH_2CH_2CH_2CH_2CH_2CH_2CH_2CH_2CH_2CH_2CH_2CH_2CH_2CH_2\overset{\overset{\displaystyle O}{\|}}{C}-O^-\ Na^+$$

(a)

Hydrocarbon end (dissolves in oils)   Ionic end (dissolves in water)

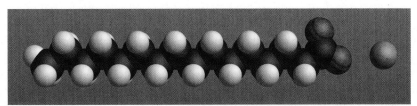

(b)

Figure 17.3 Sodium palmitate, a soap. (a) Structural formula. (b) Space-filling model.

Figure 17.4 The cleaning action of soap is visualized in this diagram of a soap micelle. A tiny oil droplet is suspended in water by having the hydrophobic hydrocarbon tails of soap molecules immersed in the oil, while their hydrophilic ionic heads extend into the water. Attraction between the water and the ionic ends of the soap molecules carries the oil droplet into the water.

$$CH_3CH_2CH_2CH_2CH_2CH_2CH_2CH_2CH_2CH_2CH_2COO^-\,Na^+\ +\ H^+\ \longrightarrow$$

A soap   An acid

$$CH_3CH_2CH_2CH_2CH_2CH_2CH_2CH_2CH_2CH_2CH_2COOH\ +\ Na^+$$

A fatty acid

(a)

Figure 17.5 Soap has two major disadvantages. (a) Acids convert soap anions to insoluble fatty acids. (b) Hard water cations such as Ca^{2+} precipitate soap as insoluble curds. Neither product has detergent action.

$$2\,CH_3CH_2CH_2CH_2CH_2CH_2CH_2CH_2CH_2CH_2CH_2COO^-\ +\ Ca^{2+}\ \longrightarrow$$

Soap anion

$$(CH_3CH_2CH_2CH_2CH_2CH_2CH_2CH_2CH_2CH_2CH_2COO^-)_2\,Ca^{2+}$$

Bathtub ring
(insoluble)

(b)

Disadvantages of Soap

For cleaning clothes and for many other purposes, soap has been largely replaced by synthetic detergents because soaps have certain shortcomings. In acidic solutions, soaps are converted to free fatty acids, which do not exhibit detergent action. What is worse, they are insoluble in water and separate as a greasy scum (Figure 17.5a).

A more serious disadvantage of soap is that it doesn't work well in hard water (Figure 17.6). Hard water is water that contains certain metal ions, particularly magnesium, calcium, and iron ions. Soap anions react with these metal ions to form greasy, insoluble curds (Figure 17.5b). These deposits make up the familiar ring around the bathtub. They leave freshly washed hair sticky and are responsible for "tattletale gray" in the family laundry.

Figure 17.6 Sudsing quality of hard water versus soft water. *From left*: Detergent in hard water, soap in hard water, detergent in soft water, and soap in soft water. Note that the sudsing of the detergent is about the same in both hard and soft water. Note also that there is little sudsing and that insoluble material is formed when soap is used in hard water.

Water Softeners

To aid the action of soaps, various water-softening agents and devices are used. An effective water softener is washing soda, sodium carbonate ($Na_2CO_3 \cdot 10\,H_2O$). The carbonate ion makes the water basic (preventing the precipitation of fatty acids) by reacting with water to raise the pH.

$$CO_3^{2-}(aq) + H_2O(l) \longrightarrow HCO_3^-(aq) + OH^-(aq)$$

The carbonate ion also reacts with the ions that cause hard water and removes them as insoluble salts.

$$Mg^{2+}(aq) + CO_3^{2-}(aq) \longrightarrow MgCO_3(s)$$
$$Ca^{2+}(aq) + CO_3^{2-}(aq) \longrightarrow CaCO_3(s)$$

Trisodium phosphate (Na_3PO_4) is another water-softening agent. Like washing soda, it makes water basic and precipitates calcium and magnesium ions.

$$PO_4^{3-}(aq) + H_2O(l) \longrightarrow HPO_4^{2-}(aq) + OH^-(aq)$$
$$2\,PO_4^{3-}(aq) + 3\,Mg^{2+}(aq) \longrightarrow Mg_3(PO_4)_2(s)$$

Phosphates also seem to aid in the cleaning process in some other way that is not yet well understood.

Water-softening tanks are also available for use in homes and businesses. These tanks contain an insoluble polymeric material that attracts and holds calcium, magnesium, and iron ions to its surface, replacing them with sodium ions and thus softening the water (Figure 17.7). After a period of use, the polymer becomes saturated and must be discarded or regenerated.

Before leaving the subject of soap, let's mention that soap has some advantages. It is an excellent cleanser in soft water, it is relatively nontoxic, it is derived from renewable resources (animal fats and vegetable oils), and it is biodegradable.

17.3 Synthetic Detergents

A second technological approach to the problems associated with soap was to develop a new synthetic detergent. Molecules of synthetic detergents were enough like those of soap to have the same cleaning action but different enough to resist the effects of

---

Water softeners are used to remove ions of calcium, magnesium, and iron, the ions that make water "hard."

Recall from Chapter 7 that a substance that produces hydroxide ions in water is a base. Carbonate and phosphate ions are bases because they react with water to form hydroxide ions. Bases are also defined as proton acceptors, and carbonate and phosphate ions accept protons from water molecules.

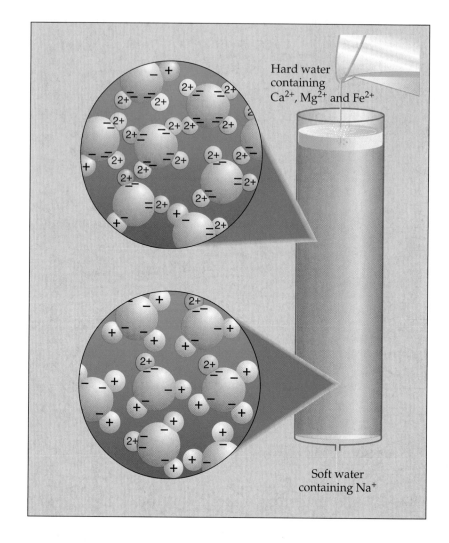

Figure 17.7 Water softening by ion exchange. Doubly charged cations (gray)—Ca^{2+}, Mg^{2+}, and Fe^{2+}—replace Na^+ (orange) on the negatively charged surfaces of the ion exchange particles (top). By the time the water reaches the bottom of the column, it has been softened. The resin there has mainly Na^+ ions attached.

[Web Reference 2] A wealth of information from the Soap and Detergent Association of North America.

acids and hard water. The raw materials for soap manufacture were scarce and expensive during and immediately after World War II. The synthetic detergent industry developed rapidly in the postwar period.

ABS Detergents: Nonbiodegradable

Within a few years, cheap synthetic detergents, made from petroleum products, were widely available. Alkylbenzenesulfonate (ABS) detergents were made from propylene ($CH_2 = CHCH_3$), benzene, and sulfuric acid. The resulting sulfonic acid (RSO_3H) was neutralized with a base, usually sodium carbonate, to yield the final product.

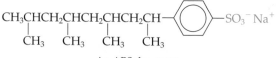

An ABS detergent

Sales of ABS detergents soared. For a decade or more, nearly everyone was happy, but suds began to accumulate in sewage treatment plants. Foam piled high in rivers (Figure 17.8), and in some areas a head of foam even appeared on drinking water (Figure 17.9). The branched-chain structure of ABS molecules was not readily broken

Figure 17.8 Foaming rivers were common during the period when ABS detergents were used. The problem was solved in the United States by replacing ABS detergents with biodegradable LAS detergents. This photograph shows detergent foam on the Bogota River in Colombia.

Figure 17.9 A glass of suds, straight from the tap. The water comes from a groundwater source contaminated with ABS detergents.

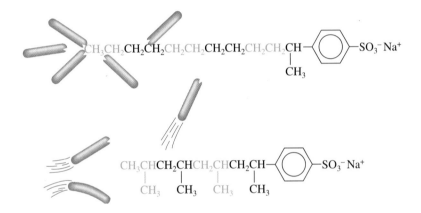

Figure 17.10 Microorganisms can readily metabolize LAS detergents, removing two carbon atoms at a time from the molecule. It takes them much longer to break down the branched chains of the ABS molecules.

down by the microorganisms in sewage treatment plants, and the groundwater supply was threatened. Public outcries caused laws to be passed and industries to change their processes. Biodegradable detergents were quickly put on the market, and non-biodegradable detergents were banned.

LAS Detergents: Biodegradable

Biodegradable detergents called linear alkylsulfonates (LAS) have linear chains of carbon atoms. Microorganisms can break down LAS molecules by producing enzymes that degrade the molecule two (and only two) carbon atoms at a time (Figures 17.10 and 17.11). The branched chain of ABS molecules blocks this enzyme action, preventing their degradation. Thus, technology has solved the problem of foaming rivers.

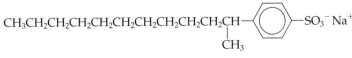

An LAS detergent

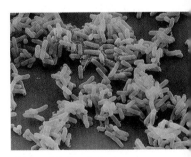

Figure 17.11 Microorganisms such as *Escherichia coli*, shown here magnified 42,500 times by scanning electron microscopy, are able to degrade LAS detergents.

The cleansing action of synthetic detergents is quite similar to that of soaps. Synthetics work better in acidic solution and in hard water, though. Their calcium and

[Web Reference 3] Detergents and other surfactants wind up in our streams and oceans. Go to this website for a look at the ecological safety of LAS surfactants from the industrial point of view.

magnesium salts, unlike those of soap, are soluble and do not separate out, even in extremely hard water (Figure 17.6). Thus, the cleansing action of a synthetic detergent is little affected by hard water.

17.4 Laundry Detergent Formulations

Detergent products used in homes and commercial laundries usually contain much more than surfactant LAS molecules. Common additives include builders, brighteners, fabric softeners, and other substances to lessen the redeposition of dirt or simply to reduce cost.

Builders

Laundry detergents are sometimes eaten by small children. Products with sodium carbonate builders are more toxic than those with phosphate builders and can actually kill children who consume them in large amounts.

Any substance added to a surfactant to increase its detergency is called a **builder**. Common builders, once widely used but now widely banned or restricted, are phosphates. An example is sodium tripolyphosphate ($Na_5P_3O_{10}$). It ties up Ca^{2+} and Mg^{2+} in soluble complexes—thus softening the water—and produces a mild alkalinity, providing a favorable environment for detergent action. We have seen (Chapter 13) how phosphates speed the eutrophication of lakes. In some areas, phosphates from detergents have been shown to contribute somewhat to this process.

[Web Reference 4] Papers and technical articles on phosphates are available.

Several state and local governments have banned the sale of detergents containing phosphates, and the detergent industry has offered a variety of replacements; the most prominent are sodium carbonate and complex aluminosilicates called *zeolites*. Sodium carbonate acts by precipitating calcium ions, thus softening the water (Section 17.3).

$$Ca^{2+}(aq) + CO_3^{2-}(aq) \longrightarrow CaCO_3(s)$$

The $CaCO_3$ precipitate can be harmful to automatic washing machines. Further, excess carbonate ions form strongly basic solutions—that is, solutions that contain an excess of OH^- ions.

$$CO_3^{2-}(aq) + H_2O(l) \longrightarrow HCO_3^-(aq) + OH^-(aq)$$

Zeolites are perhaps the most promising of the substitutes. Zeolite anions trap calcium ions by exchanging them for their own sodium ions.

$$Ca^{2+}(aq) + Na_2Al_2Si_2O_8(s) \longrightarrow 2\,Na^+(aq) + CaAl_2Si_2O_8(s)$$

Calcium ions are held in suspension by the zeolites (rather than being precipitated). Further, zeolite solutions are not strongly basic and are therefore less likely than sodium carbonate solutions to irritate the skin and eyes.

Brighteners

Many detergents contain **optical brighteners**. These compounds, called *blancophors* (or colorless dyes), absorb the invisible ultraviolet component of sunlight and reemit it as visible light at the blue end of the spectrum (Figure 17.12). The fabric appears brighter, and the blue light camouflages any yellowing. Clothes treated with an optical brightener on the surface may be dirty underneath, but they look "whiter and brighter than new" (Figure 17.13). These brighteners are also used in cosmetics, paper, soap, plastics, and other products.

Optical brighteners appear to be of little harm to humans, although they have been shown to cause skin rashes in some people. Their effect on lakes and streams is largely unknown, despite the large amounts entering our waterways. The only possible benefit of these compounds is cosmetic: The appearance of our laundry may be more pleasing.

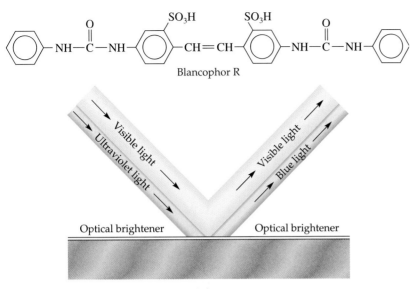

Blancophor R

Fabric

Figure 17.12 An optical brightener, such as the molecule shown here, converts invisible ultraviolet light to visible blue light, making the fabric look brighter and masking any yellowish color.

Figure 17.13 Clothes washed in a detergent formulation that has an optical brightener glow in the light from an ultraviolet lamp (black light). Paper, chalk, and other white materials often have optical brighteners added.

17.5 Liquid Laundry Detergents

Early laundry detergent preparations were powders, but liquid laundry detergent formulations now command a substantial share of the market. In general, powders are more concentrated and clean better, but many consumers find liquids more convenient. Some liquid laundry detergents are "built" with sodium carbonate or with zeolites. Unbuilt formulations are high in surfactants but contain no builders.

LAS surfactants are the cheapest surfactants in liquid laundry detergents. In unbuilt formulations, they are usually used as the sodium or triethanolamine salt, and in built varieties, they are often present as the potassium salt. Other popular surfactants for liquid formulations are alcohol ether sulfates (AES). AES molecules have a hydrocarbon portion derived from an alcohol (or alkylphenol), a polar portion derived from ethylene oxide, and a sulfate salt portion (Figure 17.14a). AES surfactants are efficient but more expensive.

CH_2——CH_2

Ethylene oxide

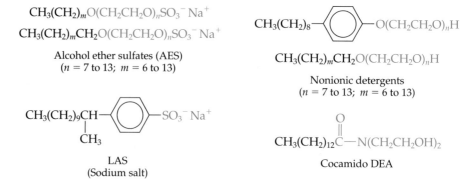

$CH_3(CH_2)_mO(CH_2CH_2O)_nSO_3^-\,Na^+$

$CH_3(CH_2)_mCH_2O(CH_2CH_2O)_nSO_3^-\,Na^+$

Alcohol ether sulfates (AES)
($n = 7$ to 13; $m = 6$ to 13)

$CH_3(CH_2)_8$—⬡—$O(CH_2CH_2O)_nH$

$CH_3(CH_2)_mCH_2O(CH_2CH_2O)_nH$

Nonionic detergents
($n = 7$ to 13; $m = 6$ to 13)

LAS
(Sodium salt)

Cocamido DEA

$CH_3(CH_2)_9CH$—⬡—$SO_3^-\,HN^+(CH_2CH_2OH)_3$
|
CH_3

LAS
(Triethanolamine salt)

Figure 17.14 Some of the kinds of substances used in various detergent formulations. Nonoxynol-9, the nonylphenol ether in which $n = 9$, is also a spermicide used in contraceptives such as the Today Sponge.

All the surfactants discussed so far, including soap, are **anionic surfactants**; the working part is an anion with a nonpolar part and an ionic end. Some liquid detergents contain **nonionic surfactants**. Examples are alcohol ethoxylates and alkylphenol ethoxylates. Because the oxygen atoms hydrogen-bond with water molecules, they make their ends of the molecules water-soluble, just like the ionic end of an anionic surfactant (Figure 17.14b). Nonionic surfactants effectively remove oily soil from fabrics. They are not as good as anionic surfactants at keeping dirt particles in suspension. Alcohol ethoxylates have the unusual property of being more soluble in cold water than in hot, which makes them particularly suitable for cold-water laundering.

Dishwashing Detergents

Liquid detergents for washing dishes by hand generally contain one or more surfactants as the only active ingredients. These include LAS as the sodium and/or the triethanolamine salt (Figure 17.14c,e). Some use nonionic surfactants such as cocamido DEA (Figure 17.14d), an amide made from a fatty acid and diethanolamine. Few dishwashing liquids contain builders; those that do usually have only small amounts. Most liquid dishwashing detergents differ significantly only in the concentration or effectiveness of the surfactant.

Detergents for automatic dishwashers are quite another matter. They often are quite caustic and should never be used for hand dishwashing. They contain sodium tripolyphosphate ($Na_5P_3O_{10}$), sodium metasilicate (Na_2SiO_3), sodium sulfate (Na_2SO_4), a chlorine bleach, and only a small amount of surfactant, usually a nonionic type. They depend mainly on their strong alkalis and the vigorous agitation of the machine for cleaning.

17.6 Quaternary Ammonium Salts: Dead Germs and Soft Fabrics

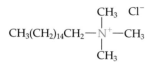

Hexadecyltrimethylammonium
chloride
(a cationic surfactant)

In addition to anionic detergents and nonionic surfactants, there is a third type, called **cationic surfactants**, in which the working part is a positive ion. The most common of these are called quaternary ammonium salts because they have four hydrocarbon groups attached to a nitrogen atom that bears a positive charge. An example of such a cationic surfactant is hexadecyltrimethylammonium chloride.

These cationics are not good detergents, but they have a degree of germicidal action. Sometimes they are used along with nonionic surfactants as cleansers and disinfectants in the food and dairy industries. Cationic surfactants are seldom used with anionic ones because the ions of opposite charge tend to clump together and precipitate from solution, destroying the detergent action of both. Only about 10% of surfactants are cationic. Anionic surfactants predominate, making up about 65%, and nonionic substances make up about 25% of surfactants.

Another kind of quaternary salt with *two* long carbon chains and two smaller groups on nitrogen is used as a fabric softener. An example is dioctadecyldimethyl-ammoniun chloride. These compounds are strongly adsorbed by the fabric, forming a film on the fabric's surface one molecule thick. The long hydrocarbon chains lubricate the fibers, imparting increased flexibility and softness to the fabric.

$$CH_3(CH_2)_{16}CH_2-\overset{\overset{\displaystyle CH_3}{|}}{\underset{\underset{\displaystyle CH_3(CH_2)_{16}CH_2}{|}}{N^+}}-CH_3 \quad Cl^-$$

Dioctadecyldimethylammonium chloride

17.7 Bleaches: Whiter Whites

Bleaches are oxidizing agents (Chapter 8). The familiar liquid laundry bleaches (such as Clorox and Purex) are all 5.25% sodium hypochlorite (NaOCl) solutions. They differ only in price. Hypochlorite bleaches release chlorine rapidly, and high concentrations of chlorine can be quite damaging to fabrics. These bleaches do not work well on polyester fabrics, often causing yellowing rather than the desired whitening.

Other bleaches are available in solid forms that release chlorine slowly in water so as to minimize damage to fabrics. Symclosene is an example of a cyanurate-type bleach.

Oxygen-releasing bleaches usually contain sodium perborate ($NaBO_2 \cdot H_2O_2$). As indicated by the formula, this compound is a complex of $NaBO_2$ and hydrogen peroxide (H_2O_2). Above 65°C, the hydrogen peroxide is liberated and acts as a bleach. It decomposes in turn to liberate oxygen.

$$2\ H_2O_2(l) \longrightarrow 2\ H_2O(l) + O_2(g)$$

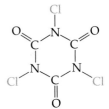

Symclosene

Borates are somewhat toxic. Perborate bleaches are less active than chlorine bleaches and require higher temperatures, higher alkalinity, and higher concentrations to do an equivalent job. They are used mainly for bleaching white, resin-treated polyester–cotton fabrics. These fabrics last much longer with oxygen bleaching than with chlorine bleaching. Also, properly used, oxygen bleaches make fabrics whiter than do chlorine bleaches.

17.8 All-Purpose Cleaning Products

Various all-purpose cleaning products are available for use on walls, floors, countertops, appliances, and other tough, durable surfaces. Those for use in water may contain surfactants, sodium carbonate, ammonia, solvent-type grease cutters, disinfectants, deodorants, and other ingredients. Some are great for certain jobs but not as good for others. They damage some surfaces but work especially well on others. Most important, they may be harmful when used improperly. Reading the labels is important.

Mixing bleach with other household chemicals can be quite dangerous. For example, mixing a hypochlorite bleach with hydrochloric acid produces poisonous chlorine gas.

$$2\ HCl(aq) + ClO^-(aq) \longrightarrow Cl^-(aq) + H_2O(l) + Cl_2(g)$$

Mixing bleach with toilet bowl cleaners that contain HC1 is especially dangerous. Most bathrooms are small and poorly ventilated, and generating chlorine in such a limited space is hazardous. Chlorine can do enormous damage to the throat and the entire respiratory tract. If the concentration of chlorine is high enough, it can kill.

Mixing bleach with ammonia is also extremely hazardous. Two of the gases produced are chloramine (NH_2Cl) and hydrazine (NH_2NH_2) both of which are violently toxic.

We repeat: *never* mix bleach with other chemicals without specific directions to do so. In fact, it is a good rule not to mix any chemicals unless you know exactly what you are doing.

Household ammonia solutions, straight from the bottle, are good for loosening baked-on grease or burned-on food. Diluted with water, they clean mirrors, windows, and other glass surfaces. Mixed with detergent, ammonia rapidly removes wax from vinyl floor coverings. Ammonia vapors are highly irritating, and this cleanser should never be used in a closed room. Ammonia should not be used on asphalt tile, wood surfaces, or aluminum because it may stain, pit, or erode these materials.

Baking soda (sodium bicarbonate, $NaHCO_3$) straight from the box, is a mild abrasive cleanser. It absorbs food odors readily, making it good for cleaning the inside of a refrigerator. Vinegar (acetic acid) cuts grease film. It should not be used on marble, because it reacts with the marble, pitting the surface.

$$\underset{\text{Marble}}{CaCO_3} + \underset{\text{Vinegar}}{2\ CH_3COOH} \longrightarrow Ca^{2+} + 2\ CH_3COO^- + CO_2 + H_2O$$

17.9 Special-Purpose Cleaners

$$\begin{array}{l} H_2C-COOH \\ \quad | \\ HOC-COOH \\ \quad | \\ H_2C-COOH \end{array}$$

Citric acid

There are many highly specialized cleaning products on the market. For metals there are various kinds of chrome cleaners, brass cleaners, copper cleaners, and silver cleaners. There are lime removers, rust removers, and grease removers. There are cleaners specifically for wood surfaces, for vinyl floor coverings, and for ceramic tile. Let us look at just a few special purpose cleaners that you would find in almost any home.

Toilet Bowl Cleaners

The buildup that forms in toilet bowls is mainly calcium carbonate ($CaCO_3$), deposited from hard water, along with discolorations due to such things as iron compounds and fungal growth. Because calcium carbonate is readily dissolved by acid, toilet bowl cleaners tend to be strongly acidic. The solid crystalline products are usually sodium bisulfate ($NaHSO_4$), and the liquid cleaners are hydrochloric acid (HCl), citric acid, or some other acidic material.

Scouring Powder

Most powdered cleansers contain an abrasive such as silica (SiO_2) for rubbing stains from hard surfaces. They also usually contain a surfactant to dissolve grease, and some feature a chlorine-releasing bleach. These cleansers are mainly intended for removing stains from porcelain tubs and sinks. Such abrasive cleansers may scratch the finish on appliances, countertops, and metal utensils. They may even scratch the surfaces of sinks, toilet bowls, and bathtubs. Dirt gets into the scratches and makes cleaning even more difficult.

Glass Cleaners

Cleaners for windowpanes and mirrors are volatile liquids that evaporate without leaving a residue. The most common glass cleaner is simply isopropyl alcohol diluted with water. Sometimes ammonia or vinegar is added for greater cleaning power.

Drain Cleaners

When kitchen drains become clogged, it is usually because the pipes have become clogged with grease. Drain cleaners often contain sodium hydroxide (NaOH), either in the solid form or as a concentrated liquid. The sodium hydroxide reacts with the water in the pipe to generate heat, which melts much of the grease. The sodium hydroxide then reacts with some of the fat, converting it to soap, which cleans out the rest of the grease in the pipe.

In some products there are also bits of aluminum metal which react with the sodium hydroxide solution to form hydrogen gas, which bubbles out of the clogged area of the drain, creating a stirring action.

Many liquid drain cleaners contain bleach (NaOCl) as well as concentrated sodium hydroxide. Shower drains are often clogged with hair; the bleach degrades the hair, helping to unplug the drain. The best drain cleaner is prevention: Don't pour grease down the kitchen sink and try to keep hair out of the drain. If a drain does get plugged, a mechanical device such as a plumber's snake is often a better choice than a chemical drain cleaner.

Oven Cleaners

Most oven cleaners also contain sodium hydroxide. Several popular products dispense it as an aerosol foam. The greasy deposits on oven walls are converted to soaps when they react with the sodium hydroxide. The resulting mixture can then be washed off with a wet sponge. (You must wear rubber gloves, of course. Sodium hydroxide is extremely caustic and very hard on the skin.)

17.10 Organic Solvents in the Home

Solvents are used in the home to remove paint, varnish, adhesives, waxes, and other materials. Petroleum solvents are also added to some all purpose cleansers as grease cutters (Figure 17.15). They dissolve grease readily but, like gasoline, are highly flammable and deadly when swallowed. The lungs become saturated with hydrocarbon vapors, fill with fluid, and fail to function.

Most organic solvents used around the home are volatile and flammable. Many have toxic fumes, and nearly all are narcotic at high concentrations. Such solvents should be used only with adequate ventilation and never used around a flame. Be sure to read—and heed—all precautions before you use any solvent. And never use gasoline for cleaning clothes; it is too hazardous in too many ways.

A troubling problem connected with household solvents is the practice of inhaling fumes to get "high." The popularity of solvent sniffing seems to be related mainly to peer pressure, especially among young teenagers. Long-term sniffing can cause permanent damage to vital organs, especially the lungs. Often there is irreversible brain damage as well. Some people have died of heart failure while sniffing glue.

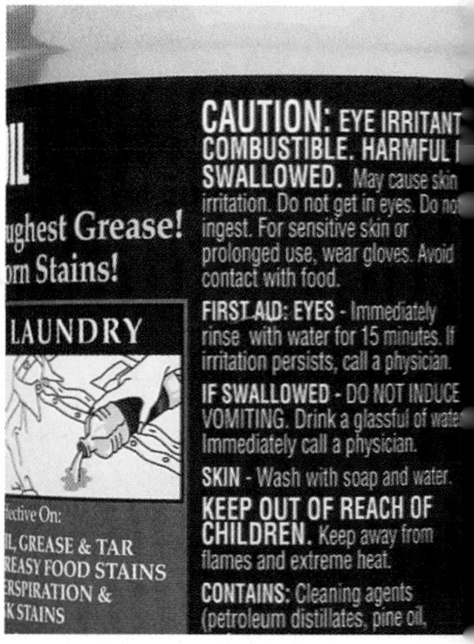

Figure 17.15 Cleansers that contain petroleum distillates warn of flammability and of the hazard of swallowing them.

17.11 Paints

Paint is a broad term that covers a wide variety of products—lacquers, enamels, varnishes, oil-base coatings, and a number of different water-base finishes. Any or all of these materials can be found in any home.

A paint contains three basic ingredients: a pigment, a binder, and a solvent. The universal pigment today is titanium dioxide (TiO_2), which has taken the place of "white lead," the poisonous white pigment that was finally banned in 1977.

Titanium dioxide is a brilliant white pigment with great stability and excellent hiding power. All ordinary paints are pigmented with titanium dioxide. For colored

White lead is basic lead carbonate, $2\,PbCO_3 \cdot Pb(OH)_2$.

paints, small amounts of colored pigments or dyes are added to the white base mixture. The binder, or film former, is the substance that binds the pigment particles together and holds them on the painted surface. In oil paints the binder is usually tung oil or linseed oil. In water-base paints it is a polymer of some kind. Most interior paints have polyvinyl acetate as the binder. Exterior water-base paints use acrylic resins as binders. Acrylic latex paints are much more resistant to rain and sunlight. The solvent is added in order to keep the paint fluid until it is applied to a surface. The solvent might be an alcohol, a hydrocarbon, an ester (or some mixture thereof), or water.

The paint may also contain additives: a drier (or activator) to make the paint dry faster, a fungicide to act as a preservative, a thickener to increase the paint's viscosity, an antiskinning agent to keep the paint from forming a skin inside the can, and perhaps a surfactant to stabilize the mixture.

17.12 Waxes

Waxes are esters (Section 9.13) of long-chain organic acids (fatty acids) with long-chain alcohols. They are produced by plants and animals mainly as protective coatings. (These compounds are not to be confused with paraffin wax, which is made up of hydrocarbons.)

Beeswax is the material from which bees build honeycombs. The following ester is a typical molecule in beeswax. Beeswax is used in such household products as candles and shoe polish.

$$CH_3(CH_2)_{24}\overset{\overset{\displaystyle O}{\|}}{C}-OCH_2(CH_2)_{32}CH_3$$

Carnauba wax is a coating that forms on the leaves of certain palm trees in Brazil. It is a mixture of esters similar to those in beeswax. It is used in making automobile wax, floor wax, and furniture polish.

Spermaceti wax, which is extracted from the head of a sperm whale, is largely cetyl palmitate.

$$CH_3(CH_2)_{14}\overset{\overset{\displaystyle O}{\|}}{C}-OCH_2(CH_2)_{14}CH_3$$

Once widely used in making cosmetics and other products, spermaceti wax is now in short supply because the sperm whale has been hunted almost to extinction.

Lanolin, the grease in sheep's wool, is also a wax. Because it forms stable emulsions with water, it is useful in making various skin creams and lotions.

Many natural waxes have been replaced by synthetic polymers such as silicones (Chapter 10), which are often cheaper and more effective.

17.13 Cosmetics: Personal Care Chemicals

Ages ago, people used materials from nature for cleansing, beautifying, and otherwise altering their appearance. Evidence indicates that 7000 years ago, Egyptians, used powdered antimony (Sb) and the green copper ore malachite as eye shadow. Egyptian pharaohs used perfumed hair oils as far back as 3500 B.C.E. Claudius Galen, a Greek physician of the second century C.E. is said to have invented cold cream. Dandy gentlemen of seventeenth-century Europe used cosmetics lavishly, often to cover the fact that they seldom bathed. Ladies of eighteenth-century Europe whitened their faces with lead carbonate ($PbCO_3$) and many died from lead poisoning.

The use of cosmetics has a long and interesting history, but nothing in the past comes close to the amounts and varieties of cosmetics used by people in the modern industrial world. Each year we spend billions of dollars on everything from hair sprays to toenail polishes, from mouthwashes to foot powders.

What is a cosmetic? The United States Food, Drug, and Cosmetic Act of 1938 defined **cosmetics** as "articles intended to be rubbed, poured, sprinkled or sprayed on, introduced into, or otherwise applied to the human body or any part thereof, for cleansing, beautifying, promoting attractiveness or altering the appearance...." Soap, although obviously used for cleansing, is specifically excluded from coverage by the law. Also excluded are substances that affect the body's structure or functions. Antiperspirants, products that reduce perspiration, are legally classified as drugs. So are antidandruff shampoos. The main difference between drugs and cosmetics is that drugs must be proven "safe and effective" before they are marketed; cosmetics generally do not have to be tested before they're marketed. Most brands of a given type of cosmetic contain the same (or quite similar) active ingredients. Thus, advertising is usually geared toward selling a name, a container, or a fragrance, rather than the actual product itself.

Skin Creams and Lotions

Skin is a complex organ that encloses our bodies, forming a barrier to keep harmful substances out and moisture and nutrients in (Figure 17.16). The outer layer of skin is called the *epidermis*. The epidermis in turn is divided into two parts: dead cells on the outside (the corneal layer), and living cells on the inside, continually replacing corneal cells, which are then sloughed off.

The corneal layer is composed mainly of a tough, fibrous protein called **keratin**. Keratin has a moisture content of about 10%. Below 10% moisture, the human skin is dry and flaky. Above 10%, conditions are ideal for the growth of harmful microorganisms. Skin is protected from loss of moisture by **sebum**, an oily secretion of the sebaccous glands. Exposure to sun and wind may leave the skin dry and scaly, and too-frequent washing also removes natural skin oils.

[Web Reference 5]
 The Cosmetic Ingredient Review (CIR) was established in 1976 by the Cosmetic, Toiletry, and Fragrance Association in an effort to make sure that the safety of ingredients used in cosmetics is thoroughly reviewed.

[Web Reference 6]
The Food and Drug Administration (FDA) presents its view of the chemicals in cosmetics.

About 8000 different chemicals are used in cosmetics. They are sold in 20,000–40,000 different combinations.

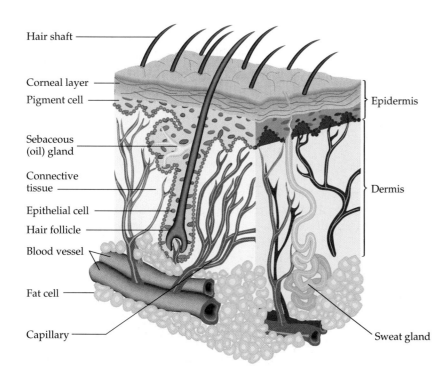

Figure 17.16 Cross section of an area of skin. Cosmetics affect only the outer corneal layer of dead cells.

Hair shaft

Corneal layer
Pigment cell

Sebaceous (oil) gland

Connective tissue

Epithelial cell

Hair follicle

Blood vessel

Fat cell

Capillary

Epidermis

Dermis

Sweat gland

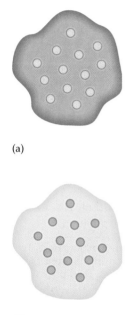

(a)

(b)

Figure 17.17 A lotion is an emulsion of an oil in water (a). A cream is an emulsion of water in an oil (b). A lotion feels cool because evaporating water removes heat from the skin. A cream feels greasy.

Some cosmetics contain *humectants* (Chapter 16) such as glycerol. Glycerol, with its three hydroxyl groups, holds water by hydrogen bonding. Emollients such as petroleum jelly keep the skin moist by forming a physical barrier that hinders evaporation of water from the skin.

Cosmetics are applied to the dead cells of the corneal layer. About $20 billion a year is spent in the United States on various preparations applied to the skin, consisting mainly of lotions and creams. A **lotion** is an emulsion of tiny oil droplets dispersed in water. A **cream** is the opposite; tiny water droplets are dispersed in oil (Figure 17.17). The essential ingredient of each is a fatty or oily substance that forms a protective film over the skin and helps hold moisture in. Typical ingredients are mineral oil and/or petroleum jelly, both of which are mixtures of alkanes obtained from petroleum. Other ingredients include natural fats and oils, perfumes, waxes, water, and emulsifiers (compounds that keep the oily portions from separating from the water). Natural materials used on the skin include lanolin, a fat obtained from sheep's wool, and olive oil. Beeswax is often added to harden the product.

Some creams have been formulated with hormones, queen bee jelly, and other strange ingredients. None of these has been found to confer any particular benefit. Creams and lotions protect the skin by providing a protective coating and by softening it. Such skin softeners are called **emollients**. Petroleum jelly (one trade name is Vaseline) or a good grade of white mineral oil (baby oil) works as well as fancy creams.

It may seem strange that gasoline, a mixture of alkanes, dries out the skin but that the higher alkanes in mineral oil and petroleum jelly soften it. Keep in mind, however, that gasoline is a thin, free-flowing liquid. It dissolves natural skin oils and carries them away. Higher alkanes are viscous, staying right on the skin and serving as emollients.

Moisturizers, which hold moisture to the skin, are usually substances such as lanolin. Collagen (the protein in connective tissue) has also been found to be an effective moisturizer in skin lotions.

Ultraviolet rays in sunlight turn light skin darker by triggering production of the pigment **melanin**. The dark melanin then protects the deeper layers of the skin from damage. Excessive exposure to ultraviolet radiation causes premature aging of the skin and leads to skin cancer. Our quest for tanned skin has resulted in an epidemic of skin cancer. Shorter-wavelength (UV-B) rays are more energetic and are especially harmful. **Sunscreen lotions** block UV-B radiation while letting through the less energetic long-wave UV-A rays that promote tanning. For many years the active ingredient in many of these preparations has been *para*-aminobenzoic acid or one of its esters. Various concentrations of ultraviolet-absorbing substances in a lotion provide the **skin protection factor (SPF)** ratings. SPF values vary from 2 to 35 or more. An SPF of 10, for example, is supposed to mean that you can stay in the sun without burning 10 times as long as you could with unprotected skin.

$$H_2N- \!\!\!\bigcirc\!\!\!-\overset{\overset{\displaystyle O}{\|}}{C}-OH$$

para-Aminobenzoic acid

Cigarette smoking also leads to premature aging of the skin. Nicotine causes constriction of the tiny blood vessels that feed the skin. Repeated constrictions over the years cause the skin to lose its elasticity and become wrinkled. The best treatment for wrinkling of the skin is prevention, accomplished largely by avoiding excessive exposure to the sun and not smoking cigarettes.

Lipstick

Lipstick is quite similar to skin creams in composition. It is made of an oil and a wax. To keep the lipstick firm, a higher proportion of wax is used than in creams. Dyes and pigments provide color. The oil is frequently castor oil. Waxes often employed are beeswax, carnauba, and candelilla. Perfumes are added to cover up the unpleasant fatty odor of the oil, and antioxidants are used to retard rancidity. Bromo acid dyes such as tetrabromofluorescein, a bluish-red compound, are responsible for the color of most modern lipsticks. Because it has little in the way of protective oils, the skin of the lips easily dries out, leading to chapped lips. With or without coloring, lipsticks do protect the lips from drying out.

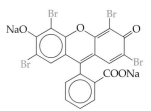

Tetrabromofluorescein

Eye Makeup

Various chemicals are used to decorate the eyes. **Mascara**, to darken eyelashes, has a base of soap, oils, fats, and waxes. Mascara is colored brown by iron oxide pigments, black by carbon (lampblack), green by chromium(III) oxide (Cr_2O_3), or blue by ultramarine (a silicate that contains some sulfide ions). A typical composition is 40% wax, 50% soap, 5% lanolin, and 5% coloring matter. Eyebrow pencils have about the same ingredients.

Eye shadow has a base of petroleum jelly with the usual fats, oils, and waxes. It is colored by dyes or made white by zinc oxide (ZnO) or titanium dioxide (TiO_2) pigments. A typical composition is 60% petroleum jelly, 10% fats and waxes, 6% lanolin, and the remainder dyes, pigments, or both.

Some people have allergic reactions to ingredients in eye makeup. A more serious problem is eye infection caused by bacterial contamination. It is recommended that eye makeup be discarded after 3 months. Mascara is of special concern because of the wand applicator. A slip of the hand and it can scratch the cornea.

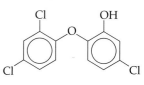

Triclosan

Deodorants and Antiperspirants

Deodorants are products that have perfume to mask body order and a germicide to kill odor-causing bacteria. Bacteria act on perspiration residue and sebum (natural body oil) to produce malodorous compounds such as short-chain fatty acids and amines. The germicide is usually a long-chain quaternary ammonium salt (Section 17.7) or a phenol such as triclosan.

$$Al_2(OH)_5Cl \cdot 2\,H_2O$$

Aluminum chlorohydrate

Antiperspirants are usually deodorants as well, but they also retard perspiration. The active antiperspirant ingredient is a variable complex of chlorides and hydroxides of aluminum and zirconium. (Aluminum chlorohydrate is a typical compound found in many antiperspirants.) Zirconium and aluminum chlorides and hydroxides function as **astringents**, which constrict the openings of the sweat glands, thus restricting the amount of perspiration that can escape. Antiperspirants can be formulated into creams or lotions, or they can be dissolved in alcohol and applied as sprays. Sweating is a natural, healthy body process, and to stop it on a routine basis is probably not wise. Regular bathing and changing of clothes make antiperspirants unnecessary for most people. But if you still feel the need for underarm protection, a simpler deodorant product should suffice.

Figure 17.18 Toothpastes are available under many brand names and in various formulations. Some contain ingredients such as baking soda (sodium bicarbonate), hydrogen peroxide, or both, but there is only marginal evidence of their effectiveness. The only essential ingredients in a toothpaste are a detergent and an abrasive.

17.14 Toothpaste: Soap with Grit and Flavor

After soap (which doesn't count, because the law says it isn't a cosmetic), toothpaste is the most important cosmetic product. Its only essential components are a detergent and an abrasive (Figure 17.18). Soap and sodium bicarbonate can do the job quite well but would be rather unpalatable. The ideal abrasive should be hard enough to clean teeth but not hard enough to damage tooth enamel. Abrasives frequently used in toothpaste are listed in Table 17.1. Some have been criticized as being too harsh.

A typical detergent is sodium dodecyl sulfate (sodium lauryl sulfate).

$$CH_3CH_2CH_2CH_2CH_2CH_2CH_2CH_2CH_2CH_2CH_2CH_2OSO_3^-\ Na^+$$

Table 17.1 ▌ Abrasives Commonly Used in Toothpaste

Name	Chemical Formula
Precipitated calcium carbonate	$CaCO_3$
Insoluble sodium metaphosphate	$(NaPO_3)_n$
Calcium hydrogen phosphate	$CaHPO_4$
Titanium dioxide	TiO_2
Tricalcium phosphate	$Ca_3(PO_4)_2$
Calcium pyrophosphate	$Ca_2P_2O_7$
Hydrated alumina	$Al_2O_3 \cdot nH_2O$
Hydrated silica	$SiO_2 \cdot nH_2O$

Table 17.2 ▌ A Typical Recipe for Toothpaste

Ingredient	Function	Amount
Precipitated calcium carbonate	Abrasive	46 g
Castile soap or sodium dodecyl sulfate	Detergent	4 g
Glycerol (glycerin)	Sweetener	20 g
Gum tragacanth or gum cellulose	Thickener	1 g
Oil of peppermint (or peppermint extract)	Flavoring	1 mL
Water	—	28 mL

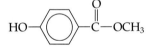

Methyl *para*-hydroxybenzoate
(Methylparaben)

$Ca_5(PO_4)_3OH$

Hydroxyapatite

$Ca_5(PO_4)_3F$

Fluorapatite

Any pharmaceutical grade of soap or detergent probably would work satisfactorily. Most toothpastes today are full of minty flavors, colors, aromas, and sweet tastes. Ingredients include sweeteners such as sorbitol, glycerol (glycerin), and saccharin (Chapter 16); flavors such as wintergreen and peppermint; thickeners such as cellulose gum and polyethylene glycols; and preservatives such as methyl *para*-hydroxybenzoate. Table 17.2 gives a typical recipe for toothpaste.

Tooth decay is caused primarily by bacteria that convert sugars to sticky dextrans or plaque and to acids such as lactic acid ($CH_3CHOHCOOH$). Acids dissolve tooth enamel. Brushing and flossing remove plaque and thus prevent decay. Decay can be minimized by eating sugars only at meals rather than in snacks and by brushing immediately after eating. Acids such as the phosphoric acid (H_3PO_4) in beer and soda pop may also erode the tooth enamel of people who consume these beverages in large amounts.

Many modern toothpastes contain fluorine compounds, such as stannous fluoride (SnF_2) shown to be effective in reducing the incidence of tooth decay. The enamel of teeth is composed mainly of hydroxyapatite. Fluoride from drinking water (or toothpaste) converts part of the enamel to fluorapatite. Fluorapatite is stronger and more resistant to decay than hydroxyapatite.

A major cause of tooth loss in adults is gum disease. Toothpaste formulations that include ingredients such as baking soda ($NaHCO_3$) and hydrogen peroxide (H_2O_2) are purported to prevent gum disease, but there is little evidence that they really do.

17.15 Perfumes, Colognes, and Aftershaves

Although **perfumes** are among the most ancient and widely used cosmetics, their chemistry is exceedingly complex. Originally, all perfumes were extracted from natural sources. Chemists have identified many of the components of perfumes and synthesized them in the laboratory, and so there are many synthetic perfumes. The best ones are probably still made from natural materials because chemists have so far been unable to identify all the many important, but minor, ingredients.

A good perfume may have a hundred or more constituents. Often the components are divided into three categories, called *notes*, based on differences in volatility. The most volatile fraction (that which vaporizes most readily) is called the **top note**. This fraction, made up of relatively small molecules, is responsible for the odor when a perfume is first applied. The **middle note** is intermediate in volatility. It is responsible for the lingering aroma after most of the top-note compounds have vaporized. The **end-note** fraction has low volatility and is made up of compounds with large molecules.

Several compounds with flowery or fruity odors are synthesized in large quantities for use in perfumes. Some are listed in Table 17.3. Odors vary with dilution. A concentrated solution may be unpleasant, yet a dilute solution of the same compound may have a pleasant aroma.

Note that fragrance molecules (Table 17.3) are fat-soluble. They dissolve in skin oils and can be retained for hours.

Table 17.3 ∎ Compounds with Flowery and Fruity Odors Used in Perfumes

Name	Structure*	Odor
Citral	$CH_3C{=}CHCH_2CH_2C{=}CHC{-}H$ (with CH_3 groups and O)	Lemon
Irone	CH_3 ring with $CH{=}CH{-}C{=}O$, CH_3	Violet
Jasmone	cyclopentenone with $-CH_2CH{=}CHCH_2CH_3$ and CH_3	Jasmine
Geraniol	$CH_3C{=}CHCH_2CH_2C{=}CHCH_2OH$ (with CH_3 groups)	Rose

*Note that each compound has only one oxygen atom for ten or more carbon atoms.

Many flowery or fruity odors are sickeningly sweet, even in dilute solutions. Compounds such as musks are often added to moderate the odor (Table 17.4). Musks and similar compounds have extremely disagreeable odors when concentrated but are often pleasant in extreme dilution.

The civet cat, from which the compound civetone is obtained, is a skunklike animal of eastern Africa. Its secretion, like that of the skunk, is a defensive weapon. The secretion from musk deer is probably a pheromone (Chapter 16) that serves as a sex attractant for the deer.

Are there sex attractants for humans? The evidence is scanty, but some perfume makers add α-androstenol to their products. α-Androstenol is a steroid that occurs naturally in human hair and urine. Some studies hint that it may act as a sex attractant for human females, but this seems more likely to be just an advertising ploy. Even if there are human sex attractants, it is doubtful that they have much influence on human behavior because we probably have overriding cultural constraints. Anyway, we seem to prefer the sex attractant of the musk deer.

A perfume usually consists of 10–25% fragrant compounds and fixatives dissolved in ethyl alcohol. **Colognes** are perfumes diluted with ethyl alcohol or an alcohol-water mixture. They are only about 10% as strong as perfumes, usually containing only about 1 or 2% of perfume essence.

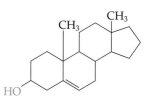

α-Androstenol

Table 17.4 ▌ Unpleasant Compounds Used to Fix Delicate Odors in Perfumes

Compound	Structure	Natural Source
Civetone	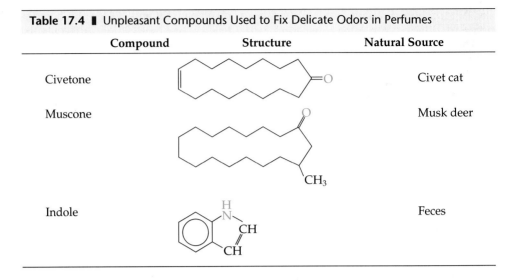	Civet cat
Muscone		Musk deer
Indole		Feces

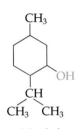

Menthol

Aftershave lotions are similar to colognes. Most are about 50–70% ethanol, the remainder being water, perfume, and food coloring. Some have menthol added for a cooling effect on the skin; others have an added emollient of some sort to soothe chapped skin.

Perfumes are the source of many allergic reactions associated with cosmetics and other consumer products. **Hypoallergenic cosmetics** are those that purport to cause fewer allergic reactions than regular products. The term has no legal meaning, but most "hypoallergenic" cosmetics do not contain perfume.

17.16 Some Hairy Chemistry

Like skin, hair is composed primarily of the fibrous protein keratin. Recall (Chapter 15) that protein molecules are made up of amino acid chains held together by four types of forces: hydrogen bonds, salt bridges, disulfide linkages, and hydrophobic interactions (Figure 15.18). Of these, hydrogen bonds and salt bridges are important to our understanding of the actions of shampoos and conditioners. Hydrogen bonds between protein chains are disrupted by water (Figure 17.19), salt bridges are destroyed by changes in pH (Figure 17.20), and disulfide linkages are broken and restored when hair is permanently waved or straightened.

Shampoo

The keratin of hair has five or six times as many disulfide linkages as the keratin of skin. When hair is washed, the keratin absorbs water and is softened and made more stretchable. The water disrupts hydrogen bonds and some of the salt bridges. Acids

Figure 17.19 In hair, adjacent protein chains are held together by hydrogen bonds that link the carbonyl group of one chain to the amide group of another (a). When hair is wet, these groups can hydrogen-bond with water rather than with each other (b), disrupting the hydrogen bonds between chains.

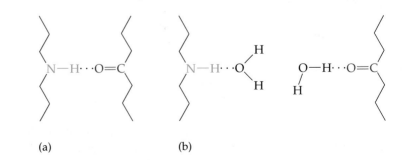

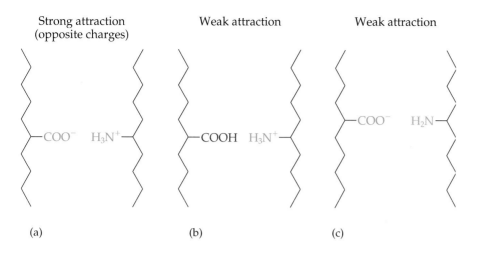

Strong attraction (opposite charges)   Weak attraction   Weak attraction

(a)   (b)   (c)

Figure 17.20 Strong electrostatic forces hold a protein chain with an ionized carboxyl group to another chain with an ionized amino function (a). A change in pH can disrupt this salt bridge. If the pH drops, the carboxyl group takes on a proton and loses its charge (b). If the pH rises, a proton is removed from the amino group and it is left uncharged (c). In either case, the forces between the protein chains are weakened considerably.

and bases are particularly disruptive to salt bridges, making control of pH important in hair care. The maximum number of salt bridges exist at a pH of 4.1.

The visible portion of hair is dead; only the root is alive. The hair shaft is lubricated by sebum. Washing the hair removes this oil and any dirt adhering to it.

Before World War II, the cleansing agent in shampoos was soap. Soap-based shampoos worked well in soft water but left a dulling film on hair in hard water. People often removed the film by using a rinse containing vinegar or lemon juice, however, such rinses are usually not needed with today's products.

Modern shampoos use a synthetic detergent as a cleansing agent. In shampoos for adults, the detergent is often an anionic type, such as sodium dodecyl sulfate, the same detergent used in many toothpastes. In shampoos used for babies and children, the detergent is often an amphoteric type that is less irritating to the eyes. Amphoteric detergents react with both acids and bases. The following molecule is an example.

Hydrocarbon chain   Ionic end

In an acidic solution, it can accept a proton on the negatively charged oxygen. In a basic solution, it can give up one of the protons from the nitrogen.

The only essential ingredient in shampoo is a detergent of some sort (Figure 17.21). What, then, is all the advertising about? You can buy shampoos that are fruit- or herb-flavored, protein-enriched, pH-balanced, or made especially for oily or dry hair. Shampoos for oily, normal, and dry hair seem to differ primarily in the concentration of the detergent. Shampoo for oily hair is more concentrated; dry hair shampoo is more dilute.

Because hair is protein, a protein-enriched shampoo does give it more body. The protein (usually keratin or collagen) coats the hair and literally glues split ends together. Protein is often added to conditioners, too. Conditioners are mainly long-chain alcohols or long-chain quaternary ammonium salts. "Quats" are similar to the compounds used in fabric softeners (Section 17.7), and they work in much the same way to coat hair fibers. Although sometimes combined with shampoo, conditioners are normally used after shampoo.

There are acidic and basic groups on the protein chains in hair. It therefore stands to reason that the acidity or basicity of a shampoo affects hair. Hair and skin are both

Figure 17.21 Shampoos are available in many forms and colors and under many brand names. The only essential ingredient in any shampoo is a detergent.

slightly acidic. Highly basic (high-pH) or strongly acidic (low-pH) shampoos would damage hair, and such products would also irritate the skin and eyes. Most shampoos, however, have pH values between 5 and 8, close to neutral or very slightly acidic.

How about all those flavors and fragrances? Ample evidence indicates that such "natural" ingredients as milk, honey, strawberries, herbs, cucumbers, and lemons add nothing to the usefulness of shampoos or other cosmetics. Why are they there? Smells sell, and there is an appeal to those interested in the back-to-nature movement. There is one hazard in the use of such fragrances: bees, mosquitoes, and other insects like fruit and flower odors, too. Using such products before going on a picnic or a hike could lead to a bee in your bonnet.

Hair Coloring

The color of hair and skin is determined by the relative amounts of two pigments: *melanin*, a brownish black pigment, and *phaeomelanin*, a red-brown pigment that colors the hair and skin of redheads. Brunettes have lots of melanin, but blondes have little of either pigment. Brunettes who would like to become blondes can do so by oxidizing the colored pigments in their hair to colorless products. Hydrogen peroxide (Chapter 8) is the oxidizing agent usually used to bleach hair.

para-Phenylenediamine

Hair dyeing is more complicated than bleaching. The color may be temporary, from water-soluble dyes that can be washed out, or more permanent, from dyes that penetrate the hair and remain there. These dyes often are used in the form of a water-soluble, often colorless, precursor that soaks into the hair and then is later oxidized by hydrogen peroxide to a colored compound.

Permanent dyes are often derivatives of an aromatic amine called *para*-phenylenediamine. Variations in color can be obtained by placing substituents on this molecule. The *para*-phenylenediamine itself produces a black color, and the derivative *para*-aminodiphenylaminesulfonic acid is used in blonde formulations. Intermediate colors can be obtained by the use of other derivatives. One derivative, *para*-methoxy-*meta*-phenylenediamine (MMPD), has been shown to be carcinogenic when fed to rats and mice, but the hazard to those who use it as hair dye is not yet known. It is interesting to note that one substitute for MMPD was its homolog, *para*-ethoxy-*meta*-phenylenediamine (EMPD). Screening revealed that EMPD causes mutations in bacteria. Such mutagens often are also carcinogens.

The FDA cannot ban a cosmetic without proof of harm, but testing for carcinogenicity would cost at least $500,000 and take 2 years. The testing of products for carcinogenicity is considered in Chapter 20.

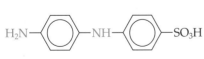

para-Aminodiphenylaminesulfonic acid   MMPD   EMPD

The chemistry of colored oxidation products is quite complex. These products probably include quinones and nitro compounds, among others, and are well-known products of the oxidation of aromatic amines. It should be noted that even permanent dyes affect only the dead outer portion of the hair shaft. New hair, as it grows from the scalp, has its natural color.

Hair treatments, such as Grecian Formula, that develop color gradually use rather simple chemistry. A solution containing colorless lead acetate [$Pb(CH_3COO)_2$] is rubbed on the hair. As it penetrates the hair shaft, Pb^{2+} ions react with sulfur atoms in the hair to form black lead sulfide (PbS). Repeated applications produce darker colors as more lead sulfide is formed.

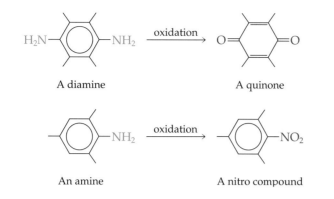

Natural hair

Wave lotion
containing
HSCH₂COOH

Neutralizer
containing H₂O₂

Waved hair

Permanent Waving: Chemistry to Curl Your Hair

The chemistry of curly hair is interesting. Hair is protein, and adjacent protein chains are held together by disulfide linkages. Permanent wave lotion contains a reducing agent such as thioglycolic acid. This wave lotion ruptures the disulfide linkages (Figure 17.22), allowing the protein chains to be pulled apart as the hair is held in a curled position on rollers. The hair is then treated with a mild oxidizing agent such as hydrogen peroxide. Disulfide linkages are formed in new positions to give shape to the hair.

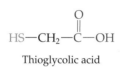

Thioglycolic acid

The same chemical process can be used to straighten naturally curly hair. The change in curliness depends only on how you arrange the hair after the disulfide bonds have been reduced and before the linkages have been restored. As in the case of permanent dyes, permanent curls grow out as new hair is formed.

Hair Sprays

Hair can be held in place by using **resins**, solid or semisolid organic materials that form a sticky film on the hair. Common resins used on hair are polyvinylpyrrolidone (PVP) and its copolymers.

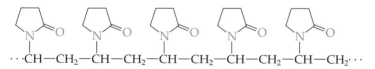

The resin is dissolved in a solvent and sprayed on the hair where the solvent evaporates. The propellants usually used are volatile hydrocarbons, which are also flammable.

Holding resins are also available as mousses. A **mousse** is simply a foam or froth. The active ingredients, like those of hair sprays, are resins such as PVP. Coloring agents and conditioners are also available as mousses. Silicone polymers are often used to impart a sheen to the hair.

Hair Removers

Chemicals that remove unwanted hair are called **depilatories**. Most of them contain a soluble sulfur compound, such as sodium sulfide or calcium thioglycolate

Figure 17.22 Permanent waving of hair is accomplished by breaking disulfide bonds between protein chains and then re-forming them in new positions.

$[Ca(CH_2SHCOO)_2]$, formulated into a cream or lotion. They are strongly basic mixtures that destroy some of the peptide bonds in the hair so that it can be washed off. Remember that skin is made of protein, too, and so any chemical that attacks the hair can also damage the skin.

Hair Restorers

For years, men have been searching for a way to make hair grow on bald spots. Women suffer from baldness, too, but seem to consider wigs more acceptable than men do.

Minoxidil was first introduced as a drug for treating high blood pressure. It acts by dilating the blood vessels. When people who were taking the drug started growing hair on various parts of their bodies, it was applied to the scalps of people who were becoming bald. Minoxidil can produce a growth of fine hair anyplace on the skin where there are hair follicles. Minoxidil is sold under the trade name Rogaine. It needs to be used on a continuous basis, and the cost approaches $1000 per year.

17.17 The Well-Informed Consumer

We have made no attempt here to tell you everything about the many chemical products you have in your home. Books have been written about some of them, and there are many volumes in the library that can help you if you want to know more. We hope, however, that the knowledge you have gained here, coupled with what you read on product labels, will make you a better-informed consumer.

When you look over the many different brands of a product at the supermarket, remember that the most expensive one is not necessarily better than the others. Consider the soap and detergent industry, which does a lot of advertising. Each laundry powder and shampoo claims to be superior to all its competitors, even though in most cases all the brands are rather similar. Advertising is expensive, and so the brands that are most widely advertised probably cost the most, too. Of course, it sometimes happens that one brand really is somewhat better than another, but the better product might be the less expensive one.

If you are unhappy with a product you have bought, tell the company that made it. The cosmetic industry gets approximately 5000 complaints from customers every year. In addition, you can complain to the FDA. In 1990 there were about 100 reports to the FDA regarding adverse effects of cosmetic products, about 25% pertaining to makeup. The others had to do mainly with skin care products, fragrances, and various kinds of hair care products.

Most cosmetics are made from inexpensive ingredients, and yet many highly advertised cosmetics have extremely high price tags. Are they worth it? This is a judgment that lies beyond the realm of chemistry. The product inside that fancy little $100 bottle with the famous label may have cost the manufacturer less than $5 to produce. On the other hand, if the product makes you look better or feel better, perhaps the price is not important. Only you can decide.

Critical Thinking Exercises

Apply knowledge that you have gained in this chapter and one or more of the FLaReS principles (Chapter 1) to evaluate the following statements or claims.

17.1 An advertisement claims that a laundry detergent "leaves clothes brighter than bright."

17.2 An advertisement claims that a "moisture-rich shampoo…restores moisture…to dry, damaged hair."

17.3 An advertisement claims that a new skin cream "restores life and vitality to skin."

Summary

1. Although soap has been known for many centuries, only during the past two centuries has it come into common use.

2. Soap is made from fat by a reaction with lye (sodium hydroxide).

3. A soap molecule has a hydrophobic hydrocarbon tail (soluble in oil or grease) and a hydrophilic polar head (soluble in water), and so it can disperse oil or grease in water.

4. In hard water, which contains calcium and magnesium ions, soap forms an insoluble scum.

5. Water softeners, such as sodium carbonate, remove as insoluble salts the ions that otherwise cause water to be hard.

6. Most synthetic detergents are salts of long-chain acids that are not natural fatty acids. Usually they are sulfuric acid derivatives.

7. When the hydrocarbon portion of a detergent molecule has a branched-chain structure, it is not easily biodegraded. Biodegradable detergents have linear hydrocarbon chains.

8. In addition to a surfactant, detergent formulations also include builders and optical brighteners.

9. Whereas most solid detergents are anionic surfactants, liquid laundry detergents usually contain nonionic surfactants.

10. Cationic surfactants (alkylammonium salts) are useful as disinfectants. Those with two long hydrocarbon chains are used as fabric softeners.

11. Bleaches are oxidizing agents. Hypochlorite bleaches release chlorine; perborate bleaches release oxygen.

12. Any organic solvents around the house should be carefully stored because most are flammable and most are toxic.

13. A paint contains three basic ingredients: a pigment, a binder, and a solvent.

14. Waxes are esters of fatty acids with long-chain alcohols.

15. The skin is made of a tough, fibrous protein called keratin.

16. Skin lotions are emulsions of oil in water; skin creams are emulsions of water in oil.

17. Eye makeup and lipstick are mainly blends of oil, waxes, and pigments.

18. Toothpaste contains a detergent, an abrasive, a thickener, flavoring, and often an added fluoride compound.

19. Perfumes can usually be split into three basic fractions: the top note (which is the most volatile), the middle note (of intermediate volatility), and the end note (of lowest volatility). Colognes are diluted perfumes.

20. Shampoos are synthetic detergents blended with perfume ingredients and sometimes mixed with protein or other conditioners to give the hair more body.

21. Dark hair contains the pigment melanin, red hair contains phaeomelanin, and blonde hair contains very little of either pigment.

22. Permanent waving of the hair uses a reducing agent to break the disulfide linkages and then an oxidizing neutralizer to reform the disulfide bonds.

23. The most expensive brand of a given product is not necessarily the best.

Key Terms

anionic surfactants   17.5
antiperspirants   17.13
astringents   17.13
bleaches   17.7
builder   17.4
cationic surfactants   17.6
colognes   17.15
cosmetics   17.13
cream   17.13
deodorants   17.13
depilatories   17.16

emollients   17.13
end note   17.15
eye shadow   17.13
hydrophilic   17.2
hydrophobic   17.2
hypoallergenic
  cosmetics   17.15
keratin   17.13
lanolin   17.12
lotion   17.13
mascara   17.13

melanin   17.13
micelle   17.2
middle note   17.15
moisturizers   17.13
mousse   17.16
nonionic surfactants   17.5
optical brighteners   17.4
paint   17.11
perfumes   17.15
resins   17.16
saponins   17.1

sebum   17.13
skin protection factor
  (SPF)   17.13
soap   17.2
sunscreen lotions   17.13
surfactant   17.2
surface-active
  agent   17.2
top note   17.15
waxes   17.12

Review Questions

1. What are saponins?

2. How do carbonate ions make water basic?

3. What is a soap?

4. What products are formed when tripalmitin reacts with lye (sodium hydroxide)?

5. List some ingredients found in toilet soaps.

6. What kind of ingredients are used in scouring soaps?

7. How does a potassium soap differ from a sodium soap?

8. Give the structural formula for each of the following.
   a. potassium stearate   b. sodium palmitate

9. How does soap (or detergent) clean a dirty surface?

10. Why does soap fail to work in acidic water?

11. What is hard water?

12. How does hard water affect the action of soap?

13. How does trisodium phosphate soften water? (List two ways.)

14. How does a water-softening tank work?

15. What are some advantages of soap?

16. What are some advantages of synthetic detergents?

17. What is an LAS detergent?

18. What is a surface-active agent (surfactant)?

19. What is a (detergent) builder?

20. How does sodium tripolyphosphate aid the cleaning action of a soap or detergent?

21. How do zeolites aid the cleaning action of a soap or detergent?

22. List two advantages of sodium carbonate as a replacement for phosphates as a builder in detergent formulations.

23. What advantages do zeolites have over carbonates as builders?

24. What is an optical brightener? How do these compounds work?

25. What is an anionic surfactant?

26. What is an nonionic detergent?

27. What ingredients are used in liquid dishwashing detergents (for hand dishwashing)?

28. How do detergents for automatic dishwashers differ from those used for hand dishwashing?

29. How do cyanurate bleaches (such as Symclosene) work?

30. How does sodium perborate work as a bleach?

31. List some uses of diluted household ammonia.

32. What safety precautions should be followed when using ammonia?

33. What sorts of surfaces should not be cleaned with ammonia? Why?

34. List two properties of baking soda that make it useful as a cleaning agent.

35. Mention a use of vinegar in cleaning.

36. Why should marble surfaces not be cleaned with vinegar?

37. List some uses of organic solvents in the home.

38. List two hazards of cleansers that contain petroleum distillates.

39. What are the main hazards of using solvents in the home?

40. Should gasoline be used as a cleaning solvent? Why or why not?

41. What are cationic surfactants?

42. Cationic surfactants are not particularly good detergents, yet they are widely used, especially in the food industry. Why?

43. Cationic surfactants are often used along with nonionic surfactants but are seldom used with anionic surfactants. Why not?

44. Give an example of a compound that acts as a fabric softener.

45. How do fabric softeners work?

46. List the essential ingredients in each of the following household products.
   a. scouring cleaner
   b. window cleaner
   c. oven cleaner
   d. toilet bowl cleaner
   e. liquid drain cleaner
   f. solid drain cleaner
   g. paint remover

47. What is the legal definition of a cosmetic?

48. Is soap a cosmetic?

49. What are the essential ingredients in a toothpaste?

50. How do fluorides strengthen tooth enamel?

51. What is the principal cause of tooth decay?

52. What are the best ways to prevent tooth decay?

53. What is the principal material of the corneal layer of the skin?

54. What is sebum?

55. What is an emollient?

56. What is the ideal moisture content of skin?

57. What are the principal components of creams and lotions?

58. What is a skin moisturizer?

59. How does lipstick differ from skin cream?

60. What materials are used to color lipsticks?

61. What is a perfume?

62. What are the three fractions of a perfume? How do these fractions differ at the molecular level?

63. What is musk? Why are musks added to perfumes?

64. What are colognes?

65. What are the principal ingredients of aftershave lotion?

66. Why is menthol added to some aftershave lotions?

67. What is a deodorant?

68. What is an antiperspirant?

69. What is the only essential ingredient of shampoos?

70. What type of detergent is used in baby shampoos? Why?

71. What chemical substances determine the color of hair and skin?

72. What chemical compound is used most often to bleach hair?

73. What is the difference between a temporary and a permanent hair dye? Why is a permanent dye not really permanent?

74. What kind of chemical reagent is used to break disulfide linkages in hair?

75. What kind of chemical reagent is used to restore disulfide linkages in hair?

76. What are resins?

Problems

Detergents: Structures and Properties

Problems 77 through 81 refer to the following structures.

$$CH_3CH_2CH_2CH_2CH_2CH_2CH_2CH_2CH_2CH_2\overset{\overset{\displaystyle CH_3}{|}}{CH}——SO_3^{-}\,Na^{+}$$

I

$$CH_3CH_2CH_2CH_2CH_2CH_2CH_2CH_2CH_2CH_2CH_2CH_2—O—\overset{\overset{\displaystyle O}{\|}}{\underset{\underset{\displaystyle O}{\|}}{S}}—O^{-}\,Na^{+}$$

II

$$CH_3CH_2CH_2CH_2CH_2CH_2CH_2CH_2CH_2CH_2—\overset{\overset{\displaystyle CH_3}{|}}{CH}——\overset{\overset{\displaystyle CH_3}{|}}{\underset{\underset{\displaystyle CH_3\ \ Cl^{-}}{|}}{N^{+}}}—CH_3$$

III

77. Which are biodegradable?

78. Which is an LAS detergent?

79. Which is sodium lauryl sulfate?

80. Which are anionic detergents?

81. Which is a cationic detergent?

Chemical Equations

82. Write the equation that shows how carbonate ion reacts with water to form a basic solution.

83. Write the equation that shows how phosphate ion reacts with water to form a basic solution.

84. Give the equation that shows how carbonate ions precipitate hard-water ions. You may use M^{2+} to represent the hard water ion.

85. Give the equation that shows how phosphate ions precipitate hard-water ions. You may use M^{2+} to represent the hard water ion.

86. Give the structures of the products formed in the following reaction.

$$
\begin{array}{l}
CH_2O—\overset{\overset{\displaystyle O}{\|}}{C}(CH_2)_{10}CH_3 \\
| \\
CH—O—\overset{\overset{\displaystyle O}{\|}}{C}(CH_2)_{10}CH_3 \quad + \quad 3\,NaOH \longrightarrow \\
| \\
CH_2O—\overset{\overset{\displaystyle O}{\|}}{C}(CH_2)_{10}CH_3
\end{array}
$$

87. Give the structures of the products formed in the following reaction.

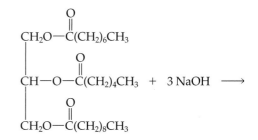

$$+ \quad 3\,NaOH \longrightarrow$$

Types of Chemical Reactions

88. What kind of chemical reaction is involved when disulfide linkages in hair are broken?

89. What kind of chemical reaction is involved when new disulfide linkages are formed in hair?

90. What kind of chemical reaction is involved when a colorless hair dye compound is converted to a colored dye?

91. What kind of chemical reaction is involved in the bleaching of hair?

Chemical Structures and Properties

Problems 92 through 95 refer to structures I through IV below.

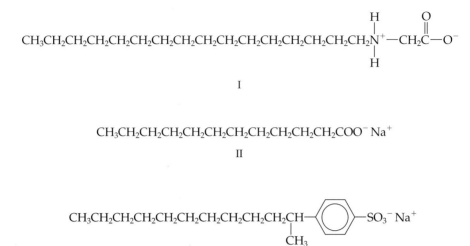

I

$$CH_3CH_2CH_2CH_2CH_2CH_2CH_2CH_2CH_2CH_2CH_2COO^- \ Na^+$$

II

III

IV

92. Which are synthetic detergents?

93. Which are soaps?

94. Which is an amphoteric detergent?

95. Which is a principal ingredient in baby shampoo?

Projects

96. The Clearwater Chemical Company has announced the development of a biodegradable detergent. What tests, if any, should be made before the detergent is marketed? Justify your recommendations.

97. Maxisuds, Inc., has announced that it has found a replacement for phosphate builders in detergents. What tests, if any, should be made before this builder is used in detergent formulations? Should Maxisuds be allowed to market the builder until it is proven harmful? Can a product ever be proven safe? Justify your position.

98. Read the labels of five brands of toothpaste. List the ingredients of each. If you can, classify each ingredient according to its function—that is, as a detergent, an abrasive, and so on. You may find a reference book such as *The Merck Index* (Reference 4) to be helpful.

99. Read the labels on five brands of lotions or creams (designed for use on the skin—face, hands, or body). List the ingredients in each brand. If you can, name the function of each ingredient.

100. Read the labels on five brands of shampoos. List the ingredients in each brand. Which ingredients are detergents? If you can, name the function of each nondetergent ingredient.

Online Projects

101. Several industry-sponsored websites about soaps have been mentioned in this chapter. Find a site sponsored by an ecological organization that talks about soap in the environment, and compare the two viewpoints.

102. One of the newer household cleaning products on the market is called Clean Shower. Search the web and see what you can find out about this interesting liquid spray.

103. A product called Febreze, an odor-removing spray appeared on the market in 1998. Search the web to see what you can find out about the composition and mode of action of this product.

References and Readings

1. "The Dirt on Detergents." *Consumer Reports*, February 2000, pp. 35–37.

2. "Dishwashing Liquids." *Consumer Reports*, February 1996, p. 45.

3. Dorrian, John. "Dissolving Household Chores." *Chem-Matters*, December 1997, pp. 13–15.

4. "Furniture Cleaners and Polishes." *Consumer Reports*, June 1993, pp. 358–360.

5. "Hair Loss: Does Anything Really Help?" *Consumer Reports*, August 1996, pp. 62–63.

6. "Hand and Bath Soaps." *Consumer Reports*, November 1995, pp. 730–733.

7. "Interior Latex Paints." *Consumer Reports*, February 1994, pp. 127–131.

8. Kiefer, David M. "The Tide Turns for Soap." *Today's Chemist at Work*, October 1996, pp. 70–74.

9. "Laundry Detergents." Consumer Reports, February 1995, pp. 92–95.

10. "Lipsticks." *Consumer Reports*, July 1995, pp. 454–457.

11. McCoy, Michael. "Soaps and Detergents." Chemical & Engineering News, 24 January 2000, pp. 37–46. A survey of the industry. One of a series of such articles published each January.

12. Morris, E.T. Fragrance: *The Story of Perfume from Cleopatra to Chanel*. New York: Scribner's, 1990.

13. Prybus, Davis, and Charles Sell, Ed. *The Chemistry of Fragrances*. Cambridge: Royal Society of Chemistry, 1999.

14. "Putting Sunscreens to the Test." *Consumer Reports*, May 1995, pp. 334–339.

15. Schamper, Tom. "Chemical Aspects of Antiperspirants and Deodorants." *Journal of Chemical Education*, March 1993, pp. 242–244.

16. "Shampoos and Conditioners: Heading Off the Hype." *Consumer Reports*, June 1992, pp. 395–403.

17. "The Skin Game." *Consumer Reports*, January 2000, pp. 38–41. Discussion about various commercial moisturizers.

18. "Straight Talk about Curly Hair." *Consumer Reports*, September 1999, pp. 49–52. What precautions to take when you perm straight hair or relax curly hair.

19. "Which Toothpaste Is Right for You?" *Consumer Reports*, August 1998, pp. 11–14.

20. Willensky, Diana. "Hair and Nails." *American Health*, November 1994, pp. 94–95.

21. Williams, D. F., and W. H. Schmitt (Eds.). *Chemystry and Technology of the Cosmetic and Toiletries Industry*. New York: Routledge, Chapman, and Hall, 1992.

Fitness and Health
Some Chemical Connections

To maintain fitness and good health, the secret lies
In good nutrition and a lot of exercise.

Exercise is essential to good health. Through chemistry, we gain a better understanding of the effects of exercise on our bodies and minds.

(a)

(b)

(c)

W e in the Western world have spent centuries developing labor-saving devices. Now many of us earn our daily bread with little physical exertion. This presents us with a different kind of problem: how to stay healthy and physically fit. We now put great effort into keeping our bodies fit through exercise and proper diet.

In our culture today, thin is in. Many former sedentary people have joined the ranks of joggers, walkers, tennis players, bodybuilders, and bikers in pursuit of the bulgeless body. About a million people per week are treated at weight-loss clubs and clinics. Countless others treat themselves with crash diets, dietary supplements, diet pills, and other programs. Diet books populate the bestseller lists. Many of these diet plans are simply fads, more likely to increase their creator's wealth than to decrease your weight. Many are grossly unbalanced and can lead to nutritional deficiencies, decreased resistance to disease, and a decline in general health. Despite the prevalence of diet plans, the percentage of the population that is obese increased from 12% in 1991 to 18% in 1998, a 50% increase in only 7 years!

Through modern science we have a better understanding of what it means to be physically fit and what it takes to get there (Figure 18.1). Athletes are larger, stronger, and faster than ever before. Quackery abounds, however. We read of diets, drugs, and exercise plans that promise health and longer life. Often these schemes require little effort on our part. It is sometimes difficult to distinguish science from nonsense. In this chapter we examine some of the ways chemistry contributes to athletic achievement and to our understanding of health and fitness. We examine muscle action, diet, electrolyte balance, drugs, and other items of concern to those interested in their own well-being.

18.1 Nutrition

A nutritious diet is essential to good health. Good nutrition includes sufficient carbohydrates, fats, and proteins, as well as all the essential minerals and vitamins, along with adequate dietary fiber and water.

Fewer Calories

For many people around the world, it is a daily struggle just to obtain enough food to survive. At the same time, in the United States and other developed nations, too much food that is too rich in fats and calories makes obesity a major health problem.

Research shows that undereating is healthier than overeating. Large studies with mice have shown that when given 40% less food than a control group chooses to eat, the mice live longer and have fewer tumors. Humans also seem to stay healthier when they eat a bit too little than when they eat too much.

Total calories consumed is important but the *distribution* of calories is even more important. Modern advice, as shown in the food guide pyramid (Figure 18.2), is that good nutrition includes generous portions of complex carbohydrates (starches), along with fruits and vegetables, modest portions of protein, and limited amounts of fats, oils, and sugars.

Less Fat but More Starch

The recommended diet contains no more than 25% of calories from fat. Unfortunately, the typical American diet is 34% fat. (For some it runs as high as 45%.) In China, people normally consume about 20% more calories than Americans do, but they are 25% thinner. The big difference is that the Chinese eat twice as much starch as Americans do, but only one-third as much fat.

Figure 18.1 Chemistry has transformed athletic performance in recent years. New materials have provided protective gear such as plastic helmets and foam padding (a). New equipment has changed the nature of most sports events. For example, pole vaulting (b) was revolutionized by the replacement of wooden poles with fiberglass-reinforced plastic ones in the 1960s. Biochemistry has brought a better understanding of the human body and the way it works (c). Chemistry has provided drugs that keep athletes healthier but also increase the potential for abuse.

 [Web Reference 1] Dietary guidelines from the U. S. Department of Agriculture.

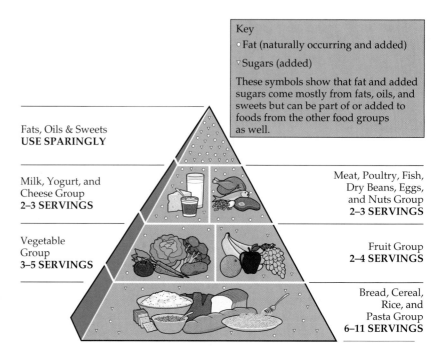

Key
- Fat (naturally occurring and added)
- Sugars (added)

These symbols show that fat and added sugars come mostly from fats, oils, and sweets but can be part of or added to foods from the other food groups as well.

Fats, Oils & Sweets
USE SPARINGLY

Milk, Yogurt, and Cheese Group
2–3 SERVINGS

Meat, Poultry, Fish, Dry Beans, Eggs, and Nuts Group
2–3 SERVINGS

Vegetable Group
3–5 SERVINGS

Fruit Group
2–4 SERVINGS

Bread, Cereal, Rice, and Pasta Group
6–11 SERVINGS

Figure 18.2 The daily food guide pyramid. Start with bread, cereal, rice, and pasta, along with fruits and vegetables; then add some servings from the milk and meat groups. Go easy on fats, oils, and sweets.

Less Protein—Especially Red Meat

Americans also consume more meat than they need. However, Americans are reducing their consumption of red meat. In 1976 the average consumption of beef per person was 95 lb/year. By 1994 it had been reduced to 70 lb per person, perhaps as a result of knowledge that a steady diet of red meat increases the rate of colon cancer.

It appears that for people with a high protein diet, weight gain is faster and weight loss is more difficult. Studies on rats indicate that when protein is strictly limited, rats can eat as much as they want with little change in weight.

Nutrition and the Athlete

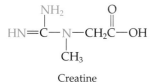

Creatine

Athletes need more calories because they expend more than the average sedentary individual. An athlete should get these extra calories as carbohydrates. Foods rich in carbohydrates, especially starches, are the preferred source of energy for the healthy body. Fatty and protein-rich foods also supply calories, but protein metabolism produces more toxic wastes that tax the liver and kidneys. The pregame steak dinner consumed by football players in the past was based on a myth that protein builds muscle. This is simply not true. Although athletes do need the **recommended daily allowance (RDA)** for protein (based on grams per kilogram of body weight), they do not need an excess. Protein consumed in amounts greater than needed for synthesis and repair of tissue only makes the athlete fatter (as a result of excessive calorie intake)—not more muscular.

Muscles are built through exercise, not through eating excess protein. When a muscle contracts against a resistance, an amino acid called *creatine* is released. Creatine stimulates production of the protein myosin, thus building more muscle tissue. If the exercise stops, the muscle begins to shrink after about 2 days. After about 2 months without exercise, muscle built through an exercise program is almost completely gone. (The muscle does *not* turn to fat, as some athletes believe. Former athletes often do get fat however, because they continue to take in the same number of calories while expending fewer.)

Some extreme-endurance athletes, such as triathletes and ultramarathoners, may need a bit more than the RDA for protein. Because nearly all Americans eat 50% more protein than they need, these athletes seldom need protein supplements. Even weightlifters and bodybuilders do not need extra protein.

18.2 Vitamins and Minerals

Vitamins are molecules that the human body needs but cannot manufacture. Unlike hormones and enzymes, which the body can synthesize, vitamins must be included in the diet. Minerals must also be present in the diet. These are all the inorganic elements that the body needs—some in sizable quantities, and some in only trace amounts.

Medical spokespersons often insist that vitamin and mineral supplements are unnecessary when the diet is well balanced and includes the RDA (or DRI) for each nutrient. However, for those who choose to take such supplements, reasonable doses of vitamin and mineral tablets and capsules certainly appear to do no harm. In fact, they can sometimes be beneficial. There are times when the body has an increased demand for vitamins and minerals, as during periods of rapid growth, during pregnancy and lactation, and during periods of trauma and recovery from disease. A study of retired people aged 65 years and older showed that a daily dose of 18 vitamins and minerals seems to strengthen the immune system. The number of infections in the study group was less than 50% of that in a control group, which received no supplements.

The RDA values for vitamins and minerals are set by the Food and Nutrition Board of the U.S. National Academy of Sciences and the National Research Council. The values are quite modest, and any well-balanced diet should have no trouble supplying them. The RDA values for vitamins are the amounts that will prevent deficiency diseases such as beriberi, pellagra, and scurvy. Some scientists believe that the *optimum* intakes of vitamins are much higher than their RDA values. In Table 18.1, the daily intakes recommended by four different scientists are compared with the RDA values.

If you eat a well-balanced diet and you are in good health, you probably do not need vitamin supplements. But if you do take them, they probably will do no harm. In small doses, vitamins are not toxic; however, in extremely large doses they might be, especially if they are fat-soluble.

> Vitamins and minerals are both discussed in Chapter 16. See Table 16.4 for a list of essential minerals and trace minerals. Table 16.5 lists most vitamins along with their molecular structures, food sources, and deficiency symptoms.

> The DRI is the *dietary reference intake*, a newer nutrition standard similar to the more familiar RDA.

Table 18.1 ▍ Recommended Daily Adult Intakes of Vitamins

Vitamin	RDA	Williams*	Allen[†]	Leibovitz[‡]	Pauling[§]
Vitamin C	60 mg	2500 mg	1500 mg	2500 mg	1000–18,000 mg
Vitamin E	10 IU	400 IU	600 IU	300 IU	800 IU
Vitamin A	5000 IU	15,000 IU	15,000 IU	20,000 IU	20,000–40,000 IU
Vitamin K	None	100 mg	None	None	None
Vitamin D	400 IU	400 IU	300 IU	800 IU	800 IU
Thiamine, B_1	1.5 mg	20 mg	300 mg	100 mg	50–100 mg
Riboflavin, B_2	1.7 mg	20 mg	200 mg	100 mg	50–100 mg
Niacinamide, B_3	18 mg	200 mg	750 mg	300 mg	300–600 mg
Pyridoxine, B_6	2.2 mg	30 mg	350 mg	100 mg	50–100 mg
Cyanocobalamine, B_{12}	3 mg	90 mg	1000 mg	100 mg	100–200 mg
Folacin	400 mg	400 mg	400 mg	400 mg	400–800 mg
Pantothenic acid	None	150 mg	500 mg	200 mg	100–200 mg

*Roger J. Williams, 1975.

[†] Mary B. Allen, 1981.

[‡] Brian Leibovitz, 1984.

[§] Linus Pauling, 1986.

Vitamin A

Vitamin A is essential for good vision, bone development, and skin maintenance. There is also some evidence that it may confer resistance to certain kinds of cancer. This may help to explain the anticancer activity that has been noted for cruciferous vegetables such as broccoli, cauliflower, brussels sprouts, and cabbage. Most cruciferous vegetables are rich in β-carotene, which is converted to vitamin A in the body. Of course, vitamin A is a fat-soluble vitamin stored in the fatty tissues of the body, especially in the liver, and so large doses can be toxic.

Vitamin B Complex

Dorothy Crowfoot Hodgkin (1910–1994) used X-rays to determine the structures of several important organic compounds including penicillin and cholesterol. She was awarded the 1964 Nobel prize in chemistry for her work on the structure of vitamin B_{12}.

Marie Curie had received another Nobel prize in 1903, but it was in physics.

 [Web Reference 2] Learn more about folic acid at this website.

People with low levels of vitamin C are more likely to develop cataracts and glaucoma, as well as gingivitis and periodontal disease.

There are at least eight members of the vitamin B family (Table 16.5). Because they are all water-soluble, any excess intake is excreted in the urine. There appears to be no toxicity connected with B vitamins, with the possible exception of vitamin B_6, which apparently can cause neurological damage in some people if taken in extremely large daily doses. As a group, B vitamins are important in maintaining the skin and the nervous system.

Niacin (vitamin B_3) in addition to preventing the skin lesions of pellagra, offers some relief from arthritis. It also helps in lowering the blood cholesterol level. Very large daily doses (5–30 g) have been taken for years by schizophrenic patients without any toxic effects.

Vitamin B_6, pyridoxine, has been found to help people with arthritis by shrinking the synovial membranes that line the joints. Vitamin B_6 is known to be a coenzyme for more than 100 different enzymes.

Vitamin B_{12} is not found in plants, and vegetarians are apt to be deficient in this vitamin. A deficiency of vitamin B_{12} can lead to pernicious anemia. Vitamin B_{12}, also called cyanocobalamine, is a very complicated molecule. (See the molecular structure given in Table 16.5.) The structure was finally determined by X-ray crystallography. Dorothy Hodgkin of Oxford University received the 1964 Nobel prize in chemistry for this work. (Only three Nobel prizes in chemistry have ever been awarded to women. The other two went to Marie Curie in 1911 and to her daughter, Irène, in 1935.)

One of the highly important B complex vitamins is folic acid, which is critical in the development of the nervous system of a fetus. Its presence in a pregnant woman's diet prevents spina bifida in her baby. Folic acid also helps prevent cardiovascular disease.

Vitamin C

Vitamin C is ascorbic acid, the component in citrus fruits that acts as an antiscurvy factor. However, vitamin C does much more than that. It promotes the healing of wounds and burns, and it has demonstrated an ability to heal gastric ulcers. It also seems to play an important role in maintaining the body's collagen supply. Like vitamin E, it is an antioxidant, and these two vitamins, along with β-carotene, have been making headlines because of their antioxidant activity. Antioxidants often act as anticarcinogens.

There is mounting evidence that vitamin C is essential for efficient functioning of the immune system. Interferons have been recognized as agents in the immune system. (Interferons are large molecules formed by the action of viruses on their host cells. By producing an interferon, a virus can interfere with the growth of another virus.) It has been demonstrated that an increased level of vitamin C increases the body's production of interferons.

Linus Pauling, Vitamin C, and the Common Cold

In 1970 Linus Pauling suggested that ascorbic acid, vitamin C, could be used as a weapon against the common cold. He said that daily doses of 1000–18,000 mg, depending on the person and the circumstances, could prevent colds or lessen their severity. Pauling's claims were greeted with skepticism—and even ridicule—by many in the scientific and medical communities. However, in recent years his work has received support from medical research. Public support has always been strong; sales of vitamin C have soared since he made his first announcement.

According to Pauling, the RDA of 60 mg of vitamin C is enough to prevent scurvy, but vitamin C does much more than prevent scurvy. He believed that the optimum intake of vitamin C was much higher than the RDA and suggested that it probably varies from one person to another. He pointed out that most species manufacture their own ascorbic acid. He believed that primates once had this ability but lost it somewhere along the evolutionary trail, so that now ascorbic acid is a vitamin (a molecule the human body needs but cannot make).

Pauling contended that vitamin C not only helps prevent (or lessen the effects of) the common cold but also has value in preventing and treating influenza, as well as several other viral diseases. Vitamin C, he believed, has many functions, including strengthening of the immune system

For whatever reasons, whether to prevent colds, to protect against cancer, or simply to maintain an optimum state of health, many people now take vitamin C supplements on a regular basis. They rarely take the larger megadoses that Pauling recommended, but they do continue to take supplements year after year. If vitamin C is not helping all these people to stay healthier, at least they seem to believe that it is.

Linus Pauling (1901–1994), winner of two Nobel prizes (for chemistry in 1954 and for peace in 1962), advocated massive doses of vitamin C to prevent the common cold.

Vitamin D

Vitamin D is a steroid-type vitamin that protects children against rickets. It promotes the absorption of calcium and phosphorus from foods to produce and maintain healthy bones. However, there is less need for vitamin D in adults. In fact, too much vitamin D can lead to excessive calcium and phosphorus absorption, with subsequent formation of calcium deposits in various soft body tissues including those of the heart. The RDA for vitamin D is 400 international units (IU), and you probably should not exceed that amount of this fat-soluble vitamin.

Vitamin E

Vitamin E is a mixture of tocopherols (Table 16.5). It seems to have considerable value in maintaining the cardiovascular system. Vitamin E has been used to treat coronary heart disease, angina, rheumatic heart disease, high blood pressure, arteriosclerosis, varicose veins, and a number of other cardiovascular problems. It is also an anticoagulant that has been useful in preventing blood clots after surgery.

Rats deprived of vitamin E become sterile. For this reason, vitamin E is sometimes called the antisterility vitamin. Vitamin E deficiency can also lead to muscular dystrophy, a disease of the skeletal muscles.

Vitamin E is an antioxidant that can inactivate free radicals. It is generally believed that much of the physiological damage from aging is a result of the production of free radicals. Indeed, vitamin E has been referred to as the *antiaging vitamin.*

LDL cholesterol is oxidized before it is deposited in arteries, and vitamin E prevents the oxidation of cholesterol. This may be the mechanism by which it helps to protect against cardiovascular disease.

There is good evidence that eating lots of fruits and vegetables protects against cancer, but there is little evidence that pills containing antioxidants do so.

A lack of vitamin E can lead to a deficiency in vitamin A. (Vitamin A can be oxidized to an inactive form when vitamin E is no longer present to act as an antioxidant.) Vitamin E also collaborates with vitamin C in protecting blood vessels and other tissues against oxidation. Vitamin E is the fat-soluble and vitamin C the water-soluble antioxidant vitamin. Oxidation of unsaturated fatty acids in the cell membrane can be prevented or reversed by vitamin E, which is itself oxidized in the process. Vitamin C can then restore vitamin E to its unoxidized form.

18.3 Body Fluids: Electrolytes

Another aspect of chemistry and nutrition that you should be aware of is the balance between fluid intake and electrolyte intake. **Electrolytes** are substances that conduct electricity when dissolved in water. Sodium ions (Na^+), potassium ions (K^+), and chloride ions (Cl^-) are the major electrolytes essential for proper cell function. Water is also an essential nutrient, a fact obvious to anyone who has been deprived of it. Modern Western people seem to prefer to meet the body's need for water by consuming soda pop, coffee, beer, fruit juice, and other beverages. Some of these products actually impair the body's use of the water in these beverages. Alcoholic drinks promote water loss by blocking action of the *antidiuretic hormone (ADH)*. Caffeine has a diuretic effect on the kidneys, promoting urine formation and consequent water loss.

The best way to replace water lost through sweat, tears, respiration, and urination is to drink water. Unfortunately, thirst is often a *delayed* response to water loss, and it may be masked by such symptoms as exhaustion, confusion, headache, and nausea (the symptoms are a result of dehydration). Body sweat contains 99% water, sodium ions (Na^+) and chloride ions (Cl^-), minute amounts of calcium ions (Ca^{2+}) and potassium ions (K^+), urea, lactic acid, and body oils. The normal American diet probably contains too much sodium chloride (NaCl). Thus, it makes the most sense to replace the water component of lost sweat with pure water itself. Commercial "sports drinks" are quite popular with both serious and weekend athletes, but because they are so concentrated (a hazard that can lead to diarrhea), they are of marginal value except for endurance athletes (Figure 18.3).

How do you know if you are drinking enough water and are in a proper state of hydration? The urine is a good indicator of hydration. When you are dehydrated, your urine gets cloudy because your kidneys are trying to conserve water so as to keep the blood volume from shrinking and to prevent shock. As dehydration worsens, muscles tire and cramp. Dizziness and fainting may follow, and brain cells may shrink, resulting in mental confusion. The heat regulatory system may also fail, causing **heat stroke**, which can be life-threatening without prompt medical attention.

Figure 18.3 The best replacement for fluids lost during exercise generally is plain water. Sports drinks help little, except when engaging in endurance activities lasting for 2 hr or more. At that point, carbohydrates in the drinks delay the onset of exhaustion by a few minutes.

18.4 Weight-Loss Diets

It is possible to lose weight through dieting. If you reduce your intake by 100 kcal/day and keep your activity constant, you will burn off a pound of adipose tissue (3500 kcal) in 35 days. Unfortunately, people are seldom so patient and they resort to more stringent diets. To achieve their goals more rapidly, they exclude certain foods and reduce the amounts of others. Such diets are harmful. Diets with fewer than 1200 kcal/day are likely to be deficient in necessary nutrients, particularly in B vitamins and iron. Further, dieting slows down metabolism. Weight lost through dieting is quickly regained when the dieter resumes old eating habits.

Example 18.1

If you ordinarily expend 2200 kcal/day and you go on a diet of 1500 kcal/day, how long will it take to diet off 1 lb of fat?

Solution

You will use 700 kcal/day more than you consume. There are 3500 kcal in 1 lb of fat, and so it will take

$$3500 \; \cancel{kcal} \times \frac{1 \; day}{700 \; \cancel{kcal}} = 5 \; days$$

(Keep in mind, however, that your weight loss will not be all fat. You will probably lose more than 1 lb, but it will be mostly water with some protein and a little glycogen.)

Exercise 18.1

A person who expends 1800 kcal/day goes on a diet of 1200 kcal/day without a change in activities. Estimate how much fat she will lose if she stays on this diet for 3 weeks.

Hunger is regulated by the hypothalamus gland, according to a popular theory. When the hypothalamus senses that the level of fatty acids circulating in the bloodstream is low, it triggers the hunger mechanism. According to the **set-point theory**, each of us has a unique level at which the hypothalamus acts. Some of us might need a higher level of body fat than others to avoid constant hunger. This is consistent with the fact that obesity seems to be inherited. It is also rather grim news for some of us who would like to lose weight to know that we can do so only by being constantly hungry. There is some good news, however. It appears that our set point can be lowered through exercise.

Crash Diets: Quick = Quack

Any weight-loss program that promises a loss of more than a pound or two a week is likely to be dangerous quackery. Most quick-weight-loss diets depend on factors other than fat metabolism to hook a prospective customer. Many recommend a **diuretic,** such as caffeine, to increase the output of urine. Weight loss is water loss, and weight is regained when the body is rehydrated.

Other such diets depend on depleting the body's stores of glycogen. On a low-carbohydrate diet the body draws on its glycogen reserves, depleting them in about 24 hr. Recall (Chapter 15) that glycogen is a polymer of glucose. Glycogen molecules have lots of hydroxyl (—OH) groups that can form hydrogen bonds to water molecules. Each pound of glycogen carries about 3 lb of water held to it by these hydrogen bonds. Depleting the pound or so of glycogen results in a weight loss of about 4 lb (1 lb glycogen + 3 lb water). No fat is lost, and the weight is quickly regained when the dieter resumes eating carbohydrates.

One pound of adipose (fatty) tissue stores about 3500 kcal of energy. If your normal energy expenditure is 2400 kcal/day, the most fat you can lose by total fasting for a day is 0.69 lb (2400 kcal/day divided by 3500 kcal/lb adipose tissue). This assumes that your body burns nothing but fat. It doesn't. Recall (Chapter 16) that the brain runs on glucose; if that glucose isn't supplied in the diet, it is obtained from protein. Any diet that restricts carbohydrate intake results in a loss of muscle mass as well as fat. When you gain the weight back (as 90% of all dieters do), you gain only fat. People who diet

A moderately active person can calculate the calories needed each day to maintain proper weight by multiplying the desired weight (in pounds) by 15 kcal/lb. To maintain a weight of 120 lb, you need (120 lb × 15 kcal/lb) = 1800 kcal per day.

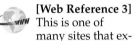
[Web Reference 3] This is one of many sites that explain the benefits of regular exercise.

[Web Reference 4] Many websites give tips on diet and exercise. This is the American Dietetic Association site.

[Web Reference 5] Here are the Unified Dietary Guidelines from the American Cancer Society, the American Dietetic Association, the American Academy of Pediatrics, the American Heart Association, and the National Institutes of Health.

[Web Reference 6] A good summary of the various dietary guidelines.

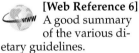
[Web Reference 7] An interesting site where you can find out how many calories you need per day and what you can eat to maintain, lose, or gain weight.

On a weight-loss diet (without exercise), about 65% of the weight lost is fat and about 11% is protein (muscle tissue). The rest is water and a little glycogen.

and then gain back the lost weight replace metabolically active tissue (muscle) with inactive fat. Weight loss becomes harder with each subsequent attempted diet.

Many crash diets are also deficient in minerals such as iron, calcium, and potassium. A deficiency of these minerals can interrupt the smooth function of nerve impulse transmission to muscle, which impairs athletic performance. Impulse transmission to vital organs may also be impaired in cases of severe restriction. Several deaths from cardiac arrest have resulted from variations of the "liquid protein diet."

18.5 Exercise for Weight Loss

Studies consistently show that people who exercise regularly live longer. They are sick less often and have fewer signs of depression. They can move faster, and they have stronger bones and muscles. There are dozens of good reasons for regular exercise, but in many cases people begin exercise programs for one simple reason. They want to lose weight.

People who do not increase their food intake when they begin an exercise program lose weight. Contrary to a common myth, exercise (up to 1 hr/day) does not cause an increase in appetite. Most of the weight loss from exercise results from an increase in metabolic rate during the activity, but the increased metabolic rate continues for several hours after completion of the exercise. Exercise helps us maintain both fitness and thinness.

Example 18.2

A fast game of tennis burns off about 10 kcal/min. How long would you have to play to burn off 1 lb of adipose (fat) tissue?

Solution

One pound of adipose tissue stores 3500 kcal of energy. To burn it off at 10 kcal/min requires

$$3500 \text{ kcal} \times \frac{1 \text{ min}}{10 \text{ kcal}} = 350 \text{ min}$$

It takes 350 min (4 hr 50 min) to burn off 1 lb of fat, even with a fast-paced game of tennis. (You don't have to do it all in one day.)

Exercise 18.2

Walking a mile burns off about 100 kcal. How far do you have to walk to burn off 1 lb of fat?

The most sensible approach to weight loss is to adhere to a balanced low-calorie diet that meets the RDA for essential nutrients, and to engage in a reasonable, consistent individualized exercise program. In this way the principles of weight loss are applied by decreasing intake and increasing output. Everyone, athletes included, should be aware of the possible risks of dieting. A vitamin deficiency often develops slowly over months or years, but it can be corrected or avoided by taking vitamin supplements.

Weight loss or gain is based on the law of conservation of energy (Chapter 14). When we take in more calories than we use up, the excess calories are stored as fat. When we take in fewer calories than we need for our activities, our bodies burn some of the stored fat to make up for the deficit. One pound of adipose tissue requires 200 mi of blood capillaries to serve its cells. Excess fat therefore puts extra strain on the heart.

18.6 Measuring Fitness

To measure fitness is usually to measure fatness. How much fat is enough? The male body requires about 3% essential body fat, and the average female body needs 10–12%. It is not easy to measure percent body fat accurately. Skinfold calipers are quite inaccurate and measure water retention as well as fat. Weight alone does not indicate degree of fitness. A tall 180-lb man may be much more fit than a 150-lb man with a smaller frame.

One way to estimate body fat is by measuring a person's density. Mass is determined by weighing, and volume is calculated with the help of a dunk tank like the one shown in Figure 18.4. The difference between a person's weight in air and when submerged in water corresponds to the *mass* of the displaced water. When this mass is divided by the density of water and then corrected for air in the lungs, it gives a reasonable value for the person's volume. Mass divided by volume yields the person's density, although results can vary with the amount of air in the lungs. Fat is less dense (0.903 g/mL) than the water (1.000 g/mL), that makes up most of the mass of our bodies. The higher the proportion of body fat a person has, the lower the density and the more buoyant that person is in water.

A much simpler way to estimate degree of fatness is by making measurements of the waist and hips. The waist should be measured at the narrowest point and the hips should be measured where the circumference is the largest. The waist measurement should then be divided by the hip measurement. For men the ratio should be less than 1, and for women it should be 0.8 or less. (Measurements can be made in inches or centimeters, as long as the same units are used for both, since the units cancel out.)

In Europe during the Middle Ages people accused of crimes were often tried by some kind of ordeal. In trial by water, the innocent sank and the guilty floated. Suspected witches were tied hand to foot and thrown into the water. The guilty floated, and were fished out, dried off, and burned at the stake. Body density was *really* important back in those days!

Body Mass Index

Body mass index (BMI) is a commonly used measure of fatness. It is defined as weight (in kilograms) divided by the square of the height (in meters). For a person who is 1.8 m tall and weighs 82 kg, the body mass index is $82/(1.8)^2 = 25$. Average BMI values for adults are between 20 and 25. Values greater than 27 may be associated with obesity as well as other related problems such as high blood pressure and diabetes. When measurements are in pounds and inches, the equation is

$$\text{BMI} = \frac{705 \times \text{body weight (lb)}}{[\text{height (in)}]^2}$$

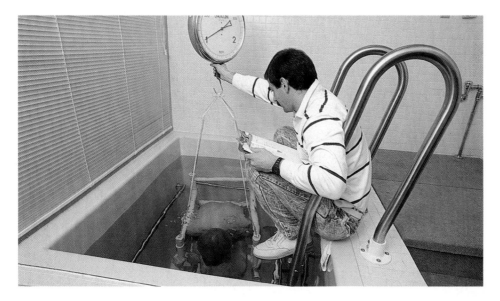

Figure 18.4 When submerged in water, a body displaces its own volume of water. The difference between a person's weight in air and when submerged in water corresponds to the *mass* of the displaced water. Since the density of water is 1 g/mL, the mass of water displaced (in grams) equals the person's volume (in milliliters).

In other words, a person who is 5 ft 10 in. tall (70 in.) and weighs 180 lb has a BMI of

$$\frac{705 \times 180}{70 \times 70} = 26$$

which is just a little on the high side.

Example 18.3

What is the BMI for a person who is 6 ft 2 in. tall and weighs 230 lb?

Solution
The person's height is $6 \times 12 + 2 = 72 + 2 = 74$ in

$$\text{BMI} = \frac{705 \times 230}{74 \times 74} = 29$$

The body mass index is 29.

Exercise 18.3
What is the BMI for a person who is 5 ft tall and weighs 120 lb?

18.7 Some Chemistry of Muscles

The human body has about 600 muscles. Exercise makes these muscles larger, more flexible, and more efficient in their use of oxygen. Exercise helps to protect against heart disease because the heart is an organ comprised mainly of muscle. A strong heart is a healthy heart. With regular exercise, resting pulse and blood pressure usually decline. After several months of an effective exercise program, pulse and blood pressure remain lower even *during* exercise. The net result called the **training effect,** is that a person who exercises regularly is able to do more physical work with less strain.

People who expand their capacity to do more physical work under less strain often begin to think of doing more—faster and with more agility and accuracy—of whatever they do. These people become athletes. Exercise is an art, but it is also increasingly becoming a science—a science in which chemistry plays a vital role.

Energy for Muscle Contraction: ATP

When cells metabolize glucose or fatty acids, only part of the chemical energy in these substances is converted to heat. Some of it is stored in the high-energy phosphate bonds of adenosine triphosphate (ATP) molecules.

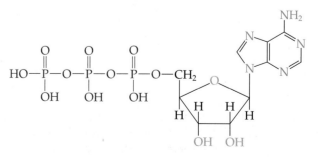

Adenosine triphosphate

The stimulation of muscles causes them to contract. This contraction is work, and it requires energy which the muscles get from the molecules of ATP. The energy stored in ATP powers the physical movement of muscle tissue. Two proteins, actin and myosin, play important roles in this process. Together they form a loose complex called *actomyosin,* the contractile protein of which muscles are made (Figure 18.5). When ATP is added to actomyosin in the laboratory, the protein fibers contract. It seems likely that the same process occurs in vivo—that is, in the muscles of living animals. Not only does myosin serve as part of the structural complex in muscles, it also acts as an enzyme for the removal of a phosphate group from ATP. Thus, it is directly involved in liberating the energy required for the contraction.

In a resting person, muscle activity (including that of the heart muscle) accounts for only about 15–30% of the energy requirements of the body. Other activities, such as cell repair, transmission of nerve impulses, and maintenance of body temperature, account for the remaining energy needs. During intense physical activity, the energy requirements of muscle may be more than 200 times the resting level.

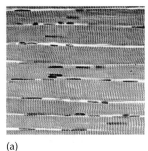
(a)

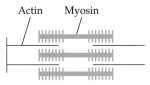

Aerobic Exercise: Plenty of Oxygen

The ATP in muscle tissue is sufficient for activities lasting for at most a few seconds. Fortunately, muscles have a more extensive energy supply in the form of **glycogen,** a starchlike stored form of dietary carbohydrate that has been ingested, digested to glucose, and absorbed. It is stored as glycogen when not needed immediately for energy.

When muscle contraction begins, glycogen is converted to pyruvic acid by muscle cells in a series of steps.

$$(C_6H_{11}O_5)_n \longrightarrow 2n \; CH_3\overset{\displaystyle O}{\overset{\displaystyle \|}{C}}-\overset{\displaystyle O}{\overset{\displaystyle \|}{C}}OH$$

Glycogen   Pyruvic acid

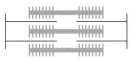
(b)

Then, if sufficient oxygen and other factors are readily available, the pyruvic acid is oxidized to carbon dioxide and water in another series of steps.

$$2 \; CH_3\overset{\displaystyle O}{\overset{\displaystyle \|}{C}}-\overset{\displaystyle O}{\overset{\displaystyle \|}{C}}OH \; + \; 5 O_2 \longrightarrow 6 CO_2 \; + \; 4 H_2O$$

Muscle contractions that occur under these circumstances—that is, in the presence of oxygen—constitute **aerobic exercise** (Figure 18.6).

Figure 18.5 Skeletal muscle tissue has a banded or *striated* appearance, shown here magnified 180 times (a). The diagram of the actomyosin complex in muscle (b) shows extended muscle fibers (top), resting fibers (middle), and partially contracted fibers (bottom).

Anaerobic Exercise and Oxygen Debt

When sufficient oxygen is not available, pyruvic acid is reduced to lactic acid.

$$CH_3COCOOH + [2 H] \longrightarrow CH_3CHOHCOOH$$

If this **anaerobic exercise** persists, an excess of lactic acid builds up in the muscle cells. This lactic acid ionizes, forming lactate ions and hydronium ions.

$$CH_3CHOHCOOH + H_2O \longrightarrow CH_3CHOHCOO^- + H_3O^+$$

The pH in the cells drops, inhibiting an enzyme (Section 15.9) essential to muscle function. At the **lactate threshold,** lactic acid concentration builds up rapidly in the blood and muscles, and the athlete suffers severe muscle fatigue. Well-trained athletes can continue for a while, but most people quit. At this point the **oxygen debt** is

Figure 18.6 Aerobic exercise is performed at a pace that allows us to get enough oxygen to our muscle cells to oxidize pyruvic acid to carbon dioxide and water. Aerobic dance is a popular form of aerobic exercise.

Myoglobin is the heme-containing protein in muscle that stores oxygen (similar to the way hemoglobin stores and transports oxygen in the blood).

Figure 18.7 After a race, the metabolism that fueled the effort continues. The athlete gulps air to repay her oxygen debt.

repaid (Figure 18.7). After exercise ends, the cells' demand for oxygen decreases, making more oxygen available to oxidize the lactic acid resulting from anaerobic metabolism back to pyruvic acid. This acid is then converted to carbon dioxide, water, and energy.

Athletes usually emphasize one type of training (anaerobic or aerobic) over the other. For example, an athlete training for a 60-m dash does mainly anaerobic work, but one planning to run a 10-km race does mainly aerobic training. Sprinting and weightlifting are largely anaerobic activities; a marathon run is largely aerobic. During a marathon, athletes must set a pace to run for more than 2 hr. Their muscle cells depend on slow, steady aerobic conversion of carbohydrates to energy. During anaerobic activities, however, muscle cells use almost no oxygen. Rather, they need the quick energy provided by anaerobic metabolism.

After glycogen stores are depleted, muscle cells can switch over to fat metabolism. Fats are the main source of energy for sustained activity of low or moderate intensity, such as the last part of a marathon run (42 km).

Muscle Fibers: Twitch Kind Do You Have?

The quality and type of muscle fibers that an athlete has affect his or her athletic performance. The quality of machinery plays a role in how work is done. For example, we can remove snow from a driveway in more than one way. We can eliminate it in 10 min with a snowblower, or we can spend 50 min shoveling. Muscles are tools with which we do work. They are classified according to the speed and effort required to accomplish this work. The two classes of muscle fibers are **fast-twitch fibers** (those stronger and larger and most suited for anaerobic activity) and **slow-twitch fibers** (those best for aerobic work). Table 18.2 lists some characteristics of these two types of muscle fibers.

Slow-Twitch Fibers: Endurance Activities

Type I (slow-twitch) fibers are called on during activity of light or moderate intensity. The respiratory capacity of these fibers is high, which means that they can provide a large amount of energy via aerobic pathways. The myoglobin level is also high. Aerobic oxidation requires oxygen, and muscle tissue rich in slow-twitch fibers is geared to supplying high levels of oxygen.

The capacity of type I muscle fibers to use glycogen is low. This tissue is not geared to anaerobic generation of energy and does not require the hydrolysis of glycogen. The catalytic activity of the actomyosin complex is low. Remember that actomyosin not only is the structural unit in muscle that actually undergoes contraction but also is responsible for catalyzing the hydrolysis of ATP to provide energy for the contraction. Low catalytic activity means that the energy is parceled out more slowly, which is not good if you want to lift 200 kg but is perfect for a 15-km run.

Fast-Twitch Fibers: Bursts of Power

Type IIB (fast-twitch) fibers have characteristics opposite those of type I fibers. Low respiratory capacity and low myoglobin levels argue against aerobic oxidation. A high capacity for glycogen use and a high catalytic activity of actomyosin allow tissue rich in fast-twitch fibers to generate ATP rapidly and also to hydrolyze this ATP rapidly during intense muscle activity. Thus, this type of muscle tissue provides the capacity to do short bursts of vigorous work. We say "bursts" because this type of muscle fatigues relatively quickly. A period of recovery, during which lactic acid is cleared from the muscle, is required between brief periods of activity.

Table 18.2 ▌ A Comparison of Two Types of Muscle Fibers

Characteristic	Type I	Type IIB*
Category	Slow twitch	Fast twitch
Color	Red	White
Respiratory capacity	High	Low
Myoglobin level	High	Low
Catalytic activity of actomyosin	Low	High
Capacity for glycogen use	Low	High

*A type IIA fiber exists that resembles type I in some respects and type IIB in others. We will discuss only the two types described in this table.

Carbohydrate Loading

Athletes, it seems, will try almost anything to gain a competitive edge. Those engaged in endurance activities, such as long-distance running, often try a technique called *carbohydrate loading*. Muscles run on glycogen as the preferred fuel, but the human body can store only a limited amount—perhaps about 500 g in a 70-kg person.

In carbohydrate loading, the athlete cuts back on training a few days before competition and eats a diet high in carbohydrates. Under this regimen, the body stores more glycogen than usual. The benefits of carbohydrate loading, even for top athletes, are marginal. For casual athletes, the technique is probably ridiculous.

Building Muscles

Endurance exercise increases myoglobin levels in skeletal muscles. This provides for faster oxygen transport and increased respiratory capacity (Figure 18.8), and these changes are apparent within 2 weeks. Endurance training does not necessarily increase the size of muscles. If you want larger muscles, try weightlifting. Weight training (Figure 18.9) develops fast-twitch muscle fibers. These fibers increase in size and strength with repeated anaerobic exercise. Weight training does *not* increase respiratory capacity, however.

Muscle-fiber type seems to be inherited. Research shows that world-class marathon runners may possess up to 80–90% slow-twitch fibers, compared with championship sprinters who may have up to 70% fast-twitch muscle fibers. Some exceptions have been noted, however. Thus, factors such as training and body composition are important in athletics. Other factors are equally important, including nutrition, fluid and electrolyte balance, and drug use or misuse.

18.8 Drugs and the Athlete

Muscle chemistry and metabolism, aerobic and anaerobic exercise, and nutrition and fluid balance all relate to normal internal processes. It seems unlikely that properly fed, well-trained athletes can safely improve their performance with magic, superstition, or drugs. Many athletes, however, are not satisfied with the idea that hard work and proper nutrition are the best answers to improved performance. They turn instead to dangerous drugs, such as the stimulants amphetamine and cocaine, and to anabolic steroids to build muscles.

Figure 18.8 The New York City marathon tests the aerobic capacity of muscle.

Figure 18.9 Strength exercises build muscle mass but do not increase respiratory capacity of muscles.

Restorative Drugs

Athletes also use drugs to alleviate pain or soreness that results from overtraining and to treat injuries. These **restorative drugs** include analgesics (painkillers), such as aspirin and acetaminophen (Chapter 19), and anti-inflammatory drugs, such as aspirin, ibuprofen, ketoprofen, and cortisone. Cortisone derivatives are often injected to reduce swelling in damaged joints and tissues. Relief is often transitory, however, and side effects are often severe. Prolonged use of these drugs can cause fluid retention, hypertension, ulcers, disturbance of the sex hormone balance, and other problems. Another substance, one that can be bought over the counter and applied externally, is methyl salicylate, or oil of wintergreen. This substance causes a mild burning sensation when applied to the skin, thus serving as a counterirritant for sore muscles. An aspirin derivative, it also acts as an analgesic.

Stimulant Drugs

Some athletes use *stimulant drugs* (Chapter 19) to try to improve performance. The caffeine in a strong cup of coffee may be of some benefit. Caffeine triggers the release of fatty acids. Metabolism of these compounds may conserve glycogen, but any such effect is small. Caffeine also increases the heart rate, speeds metabolism, and increases urine output. The latter could lead to dehydration.

Other athletes resort to stronger stimulants, including amphetamines and cocaine. A multipurpose drug, cocaine was first used in the late 1800s as a local anesthetic to deaden pain during dental work or minor surgery. It is also a powerful stimulant, which is the reason for its current popularity. Cocaine stimulates the central nervous system, increasing alertness, respiration, blood pressure, muscle tension, and heart rate. Both stimulants mask symptoms of fatigue and give an athlete a sense of increased stamina. The cocaine intoxication deaths of several thousand people, including prominent professional athletes, brought this problem to public attention. Other athletes have been arrested, suspended, and banned for cocaine use.

[Web Reference 8] An extensive article entitled "Drugs Athletes Use to Enhance Performance."

[Web Reference 9] Some sites dealing with anabolic steroids and their dangers.

Anabolic Steroids

Many strenuous athletic performances depend on well-developed muscles. Muscle mass depends on the level of the male hormone *testosterone* (Chapter 19). When boys reach puberty, testosterone levels rise, and the boys become more muscular if they exercise. Men generally have larger muscles than women because they have more testosterone.

Testosterone and some of its semisynthetic derivatives are taken by athletes in an attempt to build muscle mass more rapidly. Steroid hormones used to increase muscle mass are called **anabolic steroids.** These chemicals aid in the building (anabolism) of body proteins and thus of muscle tissue.

There are no good controlled studies demonstrating the effectiveness of anabolic steroids. They do seem to work—at least for some people—but the side effects are many. In males, side effects include testicular atrophy and loss of function, impotence, acne, liver damage, edema (swelling), elevated cholesterol levels, and growth of breasts. Liver cancer is now showing up at an alarming rate in athletes who began using steroids in the 1960s.

Anabolic steroids act as male hormones (androgens), making women more masculine. They help women build larger muscles, but they also result in balding, extra body hair, a deep voice, and menstrual irregularities. What price will an athlete pay for improved performance?

Drugs, Athletic Performance, and Drug Screening

The use of drugs to increase athletic performance is not only illegal and physically dangerous but it can also be psychologically damaging. Stimulant drugs give the user a false sense of confidence by affecting the person's ego, giving the delusion of invincibility. Such "greatness" often ends as the competition begins, however, because the stimulant effect is short-lived (it lasts for only half an hour to an hour for cocaine). Exhaustion and extreme depression usually follow. More stimulants are needed to combat these "down" feelings, and the user then runs the risk of addiction or overdose.

Because drug use for the enhancement or improvement of performance is illegal, chemists have another role to play in sports: screening blood and urine samples for illegal drugs. Using sophisticated instruments, chemists can detect minute amounts of illegal drugs. Drug testing has been used at the Olympic Games since 1968, and it is rapidly becoming standard practice for athletes in college and professional sports.

Drugs are used at great risk to the athlete and with little evidence of benefit in most cases. The use of illegal drugs violates the spirit of fair athletic competition. Let us work toward a world in which events and medals are won by drug-free athletes who are highly motivated, well nourished, and well trained.

18.9 Exercise and the Brain

Hard training improves athletic performance. Few chemical substances help very much. Our own bodies, however, manufacture a variety of substances that do help. Examples are the pain-relieving **endorphins** (Section 19.17).

After vigorous activity, athletes have an increased level of endorphins in their blood. Stimulated by exercise, deep sensory nerves release endorphins to block the pain message. Exercise extended over a long period of time, such as distance running, causes in an athlete many of the same symptoms experienced by opiate users. Runners get a euphoric high during or after a hard run. Unfortunately, both natural endorphins and their analogs, such as morphine, seem to be addictive. Athletes, especially runners, tend to suffer from withdrawal: They feel bad when they don't get the vigorous exercise of a long, hard run. This can be positive because exercise is important to the maintenance of good health. It can be negative if the person exercises when injured or to the exclusion of family or work obligations. Athletes also seem to develop a tolerance to their own endorphins: They have to run farther and farther to get that euphoric feeling.

Good physical condition, daily vigorous exercise, and experiencing a natural, cheap, legal high seem safer than resorting to illegal drugs. You may get hooked on exercise, but this seems far better than getting hooked on narcotics.

Exercise also directly affects the brain by increasing its blood supply and by producing **neurotrophins,** substances that enhance the growth of brain cells. Exercises that involve complex motions, such as dance movements, lead to an increase in the connections between brain cells. Both our muscles and our brains work better when we exercise regularly.

18.10 No Smoking

If you want to stay healthy and fit, one thing you certainly should *not* do is use tobacco. According to the U.S. Centers for Disease Control and Prevention, cigarette smoking is associated with death from 24 different diseases, including heart disease, stroke, and lung diseases such as cancer, emphysema, and pneumonia. Lung cancer is not the only malignancy caused by cigarette smoking. Cancers of the pancreas, bladder, breast, kidney, and cervix are also associated with smoking.

We fear cancer, but the major killers of smokers are heart attack and stroke. Cigarette smokers have a 70% higher risk of heart attack than nonsmokers. When they are also overweight, with high blood pressure and high cholesterol, their risk is 200% higher. At every age the death rate is higher among smokers than among nonsmokers.

Women who smoke during pregnancy have more stillborn babies, more premature babies, and more babies who die before they are a month old. Women who take birth control pills and also smoke have an abnormally high risk of stroke. Women who smoke are advised to use some other method of birth control.

One of the gases in cigarette smoke is carbon monoxide, which ties up about 8% of the hemoglobin in the blood. Therefore, smokers breathe less efficiently. They work harder to breathe, but less oxygen reaches their cells. It is no wonder that fatigue is a common problem for smokers.

In the linings of the lungs there are cilia (tiny, hairlike processes) that help to sweep out particulate matter breathed into the lungs. But in smokers the tar from cigarette smoke forms a sticky coating that keeps the cilia from moving freely. The reduced action of the cilia may explain the smoker's cough. Pathologists claim that during an autopsy it is usually easy to tell if the body belonged to a smoker. Lungs are normally pink; a smoker's lungs are black.

Unfortunately, cigarette smoke does not harm only the smoker. Many nonsmokers are exposed to secondhand smoke, which causes an estimated 3000 premature deaths each year in the United States. These nonsmokers often suffer the same health problems as smokers. Children of smokers are especially vulnerable; for example, they miss twice as much school time because of respiratory infections as do children of nonsmokers.

In recent years there have been efforts to lessen the indoor pollution caused by smoking. Many restaurants and government buildings have been declared nonsmoking areas, and Congress may outlaw smoking in *all* public places. Smoking has never been a smart or a healthy thing to do, but now it isn't even fashionable.

18.11 Chemistry of Sports Materials

The athlete's world is shaped by chemicals. The ability to perform is a reflection of biochemistry within the athlete. Chemicals taken by athletes may hinder or enhance their performance. But there is yet another chemical dimension to the sports world:

Two million people die each year in the United States, about 1 million of them prematurely. Tobacco use contributes to about half of these early deaths. Each day 6000 people try cigarettes for the first time. Half will become addicted, and half of these will die prematurely.

 [Web Reference 10] Several smoking-related sites of the Centers for Disease Control and Prevention.

More than 4000 chemical compounds have been identified in cigarette smoke, including more than 40 carcinogens.

A study of military veterans, conducted at Harvard University, concluded that smokers are 4.3 times as likely as nonsmokers to develop Alzheimer's disease.

 [Web Reference 11] A report on the irreversible effects of smoking.

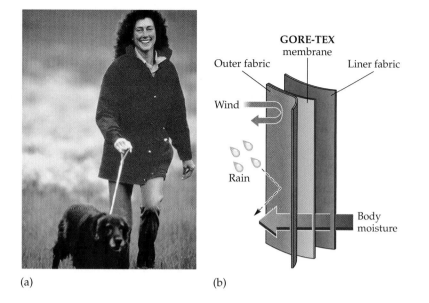

(a)   (b)

Figure 18.10 Imaginative use of fibers and other materials makes it possible to exercise in relative comfort even in inclement weather (a). Gore-Tex fabrics (b) repel wind and raindrops but allow body moisture to escape.

the chemicals involved in providing athletic clothing and equipment (Table 18.3). Athletic clothing depends on chemicals for its shape and protective qualities.

Swimsuits, ski pants, and elastic supports stretch because of synthetic fibers. Joggers run in thunderstorms and windstorms in Gore-Tex suits without getting cold and wet. Gore-Tex is a thin, membranous material made by stretching fibers of polytetrafluoroethylene (Teflon). The material has billions of tiny holes that are too small to pass drops of water but readily pass water vapor. Raindrops falling on the outside are held together by surface tension and run off rather than penetrating the fabric. Water vapor from sweating skin, however, can pass from the warm area inside the suit where vapor concentration is high to the cooler area outside the suit where the vapor concentration is lower. The runner stays relatively dry. (Figure 18.10).

Joggers are not the only athletes pampered by equipment made of materials developed by the chemical industry. Football, hockey, and baseball players are protected

Table 18.3 ▌ Some Sports and Recreational Items Made From Petroleum

Golf balls	Whistles	Tote bags	Hockey pucks
Wet suits	Motorcycle helmets	Darts	Ice buckets
Parachutes	Dune-buggy bodies	Tents	Fishing nets
Card tables	Checkers	Stadium cushions	Hiking boots
Golf-cart bodies	Chess boards	Fingerpaints	Frisbees
Warmup suits	Shorts	Foul-weather gear	Fishing boots
Ping-pong paddles	Tennis shoes	Foot pads	Diving masks
Rafts	Paddles	Visors	Guitar picks
Uniforms	Decoys	Swimming-pool liners	Beach balls
Phonographs	Volley balls	Vinyl tops for cars	Sunglasses
Racks	Sleeping bags	Ice chests	Dog leashes
Track shoes	Sports-car bodies	Life jackets	Dice
Dominoes	Tennis balls	Audio tapes	Pole-vaulting poles
Windbreakers	Reclining chairs	Model planes	Aquariums
Sails	Insect repellents	In-line skates	Golf club impact surfaces

by polycarbonate helmets and protective pads of synthetic foamed rubbers. Brightly dyed nylon uniforms with synthetic colors add to the glamour of amateur and professional teams. Many sports events are played on artificial turf, a carpet of nylon or polypropylene with an underpad of synthetic foamed polymers. Stadiums are protected from the weather by Teflon covers reinforced with glass fibers and held aloft by air pressure. Baseball or football on artificial turf in an air-conditioned, enclosed stadium is quite different from a game played outside on grass (Figure 18.11), and perhaps in poor weather. Balls bounce differently and players wear shoes especially designed to cope with the artificial surface.

The dramatic effect of new materials is perhaps best illustrated in pole vaulting. Between 1940 and 1960 the world-record height for this event increased only 23.5 cm. After the development of plastic poles reinforced with glass fibers, however, the record rapidly rose another 100 cm. Other sports have been affected similarly, if less dramatically. From the soles of sports shoes and the wax used on cross-country skis, to titanium golf clubs and tennis sweaters that stretch and yet retain their shape, the sports world is thoroughly immersed in chemicals.

Figure 18.11 Baseball played on an artificial surface inside a domed stadium is quite different from baseball played outside on grass.

Critical Thinking Exercises

Apply knowledge that you have gained in this chapter and one or more of the FLaReS principles (Chapter 1) to evaluate the following statements or claims.

18.1 An advertisement in an American sports magazine claims that an amino acid supplement helps build muscle mass.

18.2 An athlete claims that a certain brand of shoe helps him jump higher than any other kind of shoe.

18.3 An athlete claims that a piece of tape with a certain shape applied across the bridge of the nose helps her breathe better and improves her athletic performance.

18.4 An advertisement for a weight-loss pill claims that you can "Eat all you want and still lose weight."

Summary

1. Good health and fitness require a nutritious diet and regular exercise.

2. A good diet consists largely of complex carbohydrates (starches) with plenty of fruits and vegetables, modest amounts of dairy and meat products, and very little fat and sugar.

3. Most Americans should eat less fat and less protein but more carbohydrates.

4. Athletes need more calories than sedentary people, and they should get them in the form of carbohydrates.

5. Quick-weight-loss diets do not usually lead to permanent weight loss.

6. If you begin an exercise program and do not increase your food intake, you will lose weight.

7. Each pound of excess adipose tissue requires 200 extra miles of blood capillaries.

8. The body mass index (a measure of fatness) can be calculated from a person's height and weight.

9. Linus Pauling and others have suggested daily vitamin intakes much higher than the recommended daily allowance (RDA) values.

10. Vitamin C, vitamin E, and β-carotene are antioxidants, which help to prevent damage by free radicals.

11. If you wish to maintain a state of optimum health, do not smoke.

12. Exercise strengthens muscles, prolongs life, and lowers the incidence of disease.

13. Aerobic exercise involves muscle contraction in the presence of oxygen. A long, steady run is aerobic exercise.

14. Anaerobic exercise involves quick bursts of energy. A 60-m dash is anaerobic exercise.

15. There are two classes of muscle fibers, fast-twitch (which are larger and stronger and good for rapid bursts of power) and slow-twitch (which are geared for steady exercise of long endurance).

16. Weight training develops fast-twitch muscles but does not increase respiratory capacity.

17. Long-distance running develops slow-twitch muscles and increases respiratory capacity.

18. Carbohydrate loading is done before endurance activities in order to maximize glycogen storage, but the benefits are marginal.

19. Water is the best drink to prevent dehydration.

20. Athletes sometimes use restorative drugs, such as aspirin, and stimulant drugs, such as caffeine and cocaine.

21. Anabolic steroids increase muscle mass, but they have many bad side effects, including impotence and liver cancer.

22. It appears that athletes can become addicted to their own endorphins.

23. From modern sports clothing and equipment to artificial turf, chemistry has contributed greatly to the world of sports.

Key Terms

aerobic exercise 18.7	endorphins 18.9	neurotrophins 18.9	set-point theory 18.4
anabolic steroids 18.8	fast-twitch fibers 18.7	oxygen debt 18.7	slow-twitch fibers 18.7
anaerobic exercise 18.7	glycogen 18.7	recommended daily	training effect 18.7
diuretic 18.4	heat stroke 18.3	allowance (RDA) 18.2	
electrolytes 18.3	lactate threshold 18.7	restorative drugs 18.8	

Review Questions

1. According to the food guide pyramid, what foodstuff should make up the largest part of your diet?

2. What foodstuffs should be eaten only sparingly?

3. The recommended diet contains what percent of its calories as fat?

4. The typical American diet contains what percent fat?

5. Mention two good reasons for not going on a high-protein diet.

6. Why have many people cut back on their consumption of red meat?

7. Do athletes need more protein than other people?

8. Muscles are made of protein. Does eating more protein result in larger muscles?

9. Do you think it is likely that obesity is inherited?

10. How might you lose 5 lb in just 1 week?

11. What are recommended daily allowance (RDA) values?

12. Do we ever need more of a given vitamin than its RDA?

13. What are some of the benefits of vitamin A?

14. In addition to preventing pellagra, what does vitamin B_3 (niacin) do?

15. What kinds of problems are helped by vitamin B_6 (pyridoxine)?

16. What benefit is derived from vitamin B_{12}?

17. Is taking a daily megadose of vitamin D a good idea?

18. What are the benefits of vitamin D?

19. Vitamins C and E are antioxidants. What does this mean?

20. Mention several health problems that vitamin E can be used to treat.

21. What are some of the health problems related to smoking?

22. Why is it important for pregnant women not to smoke?

23. List three ways that chemistry has had an impact on sports.

24. How many muscles do humans have?

25. Describe the training effect.

26. What is the immediate source of energy for muscle contraction?

27. What two proteins make up the actomyosin protein complex?

28. What are the two functions of the actomyosin protein complex?

29. What is aerobic exercise?

30. What is anaerobic exercise?

31. Which type of metabolism (aerobic or anaerobic) is primarily responsible for providing energy for intense bursts of vigorous activity?

32. Which type of metabolism (aerobic or anaerobic) is primarily responsible for providing energy for prolonged low levels of activity?

33. What is meant by oxygen debt?

34. What is the lactate threshold?

35. Identify type I and type IIB muscle fibers as
    a. fast twitch or slow twitch.
    b. suited to aerobic oxidation or to anaerobic use of glycogen.

36. Explain why high levels of myoglobin are appropriate for muscle tissue geared to aerobic oxidation.

37. Why does the high catalytic activity of actomyosin in type IIB fibers suggest that these are the muscle fibers engaged in brief, intense physical activity?

38. Why can the muscle tissue that uses anaerobic glycogen metabolism for its primary source of energy be called on only for brief periods of intense activity?

39. Which type of muscle fiber is most affected by endurance training exercises? What changes occur in the muscle tissue?

40. Birds use large, well-developed breast muscles for flying. Pheasants can fly 80 km/hr, but only for short distances. Great blue herons can fly only about 35 km/hr but can cruise great distances. What kind of fibers do each have in their breast muscles?

41. How do the nutritional needs of an athlete differ from those of a sedentary individual? How is this extra need best met?

42. Muscle is protein. Does an athlete need extra protein (above the RDA) to build muscles? What is the only way to build muscles?

43. Describe the biochemical process by which muscles are built.

44. List two ways to determine percent body fat. Describe a limitation of each method.

45. List some problems that result from low-calorie diets.

46. How many calories of energy are stored in 1 lb of adipose tissue?

47. Why does excess body fat put a strain on the heart?

48. Describe the set-point theory of body weight.

49. Why does a diet that restricts carbohydrate intake lead to loss of muscle mass as well as fat?

50. List two ways in which fad diets lead to "quick weight loss." Why is this weight rapidly regained?

51. How much glycogen can the average human body store?

52. What is carbohydrate loading? How does it aid an athlete in an endurance event?

53. List the three major electrolytes essential for proper cellular function.

54. What is a diuretic?

55. Describe the function of antidiuretic hormone (ADH).

56. What fluid is best for replacing water lost during exercise?

57. Why are beer and cola drinks not recommended for fluid replacement?

58. Is thirst a good indicator of dehydration? Are you always thirsty when dehydrated?

59. How is the appearance of urine related to dehydration?

60. What is heat stroke?

61. What are restorative drugs?

62. How is cortisone used in sports medicine? What are some of the side effects of its use?

63. What are anabolic steroids?

64. List some side effects of the use of anabolic steroids in males.

65. What are the effects of anabolic steroids on females?

66. Does cocaine enhance athletic performance? Explain fully.

67. What happens when the effect of cocaine wears off?

68. What is the role of chemists in the control of drug use by athletes?

69. Give a biochemical explanation of the "runner's high."

70. List the evidence indicating that long-distance running is addictive.

71. Give a biochemical explanation of addiction to long-distance running.

72. How does Gore-Tex clothing work? What is it made of?

Problems

Body Mass Index

73. What is the body mass index for a 6-ft baseball player who weighs175 lb?

74. What is the body mass index for a 6-ft 6-in. basketball player who weighs 240 lb?

Diet and Exercise

75. A giant double-decker hamburger provides 600 kcal of energy. How long would you have to walk to burn off this energy if 1 hr of walking uses about 300 kcal?

76. One kilogram of fat tissue stores about 7700 kcal of energy. An average person burns about 40 kcal in walking 1 km. If such a person walks 5 km/day, how much fat will be burned in 1 year?

77. How long would you have to run to burn off the 110 kcal in one glass of beer if 1 hr of running burns off 1100 kcal?

78. How far would you have to run to burn off 5 kg of fat if running burns off 100 kcal/km? (Assume that there are 7700 kcal in 1 kg of fat.)

79. The RDA for protein is about 0.8 g/kg body weight. How much protein is required each day by a 125-kg football player?

80. How much protein is required each day by a 50-kg gymnast (See Problem 79)?

81. A 70-kg man can store about 2000 kcal as glycogen. How far can such a man run on this stored starch if he expends 100 kcal/km while running and the glycogen is his only source of energy?

82. A 70-kg man can store about 100,000 kcal of energy as fat. How far can such a man run on this stored fat if he expends 100 kcal/km while running and the fat is his only source of energy?

Additional Problems

83. If you are moderately active and want to maintain a weight of 150 lb, about how many calories do you need each day?
84. An athlete can run a 400-m race in 45 s. Her maximum oxygen intake is 4 L/min, yet working muscles at their maximum exertion require about 0.2 L of oxygen gas per minute for each kilogram of body weight. If an athlete weighs 50 kg, what oxygen debt will she incur?
85. Fat tissue has a density of about 0.9 g/mL, and lean tissue a density of about 1.1 g/mL. Calculate the density of a person with a body volume of 80 L who weighs 85 kg. Is the person fat or lean?

Projects

86. List as many synthetic materials as you can that are used in each of the following sports.
    a. baseball
    b. basketball
    c. football
    d. ice hockey
    e. golf
    f. tennis
    g. swimming
    h. sailing
    i. skiing
    j. track and field
    k. long-distance running
    l. cold-weather running
87. Only three Nobel prizes in chemistry have been awarded to women. One of these women was Dorothy Crowfoot Hodgkin. Write a brief essay about her based on information sources from the library or the internet.

Online Projects

88. Write a report on the various drugs that athletes use to enhance their performance, using information available on the internet.
89. Search the web for information on the health effects of smoking, and make your own list of the many reasons why people should not smoke.
90. See how much can you learn about the "obesity gene" by searching the web.
91. Using the internet, see if you can find an organization or corporation that might disagree with the Unified Dietary Guidelines.

References and Readings

1. Brink, Susan. "Smart Moves." *U.S. News & World Report*, 15 May 1995. Aerobic exercise shapes up the brain as well as the body.
2. "Do You Know How to Lose Weight?" *American Health*, November 1994, pp. 29–31.
3. Ember, Lois R. "The Nicotine Connection." *Chemical and Engineering News*, 28 November 1994, pp. 8–18. Discusses the chemistry of nicotine addiction.
4. "Hypertension: What Works?" *Consumer Reports*, May 1999, pp. 60–62. Some drugs help control high blood pressure, but a new diet is often enough.
5. Jackson, Donald Dale. "The Art of Wishful Thinking Has Made a Lot of People Rich." *Smithsonian*, November 1994, pp. 146–156. Includes a discussion of fad diets.
6. Layman, Donald K. (Ed.) *Nutrition and Aerobic Exercise.* Washington, DC: American Chemical Society, 1986.
7. Liebman, Bonnie. "The Changing American Diet." *Nutrition Action Healthletter*, June 1995, pp. 8–9. We eat more sugar and fat and drink more soda pop than ever.
8. Nossal, Sir Gustav J. V., "Life, Death, and the Immune System." *Scientific American Reviews in Science*, 1999, pp. 35–43.
9. "Saving Our Children from Tobacco." *FDA Consumer*, October 1996, pp. 7–8. Each day 3000 young people in the United States become regular smokers, and 1000 of them will die prematurely from smoking-related diseases.
10. "Secondhand Smoke: Is It a Hazard?" *Consumer Reports*, January 1995, pp. 27–33.
11. "ShapingUp." *Consumer Reports*, January 1998, pp. 20–29.
12. Smart, Joanne McAllister. "Antioxidants for the Uninitiated." *Vegetarian Times*, April 1994, p. 26.
13. Smith, Trevor. "Chemistry, Exercise, and Weight Control." *Today's Chemist*, February 1990, pp. 10–11, 21.
14. Smith, Trevor. "Chemistry in Sports Medicine." Today's Chemist, October 1990, pp. 14–17.
15. Smith, Trevor. "Exercise and Weight Control—Another Look." *Today's Chemist at Work*, October 1995, pp. 47–52.
16. Smith, Trevor. "Muscles: Use 'Em or Lose 'Em." *Today's Chemist at Work*, November 1995, pp. 61–63.
17. "Smoking Leaves Fingerprints on DNA." *Science News*, 2 November 1996, p. 284. The definitive link between cigarette smoking and lung cancer is described.
18. Travis, J. "It Takes Nerve to Make Muscles Bond." *Science News*, 25 May 1996, p. 327.
19. Upton, Arthur C., and Eden Graber. *Staying Healthy in a Risky Environment.* New York: Simon and Schuster, 1993.
20. Waterhouse, Andrew L. "Wine and Heart Disease." *Chemistry & Industry*, 1 May 1995, pp. 338–341. Antioxidants in wine may protect against heart disease.

MediaLab

Fitness or Overexertion?

Keeping physically fit—what does this mean? Experts agree that a combination of the right exercise and good nutrition is essential for fitness, but how is the goal defined? Does it mean being slender, being muscular, or simply having the recommended weight for your body build? Or should fitness be described as the absence of illness or having the energy to do what you want? Is fitness the same for a child as for an adult?

Sometimes the quest for a "perfect" body goes beyond fitness and health. The goal of developing a slender body, endurance, or strong muscles is distorted into a fixation on bodybuilding or a push to train and train without listening to the body's needs for rest or relaxation. Some persons become "addicted" to exercise, pushing their bodies to exhaustion.

In these *Web Investigations*, you will look at how fitness is defined by different people and try to determine where the boundary lies between fitness and overexertion or obsession. In *Communicate Your Results*, you will discuss fitness and try to describe what it means for different persons. You will explain why a balance is important: between exercise and rest and between aerobic and strength-building exercises.

WEB INVESTIGATION

Investigation 1
What Is Fitness?

No single set of body standards applies to all people. Fitness for a runner of slender build differs from that for a computer consultant with a large frame. Fitness for a teenager differs from that for an elderly man who just wants to keep active. Are there, however, some fitness components that apply to all? To look at a good definition of fitness, select Keyword **FIT**, and go to "Physical Functional Capacity." Read as much as you like about each of the five listed components of fitness. A briefer version of these components may be found on a page for seniors at Keyword **FITNESS**. For a different view of fitness, look at a bodybuilding site: Select Keyword **BODY**.

Roofing requires fitness!

read about the physiology of aging. Aging is inevitable, but many of these changes can be slowed or mitigated through exercise. Select Keyword **AGING** for recommendations aimed at the elderly, and Keyword **BENEFITS** for a list of the benefits of exercising.

Investigation 2
Do Not Go Gentle: Aging and Exercise

Many changes take place as we age: metabolism slows, bones lose mass, and joints stiffen. Select Keyword **AGING** to

Investigation 3
Exercise: What's Right for You? How Much Is Too Much?

The amount and type of exercise you need depends on your age, health, and realistic goals. Each person should choose a form that can be incorporated easily into daily life. Most people will never be marathon winners—no matter how much they train—but they can run for enjoyment and fitness. Select Keyword **RECOMMENDATIONS** for some exercise guidelines, and Keyword **DAILY** for hints about physical activity in daily life.

Investigation 4
Overtraining and Stress Injuries

Experts from sports medicine physicians to college coaches preach the importance of rest: days off from training, rest intervals between repetitions of running or weightlifting, and sleep! Select Keyword **OVERTRAINING** for a discussion

of the overtraining syndrome. At Keyword **REST**, you'll find information about incorporating rest intervals into training regimes for various sports.

Investigation 5
The Mind–Body Connection: Visualization and Imagery in Sports

Visualization—imagining successful accomplishment—is increasingly used to aid concentration and competitiveness in sports. Select Keyword **VISUALIZATION** for a description of the role of visualization in sports. Then select Keyword **PSYCHOLOGY** for a wealth of information about the psychology of sports. Select either "Imagery & Simulation" or "Focus & Flow" from this site.

Rock climbers must be fit and safety conscious.

COMMUNICATE YOUR RESULTS

Exercise 1
Defining Fitness

Write a one- to two-page paper about the five attributes of fitness described in *Web Investigation 1*, expanding on each and telling why it is important. (You may modify the list if you describe your reasoning.)

Exercise 2
Everyday Fitness: Not Just a Weekend Athlete

In one to two pages, discuss the fitness—and fitness attributes—required for different sports and types of work. Possible approaches might be: What is needed for each of the people in the two pictures to perform well? Does a person have to work out to be fit? Is it possible to achieve and maintain fitness during work activities? What types of work and recreational activities are most conducive to maintenance of fitness after retirement?

Exercise 3
Is It All Chemical?

The biochemistry and physiology of exercise are active areas of research, particularly in relation to the aging process. As more people live longer, it becomes vital to address the issue of fitness. Look at the benefits of exercise described in *Web Investigation 2* and decide which of these are based at least in part on the biochemistry of the body. After listing them, choose one and write a one-page appeal for more research in this area.

Exercise 4
Finding the Boundaries: What Is Extreme?

Write a brief summary of overtraining. Then, using the guidelines in the *Web Investigations*, propose a training regimen for the sport of your choice.

19 Drugs
Chemical Cures, Comforts, and Cautions

Some calm us; some keep us alert; some cure us when we're ill.
For almost every malady we have some kind of pill.

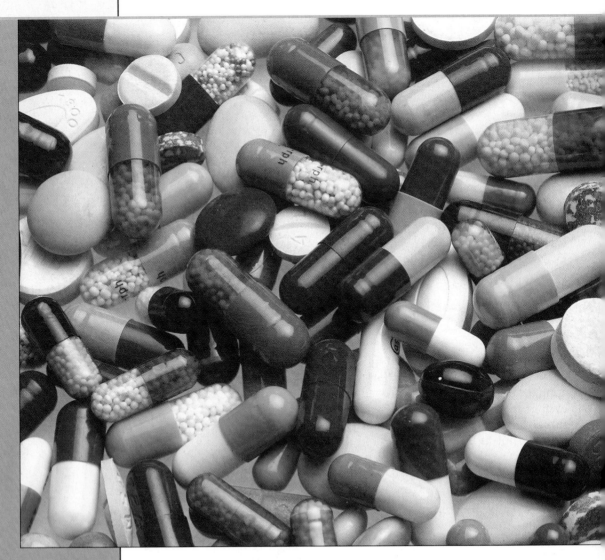

Modern chemistry provides medicines for many different ailments, from headaches and infectious diseases to cancer and mental illness.

P eople today use many different kinds of drugs. We use them to fight infections, relieve pain, stay awake, get to sleep, prevent conception, calm our anxieties, lower blood pressure, treat arthritis, and do any number of other things.

Ours is not the first society to use drugs. The use of medicines to try to relieve pain and cure illnesses dates to prehistoric times. Most societies used alcohol, and the use of marijuana goes back to at least 3000 B.C.E. The narcotic effect of the opium poppy was known to the Greeks in the third century B.C.E (Figure 19.1), and the Indians of the Andes Mountains have long chewed leaves of the coca plant because of the stimulating effect of the cocaine.

In 1904 Paul Ehrlich (1854–1915), a German chemist, realized that certain chemicals were more toxic to disease organisms than to human cells and could therefore be used to control or cure infectious diseases. He coined the term **chemotherapy** (from "chemical therapy"). Ehrlich found that certain dyes used to stain bacteria to make them more visible under a microscope could also be used to kill the bacteria. He used dyes against the organism that causes African sleeping sickness and prepared an arsenic compound effective against the organism that causes syphilis. Ehrlich was awarded the Nobel prize in 1908.

The use of drugs is not new, but never before has there been such a vast array of them. About 3500 specific prescription pharmaceuticals are on the market in the United States. Many are sold under several different brand names, resulting in a total of about 25,000 prescription medicines. There are only about 700 medicines available over the counter, but because of a plethora of brand names and various combinations, there are 300,000 such products on the market. Some of these drugs are vital to the lives and good health of the people who take them. Others are mainly recreational and can be quite destructive. Many of these recreational drugs are addictive, and most are also illegal. In this chapter we look at some of the many drugs that are available today.

19.1 Pain Relievers: Aspirin

Aspirin is the most widely used drug in the world. Its history goes back at least to the time of Hippocrates in ancient Greece because he knew that sick people could ease their pain and lower their fever by chewing willow bark. It was not until 1827 that chemists identified the active ingredient as *salicin*, a derivative of salicylic acid (Chapter 9). Salicylic acid was a very effective pain reliever and fever reducer, but it also caused great stomach distress and bleeding. Aspirin itself itself did not come into use until 1893, when the Baeyer company in Germany made the acetyl ester of salicylic acid. Acetylsalicylic acid (aspirin) had all the pain-relieving and fever-lowering properties of salicylic acid, but it was much gentler to the stomach. Aspirin soon became Baeyer's primary product, which it remains today. Aspirin is still the most widely known **analgesic** (pain reliever) and **antipyretic** (fever reducer) in the world.

Over 80 billion aspirin tablets, under a variety of trade names and store brands, are consumed annually in the United States, amounting to 14 million kg/year. Each aspirin tablet typically contains 325 mg of acetylsalicylic acid held together with an inert binder (usually starch). "Extra strength" formulations usually contain 500 mg. "Buffered" aspirin contains added antacids to prevent the stomach irritation that sometimes occurs when aspirin is taken on an empty stomach.

Aspirin: Pain Relief and Anti-Inflammatory Action

Aspirin relieves minor aches and pains and suppresses inflammation. It works by inhibiting the production of *prostaglandins*, compounds involved in sending pain messages to the brain. Aspirin doesn't cure whatever is causing the pain; it merely kills the messenger.

Figure 19.1 Some cults in ancient Greece worshipped opium. This Greek coin from that period shows an opium poppy capsule.

Drug industry sales in the United States in 1995 were $93.9 billion, about three-fourths for prescription drugs.

[Web Reference 1] Information about colds can be found at the Common Cold Center of Cardiff University in the United Kingdom.

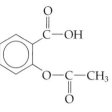

Acetylsalicylic acid (Aspirin)

Prostaglandins are pro-
duced in all body tissues
and help to regulate many
biochemical processes.
They are discussed in Section 19.6.

Inflammation is caused by an overproduction of prostaglandin derivatives. The **anti-inflammatory** action of aspirin results from its inhibition of prostaglandin synthesis. Aspirin is often the initial drug of choice for treatment of arthritis, a disease characterized by the inflammation of joints and connective tissues.

Aspirin As an Anticoagulant: Heart Attack and Stroke Prevention

Aspirin also acts as an **anticoagulant**; it inhibits the clotting of blood. Studies indicate that some intestinal bleeding occurs every time aspirin is ingested. The blood loss is usually minor (0.5–2.0 mL per two-tablet dose), but it may be substantial in some cases. Aspirin should not be used by people facing surgery, childbirth, or some other hazard involving the possible loss of blood (nonuse should start a week before the event). On the other hand, small daily doses seem to lower the risk of coronary heart attack and stroke, presumably by the same anticoagulant action that causes bleeding in the stomach.

Aspirin and Fever Reduction

Pyrogens that do not work
through prostaglandins
are not affected by aspirin.

Fevers are induced by substances called *pyrogens*. These compounds are produced by and released from leukocytes and other circulating cells. Pyrogens usually use prostaglandins as secondary mediators. Fevers therefore can be reduced by aspirin and other prostaglandin inhibitors.

Our bodies try to fight off infections by elevating the temperature. Mild fevers (those below 39°C or 102°F) are usually best left untreated. High fevers, however, can cause brain damage and require immediate treatment.

Limitations of Aspirin

Aspirin is not effective for severe pain, as from a migraine headache. Prolonged use of aspirin, as in the case of arthritic pain, can lead to gastrointestinal disorders. Like all drugs, aspirin is somewhat toxic. It is the drug most often involved in the accidental poisoning of children. The toxicities of aspirin and other drugs are listed in Table 19.1 (Also see Section 20.13.)

Still another hazard associated with aspirin use is allergic reaction. In some people, an allergy to aspirin can cause a skin rash, an asthmatic attack, or even loss of

Table 19.1 ▪ Acute Toxicities of Chemicals Presently or Formerly Used in Over-the-Counter Drugs

Chemical Compound	LD_{50}*
Acetaminophen	338[†]
Acetanilide	800
Aspirin	1500
Caffeine	335 (246)
Diphenhydramine	500
Ibuprofen	(1050)
Ketoprofen	101
Methyl salicylate	887
Naproxen	534
Phenacetin	1650

*LD_{50} values are for oral administration of the drug to rats in milligrams per kilogram of body weight. Values in parentheses are for male rats.

[†] Orally in mice.

Source: Susan Budavari (Ed.), *The Merck Index*, 12th edition. Rahway, NJ: Merck and Co., 1996.

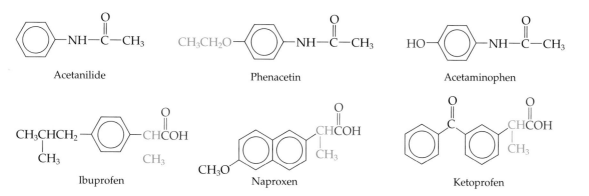

Figure 19.2   Some aspirin substitutes. Phenacetin and acetaminophen are derivatives of acetanilide. Ibuprofen (Advil, Nuprin), naproxen (Aleve), and ketoprofen (Orudis KT) have long been available by prescription but are now available over the counter; all three are derivatives of propionic acid. Phenacetin and acetanilide are no longer used.

consciousness. Some doctors claim that an allergic reaction can be delayed 3–5 h, and so the victim may not associate the reaction with aspirin. Susceptible individuals must be careful to avoid taking aspirin by itself or in any combination with other drugs.

Aspirin Substitutes

Some people who are allergic to aspirin or susceptible to bleeding can safely take substitute medicines (Figure 19.2). The most common is acetaminophen. This compound gives relief of pain and reduction of fever comparable to the action of aspirin. Unlike aspirin, however, it is not effective against inflammation, and it doesn't induce bleeding. Thus, it is of limited use to people with arthritis, but it is often used to relieve the pain that follows minor surgery. Acetaminophen usually costs more than aspirin, especially in the form of highly advertised brands such as Tylenol, Panadol, and Anacin-3. Regular acetaminophen tablets are 325 mg, and extra strength forms are 500 mg. Overuse of acetaminophen has been linked to liver and kidney damage.

Reye's Syndrome

The use of aspirin to treat fevers in children suffering from the flu or chickenpox is associated with Reye's syndrome. This syndrome is characterized by vomiting, lethargy, confusion, and irritability. Fatty degeneration of the liver and other organs can lead to death unless treatment is begun promptly. Just how aspirin is involved in the onset of Reye's syndrome is not known, but the correlation is quite strong. Aspirin products carry a warning indicating that they should not be used to treat children with fevers.

Ibuprofen, naproxen, and ketoprofen are anti-inflammatory drugs that for years were available only by prescription but now are available as over-the-counter medications. Ibuprofen is sold under brand names such as Advil, Nuprin, and Motrin IB. It is also available in generic form. The usual nonprescription dosage is 200 mg. Ibuprofen is perhaps superior to aspirin in its action against inflammation. It also relieves mild pain and reduces fevers but is unlikely to be any better than aspirin for these purposes. It is usually much more expensive than aspirin, and many people who are sensitive to aspirin are also sensitive to ibuprofen.

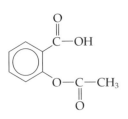

Acetylsalicylic acid
(Aspirin)

Ketoprofen (Orudis KT) has a ketone function, and naproxen (Naprosyn) is a derivative of naphthalene. Both are similar to ibuprofen in their action.

Combination Pain Relievers

Many of the analgesics we buy are combinations of aspirin and other drugs. For many years, the most familiar combination was aspirin, phenacetin, and caffeine (APC), available under various trade names. Phenacetin has about the same effectiveness as aspirin in reducing fever and relieving minor aches and pains. However, it has been implicated in damage to the kidneys, in blood abnormalities, and as a likely carcinogen. The FDA banned further use of phenacetin in 1983.

Anacin, which (along with Empirin and Excedrin) was once an APC formulation, now contains only aspirin and caffeine. Caffeine is a mild stimulant found in coffee, tea, and cola syrup. There is no reliable evidence that caffeine significantly enhances the effect of aspirin. (In fact, evidence indicates that for fever reduction, caffeine *counteracts* the effect of aspirin. Combinations containing caffeine are therefore *less effective* than plain aspirin for this use.) Excedrin also contains caffeine, along with aspirin and acetaminophen. Many combination pain relievers are available, but for most people, plain aspirin is the cheapest, safest, and most effective product.

19.2 Chemistry and the Common Cold

We spend a lot of money for cold remedies, but as yet no real cure for the common cold has been developed. Colds are caused by as many as 200 related viruses, and no antibiotic or other drug is effective against them. Most cold remedies treat just the symptoms. Included in the arsenal of weapons used against colds are antihistamines, cough suppressants, expectorants, bronchodilators, anticholinergics, and nasal decongestants. (Figure 19.3 gives chemical structures for several compounds used in cold medicines.) An FDA advisory panel reviews these compounds periodically for safety and effectiveness.

An *anticholinergic* drug is one that acts on nerves that use acetylcholine as a neurotransmitter (Section 19.10).

Antihistamines

Antihistamines, such as diphenhydramine, chlorpheniramine, and promethazine (Figure 19.3a), relieve the symptoms of allergies: sneezing, itchy eyes, and runny nose. Antihistamines are not effective at curing colds, although they can temporarily relieve some of the symptoms, especially those caused by allergies. When an **allergen** (a substance that triggers an allergic reaction) binds to the surfaces of certain cells, it triggers the release of histamine, which causes the redness, swelling, and itching associated with allergies. Antihistamines block the release of histamine, but most also enter the brain and act on the cells controlling sleep, thus making patients drowsy. There are prescription drugs, such as cetirizine (Zyrtec), loratadine (Claritin), and fexofenadine (Allegra), that inhibit the release of histamine but cannot enter the brain and so do not cause drowsiness.

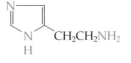

Histamine

Cough Suppressants

The FDA panel has found only three compounds to be effective as *cough suppressants (antitussives)*. Two of them, codeine and dextromethorphan (Figure 19.3b), are narcotics (Section 19.15). The other, diphenhydramine, is an antihistamine. Should a cough really be suppressed? Not usually, because the respiratory tract uses the cough mechanism to rid itself of congestion. When a cough is dry or interferes with needed rest, however, temporary cough suppression may be advisable.

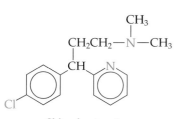

Chlorpheniramine

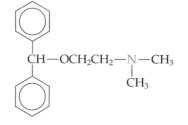

Diphenhydramine

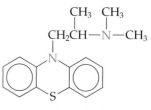

Promethazine

(a) Antihistamines

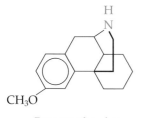

Dextromethorphan

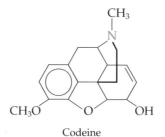

Codeine

(b) Cough suppressants

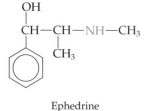

Ephedrine

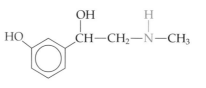

Phenylephrine

(c) Nasal decongestants

Figure 19.3   Some chemicals used to treat colds: antihistamines (a), cough suppressants (b), and some nasal decongestants (c). Diphenhydramine, an antihistamine, is also used as a cough suppressant. All seven compounds are amines. They are used in medicine as hydrochloride salts because the salts are more water-soluble than the uncombined amines (free bases).

Expectorants and Decongestants

Expectorants help bring up mucus from the bronchial passages. Only guaifenesin appears to be a safe and effective expectorant, and its effectiveness is only marginal.

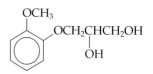

Guaifenesin

   Nasal decongestants seem to be safe and effective for occasional use, although repeated use leads to a rebound effect in which the nasal passages swell and make congestion seem worse. Examples of nasal decongestants are phenylephrine and ephedrine (Figure 19.3c) and their salts.

   How should you treat a common cold? First, you should drink plenty of liquids and get lots of rest. Then, you might want to increase your intake of vitamin C (Section 18.2). At least 20 studies have shown that vitamin C can reduce the length of a cold or lessen its severity. When you are in good physical condition and have a strong immune system, colds seem to strike less often. Frequent washing of the hands definitely seems to help, too. Many people claim to get cold relief by using some favorite over-the-counter drug. And, of course, there are those who still swear by chicken soup!

19.3 Antibacterial Drugs

Less than a century ago, infectious diseases were the principal cause of death in the United States (Figure 19.4); today only one of these infectious disease categories (pneumonia and influenza) remains among the ten leading causes of death. However, the list now includes a new infectious disease category: a viral infection that causes AIDS (acquired immune deficiency syndrome). Many diseases have been brought under control by the use of *antibacterial drugs*. The first of these were sulfa drugs, the

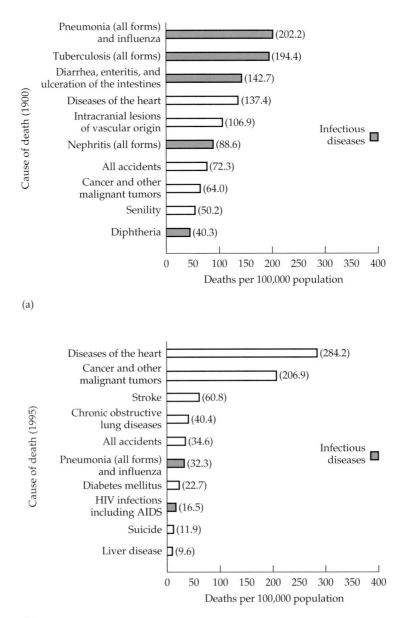

Figure 19.4 (a) In 1900 half of the ten leading causes of death in the United States were infectious diseases. By 1980 the only infectious disease category on the list was pneumonia and influenza, but AIDS was added to the list in the 1980s. (b) The leading causes of death in 1995 were so-called lifestyle diseases: heart disease, stroke, and diabetes are related in part to our diets; cancer and lung diseases to cigarette smoking; and liver disease, accidents, and suicide to excessive consumption of alcohol.

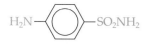

Sulfanilamide

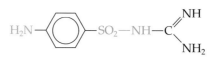

Sulfaguanidine

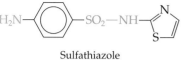

Sulfathiazole

para-Aminobenzoic acid

Figure 19.5 Sulfanilamide and two other sulfa drugs. *para*-Aminobenzoic acid has a similar structure.

prototype of which was discovered in 1935 by the German chemist Gerhard Domagk (1895–1964). Sulfa drugs were used extensively during World War II to prevent wound infections. Many soldiers lived who would have died in earlier wars.

Sulfa Drugs

Sulfanilamide, the simplest sulfa drug, was one of the first to have its action understood at the molecular level. Its effectiveness is based on a case of mistaken identity. Bacteria need *para*-aminobenzoic acid (PABA) to make folic acid, which is essential for the formation of certain compounds the bacteria require for proper growth. But bacterial enzymes can't tell the difference between sulfanilamide and PABA because the substances are so similar in structure (Figure 19.5). When sulfanilamide is applied to an infection in large amounts, bacteria incorporate it into pseudofolic acid molecules that cannot act normally; hence, the bacteria cease to grow. Of the thousands of sulfanilamide analogs that have been developed and tested, only a few are used today. The structures of two common ones are given in Figure 19.5. Some sulfa drugs tend to cause kidney damage or other problems.

Penicillins

The next important discovery was that of penicillin, an antibiotic. **Antibiotics** are soluble substances (derived from molds or bacteria) that inhibit the growth of other microorganisms (Figure 19.6). Penicillin was first discovered in 1928 but was not tried on humans until 1941. Alexander Fleming, a Scottish microbiologist then working at the University of London, first observed the antibacterial action of a mold, *Penicillium notatum*. Fleming was studying an infectious bacterium, *Staphylococcus aureus*, and one of his cultures became contaminated with a blue mold. Contaminated cultures generally are useless. Most investigators probably would have destroyed the culture and started over, but Fleming noted that bacterial colonies had been destroyed in the vicinity of the mold.

Fleming was able to make crude extracts of the active substance. This material, later called *penicillin*, was further purified and improved by Howard Florey (1898–1968), an Australian, and Ernst Boris Chain (1906–1979), a refugee from Nazi Germany, both working at Oxford University. Fleming, Florey, and Chain shared the 1945 Nobel prize in physiology and medicine for their work on penicillin.

It soon became apparent that penicillin was not a single compound but a group of related compounds. Chemists recognized that by designing molecules with different structures they could change the properties of the drugs. The penicillins that resulted from these experiments vary in effectiveness. Some must be injected; others can be taken orally. Bacteria resistant to one penicillin may be killed by another. The structures of several common penicillins are shown in Figure 19.7.

The mode of action of penicillin has been unraveled. In certain bacteria, the cell walls are made up of mucoproteins, polymers in which amino sugars are combined with protein molecules. Penicillin prevents cross-linking between these large molecules and thereby prevents the synthesis of cell walls in bacteria. Cells of higher animals do not have mucoprotein walls. Instead, they have external membranes that

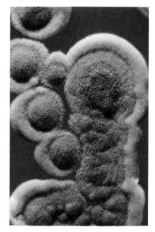

Figure 19.6 Penicillin molds. These symmetrical colonies of mold are *Penicillium chrysogenum*, a mutant form of which now produces nearly all commercial penicillin.

Alexander Fleming (1881–1955), discoverer of penicillin, is shown here on a 1981 Hungarian stamp.

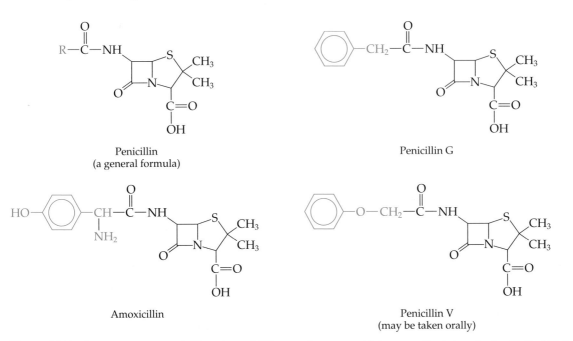

Penicillin
(a general formula)

Penicillin G

Amoxicillin

Penicillin V
(may be taken orally)

Figure 19.7 Some common penicillins. Amoxicillin was the most widely prescribed drug in the United States in 1988, but the development of resistant bacterial strains has caused a dramatic decline in its use.

differ in composition from the cell walls of bacteria and are not affected by penicillin. Thus, penicillin can destroy bacteria without harming human cells. Many people—perhaps as many as 5% of the population—are allergic to penicillin, however.

During their early history, antibiotics came to be known as miracle drugs. The number of deaths from blood poisoning, pneumonia, and other infectious diseases was reduced substantially by the use of antibiotics. Prior to 1941 a person with a major bacterial infection almost always died. Today, such deaths are rare except for those ill with other conditions such as AIDS. Six decades ago, pneumonia was a dread killer of people of all ages. Today, it kills mainly the elderly and those with AIDS. Antibiotics have indeed worked miracles in our time, but even miracle drugs are not without problems. It wasn't long after these drugs were first used that disease organisms began to develop strains resistant to them. As long as erythromycin (an antibiotic obtained from *Streptomyres erythreus*) had only limited use, all strains of staphylococci could be handled readily by the drug, but after it was put into extensive use, resistant strains began to appear. Staph infections are now a serious problem in hospitals. People who go to a hospital to be cured sometimes develop serious bacterial infections instead. Some even die of staph infections picked up at hospitals.

> Drug-resistant tuberculosis now accounts for one out of every seven new cases. (Many of those infected with these resistant organisms have already died.)

Cephalosporins

In a race to stay ahead of resistant strains of bacteria, scientists continue to seek new antibiotics. Penicillins now have been partially displaced by related compounds called *cephalosporins*. Perhaps the most notable of these is cephalexin (Keflex). Unfortunately, some strains of bacteria have already shown resistance to cephalexin.

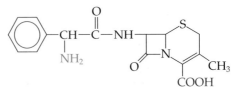

Cephalexin

Tetracyclines

Another important development in the field of antibiotics was the discovery of a group of compounds called *tetracyclines*. The first of these, aureomycin, was isolated in 1948 by Benjamin Duggor (1872–1956) from *Streptomyces aureofaciens*. Scientists at Pfizer Laboratories isolated terramycin from *Streptomyces rimosus* in 1950, and both drugs were later found to be derivatives of tetracycline, a compound now obtained from *Streptomyces viridifaciens*. All three compounds (Figure 19.8) are **broad-spectrum antibiotics**, so called because they are effective against a wide variety of bacteria.

Tetracyclines bind to bacterial ribosomes, inhibiting bacterial protein synthesis and blocking growth of the bacteria. Tetracyclines do not bind to mammalian ribosomes and thus do not affect protein synthesis in host cells. Several microorganisms have developed strains resistant to tetracyclines.

When given to young children, tetracyclines can cause the discoloration of permanent teeth, even though the teeth may not appear until several years later. This probably results from the interaction of tetracyclines with calcium during the period of tooth development. Calcium ions in milk and other foods combine with hydroxyl groups on tetracycline molecules.

Fluoroquinolones

Fluoroquinolones are an important class of antibiotics with activity against a broad spectrum of bacteria and a low incidence of side effects. They are especially useful against bacteria with penicillin resistance. They act by inhibiting bacterial DNA replication through interference with the action of an enzyme called DNA gyrase or topoisomerase. Because their mechanism of action is different from that of other antibiotics, it was hoped that resistance to these drugs would be slow to emerge. However, resistance to fluoroquinolones is already being seen. Ciprofloxacin (Cipro) is the leading representative of this class.

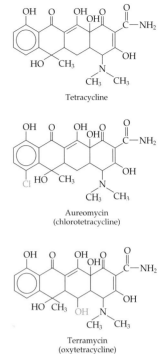

Tetracycline

Aureomycin
(chlorotetracycline)

Terramycin
(oxytetracycline)

Figure 19.8 Three tetracycline antibiotics.

Structure–Function Relationships

Fluoroquinolines are of particular interest to chemists and pharmacologists because many structural features are correlated with their activity, making possible a rational, logical approach to synthesis of new drugs in this class.

The basic skeletal structure for fluoroquinolines is shown on the right. The fluorine atom at position 6 is essential for broad antimicrobial activity when taken orally. Effectiveness depends on a carboxyl group at position 3 and a carbonyl oxygen at position 4. These are responsible for binding to the DNA complex. A nitrogen-containing ring structure at position 7 or a methoxy group at position 8 broadens the range of bacteria affected. If position 7 has a bulky substituent, the drug has fewer side effects on the nervous system.

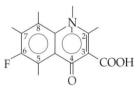

[Web Reference 2] The important issue of increasing bacterial resistance to antibiotics.

19.4 Viruses and Antiviral Drugs

For most of us, antibiotics have taken the terror out of bacterial infections such as pneumonia and diphtheria. We worry about resistant strains of bacteria, but we hope these problems are not insurmountable. Viral diseases cannot be cured by antibiotics, however, and viral infections including colds, herpes, and **acquired immune deficiency syndrome (AIDS)** still plague us. Some viral infections—such as poliomyelitis, mumps, measles, and smallpox—can be prevented by vaccination. Influenza vaccines are quite effective against common recurrent strains, but there are many different strains of flu viruses, and new ones appear periodically.

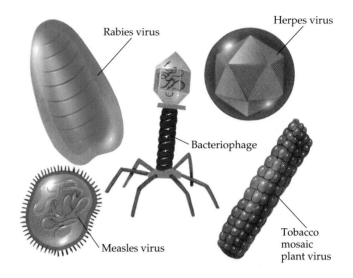

Figure 19.9 Viruses come in a variety of shapes, which are determined by their protein coats.

DNA Viruses and RNA Viruses

Viruses are composed of nucleic acids and proteins (Figure 19.9). They have an external coat with a repetitive pattern of protein molecules. Some coats also include a lipid membrane. Others have sugar–protein combinations called glycoproteins. The genetic material of a virus is either DNA or RNA. A *DNA virus* enters a host cell where the DNA is replicated, and it directs the host cell to produce viral proteins. The viral proteins and viral DNA assemble into new viruses that are released by the host cell. These new viruses can then invade other cells and continue the process.

Most *RNA viruses* use their nucleic acids in much the same way. The virus penetrates a host cell where the RNA strands are replicated, and they induce the synthesis of viral proteins. The new RNA strands and viral proteins are then assembled into new viruses. Some RNA viruses, called **retroviruses**, synthesize DNA in the host cell. This process is the opposite of the transcription of a DNA code into RNA (Chapter 15) that normally occurs in cells. The synthesis of DNA from an RNA template is catalyzed by an enzyme called *reverse transcriptase*. The human immunodeficiency virus (HIV) that causes AIDS is a retrovirus. HIV invades and eventually destroys T cells, white blood cells that normally help protect the body from infections. With the T cells destroyed, the AIDS victim often succumbs to pneumonia or some other infection.

Antiviral Drugs

In recent years, scientists have developed drugs that are effective against some viruses. Structures of some familiar ones are given in Figure 19.10. Amantadine helps prevent some influenza A infections. Acyclovir (Zovirax) controls flareups of herpes viruses that cause genital sores, chickenpox, shingles, mononucleosis, and cold sores.

Three important kinds of drugs are being used against AIDS. (1) Azidothymidine (AZT or Zidovudine) is a nucleoside analog that substitutes for thymidine. (3TC, ddI, ddC, d4T, and abacavir are other drugs that substitute for a nucleoside in the virus DNA, thus crippling it.) AZT slows down the AIDS virus, but it does not cure the disease. (2) Reverse transcriptase inhibitors stop the reverse transcriptase from working properly to make more virus. (3) Protease inhibitors, a new class of anti-HIV drugs, block the enzyme protease so that new copies of the virus can't infect new cells. Protease inhibitors include amprenavir (Agenerase), saquinavir (Fortovase), ritonavir (Norvir), indinavir (Crixivan), and nelfinavir (Viracept). Combinations of these drugs seem to be more effective than any one of them alone.

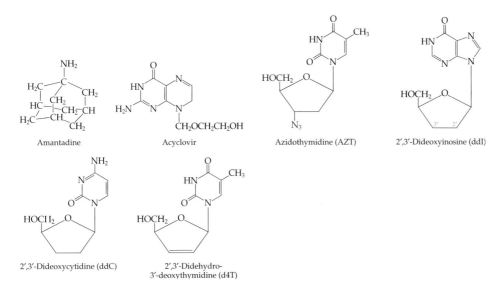

Figure 19.10 Some antiviral drugs. Amantadine is completely synthetic; the others are derived from the purine and pyrimidine bases of nucleic acids.

Unfortunately, they are too toxic to use in quantities sufficient to stop virus replication completely.

[Web Reference 3] Some of the many, many websites dealing with AIDS and its treatment.

Basic Research and Drug Development

Scientists had to understand the normal biochemistry of cells before they could develop drugs to treat abnormal conditions caused by the invasion of viruses. Gertrude Elion and George Hitchings of Burroughs Wellcome Research Laboratories in North Carolina and James Black of Kings College in London did the basic biochemical research that led to the development of antiviral drugs (and many anticancer drugs) (Section 19.7). They determined the shapes of cell membrane receptors, and they learned how normal cells work. They and other scientists were then able to design drugs to block receptors in infected cells. Elion, Hitchings, and Black shared the 1988 Nobel prize in physiology and medicine for this work.

Today scientists use powerful computers to design molecules to fit receptors. Drug design was often a hit-or-miss procedure in its early decades. It is now becoming a more precise science.

19.5 Chemicals Against Cancer

Chemists have designed molecules to relieve headaches, cure infectious diseases, and prevent conception. Why can't they do something about cancer? They have done a lot, but much remains to be done. The main problem is that the drugs that kill cancer cells can damage normal cells as well. The primary aim of cancer research is to find a way to kill cancerous tissue without killing normal cells. Treatment with drugs, radiation, and surgery has led to a high rate of cure for some kinds of cancer (for example, one form of skin cancer). For other types, such as lung cancer, the rate of cure is still quite low. Dozens of different chemical substances are used widely in the treatment of cancer, and this number will no doubt increase rapidly as our understanding of basic cell chemistry increases.

Chemotherapy affects any body cells that undergo rapid replacement, not just cancer cells. Included are the cells that line the digestive tract and those that produce hair. Side effects of chemotherapy therefore include nausea and loss of hair. Eventually normal cells are affected to such a degree that treatment must be discontinued.

Antimetabolites: Inhibition of Nucleic Acid Synthesis

In cancer chemotherapy, **antimetabolites** are usually compounds that inhibit the synthesis of nucleic acids. Rapidly dividing cells, characteristic of cancer, require large quantities of DNA. Anticancer metabolites block DNA synthesis and therefore

Cisplatin

Transplatin

block the increase in the number of cancer cells. Because cancer cells are undergoing rapid growth and cell division, they are generally affected to a greater extent than normal cells.

One widely used anticancer drug is cisplatin, a platinum-containing compound. Cisplatin binds to DNA and blocks its replication.

Two other prominent antimetabolites are 5-fluorouracil and its deoxyribose derivative, 5-fluorodeoxyuridine. In the body, both these compounds can be incorporated into a phosphate–sugar base unit called a *nucleotide* (Chapter 15). The fluorine-containing nucleotide inhibits the formation of thymine-containing nucleotides required for DNA synthesis. Thus, both compounds slow the division of cancer cells; they have been employed against a variety of cancers, especially those of the breast and the digestive tract.

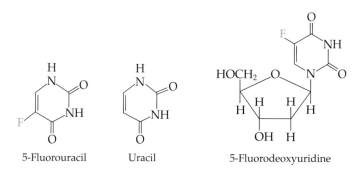

5-Fluorouracil   Uracil   5-Fluorodeoxyuridine

Another common antimetabolite is 6-mercaptopurine. This compound can substitute for adenine in a nucleotide. The pseudonucleotide then inhibits the synthesis of nucleotides incorporating adenine and guanine. Hence, DNA synthesis and cell division are slowed. 6-Mercaptopurine has been used in the treatment of leukemia.

The antimetabolite methotrexate acts in a somewhat different manner. Note the similarity between its structure and that of folic acid. Like the pseudofolic acid formed from sulfanilamide, methotrexate competes successfully with folic acid for an enzyme but cannot perform the growth-enhancing function of folic acid. Again, cell division is slowed and cancer growth is retarded. Methotrexate is used frequently against leukemia.

6-Mercaptopurine

Adenine

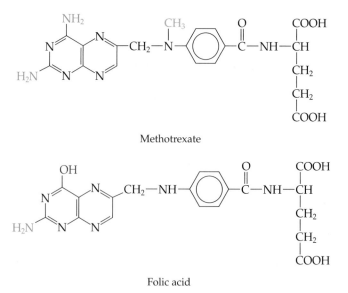

Methotrexate

Folic acid

Alkylating Agents: Turning War Gases on Cancer

Alkylating agents are highly reactive compounds that can transfer alkyl groups to compounds of biological importance. These foreign alkyl groups then block the usual action of the biological molecules. Some alkylating agents are used against cancer. Typical among these are nitrogen mustards, compounds that arose out of chemical warfare research.

The original mustard "gas" was a sulfur-containing blister agent used during World War I. Contact with either the liquid or the vapor causes blisters that are painful and slow to heal. It is easily detected, however, by its garlic or horseradish odor. Mustard gas is denoted by the military symbol H.

$$Cl-CH_2CH_2-S-CH_2CH_2-Cl$$

Mustard gas

H

Nitrogen mustards (symbol HN) were developed about 1935. Though not quite as effective overall as mustard gas, nitrogen mustards produce greater eye damage and don't have an obvious odor. Structurally, nitrogen mustards are chlorinated amines.

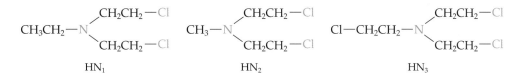

To our pleasant surprise, nitrogen mustards have been found to be effective in the treatment of a formerly fatal type of skin cancer. Complete remission is often obtained by bathing patients in a solution of nitrogen mustards. Knowledge gained through science is neither good nor evil. The same knowledge can be used either for our benefit or for our destruction.

The nitrogen mustard of choice for cancer therapy nowadays is a compound called cyclophosphamide (Cytoxan).

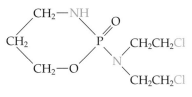

Cyclophosphamide

Chemical Warfare

Chemical agents were used extensively during World War I. More than 30 such substances were employed, killing 91,000 and wounding 1.2 million (many of them for life). Fritz Haber supervised the release of chlorine gas in the first attack by the Germans. Adolf Hitler was among those wounded when the British retaliated with phosgene a few days later. Haber considered gas warfare to be "a higher form of killing." Fortunately, his views have not become widely accepted. The use of chemical warfare agents was largely avoided during World War II. They have been used in smaller wars, however, most recently by the Iraqi government against rebels.

Alkylating agents can cause cancer as well as cure it. For example, the nitrogen mustard HN_2 causes lung, mammary, and liver tumors when injected into mice; yet it can be used with some success in the management of certain human tumors. There is still a lot of mystery—and seeming contradiction—regarding the causes and cures of cancer.

Miscellaneous Anticancer Agents

There are many other anticancer agents. Alkaloids from vinca plants have been shown to be effective against leukemia and Hodgkin's disease. Paclitaxel (Taxol), obtained from the Pacific yew tree, is effective against cancers of the breast, ovary, and cervix. Several antibiotics have been found to kill cancer cells as well as bacteria. Actinomycin, a mixture of complex compounds obtained from the molds *Streptomyces antibioticus* and *S. parvus*, is used against Hodgkin's disease and other types of cancer. It is quite effective but extremely toxic. Actinomycin acts by binding to the double helix of DNA, thus blocking the replication of RNA on the DNA template. Protein synthesis is inhibited.

Sex hormones can be used against cancers of the reproductive system. For example, the female hormones estradiol (a natural hormone) and diethyl stilbestrol (DES, a synthetic hormone) can be used to treat cancer of the prostate gland. Conversely, male hormones such as testosterone can be used against breast cancer. Such treatment often brings about a temporary cessation—or even a regression—in the growth of cancer cells.

The food additive BHT (Chapter 16) has been shown to be anticarcinogenic in tests involving laboratory animals. Some people involved in cancer research speculate that the use of this additive as a preservative in foods may account in part for the declining rate of stomach cancer in the United States.

Chemotherapy is only part of the treatment for cancer. Surgical removal of tumors and radiation treatment remain major weapons in the war on cancer. It is unlikely that a single agent will be found to cure all cancers. Perhaps a greater hope lies in the prevention of cancer. Much active research is underway on the mechanisms of carcinogenesis. The more we learn about what causes cancer, the better equipped we will be in trying to prevent it.

19.6 Hormones: The Regulators

Before we can discuss the next group of drugs, we must take a brief look at the human endocrine system and some of the chemical compounds, called **hormones,** that this system manufactures. Hormones are chemical messengers produced in the endocrine glands (Figure 19.11). Those released in one part of the body send signals for profound

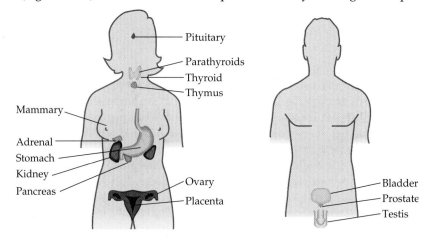

Figure 19.11 The approximate locations of the endocrine glands in the human body.

Table 19.2 ▮ Some Human Hormones and Their Physiological Effect

Name	Gland and Tissue	Chemical Nature	Effect
Various releasing and inhibitory factors	Hypothalamus	Peptide	Triggers or inhibits release of pituitary hormones
Human growth hormone (HGH)	Pituitary, anterior lobe	Protein	Controls general body growth; controls bone growth
Thyroid-stimulating hormone (TSH)	Pituitary, anterior lobe	Protein	Stimulates growth of the thyroid gland and production of thyroxin
Adrenocorticotrophic hormone (ACTH)	Pituitary, anterior lobe	Protein	Stimulates growth of the adrenal cortex and production of cortical hormones
Follicle-stimulating hormone (FSH)	Pituitary, anterior lobe	Protein	Stimulates growth of follicles in ovaries of females and of sperm cells in testes of males
Luteinizing hormone (LH)	Pituitary, anterior lobe	Protein	Controls production and release of estrogens and progesterone from ovaries and testosterone from testes
Prolactin	Pituitary, anterior lobe	Protein	Maintains the production of estrogens and progesterone; stimulates the formation of milk
Vasopressin	Pituitary, posterior lobe	Peptide	Stimulates contractions of smooth muscle; regulates water uptake by the kidneys
Oxytocin	Pituitary, posterior lobe	Peptide	Stimulates contraction of the smooth muscle of the uterus; stimulates secretion of milk
Parathyroid	Parathyroid	Protein	Controls metabolism of phosphorus and calcium
Thyroxine	Thyroid	Amino acid derivative	Increases rate of cellular metabolism
Insulin	Pancreas, beta cells	Protein	Increases cell usage of glucose; increases glycogen storage
Glucagon	Pancreas, alpha cells	Protein	Stimulates conversion of liver glycogen to glucose
Cortisol	Adrenal gland, cortex	Steroid	Stimulates conversion of proteins to carbohydrates
Aldosterone	Adrenal gland, cortex	Steroid	Regulates salt metabolism; stimulates kidneys to retain Na^+ and excrete K^+
Epinephrine (adrenaline)	Adrenal gland, medulla	Amino acid derivative	Stimulates a variety of mechanisms to prepare the body for emergency action including the conversion of glycogen to glucose
Norepinephrine (noradrenaline)	Adrenal gland, medulla	Amino acid derivative	Stimulates sympathetic nervous system; constricts blood vessels; stimulates other glands
Estradiol	Ovary, follicle	Steroid	Stimulates female sex characteristics; regulates changes during menstrual cycle
Progesterone	Ovary, corpus luteum	Steroid	Regulates menstrual cycle; maintains pregnancy
Testosterone	Testis	Steroid	Stimulates and maintains male sex characteristics

physiological changes in other parts of the body. By causing reactions to speed up or slow down, hormones control growth, metabolism, reproduction, and many other functions of body and mind. A number of important human hormones and their physiological effects are listed in Table 19.2.

Prostaglandins: Hormone Mediators

Closely related to hormones are **prostaglandins**. People engaged in medical research are excited about the possibilities of drugs related to prostaglandins, and thousands of papers are published each year on these compounds. Our bodies synthesize prostaglandins from arachidonic acid, a fatty acid with 20 carbon atoms

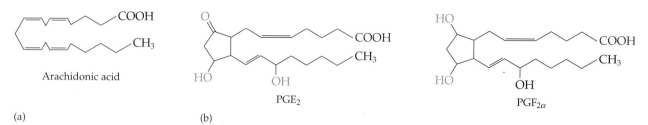

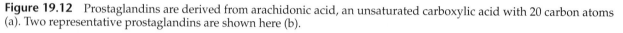

Figure 19.12 Prostaglandins are derived from arachidonic acid, an unsaturated carboxylic acid with 20 carbon atoms (a). Two representative prostaglandins are shown here (b).

Percy Lavon Julian (1899–1975) was involved in the synthesis of a variety of steroids, including physostigmine, a drug used to treat glaucoma. Untreated, glaucoma causes blindness. Julian is shown here on a 1992 U.S. postage stamp.

(Figure 19.12). There are six primary prostaglandins, and many others have been identified. These compounds are widely distributed throughout the body at extremely low concentrations. They are among the most potent biological chemicals, and extremely small doses can elicit marked changes.

Prostaglandins act as mediators of hormone action. They regulate such things as smooth muscle activity, blood flow, and secretion of various substances. This range of physiological activity has led to the synthesis of hundreds of prostaglandins and their analogs. Two are now in use in the United States to induce labor. Others can be used clinically to lower or increase blood pressure, to inhibit stomach secretions, to relieve nasal congestion, to provide relief from asthma, and to prevent formation of the blood clots associated with heart attacks and strokes.

$PGF_{2\alpha}$ is used in cattle breeding. A prize cow is treated with a hormone and with $PGF_{2\alpha}$ to induce the release of many ova. The ova are then fertilized with sperm from a champion bull, and the developing embryos are implanted in less valuable cows. This procedure enables a farmer to get several calves a year from one outstanding cow.

Dozens of prostaglandins and related compounds are now under investigation as possible drugs. They hold great promise for a variety of purposes.

Steroids

Steroids all have the same skeletal four-ring structure (Figure 19.13). Not all steroids are hormones. Cholesterol is a common component of all animal tissues. About 10%

Figure 19.13 All steroids have the same basic four-ring structure (a). Cholesterol (b) is an essential component of all animal cells. Cortisone and prednisone (c) are anti-inflammatory substances.

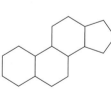

Steroid skeletal structure

(a)

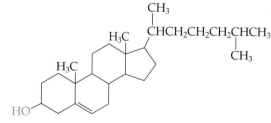

Cholesterol

(b)

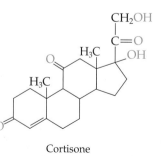

Cortisone

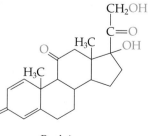

Prednisone

(c)

of the brain is cholesterol, which makes up a vital part of the membranes of nerve cells. Cholesterol is a major component of certain types of gallstones and is also found in deposits in hardened arteries.

Another steroid of some interest is the adrenal hormone cortisone. Applied topically or injected, cortisone acts in the body to reduce inflammation. It was once widely used in the treatment of arthritis but has been largely replaced by the related compound prednisone, which is effective in much smaller doses and has fewer side effects.

There are various kinds of drugs, both natural and synthetic, that are based on steroids. Some of these are the anabolic steroids (Section 18.8) taken illegally by many athletes to improve their performance. Some are anti-inflammatory drugs used to treat such conditions as arthritis, bronchial asthma, dermatitis, and eye infections. And some are sex hormones used in formulating birth control pills (Section 19.7).

Serendipity often plays a role in the discovery of new drugs. Percy Julian was a chemist at the Glidden Paint Company doing research on soybeans when his work led to the development of new steroid-based drugs. It often happens that new drugs are discovered by chemists who are not really looking for them.

Sex Hormones

Closely related in structure to cholesterol, cortisone, and prednisone are sex hormones (Figure 19.14). It is interesting to note that male sex hormones differ only slightly in structure from female sex hormones. In fact, the female hormone progesterone can be converted to the male hormone testosterone by a simple biochemical reaction. The physiological actions of these structurally similar compounds, however, are markedly different.

Male sex hormones, called **androgens**, are secreted by the testes. These hormones are responsible for development of the sex organs and for secondary sexual characteristics, such as voice and hair distribution. The most important male hormone is testosterone.

There are two important groups of female sex hormones. **Estrogens** are produced mainly in the ovaries. They control female sexual functions, such as the menstrual cycle, the development of breasts, and other secondary sexual characteristics. Two important estrogens are estradiol and estrone. Another female sex hormone is progesterone, which prepares the uterus for pregnancy and prevents the further release of eggs from the ovaries during pregnancy.

Sex hormones—both natural and synthetic—are sometimes used therapeutically. For example, a woman who has passed menopause may be given hormones to compensate for those no longer being produced by her ovaries. Some of the earliest compounds employed in cancer chemotherapy were sex hormones (Section 19.5).

19.7 Chemistry and Social Revolution: The Pill

When administered by injection, progesterone serves as an effective birth control drug. This knowledge led to attempts by chemists to design a contraceptive that would be effective when taken orally.

The structure of progesterone was determined in 1934 by Adolf Butenandt, who received the Nobel prize in chemistry in 1939 for his work on sex hormones. In 1938 Hans Inhoffen synthesized the first oral contraceptive, ethisterone, which has a triple-bonded ethynyl group ($-C\equiv CH$). However, it had to be taken in large doses to be effective and was not widely used.

The next breakthrough came in 1951, when Carl Djerassi synthesized 19-nor-progesterone, progesterone with one of its methyl groups missing. This compound was four to eight times as effective as progesterone as a birth control agent. Like

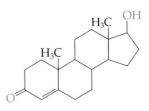

Testosterone

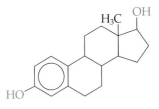

Estradiol

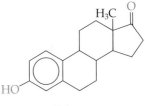

Estrone

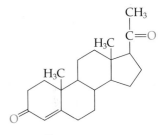

Progesterone

Figure 19.14 The principal sex hormones. Testosterone is the main male sex hormone (androgen). Estradiol and estrone are female sex hormones (estrogens). Progesterone, also produced by females, is essential to the maintenance of pregnancy.

[Web Reference 4] A primer on women's hormones.

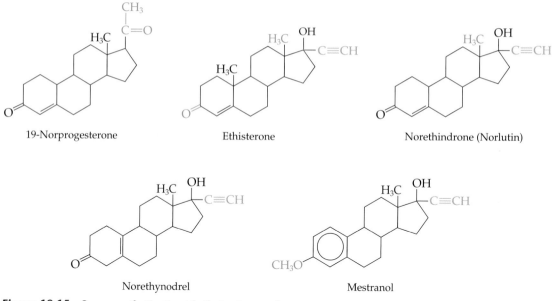

19-Norprogesterone   Ethisterone   Norethindrone (Norlutin)

Norethynodrel   Mestranol

Figure 19.15   Some synthetic steroids that act as sex hormones.

Carl Djerassi (1923–), professor of chemistry at Stan-
ford University and
president of Zoecon Cor-
poration, Palo Alto, California, synthesized
Norlutin, a progestin, in
1951. He was awarded a
patent in 1956.

progesterone itself, it had to be given by injection, which was undesirable. Djerassi then put it all together. Removal of a methyl group made the drug more effective,
and the ethynyl group allowed oral administration. He then synthesized
17α-ethynyl-19-nortestosterone, better known by the trade name Norlutin. This
drug proved effective when it was taken in small doses. At about the same time,
Frank Colton synthesized norethynodrel. Note that the two substances differ only
in the position of a double bond (Figure 19.15). Colton's employer, G. D. Searle,
produced the first approved contraceptive, Enovid, in 1960. The pill contained
9.85 mg of norethynodrel and 150 mg of mestranol. Norethynodrel, norethindrone,
and related compounds are called **progestins** because they mimic the action of progesterone. Mestranol is a synthetic estrogen added to regulate the menstrual cycle.
Progestin acts by establishing a state of false pregnancy (Figure 19.16). A woman
does not ovulate when she is pregnant (or in the state of false pregnancy established by the progestin), and because she does not ovulate, she cannot conceive.

Chemical Treasure in the Rainforest

In 1940 progesterone was one of the most expensive compounds in the world. It
was obtained from animals, with great difficulty, and sold for $200 per gram. A
chemist named Russell Marker wondered if progesterone could be made from
some kind of plant steroid. He took a steroid called *diosgenin*, which occurs in
small amounts in lily roots, and converted it to progesterone in the laboratory.
   He then wondered whether there might be a more abundant source of this
plant steroid. He examined more than 400 plants from all over the world and finally identified a Mexican vine, a kind of wild yam, that contained a sizable
amount of diosgenin in its roots. He eventually left his faculty post at Penn State
University to look for more of these vines in the tropical forests of Mexico. Eventually he found a profuse growth of them and within a few months synthesized

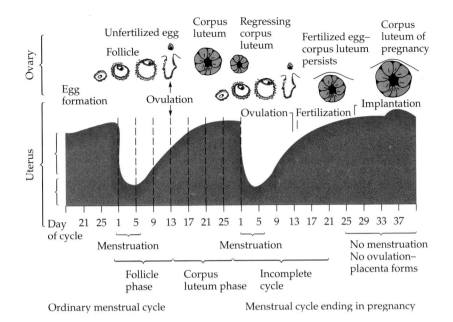

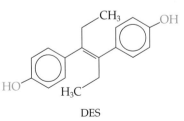

Figure 19.16 Changes in the ovary and uterus during the menstrual cycle. Pregnancy—or the pseudopregnancy caused by the birth control pill—prevents ovulation.

almost a kilogram of progesterone. Marker had found a commercial source not only for progesterone but for all the other steroid hormones as well.

Tropical rainforests are warm, moist, and fertile areas where there are more different kinds of plants and animals than any place else on Earth. These forests are living chemical factories that contain valuable and irreplaceable natural resources. Unfortunately, the world's rainforests are rapidly disappearing.

DES: A Missed-Period Pill

DES has been used for two seemingly opposite purposes. It was employed at low doses in an attempt to help women with a history of miscarriages to maintain pregnancy, and at high doses to induce abortion. We know now that it was not effective in maintaining pregnancy, but many doctors prescribed it for that purpose during the 1950s and 1960s. DES caused vaginal cancer in some daughters and sterility in some sons of women who took it to prevent miscarriage. It is effective at high doses in inducing abortions, but unpleasant side effects are frequently noted.

Frank Colton (1923–), a chemist at G. D. Searle Co., synthesized norethynodrel, for which Searle was awarded patents in 1954 and 1955. Norethynodrel, a progestin, was used in Enovid, the first birth control pill.

DES

DES was also used for several years as a growth promoter in cattle and poultry (Chapter 16). It was banned from that use by the FDA when residues were found (as incidental additives) in the meat of these animals. How can the FDA approve the use of a substance as a drug and yet ban it as an additive in the feed of animals intended as human food? The answer lies in the way the law is written and in the way the chemical is used. The Delaney clause *requires* the banning of any food additive that causes cancer in humans or laboratory animals. DES had been shown to cause vaginal cancer; therefore it had to be banned as a food additive. There is no such requirement for drugs; DES could be approved for use in inducing abortions. It also was pointed out that if the drug were effective, there would be no offspring to have cancer.

DES is not a steroid, but the shape of the molecule shows some similarity to the structure of estradiol (Figure 19.15). By its action, DES is classified as a synthetic female hormone.

RU-486: Convenience and Controversy

Progesterone is essential for maintaining pregnancy. If its action is blocked, pregnancy cannot be established or maintained. Rousel-Uclaf, a French subsidiary of Hoescht, has developed a drug that blocks the action of progesterone. Mifepristone (RU-486), long available in France and China, has replaced a substantial number of surgical abortions. A woman who wishes to abort a pregnancy takes three mifepristone tablets, followed in a few days by an injection of a prostaglandin (Section 19.8). The lining of the uterine wall and the implanted fertilized egg are sloughed off, and the pregnancy is terminated.

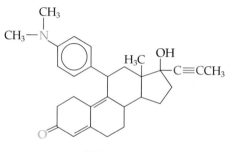

Mifepristone

RU-486 was approved in the United States in 1996. People who believe that human life begins when the ovum is fertilized by the sperm equate the use of mifepristone with abortion and oppose making the drug available. Others consider the drug just another method of birth control and see its use as being safer than a surgical abortion.

RU-486 has other medical uses, too. It can increase uterine contractions when labor has stalled during childbirth. It appears to trigger lactation in mothers and to increase milk production. It seems to slow the growth of certain types of cancer. And it is also being studied as a treatment for Cushing's syndrome, which results from the excessive production of cortisone.

Risks of Taking Birth Control Pills

Oral contraceptives have been used by millions of women in the United States since 1960. They appear to be safe in most cases, but some women experience hypertension, acne, or abnormal bleeding. These pills increase the risk of blood clotting in some women, but so does pregnancy. Blood clots can clog arteries and cause death by stroke or heart attack. The death rate associated with birth control pills is about 3 in 100,000, only one-tenth of that associated with childbirth (which is about 30 per 100,000). For smokers, however, the risks are much higher. For women over 40 who smoke 15 cigarettes per day, the risk of death from stroke or heart attack is 1 in 5000. The FDA advises all women who smoke, especially those over 40, to use some other method of contraception.

The Minipill

Because most of the side effects of oral contraceptives are associated with the estrogen component, the amount of estrogen in these pills has been greatly reduced over the years. Today's pills contain only a fraction of a milligram of estrogen. In fact, minipills are now available that contain only small amounts of progestin and no estrogen at all. Minipills are not quite as effective as the combination pills, but they have fewer side effects.

Birth Control by Implant

In 1990 the FDA approved the use of an under-the-skin implant for birth control. Levonorgestrel (Norplant) (Figure 19.17), a synthetic analog of progesterone, is incorporated into six plastic capsules which are implanted under the skin of the upper arm. Release of the progestin over time prevents pregnancy for 5 years with a failure rate of only 0.2%. It has about the same side effects as the progestin-only minipill. (Levonorgestrel is the progestin used in some ordinary birth control pills.)

A Pill for Males?

Why should females have to bear all the responsibility for contraception? Why not a pill for males? Studies have shown that many men—including a majority of younger ones—are willing to share the risks and the responsibility of contraception. Nevertheless, there are biological reasons for females to bear the burden. Women are the ones who get pregnant when contraception fails, and in females contraception has to interfere with only one monthly event—ovulation. Males produce sperm continuously.

A good deal of research has gone into the development of contraceptives for males. Testosterone and its analogs work quite well, but they have to be injected on a daily basis. Another product was an effective contraceptive, but an alcoholic drink would bring on a state of extreme nausea. Estrogens would work as male contraceptives, but they would also cause the development of female characteristics in men. A prostaglandin, PGE_2, has shown some promise as a birth control product for men. But perhaps the most successful male contraceptive so far is a protein fragment called P26h which "neutralizes the sperm's fertilizing power" according to its Montreal manufacturer.

The availability of vasectomies—simple surgical procedures that block the emission of sperm—has lessened the demand for a male contraceptive, but a major drawback is that these procedures are often irreversible.

Figure 19.17 Norplant, a progestin in controlled release plastic capsules that are implanted under the skin, prevents pregnancy for up to 5 years. Each capsule is about 2.5 cm long and the width of a matchstick.

19.8 Drugs for the Heart

Cardiovascular drugs are a highly important group of life-saving compounds. They rank first in worldwide sales of prescription drugs, comprising about 40% of retail pharmaceutical sales. They have been shown both to prolong life and to improve its quality. Your heart is a muscle that beats virtually every second of every day for as long as you live. Over the course of 24 hr it pumps more than 8000 L of blood through a total of almost 100,000 km of blood vessels.

The major diseases of the heart and blood vessels include ischemic coronary artery disease (inadequate supply of oxygen to the heart), heart arrhythmias (abnormal heartbeat), hypertension (high blood pressure), and congestive heart failure. Atherosclerosis (fatty deposit buildup in the lining of the arteries) is the primary cause of coronary artery disease, which in turn causes myocardial infarction (heart attack). Thus, most drug treatments for the heart aim to increase its supply of blood (and oxygen), to normalize its rhythm, to lower blood pressure, or to prevent accumulation of lipid plaque deposits in blood vessels.

The word *ischemic* means "lacking oxygen."

Lowering Blood Pressure

Hypertension, or high blood pressure, (usually defined as pressure above 140/90) is the most common cardiovascular disease. Since it normally does not produce symptoms, many people have hypertension without realizing it. Four of the major drug categories for lowering blood pressure are *diuretics* (which act on the kidneys to excrete more water, thus lowering the blood volume), *beta blockers* (which slow the heart rate and reduce the force of the heartbeat), *calcium channel blockers* (which are powerful vasodilators, inducing muscles around the blood vessels to relax), and *angiotensin-converting enzyme (ACE) inhibitors*. Two common beta blockers are propranolol and metaprolol (Section 19.11).

Normalizing Heart Rhythm

An *arrhythmia* is an abnormal heartbeat. Some arrhythmias exhibit no symptoms and are discovered only during a physical examination, but others can be life-threatening. The electrical properties of nerves and muscles arise from the flow of ions across cell membranes, and drugs for arrhythmia alter this flow. There are several types of such drugs, with different mechanisms of action. These drugs tend to have a very narrow margin between the therapeutic dose and a harmful amount. A change in the heart's rhythm can have serious consequences. The too-rapid heartbeat, known as *fibrillation*, is so common that defibrillator devices are now available on many airplanes and in various other public places.

Treating Coronary Artery Disease

The chest pain called *angina pectoris* is the most common symptom of coronary artery disease, in which the heart gets less oxygen than it needs. This is usually due to partial blockage of the coronary arteries by lipid-containing plaque (arteriosclerosis). When the blockage is complete, a heart attack occurs, and some of the heart muscle dies. Medical treatment for coronary artery disease usually involves dilation (widening) of the blood vessels to the heart to increase the blood flow and slowing of the heart rate to decrease its work load and its demand for oxygen. Some of the drugs used are the same as those used to lower high blood pressure (beta blockers, for example). Several organic nitro compounds, especially amyl nitrite and nitroglycerin, have a long history in the treatment of angina pectoris.

$$H_2C-O-NO_2$$
$$HC-O-NO_2$$
$$H_2C-O-NO_2$$

Nitroglycerin

$$CH_3CH_2CH_2CH_2CH_2-ONO$$

amyl nitrite

These compounds act by releasing nitric oxide, NO, which relaxes the constricted vessels that are reducing the supply of blood and oxygen to the heart.

Although many new drugs are now being used to treat heart failure, one that has been used for centuries still plays an important role. Digitalis, from the foxglove plant, was used by the ancient Egyptians and Romans and is still used to treat patients with heart failure. Digitalis is a mixture of glycosides that yield carbohydrates and steroids on hydrolysis. Hydrolysis of digitoxin (one of the digitalis glycosides) yields the steroid digitoxigenin.

19.9 Drugs and the Mind

Psychotropic drugs are those that affect the human mind. Probably the first drugs used by primitive peoples influenced the mind. Alcohol, marijuana, opium, cocaine, peyote, and other plant materials have been known for thousands of years. Generally, only one or two of these were used in any one society; however, thousands of drugs that affect the mind are readily available today. Some still come from plants, but the majority are synthetic.

[Web Reference 5] Website of the American Heart Association on heart disease and its treatment.

There is no clear distinction between drugs that affect the mind and those that affect the body. Most of the drugs we take probably affect our minds as well as our bodies. It can still be useful, however, to distinguish between drugs that act primarily on the body and drugs that affect primarily the mind.

Drugs that affect the mind are divided into three classes. **Stimulant drugs**, such as cocaine and amphetamines (Section 19.16), increase alertness, speed up mental processes, and generally elevate the mood. **Depressant drugs**, such as alcohol, most anesthetics, opiates, barbiturates, and minor tranquilizers (Sections 19.12 and 19.13), reduce the level of consciousness and the intensity of reactions to environmental stimuli. In general, depressants dull emotional responses. **Hallucinogenic drugs**, such as lysergic acid diethylamide (LSD) and marijuana (Sections 19.17 and 19.18), are often called "mindbenders." They alter qualitatively the way we perceive things. We will examine some of the major drugs in each category, but first let's look at the chemistry of the nervous system and explore how nerve cells work.

19.10 Some Chemistry of the Nervous System

The nervous system is made up of about 12 billion **neurons** (nerve cells) with 10^{13} connections between them. The brain operates with a power output of about 25 W and has the capacity to handle about 10 trillion bits of information. Nerve cells vary a great deal in shape and size. One type is shown in Figure 19.18. The essential parts of each cell are the cell body, the axon, and the dendrites. We discuss here only the nerves that make up the involuntary (autonomic) nervous system. These nerves carry messages between organs and glands that act involuntarily (such as the heart, the digestive organs, and the lungs) and the brain and spinal column.

Although the axons on a given nerve cell may be up to 60 cm long, there is no continuous pathway from an organ to the central nervous system. Messages must be transmitted across tiny, fluid-filled gaps, or **synapses** (Figure 19.19). When an electrical signal from the brain reaches the end of an axon, specific chemicals (called **neurotransmitters**) that carry the impulse across the synapse to the next cell are liberated. There are perhaps a hundred or so neurotransmitters, and each has a specific function. Messages are carried to other nerve cells, to muscles, and to the endocrine glands (such as the adrenal glands). Each neurotransmitter fits a specific receptor site on the receptor cell (Figure 19.20). Many drugs (and some poisons) act by mimicking the

[Web Reference 6]
www Fact sheets about various aspects of mental health and drugs

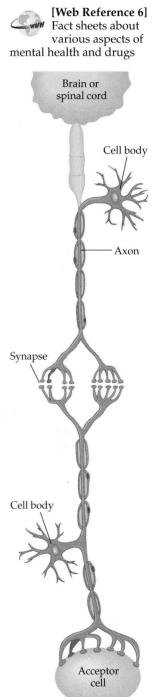

Figure 19.20 Diagram of the pathway by which messages are transmitted to (and from) a receptor cell in a gland or an organ from (and to) the central nervous system.

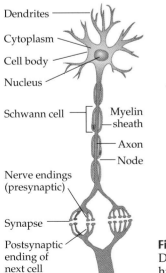

Figure 19.18 Diagram of a human nerve cell.

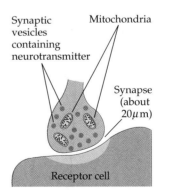

Figure 19.19 Diagram of a synapse. When an electrical signal reaches the presynaptic nerve ending, neurotransmitter molecules are released from the vesicles. They migrate across the synapse to the receptor cell where they fit specific receptor sites.

action of the neurotransmitter. Others act by blocking the receptor and preventing the neurotransmitter from acting on it. Several neurotransmitters are amines (Chapter 9), as are some of the drugs that affect the chemistry of our brains.

19.11 Brain Amines: Depression and Mania

We all have our ups and downs in life. These moods probably result from multiple causes, but it is likely that a variety of chemical compounds formed in the brain are involved. Before we consider these ups and downs, however, let's take a look at epinephrine, an amine formed in the adrenal glands.

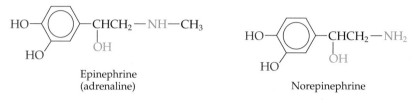

Epinephrine
(adrenaline)

Norepinephrine

[Web Reference 7] Some excellent articles on depression and the neurobiology involved.

Commonly called adrenaline, epinephrine is secreted by the adrenal glands. A tiny amount of epinephrine causes a great increase in blood pressure. When a person is under stress or is frightened, the flow of adrenaline prepares the body for fight or flight. Because culturally imposed inhibitions prevent fighting or fleeing in most modern situations, the adrenaline-induced supercharge is not used. This sort of frustration has been implicated in some forms of mental illness.

A Biochemical Theory of Mental Illness

Biochemical theories of mental illness involve brain amines. One is norepinephrine (NE), a relative of epinephrine. NE is a neurotransmitter formed in the brain. When produced in excess, NE causes a person to be elated—perhaps even hyperactive. In large excess, NE induces a manic state. A deficiency of NE, on the other hand, can cause depression.

Levels of 5-HIAA are also low in murderers and other violent offenders. However, they are higher than normal in people with obsessive–compulsive disorders, sociopaths, and people with guilt complexes.

Another brain amine is serotonin, also a neurotransmitter. Serotonin is involved in sleep, sensory perception, and regulation of body temperature. A metabolite of serotonin, 5-hydroxyindoleacetic acid (5-HIAA), is found in unusually *low* levels in the spinal fluid of violent suicide victims. This indicates that abnormal serotonin metabolism plays a role in depression. Recent research suggests that a reduced flow of serotonin through the synapses in the frontal lobe of the brain causes depression.

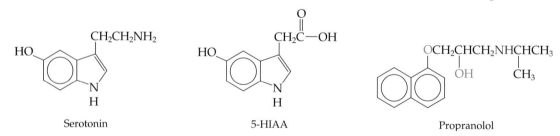

Serotonin

5-HIAA

Propranolol

Brain Amine Agonists and Antagonists in Medicine

Our cells have at least six different receptors that are activated by NE and related compounds. NE agonists (drugs that enhance or mimic its action) are stimulants. NE antagonists (drugs that block the action of NE) slow down various processes. Drugs called beta blockers reduce the stimulant action of epinephrine and NE on various kinds of cells. Propranolol (Inderal) is used to treat cardiac arrhythmias, angina, and hypertension by slightly lessening the force of the heartbeat. Unfortunately, it also causes lethargy and depression. Metoprolol (Lopressor) acts selectively on the cells of the heart. It can be used by hypertensive patients who have asthma because it does not act on receptors in the bronchi.

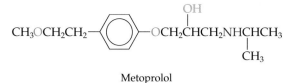

Metoprolol

Serotonin agonists are used to treat depression and anxiety. Agonist drugs are being used experimentally to treat obsessive–compulsive disorder. Serotonin antagonists are used to treat migraine headaches and to relieve the nausea caused by cancer chemotherapy.

Brain Amines and Diet: You Feel What You Eat

Richard Wurtman of the Massachusetts Institute of Technology has found a relationship between diet and serotonin levels in the brain. Serotonin is produced in the body from the amino acid tryptophan (Figure 19.21). Wurtman found that diets high in carbohydrates lead to high levels of serotonin. High levels of protein lower the serotonin concentration.

Norepinephrine also is synthesized in the body from an amino acid; it is derived from tyrosine. The synthesis is complex and proceeds through several intermediates (Figure 19.22). Because tyrosine is also a component of our diets, it may well be that our mental state depends to a fair degree on our diet.

Nearly one out of every ten people in the United States suffers from mental illness. Over half the patients in hospitals are there because of mental problems. When the biochemistry of the brain is more fully understood, mental illness may be cured (or at least alleviated) by the administration of drugs.

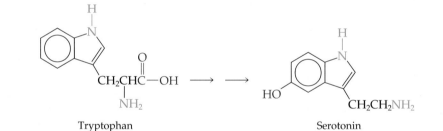

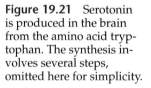

Tryptophan   Serotonin

Figure 19.21 Serotonin is produced in the brain from the amino acid tryptophan. The synthesis involves several steps, omitted here for simplicity.

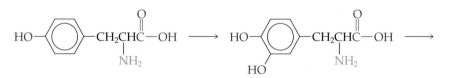

Tyrosine   Dopa

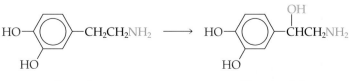

Dopamine   Norepinephrine

Figure 19.22 The biosynthesis of norepinephrine from tyrosine. L-Dopa, the left-handed form of the compound (see the box feature in Section 19.16), is used to treat Parkinson's disease, which is characterized by rigidity and stiffness of muscles. Parkinson's results from inadequate dopamine production, but dopamine cannot be used directly to treat the disease because it is not absorbed into the brain. Dopamine has been used to treat hypertension. It also is a neurotransmitter; schizophrenia has been attributed to an overabundance of dopamine in nerve cells.

β-Phenylethylamine

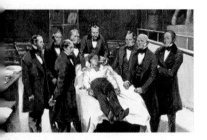

Phenylacetic acid

Love: A Chemical Connection

The notion that love might be chemical in origin is an unsettling one. However, it is possible that the emotions that trigger romantic relationships are governed in part by a chemical called β-phenylethylamine (PEA). PEA functions as a neurotransmitter in the human brain. It appears to create excited, alert feelings and moods. Increased levels of PEA produce a "high" feeling identical to the feeling that people describe as "being in love." Not surprisingly, the chemical structure of PEA resembles that of norepinephrine.

How much PEA does it take to get back that old feeling? Levels of PEA in the brain can be estimated by measuring levels of its metabolite, phenylacetic acid, in the urine. Low levels of urinary phenylacetic acid correlate with depression. This has prompted researchers to investigate factors that increase PEA levels in the brain. There are no food sources of PEA, but protein-rich foods contain phenylalanine (Chapter 16), an amino acid precursor of PEA. Perhaps a steak dinner is a way to your true love's heart after all.

Toxic Gases and the Learning Process

Gases such as carbon monoxide (CO), nitric oxide (NO), and hydrogen sulfide (H_2S) have long been known for their toxic effects. Imagine the surprise when scientists discovered that brain cells make NO and use it for communication.

A living cell constantly senses its environment. It changes in response to information flowing into and out of the cell. Most chemical messengers are complex substances such as norepinephrine and serotonin. But NO and possibly CO and H_2S are also chemical messengers. NO is formed in cells from arginine, a nitrogen-rich amino acid (Chapter 15), in a reaction catalyzed by an enzyme. NO kills invading microorganisms, probably by deactivating iron-containing enzymes in much the same way that CO destroys the oxygen-carrying capacity of hemoglobin (Chapter 12). NO also helps to regulate blood pressure and is involved in the formation of long-term memories.

CO is formed in cells by the enzyme-catalyzed oxidation of heme. Like NO, it seems to be involved in the long-term potentiation of learning. H_2S has been found in the brain cells of rats, but there is as yet no proof that the cells make it. There is evidence that H_2S stimulates certain receptors that strengthen the connections between brain cells, an indication of long-term learning. Perhaps learning can be a gas, after all.

Figure 19.23 This painting shows an operation in Boston in 1846 during which ether was used as an anesthetic.

19.12 Anesthetics

Anesthetics are the ultimate depressants. A **general anesthetic** acts on the brain to produce unconsciousness, along with a general insensitivity to feeling or pain. Diethyl ether was the first general anesthetic (Figure 19.23). It was introduced into surgical practice in 1846 by a Boston dentist, William Morton. Inhalation of ether vapor produces unconsciousness by depressing the activity of the central nervous system. Ether is relatively safe, since there is a fairly wide gap between the dose that produces an effective level of anesthesia and the lethal dose. Its disadvantages are high flammability and an undesirable side effect, nausea.

Nitrous oxide, or laughing gas (N_2O), was tried by Morton without success before he tried ether. Nitrous oxide was discovered by Joseph Priestley in 1772, and its narcotic effect was soon noted. Mixed with oxygen, nitrous oxide finds some use in modern anesthesia. It is quick-acting but not very potent. Concentrations of 50% or greater must be used for it to be effective. When nitrous oxide is mixed with ordinary air instead of oxygen, not enough oxygen gets into the patient's blood and permanent brain damage can result.

Chloroform ($CHCl_3$) was introduced as a general anesthetic in 1847. It quickly became popular after Queen Victoria gave birth to her eighth child while anesthetized by chloroform in 1853, and it was used widely for years. It is nonflammable and produces effective anesthesia, but it has serious drawbacks. For one, it has a narrow safety margin; the effective dose is close to the lethal dose. It also causes liver damage, and it must be protected from oxygen during storage to prevent the formation of deadly phosgene gas.

Modern Anesthesia

Modern anesthetics include fluorine-containing compounds such as halothane, enflurane, and methoxyflurane (Figure 19.24). These compounds are nonflammable and relatively safe for patients. Their safety—particularly that of halothane—for operating room personnel, however, has been questioned. For example, female operating room workers suffer a higher rate of miscarriages than women in the general population.

Modern surgical practice has moved away from the use of a single anesthetic. Generally, a patient is given an intravenous anesthetic such as thiopental (Section 19.13) to produce unconsciousness. A gaseous anesthetic is then administered to provide insensitivity to pain and to keep the patient unconscious. A relaxant such as curare may also be employed. Curare and related compounds produce profound relaxation; thus, only light anesthesia is required, avoiding the hazards of deep anesthesia.

The potency of an anesthetic is related to its solubility in fat. General anesthetics seem to work by dissolving in the fatlike membranes of nerve cells, changing their permeability and depressing the conductivity of the neurons.

Solvent Sniffing: Self-Administered Anesthesia

Nearly all gaseous and volatile liquid organic compounds exhibit anesthetic action. It is this anesthetic action of organic solvents that leads to their abuse. The sniffing of glue solvents, gasoline, aerosol propellants, and other inhalants is perhaps the deadliest form of drug abuse. The dose required for intoxication is not far from that which will stop the heart. And it is difficult to measure the dose inhaled from the plastic or paper bag normally used in glue sniffing. Also, as with nitrous oxide, sublethal doses can cause permanent brain damage by cutting down the oxygen supply to the brain.

Local Anesthetics

A **local anesthetic** renders one part of the body insensitive to pain, but the patient remains conscious. For dental work and minor surgery, it is usually desirable to deaden the pain in only one part of the body. The first local anesthetic to be used successfully was cocaine, a drug first isolated in 1860 from the leaves of the coca plant (Figure 19.25). Its structure was determined by Richard Willstätter in 1898. Even before Willsätter's work, there were attempts to develop synthetic compounds with similar properties. Cocaine is a powerful stimulant. Its abuse is discussed later in this chapter.

Nitrous oxide
(dinitrogen monoxide)

Chloroform
(trichloromethane)

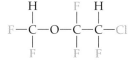

Halothane

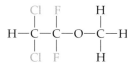

Enflurane

Methoxyflurane

Figure 19.24 Three modern general anesthetics. Note that enflurane and methoxyflurane are ethers.

Figure 19.25 Coca leaves.

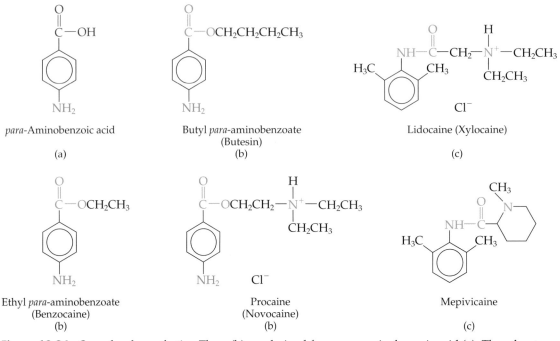

Figure 19.26 Some local anesthetics. Three (b) are derived from *para*-aminobenzoic acid (a). The other two (c) have amide functional groups. All are often used in the form of the hydrochloride salt, which is more soluble in water than the free base.

Certain esters of *para*-aminobenzoic acid (PABA) act as local anesthetics (Figure 19.26). The ethyl and butyl esters are used to relieve the pain of burns and open wounds. These esters are applied as ointments, usually in the form of picrate salts.

More powerful in their anesthetic effects are a series of derivatives with a second nitrogen atom in the alkyl group of the ester. Perhaps the best known of these is procaine (Novocaine), first synthesized in 1905 by Alfred Einhorn (1865–1917), who had worked with Willstätter on the structure of cocaine. Procaine can be injected as a local anesthetic or injected into the spinal column to deaden the entire lower portion of the body. Local anesthetics work by blocking nerve impulses to the brain. When the block involves the spinal cord, messages of pain from the lower parts of the body are prevented from reaching the brain.

The local anesthetic of choice nowadays is often lidocaine or mepivicaine. Both compounds are highly effective but have fairly low toxicity (Table 19.3). Lidocaine and mepivicaine are not derivatives of PABA, but they share some structural features with compounds that are.

Dissociative Anesthetics: Ketamine and PCP

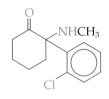

Ketamine

Phencyclidine
(PCP)

Ketamine, an intravenous anesthetic, is called a **dissociative anesthetic**: it induces hallucinations similar to those reported by people who have had near-death experiences. They seem to remember observing their rescuers from a vantage point above the scene, or moving through a dark tunnel toward a bright light. Unlike thiopental (Section 19.14), ketamine seems to affect associative pathways before it reaches the brainstem.

Little is known about the action of ketamine at the molecular level. If it acts by binding to receptors in the body, we can assume that our bodies also produce chemicals that fit these receptors. These compounds may be synthesized or released only in extreme circumstances—such as in near-death experiences.

Table 19.3 ▮ Toxicities of Various Drugs*

Drug	LD$_{50}$ (mg/kg)	Method of Administration	Experimental Animals
Local anesthetics			
Lidocaine	292	Oral	Mice
Procaine	45	Intravenous	Mice
Cocaine	17.5	Intravenous	Rats
Barbiturates			
Pentobarbital	118	Oral	Rats
Phenobarbital	660	Oral	Rats
Thiopental sodium	149	Intraperitoneal	Mice
Narcotics			
Morphine	500	Subcutaneous	Mice
Heroin (diacetylmorphine)	21.8	Intravenous	Mice
Meperidine	170	Oral	Rats
Stimulants			
Caffeine	355	Oral	Rats
Nicotine	230	Oral	Mice
	0.3	Intravenous	Mice
Amphetamine	180	Subcutaneous	Rats
Methamphetamine	70	Intraperitoneal	Mice
Mescaline	370	Intraperitoneal	Rats

*Comparisons of toxicities in different animals—and extrapolation to humans—are at best crude approximations. The method of administration can have a profound effect on the observed toxicity.

Source: Susan Budavari (Ed.). *The Merck Index*, 12th edition. Rahway, NJ: Merck and Co., 1996.

Closely related to ketamine is phencyclidine (PCP), known on the street as "angel dust." PCP is soluble in fat and has no appreciable water solubility. It is stored in fatty tissue and released when the fat is metabolized; this accounts for the "flashbacks" commonly experienced by users.

PCP is common on the illegal drug scene. It is cheap and easily prepared. It was tested and declared to be too dangerous for humans, but it found use as an animal tranquilizer. Many users experience bad "trips" with PCP, and about 1 in 1000 develops a severe form of schizophrenia. Laboratory tests show that PCP depresses the immune system, which could lead to an increased risk of infection. Despite these well-known problems, every few years a new group of young people appear on the scene ready to be victimized by this hog tranquilizer.

19.13 Depressants

Alcohol

Ethyl alcohol was probably the first synthetic chemical to be made by humans. We know that as early as 3700 B.C. the Egyptians fermented fruit to make wine and the Babylonians made beer from barley. People have been fermenting fruits and grains ever since, but relatively pure alcohol was not available until the invention of distillation during the Middle Ages.

Many people think that they are being stimulated when they take a drink of alcohol, but alcohol is actually a depressant. It slows down both physical and mental activity. However, there does seem to be a positive side to drinking alcohol. Longevity studies indicate that those who use alcohol only moderately (no more than a drink or two a day) live longer than nondrinkers. This is probably a result of the relaxing effect of the alcohol. Heavier drinking, however, can cause many health problems and shortens the life span.

We discussed ethyl alcohol in some detail in Section 9.8, but we discuss it again in this chapter on drugs because alcohol is by far the most used and abused drug in the world.

About 100 million people in the United States drink alcohol, and at least 10 million of them are alcoholics. Driving after drinking alcohol is a leading cause of accidental death in the United States, and for the under-35 age group, it is the leading cause of death. Alcoholism is the third leading health problem in the United States, right below cancer and heart disease. The suicide rate for alcoholics is 58 times greater than for other people.

Alcohol plays a major role in about half of all highway fatalities, and more than half of all arrests are related to alcohol abuse. It is also a significant factor in cases involving battered children. Alcohol is also a potent teratogen; it causes fetal alcohol syndrome (Section 9.8). A baby born to an alcoholic mother might have an abnormally small head and defects of the arms, legs, face, or other body parts. The greater the mother's use of alcohol, the more serious the birth defects of her child are apt to be. In the United States, 4000–5000 babies are born each year with birth defects due to alcohol.

[Web Reference 8] Information about the effects of alcohol, including its possible benefits for some people

Some Biochemistry of Ethanol

Alcoholic beverages are generally high in calories. Pure ethanol furnishes about 7 kcal/g. Ethanol is metabolized by oxidation to acetaldehyde at a rate of about 1 oz (28 mL)/hr, and a buildup of ethanol concentration in the blood caused by drinking more than this amount results in intoxication.

Oxidation of ethanol uses up a substance called NAD^+, the oxidized form of nicotinamide adenine dinucleotide, represented in the reduced form as NADH, where the H is a hydrogen atom. Because NAD^+ is usually employed in the oxidation of fats, when one drinks too much alcohol, the fat is deposited—mainly in the abdomen—rather than metabolized. This results in the familiar "beer gut" of many heavy drinkers.

Just how ethanol intoxicates is still somewhat of a mystery. Researchers have found that ethanol disrupts a receptor for *gamma*-aminobutyric acid (GABA), a substance the body uses to shut off nerve cell activity. Ethanol thus disrupts the brain's control over muscle activity, causing the drunk to stagger and fall. (The action of drugs on nerve cell receptors is discussed in Section 19.10.)

There are several theories about the cause of alcoholism. Studies on twins raised separately show that there is undoubtedly a genetic component. Biochemically, acetaldehyde is known to react with serotonin (Section 19.11) to produce a compound that acts on opiate receptors (Section 19.14) to produce an effect much like that of morphine or heroin. We still have much to learn about this ancient drug.

[Web Reference 9] A fact sheet from the March of Dimes about alcohol and pregnancy.

Barbiturates

As a family of related compounds, barbiturates display a wide variety of properties. They can be employed to produce mild sedation, deep sleep, or even death.

Barbituric acid was first synthesized in 1864 by Adolph von Baeyer. He made it from urea, which occurs in urine, and malonic acid, which occurs in apples. The term *barbiturates*, according to Willstätter, came about because at the time of the discovery von Baeyer was infatuated with a girl named Barbara. The word comes from "Barbara" and "urea."

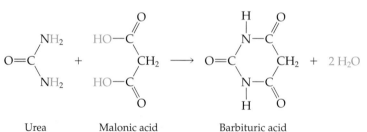

Urea   Malonic acid   Barbituric acid

Synergism: Barbiturates and Alcohol

Barbiturates are especially dangerous when ingested along with ethyl alcohol. This combination produces an effect much more drastic than the sum of the effects of two depressants. The effects of barbiturates are enhanced by a factor of up to 200 when they are taken with alcoholic beverages. This effect in which one drug enhances the action of another is called a **synergistic effect**. Synergistic effects can be deadly, and they are not limited to alcohol–barbiturate combinations. Two drugs should never be taken at the same time without competent medical supervision.

Barbiturates are extremely addictive. Habitual use leads to the development of a tolerance to the drugs, and ever-larger doses are required to produce the same degree of intoxication. Barbiturates are legally available by prescription only, but they are also part of the illegal drug scene. They are known as "downers" because of their depressant, sleep-inducing effects.

The side effects of barbiturates are similar to those of alcohol. Abuse leads to hangovers, drowsiness, dizziness, and headaches. Withdrawal symptoms are often severe, accompanied by convulsions and delirium. In fact some medical authorities now say that withdrawal from barbiturates is more likely to cause death than withdrawal from heroin.

Barbiturates are cyclic amides. Notice, however, that the barbiturate ring resembles that of thymine, one of the bases found in nucleic acids. Evidence indicates that barbiturates may act by substituting for thymine (or cytosine or uracil) in nucleic acids, thus interfering with protein synthesis.

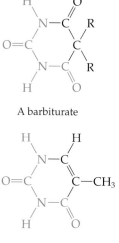

A barbiturate

Thymine

Barbiturates as Medicines

Several thousand barbiturates have been synthesized over the years, but only a few have found widespread use in medicine (Figure 19.27). Pentobarbital (Nembutal) is employed as a short-acting hypnotic drug. Before the discovery of modern tranquilizers, it was used widely to calm anxiety and other disorders of psychic origin.

Phenobarbital (Luminal) is a long-acting drug. It, too, is a hypnotic and can be used as a sedative. Phenobarbital is employed widely as an anticonvulsant for epileptics and brain-damaged people. Thiopental (Pentothal), a compound that differs from pentobarbital only in that an oxygen atom on the ring has been replaced by a sulfur atom, is used widely as an anesthetic.

Barbiturates were once used in small doses as sedatives, the dosage generally being a few milligrams. In larger dosages (about 100 mg), barbiturates induce sleep. They were once the sleeping pills of choice but now have been replaced by drugs such as triazolam (Section 19.20). Barbiturates are quite toxic, the lethal dose being about 1500 mg (1.5 g). There is also potential for accidental overdose. After ingesting a couple of tablets, a person becomes groggy, and if unable to remember whether he or she took the sleeping pills, may take more.

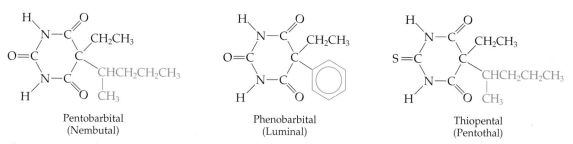

Pentobarbital
(Nembutal)

Phenobarbital
(Luminal)

Thiopental
(Pentothal)

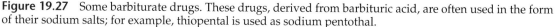

Figure 19.27 Some barbiturate drugs. These drugs, derived from barbituric acid, are often used in the form of their sodium salts; for example, thiopental is used as sodium pentothal.

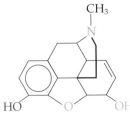

Morphine

19.14 Narcotics

Narcotics are drugs that produce narcosis (stupor or general anesthesia) and relief of pain (analgesia). Many drugs produce these effects, but in the United States only those that are also *addictive* are legally classified as narcotics. Their use is regulated by federal law.

Opium and Morphine

Opium is the dried, resinous juice of the unripe seeds of the oriental poppy (*Papaver somniferum*) (Figure 19.28). It is a complex mixture of about 20 nitrogen-containing organic bases (alkaloids), sugars, resins, waxes, and water. The principal alkaloid, morphine, makes up about 10% of the weight of raw opium. Opium was used in many patent medicines during the nineteenth century, such as Mrs. Winslow's Soothing Syrup (Figure 19.29).

Morphine was first isolated in 1805 by Friedrich Sertürner, a German pharmacist. With the invention of the hypodermic syringe (Figure 19.30) in the 1850s, a new method of administration became available. Injection of morphine directly into the

Figure 19.28 Opium poppy flower (a) and seed pod (b). The scratches on the seed pod exude the resinous juice that becomes opium when it dries.

(a)

(b)

Figure 19.29 Trade card advertising Mrs. Winslow's Soothing Syrup.

bloodstream was more effective for the relief of pain, but this method also seriously escalated the problem of addiction. Morphine was used widely during the American Civil War (1861–1865). It was effective for the relief of pain caused by battle wounds. One side effect of morphine use is constipation. Noting this, soldiers came to use morphine as a treatment for that other common malady of men on the battlefront— dysentery. During their wartime service, more than 100,000 soldiers became addicted to morphine. The affliction was so common among veterans that it came to be known as "soldier's disease."

Morphine and other narcotics were placed under control of the federal government by the Harrison Act of 1914. Morphine is still used by prescription for the relief of severe pain. It also induces lethargy, drowsiness, confusion, euphoria, chronic constipation, and depression of the respiratory system. Morphine is addictive, and this becomes a problem when it is administered in amounts greater than the prescribed dose or for a period longer than the prescribed time.

Figure 19.30 An early hypodermic syringe.

Codeine and Heroin

Slight changes in the molecular architecture of morphine produce altered physiological properties. Replacement of one of the —OH groups by an —OCH$_3$ group produces codeine. Actually, codeine is present in opium to an extent of about 0.5%. It is usually synthesized, however, by methylating the more abundant morphine molecules. About 55,000 kg of codeine is produced each year in the United States, enough for 16 doses of 15 mg for every person in the country.

Codeine is similar to morphine in its physiological action, except that it is less potent and has less tendency to induce sleep. It is also thought to be less addictive. In amounts of less than 2.2 mg/mL, codeine is exempt from the stringent narcotics regulations and is used in a few "controlled-substance" cough syrups.

In the laboratory, the reaction of morphine with acetic anhydride (acetic anhydride is derived from acetic acid by the removal of water) produces heroin. This morphine derivative was first prepared by chemists at the Baeyer Company of Germany in 1874. It received little attention until 1890, when it was proposed as an antidote for morphine addiction. Shortly thereafter, Baeyer widely advertised heroin as a sedative for coughs, often in the same ads describing aspirin (Figure 19.31). It soon was found, however, that heroin induced addiction more quickly than morphine and that heroin addiction was harder to cure.

The physiological action of heroin is similar to that of morphine. Heroin is less polar than morphine; it enters the fatty tissues of the brain more rapidly and seems to produce a stronger feeling of euphoria than morphine. Heroin is not legal in the United States, even by prescription. It has, however, been advocated for use in the relief of pain in terminal cancer patients and has been so used in Britain.

Figure 19.31 Heroin was regarded as a safe medicine in 1900; it was widely used as a sedative for coughs. It was also thought to be a nonaddictive substitute for morphine.

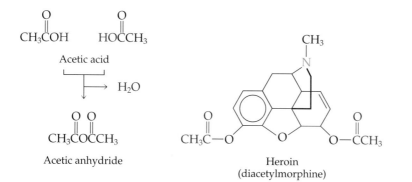

Addiction probably has three components: emotional dependence, physical dependence, and tolerance. Psychological dependence is evident in the uncontrollable desire for the drug. Physical dependence is shown by acute withdrawal symptoms such as convulsions. Tolerance for the drug is evidenced by the increasing dosages required to produce in the addict the same degree of narcosis and analgesia.

Deaths from heroin are usually attributed to overdoses, but the situation is not always clear. The problem seems to be a matter of quality control. Street drugs vary considerably in potency, and a particular dose can be much higher in heroin than expected.

Synthetic Narcotics

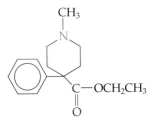

Meperidine
(Demerol)

Much research has gone into developing a drug that would be as effective as morphine for the relief of pain but would not be addictive. Perhaps the best known synthetic narcotic is meperidine (Demerol). Meperidine is somewhat less effective than morphine, but it has the advantage that it does not cause nausea. Repeated use, unfortunately, does lead to addiction.

Another synthetic narcotic is methadone, a drug widely used to treat heroin addiction. Like heroin, methadone is highly addictive. However, when taken orally, it does not induce the sleepy stupor characteristic of heroin intoxication. Unlike a heroin addict, a person on methadone maintenance is usually able to hold a productive job. Methadone is available free in clinics. If an addict who has been taking methadone reverts to heroin, the methadone in his or her system effectively blocks the euphoric rush normally given by heroin and so reduces the addict's temptation to use heroin.

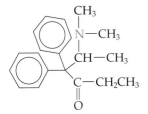

Methadone

When injected into the body, methadone produces an effect similar to that of heroin, and methadone has been diverted for illegal use in this manner. An addict on methadone is still an addict. All the problems of tolerance (and cross-tolerance with heroin and morphine) still exist.

More Morphine Analogs: Agonists and Antagonists

Chemists have synthesized thousands of morphine analogs, but only a few have shown significant analgesic activity, and most are addictive. Morphine acts by binding to receptors in the brain. Molecules that have morphine-like action are called **agonists**. Morphine **antagonists** are drugs that block the action of morphine by blocking the receptors. Some molecules have both agonist and antagonist effects. An example is pentazocine (Talwin), which is less addictive than morphine and is effective for the relief of pain.

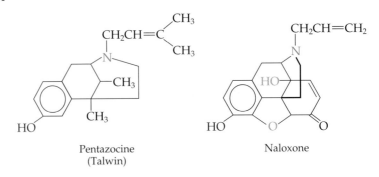

Pentazocine
(Talwin)

Naloxone

Pure antagonists such as naloxone are of value in treating opiate addicts. An addict who has overdosed can be brought back from death's door by an injection of naloxone.

Natural Opiates

Morphine acts by binding to specific receptor sites in the brain. These morphine receptors were first demonstrated in 1973 by Solomon Snyder and Candace Pert at Johns Hopkins University School of Medicine.

Why should the human brain have receptors for a plant-derived drug such as morphine? There seemed to be no good reason, and so several investigators started a search for morphine-like substances produced by the human body. Not one, but several such substances, called **endorphins** (endogenous morphines), were soon found. Each was a short peptide chain composed of amino acid units. Those with five amino acid units are called *enkephalins*. There are two enkephalins, and they differ only in the amino acid at the end of the chain. *Leu*-enkephalin has the sequence Tyr-Gly-Gly-Phe-Leu, and *Met*-enkephalin is Tyr-Gly-Gly-Phe-Met. Other substances with chains of 30 amino acids also were found.

Some enkephalins have been synthesized and shown to be potent pain relievers. Their use in medicine is quite limited, however, because after being injected they are rapidly broken down by the enzymes that hydrolyze proteins. Researchers have sought to make analogs more resistant to hydrolysis that can be employed as morphine substitutes for the relief of pain. Unfortunately, both natural enkephalins and their analogs, such as morphine, seem to be addictive.

It appears that endorphins are released as a response to pain deep in the body. Some evidence indicates that acupuncture anesthetizes somewhat by stimulating the release of brain "opiates." The long needles stimulate deep sensory nerves that cause the release of peptides that then block the pain signals.

Endorphin release has also been used to explain other phenomena once thought to be largely psychological. A soldier, wounded in battle, feels no pain until the skirmish is over. His body has secreted its own painkiller.

[Web Reference 10] Some excellent fact sheets on various substances, including information about their short- and long-term effects, withdrawal symptoms, and potential for abuse.

19.15 Antianxiety Agents

The hectic pace of life in the modern world causes some people to seek rest and relaxation in chemicals. Ethyl alcohol is undoubtedly the most widely used tranquilizer. The drink before dinner—to "unwind" from the tensions of the day—is very much a part of the American way of life. Many people, however, seek relief in other chemical forms.

One class of widely used antianxiety drugs are the benzodiazepines, compounds that feature seven-member heterocyclic rings. Of these, perhaps the best known is diazepam (Valium). Antianxiety agents make people feel better simply by making them feel dull and insensitive, but they do not solve any of the underlying problems that cause anxiety.

Certain benzodiazepines are also used to treat insomnia. Triazolam (Halcion) is a familiar example. It does help a person get to sleep, but the sleep is not restful and the drug does not really cure insomnia. Furthermore, Halcion and similar drugs have been banned by the British because of their dangerous side effects.

Like most other mind-altering drugs, benzodiazepines act by binding to specific receptors. Presumably our bodies produce compounds that also fit these receptors, although no such compound has yet been found. Rather, scientists have discovered compounds, called *beta*-carbolines, that act on the brain's anxiety receptors to produce terror. There is still so much to learn about the chemistry of the brain.

In any case, what price tranquility? After 20 years of use, benzodiazepines were found to be addictive. People trying to go off these drugs after prolonged use experience painful withdrawal.

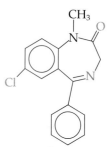

Diazepam
(Valium)

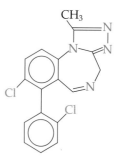

Triazolam
(Halcion)

Antianxiety agents are sometimes called *minor tranquilizers*.

Antipsychotic agents are often referred to as *major tranquilizers.*

Antipsychotic Agents

For centuries, the people of India used the snakeroot plant, *Rauwolfia serpentina*, to treat a variety of ailments including fever, snakebite, and other poisonings, and—most importantly—to treat maniacal forms of mental illness. Western scientists became interested in the plant near the middle of the twentieth century—after dismissing such remedies as quackery for generations.

In 1952 rauwolfia was introduced into American medical practice as an antihypertensive (blood-pressure-reducing) agent by Robert Wilkins of Massachusetts General Hospital. The same year, Emil Schlittler isolated an active alkaloid from rauwolfia in Switzerland and named it reserpine. Reserpine was found not only to reduce blood pressure but also to bring about sedation. The latter finding attracted the interest of psychiatrists, who found it so effective that by 1953 it had replaced electroshock therapy for 90% of psychotic patients.

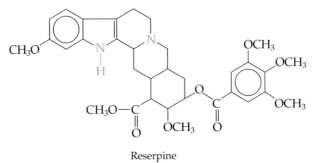

Reserpine

Phenothiazines

Also in 1952 chlorpromazine (Thorazine) was administered as a tranquilizer to psychotic patients in the United States. The drug had been tested in France as an antihistamine, and medical workers there had noted that it calmed mentally ill patients being treated for allergies. Chlorpromazine was found to be quite effective in controlling the symptoms of schizophrenia and truly revolutionized mental illness therapy.

Chlorpromazine is one of a group of related compounds called phenothiazines. Several of these compounds are used in medicine. Promazine (chlorpromazine without the chlorine atom) is also a tranquilizer, but it is much less potent than chlorpromazine.

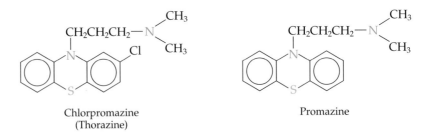

Chlorpromazine
(Thorazine)

Promazine

Phenothiazines are dopamine antagonists. Dopamine (Figure 19.22) is important in the control of detailed motion (such as grasping small objects), in memory and emotions, and in exciting the cells of the brain. Some researchers think schizophrenic patients produce too much dopamine, whereas others think that they have too many dopamine receptors. In either case, blocking the action of dopamine relieves the symptoms of schizophrenia.

Antipsychotic drugs have been one of the real triumphs of chemical research. They have served to reduce greatly the number of patients confined to mental hospitals by controlling the symptoms of schizophrenia to the extent that 95% of all schizophrenics no longer need hospitalization.

[**Web Reference 11**] An interesting, though lengthy, review of treatments for schizophrenia.

Antidepressants

It is interesting to note that slight changes in the structure of drugs can result in profound changes in their properties. Replacing the sulfur atom of promazine with a —CH_2CH_2— group produces imipramine (Tofranil), a compound that is not a tranquilizer but rather an antidepressant. Another common tricyclic (three-ring) antidepressant drug is amitriptylene (Elavil), in which the ring nitrogen atom also is replaced by a carbon atom. Although tricyclic antidepressants have been available since the 1950s, they have been only mildly successful. First, there is a narrow range in which the dose is both safe and effective. Low doses have little effect, and higher doses quickly become toxic. Second, there are undesirable side effects, such as nausea, headache, dizziness, loss of appetite, and grogginess, as well as more serious problems such as jaundice and high blood pressure.

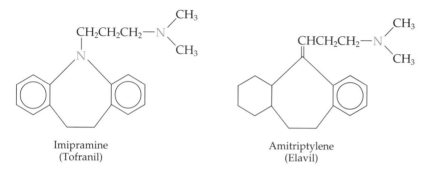

Imipramine
(Tofranil)

Amitriptylene
(Elavil)

Newer antidepressants are now available, including fluoxetine (Prozac). Doctors prescribe Prozac to help people cope with gambling problems, obesity, fear of public speaking, and premenstrual syndrome (PMS). Sometimes they prescribe it for healthy people just to help them loosen up a little or to develop a more cheerful disposition. This drug works by enhancing the effect of serotonin, blocking its reabsorption by cells. It seems to be safer than the older antidepressants and more easily tolerated.

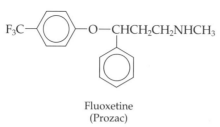

Fluoxetine
(Prozac)

19.16 Stimulant Drugs

Among the more widely known stimulant drugs are a variety of synthetic amines related to β-phenylethylamine (Figure 19.32). Note the similarity of these molecules to epinephrine and norepinephrine; all are derived from the basic phenylethylamine structure. Amphetamines probably act as stimulants by mimicking natural brain amines.

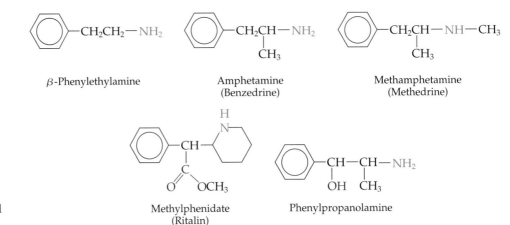

Figure 19.32
β-Phenylethylamine and related compounds.

Amphetamines

Amphetamine and methamphetamine have been widely abused. Amphetamine was once extensively used for weight reduction and it has also been employed in treating mild depression and narcolepsy, a rare form of sleeping sickness. Amphetamine induces excitability, restlessness, tremors, insomnia, dilated pupils, increased pulse rate and blood pressure, hallucinations, and psychoses. It is no longer recommended for weight reduction. It was generally found that any weight loss achieved was only temporary. The greatest problem, however, was the diversion of vast quantities of amphetamines into the illegal drug market. Amphetamines are inexpensive.

Methamphetamine has a more pronounced psychological effect than amphetamine. Generally, the "speed" that abusers inject into their veins is methamphetamine. Such injections are said to give the abuser a euphoric rush, at least initially. Shooting methamphetamine is quite dangerous, however, because the drug is rather toxic (Table 19.3).

Another amphetamine derivative, phenylpropanolamine, is widely used as an over-the-counter appetite suppressant. Like its relatives, this compound is a stimulant. Studies show that it is at best marginally effective as a diet aid, and it poses a threat to people with hypertension. Nevertheless, sales of phenylpropanolamine reach 1 billion tablets at $150 million each year.

One controversial use of amphetamines is their employment in the treatment of attention deficit disorder (ADD) in children. The drug of choice is often methylphenidate (Ritalin). Although it is a stimulant, the drug seems to calm children who otherwise can't sit still. This use has been criticized as "leading to drug abuse" and as "solving the teacher's problem, not the kid's."

Like other amine drugs, amphetamines are usually used in the form of hydrochloride salts. A freebase form of methamphetamine (like crack cocaine) is used for smoking. It is called "ice" because it is a clear, crystalline solid that looks like frozen water.

Right- and Left-Handed Molecules

Dextroamphetamine (Dexedrine) is another stimulant drug that has been abused widely. It is related to amphetamine in a very subtle way. Actually, amphetamine is a mixture of two isomers. These isomers have the same atoms and groups of atoms, but the relative spatial orientations of these atoms and groups differ in the two isomers. One isomer is the mirror image of the other.

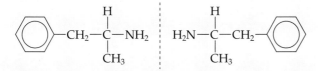

These isomers, called *enantiomers*, are not superimposable but are related to one another in much the same way that your right hand is related to your left. You can't fit a right-hand glove on your left hand, or vice versa. These mirror-image isomers fit enzymes differently; thus they have different effects. The **dextro isomer** (right-handed) is a stronger stimulant than the **levo isomer** (left-handed). Dexedrine is the trade name for the pure dextro isomer. Benzedrine is the trade name for a mixture of the two isomers in equal amounts. Dexedrine is two to four times as active as Benzedrine.

Another example of this phenomenon is Lipitor, a statin (lipid-lowering) drug widely used to lower blood cholesterol. Initial research in 1989 was promising but not outstanding. Later it was found that only the left-handed isomer is active and that it is much more effective when separated from its right-handed form.

Mirror-image isomerism is quite common in organic chemicals of biological importance. Any molecule that has a carbon atom with four different kinds of atoms or groups attached to it can exist as mirror-image isomers. More than 40% of current drugs have dextro and levo isomers, and most of them are sold as mixtures of the isomers. The development of drugs that contain only one enantiomer is becoming increasingly important in the pharmaceutical industry.

Cocaine

Cocaine, first used as a local anesthetic (Section 19.13), also serves as a powerful stimulant. The drug is obtained from the leaves of a shrub that grows almost exclusively on the eastern slopes of the Andes Mountains. Many of the Indians living in and around the area of cultivation chew coca leaves—mixed with lime and ashes—because of their stimulant effect. Cocaine used to arrive in the United States as the salt, cocaine hydrochloride, but now much of it comes in the form of broken lumps of the free base, a form called *crack cocaine*.

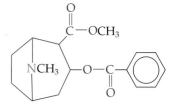

Cocaine

Cocaine hydrochloride is readily absorbed through the watery mucous membrane of the nose, and this is the form used by those who snort cocaine. Those who smoke cocaine use the free base (crack), which readily vaporizes at the temperature of a burning cigarette. When smoked, cocaine reaches the brain in 15 s. It acts by preventing the neurotransmitter dopamine from being reabsorbed after it is released by nerve cells. High levels of dopamine are therefore available to stimulate the pleasure centers of the brain. After the binge, dopamine is depleted in less than an hour, leaving the user in a pleasureless state and (often) craving more cocaine.

The use of cocaine increases stamina and reduces fatigue, but the effect is short-lived. Stimulation is followed by depression. Once quite expensive and limited to use mainly by the wealthy, cocaine is now available in cheap, potent forms. Hundreds, including several well-known athletes, have died from cocaine overdose.

Caffeine: Coffee, Tea, or Cola

Coffee and tea and some soft drinks contain the mild stimulant caffeine. An effective dose of caffeine is about 200 mg, corresponding to about two cups of strong coffee

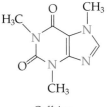

Caffeine

Each year in the United States a million kilograms of caffeine is added to food, mostly to soda pop.

or tea. Caffeine is also available in tablet form as a stay-awake or keep-alert type of drug. The best known brands are probably No-Doz and Vivarin. No-Doz contains about 100 mg of caffeine per tablet; each Vivarin tablet has 200 mg.

Is caffeine addictive? The "morning grouch" syndrome suggests that it is mildly so. There is also evidence that caffeine may be involved in chromosome damage. To be safe, people in their childbearing years should avoid large quantities of caffeine. Overall, the hazards of caffeine ingestion seem to be slight.

Nicotine: Going Up in Smoke

Nicotine

Another common stimulant is nicotine. This drug is taken by smoking or chewing tobacco. Nicotine is highly toxic to animals (Table 19.3) and has been used in agriculture as a contact insecticide. It is especially deadly when injected; the lethal dose for a human is estimated to be about 50 mg. Nicotine seems to have a rather transient effect as a stimulant. This initial response is followed by depression, but smokers generally keep a near-constant level of nicotine in their bloodstream by indulging frequently.

Is nicotine addictive? Casual observation of a person trying to quit smoking seems to indicate that it is. Consider the 1972 memorandum from a Philip Morris scientist who noted that "no one has ever become a cigarette smoker by smoking cigarettes without nicotine." He suggested that the company "think of the cigarette as a dispenser for a dose unit of nicotine."

Clonidine, a drug used to treat high blood pressure, reduces nicotine withdrawal symptoms.

It is interesting that teenagers who smoke are much more likely to use hard drugs than those who are nonsmokers. Among heavy smokers from 12 to 17 years old, four out of five are also drug users (Table 19.4).

Table 19.4 ▪ Use of Drugs Among Smokers and Nonsmokers Aged 12 to 17

Drug	Use Among People Who Never Smoked (%)	Use Among All Smokers (%)	Use Among Heavy Smokers* (%)
Heroin	0.1	1.2	5.1
Cocaine	0.3	15.4	31.7
Crack	0.1	5.7	11.1
Marijuana	2.5	57.5	68.2
Any illicit substance†	7.2	67.7	78.9

*Smoke more than one pack a day.

†Includes marijuana, cocaine, crack, heroin, PCP, hallucinogens, and, except for medical use, inhalants, sedatives, tranquilizers, stimulants, analgesics, and steroids.

Source: Center on Addiction and Substance Abuse at Columbia University. Data derived from the national household survey on drug abuse by the National Institute of Drug Abuse, 1991.

19.17 The Mindbenders: LSD

The third major class of drugs are popularly called mindbenders because they qualitatively change the way we perceive things. Probably the most powerful of these drugs is LSD. The physiological properties of this compound were discovered quite accidentally by Albert Hofmann in 1943. Hofmann, a chemist in Switzerland, unintentionally ingested some LSD. He later took 250 mg, which he considered a small dose, to verify that LSD had caused the symptoms he had experienced. Hofmann had a rough time for the next few hours, experiencing such symptoms as visual disturbances and schizophrenic behavior.

Lysergic acid diethylamide
(LSD)

Several medical drugs are obtained from the ergot fungus. Because ergotamine shrinks blood vessels in the brain, it is used to treat migraine headaches. Ergonovine induces uterine contractions and can reduce bleeding after childbirth. Like LSD, both these compounds are lysergic acid amides.

Lysergic acid is obtained from ergot, a fungus that grows on rye. It is converted to the diethylamide by treatment with thionyl chloride ($SOCl_2$) followed by diethylamine. Note that part of the LSD structure (color) resembles that of serotonin. LSD seems to act as a serotonin agonist.

LSD is a potent drug, as indicated by the small amount required for a person to experience its fantastic effects. The usual dose is probably about 10–100 μg. No wonder Hofmann had a bad time with 250 mg. To give you an idea of how small 10 μg is, let's compare that amount of LSD with the amount of aspirin in one tablet—one aspirin tablet contains 325,000 μg of aspirin.

19.18 Marijuana: Some Chemistry of Cannabis

Books have been written about marijuana, yet all we know for certain about this drug would fill only a few pages. *Cannabis sativa* (Figure 19.33) has long been useful. The stems yield tough fibers for making ropes. *Cannabis* has been used as a drug in tribal religious rituals and also has a long history as a medicine, particularly in India. In the United States, marijuana is second only to alcohol in popularity as an intoxicant.

The term **marijuana** refers to a preparation made by gathering the leaves, flowers, seeds, and small stems of the plant (Figure 19.34), which are generally dried and smoked. They contain various chemical substances, many of them still unidentified. The principal active ingredient, however, is tetrahydrocannabinol (THC). Although there are several active cannabinoids in marijuana, only one is shown here.

Marijuana plants vary considerably in THC potency, depending on their genetic variety. Wild plants native to the United States have a low THC content, usually about 0.1%, but some marijuana sold in North America now has a THC content approaching 6%.

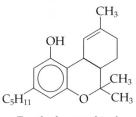

Tetrahydrocannabinol
(THC)

In 1996 voters in California and Arizona approved the medical use of marijuana and some other drugs. Still illegal under federal law, synthetic THC is legal by prescription for this use.

Figure 19.33 The marijuana plant.

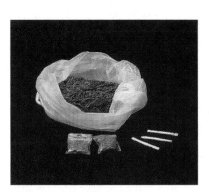

Figure 19.34 Retail forms of marijuana.

The gene for a THC receptor has been cloned. These receptors are found in movement control centers in the brain, explaining the loss of coordination seen in those intoxicated by the drug. Memory and cognition areas of the brain are also rich in THC receptors, which explains why marijuana users do poorly on tests. Few receptors are present in the brainstem where breathing and heartbeat are controlled, however, which is probably why it is hard to get a lethal dose of pure marijuana.

 [Web Reference 12] A discussion of the medical uses of marijuana.

If THC binds to receptors in the brain, the brain must produce a THC-like substance. This substance is thought to be anandamide, derived from arachidonic acid, which is the precursor of prostaglandins.

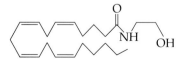

Anandamide

Effects of Marijuana

The effects of marijuana are difficult to measure, partly because of the variable amount of THC in different samples. A variety of standard potency is now grown for controlled clinical studies. With this product, some of the effects of marijuana can be measured in reproducible experiments. Smoking *Cannabis* increases the pulse rate, distorts the sense of time, and impairs some complex motor functions. Other possible effects include a euphoric floating sensation, a feeling of anxiety, a heightened enjoyment of food, and a false impression of brilliance. Although studies have shown no mind-expanding effects, users sometimes experience hallucinations.

The long-term effects of marijuana use are more difficult to evaluate. There is little objective evidence that smoking marijuana leads to the use of harder drugs. (More heroin addicts start drug use with alcohol than with marijuana.) However, there is some evidence that long-term marijuana use causes brain damage. People who use marijuana heavily are often lazy, passive, and mentally sluggish, but it is difficult to prove that marijuana is the cause. Even if it is, the damage is less extensive than that caused by heavy use of alcohol.

Some people claim that excessive use of marijuana leads to psychoses. It has long been known to induce short-term psychotic episodes in those already predisposed and in others who take excessive amounts. Long-term psychoses, however, occur at the same rate among regular marijuana users as among members of the general population.

An interesting report from Harvard Medical School concerns several young men being treated for gynecomastia (enlarged breasts). All were heavy marijuana users. Their breasts also discharged a white milky liquid. The painful swelling receded in three of the men after they stopped smoking *Cannabis*, but three others needed surgery. Their doctors are convinced that marijuana contains a "feminizing ingredient." There is a slight structural similarity between THC (this page) and female hormones (Figure 19.14). Some studies indicate that THC binds weakly to estrogen receptors.

Despite its apparent feminizing properties, THC in both high and low doses causes an initial rise in testosterone levels in men. With high doses, however, the slight increase is followed by a rapid fall to below-normal levels.

Chemists and Marijuana

What do chemists have to do with all this? Well, they have isolated the active components and synthesized them, and they can measure the THC content of marijuana. They can monitor THC in the bloodstream and identify the products of its breakdown. However, they cannot tell how it changes body chemistry or what its long-term effects are.

Perhaps the most significant harm from marijuana comes through its impairment of complex motor functions—such as those used in driving an automobile. Incidentally, chemists have developed a THC detector similar to the one used to test blood alcohol. Unlike alcohol, THC persists in the bloodstream for several days because it is fat-soluble. The products of its breakdown remain in the blood for as long as 8 days.

Marijuana does have some legitimate medical uses. It reduces eye pressure in people who have glaucoma. If not treated, this increasing pressure eventually causes blindness. Marijuana also relieves the nausea that afflicts cancer patients undergoing radiation treatment and chemotherapy.

19.19 Drug Problems

Our primary drug problem is the existence of illegal drugs. When drugs are illegal, their marketing is covert, with sellers failing to report their income or pay taxes on it. This makes the illicit drug business enormously lucrative. Annual revenues for drug dealers amount to at least $400 billion, about half of which comes from the United States. Worldwide, people spend more money for illegal drugs than for food.

So much unlawful wealth makes illegal drugs a dangerous business in which crimes are frequent, murders are common, and corrupt politicians abound. **Drug abuse** (using drugs for their intoxicating effects) is an extremely serious problem, not only for the abusers but for society in general.

The illegal drug user is the biggest loser. Street drugs are expensive, and most are addictive. Some addicted users have to steal in order to pay for drugs and eventually end up in jail. In the workplace 50–80% of all accidents and personal injuries are drug-related. Drug users are absent from work 2.5 times as often as nonusers, and they are 5 times as likely to file claims for compensation. They are a drain on their employers' income, and they often have trouble keeping a job.

One recurring problem is that illegal drugs are not always what they are supposed to be. Buyers simply have to trust the information sellers give them about the identity and quality of the products they buy. Even with marijuana, a readily identified weed, there are problems of quality control. A nonpotent variety is often used to dilute potent material. On occasion a nonpotent variety is laced with other drugs—some highly toxic. In fact, crime labs have generally found nearly two-thirds of all drugs (other than marijuana) brought in for analysis to be something other than what the dealers said they were.

Of course, there can be difficulties even with legal drugs that have been approved and tested. Problems range from faulty prescriptions written by physicians, to pharmacy errors in filling prescriptions, to patient mistakes in taking medications. Some patients have allergies to particular drugs, which they forget to mention. Many drugs have undesirable side effects that are worse than the condition being treated. There are drugs (both prescription and over-the-counter) that have such similar names that they are easily confused. **Drug misuse** (for example, using penicillin, which has no effect on viruses, to treat a viral infection) is all too common. (Such overuse of penicillin has led to strains of bacteria that are now immune to penicillin.) Another problem is that the cost of many drugs makes them unaffordable for the people who need them. Pharmaceutical research and development is expensive, and companies must often charge high prices for the products they develop, at least for a while. After the patent rights expire, other companies can make "generic" versions of the same products at lower prices because they have incurred no research costs. Generic drugs are usually chemically identical to the original products, but there have been cases in which fraudulent generic drugs have caused much distress.

Chemistry has provided many drugs of enormous benefit to society, but sometimes these drugs can create problems.

19.20 The Placebo Effect

An interesting phenomenon associated with the testing of drugs is the placebo effect. A **placebo** is an inactive substance given in the form of medication to a patient who thinks it is the real thing.

It is common practice when evaluating a new drug to administer it to one group of patients and to give placebos to a similar "control" group. The patients have no idea whether they are receiving the real drug or the placebo and generally believe that they are really taking the new drug. Doctors treating the patients usually do not know who is receiving the drug and who is getting the placebo until the study has been completed.

Sometimes people who think they are receiving a certain drug expect positive results and actually experience such results, even though they have not been given the actual drug. In one study of a new tranquilizer, 40% of the patients who had been given placebos reported that they felt much better and rated the drug as highly effective.

The placebo effect is good evidence of the fact that there is a strong connection between the mind and the body. Medical treatment of the human body should not ignore the enormous potential of the human mind.

Critical Thinking Exercises

Apply knowledge that you have gained in this chapter and one or more of the FLaReS principles (Chapter 1) to evaluate the following statements or claims.

19.1 An actor in a television advertisement states, "For aches and pain I take Tylenol, but for headaches I take Excedrin. Excedrin works better for headaches."

19.2 An advertisement states that "Mr. Sinibaldi [had] high blood pressure [that persisted through] 15 years and five doctors … in spite of taking up to four different drugs at once. …

Using a combination of diet and garum amoricum … brought Mr. Sinibaldi's blood pressure down … in just 14 days! … [Garum amoricum] could also rescue you … from high blood pressure … headaches … constipation … [and 20 other maladies]."

19.3 A nurse claims to be able to cure disease by "therapeutic touch," a technique which she says allows her to sense "human energy fields" by simply moving her hands above the patient's body.

Summary

1. Drugs are substances used to relieve pain, to treat an illness, or to improve one's state of health or well-being.

2. Analgesics are pain relievers, aspirin being the foremost example.

3. Drugs used to treat colds include antihistamines, cough suppressants (antitussives), and nasal decongestants.

4. The first antibacterial drug was sulfanilamide.

5. Antibiotics (derived from mold or bacteria) include penicillins, cephalosporins, and tetracyclines.

6. Viral diseases, such as measles, mumps, and smallpox, are usually prevented by vaccination, but vaccines are not available for all viral diseases.

7. Herpes and AIDS are both viral infections that have no cure, but several antiviral drugs have been developed to treat them.

8. Anticancer drugs are mainly antimetabolites, which interfere with DNA synthesis, thereby slowing down the growth of cancers.

9. "Birth control" pills are mixtures of an estrogen and a progestin, which act by creating a state of pseudopregnancy during which a woman does not ovulate and therefore cannot conceive.

10. Psychotropic drugs are drugs that affect the mind. They can be divided into three classes: depressants, stimulants, and hallucinogenics.

11. Anesthetics are depressants used during surgery to produce unconsciousness or local insensitivity to pain.

12. Narcotics are drugs that relieve pain and can produce anesthesia, but they are also addictive. Morphine is an example.

13. Natural painkillers called endorphins are produced within the body.

14. Tranquilizers are depressants that many people use to relax.

15. Stimulant drugs include amphetamines.

16. Many drugs exist in mirror image (*dextro* and *levo*) forms. Usually one form is more active than the other, and sometimes the other form is actually harmful.

17. Caffeine in coffee, tea, and soft drinks and nicotine in tobacco are common legal stimulants.

18. Although cocaine was first used as an anesthetic, it is also a powerful stimulant.

19. The best known among the psychedelic or mindbending drugs are LSD and marijuana.

20. Not only do street drugs have a negative effect on one's health, but they are also likely not to be what they are supposed to be. Quality control does not exist in the illegal drug market.

21. The placebo effect occurs when people experience the effects of having taken a drug when they believe that they have taken it, even though they actually have not.

Key Terms

Review Questions

1. What is an analgesic?
2. What is an antipyretic?
3. What is the chemical name for aspirin?
4. How does aspirin suppress inflammation?
5. In what ways do aspirin and acetaminophen act similarly?
6. Why is aspirin chosen over acetaminophen for the treatment of arthritis?
7. Why is acetaminophen chosen over aspirin for the relief of pain associated with surgical procedures?
8. What is ibuprofen?
9. What is a placebo?
10. What are sulfa drugs? How do they work?
11. What are antibiotics?
12. How can a drug such as penicillin kill bacterial cells without killing human cells?
13. Tetracyclines are broad-spectrum antibiotics. What does this mean?
14. What is a hormone?
15. What is cortisone?
16. What is an androgen?
17. What is an estrogen?
18. How do birth control pills work?
19. What is DES? What problems have been associated with its use?
20. How does mifepristone (RU-486) work?
21. What are prostaglandins?
22. List two major classes of anticancer drugs.
23. What are psychotropic drugs?
24. What is the difference between drug misuse and drug abuse? Give an example of each.
25. List the three classes of drugs that affect the mind.
26. What are the effects of a stimulant drug?
27. What are the effects of a depressant drug?
28. What are the effects of a hallucinogenic drug?
29. What is a general anesthetic?

30. What is a local anesthetic?
31. Which of these anesthetics are dangerous because of flammability?
    a. diethyl ether   b. halothane
    c. chloroform
32. For each of the following anesthetics, describe a disadvantage associated with its use. Do not include flammability.
    a. nitrous oxide   b. halothane
    c. diethyl ether
33. What application do curare and related substances have in modern anesthesiology?
34. Describe two medical uses of barbiturates and the relative dosage level of each.
35. What are the hazards of long-term barbiturate use?
36. Describe a synergistic reaction involving drugs.
37. Name two dissociative anesthetics.
38. What are narcotics?
39. What is opium?
40. How does codeine differ from morphine in its physiological effects? In what ways are the two drugs similar?
41. How does methadone maintenance work?
42. How are endorphins related to each of the following?
    a. the anesthetic effect of acupuncture
    b. the absence of pain in a badly wounded soldier
43. What is an agonist?
44. What is an antagonist?
45. Define or identify each of the following.
    a. neuron   b. synapse
    c. neurotransmitter
46. How may our mental state be related to our diet?
47. Are tranquilizers a cure for schizophrenia?
48. What are some of the problems involved in the clinical evaluation of LSD?
49. What are some of the problems involved in the clinical evaluation of marijuana?

Problems

Chemical Structures and Properties

50. What two functional groups are present in an acetaminophen molecule?
51. What two functional groups are present in an aspirin (acetylsalicylic acid) molecule?
52. Terfenadine (Seldane) has the chemical formula $C_{32}H_{41}NO_2$. Is the compound water-soluble or fat-soluble? Explain your answer.
53. Ephedrine hydrochloride has the chemical formula $C_{10}H_{16}NO^+Cl^-$. Is the compound water-soluble or fat-soluble? Explain your answer.
54. From what carboxylic acid are many local anesthetics derived?

55. What structural feature characterizes the more powerful local anesthetics?
56. What is the basic structure common to all barbiturate molecules? How is this structure modified to change the properties of individual barbiturate drugs?

Drug Toxicities

57. When administered intravenously to rats, procaine and cocaine have LD_{50} values of 50 mg/kg and 17.5 mg/kg, respectively. Which is more toxic? Explain.
58. If the minimum lethal dose (MLD) of amphetamine is 5 mg/kg, what the MLD for a 70-kg person? Can toxicity studies on animals always be extrapolated to humans?

Additional Problems

59. What structural feature makes a birth control steroid effective orally?

60. Which of the following compounds are classified as steroids?
  a. tristearin   **b.** cholesterol
  c. testosterone   **d.** prostaglandins

61. Give the structure for *para*-aminobenzoic acid.

62. Give the structure for sulfanilamide.

63. What structural feature is shared by all tetracyclines?

64. What structural feature is shared by all steroids?

65. When administered intravenously to rats, the LD_{50} of procaine is 50 mg/kg body weight, and that of nicotine, 1.0 mg/kg body weight. Which drug is more toxic? What is the approximate lethal dose of each for a 50-kg human?

Projects

66. Following are the generic names of several of the most widely prescribed drugs. Use *The Merck Index* or a similar reference to determine (if possible) the chemical structure, medical use, toxicity, and side effects of each.
  a. furosemide   **b.** methyldopa
  c. hydrochlorothiazide **d.** digoxin
  e. triazolam   **f.** amoxicillin
  g. metoclopramide

67. Examine the labels of three over-the-counter sleeping pills (such as Nytol, Sominex, and Sleep-eze). Make a list of the ingredients in each. Look up the properties (medical uses, dosages, side effects, toxicities) of each in a reference work such as *The Merck Index*.

68. Do a cost analysis of five brands of plain aspirin. Calculate the cost per gram of each.

69. Compare the cost per gram of an extra-strength aspirin formulation with that of plain aspirin.

70. Discuss some of the problems involved in proving a drug safe or proving it harmful. Which is easier? Why?

Online Projects

71. Do an internet search to learn more about drug trials so that you can answer the following questions.
  a. Why might a person who is gravely ill not want to participate in a placebo-controlled drug study?
  b. What are the advantages and disadvantages of participating in such a drug study?
  c. Why is it important that studies use placebos as well as the test drug?
  d. Sometimes the control group receives the standard treatment for the disease instead of a placebo. Why does this happen?

72. Search the web for information on any new drugs for treating one of the following diseases.
  a. pneumonia   **b.** arthritis
  c. cancer   **d.** diabetes

References and Readings

1. Borman, Stu. "Enzyme Structures Could Improve Pain Relievers." *Chemical & Engineering News*, 18 November 1996, p. 32. Anti-inflammatory drugs such as aspirin act by blocking enzymes.

2. Christiansen, Damaris. "Designer Estrogens: Getting All the Benefits, Few of the Risks." *Science News*, 16 October 1999, pp. 252–254.

3. Drews, Jurgen. *In Quest of Tomorrow's Medicines*. (David Kramer, trans.) New York: Springer-Verlag, 1999.

4. Ember, Lois R. "The Nicotine Connection." *Chemical & Engineering News*, 28 November 1994, pp. 8–18.

5. Hales, Dianna, and Robert E. Hales. "The Brain's Power to Heal." *Parade Magazine*, 21 November 1999, pp. 10–13.

6. Hellman, Samuel, and Everett E. Vokes. "Advancing Current Treatments for Cancer." *Scientific American*, September 1996, pp. 118–123. Part of a special issue on cancer.

7. "Herbal Rx: The Promise and Pitfalls." *Consumer Reports*, March 1999, pp. 44–48. Gives a list of popular herbal supplements along with their purported benefits, evidence of efficacy, and possible side effects.

8. Jackson, Ian. "Pharmaceutical Revolution: The Pill Generation." *Today's Chemist*, October 1990, pp. 17–20.

9. Julien, Robert M. *A Primer of Drug Action*, 5th edition. San Francisco: W. H. Freeman, 1989.

10. Landau, Ralph. *Pharmaceutical Innovation*. Philadelphia: Chemical Heritage Foundation, 1999.

11. Leland, John. "Herbal Warning." *Newsweek*, 6 May 1996, pp. 60–67. Many herbs are toxic.

12. Lemonick, Michael D. "Glimpses of the Mind." *Time*, 17 July 1995, pp. 44–52.

13. Lipman, Marvin M., Michael Leff, and the editors of *Consumer Reports. The Best of Health*. New York: Consumer Reports, 1998. Answers to questions that people might want to ask their doctor.

14. Liska, Ken. *Drugs and the Human Body: With Implications for Society*, 6th edition. Upper Saddle River, NJ: Prentice Hall, 2000.

15. *The Merck Manual*, 17th edition, Rahway, NJ: Merck & Co., 1999. Professional medical information regarding a large number of diseases and syndromes.

16. Modeland, Vern. "Modern Anesthesia: Going Under Safely." *FDA Consumer*, December 1989–January 1990, pp. 13–17.

17. Nordenberg, Tamar. "Now Available Without a Prescription." *FDA Consumer*, November 1996, pp. 6–11. Several drugs, including pain relievers and nicotine patches, long available only by prescription, can now be purchased over the counter.

18. *Physicians' Desk Reference*, 54th edition. Oradell, NJ: Medical Economics Company, 2000.

19. Podolsky, Doug. "Birth Control." *US. News & World Report*, 24 December 1990, pp. 58–65.

20. Prendergast, Alan. "Beyond the Pill." *American Health*, October 1990, pp. 37–44. Discusses contraceptives for males.

21. "Prescription and Over-the-Counter Drugs." Pleasantville, NY: *Reader's Digest*, 1998. Information on usage, dosage, precautions, interactions, and side effects of a large number of drugs.

22. Rubin, Rita. "The War on Cancer." *U.S. News & World Report*, 5 February 1996, pp. 54–77.

23. Stachulski, Andrew V., and Martin S. Lennard. "Drug Metabolism: The Body's Defense Against Chemical Attack." *Journal of Chemical Education*, March 2000, pp. 349–353.

24. Stinson, Stephen C. "Drug Firms Restock Antibacterial Arsenal." *Chemical & Engineering News*, 23 September 1996, pp. 75–100. Bacteria are resistant to many present-day drugs, spurring pharmaceutical firms to seek new types.

25. Weissmann, Gerald. "Aspirin." *Scientific American*, January 1991, pp. 84–90. Its action is even more complicated than we thought.

Poisons
Chemical Toxicology

Sometimes one drop is all it takes to make one comatose.
The greater the toxicity, the less the lethal dose.

The globe fish (*Spheroides rubripes*), a variety of puffer fish, is considered a great delicacy in Japan. The ovaries and liver of this and related species produce a poison called tetrodotoxin. This toxin has an LD_{50} (Section 20.7) of 10 μg/kg body weight (intraperitoneally in mice), making it one of the most toxic substances known. Hundreds of Japanese have died from eating improperly prepared puffer fish.

W hen the Greek philosopher Socrates was accused of corrupting the youth of Athens in 399 B.C.E, he was given the choice of exile or death. He chose death and implemented his decision by drinking a cup of hemlock.

Poisons from plant and animal sources were well known in the ancient world. Both snake and insect venoms and plant alkaloids were used. Today, many aboriginal tribes still employ a variety of poisons in hunting and warfare. Curare, used by certain South American tribes, is one of the more notorious examples.

Although poisonous substances have been known and used for centuries, it is only within the last 150 years that scientists have learned the nature of the chemicals that are the active components of many of them. The hemlock taken by Socrates was probably prepared from the fully-grown but unripe fruit of *Conium maculatum* (poison hemlock). Usually, the fruit is dried carefully and then brewed into a "tea." Hemlock contains several alkaloids, but the principal one is coniine. This drug causes nausea, weakness, paralysis, and—as in the case of Socrates—death.

Poisons have always been with us, but our knowledge of them is greater than ever before. Industrial accidents, such as that at Bhopal, India, in 1984 have made the public acutely aware of problems caused by toxic substances. At Bhopal, the accidental release of methyl isocyanate ($CH_3N{=}C{=}O$), an intermediate in the synthesis of carbamate insecticides (Chapter 16), killed more than 2000 people and injured

Jacques Louis David's painting *The Death of Socrates* (1787) shows Socrates about to drink a cup of hemlock to carry out the death sentence decreed by the rulers of Athens.

Conium maculatum
(poison hemlock)

Coniine

Poison hemlock (*Conium maculatum*), the plant from which the hemlock taken by Socrates was probably prepared. The principal alkaloid in poison hemlock is coniine.

601

countless others. People are also concerned about long-term exposure to toxic substances in the air, in their drinking water, and in their food. Chemists can detect exceedingly tiny quantities of such substances; however, it is still quite difficult to determine the effects of these trace amounts of toxic materials on human health.

In this chapter, we consider substances that are poisonous or otherwise injurious. **Toxicology** is the branch of pharmacology (Chapter 19) that deals with the effects of poisons, their identification or detection, and the development and use of antidotes.

20.1 All Things Are Poisons

What is a poison? Perhaps a better question would be; How much is a poison? A substance may be harmless—or even a necessary nutrient—in one amount, and injurious—or even deadly—in another. Even common substances such as salt and sugar can be poisonous when eaten in abnormally large amounts. Too much sugar—candy or sweets—can give a child a stomachache. Too much salt—sodium chloride—can induce vomiting. There have even been cases of fatal poisoning when salt was accidentally substituted for lactose (milk sugar) in formulas for infants. Some substances are obviously more toxic than others, however. It would take a massive dose of salt to kill the average healthy adult, whereas only a few micrograms of certain nerve poisons can be fatal. Toxicity depends on the chemical nature of the substance.

Also, individuals may respond differently to the same chemical. To cite an extreme case, a few grams of sugar would cause no acute symptoms in a normal person but might be dangerous to a diabetic. Excessive amounts of salt could be serious for a person with edema (swelling due to excessive amounts of fluid in the tissues). Many drugs, including penicillin, can cause allergic reactions in a few individuals.

Still another complicating factor is that chemicals behave differently when administered in different ways. Recall (Table 19.3.) that nicotine is more than 50 times as toxic when given intravenously as when taken orally. Good, fresh water is delightful when taken orally, but water can be deadly when inhaled in sufficient quantity. Even closely related animal species can react differently to a given chemical; this complicates toxicity testing.

Recall from Section 16.5 that vitamin A is essential to health but that it can build to toxic levels in the body's fatty tissues.

Poisons Around the House

In Chapter 17, we noted that many household chemicals are poisonous. Drain cleaners, oven cleaners, and toilet bowl cleaners are highly corrosive, and some insecticides and rodenticides are quite toxic. Laundry bleach and ammonia are both toxic, and when they are mixed together they produce deadly gases.

Even seemingly harmless products around the house can be dangerous if a young child happens to drink or eat them. A bottle of cough syrup can trigger an emergency visit to the hospital. The number of deaths of children who ate their parents' aspirin was the impetus for the required use of "childproof" containers for hazardous substances.

Poisons in the Garden

Some of the toxic products used on farms (Chapter 16) are also used in home gardens. But herbicides and insecticides are not the only poisons found in a garden. Sometimes the plants themselves are toxic.

Poisons sometimes come in pretty packages. Iris are beautiful, and so are azaleas and hydrangeas, but all these popular perennials are poisonous. Holly berries, wisteria seeds, and the leaves and berries of privet hedges are also among the more poisonous products of the home garden.

Even the healthy-looking green plants that beautify the inside of a house can be toxic. One poisonous houseplant, philodendron, is probably the most widely cultivated of all indoor plants.

In our discussion here, we cover only a few of the many toxic substances. We organize them into groups with similar effects, giving priority to those that are more likely to be encountered in everyday life.

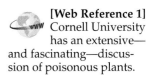

[Web Reference 1]
Cornell University has an extensive—and fascinating—discussion of poisonous plants.

20.2 Corrosive Poisons: A Closer Look

In Chapter 7, we mentioned the corrosive effects of strong acids and strong bases on human tissue. These chemicals indiscriminately destroy living cells. Corrosive chemicals in lesser concentrations exert more subtle effects.

Strong Acids and Bases

Both acids and bases, even in dilute solutions, catalyze the hydrolysis of protein molecules in living cells. These reactions involve breaking of the amide (peptide) linkages in these molecules.

Protein molecule $+$ H_2O $\xrightarrow{H^+ \text{ or } OH^-}$

Fragments

Generally, the fragments are not able to carry out the functions of the original protein. In cases of severe exposure, fragmentation continues until the tissue is completely destroyed.

Acids in the lungs are particularly destructive. In Chapter 12, we saw how sulfuric acid is formed when sulfur-containing coal is burned. Acids are also formed when plastics and other wastes are burned. These acid pollutants cause the breakdown of lung tissue (Figure 20.1).

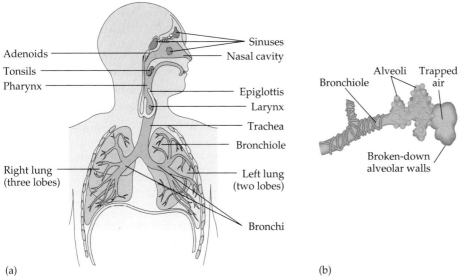

(a)

(b)

Figure 20.1 The respiratory system (a). Air flows through the pharynx (throat), larynx (voice box), and trachea (windpipe), into the bronchi and bronchioles (bronchial tubes), and finally into the alveoli (air sacs). In pulmonary emphysema (b), the alveolar walls deteriorate and lose elasticity. They can no longer expel the old, carbon dioxide-rich air and therefore cannot take in fresh, oxygen-rich air.

Oxidizing Agents

Other air pollutants also damage living cells. Ozone, peroxyacetyl nitrate (PAN), and the other oxidizing components of photochemical smog probably do their main damage through the deactivation of enzymes. The active sites of enzymes often incorporate the sulfur-containing amino acids cysteine and methionine. Cysteine is readily oxidized by ozone to cysteic acid.

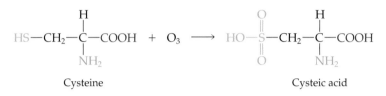

Cysteine   Cysteic acid

Methionine is oxidized to methionine sulfoxide.

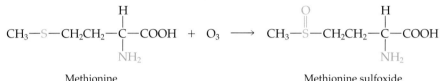

Methionine   Methionine sulfoxide

Still another amino acid, tryptophan, is known to react with ozone. Tryptophan, which does not contain sulfur, undergoes a ring-opening oxidation at the double bond.

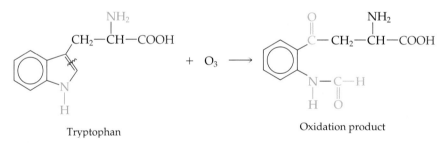

Tryptophan   Oxidation product

No doubt oxidizing agents can break bonds in many other chemical substances in a cell. Such powerful agents as ozone are more likely to make an indiscriminate attack than to react in a highly specific way.

20.3 Poisons Affecting Oxygen Transport and Oxidative Processes

Certain chemical substances prevent cellular oxidation of metabolites by blocking the transport of oxygen in the bloodstream or by interfering with oxidative processes in the cells. These chemicals act on the iron atoms in complex protein molecules.

Blood Agents

Probably the best known of these metabolic poisons is carbon monoxide. Recall that this gas binds tightly to the iron atom in hemoglobin, blocking the transport of oxygen (Chapter 12).

Nitrates, which occur in dangerous amounts in the groundwater in some agricultural areas (Chapter 13), also diminish the ability of hemoglobin to carry oxygen. Microorganisms in the digestive tract reduce nitrates to nitrites. We can write the process as a reduction half-reaction (Chapter 8).

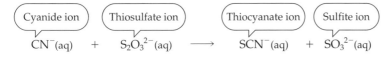

$$2\,H^+(aq) \quad + \quad NO_3^-(aq) \quad + \quad 2e^- \quad \longrightarrow \quad NO_2^-(aq) \quad + \quad H_2O$$

The nitrite ions oxidize the iron atoms in hemoglobin from Fe^{2+} to Fe^{3+}. The resulting compound, called *methemoglobin*, is incapable of carrying oxygen. The consequent oxygen deficiency disease is called *methemoglobinemia*. In infants, this disease is known as the blue-baby syndrome.

Cyanides: Agents of Death

Cyanides are among the most notorious poisons in both fact and fiction. They act quickly, and it takes only a tiny amount to kill. The average fatal dose is only 50 or 60 mg of gaseous hydrogen cyanide ($H—C\equiv N$) or of a solid salt containing the cyanide ion ($C\equiv N^-$). Specially trained experts use HCN gas to exterminate insects and rodents in the holds of ships, in warehouses, in railway cars, and on citrus and other fruit trees. Sodium cyanide (NaCN) is employed to extract gold and silver from ores. It is also used in electroplating baths. Hydrogen cyanide is generated from the sodium salt by treatment with an acid.

$$NaCN(s) + H_2SO_4(aq) \longrightarrow HCN(g) + NaHSO_4(aq)$$

Unlike carbon monoxide, cyanide does not react with hemoglobin. Instead, it blocks the oxidation of glucose inside the cell by forming a stable complex with iron(III) ions in oxidative enzymes called *cytochrome oxidases*. These enzymes normally act by providing electrons for the reduction of oxygen in the cell. Cyanide blocks this action and brings an abrupt end to cellular respiration, causing death in minutes.

Any antidote for cyanide poisoning must be administered quickly. Providing 100% oxygen to support respiration can sometimes help. Sodium nitrite is often given intravenously to oxidize iron atoms in the enzymes back to the active Fe^{3+} form. Sodium thiosulfate ("hypo" used in developing photographic film) is then used if time permits. The thiosulfate ion transfers a sulfur atom to the cyanide ion, converting it to the relatively innocuous thiocyanate ion.

Cyanide ion   Thiosulfate ion   Thiocyanate ion   Sulfite ion

$$CN^-(aq) \quad + \quad S_2O_3^{2-}(aq) \quad \longrightarrow \quad SCN^-(aq) \quad + \quad SO_3^{2-}(aq)$$

Unfortunately, few victims of cyanide poisoning survive long enough to be treated.

20.4 Make Your Own Poison: Fluoroacetic Acid

Although the body generally acts to detoxify poisons, there are notable exceptions in which it converts an essentially harmless chemical to a deadly poison. Fluoroacetic acid (FCH_2COOH) is one such compound.

Body cells use acetic acid to produce citric acid, which is then broken down in a series of steps, most of which release energy. When fluoroacetic acid is ingested, it is incorporated into fluorocitric acid. The latter effectively blocks the citric acid cycle by tying up the enzyme that acts on citric acid. Thus, the energy-producing mechanism of the cell is shut off and death comes quickly.

Sodium fluoroacetate ($FCH_2COO^-Na^+$), also known as Compound 1080, is used to poison rats and predatory animals. It is not selective, and thus it is dangerous to humans, pets, and other animals. Ranchers use sodium fluoroacetate to poison coyotes, eagles, and other animals suspected of preying on sheep and cattle. Such use

Hemoglobin is bright red and is responsible for the red color of blood. Methemoglobin is brown. During cooking, red meat turns brown because of the oxidation of hemoglobin to methemoglobin. Dried bloodstains turn brown for the same reason.

Deaths caused by house fires are often actually caused by toxic gases. Smoldering fires produce carbon monoxide, but hydrogen cyanide (formed by burning plastics and fabrics, Chapter 10) is also important. A medical journal, *The Lancet* (20 August 1988, p. 457), recommends that fire department rescue crews carry spring-loaded hypodermic syringes filled with sodium thiosulfate solution for emergency treatment of victims of smoke inhalation.

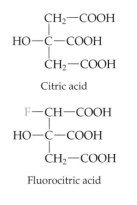

Citric acid

Fluorocitric acid

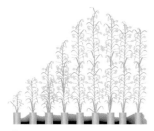

Figure 20.2 The effect of copper ions on the height of oat seedlings. From left to right, the concentrations of Cu^{2+} are 0, 3, 6, 10, 20, 100, 500, 2000, and 3000 $\mu g/L$. The plants on the left show varying degrees of deficiency; those on the right show copper ion toxicity. The optimum level of Cu^{2+} for oat seedlings is therefore about 100 $\mu g/L$.

drove eagles nearly to extinction, and as a result, the poisoning of predators has been banned on federal land.

Fluoroacetic acid occurs in nature in an extremely poisonous plant called *gifblaar* that grows in South Africa. Natives have used this plant to poison the tips of their arrows.

20.5 Heavy Metal Poisons

People have long used a variety of metals in industry and in agriculture and around the home. Most metals and their compounds show some toxicity when ingested in large amounts. Even essential mineral nutrients can be toxic when taken in excessive amounts (Figure 20.2). In many cases, too little of a metal ion (a deficiency) can be as dangerous as too much (toxicity). For example, the average adult requires 10–18 mg of iron every day. When less is taken in, the person suffers from anemia. Yet an overdose can cause vomiting, diarrhea, shock, coma, and even death. As few as 10–15 tablets containing 325 mg each of iron (as $FeSO_4$) have been fatal to children.

We don't know exactly how iron poisoning works. Heavy metals—those near the bottom of the periodic table—exert their action primarily by inactivating enzymes. Heavy metal ions react with hydrogen sulfide to form insoluble sulfides.

$$Pb^{2+}(aq) + H_2S(g) \longrightarrow PbS(s) + 2\,H^+(aq)$$

$$Hg^{2+}(aq) + H_2S(g) \longrightarrow HgS(s) + 2\,H^+(aq)$$

Most enzymes have amino acids with sulfhydryl (—SH) groups. Heavy metal ions tie up these groups, rendering the enzymes inactive (Figure 20.3).

Figure 20.3 Mercury poisoning. Enzymes catalyze reactions by binding a reactant molecule at the active site (Chapter 15). Mercury ions react with sulfhydryl groups to change the shape of the enzyme and destroy the active site.

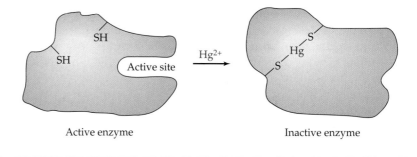

Active enzyme   Inactive enzyme

Arsenic Poisoning

Arsenic is not a metal, but it has some metallic properties. In commercial poisons, arsenic is usually found as arsenate (AsO_4^{3-}) or arsenite (AsO_3^{3-}) ions. Like heavy metal ions, these ions render enzymes inactive by tying up sulfhydryl groups.

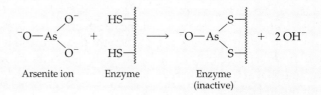

Arsenite ion   Enzyme   Enzyme
                               (inactive)

Organic compounds containing arsenic are well-known. One such compound, arsphenamine, was the first antibacterial agent and was once used widely in the treatment of syphilis.

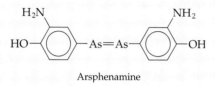

Arsphenamine

Another arsenic compound, lewisite, first synthesized by (and named for) W. Lee Lewis, was developed as a blister agent for use in chemical warfare. The United States started large-scale production of lewisite in 1918, but fortunately World War I came to an end before this gas could be employed.

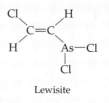

Lewisite

Quicksilver → Slow Death

Mercury (Hg) is a most unusual metal. It is the only common metal that is a liquid at room temperature. People have long been fascinated by this bright, silvery, dense liquid, formerly known as quicksilver. Children sometimes play with the mercury from a broken thermometer. This metal has many uses. Dentists use it to make amalgams for filling teeth, and laboratory workers employ mercury and its compounds in a variety of ways. Farmers use seeds treated with compounds of mercury.

Mercury presents a hazard to those who work with it because its vapor is hazardous. An open container of mercury or a spill on the floor can release enough mercury vapor into the air to exceed the established maximum safe level by a factor of 200. Because mercury is a cumulative poison (it takes the body about 70 days to rid itself of *half* of a given dose), chronic poisoning is a threat to those continually exposed.

Fortunately, there are antidotes for mercury poisoning. British scientists, searching for an antidote for the arsenic-containing war gas lewisite, came up with a compound effective for heavy metal poisoning as well. The compound, a derivative of glycerol, is called British anti-lewisite (BAL). It acts by chelating (from the Greek *chela* meaning "claw") Hg^{2+} ions. Thus tied up, the mercury cannot attack vital enzymes.

Now for the bad news. The effects of mercury poisoning may not show up for several weeks. By the time the symptoms—loss of equilibrium, sight, feeling, and hearing—are recognizable, extensive damage has already been done to the brain and the nervous system. Such damage is largely irreversible. The BAL antidote is effective only when a person knows that he or she has been poisoned and seeks treatment right away.

Metallic mercury does not seem to be very toxic when ingested (swallowed). Most of it passes through the system unchanged. Indeed, there are numerous reports of mercury being given orally in the eighteenth and nineteenth centuries as a remedy for obstruction of the bowels. Doses varied from a few ounces to a pound or more.

However, mercury vapor is quite hazardous when inhaled, particularly when exposure takes place over a long period of time. Such chronic exposure usually occurs with regular occupational use of the metal, such as in mining or extraction. By some as yet unknown mechanism, the body converts the inhaled mercury, to Hg^{2+} ions. All compounds of mercury, except those that are essentially insoluble in water, are poisonous no matter how they are administered.

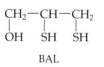

BAL

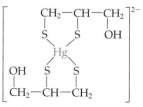

Mercury atom chelated by two BAL molecules

 **[Web Reference 2]** One of the environmental tragedies of the twentieth century was the poisoning of thousands of Japanese by industrial waste mercury dumped into Minamata Bay. You can also find an account of the detoxification/restoration effort.

Lead in the Environment

Compounds of lead are widespread in the environment, reflecting the many uses we have for this soft, dense, corrosion-resistant metal and its compounds. Lead (as Pb^{2+}) is present in many foods, generally in concentrations of less than 0.3 ppm. Lead (again as Pb^{2+}) also gets into our drinking water (up to 0.1 ppm) from lead-sealed pipes. We used to breathe lead compounds, mainly from automobiles that burned leaded gasoline, but this exposure has decreased dramatically as we have switched to unleaded gasoline.

Lead compounds are quite toxic. Metallic lead is generally converted to Pb^{2+} in the body. So, with all that lead, why aren't we dead? We can excrete about 2 mg of lead per day. Our intake from air, food, and water is generally less than that, and so we generally do not accumulate toxic levels. If intake exceeds excretion, however, lead builds up in the body and chronic irreversible lead poisoning results.

Lead poisoning is a major problem in children, particularly in areas with old, run-down buildings. Some children develop a craving that causes them to eat unusual things, and children with this syndrome (called *pica*) eat chips of peeling lead-based paints. These children probably also pick up lead compounds from the streets, where they were deposited by automobile exhausts. They also ingest lead from food and other sources. In all, thousands of children suffer from lead poisoning each year. Such poisoning often leads to mental retardation and neurological disorders caused by damage to the brain and nervous system.

Lead poisoning is usually treated with a combination of BAL and another chelating agent called EDTA ethylenediaminetetraacetic acid (EDTA).

$$HOOC-CH_2 \diagdown \quad \quad \quad \quad CH_2-COOH$$
$$N-CH_2CH_2-N$$
$$HOOC-CH_2 \diagup \quad \quad \quad \quad CH_2-COOH$$

EDTA

The calcium salt of EDTA is administered intravenously. In the body, calcium ions are displaced by lead ions, which the chelate binds more tightly.

$$CaEDTA^{2-} + Pb^{2+} \longrightarrow PbEDTA^{2-} + Ca^{2+}$$

The lead–EDTA complex is then excreted.

As in mercury poisoning, the neurological damage done by lead compounds is essentially irreversible. Treatment must be begun early to be effective.

Cadmium: The "Ouch-Ouch" Disease

Cadmium is used in alloys, in the electronics industry, in nickel–cadmium rechargeable batteries, and in many other applications. Like mercury and lead, cadmium (as Cd^{2+} ions) has caused major catastrophes. Cadmium poisoning leads to loss of calcium ions (Ca^{2+}) from the bones, leaving them brittle and easily broken. It also causes severe abdominal pain, vomiting, diarrhea, and a choking sensation.

The most notable cases of cadmium poisoning occurred along the upper Zintsu River in Japan, where cadmium ions entered the water in milling wastes from a mine. Downstream, farm families used the water to irrigate their rice fields and for drinking, cooking, and other household purposes. Soon these farm folk began to suffer from a strange, painful malady that became known as *itai-itai*, the "ouch-ouch" disease. Over 200 people died, and thousands were disabled.

20.6 More Chemistry of the Nervous System

Some poisons—among them the most toxic substances known—act on the nervous system. Signals are shuttled across synapses between cells by chemical substances

The EPA estimates that lead poisoning annually contributes to 123,000 cases of hypertension and 680,000 miscarriages and retards the growth of 7000 children. These problems add $635 million to health care costs annually.

[Web Reference 3] Extensive information about lead poisoning—with some surprising sources.

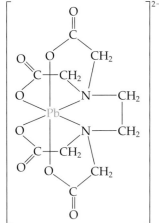

Lead–EDTA complex

called *neurotransmitters* (Chapter 19). Neurotoxins—and some drugs—can disrupt the action of a neurotransmitter in several ways: by interfering with its synthesis or transport, by occupying the transmitter's receptor site, or by blocking degradation of the transmitter.

One such chemical messenger is acetylcholine. It activates the next cell by fitting a specific receptor and thus changing the permeability of the cell membrane to certain ions. Once acetylcholine has carried the impulse across the synapse, it is rapidly hydrolyzed to acetic acid and relatively inactive choline. This reaction is catalyzed by an enzyme, acetylcholinesterase.

People with bipolar (manic-depressive) disorder are overly sensitive to acetylcholine; they seem to have too many receptors, which may account for their wild swings in mood.

$$\underset{\text{Acetylcholine}}{CH_3\overset{O}{\overset{\|}{C}}OCH_2CH_2\overset{CH_3}{\underset{CH_3}{\overset{|}{\underset{|}{N^+}}}}-CH_3} + H_2O \xrightarrow{\text{acetylcholinesterase}} \underset{\substack{\text{Acetic} \\ \text{acid}}}{CH_3\overset{O}{\overset{\|}{C}}-OH} + \underset{\text{Choline}}{HOCH_2CH_2\overset{CH_3}{\underset{CH_3}{\overset{|}{\underset{|}{N^+}}}}-CH_3}$$

The receptor cell releases the hydrolysis products and is then ready to receive further impulses. Other enzymes, such as acetylase, convert the acetic acid and choline back to acetylcholine, completing the cycle (Figure 20.4).

People with Alzheimer's disease are deficient in the enzyme acetylase. They produce too little acetylcholine for proper brain function.

Nerve Poisons: Stopping the Acetylcholine Cycle

Various substances disrupt the acetylcholine cycle at three different points. First, botulin, the deadly toxin produced by *Clostridium botulinum* (an anaerobic bacterium) found in improperly processed canned food (Chapter 16), blocks the synthesis of acetylcholine. With no messenger formed, no messages are carried. Paralysis sets in and death occurs, usually from respiratory failure.

Second, curare, atropine, and some local anesthetics (Chapter 19) act by blocking receptor sites. In this case, the message is sent but not received. In the case of local anesthetics, this can be good for pain relief in a limited area, but these drugs, too, can be lethal in sufficient quantity.

Third, anticholinesterase poisons, inhibit the enzyme cholinesterase (Figure 20.5). Organic phosphorus insecticides (Chapter 16) are well-known nerve poisons. The phosphorus–oxygen linkage is thought to bond tightly to acetylcholinesterase, blocking the breakdown of acetylcholine. Acetylcholine therefore builds up, causing receptor nerves to fire repeatedly. This overstimulates the muscles, glands, and organs. The heart beats wildly and irregularly, and the victim goes into convulsions and dies quickly.

Atropine

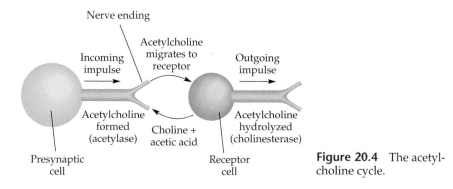

Nerve ending

Incoming impulse

Acetylcholine migrates to receptor

Outgoing impulse

Acetylcholine formed (acetylase)

Choline + acetic acid

Acetylcholine hydrolyzed (cholinesterase)

Presynaptic cell

Receptor cell

Figure 20.4 The acetylcholine cycle.

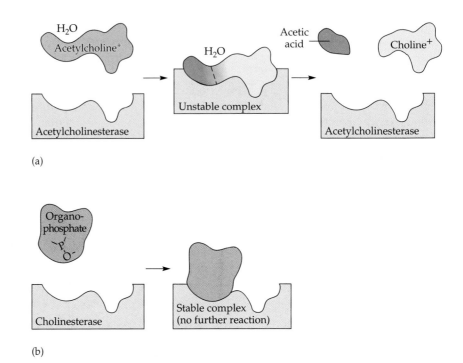

Figure 20.5 (a) Acetylcholinesterase catalyzes the hydrolysis of acetylcholine to acetic acid and choline. (b) An organophosphate ties up acetylcholinesterase, preventing it from breaking down acetylcholine.

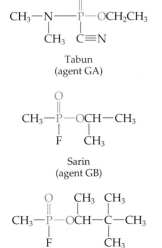

Tabun
(agent GA)

Sarin
(agent GB)

Soman
(agent GD)

Figure 20.6 Three organophosphorus compounds used as nerve poisons in chemical warfare. Compare these structures with those of the insecticides malathion and parathion. Sarin was used in the 1995 terrorist attack on the Tokyo subway that killed 12 people and sickened thousands.

Organophosphorus Compounds: Insecticides and Weapons of War

While doing research on organophosphorus compounds as possible insecticides during World War II, German scientists discovered some extremely toxic compounds with a frightening potential for use in warfare. The Russians captured a German plant that manufactured a compound called tabun (designated agent GA by the U.S. Army), a compound with a fruity odor. The Soviets dismantled the factory and moved it to Russia; however, the U.S. Army captured some of the stock.

The United States has developed other nerve poisons (Figure 20.6). One is called sarin (agent GB) and is four times as toxic as tabun. It also has the "advantage" of being odorless. Another organophosphorus nerve poison is soman (agent GD). It is moderately persistent, whereas tabun and sarin are generally nonpersistent.

Note again the variation in structure from one compound to the next. The approach in developing chemical warfare agents is much the same as the approach in developing drugs—find one that works and then synthesize and test structural variations.

Nerve poisons are among the most toxic synthetic chemicals known (but are not nearly as toxic as the natural toxin botulin). They are inhaled or absorbed through the skin, causing a complete loss of muscular coordination and subsequent death by cessation of breathing. The usual antidote is atropine injection and artificial respiration. Without an antidote, death may occur in 2–10 min.

The insecticides malathion and parathion (Chapter 16) are quite similar to the warfare nerve agents but are somewhat less toxic. These phosphorus-based insecticides and others like them should be used with extreme caution. Even relatively safe (to mammals) chlorinated hydrocarbon pesticides act as nerve poisons. Acute DDT poisoning causes tremors, convulsions, and cardiac or respiratory failure. Chronic exposure to DDT leads to degeneration of the central nervous system. Other chlorinated compounds, such as the PCBs (Chapter 10), act in a similar manner.

Nerve poisons have helped us gain an understanding of the chemistry of the nervous system. This knowledge enables scientists to design antidotes for nerve poisons, to understand diseases of the nervous system better, and to progress in the development of drugs against pain and diseases such as Alzheimer's disease and Parkinson's disease.

20.7 The Lethal Dose

What do we mean by "most toxic"? Some poisonous substances are much more poisonous than others. In order to quantify toxicity, scientists use the term LD_{50} (Lethal Dose for 50%) to indicate a dosage that kills 50% of a population of test animals. This statistic is used because some animals are unusually strong and can survive fairly large doses of a given poison, whereas others are more susceptible and die from much smaller doses. The dose that kills 50% is the average lethal dose.

Usually LD_{50} values are given in terms of mass of poison per unit body weight of the test animal. It is assumed that the material is given orally unless specified otherwise. (Intravenous values can be quite different from those measured with oral ingestion.) Although LD_{50} values are useful in comparing the relative toxicities of various substances, they are not necessarily the same values that would be measured for humans.

The substances listed in Table 20.1 are not usually thought of as poisons, yet there is a dose of each one that can kill half of a given population of test animals. The larger the LD_{50} value, the less toxic the substance. Nicotine and caffeine are the most toxic substances listed in Table 20.1 because they have the smallest LD_{50} values.

The substances in Table 20.2 are highly lethal poisons. Notice that the doses are given in milligrams per kilogram of body weight, whereas the values in Table 20.1 are in grams per kilogram. The most highly poisonous substance of all is botulin toxin.

Table 20.1 ■ LD_{50} Values for Some Common Substances

Substance	Test Animal	LD_{50} (g/kg)
Ethyl alcohol	Rat	10.3
Vitamin B_1	Mouse	8.2
Sodium chloride	Rat	3.75
Aspirin	Mouse	1.5
Acetaminophen	Mouse	0.34
Nicotine	Mouse	0.23
Caffeine	Mouse	0.13

Table 20.2 ■ Estimated LD_{50} Values for Some Highly Lethal Poisons

Substance	LD_{50} (mg/kg)
Sodium cyanide (NaCN)	15
Arsenic trioxide (As_2O_3)	15
Aflatoxin B	10
Rotenone	3
Strychnine	0.5
Muscarine	0.2
Tetanus toxin	0.000005
Botulin toxin	0.0000003

Botulin is the most toxic substance known. It has a minimum lethal dose of 0.0003 μg/kg body weight in mice. Yet botulin, in a commercial formulation called BoTox, is used in medicine to treat intractable muscle spasms. When a muscle contracts over a long period of time without relaxing, it can cause severe pain. Tiny amounts of BoTox can be injected into the muscle where it binds to receptors on nerve endings. This stops the release of acetylcholine, and the signal for the muscle to contract is blocked. The muscle fibers that were pulling too hard are paralyzed, providing relief from the spasm.

BoTox is also used to treat afflictions such as stuttering, crossed eyes, and Parkinson's disease. In the case of crossed eyes, for example, the optic muscles cause the eyeballs to be drawn to focus inward. Treated with botulin, the muscles are hindered from doing so, and repeated treatments correct the problem.

[Web Reference 5] The Agency for Toxic Substances and Disease Registry (of the Centers for Disease Control and Prevention) provides extensive information about poisonous chemicals, including molecular models. Emphasis is on contaminants found at hazardous waste sites.

What Kills You? What Makes You Sick?

The media are filled with reports of "toxic chemicals." You might think that chemicals are a leading cause of death and injury. Some chemicals are quite toxic. Misused, they can make you sick or even kill you. However, when you look once more at Figure 19.5, it is difficult to associate any of the leading causes of death directly with chemicals. In developed countries, most of the causes of premature death are related to lifestyle. Nearly half are due to cardiovascular diseases, and another 20% or so are due to cancer. In these countries, it is "sociologicals," not "chemicals," that kill us.

Accidents and suicides often involve automobile crashes and guns. More than 20,000 people are murdered each year in the United States, most with guns, knives, or clubs of some sort. It is the physical force of a moving projectile that does the damage—these "physicals" kill us.

Worldwide, the main causes of premature death are infectious and parasitic diseases, "biologicals" that account for nearly one-third of the total. Three million children under the age of 5 die each year of diarrhea caused by drinking water contaminated with microorganisms, and 2,700,000 people die each year of tuberculosis, 1,200,000 of measles, and 2,000,000 of malaria. Even in the United States, pneumonia, influenza, and AIDS—caused by bacteria and viruses—are major causes of death. Food poisoning, caused by *Escherichia coli*, *Cyclospora*, *Listeria*, *Salmonella*, and other microorganisms sicken thousands each year. Chicken meat and eggs are frequently contaminated with *Salmonella*. *Escherichia coli* is a common inhabitant of cow intestines; people who eat beef that is carelessly slaughtered or foods exposed to water contaminated with cow manure are often infected. In 1996 a strain of *E. coli*, called *E. coli* O157:H7, made 9500 people ill and killed 11 people in Japan.

In all these cases, of course, chemistry is involved at some level. We can identify the toxins produced by bacteria and work out the molecular mechanisms by which viruses invade and destroy cells. Nicotine is the chemical—a natural one, but a chemical no less—that makes tobacco addictive. Perfectly natural ethanol is the most dangerous chemical in alcoholic beverages. Chemicals can be dangerous, but with care we can use them to our advantage. As far as being threats to our lives and health, though, chemicals are far down the list. We are even more likely to die of "geologicals"—earthquakes, floods, storms, and extremely hot or cold weather.

20.8 Your Liver: A Detox Facility

The human body can handle moderate amounts of some poisons. The liver is able to detoxify some compounds by oxidation, reduction, or coupling with amino acids or other normal body chemicals.

Perhaps the most common route is oxidation. Ethanol (Chapter 9) is detoxified by oxidation to acetaldehyde, which in turn is oxidized to acetic acid and then to carbon dioxide and water.

An antidote for methyl alcohol poisoning is ethyl alcohol. Ethanol is administered intravenously in an attempt to "load up" the liver enzymes with ethanol and thus block the oxidation of methanol until the compound can be excreted.

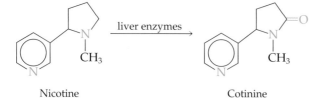

Highly toxic nicotine from tobacco is detoxified by oxidation to cotinine.

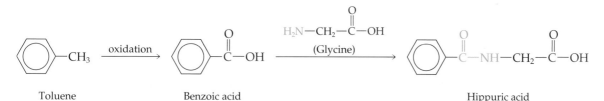

Nicotine   Cotinine

Cotinine is less toxic than nicotine. The added oxygen atom also makes cotinine more water-soluble, and thus more readily excreted in the urine, than nicotine.

The liver is equipped with a system of enzymes, called P-450, that oxidize fat-soluble substances (which are otherwise likely to be retained in the body) into water-soluble ones that are readily excreted. It can also conjugate compounds with amino acids. For example, toluene is essentially insoluble in water. P-450 enzymes oxidize toluene to more soluble benzoic acid. The latter is then coupled with the amino acid glycine to form hippuric acid, which is even more soluble and is readily excreted.

Toluene   Benzoic acid   Hippuric acid

It should be pointed out that liver enzymes simply oxidize, reduce, or conjugate. The end product is not always less toxic. For example, methanol is oxidized to a more toxic form, formaldehyde. Formaldehyde reacts with proteins in the cells to cause blindness, convulsions, respiratory failure, and death.

The same enzymes that oxidize alcohols deactivate the male hormone testosterone. Buildup of these enzymes in a chronic alcoholic leads to a more rapid destruction of testosterone. Thus, we have the mechanism for alcoholic impotence, one of the well-known characteristics of the disease.

Benzene, because of its general inertness in the body, is not acted on until it reaches the liver. There it is slowly oxidized to an epoxide.

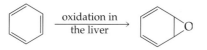

The epoxide is a highly reactive molecule that can attack certain key proteins. The damage done by this epoxide sometimes results in leukemia.

Carbon tetrachloride (CCl_4), is also quite inert in the body. But when it reaches the liver, it is converted to the reactive trichloromethyl free radical ($Cl_3C\cdot$) which in turn attacks the unsaturated fatty acids in the body. This action can trigger cancer.

Getting Rid of "Toxins"

Two thousand years ago Greek physicians used the theory of the four humors—blood, phlegm, black bile, and yellow bile—to diagnose illnesses. They held that people in good health had a balance of these fluids and that too much or too little of one or more of them caused disease. For a patient short of a humor, the treatment included bed rest, a change in diet, or medicines. For those with too much of a humor, the practice included sweating, purging, and bloodletting. In some circles today, people believe that sweating gets rid of "toxins" and that laxatives somehow purify the body. Erroneous beliefs often persist for centuries.

20.9 Chemical Carcinogens: Slow Poisons

Carcinogens cause the growth of tumors. A tumor is an abnormal growth of new tissue and can be either benign or malignant. *Benign tumors* are characterized by slow growth; they often regress spontaneously, and they do not invade neighboring tissues. *Malignant tumors*, often called cancers, can grow slowly or rapidly, but their growth is generally irreversible. Malignant growths invade and destroy neighboring tissues. Actually, the term "cancer" does not apply to a single disease. It is a catchall term used to describe 200 different afflictions, many of which are not even closely related to each other.

What Causes Cancer?

Most people seem to believe that chemicals are a major cause of cancer, but the facts indicate otherwise. Like overall deaths (Chapter 19), most cancers are caused by lifestyle factors (Figure 20.7). Nearly two-thirds of all cancer deaths in the United States are linked to tobacco, diet, or a lack of exercise with resulting obesity. Even

Figure 20.7 Ranking cancer risks. About 30% of all cancer deaths are attributed to tobacco. Another 30% or so are caused by a diet high in fat and calories and low in fruits and vegetables. Factors such as environmental pollution and food additives rank high in public perception of risks but are only relatively minor contributors to cancer deaths. (Based on data from Harvard University School of Public Health.)

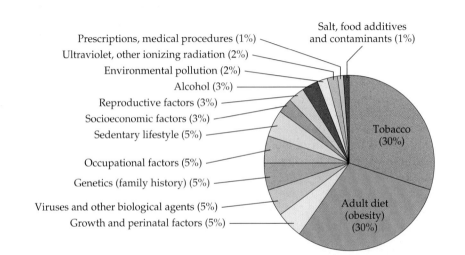

Salt, food additives and contaminants (1%)
Prescriptions, medical procedures (1%)
Ultraviolet, other ionizing radiation (2%)
Environmental pollution (2%)
Alcohol (3%)
Reproductive factors (3%)
Socioeconomic factors (3%)
Sedentary lifestyle (5%)
Occupational factors (5%)
Genetics (family history) (5%)
Viruses and other biological agents (5%)
Growth and perinatal factors (5%)
Tobacco (30%)
Adult diet (obesity) (30%)

among the chemicals that are most suspect, such as pesticides, many cannot be shown to be carcinogenic. Only about 30 chemical compounds have been identified as human carcinogens. Another 300 or so have been shown to cause cancer in laboratory animals, and some of these 300 are widely used.

Few of the known carcinogens are synthetic chemicals. Some, such as safrole in sassafras and aflatoxins produced by molds on foods, occur naturally. Some scientists estimate that 99.99% of all carcinogens that we ingest are natural ones. Plants produce compounds to protect themselves from fungi, insects, and higher animals, including humans. Some of these compounds are carcinogens found in mushrooms, basil, celery, figs, mustard, pepper, fennel, parsnips, and citrus oils—almost every place that a curious chemist looks. Carcinogens are also produced during cooking and as products of normal metabolism. Because carcinogens are so widespread, we must have some way of protecting ourselves from them.

There are about 400,000 cancer deaths each year in the United States. Of these about 150,000 are related to cigarette smoking, and another 150,000 are related to diet.

How Cancers Develop

How do chemicals and physical factors cause cancer? Their mechanisms of action are probably quite varied. Some carcinogens chemically modify DNA, thus scrambling the code for replication and for the synthesis of proteins. For example, aflatoxin B is known to bind to guanine residues in DNA. Just how this initiates cancer, however, has not yet been determined.

Genetics plays a role in the development of many forms of cancer. Certain genes, called *oncogenes*, seem to trigger or sustain the processes that convert normal cells to cancerous ones. Oncogenes arise from ordinary genes that regulate cell growth and cell division. Chemical carcinogens, radiation, or perhaps some viruses can activate oncogenes. It seems that more than one oncogene must be turned on, perhaps at different stages of the process, before a cancer develops. We also have *suppressor genes* that ordinarily prevent the development of cancers. These genes must be inactivated before a cancer develops. Suppressor gene inactivation can occur through mutation, alteration, or loss. In all, several mutations may be required in a cell before it turns cancerous.

One kind of breast cancer, a hereditary form that usually develops before age 50, results from a mutation in a tumor-suppressor gene called *BRCA1*. Women who have a mutant form of BRCA1 have an 82% risk of developing breast cancer, compared with most women in the United States, who have about a 10% risk. One hope of genetic engineering is that suppressor genes can be produced and used in therapy.

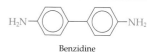

Benzidine

Chemical Carcinogens

A variety of widely different chemical compounds are carcinogenic. We cannot cover all types of carcinogens here and so we will concentrate on a few major classes.

Polycyclic aromatic hydrocarbons (Chapter 9), of which 3,4-benzpyrene is perhaps the best known, are among the more notorious carcinogens. These hydrocarbons are formed during the incomplete burning of nearly any organic material. They have been found in charcoal-grilled meats, cigarette smoke, automobile exhausts, coffee, burnt sugar, and many other materials. Not all polycyclic aromatic hydrocarbons are carcinogenic. There are strong correlations between carcinogenicity and certain molecular sizes and shapes.

Aromatic amines make up another important class of carcinogens. Two prominent ones are β-naphthylamine (Chapter 16) and benzidine. These compounds, once widely used in the dye industry, were responsible for a high incidence of bladder cancer among workers whose jobs brought them into prolonged contact with these substances.

Not all carcinogens are aromatic. Two prominent aliphatic (nonaromatic) ones are dimethylnitrosamine (Chapter 16) and vinyl chloride (Chapter 10). Others include three- and four-membered heterocyclic rings containing nitrogen or oxygen (Figure 20.8). The epoxides and derivatives of ethyleneimine are examples. Others are cyclic esters called lactones.

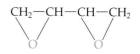

Bis(epoxy)butane

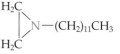

N-Laurylethyleneimine

β-Propiolactone

Figure 20.8   Three small-ring heterocyclic carcinogens.

[Web Reference 6]
A Report on Carcinogens by National Institute of Environmental Health Sciences.

Keep in mind that this list is not all-inclusive. Rather, its purpose is to give you an idea of the kinds of compounds that have tumor-inducing properties.

How Cigarette Smoking Causes Cancer

The association between cigarette smoking and cancer has been known for decades, but the precise mechanism was not determined until 1996. The research scientists who found the link focused on a tumor-suppressor gene called *P53*, a gene that is mutated in about 60% of all lung cancers, and a metabolite of benzpyrene, a carcinogen found in tobacco smoke.

In the body, benzpyrene is oxidized to an active carcinogen (an epoxide) which binds to specific nucleotides in the gene—sites called "hot spots"—where mutations frequently occur. The *P53* gene is mutated in about 60% of all lung cancers, and it seems quite likely that the benzpyrene metabolite causes many of these mutations.

Anticarcinogens

If the food we eat has many natural carcinogens, why don't we all get cancer? Other substances in our food act as **anticarcinogens**. Antioxidant vitamins are believed to protect against some forms of cancer, and the food additive butylated hydroxytoluene (BHT) may give protection against stomach cancer. Certain vitamins also have been shown to have anticarcinogenic effects.

Vitamins that are antioxidants (vitamin C, vitamin E, and β-carotene, a precursor of vitamin A) seem to exhibit the strongest anticancer properties. A diet rich in cruciferous vegetables (cabbage, broccoli, brussels sprouts, kale, and cauliflower) has been shown to reduce the incidence of cancer both in animals and in human population groups.

A number of studies with vitamins A, C, and E, used separately or in combination, seem to confirm that each of these vitamins has some ability to lower the incidence of cancer. There are probably many other anticarcinogens in our food that have not yet been identified.

Mutant Induced Revertants

Spontaneous Revertants

The Ames test. The top petri dish, to which a mutagenic chemical has been added, shows the growth of several colonies of *Salmonella* bacteria, indicating that mutations have occurred. The bottom dish is a control that shows a few spontaneous mutations.

20.10 Testing for Carcinogens: Three Ways

How do we know that a chemical causes cancer? Obviously, we can't experiment on humans to see what happens. That leaves us with no way to prove beyond doubt that a chemical does or does not cause cancer in humans. Three ways to gain evidence against a compound are bacterial screening for mutagenesis, animal tests, and epidemiological studies.

The Ames Test: Bacterial Screening

The quickest and cheapest way to find out whether or not a substance may be carcinogenic is to use a screening test such as that developed by Bruce N. Ames of the University of California at Berkeley. The **Ames test** is a simple laboratory procedure that can be carried out in a petri dish. It assumes that most carcinogens are also mutagens, altering genes in some way. (This usually seems to be the case. About 90% of the chemicals that appear on a list of either mutagens or carcinogens are found on the other list as well.)

The Ames test uses a special strain of *Salmonella* bacteria that have been modified so that they require histidine as an essential amino acid. The bacteria are placed in an agar medium containing all nutrients except histidine. Incubating the mixture in the presence of a mutagenic chemical causes the bacteria to mutate so that they no longer require histidine and can grow like normal bacteria. Growth of bacterial colonies in the petri dish means that the chemical added was a mutagen, and probably also a carcinogen.

Animal Testing

Chemicals suspected of being carcinogens can be tested on animals. Tests involving low dosages and millions of rats would cost too much, and so tests are usually done by using large doses and 30 or so rats. An equal number of rats serve as controls, which are exposed to the same diet and environment but without the suspected carcinogen. A higher incidence of cancer in the experimental animals than in the controls indicates that the compound is carcinogenic.

Animal tests are not conclusive. Humans are not usually exposed to comparable doses; there may be a threshold below which a compound is not carcinogenic. Further, human metabolism is different from that of the test animals. A carcinogen might be active in rats but not in humans (or vice versa). There is, however, some correlation between animal tests and the occurrence of human cancers.

Epidemiological Studies

The best evidence that a substance causes cancer in humans comes from epidemiological studies. A population that has a higher than normal rate of a particular kind of cancer is studied for common factors in their background. It was studies of this sort, for example, that showed that cigarette smoking causes lung cancer, that vinyl chloride causes a rare form of liver cancer, and that asbestos causes cancer of the lining of the pleural cavity (the body cavity containing the lungs). These studies sometimes require sophisticated mathematical analyses, and there is always the chance that some other (unknown) factor is involved in the carcinogenesis.

In the 1990s, epidemiological studies were carried out to determine whether or not electromagnetic fields, such as those surrounding high-voltage power lines and certain electric appliances, could cause cancer. The results indicated that if there is an effect, it is exceedingly small.

20.11 Birth Defects: Teratogens

Another group of toxic chemicals are those that cause birth defects. These substances are called **teratogens**. Perhaps the most notable teratogen is the tranquilizer thalidomide. During the late 1950s and early 1960s, thalidomide was considered so safe, based on laboratory studies, that it was often prescribed for pregnant women. In Germany it was available without a prescription. It took several years for the human population to provide evidence that laboratory animals had not provided. The drug had a disastrous effect on developing human embryos. About 12,000 women who had taken the drug during the first 12 weeks of pregnancy had babies who suffered from *phocomelia*, a condition characterized by shortened or absent arms and legs and other physical defects. The drug was used widely in Germany and Great Britain, and these two countries bore the brunt of the tragedy. The United States escaped relatively

Animal studies cost about $1 million each and take 2 years, and some people are opposed to using animals in this way.

There is only a 70% correlation between the carcinogenesis of a chemical in rats and that in mice. The correlation between carcinogenesis in either rodent and that in humans is probably less.

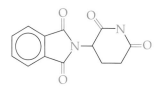

Thalidomide

Frances O. Kelsey, an FDA official, refused to approve thalidomide for marketing in the United States. Her action prevented many mothers from having to face the tragedy of giving birth to a "thalidomide baby." She is shown here receiving the Distinguished Federal Civilian Service Award from President John F. Kennedy in 1962.

Thalidomide is making a comeback. It has a unique anti-inflammatory action against a debilitating skin condition that often occurs with leprosy. The FDA approved thalidomide in 1998 for treating this condition, and strict restrictions are in place to prevent its use by pregnant women. Thalidomide is also being studied for use in the treatment of wasting, the severe weight-loss condition that often accompanies AIDS. Scientists are also working to develop new thalidomide derivatives to treat rheumatoid arthritis, AIDS, and other diseases. By modifying the structure of the molecule, they hope to enhance its beneficial effects while reducing its harmful properties.

unscathed because Frances O. Kelsey of the FDA believed there was evidence to doubt the drug's safety and therefore did not approve it for use in the United States.

Other chemicals that act as teratogens include isotretinoin (Accutane), a prescription medication approved for use in treating severe acne. When taken by women during the first trimester of pregnancy, it can cause multiple major malformations. Educational materials provided to physicians and to patients have prevented all but a few tragedies associated with isotretinoin.

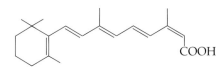

Isotretinoin
(13-*cis*-retinoic acid)

By far the most hazardous teratogen, in terms of the number of babies born with birth defects, is ethyl alcohol (Chapter 19), which causes fetal alcohol syndrome.

20.12 Hazardous Wastes

Toxic substances, carcinogens, teratogens—the public has become increasingly concerned in recent years about issues involving hazardous wastes in the environment. Problems created by chemical dumps have made household words out of Love Canal in New York and "Valley of the Drums" in Kentucky. Although often overblown in the news media, serious problems do exist. Hazardous wastes can cause fires or explosions. They can pollute the air, and they can contaminate our food and water. Occasionally they poison by direct contact. As long as we want the products that our industries produce, however, we will have to deal with the problems of hazardous wastes (Table 20.3).

 [Web Reference 7] Gives an idea of the magnitude of the problem of hazardous waste.

The first step in dealing with any problem is to understand what the problem is. A **hazardous waste** is one that can cause or contribute to death or illness or that threatens human health or the environment when improperly managed. For convenience, hazardous wastes are divided into four types: reactive, flammable, toxic, and corrosive.

Reactive wastes tend to react spontaneously or to react vigorously with air or water. They can generate toxic gases, such as hydrogen cyanide (HCN) or hydrogen sulfide (H_2S), or explode when exposed to shock or heat. Explosives such as trinitrotoluene (TNT) and nitroglycerin obviously are reactive wastes. Another example

---

Table 20.3 ▌ Industrial Products and Hazardous Waste By-Products

Product	Associated Waste
Plastics	Organic chlorine compounds
Pesticides	Organic chlorine compounds, organophosphate compounds
Medicines	Organic solvents and residues, heavy metals (for example, mercury and zinc)
Paints	Heavy metals (for example, lead), pigments, solvents, organic residues
Oil, gasoline	Oil, phenols and other organic compounds, heavy metals, ammonium salts, acids, strong bases
Metals	Heavy metals, fluorides, cyanides, acid and alkaline cleaners, solvents, pigments, abrasives, plating salts, oils, phenols
Leather	Heavy metals, organic solvents
Textiles	Heavy metals, dyes, organic chlorine compounds, solvents

(a)

(b)

(a) A waste dump in 1970 at Malkins Bank, Cheshire, England, with drums leaking chemical wastes. (b) The same site, cleaned up and restored, is now a municipal golf course.

is sodium metal. Wastes containing sodium caused explosions at Malkins Bank in Great Britain. Sodium reacted with water to form hydrogen gas.

$$2\,Na(s) + H_2O(l) \longrightarrow 2\,NaOH(aq) + H_2(g)$$

The hydrogen then exploded when ignited in air.

$$2\,H_2(g) + O_2(g) \longrightarrow 2\,H_2O(g)$$

Reactive wastes usually can be deactivated before disposal. Sodium can be treated with isopropyl alcohol, with which it reacts slowly, rather than being dumped without treatment.

Flammable wastes are those that burn readily on ignition, presenting a fire hazard. An example is hexane, a hydrocarbon solvent. In one case, hexane (presumably dumped accidentally by Ralston-Purina) was ignited in the sewers of Louisville, Kentucky.

$$2\,C_6H_{14}(l) + 19\,O_2(g) \longrightarrow 12\,CO_2(g) + 14\,H_2O(g)$$

Explosions ripped up several blocks of streets. Flammable wastes can usually be burned safely in incinerators rather than discarded.

Toxic wastes are those that contain or release toxic substances in quantities sufficient to pose a hazard to human health or to the environment. Most of the toxic substances discussed in this chapter would qualify as toxic wastes if they were improperly dumped in the environment. Some toxic wastes can be incinerated safely. PCBs (Chapter 10), for example, are broken down at high temperatures to carbon dioxide, water, and hydrogen chloride. If the hydrogen chloride is removed by scrubbing or is safely diluted and dispersed, this is a satisfactory way to dispose of PCBs. Some toxic wastes cannot be incinerated, however, and must be contained and monitored for years.

Corrosive wastes are those that require special containers because they corrode conventional container materials. Acids can't be stored in steel drums because they react with and dissolve the iron.

$$Fe(s) + 2\,H^+(aq) \longrightarrow Fe^{2+}(aq) + H_2(g)$$

Acid wastes can be neutralized (Chapter 7) before disposal, and lime (CaO) serves as cheap base for neutralization.

$$2\,H^+(aq) + CaO(s) \longrightarrow Ca^{2+}(aq) + H_2O(l)$$

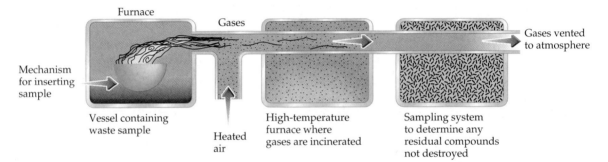

Figure 20.9   A schematic diagram of an incinerator for hazardous wastes.

The best way to handle hazardous wastes is not to produce them in the first place. Many industries have modified manufacturing processes to minimize their wastes, and some wastes can be reprocessed to recover energy or materials. Hydrocarbon solvents such as hexane can be purified and reused or burned as fuels. Sometimes one industry's waste can become raw material for another industry. For example, waste nitric acid from the metals industry can be converted to fertilizer. Finally, if a hazardous waste cannot be used or incinerated or treated to render it less hazardous, it must be stored in a secure landfill. Unfortunately, landfills often leak, contaminating the groundwater. We clean up one toxic waste dump and move the materials to another, playing a rather macabre shell game. The best technology at present for treating organic wastes, including chlorinated compounds, is incineration (Figure 20.9). At 1260°C, 99.9999+% destruction is achieved.

Scientists have identified microorganisms that degrade hydrocarbons such as those in gasoline. Other bacteria, when provided with proper nutrients, can degrade chlorinated hydrocarbons. Certain bacteria in the soil can even break down TNT and nitroglycerin. Through genetic engineering, scientists are developing new strains of bacteria that can decompose a great variety of wastes. Perhaps biodegradation will be the way of the future.

The trouble with incineration is finding a place to build the incinerator. It seems that no one wants an incinerator in his or her own backyard.

 [Web Reference 8] The EPA website stresses waste minimization—reducing waste at its source, before it is even generated—combined with environmentally sound recycling.

20.13 What Price Poisons?

We use so many poisons in and around our homes and workplaces that accidents are bound to happen. Poison control centers have been established in cities to help physicians deal with emergency poisonings. Are our insecticides, drugs, cleansers, and other chemicals worth the price we pay in terms of accidental poisonings? That is for you to decide. Generally, it is the misuse of these chemicals that leads to tragedy.

Perhaps it is easy to be negative about chemists and chemistry when you think of such horrors as nerve gases, carcinogens, and teratogens. But keep in mind that many toxic chemicals are of enormous benefit to us and that they can be used safely despite their hazardous nature. The plastics industry was able to control vinyl chloride emissions once the hazard was known. We are still able to have valuable vinyl plastics even though the vinyl chloride from which they are made causes cancer.

Increasingly, we have to decide whether the benefits we gain from hazardous substances are worth the risks we assume by using them. Many issues involving toxic chemicals are emotional; most of the decisions regarding them are political; but possible solutions to such problems lie mainly in the field of chemistry. We hope that the chemistry you have learned here will help you make intelligent decisions. Most of all, we hope that you will continue to learn more about chemistry throughout the rest of your life, because chemistry affects nearly everything you do. We wish you success and happiness, and may the joy of learning go with you always.

Critical Thinking Exercises

Apply knowledge that you have gained in this chapter and one or more of the FLaReS principles (Chapter 1) to evaluate the following statements or claims.

20.1 An advertisement implies that rotenone is safer than competitive pesticides because it is obtained from natural sources.

20.2 A book claims that all chlorine compounds are toxic and should be banned.

20.3 A natural foods enthusiast contends that no food that has a carcinogen in it should be allowed on the market.

Summary

1. Toxicology is the branch of pharmacology that deals with poisons.

2. "The dose makes the poison." Anything can be a poison if the dose is large enough.

3. Strong acids and strong bases are toxic because they are corrosive.

4. Substances such as ozone are toxic because they are strong oxidizing agents.

5. Carbon monoxide and nitrite ions are toxic because they interfere with the transport of oxygen by the blood.

6. Cyanide is a poison because it shuts down cell respiration.

7. Fluoroacetic acid is a poison because it shuts down the citric acid cycle.

8. Heavy metal poisons, such as lead and mercury, inactivate enzymes by tying up their —SH groups.

9. Nerve poisons, such as organophosphates, interfere with the acetylcholine cycle.

10. Carcinogens are slow poisons that trigger the growth of malignant tumors.

11. Teratogens are chemicals of many different types that cause birth defects.

12. Hazardous waste includes all kinds of industrial products and by-products that can cause illness or death.

Key Terms

Ames test   20.10
anticarcinogens   20.9
carcinogens   20.9

corrosive wastes   20.12
flammable wastes   20.12
hazardous wastes   20.12

reactive wastes   20.12
teratogens   20.11
toxicology   (page 602)

toxic wastes   20.12

Review Questions

1. What is toxicology? How is it related to pharmacology?
2. Is sodium chloride (table salt) poisonous? Explain your answer fully.
3. Give an example that shows how the toxicity of a substance depends on the route of administration.
4. Give an example that shows how a substance may be more harmful to one person than to another.
5. How does fluoroacetic acid exert its toxic effect?

6. List three corrosive poisons.
7. List some sources of mercury poisoning and of lead poisoning.
8. What is the single leading cause of cancer?
9. Name several natural carcinogens.
10. What is a tumor? How are benign and malignant tumors different?

Problems

Corrosive Poisons

11. How do dilute solutions of acids and bases damage living cells?
12. How does ozone damage living cells?

Poisons Affecting Oxygen Transport and Oxidative Processes

13. What is a blood agent? List two such poisons.
14. What is methemoglobin?

15. How do cyanides exert their toxic effect?
16. How does sodium thiosulfate act as an antidote for cyanide poisoning?

Heavy Metal Poisons

17. Iron (as Fe^{2+}) is a necessary nutrient. What are the effects of too little Fe^{2+}? Of too much?
18. How does mercury (as Hg^{2+}) exert its toxic effect?
19. How does BAL act as an antidote for mercury poisoning?

20. How does EDTA act as an antidote for lead poisoning?
21. How does cadmium (as Cd^{2+}) exert its toxic effect?
22. What is *itai-itai* disease?

Chemistry of the Nervous System

23. What is acetylcholine? Describe its action.
24. How do each of the following affect the acetylcholine cycle?
    a. botulin   b. curare
    c. atropine
25. What are anticholinesterase poisons? How do they act on the acetylcholine cycle?
26. List three nerve poisons developed for use in warfare.
27. How does atropine act as an antidote for poisoning by organophosphorus compounds?
28. Describe a use of botulin in medicine.

Detoxification

29. What is the P-450 system? What is its function?
30. How does the liver detoxify ethanol?
31. List two ways that the conversion of nicotine to cotinine in the liver lessens the risk of nicotine poisoning.
32. List two steps in the detoxification of ingested toluene. What is the effect of these steps?
33. Does the P-450 system always detoxify foreign substances?
34. How does ethanol work as an antidote for methanol poisoning?

Lethal Dose

35. The LD_{50} for ketoprofen (Orudis KT), orally in rats, is 101 mg/kg body weight. If this value could be extrapolated to humans, what would be the lethal dose for a 23-kg child?
36. The LD_{50} for methyl isocyanate, the substance that caused the Bhopal tragedy, orally in rats, is 140 mg/kg body weight. Estimate the lethal dose for a 125-lb human.

Chemical Carcinogens

37. What are oncogenes? How are they involved in the development of cancer?
38. What are suppressor genes? How are they involved in the development of cancer?
39. List some conditions under which polycyclic hydrocarbons are formed.
40. Name two aromatic amines that are carcinogens.
41. Describe the mechanism by which benzpyrene causes lung cancer in cigarette smokers.
42. Name two aliphatic carcinogens.
43. List some of the limitations involved in testing compounds for carcinogenicity by using laboratory animals.
44. What is an epidemiological study? Can such a study prove absolutely that a compound causes cancer?

Mutagens and Teratogens

45. What is a mutagen? A teratogen?
46. Describe the Ames test for mutagenicity. What are its limitations as a screening test for carcinogens?

Hazardous Wastes

47. What is a hazardous waste?
48. Define and give an example of a reactive waste.
49. Define and give an example of a flammable waste.
50. What is a corrosive waste?
51. Why can't waste acids be stored in steel drums?
52. What is the best method for the disposal of toxic organic wastes? Justify fully the method you select.

Additional Problems

53. Prowling around an old dump, Murgatroyd B. Muckraker finds a metal container filled with 12 oz of liquid. The label indicates that the can contains a carcinogen at a concentration of 8.2 mg/fluid ounce. On further investigation, he finds that 20 billion such containers were once filled and distributed in the United States each year.
    a. How many milligrams of the carcinogen are in each can?
    b. How many metric tons of the carcinogen were distributed in this manner each year?
    c. What should have been done about this problem?
54. Botulin toxin has an estimated LD_{50} of 300 pg/kg body weight and an estimated molar mass of 150,000 g/mol. About how many molecules comprise a lethal dose for a 55-kg human?

Projects

55. Choose a disposal method for each of the following wastes. Be as specific as possible and justify your choice.
    a. hydrochloric acid contaminated with iron salts
    b. picric acid (an explosive)
    c. soybean oil contaminated with PCBs
56. Should carcinogens that occur naturally in foods be subjected to the same tests used to evaluate synthetic pesticides? Why or why not?
57. Should substances be tested for toxicity or carcinogenicity on laboratory animals? On people? Explain your answer fully.
58. Are nerve gases less humane than bullets in warfare?

Online Projects

59. Most tooth fillings consist of amalgams containing mercury, and the safety of these amalgams is a recurring controversy. Using your favorite search methods, find some websites dealing with this issue and write a brief analysis of the opposing points of view. (*Hint:* Try the website of the American Dental Association, and then try an alternative health website.)

60. Possibly using http://www.idrc.ca/books/reports/1997/19-01e.html as a starting point, look at the effects of mercury on the environment and do a risk–benefit analysis of mercury use.

References and Readings

1. Baker, Scott R., and Chris F. Wilkinson (Eds.). *The Effects of Pesticides on Human Health*. Princeton, NJ: Princeton Scientific, 1990.

2. Bellafante, Ginia. "Minimizing Household Hazardous Waste." *Garbage*, March–April 1990, pp. 44–48.

3. Bernarde, Melvin A. *Our Precarious Habitat: Fifteen Years Later*. New York: Wiley, 1989. Chapter 13 discusses hazardous wastes.

4. Bower, B. "Excess Lead Linked to Boys' Delinquency." *Science News*, 10 February 1996, p. 86.

5. "Cancer Quantified." *U.S. News & World Report*, 2 December 1996, p. 19. Lists factors that cause cancer.

6. Cole, Leonard A. "The Specter of Biological Weapons." *Scientific American*, December, 1996, pp. 60–65.

7. Hoffmann, Franz. "Fighting Cancer with Good Taste (and a Good Diet)." *Journal of College Science Teaching*, December 1996–January 1997, pp. 197–200. Discusses chemicals in cabbage and related plants.

8. Manahan, Stanley E. *Toxicological Chemistry*. Boca Raton, FL: CRC Press, 1992.

9. Marshall, Eliot. "Solving Louisville's Friday-the-Thirteenth Explosion." *Science*, 27 March 1981, p. 1405.

10. McGinn, Anne Platt. "Phasing Out Persistent Organic Pollutants," in Lester R. Brown, Christopher Flavin, and Hilary French (Eds.), *State of the World 2000*. New York: W.W. Norton, 2000.

11. Parshall, George W. "Chemical Warfare Agents: How Do We Rid Ourselves of Them?" *The Chemist*, May–June 1995, pp. 5–7. We need a safe way to destroy weapons no longer needed.

12. Perera, Frederica P. "Uncovering New Clues to Cancer Risk." *Scientific American*, May 1996, pp. 54–62.

13. Rodricks, Joseph V. *Calculated Risks: Understanding the Toxicity and Human Health Risks of Chemicals in Our Environment*. New York: Cambridge University Press, 1992.

14. Travis, J. "Putting a Tumor Suppressor Back to Work." *Science News*, 31 August 1996, p. 134.

15. Wheelwright, Jeff. "The Berry and the Poison." *Smithsonian*, December 1996, pp. 40–50. Methyl bromide is very toxic, but it saves farmers an estimated $1.5 billion a year. It is being banned because it attacks the ozone layer.

Appendix A
A Review of Measurement and Mathematics

Accurate measurements are essential to science. Measurements can be made in a variety of units, but most scientists use the International System of Measurement (SI). The data they record often has to be converted from one kind of unit to another and otherwise manipulated mathematically. In this appendix, we extend the discussion of metric measurement that we began in Chapter 1 and review some of the mathematics that you may find useful in this course.

A.1 The International System of Measurement

As noted in Chapter 1, the standard unit of length in the International System of Measurement is the *meter*. This distance was once meant to be 0.0000001 of Earth's quadrant—that is, of the distance from the North Pole to the equator measured along a meridian. The quadrant proved difficult to measure accurately, and today the meter is defined precisely as the distance light travels in a vacuum during 1/299,792,458 of a second.

The primary unit of mass is the *kilogram* (1 kg = 1000 g). It is based on a standard platinum–iridium bar kept at the International Bureau of Weights and Measures. The *gram* is a more convenient unit for many chemical operations.

The derived SI unit of volume is the *cubic meter*. The units more frequently employed in chemistry, however, are the *liter* $(1 \text{ L} = 0.001 \text{ m}^3)$ and the *milliliter* $(1 \text{ mL} = 0.001 \text{ L})$. Other SI units of length, mass, and volume are derived from these basic units. Table A.1 lists additional metric units of length, mass, and volume and illustrates the use of prefixes.

Table A.1 ∎ Some Metric Units of Length, Mass, and Volume

Length

1 kilometer (km)	= 1000 meters (m)
1 meter (m)	= 100 centimeters (cm)
1 centimeter (cm)	= 10 millimeters (mm)
1 millimeter (mm)	= 1000 micrometers (μm)

Mass

1 kilogram (kg)	= 1000 grams (g)
1 gram (g)	= 1000 milligrams (mg)
1 milligram (mg)	= 1000 micrograms (μg)

Volume

1 liter (L)	= 1000 milliliters (mL)
1 milliliter (mL)	= 1000 microliters (μL)
1 milliliter (mL)	= 1 cubic centimeter (cm^3)

A.2 Exponential (Scientific) Notation

Scientists often use numbers that are so large or so small that they boggle the mind. For example, light travels at 300,000,000 m/s. There are 602,200,000,000,000,000,000,000 carbon atoms in 12.01 g of carbon. On the small side, the diameter of an atom is about 0.0000000001 m, and the diameter of an atomic nucleus is about 0.000000000000001 m. We find it difficult to keep track of the zeros in such quantities. Scientists find it convenient to express such numbers in exponential notation.

A number is in *exponential notation*—often called *scientific notation*—when it is written as the product of a coefficient (usually with a value between 1 and 10) and a power of 10. Two examples are

$$4.18 \times 10^3 \quad \text{and} \quad 6.57 \times 10^{-4}$$

Expressing numbers in exponential form generally serves two purposes.

1. We can write very large or very small numbers in a minimum of printed space and with a reduced chance of typographical error.

2. We can convey explicit information about the precision of measurements: The number of significant figures (Section A.4) in a measured quantity is stated unambiguously.

In the expression 10^n, n is the exponent of 10, and the number 10 is said to be raised to the nth power. If n is a *positive quantity*, 10^n has a value *greater than 1*. If n is a *negative quantity*, 10^n has a value *less than 1*. We are particularly interested in cases where n is an integer. For example,

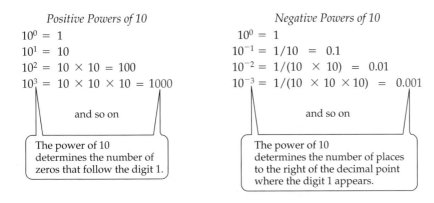

Positive Powers of 10

$10^0 = 1$
$10^1 = 10$
$10^2 = 10 \times 10 = 100$
$10^3 = 10 \times 10 \times 10 = 1000$

and so on

The power of 10 determines the number of zeros that follow the digit 1.

Negative Powers of 10

$10^0 = 1$
$10^{-1} = 1/10 = 0.1$
$10^{-2} = 1/(10 \times 10) = 0.01$
$10^{-3} = 1/(10 \times 10 \times 10) = 0.001$

and so on

The power of 10 determines the number of places to the right of the decimal point where the digit 1 appears.

We express 612,000 in exponential form as

$$612,000 = 6.12 \times 100,000 = 6.12 \times 10^5$$

We express 0.000505 in exponential form as

$$0.000505 = 5.05 \times 0.0001 = 5.05 \times 10^{-4}$$

We can use a more direct approach to converting numbers to the exponential form.

- Count the number of places a decimal point must be moved to produce a coefficient having a value between 1 and 10.
- The number of places counted then becomes the power of 10.

- The power of 10 is *positive* if the decimal point is moved to the *left*.

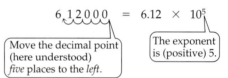

To convert a number from exponential form to the conventional form, move the decimal point in the opposite direction.

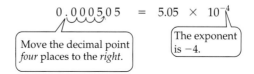

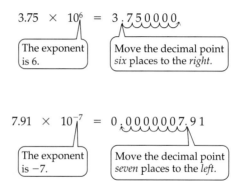

The keystrokes on your calculator may be different from those shown here. Check the instructions in the manual that came with the calculator.

It is easy to handle exponential numbers on most calculators. A typical procedure is to enter the number, followed by the key [EXP]. The keystrokes required for the number 2.85×10^7 are [2] [.] [8] [5] [EXP] [7], and the result displayed is 2.85^{07}.

For the number 1.67×10^{-5}, the keystrokes are [1] [.] [6] [7] [EXP] [5] [±], and the result displayed is 1.67^{-05}. Many calculators can be set to convert all numbers and calculated results to the exponential form, regardless of the form in which the numbers are entered. Generally, a calculator can also be set to display a fixed number of significant figures in the results.

Addition and Subtraction

To add or subtract numbers in exponential notation, it is necessary to express each quantity as the *same power of 10*. In calculations, this treats the power of 10 in the same way as a unit—it is simply "carried along." In the following, each quantity is expressed with the power 10^{-3}.

$$(3.22 \times 10^{-3}) + (7.3 \times 10^{-4}) - (4.8 \times 10^{-4}) = (3.22 \times 10^{-3}) + (0.73 \times 10^{-3}) - (0.48 \times 10^{-3})$$

$$= (3.22 + 0.73 - 0.48) \times 10^{-3}$$

$$= 3.47 \times 10^{-3}$$

Multiplication and Division

To multiply numbers expressed in exponential form, *multiply* all coefficients to obtain the coefficient of the result and *add* all exponents to obtain the power of 10 in the result.

Rewrite in exponential form.

$$0.0803 \times 0.0077 \times 455 = (8.03 \times 10^{-2}) \times (7.7 \times 10^{-3}) \times (4.55 \times 10^{2})$$

Multiply the coefficients.   Add the exponents.

$$= (8.03 \times 7.7 \times 4.55) \times 10^{(-2-3+2)}$$

$$= (2.8 \times 10^{2}) \times 10^{-3} = 2.8 \times 10^{-1}$$

Generally, most calculators perform these operations automatically, and no intermediate results need be recorded.

To divide two numbers in exponential form, *divide* the coefficients to obtain the coefficient of the result and *subtract* the exponent in the denominator from the exponent in the numerator to obtain the power of 10. In the example below, multiplication and division are combined. First, the rule for multiplication is applied to the numerator and to the denominator, and then the rule for division is used.

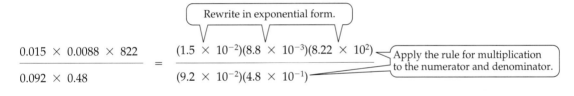

Rewrite in exponential form.

$$\frac{0.015 \times 0.0088 \times 822}{0.092 \times 0.48} = \frac{(1.5 \times 10^{-2})(8.8 \times 10^{-3})(8.22 \times 10^{2})}{(9.2 \times 10^{-2})(4.8 \times 10^{-1})}$$

Apply the rule for multiplication to the numerator and denominator.

Apply the rule for division.

$$= \frac{1.1 \times 10^{-1}}{4.4 \times 10^{-2}} = 0.25 \times 10^{-1-(-2)} = 0.25 \times 10^{1}$$

$$= 2.5 \times 10^{-1} \times 10^{1} = 2.5 \times 10^{0} = 2.5$$

Raising a Number to a Power and Extracting the Root of an Exponential Number

To raise an exponential number to a given power, raise the coefficient to that power and multiply the exponent by that power. For example, we can cube a number (that is, raise it to the *third* power) in the following manner.

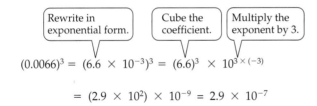

Rewrite in exponential form.   Cube the coefficient.   Multiply the exponent by 3.

$$(0.0066)^{3} = (6.6 \times 10^{-3})^{3} = (6.6)^{3} \times 10^{3 \times (-3)}$$

$$= (2.9 \times 10^{2}) \times 10^{-9} = 2.9 \times 10^{-7}$$

To extract the root of an exponential number, we raise the number to a fractional power—one-half power for a square root, one-third power for a cube root, and so on. Most calculators have keys designed for extracting square roots and cube roots. Thus, to extract the square root of 1.57×10^{-5}, enter the number 1.57×10^{-5} into a calculator, and use the $\left[\sqrt{\ }\right]$ key.

$$\sqrt{1.57 \times 10^{-5}} = 3.96 \times 10^{-3}$$

Some calculators allow you to extract roots by keying the root in as a fractional exponent. For example, we can take a cube root as follows.

$$(2.75 \times 10^{-9})^{1/3} = 1.40 \times 10^{-3}$$

Practice Problems

1. Express each of the following numbers in exponential form.
   a. 0.000017   **b.** 19,000,000
   c. 0.0034   **d.** 96,500

2. Express each of the following numbers in exponential form.
   a. 4,500,000   **b.** 0.000108
   c. 0.0341   **d.** 406,000

3. Carry out the following operations. Express the answers in exponential form.

   a. $(4.5 \times 10^{13})(1.9 \times 10^{-5})$ **b.** $\dfrac{9.3 \times 10^{9}}{3.7 \times 10^{27}}$

   c. $\dfrac{4.3 \times 10^{-7}}{7.6 \times 10^{22}}$ **d.** $\sqrt{1.7 \times 10^{-5}}$

4. Carry out the following operations. Express the answers in exponential form.

   a. $(6.2 \times 10^{-5})(4.1 \times 10^{-12})$ **b.** $(2.1 \times 10^{-6})^{2}$

   c. $\dfrac{2.1 \times 10^{5}}{9.8 \times 10^{-7}}$ **d.** $\dfrac{4.6 \times 10^{-12}}{2.1 \times 10^{3}}$

A.3 Unit Conversions

We live in a world in which almost all other countries use metric measurements. Suppose that an American applies for a job in another country and the job application asks for her height in centimeters. She knows her height in customary units is 66.0 in. Her actual height is the same, whether we express it in inches, feet, centimeters, or millimeters. Thus, when we measure something in one unit and then convert it to another one, we must not change the measured quantity in any fundamental way. We can use equivalencies such as those in Table A.2 to derive conversion factors.

Table A.2 ▌ Some Conversions Between Common and Metric Units

Length

1 mile (mi)	= 1.61 kilometers (km)
1 yard (yd)	= 0.914 meter (m)
1 inch (in.)	= 2.54 centimeters (cm)

Mass

1 pound (lb)	= 454 grams (g)
1 ounce (oz)	= 28.4 grams (g)
1 pound (lb)	= 0.454 kilogram (kg)

Volume

1 U.S. quart (qt)	= 0.946 liter (L)
1 U.S. pint (pt)	= 0.473 liter (L)
1 fluid ounce (fl oz)	= 29.6 milliliters (mL)
1 gallon (gal)	= 3.78 liters (L)

In mathematics, multiplying a quantity by 1 does not change its value. We can therefore use a factor equivalent to 1 to convert between inches and centimeters. We find our factor in the definition of the inch in Table A.2.

$$1 \text{ in.} = 2.54 \text{ cm}$$

Notice that if we divide both sides of this equation by 1 in., we obtain a ratio of quantities that is equal to 1.

$$1 = \frac{1 \text{ in.}}{1 \text{ in.}} = \frac{2.54 \text{ cm}}{1 \text{ in.}}$$

If we divide both sides of the equation by 2.54 cm, we also obtain a ratio of quantities that is equal to 1.

$$\frac{1 \text{ in.}}{2.54 \text{ cm}} = \frac{2.54 \text{ cm}}{2.54 \text{ cm}} = 1$$

The two ratios, one shown in red and the other in blue, are conversion factors. A *conversion factor* is a ratio of terms, equivalent to the number 1, used to change the unit in which a quantity is expressed. We call this process the *unit conversion method* of problem solving. Since we set up problems by examining the *dimensions* associated with the given quantity, those associated with the desired result, and those needed in the conversion factors, the process is sometimes called *dimensional analysis*.

What happens when we multiply a known quantity by a conversion factor? The original unit cancels out and is replaced by the desired unit. Thus, our general approach to using conversion factors is

Desired quantity and unit = given quantity and unit × conversion factors

Now let's return to the question about the woman's height, a measured quantity of 66.0 in. To get an answer in centimeters, we must use the appropriate conversion factor (in red). Note that the desired unit (cm) is in the numerator and the unit to be replaced (in.) is in the denominator. Thus we can cancel the unit in. so that only the unit cm remains.

$$66.0 \text{ in.} \times \frac{2.54 \text{ cm}}{1 \text{ in.}} = 168 \text{ cm (rounded off from 167.64)}$$

You can see why the other conversion factor (in blue) won't work. It gives a nonsensical unit.

$$66.0 \text{ in.} \times \frac{1 \text{ in.}}{2.54 \text{ cm}} = \frac{26.0 \text{ in.}^2}{\text{cm}}$$

Following are some examples and exercises to give you some practice in this method of problem solving.

Example A.1

a. Convert 0.742 kg to grams.
b. Convert 0.615 lb to ounces.
c. Convert 135 lb to kilograms.

Solution
a. This is a conversion from one metric unit to another. We simply use a knowledge of prefixes to convert from meters to millimeters.

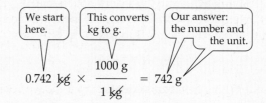

$$0.742 \; \cancel{kg} \times \frac{1000 \; g}{1 \; \cancel{kg}} = 742 \; g$$

b. This is a conversion from one common unit to another. Here we use the fact that 1 lb = 16 oz to convert from pounds to ounces. Then we proceed as in part (a), arranging the conversion factor to cancel the unit lb.

$$0.615 \; \cancel{lb} \times \frac{16 \; oz}{1 \; \cancel{lb}} = 9.48 \; oz$$

c. This is a conversion from a common unit to a metric unit. We need data from Table A.2 to convert from pounds to kilograms. Then we proceed as in part (b), arranging the conversion factor to cancel the unit lb.

$$135 \; \cancel{lb} \times \frac{1 \; kg}{2.205 \; \cancel{lb}} = 61.2 \; kg$$

(We give our answers with the proper number of significant figures. If you do not understand significant figures and wish to do so, they are discussed in Section A.4. In this text we simply use three significant figures for most calculations.)

Exercise A.1
a. Convert 16.3 mg to grams.   **b.** Convert 24.5 oz to pounds.
c. Convert 12.5 fl oz to milliliters.

Quite often, to get the desired unit, we must use more than one conversion factor. We can do so by arranging all the necessary conversion factors in a single setup that yields the final answer in the desired unit.

Example A.2
What is the length, in millimeters, of a 3.25-ft piece of tubing?

Solution
No relationship between feet and millimeters is given in Table A.2, and so we need more than one conversion factor. We can think of the problem as a series of three conversions.
1. Use the fact that 1 ft = 12 in. to convert from feet to inches.
2. Use data from Table A.2 to convert from inches to centimeters.
3. Use a knowledge of prefixes to convert from centimeters to millimeters.

We could solve this problem in three distinct steps by making one conversion in each step, but it is just as easy to combine three conversion factors into a single setup. Then we proceed as indicated below.

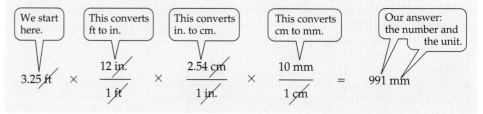

Exercise A.2
Carry out the following conversions.
a. 90.3 mm to meters
b. 729.9 ft to kilometers
c. 1.17 gal to fluid ounces

Sometimes we need to convert two or more units in the measured quantity. We can do this too by arranging all the necessary conversion factors in a single setup so that the starting units cancel and the conversions yield the final answer in the desired units.

Example A.3

A saline solution has 1.00 lb of salt in 1.00 gal of solution. Calculate the quantities in grams per liter of solution.

Solution

First let's identify the measured quantities. They can be expressed in the form of a ratio of mass of salt in pounds to a volume in gallons.

$$\frac{1 \text{ lb (salt)}}{1 \text{ gal (solution)}}$$

This ratio must be converted to one expressed in grams per liter. We must convert from pounds to grams in the numerator and from gallons to liters in the denominator. The following set of equivalent values from Table A.2 can be used to formulate conversion factors.

$$1 \text{ lb} = 453.6 \text{ g} \qquad 1 \text{ gal} = 3.785 \text{ L}$$

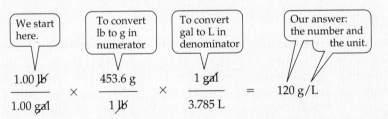

Note that we could also do the conversions in the numerator and denominator separately and then divide the numerator by the denominator.

Numerator: $\qquad 100 \text{ lb} \times \dfrac{453.6 \text{ g}}{1 \text{ lb}} = 454 \text{ g}$

Denominator: $\qquad 100 \text{ gal} \times \dfrac{3.785 \text{ L}}{1 \text{ gal}} = 3.785 \text{ L}$

Division: $\qquad \dfrac{454 \text{ g}}{3.785 \text{ L}} = 120 \text{ g/L}$

Both methods give the same answer to three significant figures.

Exercise A.3

Carry out the following conversions.
a. 88.0 km/h to meters per second
b. 1.22 ft/s to kilometers per hour
c. 4.07 g/L to ounces per quart

A.4 Precision, Accuracy, and Significant Figures

Counting usually gives exact numbers. For example, we can count exactly 24 students in a room. Measurements, on the other hand, are subject to error. One source of error is in the measuring instruments themselves. For example, an incorrectly calibrated thermometer may consistently yield a result that is 0.2°C too low. Other errors may result from the experimenter's lack of skill or care in using measuring instruments.

Precision and Accuracy

(a) Low accuracy   (b) Low accuracy
Low precision   High precision

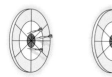

(c) High accuracy (d) High accuracy
Low precision   High precision

Comparing precision and accuracy: a dartboard analogy. (a) The darts are both scattered (low precision) and off-center (low accuracy). (b) The darts are in a tight cluster (high precision) but still off-center (low accuracy). (c) The darts are somewhat scattered (low precision) but evenly distributed about the center (high accuracy). (d) The darts are in a tight cluster (high precision) and well centered (high accuracy).

Suppose you were one of five students asked to measure a person's height using a meter stick marked off in millimeters. The five measurements are recorded in Table A.3. The **precision** of a set of measurements refers to how closely individual measurements agree with one another. We say the precision is good if each of the measurements is close to the average, and poor if there is a wide deviation from the average value. How would you describe the precision of the data in Table A.3? Examine the individual data, note the average value, and determine how much the individual data differ from the average. Because the maximum deviation from the average value is 0.003 m, the precision is quite good.

The **accuracy** of a set of measurements refers to the closeness of the average of the set to the "correct" or most probable value. Measurements of high precision are more likely to be accurate than are those of poor precision, but even highly precise measurements are sometimes inaccurate. For example, what if the meter sticks used to obtain the data in Table A.3 were actually 1005 mm long but still had 1000-mm markings? The accuracy of the measurements would be rather poor, even though the precision would remain good.

Sampling Errors

No matter how accurate an analysis, it will not mean much unless it is performed on valid, representative samples. Consider determining the level of glucose in the blood of a patient. High glucose levels are associated with diabetes, a debilitating disease. Results vary, depending on several factors such as the time of day and what and when the person last ate. Glucose levels are much higher soon after a meal high in sugars. They also depend on other factors, such as stress. Medical doctors usually take blood for analysis after a night of fasting. These fasting glucose levels tend to be more reliable, but physicians often repeat the analysis when high or otherwise suspicious results are obtained. Therefore, there is no one true value for the level of glucose in a person's blood. Repeated samplings provide an average level. Similar results from several measurements give much more confidence in the findings. For example, a diagnosis of diabetes is usually based on two consecutive fasting blood glucose levels above 120 mg/dL.

Table A.3 ▌ Five Measurements of a Person's Height

Student	Height, m
1	1.827
2	1.824
3	1.826
4	1.828
5	1.829
Average	1.827

Significant Figures

Look again at Table A.3. Notice that the five measurements of height agree in the first three digits (1.82); they differ only in the fourth digit. We say that the fourth digit is uncertain. All digits known with certainty, plus the first uncertain one, are called **significant figures**. The precision of a measurement is reflected in the number of significant figures—the more significant figures, the more precise the measurement. The measurements in Table A.3 have four significant figures. In other words, we are quite sure that the person's height is between 1.82 m and 1.83 m. Our best estimate of the average value, including the uncertain digit, is 1.827 m.

The number 1.827 has four digits; we say it has four significant figures. In any properly reported measurement, all nonzero digits are significant. Zeros, however, may or may not be significant because they can be used in two ways: as a part of the measured value or to position a decimal point.

- Zeros between two other significant digits are significant. *Examples*: 4807 (four significant figures); 70.004 (five).
- We use a lone zero preceding a decimal point for aesthetic purposes; it is not significant. *Example*: 0.352 (three significant figures).
- Zeros that precede the first nonzero digit are also not significant. *Examples*: 0.000819 (three significant figures); 0.03307 (four).
- Zeros at the end of a number are significant if they are to the right of the decimal point. *Examples*: 0.2000 (four significant figures); 0.050120 (five).

We can summarize these four situations with a general rule: When we read a number from left to right all the digits starting with the first nonzero digit are significant. Numbers without a decimal point that end in zeros are a special case, however.

- Zeros at the end of a number may or may not be significant if the number is written without a decimal point. *Example*: 700. We do not know whether the number 700 was measured to the nearest unit, ten, or hundred. To avoid this confusion, we can use exponential notation (Section A.2). In exponential notation, 700 is recorded as 7×10^2 or 7.0×10^2 or 7.00×10^2 to indicate one, two, or three significant figures, respectively. The only significant digits are those in the coefficient, not in the power of 10.

We use significant figures only with measurements—quantities subject to error. The concept does not apply to a quantity that is

1. inherently an integer, such as 3 sides to a triangle or 12 items in a dozen,
2. inherently a fraction, such as the radius of a circle equals one-half the diameter,
3. obtained by an accurate count, such as 18 students in a class,
4. a defined quantity, such as 1 km = 1000 m.

In these contexts, the numbers 3, 12, 1/2, 18, and 1000 can have as many significant figures as we want. More properly, we say that each is an exact value.

Significant Figures in Calculations: Multiplication and Division

If we measure a sheet of notepaper and find it to be 14.5 cm wide and 21.7 cm long, we can find the area of the paper by multiplying the two quantities. A calculator gives the answer as 314.65. Can we conclude that the area is 314.65 cm²? That is, can we know the area to the nearest hundredth of a square centimeter when we know the width and length only to the nearest tenth of a centimeter? It just doesn't seem reasonable—and it isn't. A calculated quantity can be no more precise than the data used in the calculation, and the reported result should reflect this fact.

A strict application of this principle involves a fairly complicated statistical analysis that we will not attempt here, but we can do a pretty good job through a practical rule involving significant figures.

> *In multiplication and division, the reported result should have no more significant figures than the factor with the fewest significant figures.*

In other words, a calculation is only as precise as the least precise measurement that enters into the calculation.

To obtain a numerical answer with the proper number of significant figures often requires that we round off numbers. In rounding, we drop all digits that are not significant and, if necessary, adjust the last reported digit. We use the following rules in rounding.

- If the leftmost digit to be dropped is less than 5, leave the final digit unchanged. *Example*: If we need four significant figures, 69.744 rounds to 69.74, and to 69.7 if we need three significant figures.
- If the leftmost digit to be dropped is greater than 5, increase the final digit by 1. *Example*: 538.76 rounds to 538.8 if we need four significant figures. Similarly, 74.397 rounds to 74.40 if we need four significant figures, and to 74.4 if we need three.
- If the leftmost digit to be dropped is exactly 5, we round up if the preceding digit is odd, and down if the preceding digit is even. *Example*: If we need three significant figures in each case, 4.735 rounds to 4.74, and 5.625 rounds to 5.62.

Example A.4

What is the area, in square centimeters, of a rectangular gauze bandage that is 2.54 cm wide and 12.42 cm long? Use the correct number of significant figures in your answer.

Solution

The area of a rectangle is the product of its length and width. In the result, we can show only as many significant figures as there are in the least precisely stated dimension, the width, which has three significant figures.

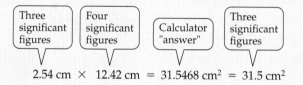

$$2.54 \text{ cm} \times 12.42 \text{ cm} = 31.5468 \text{ cm}^2 = 31.5 \text{ cm}^2$$

We use the rules for rounding off numbers as the basis for dropping the digits 468.

Exercise A.4

Calculate the volume, in cubic meters, of a rectangular block of foamed plastic that is 1.827 m long, 1.04 m wide, and 0.064 m thick. Use the correct number of significant figures.

Example A.5

For a laboratory experiment, a teacher wants to divide all of a 226.8-g sample of glucose equally among the 18 members of her class. How many grams of glucose should each student receive?

Solution

Here we need to recognize that the number 18 is a counted number, that is, an exact number that is not subject to significant figure rules. The answer should carry four significant figures, the same as in 226.8 g.

$$\frac{226.8 \text{ g}}{18} = 12.60 \text{ g}$$

In this calculation a calculator displays the result 12.6. We add the digit 0 to emphasize that the result is precise to four significant figures.

Exercise A.5

A dozen eggs has a mass of 681 g. What is the average mass of one of the eggs, expressed with the appropriate number of significant figures?

Significant Figures in Calculations: Addition and Subtraction

In addition or subtraction, we are concerned not with the number of significant figures but with the number of digits to the right of the decimal point. When we add or subtract quantities with varying numbers of digits to the right of the decimal point, we need to note the one with the fewest such digits. The result should contain the same number of digits to the right of its decimal point. For example, if you are adding several masses and one of them is measured only to the nearest gram, the total mass cannot be stated to the nearest milligram no matter how precise the other measurements are.

   We apply this idea in Example A.6. Note that in a calculation involving several steps, we need round off only the final result.

Example A.6

Perform the following calculation and round off the answer to the correct number of significant figures.

$$2.146 \text{ g} + 72.1 \text{ g} - 9.1434 \text{ g} = \text{?}$$

Solution

In this calculation, we add two numbers and subtract a third from the sum of the first two.

$$
\begin{array}{ll}
2.146 \text{ g} & \leftarrow \text{three decimal places} \\
+ \ 72.1 \text{ g} & \leftarrow \text{one decimal place} \\
\hline
74.246 \text{ g} & \\
\\
- \ 9.1434 \text{ g} & \leftarrow \text{four decimal places} \\
\hline
65.1026 \text{ g} = 65.1 \text{ g} & \leftarrow \text{one decimal place}
\end{array}
$$

Note that we do not round off the intermediate result (74.246). When using a calculator, we generally don't need to write down an intermediate result.

Exercise A.6

Perform the indicated operations and give answers with the proper number of significant figures. Note that in addition and subtraction all terms must be expressed in the same unit.

a. 48.2 m + 3.82 m + 48.4394 m

b. 148 g + 2.39 g + 0.0124 g

c. 15.436 L + 5.3 L − 6.24 L − 8.177 L

d. (51.5 m + 2.67 m) × (33.42 m − 0.124 m)

e. $\dfrac{125.1 \text{ g} - 1.22 \text{ g}}{52.5 \text{ mL} + 0.63 \text{ mL}}$

f. $\dfrac{0.307 \text{ g} - 14.2 \text{ mg} - 3.52 \text{ mg}}{(1.22 \text{ cm} - 0.28 \text{ mm}) \times 0.752 \text{ cm} \times 0.51 \text{ cm}}$

A.5 Calculations Involving Temperature and Heat

We sometimes need to make conversions from one temperature scale to another or from one energy unit to another.

Temperature

On the Fahrenheit scale, the freezing point of water is 32°F and the boiling point is 212°F, whereas on the Celsius scale the freezing point of water is 0°C and the boiling point is 100°C. A 10° temperature interval on the Celsius scale therefore equals an 18° interval on the Fahrenheit scale. From these facts we can derive two equations that relate temperatures on the two scales. One of these requires multiplying the degrees of Celsius temperature by the factor 1.8 (that is, 18/10) to obtain the degrees of Fahrenheit temperature, followed by adding 32 to account for the fact that 0°C = 32°F.

$$°F = 1.8(°C) + 32$$

In the other equation, we subtract 32 from the Fahrenheit temperature to get the number of degrees Fahrenheit above the freezing point of water. Then this quantity is divided by 1.8.

$$°C = \frac{°F - 32}{1.8}$$

Example A.7 illustrates a practical situation where conversion between Celsius and Fahrenheit temperatures is necessary.

Example A.7

At home you keep your room thermostat set at 68°F. When traveling, you have a room with a thermostat that uses the Celsius scale. What Celsius temperature will give you the same temperature as at home?

Solution

$$°C = \frac{°F - 32}{1.8}$$

$$= \frac{68 - 32}{1.8} = 20°C$$

Exercise A.7

a. Convert 85.0°C to degrees Fahrenheit.
b. Convert −12.2°C to degrees Fahrenheit.
c. Convert 355°F to degrees Celsius.   **d.** Convert −20.8°F to degrees Celsius.

Heat

Scientists usually measure quantities of heat energy in *joules*. The joule is the SI unit of heat. A joule (J) is the work done by a force of 1 newton[1] acting over a distance of 1 meter. For the most part, we will use the more familiar calorie.

$$1 \text{ cal} = 4.184 \text{ J}$$

$$1 \text{ kcal} = 1000 \text{ cal} = 4184 \text{ J}$$

---

[1] A *newton* (N) is the basic SI unit of force. A newton is the force required to give a 1-kg mass an acceleration of 1 m/s². That is, 1 N = 1 kg · m/s². Therefore, 1 J = 1 N · m = 1 kg · m²/s².

A calorie (cal) is the amount of heat required to raise the temperature of 1 g of water 1°C. (This quantity varies slightly with temperature; a calorie is defined more precisely as the amount of heat required to raise the temperature of 1 g of water from 14.5° to 15.5°C).

Some substances gain or lose heat more readily than others (Table A.4). The **specific heat** of a substance is the amount of heat required to raise the temperature of 1 g of the substance 1°C. The definition of a calorie indicates that the specific heat of water is 1.00 cal/(g · °C). In SI, the specific heat of water is 4.184 J/(g · °C). Water is a poor conductor of heat, whereas metals are good heat conductors. A sample of water must absorb much more heat to raise its temperature than a metal sample of similar mass. The metal sample may become red hot after absorbing a quantity of heat that makes the water sample only lukewarm.

Table A.4 ▪ Specific Heats of Selected Substances

Substance	Specific Heat	
	cal/g · °C	J/g · °C
Aluminum (Al)	0.216	0.902
Copper (Cu)	0.0921	0.385
Ethyl alcohol (C_2H_5OH)	0.588	2.46
Iron (Fe)	0.106	0.443
Ethylene glycol (HOC_2H_4OH)	0.561	2.35
Magnesium (Mg)	0.245	1.025
Mercury (Hg)	0.0332	0.139
Sulfur (S)	0.169	0.706
Water (H_2O)	1.000	4.182

We can use the following equation, in which ΔT is the change in temperature (in either °C or kelvins), to calculate the quantity of heat absorbed or released by a system.

$$\text{Heat absorbed or released} = \text{mass} \times \text{specific heat} \times \Delta T$$

Example A.8

How much heat, in calories, kilocalories, and kilojoules, does it take to raise the temperature of 225 g of water from 25.0° to 100.0°C?

Solution

Let's list the quantities we need for the calculations.

Mass of water	= 225 g	
Specific heat of water	= 1.00 cal/(g · °C)	= 4.184 J/(g · °C)
ΔT	= (100.0 − 25.0)°C	= 75.0°C

Then we use the equation

Heat absorbed
$$= \text{mass} \times \text{specific heat} \times \Delta T$$
$$= 225 \text{ g} \times 1.00 \text{ cal/(g · °C)} \times 75.0°C$$
$$= 16,900 \text{ cal}$$

We can then convert the unit cal to the units kcal and kJ.

$$16.900 \text{ cal} \times \frac{1 \text{ kcal}}{1000 \text{ cal}} = 16.9 \text{ kcal}$$

$$16.9 \text{ kcal} \times \frac{4.184 \text{ kJ}}{1 \text{ kcal}} = 70.7 \text{ kJ}$$

Exercise A.8
How much heat, in calories, kilocalories, and kilojoules, is released by 975 g of water as it cools from 100.0° to 18.0°C?

The "calorie" used for measuring the energy content of foods is actually a *kilocalorie* (kcal) or food Calorie (Cal). A dieter might be aware that a banana split contains 1500 "calories." If the same dieter realized that this was really 1,500,000 cal, giving up the banana split might be easier.

Table A.5 ▌ Some Conversion Units for Energy

1 calorie (cal) = 4.184 joules (J)
1 British thermal unit (Btu) = 1055 joules (J) = 252 calories (cal)
1 food Calorie = 1 kilocalorie (kcal) = 1000 calories (cal) = 4184 joules (J)

Practice Problems

You may need data from Table A.4 and/or Table A.5 for some of these problems.

1. Make the following conversions.
   a. 25°C to kelvins   b. 273 K to°C
2. Make the following conversions.
   a. 301 K to°C   b. 473°C to kelvins
3. How many calories are there in 0.82 kcal?
4. How many kilocalories are there in 65,500 cal?
5. For each of the following, indicate which is the larger unit.
   a. °C or °F   b. cal or Cal
6. Order the following temperatures from coldest to hottest: 0 K, 0°C, and 0°F.
7. Carry out the following conversions.
   a. 37.0°C to degrees Fahrenheit
   b. 5.5°F to degrees Celsius
   c. 273°C to degrees Fahrenheit

8. Carry out the following conversions.
   a. 98.2°F to degrees Celsius
   b. 2175°C to degrees Fahrenheit
   c. 25.0°F to degrees Celsius
9. Carry out the following conversions.
   a. 0.741 kcal to joules   b. 8.63 kJ to calories
10. Carry out the following conversions.
   a. 1.36 kcal to kilojoules   b. 345 cal to joules
   c. 873 kJ to kilocalories
11. How many calories are required to raise the temperature of 50.0 g of water from 20.0° to 50.0°C?
12. How many kilojoules are required to raise the temperature of 131 g of water from 15.0° to 95.0°C?

Appendix B
Glossary

acid A substance that, when added to water, produces an excess of hydronium ion; a proton donor.

acid–base indicator A substance that is one color in acid and another color in base.

acidic anhydride A substance, usually a nonmetal oxide, that forms an acid on addition of water.

acid rain Rain having a pH less than 5.6.

acquired immune deficiency syndrome (AIDS) A disease caused by a retrovirus (HIV) that weakens the immune system.

activated sludge method A combination of primary and secondary sewage treatment methods in which some sludge is recycled.

activation energy The minimum quantity of energy that must be available before a chemical reaction can take place.

active site The spot on a molecule at which reaction occurs.

addition polymerization A polymerization reaction in which all the atoms of the monomers are incorporated in the polymer.

addition reaction A reaction in which the single product contains all the atoms of two reactant molecules.

adipose tissue Connective tissue where fat is stored.

advanced sewage treatment Sewage treatment designed to remove phosphates, nitrates, and other soluble impurities.

aerobic oxidation An oxidation process occurring in the presence of oxygen.

aerobic exercise Exercise in which muscle contractions occur in the presence of oxygen.

aerosol Particles of 1 μm diameter, or less, dispersed in air.

aflatoxins Compounds produced by molds growing on stored peanuts and grains.

Agent Orange A combination of 2,4-D and 2,4,5-T used extensively in Vietnam to remove enemy cover and destroy crops that maintained enemy armies.

agonist A molecule that fits and activates a specific receptor.

alchemy A mystical blend of chemistry, magic, and religion that flourished in Europe during the Middle Ages (C.E. 500 to 1500).

alcohol (ROH) A compound composed of an alkyl group and a hydroxyl group.

aldehyde (RCHO) An organic molecule with a carbonyl group that has at least one hydrogen atom attached to the carbonyl carbon.

alkali metal A metal in Group 1A of the periodic table.

alkaline earth metal An element in Group 2A of the periodic table.

alkaloid A physiologically active nitrogen-containing organic compound obtained from plants.

alkalosis A physiological condition in which the pH of the blood rises to a life-threatening level.

alkane A hydrocarbon with only single bonds; a saturated hydrocarbon.

alkene A hydrocarbon containing one or more double bonds.

alkyl group (R—) The group of atoms that results when a hydrogen atom is removed from an alkane.

alkyne A hydrocarbon containing one or more triple bonds.

allergen A substance that triggers an allergic reaction.

allotropes Different forms of the same element in the same physical state.

alloy A mixture of two or more elements, at least one of which is a metal; an alloy has metallic properties.

alpha helix A secondary structure of a protein molecule in which the molecule has a spiral arrangement.

alpha (α) particle A cluster of two protons and two neutrons; a helium nucleus.

Ames test A laboratory test for mutagens, which are usually also carcinogens.

amide An organic compound having the functional group CON in which the carbon is double-bonded to the oxygen and single-bonded to the nitrogen.

amine A compound that contains the elements carbon, hydrogen, and nitrogen; derived from ammonia by replacing one, two, or three of the hydrogen atoms with one, two, or three alkyl group(s).

amino acid An organic compound that contains both an amino group and a carboxylic acid group; amino acids combine to produce proteins.

amino group ($-NH_2$) A substituent group comprised of a nitrogen atom bonded to two hydrogen atoms.

anabolic steroid A drug that aids in the building (anabolism) of body proteins and thus of muscle tissue.

anabolism The buildup of body tissues from simpler molecules.

anaerobic decay Decomposition in the absence of oxygen.

anaerobic exercise Exercise involving muscle contractions without sufficient amounts of oxygen.

analgesic A pain reliever.

androgen A male sex hormone.

anesthetic A substance that causes loss of feeling and pain.

anion A negatively charged ion.

anionic surfactant A surfactant with a hydrocarbon tail and a water-soluble head that bears a negative charge.

anode The electrode at which oxidation occurs.

antagonist A drug that blocks the action of an agonist by blocking the receptors.

antibiotic A soluble substance, produced by a mold or bacterium, that inhibits growth of other microorganisms.

anticarcinogen A substance that inhibits the formation of cancer.

anticholinergic A drug that acts on nerves using acetylcholine as a neurotransmitter.

anticoagulant A substance that inhibits the clotting of blood.

anticodon The sequence of three adjacent nucleotides in a tRNA molecule that is complementary to a codon on mRNA.

antihistamine A substance that relieves the symptoms of allergies: sneezing, itchy eyes, and runny nose.

anti-inflammatory A substance that inhibits inflammation.

antimetabolite A compound that inhibits the synthesis of nucleic acids.

antioxidant A chemical that is so easily oxidized that it protects other substances from oxidation; a reducing agent.

antiperspirant A formulation that retards perspiration by constricting the openings of sweat glands.

antipyretic A fever-reducing substance.

apoenzyme The pure protein part of an enzyme.

applied research An investigation aimed at the solution of a particular problem.

aqueous solution A solution in which the solvent is water.

arithmetic growth A process in which a constant amount is added during each growth period.

aromatic compound A compound that has special bonding and properties like those of benzene.

asbestos A group of related fibrous silicates.

astringent A substance that constricts the openings of the sweat glands, thus reducing the amount of perspiration that escapes.

atmosphere The gaseous mass surrounding Earth; a unit of pressure equal to 760 mmHg.

atmospheric inversion A warm layer of air above a cool, stagnant lower layer.

atom The smallest characteristic particle of an element.

atomic mass unit The unit of relative atomic weights, 1/12 the mass of a carbon 12 atom.

atomic number (Z) The number of protons in the nucleus of an atom of an element.

atomic theory A model that explains the law of multiple proportions and the law of constant composition by stating that all elements are composed of atoms.

Avogadro's hypothesis Equal volumes of gases, regardless of their compositions, contain equal numbers of molecules under the same conditions of temperature and pressure.

Avogadro's number The number of atoms in exactly 12 g of pure carbon-12; 6.02×10^{23}.

background radiation Ever-present radiation from cosmic rays and from natural radioactive isotopes in air, water, soil, and rocks.

base A substance that, when added to water, produces an excess of hydroxide ions; a proton acceptor.

base triplet The sequence of three bases on a tRNA molecule that determine which amino acid it can carry.

basic anhydride A substance, usually a metal oxide, that forms a base on addition of water.

basic research The search for knowledge for its own sake.

battery A series of electrochemical cells.

beta (β) particle An electron emitted by a radioactive atom.

binary compound A compound consisting of two elements.

binding energy Energy derived from the conversion of mass to energy when neutrons and protons are combined to form nuclei.

biochemical oxygen demand (BOD) The quantity of oxygen required by microorganisms to remove organic matter from water.

biochemistry A study of the chemistry of living systems.

biomass The total mass of plants and animals; in energy studies, usually means plant material used as a source of fuel.

bitumen A hydrocarbon mixture obtained from tar sands by heating.

bleach A substance used to remove unwanted color from fabrics, hair, or other materials.

blood sugar Glucose, a simple sugar, circulated in the bloodstream.

bonding pair (BP) A pair of electrons that comprises a chemical bond.

Boyle's law For a given mass of gas at constant temperature, the volume varies inversely with the pressure.

breeder reactor A nuclear reactor that converts nonfissile isotopes to fissile isotopes.

broad-spectrum antibiotic An antibiotic that is effective against a wide variety of microorganisms.

bronze An alloy of copper and tin.

buffer A compound that reacts with either acid or base to keep the pH of a solution essentially constant.

builder (in detergent formulations) Any substance added to a surfactant to increase its detergency.

calorie (cal) The amount of heat required to raise the temperature of 1 g of water 1°C.

carbohydrate A compound consisting of carbon, hydrogen, and oxygen; a starch or sugar.

carbon-14 dating A technique for determining the age of artifacts based on the half-life of carbon-14.

carbonyl group (—CO—) A carbon atom double-bonded to an oxygen atom.

carboxyl group (—COOH) A carbon atom double-bonded to one oxygen atom and singly bonded to a second oxygen atom which in turn is bonded to a hydrogen atom; the functional group of carboxylic acids.

carboxylic acid (RCOOH) An organic compound that contains the —COOH functional group.

carcinogen A substance or physical entity that produces tumors.

catabolism The metabolic process in which complex compounds are broken down into simpler substances.

catalyst A substance that increases the rate of a chemical reaction without itself being used up.

catalytic converter A device containing catalysts for oxidizing carbon monoxide and hydrocarbons to carbon dioxide and reduction of nitrogen oxides to nitrogen gas.

catalytic reforming A process that converts low-octane alkanes to high-octane aromatic compounds.

cathode The electrode at which reduction occurs.

cathode ray A stream of high-speed electrons.

cation A positively charged ion.

cationic surfactant A surfactant with a hydrocarbon tail and a water-soluble head that bears a positive charge.

celluloid Cellulose nitrate, a synthetic material derived from natural cellulose by reaction with nitric acid.

cellulose A polymer comprised of glucose units joined through beta linkages.

Celsius scale A temperature scale on which water freezes at 0° and boils at 100°.

cement A mixture of lime, clay, and water that hardens like stone when it dries.

ceramic A hard, solid product made from clay or similar materials.

chain reaction A self-sustaining change in which one or more products of one event cause one or more new events.

Charles' law For a given mass of gas at constant pressure, the volume varies directly with the absolute temperature.

chemical bond The force of attraction that holds atoms together in compounds.

chemical change A change in chemical composition.

chemical equation A before-and-after description in which chemical formulas and coefficients represent a chemical reaction.

chemical property Characteristic of a substance that describes any of the ways by which the substance might react so as to change its composition.

chemical symbol An abbreviation, consisting of one or two letters, that stands for an element.

chemistry The study of matter and the changes it undergoes.

chemotherapy The use of chemicals to control or cure diseases.

coal A fossilized black rock composed mainly of carbon.

codon A sequence of three adjacent nucleotides in mRNA that specifies one amino acid.

coenzyme An organic molecule (often a vitamin) that combines with an apoenzyme to make a complete, functioning enzyme.

cofactor An inorganic component that combines with an apoenzyme to make a complete, functioning enzyme.

cologne A diluted perfume.

compound A pure substance made up of two or more elements combined in fixed proportions.

concentrated solution A solution that has a relatively large amount of solvent per unit volume of solution.

condensation The reverse of vaporization; a change from the gaseous state to the liquid state.

condensation polymerization A reaction in which not all the atoms in the starting monomers are incorporated in the polymer because water is split out as the polymer is formed.

condensed structural formula An organic chemical formula that shows the atoms of hydrogen right next to the carbon atoms to which they are attached.

copolymer A polymer formed by the combination of two or more different monomer units.

corrosive waste A waste that requires a special container because it corrodes conventional container materials.

cosmetics Substances defined in the 1938 U.S. Food, Drug, and Cosmetic Act as "articles intended to be rubbed, poured, sprinkled or sprayed on, introduced into, or otherwise applied to the human body or any part thereof, for cleaning, beautifying, promoting attractiveness or altering the appearance...."

cosmic rays Extremely high-energy rays from outer space.

covalent bond A bond formed by a shared pair of electrons.

cream An emulsion of tiny water droplets in oil.

critical mass The mass of an isotope above which a self-sustaining chain reaction can occur.

crystal A solid with plane surfaces at definite angles.

cyclic hydrocarbon A ring-containing hydrocarbon.

daughter isotopes Isotopes formed by the radioactive decay of another isotope.

defoliation Premature removal of leaves from plants.

Delaney Amendment A 1958 amendment to the U.S. Food, Drug, and Cosmetic Act that automatically bans from food any chemical shown to induce cancer in laboratory animals.

density The quantity of mass per unit volume.

deodorant A product that uses perfume to mask body odor; some claim to prevent body odor by killing odor-causing bacteria.

deoxyribonucleic acid (DNA) The type of nucleic acid found primarily in the nuclei of cells.

depilatory A hair remover.

depressant drug A drug that slows down both physical and mental activity.

deuterium An isotope of hydrogen with a proton and a neutron in the nucleus (mass of 2 u).

dextro isomer A "right-handed" isomer.

dietary mineral A mineral required in the diet for proper health and well-being.

dilute solution A solution that has a relatively small amount of solvent per unit volume of solution.

dioxins Chlorinated cyclic compounds produced by burning wastes containing chlorinated compounds; once found as contaminants in herbicides.

dipole A molecule that has a positive end and a negative end.

dipole interactions The attractive forces that exist among polar covalent molecules.

disaccharide A sugar that on hydrolysis yields two monosaccharide molecules per molecule of disaccharide.

dispersion forces The momentary, usually weak, attractive forces between molecules.

dissociative anesthetic A substance that causes gross personality disorders, including hallucinations similar to those in near-death experiences.

dissolved oxygen Oxygen dissolved in water; a measure of that water's ability to support fish and other aquatic life.

disulfide linkage A covalent linkage through two sulfur atoms.

diuretic A substance that increases the body's output of urine.

double bond Two shared pairs of electrons.

doubling time The time it takes a population to double in size.

drug abuse The use of a drug for its intoxicating effect.

drug misuse The use of a drug for a purpose other than its intended use.

elastomer A synthetic polymer with rubberlike properties.

electrochemical cell A device that produces electricity by means of a chemical reaction.

electrode A carbon rod or metal strip inserted into an electrolytic or electrochemical cell.

electrolysis The process of splitting a compound by means of electricity.

electrolyte A compound that, in water solution, conducts an electric current.

electron The subatomic particle that bears a unit of negative charge.

electron capture (EC) A type of radioactive decay in which a nucleus absorbs an electron from the first or second electronic shell.

electron configuration The arrangement of an atom's electrons in space.

electron-dot (Lewis) formula The structural formula of a molecule in which valence electrons of all the atoms are indicated by dots.

electron-dot (Lewis) symbol The symbol of an element surrounded by dots representing the atom's outermost electrons.

electronegativity The ability of an atom to attract electron density toward itself when joined to another atom by a chemical bond.

electrostatic precipitator A device that removes particulate matter from smokestack gases by forming an electric charge on the particles, which are then removed by attraction to a surface of opposite charge.

element A fundamental substance in which all atoms have the same number of protons.

emollient An oil or grease used as a skin softener.

emulsion A suspension of submicroscopic particles of fat or oil in water.

end note The portion of perfume that has low volatility; composed of large molecules.

endorphins Naturally occurring peptides that bond to the same receptor sites as opiate drugs.

endothermic reaction A chemical reaction that absorbs energy from its surroundings.

energy The capacity for doing work.

energy levels The specific, quantized energy levels that an electron can have in an atom.

enrichment (food) Replacement of nutrients lost from a food during processing.

enrichment (isotope) The process by which the proportion of one isotope of an element is increased relative to those of the others.

entropy A measure of the randomness of a system.

enzyme A biological catalyst.

essential amino acid An amino acid not produced in the body that must be included in the human diet.

ester (RCOOR′) A compound derived from a carboxylic acid and an alcohol; the —OH of the acid is replaced by an —OR group.

estrogen A female sex hormone.

ether (ROR′) A molecule with two hydrocarbon groups attached to the same oxygen atom.

eutrophication The excessive growth of plants in a body of water that causes some of the plants to die because of a lack of light; the water becomes choked with vegetation, depleted of oxygen, and useless as a fish habitat or for recreation.

excited state That state in which an atom is supplied energy and an electron is moved from a lower to a higher energy level.

exothermic reaction A chemical reaction that releases heat to the surroundings.

eye shadow A product composed of a base of petroleum jelly with fats, oils, and waxes and colored by dyes or by zinc oxide or titanium dioxide pigments; used to color eyelids.

fast-twitch fibers The stronger, larger kind of muscle fibers that are suited for anaerobic work.

fat A compound formed by the reaction of glycerol with three fatty acid units; a triglyceride.

fat depots Storage places for fats in the body.

fatty acid A carboxylic acid that contains 4 to 20 or more carbon atoms in a chain.

first law of thermodynamics Energy is neither created nor destroyed.

flammable waste A waste that burns readily on ignition, presenting a fire hazard.

food additive Any substance other than basic foodstuffs that is present in food as a result of some aspect of production, processing, packaging, or storage.

formula mass The sum of the atomic masses of a chemical compound as indicated by the formula, expressed in atomic mass units (u).

fossil fuels Coal, petroleum, and natural gas.

free radical A reactive neutral chemical species that contains an unpaired electron.

freezing The reverse of melting; a change from the liquid to the solid state.

fuel A substance that burns readily with the release of significant amounts of energy.

fuel cell A device that produces electricity directly from fuels and oxygen.

functional group The atom or group of atoms that confers characteristic properties on a molecule.

fundamental particle An electron, proton, or neutron.

gamma (γ) rays Rays similar to X-rays that are emitted from radioactive substances; have higher energy and are more penetrating than X-rays.

gas The state of matter in which the substance maintains neither shape nor volume.

gasoline The fraction of petroleum containing C_5 to C_{12} hydrocarbons, mainly alkanes, used as automotive fuel.

gene The segment of a nucleic acid molecule that contains the information necessary to produce a protein; the smallest unit of hereditary information.

general anesthetic A depressant that acts on the brain to produce unconsciousness as well as insensitivity to pain.

geometric growth A doubling in number for each growth period.

geothermal energy Energy derived from the heat of Earth's interior.

glass A noncrystalline material obtained by melting sand with soda, lime, and various other metal oxides.

globular protein A protein whose molecules fold into roughly spherical or ovoid shapes that can be dispersed in water.

glycogen A polymer of glucose with alpha linkages and branched chains; a storage form of starch in animals.

GRAS list A list, established by the U.S. Congress in 1958, of food additives generally recognized as safe.

greenhouse effect The retention of the sun's heat energy by Earth as a result of excess carbon dioxide or other substances in the atmosphere; causes an increase in Earth's atmospheric and surface temperatures.

ground state The state of an atom in which all electrons are in the lowest possible energy levels.

group A vertical column in the periodic table; a family of elements.

half-life The amount of time required for one-half the radioactive nuclei in a sample to decay.

hallucinogenic drug A drug that produces visions and sensations that are not part of reality.

halogen An element in Group 7A of the periodic table.

hard water Water containing ions of calcium, magnesium, and/or iron.

hazardous waste A waste that, when improperly managed, can cause or contribute to death or illness or threaten human health or the environment.

heat A measure of a quantity of energy; a measure of how much energy a sample contains.

heat capacity (of a substance) The quantity of heat needed to change the temperature of the substance by 1°C.

heat of vaporization (of a substance) The amount of heat involved in the evaporation or condensation of 1 g of the substance.

heat stroke A failure of the body's heat regulatory system; unless the victim is treated promptly, the rapid rise in body temperature will cause brain damage or death.

herbicide A material used to kill plants.

heterocyclic compound A cyclic compound in which one or more atoms in the ring is not carbon.

homogeneous The same throughout; property of a sample with the same composition in all parts.

homologous series A series of compounds in which adjacent members of the series differ by a fixed unit of structure.

hormone A chemical messenger secreted into the blood by an endocrine gland.

humectant A moistening agent.

hydrocarbon An organic compound that contains only carbon and hydrogen.

hydrogen bomb A bomb based on the nuclear fusion of isotopes of hydrogen.

hydrogen bond A type of intermolecular force in which a hydrogen atom covalently bonded in one molecule is simultaneously attracted to a nonmetal atom in a neighboring molecule. Both the atom to which the hydrogen atom is bonded and the one to which it is attracted must be small atoms of high electronegativity, usually N, O, or F.

hydrophilic Attracted to polar solvents such as water.

hydrophobic Not attracted to water; associated with other nonpolar entities.

hydrolysis The reaction of a substance with water; literally, a splitting by water.

hypoallergenic cosmetics Cosmetics claimed to cause fewer allergic reactions than regular products.

hypothesis A guess that can be tested by experiment.

ideal gas law The volume of a gas is proportional to the amount of gas and its Kelvin temperature and inversely proportional to its pressure.

induced radioactivity Radioactivity caused by bombarding a stable isotope with radioactive particles, forming a radioactive isotope.

industrial smog Polluted air associated with industrial activities, usually characterized by sulfur oxides and particulate matter.

inorganic chemistry The study of the compounds of all elements other than carbon.

insecticide A substance that kills insects.

iodine number The number of grams of iodine that will be consumed by 100 g of fat or oil; an indication of the degree of unsaturation.

ion A charged atom or group of atoms.

ionic bond The chemical bond that results when electrons are transferred from a metal to a nonmetal; the electrostatic attraction between ions of opposite charge.

ionizing radiation Radiation that produces ions as it passes through matter.

isomers Compounds that have the same molecular formula but different structural formulas and properties.

isotopes Atoms that have the same number of protons but different numbers of neutrons.

joule (J) The SI unit of energy (1 J = 0.239 cal).

juvenile hormone A hormone that controls the rate of development of the young; used to prevent insects from maturing.

kelvin (K) The SI unit of temperature. Zero on the Kelvin scale is absolute zero.

keratin The tough, fibrous protein that comprises most of the outermost layer of the epidermis.

kerogen The complex material found in oil shale; has an approximate composition of $(C_6H_8O)_n$, where n is a large number.

ketone (R—CO—R′) An organic compound with a carbonyl group between two carbon atoms.

kilocalorie A unit of energy equal to 1000 cal.

kilogram (kg) The SI unit of mass, a quantity equal to about 2.2 lb.

kinetic energy The energy of motion.

kinetic–molecular theory A model that uses the motion of molecules to explain the behavior of the three states of matter.

lactate threshold The upper limit of lactic acid concentration in muscle tissue above which the muscle is too fatigued to contract.

lanolin A natural wax obtained from sheep's wool.

law of combining volumes The volumes of gaseous reactants and products are in a small whole-number ratio when all measurements are made at the same temperature and pressure.

law of conservation of energy The amount of energy within the universe is constant; energy cannot be created or destroyed, only transformed.

law of conservation of mass Matter is neither created nor destroyed during a chemical change.

law of definite proportions A compound always contains elements in certain definite proportions, never in any other combination; also called the law of constant composition.

law of multiple proportions Elements may combine in more than one proportion to form more than one compound—for example, CO and CO_2.

LD$_{50}$ The dosage that would be lethal to 50% of a population of test animals.

levo isomer A "left-handed" isomer.

limiting reagent The reactant that is used up first in a reaction, after which the reaction ceases no matter how much remains of the other reactants.

line spectrum The pattern of colored lines emitted by an element.

lipid A substance from animal or plant cells that is soluble in nonpolar solvents and insoluble in water.

lipoprotein A protein combined with a lipid, such as a triglyceride or cholesterol.

liquid The state of matter in which the substance assumes the shape of its container, flows readily, and maintains a fairly constant volume.

liter (L) A unit of volume equal to a cubic decimeter.

local anesthetic A substance that renders part of the body insensitive to pain while leaving the patient conscious.

lotion An emulsion of submicroscopic fat or oil droplets dispersed in water.

main group element An element in the A groups of the periodic table (customary U.S. arrangement) and in Groups 1, 2, and 13 to 18 in the periodic table recommended by IUPAC.

marijuana A preparation made from the leaves, flowers, seeds, and small stems of the *Cannabis* plant.

mascara A product composed of a base of soap, oils, fats, and waxes and colored by iron oxide pigments, carbon, chromium oxide, or ultramarine; used to darken eyelashes.

mass A measure of the quantity of matter.

mass–energy equation Einstein's equation $E = mc^2$, in which E is energy, m is mass, and c is the speed of light.

mass number The nucleon number, the sum of the numbers of protons and of neutrons in the nucleus of an atom.

matter The stuff of which all materials are made; anything that has mass.

melanin A brownish-black pigment that determines the color of the skin and hair.

melting point The temperature at which a substance changes from the solid to the liquid state.

messenger RNA (mRNA) The type of RNA that contains the codons for a protein; travels from the nucleus of the cell to a ribosome.

metabolism The sum of all the chemical reactions by which the protoplasm of an organism grows, is maintained, obtains energy, and is degraded.

metalloid An element with properties intermediate between those of metals and those of nonmetals.

metals The group of elements to the left of the heavy, stepped, diagonal line in the periodic table.

meter (m) The SI unit of length, slightly longer than 39 in.

mica A mineral composed of SiO_4 tetrahedra arranged in a two-dimensional, sheetlike array.

micelle A spherical cluster of surfactant molecules arranged so that their hydrophilic ends all lie along the outer surface.

micronutrient A substance needed by the body in only tiny amounts.

middle note The portion of perfume intermediate in volatility; responsible for the lingering aroma after most top-note compounds have vaporized.

minerals (dietary) The inorganic substances required in the diet for good health.

mixture Matter with a variable composition.

moisturizer (skin) A substance that adds moisture to the skin or acts to retain moisture in the skin.

molarity (M) The concentration of a solution in moles of solute per liter of solution.

molar mass The formula mass of a substance expressed in grams.

molar volume The volume occupied by 1 mol of a substance under specified conditions.

mole (mol) The formula mass in grams of an element or compound; the amount of a chemical substance that contains 6.02×10^{23} formula units of the substance.

molecular mass The mass of a molecule of a substance; the sum of the atomic masses as indicated by the molecular formula, expressed in atomic mass units (u).

molecule Two or more atoms joined together by covalent bonds; the smallest fundamental unit of a molecular substance.

monomer A molecule of relatively low molecular mass. Monomers are combined to make polymers.

monosaccharide A carbohydrate that cannot be hydrolyzed into simpler sugars.

mousse A foam or froth; a hair care product composed of holding resins used to hold hair in place.

mutagen Any entity that causes changes in genes without destroying the genetic material.

narcotic A depressant, analgesic drug that induces narcosis (sleep).

natural gas A mixture of gases, mainly methane, found in many underground deposits.

neuron A nerve cell.

neurotransmitter A chemical that carries an impulse across a synapse from one nerve cell to the next.

neurotrophins Substances produced during exercise that promote the growth of brain cells.

neutralization The reaction of an acid and a base to produce a salt and water.

neutron A fundamental particle with a mass of approximately 1 u and no electric charge.

nitrogen cycle The various processes by which nitrogen is cycled among the atmosphere, soil, water, and living organisms.

noble gases Generally unreactive gases that appear in the far right column (Group 8A) of the periodic table.

nonbonding pair (NBP) A pair of electrons in the valence shell of an atom not involved in a bond; also called a lone pair.

nonionic surfactant A surfactant with a hydrocarbon tail and a polar head whose oxygen atoms attract water molecules and make the head water-soluble; bears no ionic charge.

nonmetals The group of elements to the right of the heavy, stepped, diagonal line in the periodic table.

nonpolar covalent bond A covalent bond in which there is an equal sharing of electrons.

nuclear fission The splitting of an atomic nucleus into two large fragments.

nuclear fusion The combination of two small atomic nuclei to produce one larger nucleus.

nuclear reactor A plant that produces energy by nuclear fission.

nuclear winter A period of dark, cold weather that may be caused by dust and smoke entering the atmosphere after the explosion of nuclear bombs.

nucleic acid A nucleotide polymer, DNA or RNA.

nucleon A proton or neutron in an atomic nucleus.

nucleon number (A) The total number of protons and neutrons in an atom; the mass number.

nucleotide A combination of a heterocyclic amine, a pentose sugar, and phosphoric acid; the monomer unit of nucleic acids.

nucleus Concentrated, positively charged matter at the center of an atom; composed of protons and neutrons.

octane rating The antiknock quality of a gasoline as compared with mixtures of isooctane (with a rating of 100) and heptane (with a rating of 0).

octet rule Atoms seek an arrangement that will surround them with eight electrons in the outermost energy level.

oil (fatty) A substance formed from glycerol and fatty acids; liquid at room temperature.

oil shale Fossil rock containing kerogen from which oil can be obtained at high cost by distillation.

optical brightener A compound that absorbs the invisible ultraviolet component of sunlight and reemits it as visible light at the blue end of the spectrum.

orbital A region of space in an atom occupied by one or two electrons.

organic chemistry The study of the compounds of carbon.

organic farming Farming without synthetic fertilizers or pesticides.

oxidation An increase in oxidation number; combination of an element or compound with oxygen; loss of hydrogen; loss of electrons.

oxidizing agent A substance that causes oxidation and is itself reduced.

oxygen cycle The various processes by which oxygen is cycled among the atmosphere, soil, water, and living organisms.

oxygen debt The demand for oxygen in muscle cells during anaerobic exercise.

ozone layer The layer of the stratosphere that contains ozone and shields living creatures on Earth from deadly ultraviolet radiation from the sun.

paint A surface coating that contains a pigment, a binder, and a solvent.

particulate matter (PM) A pollutant composed of solid and liquid particles of greater than molecular size.

peptide bond The amide linkage that bonds amino acids in chains of peptides, polypeptides, and proteins.

percent by mass The concentration of a solution in mass of solute divided by mass of solution with the quotient multiplied by 100%.

percent by volume The concentration of a solution in volume of solute divided by volume of solution with the quotient multiplied by 100%.

perfume A fragrant mixture of plant extracts and other chemicals dissolved in alcohol.

periodic table A systematic arrangement of the elements in columns and rows; elements in a given column have similar properties.

periods The horizontal rows of the periodic table.

pesticide A substance that kills some kind of pest (weeds, insects, rodents, etc.).

petroleum A dark, oily mixture of hydrocarbons, mainly alkanes, occurring in various deposits around the world.

pH The negative logarithm of hydronium ion concentration.

pharmacology The study of the response of living organisms to drugs.

phenol A compound with an OH group attached to a benzene ring.

pheromone A natural chemical secreted by an organism to mark a trail, send out an alarm, or attract a mate.

photochemical smog Smog created by the action of sunlight on unburned hydrocarbons and nitrogen oxides, mainly from automobiles.

photon A unit particle of energy.

photoscan A permanent visual record showing the differential uptake of a radioisotope by various tissues.

photosynthesis The chemical process used by green plants to convert solar energy into chemical energy by reducing carbon dioxide.

photovoltaic cell A solar cell; a cell that converts sunlight directly to electric energy.

physical change A change in physical state or form.

physical property A quality of a substance that can be demonstrated without changing the composition of the substance.

placebo A substance that looks and tastes like a real drug but has no active ingredients.

plasma A state of matter similar to a gas but composed of isolated electrons and nuclei rather than discrete whole atoms or molecules.

plasticizer A chemical substance added to some plastics, such as vinyl, to make them more flexible and easier to work with.

pleated sheet A secondary protein structure characterized by antiparallel molecules with a zigzag structure.

polar covalent bond A covalent bond in which more than half of the bond's negative charge is concentrated around one of the two atoms.

polar molecule A molecule that has a dipole moment.

pollutant A chemical that causes undesirable effects by being in the wrong place and/or in the wrong concentration.

polyamide A polymer that has structural units joined by amide linkages.

polyatomic ion An ion consisting of two or more atoms bonded together.

polyester A polymer made from a dicarboxylic acid and a dialcohol.

polymer A molecule with a large molecular mass; a chain formed of repeating smaller units.

polymerase chain reaction (PCR) A process that reproduces many copies of a DNA fragment.

polypeptide A polymer of amino acids, usually of lower molecular mass than a protein.

polysaccharide Carbohydrates, such as starch or cellulose, one molecule of which yields many molecules of monosaccharide(s) on hydrolysis.

polyunsaturated fat A fat containing fatty acid units with two or more carbon–carbon double bonds.

positron (β^+) A positively charged particle with the mass of an electron.

potential energy Energy by virtue of position or composition.

preemergent herbicide A herbicide that is rapidly broken down in the soil and can therefore be used to kill weed plants before crop seedlings emerge.

primary plant nutrients Nitrogen, phosphorus, and potassium.

primary sewage treatment Treatment of sewage in a plant with a holding pond intended to remove some of the sewage solids as sludge by settling.

primary structure The amino acid sequence in a protein or of nucleotides in a nucleic acid.

products Substances produced by a chemical reaction; their formulas follow the arrow in a chemical equation.

progestin A compound that mimics the action of progesterone.

prostaglandin A hormonelike compound derived from arachidonic acid that is involved in increased blood pressure, the contractions of smooth muscle, and other physiological processes.

protein An amino acid polymer.

proton The unit of positive charge in the nucleus of an atom; the hydrogen nucleus in acid–base chemistry.

psychotropic drugs Drugs that affect the mind.

purine A base with two fused rings found in nucleic acids.

pyrimidine A base with one ring found in nucleic acids.

quantum An energy packet of specific size; one photon of energy.

quartz A compound composed of SiO_4 tetrahedra arranged in a three-dimensional array.

radioactive decay The disintegration of an unstable atomic nucleus by spontaneous emission of radiation.

radioactivity The spontaneous emission of alpha, beta, or gamma rays by disintegration of the nuclei of atoms.

radioisotopes Radioactive isotopes.

reactants Starting materials or original substances in a chemical change; their formulas precede the arrow in a chemical equation.

reactive wastes Wastes that tend to react spontaneously or to react vigorously with air or water.

recommended daily allowance (RDA) The recommended level of a nutrient necessary for a balanced diet.

reducing agent A substance that causes reduction and is itself oxidized.

reduction A decrease in oxidation number; a gain of electrons; a loss of oxygen; a gain of hydrogen.

replication Copying or duplication; the process by which DNA reproduces itself.

resin A polymeric material, usually a sticky solid or semisolid organic material.

restorative drug A drug used to relieve the pain and reduce the inflammation resulting from overuse of muscles.

retrovirus An RNA virus that synthesizes DNA in the host cells.

reverse osmosis A method of pressure filtration through a semipermeable membrane; water flows from an area of high salt concentration to an area of low salt concentration.

ribonucleic acid (RNA) The form of nucleic acid found mainly in the cytoplasm, but also present in all other parts of the cell.

risk–benefit analysis An approach that estimates a desirability quotient by dividing the benefits by the risks.

rule of 72 A mathematical formula that gives the doubling time for a population growing geometrically; 72 divided by the annual rate equals the doubling time.

salt An ionic compound produced by the reaction of an acid with a base.

saponins Natural chemical compounds that produce a soapy lather.

saturated fat A fat composed of a large proportion of saturated fatty acids esterified with glycerol.

saturated hydrocarbon An alkane; a compound of carbon and hydrogen with only single bonds.

science A branch of knowledge based on the laws of nature.

scientific law A summary of experimental data; often expressed in the form of a mathematical equation.

scientific model A representation that serves to explain a scientific phenomenon.

sebum An oily secretion of the body that protects the skin from moisture loss.

second law of thermodynamics The degree of randomness in the universe increases in any spontaneous process.

secondary plant nutrients Magnesium, calcium, and sulfur.

secondary sewage treatment Passing effluent from a primary treatment plant through gravel and sand filters to aerate the water and remove suspended solids.

secondary structure The arrangement of polypeptide chains in a protein—for example, helix or pleated sheet.

set-point theory An explanation of the difficulty of losing weight by dieting that holds that each person has an individual level of circulating fatty acids below which he or she is constantly hungry.

sex attractant A substance or mixture of substances released by an organism to attract members of the opposite sex of the same species for mating.

significant figures Those measured digits that are known with certainty plus one uncertain digit.

silicone A polymer with a base chain of alternating silicon and oxygen atoms.

single bond A pair of electrons shared between two atoms.

SI units (International System of Units) A measuring system used by scientists worldwide; it is based on seven base quantities and their multiples and submultiples.

skin protection factor (SPF) The rating of a sunscreen's ability to limit the penetration of ultraviolet radiation.

slag A relatively low-melting product of the reaction of limestone with silicate impurities in iron ore.

slow-twitch fibers Muscle fibers suited for aerobic work.

smog The combination of smoke and fog; polluted air.

soap A mixture of salts (usually sodium salts) of long-chain carboxylic acids.

solar cell A device used for converting sunlight to electricity; a photoelectric cell.

solid A state of matter in which the substance maintains its shape and volume.

solute The substance that is dissolved in another substance (solvent) to form a solution; usually present in a smaller amount than the solvent.

solution A homogeneous mixture of two or more substances.

solvent The substance that dissolves another substance (solute) to form a solution; usually present in a larger amount than the solute.

specific heat (of a substance) The amount of heat required to raise the temperature of 1 g of the substance by 1°C.

standard temperature and pressure (STP) Conditions of 0°C and 1 atm pressure.

starch A polymer of glucose units joined through alpha linkages; a complex carbohydrate.

starvation The withholding of nutrition from the body, whether voluntary or involuntary.

steel An alloy of iron containing small amounts of carbon and usually containing other metals such as manganese, nickel, and chromium.

steroid A molecule that has a four-ring skeletal structure, with one cyclopentane and three cyclohexane fused rings.

stimulant drug A drug that increases alertness, speeds up mental processes, and generally elevates the mood.

stoichiometric factor A factor that relates the amounts of two substances through their coefficients in a chemical equation.

stoichiometry Quantity relationships between reactants and products in a chemical reaction.

strong acid An acid that ionizes completely in water; a powerful proton donor.

strong base A base that dissociates completely in water; a powerful proton acceptor.

structural formula A chemical formula that shows how the atoms of a molecule are arranged, to which other atom(s) they are bonded, and the kinds of bonds.

sublevel A subdivision of electron energy levels in an atom; also called a subshell.

substance A type of matter having a definite composition and fixed properties that do not vary from one sample to another.

substrate The substance that bonds to the active site of an enzyme; the substance acted on.

sublevel A division of a main energy level of an atom; a subshell.

sunscreen lotion A type of lotion that promotes tanning of the skin by blocking out short-wave ultraviolet radiation while allowing longer-wave ultraviolet radiation to pass through.

surface-active agent Any agent that stabilizes the suspension in water of a nonpolar substance such as oil.

surfactant A surface-active agent.

synapse A tiny gap between nerve fibers.

synergistic effect An effect much greater than the sum of the expected effects.

tar sands Sands that contain bitumen, a thick hydrocarbon material.

technology The sum total of processes by which humans modify the materials of nature to better satisfy their needs and wants.

temperature A measure of heat intensity, or how energetic the particles of a sample are.

teratogen A toxic substance that causes birth defects when introduced into the body of a pregnant female.

tertiary structure The folds, bends, and twists in protein or nucleic acid structure.

tetracyclines Antibacterial drugs with four fused rings.

theory A detailed explanation of the behavior of matter based on experiments; may be revised if new data warrant.

thermonuclear reactions Nuclear fusion reactions that require extremely high temperatures and pressures.

thermoplastic polymer A kind of polymer that can be heated and reshaped.

thermosetting polymer A kind of polymer that cannot be softened and remolded.

top note The portion of perfume that vaporizes most quickly; composed of relatively small molecules; responsible for odor when perfume is first applied.

toxicology The division of pharmacology that deals with the effects of poisons on the body, their identification and detection, and remedies for them.

toxic waste A waste that contains or releases poisonous substances in amounts large enough to threaten human health or the environment.

tracers Radioactive isotopes used to trace movement or locate the sites of radioactivity in physical, chemical, and biological systems.

training effect The net effect, acquired through repeated exercise, of being able to do more physical work with less strain.

transcription The process by which DNA directs the synthesis of an mRNA molecule during protein synthesis.

transfer RNA (tRNA) A small molecule that contains anticodon nucleotides; the RNA molecule that bonds to an amino acid.

transition elements Metallic elements situated in the center portion of the periodic table, in the B groups (customary United States arrangement), and in Groups 3 to 12 in the periodic table recommended by IUPAC.

translation The process by which the information contained in the codon of an mRNA molecule is converted to a protein structure.

transmutation The changing of one element into another.

triglyceride A chemical combination of glycerol with three fatty acids; also called a triacylglycerol.

triple bond The sharing of three pairs of electrons between two atoms.

tritium A radioactive isotope of hydrogen with two neutrons and one proton in the nucleus (hydrogen-3).

unsaturated hydrocarbon An alkene or alkyne; a hydrocarbon containing a double or a triple bond.

valence electrons Electrons in the outermost shell.

valence shell electron pair repulsion (VSEPR) theory A theory of chemical bonding useful in determining the shapes of molecules; it states that valence shell electron pairs locate themselves as far apart as possible.

vaporization The process by which a substance changes from the liquid to the gaseous (vapor) state.

variable A factor that changes during an experiment.

vitamin An organic compound that the body cannot produce in the quantity required for good health.

volatile organic compounds Compounds that cause pollution because they vaporize readily.

VSEPR theory See valence shell electron pair repulsion (VSEPR) theory.

vulcanization The process of making naturally soft rubber harder by reaction with sulfur.

wax An ester of a long-chain fatty acid with a long-chain alcohol.

weak acid An acid that ionizes only slightly in water; a poor proton donor.

weak base A base that ionizes only slightly in water; a poor proton acceptor.

weight A measure of the force of attraction of Earth for an object.

wet scrubber A pollution control device that uses water or solutions to remove pollutants from smokestack gases.

X-rays Radiation similar to visible light but of much higher energy and much more penetrating.

zwitterion A molecule that contains both a positive charge and a negative charge; a dipolar ion.

Appendix C

Answers

Answers are provided for all in-chapter exercises, for selected review questions, for the odd-numbered problems in the matched sets, and for selected additional problems.

NOTE: For numerical problems, your answer may vary slightly from ours because of rounding and the use of significant figures (Appendix A).

Chapter 1

1.1 **A. a.** DQ would be small for using it to treat sick farm animals.

b. DQ would be large for the person with Rocky Mountain spotted fever.

B. Society considers cheap energy from coal worth the risks. Environmentalists disagree.

1.2 **A. a.** 1.00 kg   **b.** 99 lb

    B. a. 52 kg   **b.** 480 lb

1.3 **A.** physical: a, c; chemical: b

    B. physical: b, d; chemical: a, c

1.4 elements: He, No; compounds: CuO, NO, KI

1.5 **a.** 7.24 mg **b.** 5.14 μm **c.** 1.91 ns **d.** 5.58 km

1.6 **a.** 7.45×10^{-7} m   **b.** 5.25×10^{-7} s

    c. 1.415×10^{6} m   **d.** 2.06×10^{3} kg

1.7 **A.** 1.11 g/mL   **B.** 9.3 g/cm^3

1.8 **A.** 63.2 g   **B.** 1.76 kg

1.9 **A.** 5.75 cm^3 **B.** 15.2 mL

1.10 **A.** 373 K   **B.** 195 K

1. Chemistry is the study of matter and the changes it undergoes. A chemical is a type of matter.

3. Science is testable, reproducible, explanatory, predictive, and tentative. Testability is what best distinguishes science.

5. Alchemy is a mystical chemistry that flourished during the Middle Ages.

7. It was Bacon's dream that science would solve all the world's problems, increasing happiness and prosperity.

9. Technological progress has thus far kept food production up with population growth in the developed countries.

11. A scientific law is a summary of observations and data about a given subject.

13. Some problems involve too many variables to be treated by scientific methods.

21. Chemical changes involve change in chemical composition; physical changes do not.

25. A solid has a fixed shape and a fixed volume; a liquid has a fixed volume, but takes the shape of its container; a gas has neither a fixed shape nor a fixed volume. It takes the shape and volume of its container.

27. the cubic meter; liters, milliliters, cubic centimeters

31. DQ is small.

33. DQ is uncertain.

35. matter: a, c

37. yes

41. substances: a, b; mixtures: c, d

43. a substance; the law of definite proportions

45. elements: H, He, Ca; compound: HF

47. **a.** C   **b.** Cl   **c.** Fe

49. **a.** hydrogen **b.** oxygen **c.** sodium

51. 1000 mm; 100 cm

53. 7500 mm; 460 mm

55. **a.** cm   **b.** kg   **c.** dL

61. **a.** 120 g   **b.** 21.8 g

63. **a.** 53.1 cm^3 **b.** 18.7 mL

65. the brass weight

67. yardstick (0.91 m) < snake (1.04 m) < chain (1.2 m) < board (1.9 m)

69. the shorter one

71. 1.51 g/mL

Chapter 2

2.1 **A.** 1.22 g arsenic   **B.** 63 g nitrogen

2.2 4 atoms hydrogen/1 atom carbon

1. The atomistic theory assumes that matter is made up of small unit particles that cannot be further subdivided and still be the same kind of matter.

3. atomistic: c, d; continuous: a, b, f

5. When a chemical change occurs, matter is neither created nor destroyed.

11. law of conservation of mass

13. law of multiple proportions

17. 11.00 g; law of definite proportions

19. Water is a compound of hydrogen and oxygen.

21. No. The helium atoms are small enough to pass through the pores in the balloon.

23. No. They have been converted to gases (carbon dioxide and water vapor).

25. Yes. Total mass before and after the reaction is the same, 1.2000 g.

29. No. Dalton assumed that atoms of different elements had different masses.

35. 40 g hydrogen

37. 270 g carbon

Chapter 3

3.1 52 neutrons

3.2 48 neutrons

3.3 1st and 3rd are isotopes; 4th and 5th are isotopes

3.4 32 electrons

3.5 Be 2, 2; Mg 2, 8, 2; Ca 2, 8, 8, 2; all have the same number of outer electrons.

3.6 (F) $1s^2 2s^2 2p^5$

3.7 (Cl) $1s^2 2s^2 2p^6 3s^2 3p^5$

3.8 (Rb) $5s^1$ (Se) $4s^2 4p^4$

5. Isotopes are atoms of the same element with different masses (different numbers of neutrons).

7. A and D are isotopes; B and C are isotopes.

15. electrons

17. 10 electrons

19. argon

21. number of protons and neutrons (equal to the mass number)

23. 32 electrons

29. emitted

31. 6 electrons

33. Group 7 B. It is a transition metal.

35. the atom in which an electron moves from the first to the third energy level

37. **a.** 2 **b.** 11 **c.** 17
    d. 8 **e.** 12 **f.** 16

39. The "neutral" atom has 4 electrons but only 3 protons.

43. Metals are shiny, malleable, and ductile. They form positive ions and conduct electricity.

45. metal: chromium; nonmetals: sulfur and iodine

47. **a.** Cl, period 3 **b.** Os, period 6 **c.** H, period 1

49. lack of chemical reactivity

51. K is an alkali metal.

53. Ti and Tc are transition metals

57. $^{8}_{5}B$

59. $^{125}_{53}I$

61. **a.** 30 protons, 32 neutrons
    b. 94 protons, 147 neutrons
    c. 43 protons, 56 neutrons
    d. 36 protons, 45 neutrons

63. two; spherical; one

67. **a.** Be **b.** N **c.** Al

69. **a.** The $3s$ orbitals have filled up before the $2p$ orbitals.
    b. There are only 2 electrons in the "filled" $2p$ sublevel.
    c. There is no $2d$ energy sublevel.

71. Fluorine and chlorine have the same number of electrons in their outermost shells, but chlorine has one more electron shell than fluorine.

73. Sulfur has 4 $3p$ electrons, whereas chlorine has 5.

75. argon

Chapter 4

4.1 **a.** $^{250}_{100}Fm \longrightarrow {}^{4}_{2}He + {}^{246}_{98}Cf$

 b. $^{85}_{34}Se \longrightarrow {}^{0}_{-1}e + {}^{85}_{35}Br$

 c. $^{188}_{79}Au \longrightarrow {}^{0}_{+1}e + {}^{188}_{78}Pt$

 d. $^{37}_{18}Ar + {}^{0}_{-1}e \longrightarrow {}^{37}_{17}Cl$

4.2 0.03825 mg

4.3 0.5 mg

4.4 22,920 y

4.5 a neutron

4.6 silicon-30

7. $^{83}_{35}Br$

11. b and c

13. 97

17. Atomic number increases by 1; nucleon number does not change.

19. Both the atomic number and nucleon number are decreased by 1.

21. no changes

27. alpha particles

29. Alpha particles are more massive and carry a greater charge than beta particles.

31. Use shielding; move farther away from the source.

35. medical X-rays

39. $^{31}_{16}S \longrightarrow {}^{31}_{15}P + {}^{0}_{+1}e$

41. $^{21}_{12}Mg + {}^{20}_{11}Na + {}^{1}_{1}H$

43. $^{225}_{90}Th \longrightarrow {}^{221}_{88}Ra + {}^{4}_{2}He$

45. **a.** $^{179}_{79}Au \longrightarrow {}^{175}_{77}Ir + {}^{4}_{2}He$
    b. $^{23}_{10}Ne \longrightarrow {}^{23}_{11}Na + {}^{0}_{-1}e$

47. $^{24}_{11}Na$

49. $^{215}_{85}At$

51. 1500; 750

53. 6.25 mg

55. 5730 yr

57. 5730 yr; 11,460 yr

59. $^{18}_{8}O$

61. atomic number 105; nucleon number 258

63. 420 cm^3

Chapter 5

5.1 **a.** $:\overset{..}{\underset{..}{Ar}}:$ **b.** $\cdot Ca \cdot$ **c.** $:\overset{..}{\underset{..}{F}}\cdot$

d. $:\overset{..}{N}\cdot$ **e.** $K\cdot$ **f.** $:\overset{..}{\underset{.}{S}}\cdot$

5.2 $Li\cdot + :\overset{..}{\underset{..}{F}}\cdot \longrightarrow Li^+ + :\overset{..}{\underset{..}{F}}:^-$

5.3 $2\cdot Al\cdot + 3:\overset{..}{\underset{.}{O}}\cdot \longrightarrow 2\,Al^{3+} + 3:\overset{..}{\underset{..}{O}}:^{2-}$

5.4 CaF_2

5.5 K_2O

5.6 Ca_3N_2

5.7 CaS

5.8 calcium fluoride

5.9 copper(II) bromide

5.10 bromine trifluoride; bromine pentafluoride

5.11 N_2O_5

5.12 P_4Se_3

5.13 **a.** $:\overset{..}{\underset{..}{Br}}\cdot + :\overset{..}{\underset{..}{Br}}\cdot \longrightarrow :\overset{..}{\underset{..}{Br}}:\overset{..}{\underset{..}{Br}}:$

b. $H\cdot + :\overset{..}{\underset{..}{Br}}\cdot \longrightarrow H:\overset{..}{\underset{..}{Br}}:$

c. $:\overset{..}{\underset{..}{I}}\cdot + :\overset{..}{\underset{..}{Cl}}\cdot \longrightarrow :\overset{..}{\underset{..}{I}}:\overset{..}{\underset{..}{Cl}}:$

5.14 $Ca(CH_3CO_2)_2$

5.15 $CaHPO_4$

5.16 calcium carbonate

5.17 potassium dichromate

5.18 iron(III) carbonate

5.19
```
      H  H
   H:C:C:Cl:
      ..  ..
      H  H
```

5.20 $:\overset{..}{\underset{..}{F}}:\overset{..}{\underset{..}{O}}:\overset{..}{\underset{..}{F}}:$

5.21
$$\left[\begin{array}{c} H \\ H:\overset{..}{P}:H \\ H \end{array} \right]^+$$

5.22 $:\overset{..}{O}::N:\overset{..}{\underset{..}{O}}:\overset{..}{\underset{..}{F}}:$

5.23 linear

5.24 pyramidal

1. the noble gases

3. Sodium metal is quite reactive; sodium ions are quite stable.

7. **a.** $\cdot\overset{.}{C}\cdot$ **b.** $K\cdot$ **c.** $\cdot Mg\cdot$

d. $:\overset{..}{\underset{..}{Cl}}\cdot$ **e.** $:\overset{..}{N}\cdot$

15. $:\overset{..}{\underset{..}{I}}\cdot + \cdot\overset{..}{\underset{..}{I}}: \longrightarrow :\overset{..}{\underset{..}{I}}:\overset{..}{\underset{..}{I}}:$

17. a. 3+ **b.** 2− **c.** 1+ **d.** 1−

19. ClO_2

21. AlP; Mg_3P_2

27. ionic: a, c; polar covalent: b

29. nonpolar covalent: Br_2, F_2; polar covalent: HCl

31. polar: all

35. polar

37. C, N, O

39. molecules with odd number of electrons (NO_2),those with incomplete octets (BF_3), and those with expanded octets (SF_6)

41. dispersion forces (N_2), dipole interactions (HCl), hydrogen bonds (H_2O), and interionic forces (NaCl)

43. a. melting **b.** vaporization

47. a. sodium ion **b.** magnesium ion

c. aluminum ion **d.** chloride ion

e. oxide ion **f.** nitride ion

49. a. iron (III) ion **b.** copper (II) ion

c. silver ion

51. a. Br **b.** Ca^{2+} **c.** K^+ **d.** Fe^{2+}

53. a. nitrate **b.** sulfate

c. dihydrogen phosphate **d.** hydrogen carbonate

55. a. PO_4^{3-} **b.** HCO_3^- **c.** $Cr_2O_7^{2-}$ **d.** $C_2O_4^{2-}$

57. a. $\cdot Ca\cdot$ and Ca^{2+}

b. $:\overset{..}{\underset{.}{S}}\cdot$ and $:\overset{..}{\underset{..}{S}}:^{2-}$

c. $Rb\cdot$ and Rb^+

d. $:\overset{.}{P}\cdot$ and $:\overset{..}{\underset{..}{P}}:^{3-}$

59. a. Ca 2, 8, 8, 2 Ca^{2+} 2, 8, 8
b. S 2, 8, 6, S^{2-} 2, 8, 8
c. Rb 2, 8, 18, 8, 1 Rb^+ 2, 8, 18, 8
d. P 2, 8, 5 P^{3-} 2, 8, 8

61. **a.** Na^+ $:\overset{..}{\underset{..}{F}}:^-$ **b.** K^+ $:\overset{..}{\underset{..}{Cl}}:^-$

c. K^+ $:\overset{..}{\underset{..}{F}}:^-$

63. **a.** $3\,Na^+$ $:\overset{..}{\underset{..}{N}}:^{3-}$ **b.** Al^{3+} $3 :\overset{..}{\underset{..}{Cl}}:^-$

c. $3\,Mg^{2+}$ $2 :\overset{..}{\underset{..}{N}}:^{3-}$

65. $\cdot\overset{.}{Si}\cdot + 4\,H\cdot \longrightarrow$
```
      H
   H:Si:H
      ..
      H
```

67. $:\overset{..}{N}\cdot + 3:\overset{..}{\underset{..}{Cl}}\cdot \longrightarrow$
```
      :Cl:
      ..  ..
   :N:Cl:
      ..  ..
      :Cl:
```

69. a. $\overset{\delta-}{N}-\overset{\delta+}{H}$ b. $\overset{\delta+}{C}-\overset{\delta-}{F}$ c. $C-C$

71. a. potassium chloride   b. magnesium bromide
  c. copper (II) iodide   d. calcium sulfide
  e. iron (II) chloride   f. aluminum oxide

73. a. $CaCO_3$ b. KH_2PO_4 c. $Mg(CN)_2$ d. $LiHSO_4$

77. a. sodium hydrogen sulfate
  b. aluminum hydroxide
  c. sodium carbonate
  d. potassium hydrogen carbonate
  e. ammonium nitrite
  f. calcium hydrogen sulfate

79. a. lithium carbonate   b. sodium dichromate
  c. calcium dihydrogen phosphate
  d. ammonium oxalate

81. a. OF_2 b. N_2O_5 c. PBr_3 d. S_4N_4

83. a. carbon tetrabromide   b. dichlorine heptoxide
  c. tetraphosphorus decasulfide
  d. diiodine pentoxide

91. a. linear   b. triangular

93. a. bent   b. tetrahedral

95. a. pyramidal   b. tetrahedral

97. a. [Lewis structure of PF_5]   b. [Lewis structure of $AlBr_3$]

99. a. [Lewis structure of NF_3]   b. [Lewis structure $H:C::C:H$]

  c. $H:C:::C:H$ d. [Lewis structure $H:C::O$ with H]

101. a. [structure of NF_3]   b. [structure $H_2C=CH_2$]

  c. $H-C\equiv C-H$ d. [structure $H-C=O$ with H]

103. a. [Lewis structure $S:Cl$ with Cl]   b. [Lewis structure of sulfate/H_2SO_4]

  c. $:O:Xe:O:$ with O   d. [Lewis structure of $HClO_3$]

105. a. $:C:::N:^-$ b. [Lewis structure of IO_4^-]

c. [Lewis structure of HSO_4^-]   d. [Lewis structure of PO_4^{3-}]

Chapter 6

6.1 **A.** $N_2 + 3H_2 \longrightarrow 2NH_3$
  B. $P_4 + 6H_2 \longrightarrow 4PH_3$

6.2 $C_3H_8 + 5O_2 \longrightarrow 3CO_2 + 4H_2O$

6.3 $2H_3PO_4 + 3Ca(OH)_2 \longrightarrow Ca_3(PO_4)_2 + 6H_2O$

6.4 **A.** 1.48 L CO_2 **B.** 10.8 L CO_2

6.5 a. 147.0 u   b. 98.96 u   c. 98.00 u

6.6 a. 138.2 u   b. 294.2 u   c. 342.2 u

6.7 a. 1000 g H_2O b. 0.756 g $C_4H_{10}O$
  c. 73.7 g C_6H_6

6.8 a. 0.0664 mol Fe   b. 0.776 mol H_3PO_4
  c. 2.84 mol C_4H_{10}

6.9 **A.** 1.78 g/L   **B.** 73.9 g/mol

6.10 Molecular: 2 molecules of H_2S react with 3 molecules of O_2 to form 2 molecules of SO_2 and 2 molecules of H_2O.
  Molar: 2 mol of H_2S react with 3 mol of O_2 to form 2 mol of SO_2 and 2 mol of H_2O.
  Mass: 68.2 g of H_2S react with 96.0 g of O_2 to form 128 g of SO_2 and 36.0 g of H_2O

6.11 a. 1.59 mol   b. 305 mol   c. 0.612 mol

6.12 **A.** 0.763 g O_2 **B.** a. 2130 g CO_2 b. 2350 g CO_2

6.13 **A.** 39.0 g NH_3
  B. a. 0.967 g O_2 b. 8.02 g P_4O_{10} c. 21.4 g Mg

6.14 400 mmHg

6.15 **A.** a. 1800 mL   b. 503 mmHg

6.16 **A.** a. 2.98 L   b. 234 L   **B.** $-167°C$

6.17 **A.** a. 0.0471 atm   b. 5.00 L   **B.** 97.4 K

6.18 **A.** 0.00870 M   **B.** 0.968 M

6.19 a. 9.00 M   b. 1.26 M   c. 0.274 M
  d. 0.0242 M   e. 0.123 M   f. 9.23 M

6.20 a. 673 g KOH   b. 5.61 g KOH
  c. 0.0561 g KOH d. 4.63 g KOH

6.21 **A.** 0.0297 L   **B.** 16.3 mL

6.22 46.8%

6.23 Dilute 22.3 mL of acetic acid with enough water to make 67.5 mL of solution.

6.24 2.81%

6.25 Add 15.1 g of glucose to 260 g of water.

  2. atomic mass O = 15.9994 u;
    formula mass O_2 = 31.9988 u

  6. a. 22.4 L   b. 22.4 L   c. 22.4 L

  7. a. 4.003 g   b. 2.016 g   c. 30.07 g

  19. a. 4   b. 4   c. 8   d. 6

  21. 2 Al, 12 C, 18 H, 12 O

23. balanced: a, b; not balanced: c, d

25. a. $Cl_2O_5 + H_2O \longrightarrow 2\,HClO_3$
    b. $V_2O_5 + 2\,H_2 \longrightarrow V_2O_3 + 2\,H_2O$
    c. $4\,Al + 3\,O_2 \longrightarrow 2\,Al_2O_3$
    d. $Sn + 2\,NaOH \longrightarrow Na_2SnO_2 + H_2$
    e. $PCl_5 + 4\,H_2O \longrightarrow H_3PO_4 + 5\,HCl$
    f. $Na_3P + 3\,H_2O \longrightarrow 3\,NaOH + PH_3$
    g. $Cl_2O + H_2O \longrightarrow 2\,HClO$
    h. $2\,CH_3OH + 3\,O_2 \longrightarrow 2\,CO_2 + 4\,H_2O$
    i. $3\,Zn(OH)_2 + 2\,H_3PO_4 \longrightarrow Zn_3(PO_4)_2 + 6\,H_2O$
    j. $C_3H_8 + 5\,O_2 \longrightarrow 3\,CO_2 + 4\,H_2O$

27. a. 16.0 u   b. 84.0 u   c. 352.0 u

29. a. 156.9 u   b. 178.0 u   c. 294 u   d. 342 u

31. 5.00 L $CH_4(g)$

33. a. 2.65 L $H_2O(g)$ b. 105 L $O_2(g)$

35. 1.78 g/L

37. 52.6 g/mol

39. 67.6 g/mol

41. a. 156.9 g/mol   b. 98.00 g/mol
    c. 294.2 g/mol

43. a. 0.435 g   b. 47.0 g   c. 45.0 g

45. a. 1.56 mol   b. 0.0285 mol
    c. 0.0600 mol   d. 0.0356 mol

47. a. 16.7 mol   b. 55.9 mol

49. a. 2490 g NH_3 b. 193 g H_2

51. 2.64×10^6 kg

53. a. 1090 mL   b. 2600 mmHg

55. a. 9120 L   b. 19 h

57. 117 mL

59. 433K (or 160°C)

61. a. volume decreases   b. volume decreases
    c. volume increases

63. a. temperature decreases   b. pressure decreases

65. a. 22.3 L   b. 29 atm

67. 0.0781 mol

69. a. 2.40 M   b. 0.700 M

71. a. 0.907 M   b. 1.95 M

73. a. 80.0 g   b. 7.65 g

75. 0.208 L

77. 2.05 L

79. a. 4.83%   b. 5.09%

81. a. 3.96%   b. 7.32%

83. Add 77.5 g NaCl to 697.5 g water.

85. Add 0.0400 L acetic acid to enough water to make 2.00 L of solution.

89. 0.798 g

91. 2.20 g

93. 0.0521 g

95. 160 g

Chapter 7

7.1 $HBr(aq) \longrightarrow H^+(aq) + Br^-(aq)$

7.2 $HBr(aq) + H_2O \longrightarrow H_3O^+(aq) + Br^-(aq)$

7.3 HNO_3

7.4   KOH

7.5 $Ca(OH)_2(aq) + 2\,HCl(aq) \longrightarrow CaCl_2(aq) + 2\,H_2O$

7.6   11

7.7 1.0×10^{-2} M (0.01 M)

4. hydrogen ion (H^+) [or hydronium ion (H_3O^+)]

5. hydroxide ion (OH^-)

23. acids: b, c; base: a

25. a. HCl   b. H_2SO_4 c. H_2CO_3 d. HCN

27. a. LiOH   b. $Mg(OH)_2$ c. NaOH

29. $HCl(aq) + H_2O \longrightarrow H_3O^+(aq) + Cl^-(aq)$; hydrochloric acid

31. H_2SO_4 (an acid)

32. KOH (a base)

35. strong base

37. weak acid

39. a. $HI(aq) \longrightarrow H^+(aq) + I^-(aq)$
    b. $HNO_2(aq) \longrightarrow H^+(aq) + NO_2^-(aq)$
    c. $HClO_2(aq) \longrightarrow H^+(aq) + ClO_2^-(aq)$

41. $NaOH(aq) + HCl(aq) \longrightarrow NaCl(aq) + H_2O$

43. $Ca(OH)_2(aq) + 2\,HCl(aq) \longrightarrow CaCl_2(aq) + 2\,H_2O$

45. $H_3PO_4(aq) + 3\,NaOH(aq) \longrightarrow Na_3PO_4(aq) + 3\,H_2O$

47. acidic: a, c; basic: d; neutral: b

49. 3

51. 1.0×10^{-5}

55. none

57. 12

Chapter 8

8.1   oxidation: a, b, c, d

8.2   reduction: a; oxidation: b

8.3   reduction: a; oxidation: b, c, d

8.4   a. oxidizing agent: O_2; reducing agent: Se
       b. oxidizing agent: CH_3CN; reducing agent: H_2
       c. oxidizing agent: V_2O_5; reducing agent: H_2
       d. oxidizing agent: Br_2; reducing agent: K

8.7   A. $2\,Zn + O_2 \longrightarrow 2\,ZnO$
       B. $Se + O_2 \longrightarrow SeO_2$

8.8 $2\,PbS + 3\,O_2 \longrightarrow 2\,PbO + 2\,SO_2$

13. carbon (C), hydrogen (H_2), hydroquinone [$C_6H_4(OH)_2$]

17. oxidized: a (C_2H_4O gains O); reduced: b (H_2O_2 loses O)

19. neither; both are being reduced.

21. a. oxidizing agent: O_2; reducing agent: Al
    b. oxidizing agent: O_2; reducing agent: SO_2

23. a. oxidizing agent: HCl (specifically H^+); reducing agent: Fe

25. a. reduction: $Ag^+(aq) + e^- \longrightarrow Ag(s)$;
   oxidation: $Cu(s) \longrightarrow Cu^+(aq) + 2\,e^-$
b. reduction: $2\,H^+(aq) + 2\,e^- \longrightarrow H_2(g)$;
   oxidation: $Fe(s) \longrightarrow Fe^{2+}(aq) + 2\,e^-$

27. a. oxidation: $2\,I^- \longrightarrow I_2 + 2\,e^-$;
   reduction: $Cl_2 + 2\,e^- \longrightarrow 2\,Cl^-$;
   overall: $2\,I^- + Cl_2 \longrightarrow I_2 + 2\,Cl^-$
b. reduction: $SO_2 + 2\,H_2O \longrightarrow H_2SO_4 + 2\,H^+ + 2\,e^-$;
   oxidation: $HNO_3 + H^+ + e^- \longrightarrow NO_2 + H_2O$;
   overall: $SO_2 + 2\,HNO_3 \longrightarrow 2\,NO_2 + H_2SO_4$

29. a. S is oxidized; N is reduced
b. I is oxidized; Cr is reduced

31. reduced; it gains hydrogen

33. I^- is oxidized; Cl_2 is reduced

35. reduced

37. Nitrite ion is reduced; ascorbic acid is the reducing agent.

39. a. SO_2 **b.** H_2O
   c. $CO_2 + H_2O$ **d.** $CO_2 + SO_2$

41. Indoxyl is oxidized; O_2 is the oxidizing agent.

43. to protect them from oxygen and thus prevent corrosion

47. 10,800 L air; 2300 L O_2

48. Water is oxidized; CO_2 is the oxidizing agent. CO_2 is reduced; water is the reducing agent.

Chapter 9

9.1 C_8H_{18},

$$H-\overset{\displaystyle H}{\underset{\displaystyle H}{C}}-\overset{\displaystyle H}{\underset{\displaystyle H}{C}}-\overset{\displaystyle H}{\underset{\displaystyle H}{C}}-\overset{\displaystyle H}{\underset{\displaystyle H}{C}}-\overset{\displaystyle H}{\underset{\displaystyle H}{C}}-\overset{\displaystyle H}{\underset{\displaystyle H}{C}}-\overset{\displaystyle H}{\underset{\displaystyle H}{C}}-\overset{\displaystyle H}{\underset{\displaystyle H}{C}}-H$$

$CH_3CH_2CH_2CH_2CH_2CH_2CH_2CH_3$

9.2   A.

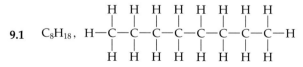

B. C_nH_{2n}

9.3 alcohol: a; ethers: b, c, e; phenol: d

9.4   A. $CH_3OCH_2CH_2CH_3$
   B. $CH_3CH_2OC(CH_3)_3$

9.5 aldehydes: b, c; ketone: a

9.6   A. a. $CH_3CH_2CH_2COOH$ **b.** $CH_3CH_2CH_2OH$

   c. $H-\overset{\displaystyle H}{\underset{\displaystyle H}{C}}-\overset{\displaystyle O}{\overset{\displaystyle \|}{C}}-H$ **d.** $H-\overset{\displaystyle H}{\underset{\displaystyle H}{C}}-\overset{\displaystyle H}{\underset{\displaystyle H}{C}}-\overset{\displaystyle O}{\overset{\displaystyle \|}{C}}-\overset{\displaystyle H}{\underset{\displaystyle H}{C}}-\overset{\displaystyle H}{\underset{\displaystyle H}{C}}-H$

   B. a. $CH_3CH_2CH_2CH_2CH_2COOH$
   b. $CH_3CH_2CH_2CH_2CH_2OH$
   c. $CH_3CH_2COCH_2CH_2CH_2CH_3$
   d. $CH_3CH_2CH_2CH_2CH_2CH_2CHO$

9.7   a. $CH_3CH_2CH_2CH_2NH_2$
   b. $CH_3CH_2NHCH_2CH_3$
   c. $CH_3NHCH_2CH_2CH_3$
   d. $CH_3\underset{\displaystyle CH_3}{\overset{\displaystyle |}{C}}HNHCH_3$

9.8   a. amine (NH)   **b.** amide (CONH)
   c. amine (N)   **d.** both (NH and $CONH_2$)

9.9 heterocyclic compounds: a, d

5. Isomers have the same molecular formula.

9. 1–4 C, gases; 5–16 C, liquids; > 18 C, solids

11. a. ethyl alcohol   **b.** isopropyl alcohol
   c. methyl alcohol

13. anesthetic; solvent

15. Aldehydes have a hydrogen atom attached to the carbonyl carbon atom.

17. esters

19. organic: a, b; inorganic: c, d

21. a. 2   **b.** 4   **c.** 7   **d.** 5

23. a. ethane   **b.** acetylene (ethyne)
   c. ethylene (ethene)

25. $CH_3CH_2CH_2CH_3$ (butane)
   $CH_3CH(CH_3)_2$ (isobutane)

27. a. CH_3CH_2- **b.** $(CH_3)_2CH-$

29. a. methanol   **b.** propyl alcohol (1-propanol)

31. <chem>benzene ring</chem>$-OH$

33. a. $CH_3-\overset{\displaystyle O}{\overset{\displaystyle \|}{C}}-H$ **b.** $H-\overset{\displaystyle O}{\overset{\displaystyle \|}{C}}-H$

35. a. formic acid (methanoic acid)
   b. propionic acid (propanoic acid)

37. a. $CH_3CH_2CH_2CH_2CH_2CH_2COOH$
   b. $CH_3CH_2CH_2CH_2CH_2CH_2CH_2CH_2CH_2COOH$

39. a. $CH_3-\overset{\displaystyle O}{\overset{\displaystyle \|}{C}}-OH$ **b.** $CH_3-\overset{\displaystyle O}{\overset{\displaystyle \|}{C}}-CH_3$

41. a. $CH_3-\overset{\displaystyle O}{\overset{\displaystyle \|}{C}}-OCH_2CH_3$

   b. $CH_3CH_2CH_2\overset{\displaystyle O}{\overset{\displaystyle \|}{C}}-OCH_3$

43. a. $CH_3CH_2NH_2$ **b.** CH_3NHCH_3

45. a. propylamine   **b.** diethylamine

47. a. same compound   **b.** same compound
   c. isomers

49. a. homologs   **b.** none of these

51. unsaturated: a; saturated: b

53. a. alkene   **b.** alkane

55. $-\overset{\overset{\displaystyle O}{\|}}{C}-$

57. the amino group

59. carboxylic acids: a, b; amine: c; amide: d

61. alcohol: c; ester: d; ether: a; carboxylic acid: b

63. 40%

65. phenol; methyl and isopropyl

67. homology

69. $2\,C_6H_6 + 15\,O_2 \longrightarrow 12\,CO_2 + 6\,H_2O$; 132 g

Chapter 10

10.1 ~CH$_2$CH~
　　　|
　　　CN

10.2 ~CH$_2$CCl$_2$CH$_2$CCl$_2$CH$_2$CCl$_2$CH$_2$CCl$_2$~

33. a. CH$_2$=CHCl　　**b.** CH$_2$=CH—⬡

35. ~CH$_2$CHClCH$_2$CHClCH$_2$CHClCH$_2$CHClCH$_2$CHCl~

37. a. ~CH$_2$CHCH$_2$CHCH$_2$CHCH$_2$CH~
　　　　　　|　　　|　　　|　　　|
　　　　　　CN　　CN　　CN　　CN

b. ~CH$_2$CH(OCOCH$_3$)CH$_2$CH(OCOCH$_3$)CH$_2$CH-(OCOCH$_3$)CH$_2$CH(OCOCH$_3$)~

c. ~CF$_2$CF$_2$CF$_2$CF$_2$CF$_2$CF$_2$CF$_2$CF$_2$~

39. CH$_2$=C—CH=CH$_2$
　　　　|
　　　　CH$_3$

41. ~NH(CH$_2$)$_4$NHC(CH$_2$)$_4$—
　　　　　　　　　　　　　　—CNH(CH$_2$)$_4$NHC(CH$_2$)$_4$C~

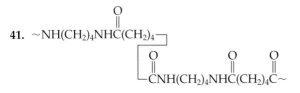

43. ~O—C⬡C—O—CH$_2$⬡CH$_2$~

45. monomer: a; repeating unit: b; polymer: c

47. acrylonitrile (CH$_2$=CH—C≡N) and styrene (CH$_2$=CH—C$_6$H$_5$)

Chapter 11

1. crust, mantle, core

11. The lithosphere is the solid portion of Earth, the hydrosphere is the watery portion, and the atmosphere is the gaseous mass surrounding Earth.

13. oxygen

15. living and once-living matter

17. the SiO$_4$ tetrahedron

19. The SiO$_4$ tetrahedra are arranged in a two-dimensional array.

21. The SiO$_4$ tetrahedra are arranged in an irregular fashion.

23. sand (silica), sodium carbonate, and limestone (calcium carbonate)

25. Cement is a mixture of lime, clay, and water that hardens like stone when it dries.

27. iron ore, limestone, and coke or coal (carbon)

29. reduction

31. An alloy is a mixture of two or more elements, at least one of which is a metal.

33. aluminum; it is difficult to obtain the metal from its ores

35. 120 million kg Al$_2$O$_3$; 260 million kg bauxite

37. TiCl$_4$ is reduced; Na is the reducing agent; Na is oxidized; TiCl$_4$ is the oxidizing agent.

39. ThO$_2$ is reduced; Ca is the reducing agent; Ca is oxidized; ThO$_2$ is the oxidizing agent.

41. AlCl$_3$ is reduced; Mn is the reducing agent; Mn is oxidized; AlCl$_3$ is the oxidizing agent.

43. 114 g

45. 0.668 g

47. 3.8 million t Zn; 0.3 million t Pb; 0.83 million t Cu

49. a cube 16 m on each side

51. 1.15 billion t

Chapter 12

1. by absorbing harmful UV radiation

5. refrigerant; blow-molding of plastic foams

6. Lead is toxic.

13. the troposphere; the stratosphere

15. water vapor: 0–4%; carbon dioxide: 365 ppm

17. by making more nitrogen compounds available to plants

19. by plants through photosynthesis

21. particulate matter, sulfur oxides, carbon monoxide

23. $S + O_2 \longrightarrow SO_2$

25. $SO_3 + H_2O \longrightarrow H_2SO_4$

27. The two interact to give an effect greater than the sum of the individual effects.

29. The lime reacts with and thus removes the SO$_2$; $CaO(s) + SO_2(g) \longrightarrow CaSO_3(s)$

31. hydrocarbons, NO$_x$, O$_3$, aldehydes, PAN

33. high temperatures; $N_2 + O_2 \longrightarrow 2NO$

34. it is broken down; $NO_2 \longrightarrow NO + O$

37. automobiles

39. A catalytic converter is a device containing catalysts for oxidizing carbon monoxide and hydrocarbons to carbon dioxide and reduction of nitrogen oxides to nitrogen gas.

41. automobiles

43. by hindering oxygen transport and thus adding to the work load of the heart

45. $O_2 + O \longrightarrow O_3$

47. increase in skin cancer

49. rain with a pH below 5.6; SO_x and NO_x are oxidized and react with water to form acids

51. acids dissolve iron;

$$Fe(s) + H_2SO_4(aq) \longrightarrow H_2(g) + FeSO_4(aq)$$

53. carbon monoxide, poorly ventilated heaters; nitrogen oxides, gas range; radon, ground

55. increased risk for lung cancer

57. The slow warming of Earth caused by gases absorbing infrared radiation.

61. 8000 μg (or 8 mg)

63. 0.19 g CO_2; 2.6%

Chapter 13

13.1 A. a. 0.1 ppb   **b.** 100 ppt
B. 89.5 ppm

1. nearly 98%

5. those that cause disease

11. The ice floats, protecting the deeper water from freezing.

13. The gasoline would not dissolve; it would float.

15. It allows water to cool the body as sweat evaporates.

17. Ice has a rigid structure with large open spaces. As ice melts, some molecules can move into those spaces.

19. dust, carbon dioxide, and sometimes nitric acid

21. water that contains Ca^{2+}, Mg^{2+}, (sometimes Fe^{2+}); it precipitates soap

23. acid precipitation; drainage from mines

25. remove sulfur from coal, scrub sulfur oxides from stack gases, reduce the use of automobiles

27. neutralize the water with a base

29. a holding pond intended to remove some of the sewage solids as sludge by settling

31. combines primary and secondary sewage treatment methods; some sludge is recycled

33. to kill pathogenic bacteria

35. hydrocarbons and chlorinated hydrocarbons

37. to carry down suspended solids

39. kills bacteria, offers residual protection; it does not kill viruses and it produces low levels of chlorinated hydrocarbons

41. by converting hydroxyapatite to harder fluorapatite

43. fertilizer; drainage from feedlots

45. $2 H^+(aq) + CaCO_3(s) \longrightarrow Ca^{2+}(aq) + CO_2(g) + H_2O$

47. a. 1 ppb   **b.** 35 ppm

51. Chlorine is reduced; it is the oxidizing agent. Sulfur dioxide is the reducing agent.

Chapter 14

14.1 A. 49,000 J   **B.** 34,600,000 J
14.2 A. 21.8 kcal   **B.** 9.48 kcal

14.3 A. 53.0 kcal   **B.** 0.0815 kcal

1. a material easily burned as a source of energy

3. wood

7. fuels: a, c

9. $C(s) + O_2(g) \longrightarrow CO_2(g)$

11. $CH_4 + 2 O_2 \longrightarrow CO_2(g) + 2 H_2O$

13. Reactions go faster at a higher temperature.

15. a reaction that gives off energy; any combustion reaction

17. 1010 kcal

19. Energy is conserved.

21. a measure of randomness (disorder); increased

23. huge reserves, available, cheap; must be mined, difficult to transport, pollutes a lot when burned unless controls are employed

25. coal, solid; petroleum, liquid; natural gas, gas

27. 37.4 years

29. 74.4 years

31. methane

33. ancient marine animals

35. by distillation, cracking, reforming; tetraethyllead is quite toxic, but a highly effective octane booster.

37. nearly 20%

39. No; the uranium is enriched to only about 3% uranium-235; to explode, it would have to be enriched to about 90%.

41. by converting nonfissile uranium-238 to fissile plutonium-239

43. Highly reactive sodium must be used as a coolant; the plutonium could be used to make bombs.

45. no fuel shortages, cleaner. Plasma is a mixture of atomic nuclei and electrons at temperatures such as that on the sun.

47. $^{238}_{90}Th \longrightarrow ^{1}_{0}n + ^{233}_{90}Th$
$^{233}_{90}Th \longrightarrow ^{0}_{-1}e + ^{233}_{91}Pa$
$^{233}_{91}Pa \longrightarrow ^{0}_{-1}e + ^{233}_{92}U$

49. $^{2}_{1}D + ^{3}_{1}T \longrightarrow ^{1}_{0}n + ^{4}_{2}He$

51. Solar energy is not concentrated. It is diffuse, making it difficult to put into useful forms.

53. plant material used as fuel

55. Fuel is fed into a fuel cell continuously, and the electrodes serve only to conduct electricity, taking no part in the reaction.

57. $C(s) + 2 H_2(g) \longrightarrow CH_4(g)$

59. $CO(g) + 2 H_2(g) \longrightarrow CH_3OH(l)$

61. 2000 watts

63. 1.83×10^6 g CO_2

65. a. neopentane   **b.** neopentane

67. a. 7.2×10^6 g   **b.** 6.4×10^6 g
c. 5.7×10^6 g

69. a. CH_4 is the lowest at 2.75 g CO_2/g fuel
b. CH_4 is the lowest at 0.055 g CO_2/kJ

Chapter 15

15.1 A. Histidylprolylvalylalanine
B. Thr-Gly-Ala-Ala-Leu

15.2 sugar: ribose; base: uracil

3. They make food by photosynthesis.
9. long-chain carboxylic acids
11. esters of glycerol with one, two, and three fatty acid residues
13. a measure of the degree of unsaturation of a fat; the carbon-to-carbon double bond
15. in every cell; muscles
17. All three contain C and H; proteins also contain N.
19. an amide function that joins two amino acid units
27. a protein with molecules that fold into a spheroid or ovoid shape
31. DNA (found mainly in the nucleus) and RNA
33. in DNA: guanine, cytosine, adenine, and thymine; in RNA: guanine, cytosine, adenine, and uracil
35. hydrogen bonding
41. mRNA; tRNA
49. Both are polymers of glucose. The acetal linkages between glucose units are alpha in starch and beta in cellulose.
51. monosaccharides: c, d
53. 51: none; 52: c, d
55. **a.** glucose **b.** galactose and glucose **c.** glucose

57.
$$CHO \quad \text{aldehyde, alcohol (hydroxyl)}$$
$$H-C-OH$$
$$HO-C-H$$
$$H-C-OH$$
$$H-C-OH$$
$$CH_2OH$$

59.
$$CHO \quad \text{aldehyde, alcohol (hydroxyl)}$$
$$H-C-OH$$
$$HO-C-H$$
$$HO-C-H$$
$$H-C-OH$$
$$CH_2OH$$

61. saturated: d, e; unsaturated: a, b, c
63. **a.** 18 **b.** 16 **c.** 18
65. corn oil; oils have more carbon-to-carbon double bonds than fats
67. no; their N-terminal and C-terminal ends are opposite

69. **a.** CH_3CHCOO^- **b.** $HOCH_2CHCOO^-$
        NH_3^+ NH_3^+

71. **a.**
$$H_3N^+CH_2\overset{\displaystyle O}{\overset{\displaystyle \|}{C}}-NH-CH-COO^-$$
$$CH_3$$

b.
$$H_3N^+CH\overset{\displaystyle O}{\overset{\displaystyle \|}{C}}-NH-CH-COO^-$$
$$CH_3 \qquad CH_2OH$$

73. nucleotide: c
75. ribose: a, b; deoxyribose: c
77. ribose; uracil
79. **a.** guanine **b.** thymine **c.** cytosine **d.** adenine
81. TTAAGC
83. AGGCTA
85. **a.** AAC **b.** CUU
87. pyrimidine: a; purine: b
89. **a.** a purine **b.** in RNA
91. lipid: b

Chapter 16

16.1 47.5%
16.2 750 kcal; 250 kcal
16.3 10%
16.4 270 mg
16.5 by 2021

1. carbohydrates, fats, proteins
3. **a.** glucose **b.** sucrose **c.** fructose
7. maltose, then glucose
9. about 4 kcal
11. energy, thermal insulation, protection of vital organs
13. more solid; more saturated
15. adipose tissue
17. 10%
23. yes
27. **a.** thyroid gland
    b. hemoglobin
    c. bones, teeth, blood clotting
    d. nucleic acids, bones, teeth
29. vitamin A: retinol; vitamin B_{12}: cyanocobalamine; vitamin C: ascorbic acid; vitamin D: calciferol; vitamin E: tocopherol
31. **a.** C **b.** D **c.** A
33. water soluble: b, c, d; fat soluble: a, e
35. fat soluble
37. glycogen
39. intentional and incidental

43. **a.** sweetener
   b. prevents rickets
   c. preservative
45. monosodium glutamate, a flavor enhancer
47. reducing agents
49. vitamin E
51. It is a dipeptide.
53. potent carcinogens found in stored peanuts and grain
59. water and carbon dioxide
61. as NO_3^- or NH_4^+
65. NH_3 without water; fertilizer
71. It interferes with calcium metabolism and formation of egg shells.
75. chemicals secreted to mark a trail, send an alarm, or attract a mate
79. causes leaves to fall off plants
83. that population growth would exceed food production
87. $6\,CO_2 + 6\,H_2O \longrightarrow C_6H_{12}O_6 + 6\,O_2$
91. 40 g
93. 32.4 g
95. 11.3%
97. 300 mg
99. 8192; geometric
103. alkene
105. ether, alkene, ester

Chapter 17

3. a salt of a long-chain carboxylic acid
11. water containing Ca^{2+}, Mg^{2+}, or Fe^{2+} ions
25. a surfactant molecule with a negative charge on its water-soluble head
41. a surfactant molecule with a positive charge on its water-soluble head
51. bacteria
53. keratin
55. skin softener
57. an oil and water
59. harder; more wax
63. sex attractant of musk deer
65. alcohol, perfume, coloring
67. product that kills bacteria that cause odor
69. detergent
71. melanin and phaeomelanin
73. Temporary dyes are water-soluble; can be washed out.
75. oxidizing agent
77. all three
79. II
81. III
83. $PO_4^{3-} + H_2O \longrightarrow HPO_4^{2-} + OH^-$

85. $2\,PO_4^{3-} + 3\,M^{2+} \longrightarrow M_3(PO_4)_2$
87.
$$\begin{array}{l} CH_2OH \\ | \\ CH{-}OH \;+\; CH_3(CH_2)_6\overset{\displaystyle O}{\overset{\|}{C}}O^-Na^+ \;+ \\ | \\ CH_2OH \end{array}$$
$$CH_3(CH_2)_4\overset{\displaystyle O}{\overset{\|}{C}}O^-Na^+ \;+\; CH_3(CH_2)_8\overset{\displaystyle O}{\overset{\|}{C}}O^-Na^+$$
89. oxidation
91. oxidation
93. II
95. I

Chapter 18

18.1 3.6 lb
18.2 35 mi
18.3 24

1. starch
3. 25%
17. No
19. They are reducing agents.
27. actin and myosin
31. anaerobic
35. Type I: slow twitch; Type II B: fast twitch
39. Type I
41. athletes need more calories; by eating more carbohydrates
49. Muscle protein is converted to glucose.
51. about 1 lb
53. Na^+, K^+, and Ca^{2+}
57. Alcohol and caffeine are diuretics.
73. 24
75. 2 hr
77. 0.1 hr (or 6 min)
79. 100 g
81. 20 km
83. 2250 kcal
85. 106 g/mL; lean

Chapter 19

1. a pain reliever
3. acetylsalicylic acid
5. Both are analgesics and antipyretics.
7. Aspirin enhances bleeding; acetaminophen does not.
9. preparation given to a patient who thinks it contains medication, although it actually does not
11. antibacterial substance produced, for example, by a mold
13. They are effective against a wide variety of bacteria.
15. a steroid produced in the adrenal cortex

17. a female sex hormone
19. diethylstilbestrol
21. hormone mediators derived from arachidonic acid
23. drugs that affect the mind
25. stimulants, depressants, and hallucinogens
29. a drug that acts on the brain to produce unconsciousness
31. diethyl ether
37. ketamine and PCP
39. dried juice of unripe seeds of the oriental poppy
43. An agonist mimics the action of a drug.
47. No
51. phenolic and carboxylic acid groups
53. water-soluble
57. Cocaine is more toxic than procaine.
59. an ethynyl group
61.

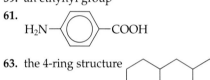

63. the 4-ring structure

Chapter 20

1. Toxicology is a branch pharmacology that deals with the effects of poisons on the body.
2. yes; in extremely large amounts
4. Sugar can be more harmful to a diabetic than to a nondiabetic.
11. by breaking down proteins, especially enzymes
13. substances that block the transport of oxygen in the bloodstream; carbon monoxide, nitrate ions
15. by blocking the oxidation of glucose, thus stopping cellular respiration
17. iron deficiency anemia; an overdose can cause vomiting, diarrhea, shock, coma, and even death
19. by chelating Hg^{2+} ions, thus enhancing their excretion
21. Cadmium poisoning leads to loss of calcium ions (Ca^{2+}) from the bones, leaving them brittle and easily broken.
23. Acetylcholine is chemical messenger that acts by activating the next cell by fitting a specific receptor and thus changing the permeability of the cell membrane to certain ions.
25. Anticholinesterase poisons inhibit the enzyme cholinesterase. They bond tightly to the enzyme, blocking the breakdown of acetylcholine and causing acetylcholine to build up and receptor nerves to fire repeatedly.
27. Atropine blocks acetylcholine receptor sites.

29. P-450 enzymes oxidize fat-soluble substances and conjugate compounds with amino acids, making them easier to excrete.
31. Cotinine is less toxic and more water-soluble than nicotine.
33. No. Some are made more toxic.
35. 2300 mg (2.3 g)
37. genes that regulate cell growth; they sustain the abnormal growth characteristic of cancer
39. burning of any organic matter; found in car exhaust fumes, cigarette smoke, and on charbroiled meats
41. Benzpyrene is oxidized to an epoxide that binds to specific nucleotides in the P53 gene where mutations frequently occur.
43. Humans are not usually exposed to doses comparable to those given test animals; human metabolism is different from that of the test animals.
45. a chemical substance that causes mutations; a chemical substance that causes birth defects
47. A waste that, when improperly managed, can cause or contribute to death or illness or threaten human health or the environment.
49. A waste that burns readily on ignition, presenting a fire hazard; hexane, gasoline.
51. They react with and dissolve the iron.
53. **a.** 98 mg **b.** 2000 t **c.** a matter of opinion

Appendix A

Practice Problems—Exponential Notation
1. **a.** 1.7×10^{-5} **b.** 1.9×10^7
   c. 3.4×10^{-3} **d.** 9.65×10^4
3. **a.** 8.6×10^8 **b.** 2.5×10^{-18}
   c. 5.7×10^{-30} **d.** 4.2×10^{-3}

Exercises
A.1 **a.** 0.0163 g **b.** 1.53 lb **c.** 370 mL
A.2 **a.** 0.0903 m **b.** 0.2224 km **c.** 149 fl oz
A.3 **a.** 24.4 m/s **b.** 1.34 km/h **c.** 0.136 oz/qt
A.4 0.12 m^3
A.5 56.8 g
A.6 **a.** 100.5 m **b.** 150 g **c.** 6.3 L
   d. 1800 m^2 **e.** 2.33 g/mL **f.** 0.634 g/cm^3
A.7 **a.** 185°F **b.** 10°F **c.** 179°C **d.** −29°C
A.8 80,000 cal; 80.0 kcal; 334 kJ

Practice Problems
1. **a.** 298 K **b.** 0°C
3. 820 cal
5. **a.** °C **b.** Cal
7. **a.** 98.6°F **b.** −14.7°C **c.** 523°F
9. **a.** 3100 J **b.** 2060 cal
11. 1500 cal

Photo Credits

Chapter 1
Page 1: Richard Johnston/Stone. Page 2 (T): Doris K. Kolb. Page 2 (M): Rachel Epstein/Stuart Kenter Associates. Page 2 (B): Doris K. Kolb. Page 3 (L): Giovanni Stradano, The Alchemist. Studiolo, Palazzo Vecchio, Florence, Italy. Scala/Art Resource, N.Y. Page 3 (R): Kunsthistorisches Museum Wien. Page 4 (T): Courtesy of the Library of Congress. Page 4 (M): UPI/Corbis. Page 8: Hank Morgan/Photo Researchers, Inc. Page 11 (T): Corbis. Page 11 (M): Rachel Fuller Brown. Page 11 (B): AP/Wide World Photos. Page 13 (T): Charles M. Duke, Jr./NASA Headquarters. Page 13 (B): Carey Van Loon. Page 16: © Kenneth Eward/BioGrafx, 2000. Page 18: © Kenneth Eward/BioGrafx, 2000. Page 19: Carey Van Loon. Page 23: Rachel Epstein/Stuart Kenter Associates. Page 32 (R): Rachel Epstein/Stuart Kenter Associates. Page 34: United States Environmental Protection Agency, Office of Pollution, Prevention, and Toxics.

Chapter 2
Page 36: IBM Research/Peter Arnold, Inc. Page 37 (T): Stamp from the private collection of Professor C. M. Lang, photography by Gary J. Shulfer, University of Wisconsin, Stevens Point. "Greece (Scott #1469);" Scott Standard Postage Stamp Catalogue, Scott Pub. Co., Sidney, Ohio. Page 37 (B): Photo Researchers, Inc. Page 38: J. L. David, Antoine Lavoisier and his wife. Oil on canvas, 1788. The Granger Collection. Page 39: Smithsonian Institution. Page 40: Stamp from the private collection of Professor C. M. Lang, photography by Gary J. Shulfer, University of Wisconsin, Stevens Point. "Hungary (Scott #2699);" Scott Standard Postage Stamp Catalogue, Scott Pub. Co., Sidney, Ohio. Page 41: John Dalton (1766-1844): color engraving, 19th century, The Granger Collection. Page 45 (T): Novosti/Science Photo Library/Photo Researchers, Inc. Page 45 (B): Stamp from the private collection of Professor C. M. Lang, photography by Gary J. Shulfer, University of Wisconsin, Stevens Point. "1957, Russia (Scott #1906) and 1969, Russia (Scott #3607);" Scott Standard Postage Stamp Catalogue, Scott Pub. Co., Sidney, Ohio. Page 47: Travelpix/FPG International LLC.

Chapter 3
Page 53: Joseph Nettis/Photo Researchers, Inc. Page 54 (L): Dr. Mitsuo Ohtsuki/Science Photo Library/Photo Researchers, Inc. Page 54 (R): Science V./IBMRI/Visuals Unlimited. Page 57: Carey Van Loon. Page 59: Burndy Library, Dibner Institute for the History of Science and Technology. Page 60 (T): Science VU/Visuals Unlimited. Page 60 (B): Stamp from the private collection of Professor C. M. Lang, photography by Gary J. Shulfer, University of Wisconsin, Stevens Point. "France #B76;" Scott Standard Postage Stamp Catalogue, Scott Pub. Co., Sidney, Ohio. Page 65 (T): Richard Megna/Fundamental Photographs. Page 65 (R): Joseph Nettis/Photo Researchers, Inc. Page 65 (B/L): David Parker/Science Photo Library/Photo Researchers, Inc. Page 65 (B/R): Stone. Page 66: Wabash Instrument Corp./Fundamental Photographs. Page 68: Stamp from the private collection of Professor C. M. Lang, photography by Gary J. Shulfer, University of Wisconsin, Stevens Point. "1963, Denmark (Scott #409);" Scott Standard Postage Stamp Catalogue, Scott Pub. Co., Sidney, Ohio.

Chapter 4
Page 79: NASA Headquarters. Page 88: Stamp from the private collection of Professor C. M. Lang, photography by Gary J. Shulfer, University of Wisconsin, Stevens Point. "New Zealand;" Scott Standard Postage Stamp Catalogue, Scott Pub. Co., Sidney, Ohio. Page 89: Corbis. Page 90 (T): Peticolas/Megna/Fundamental Photographs. Page 90 (B/L): Dyck, Anthony van (1599-1641). Saint Rosalie Interceding for the Plague-Stricken of Palermo. Oil on canvas. H. 39-1/4 in. W. 29 in. (99.7 x 73.7 cm.). The Metropolitan Museum of Art, Purchase, 1871. (71.41). Photograph © 1983 The Metropolitan Museum of Art. Page 90 (B/R): The Metropolitan Museum of Art, Purchase, 1871. (71.41). Photograph Courtesy of Brookhaven National Laboratory with permission of the Paintings Conservation Department, The Metropolitan Museum of Art. Page 92 (L): DuPont Pharmaceuticals Company. Page 92 (R): DuPont Pharmaceuticals Company. Page 93 (L): Hank Morgan/Rainbow. Page 93 (R): Dan McCoy/Rainbow. Page 97: Stamp from the private collection of Professor

C. M. Lang, photography by Gary J. Shulfer, University of Wisconsin—Stevens Point. "1979, China—People's Republic (Scott #1468);" Scott Standard Postage Stamp Catalogue, Scott Pub. Co., Sidney, Ohio. Page 98: UPI/Corbis. Page 99 (T): Stamp from the private collection of Professor C. M. Lang, photography by Gary J. Shulfer, University of Wisconsin, Stevens Point. "United States;" Scott Standard Postage Stamp Catalogue, Scott Pub. Co., Sidney, Ohio. Page 99 (B): UPI/Corbis. Page102 (L): John D. Cunningham/Visuals Unlimited. Page 102 (R): Los Alamos National Laboratory. Page 105: U.S. Navy/Science Photo Library/Photo Researchers, Inc.

Chapt er 5
Page 112: Tom Sanders/The Stock Market. Page 115: University of Pennsylvania, Van Pelt Library. Page 118 (T): Albert Copley/Visuals Unlimited. Page 118 (B): Richard Megna/Fundamental Photographs. Page 139 (T): Richard Megna/Fundamental Photographs. Page 139 (B): Richard Megna/Fundamental Photographs. Page 150: Massachusetts Institution of Technology, MIT Course 3.091. Page 151: Amethyst Galleries Inc.

Chapter 6
Page 152: Carey Van Loon. Page 156: Stamp from the private collection of Professor C. M. Lang, photography by Gary J. Shulfer, University of Wisconsin, Stevens Point. "1956, Italy (Scott #714);" Scott Standard Postage Stamp Catalogue, Scott Pub. Co., Sidney, Ohio. Page 161: Richard Megna/Fundamental Photographs. Page 162: Carey Van Loon. Page 172 (L): Richard Megna/Fundamental Photographs. Page 172 (R): Richard Megna/Fundamental Photographs. Page 177: Richard Megna/Fundamental Photographs.

Chapter 7
Page 185: Gerald Zanetti/The Stock Market. Page 187 (T): Robert Mathena/Fundamental Photographs. Page 187 (M): Richard Megna/Fundamental Photographs. Page 187 (B): Corbis. Page 192: Richard Megna/Fundamental Photographs. Page 196: Rachel Epstein/Stuart Kenter Associates. Page 197: Michael P. Gadomski/Photo Researchers, Inc. Page 202: GRID-Arendall.

Chapter 8
Page 204: Craig Hammell/The Stock Market. Page 205 (L): Carey Van Loon. Page 205 (M): Carey Van Loon. Page 205 (R): Carey Van Loon. Page 206 (L): Romilly Lockyer/The Image Bank. Page 206 (M): Derek Redfearn/The Image Bank. Page 206 (R): Chris Jones/The Stock Market. Page 210 (T): Carey Van Loon. Page 201 (B): Carey Van Loon. Page 216 (T): AP/Wide World Photos. Page 216 (B/L): Jose L. Pelaez/The Stock Market. Page 216 (T/L): Rick Smolan/Stock Boston. Page 216 (B/R): Eric Risberg/AP/Wide World Photos. Page 217: NASA Headquarters. Page 219: Carey Van Loon. Page 220 (L): Stuart Kenter Associates. Page 220 (R): Stuart Kenter Associates. Page 222 (L): Alan Pitcairn/Grant Heilman Photography, Inc. Page 222 (R): SuperStock, Inc. Page 223: Thomas R. Fletcher/Stock Boston.

Chapter 9
Page 230: Scott Camazine/Photo Researchers, Inc. Page 231: © Kenneth Eward/BioGrafx, 2000. Page 232 (T/L): Richard Megna/Fundamental Photographs. Page 232 (T/R): Richard Megna/Fundamental Photographs. Page 232 (L): Richard Megna/Fundamental Photographs. Page 234 (T): Richard Megna/Fundamental Photographs. Page 234 (L): Carey Van Loon. Page 237 (B): PhotoDisc, Inc. Page 237 (T/R): Richard Megna/Fundamental Photographs. Page 237 (B/R): Richard Megna/Fundamental Photographs. Page 239: Clive Freeman/The Royal Institution/Science Photo Library/Photo Researchers, Inc. Page 240: L. C. Clark, Jr., University of Michigan Medical School. Page 244 (T): Richard Megna/Fundamental Photographs. Page 244 (B): Richard Megna/Fundamental Photographs. Page 245: Richard Megna/Fundamental Photographs. Page 247: SIU/Photo Researchers, Inc. Page 250 (L): Richard Megna/Fundamental Photographs. Page 250 (M): Richard Megna/Fundamental Photographs.

Page 250 (R): Richard Megna/Fundamental Photographs. Page 252: Richard Megna/Fundamental Photographs. Page 264: Michel Roudnitska. Page 265: Eade Creative Services, Inc.

Chapter 10

Page 266: Chris Hartlove. Page 267: Tony Cordoza/Stone. Page 269 (L): Richard Megna/Fundamental Photographs. Page 269 (R): Richard Megna/Fundamental Photographs. Page 272: Gary A. Conner/PhotoEdit. Page 273 (T/L): Rachel Epstein/Stuart Kenter Associates. Page 273 (T/R): Corbis. Page 273 (B): Kip & Pat Peticolas/Fundamental Photographs. Page 275: Reynolds Metals Co. Page 279: Rich Chisholm/The Stock Market. Page 283 (L): Rachel Epstein/Stuart Kenter Associates. Page 283 (R): Kim Robbie/The Stock Market. Page 284 (T): Jean-Marc Barey/Photo Researchers, Inc. Page 284 (B): Amoco Fabrics & Fibers Co. Page 285: Courtesy of DuPont. Page 286 (T): Michael Mathers/Peter Arnold, Inc. Page 286 (B): Ray Pfortner/Peter Arnold, Inc. Page 292: Department of Polymer Science, University of Southern Mississippi.

Chapter 11

Page 294: Galen Rowell/Mountain Light Photography, Inc. Page 295: Guy Worthey, University of Michigan/NASA. Page 297 (L): Harold Hoffman/Photo Researchers, Inc. Page 297 (R): American Museum of Natural History. Page 298: GeoScience/Pearson Education/PH College. Page 299 (T): Chip Clark. Page 299 (B): From Asbestos and Other Fibrous Materials: Mineralogy, Crystal Chemistry, and Health Effects by H. Catherine W. Skinner, Malcolm Ross, and Clifford Frondel. New York: Oxford University Press, 1988. Page 300 (T): Mark Snyder/The Stock Market. Page 300 (B): American Museum of Natural History. Page 301: James L. Amos/Peter Arnold, Inc. Page 302 (T): Dennis O'Clair/Stone. Page 302 (B): William E. Ferguson. Page 303: Lee Boltin/American Museum of Natural History. Page 305: Breck P. Kent/Breck P. Kent. Page 306 (T): Robert Semeniuk/The Stock Market. Page 306 (B): Science VU/Visuals Unlimited. Page 308: Michael Abramson/Woodfin Camp & Associates. Page 313: The Musee de iHomme/6 Billion Human Beings Exhibit.

Chapter 12

Page 314: NASA/Johnson Space Center. Page 318 : Roger Werth/Woodfin Camp & Associates. Page 320 (T): F. Hoffmann/The Image Works. Page 320 (B): V. Leloup/Figaro/Liaison Agency, Inc. Page 321: Dr. Gerald L. Fisher/Science Photo Library/Photo Researchers, Inc. Page 323: Ellis Herwig/Stock Boston. Page 324: Jeff Greenberg/Photo Researchers, Inc. Page 328: The Goodyear Tire & Rubber Company. Page 329 (L): NASA/Goddard Space Flight Center. Page 329 (R): Susan Solomon. Page 331: Ray Pfortner/Peter Arnold, Inc. Page 336: Electric Power Research Institute. Page 339: M. C. Chamberlain/DRK Photo.

Chapter 13

Page 344: C. Haagner/Bruce Coleman Inc. Page 346 (L): Jim Zuckerman/Corbis. Page 346 (R): Michael Baytoff/Black Star. Page 347: David J. Phillip/AP/Wide World Photos. Page 349 (L): NASA Headquarters. Page 349 (R): Ray Witlin/World Bank Photo Library. Page 350 (L): Corbis Sygma. Page 350 (R): Larry Lee/Corbis. Page 351: D. Newman/Visuals Unlimited. Page 354: AP/Wide World Photos. Page 356: Kent & Donna Dannen/Photo Researchers, Inc. Page 358: Wolfgang Kaehler/Wolfgang Kaehler Photography. Page 361: James H. Karales/Peter Arnold, Inc. Page 362: National Institute of Dental Research. Page 368: U.S. Geological Survey. Page 369: U.S. Bureau of Reclimation and the Colorado River Water Users Association.

Chapter 14

Page 370: Stone. Page 371: Space Telescope Science Institute. Page 373 (L): Ed Degginger/Color-Pic, Inc. Page 373 (R): Carey B. Van Loon/Carey Van Loon. Page 374: Barry L. Runk/Grant Heilman Photography, Inc. Page 375 (L): Tom Pantages. Page 375 (R): Tom Pantages. Page 376: Kevin R. Morris/Corbis. Page 377: Continental Carbonic Products, Inc. Page 381: © The Field Museum, Neg #GEO85637C, Chicago. Photographer: John Weinstein. Page 382 (L): (14-05.01a) U.S. Department of the Interior. Page 382 (R): U.S. Department of the Interior. Page 383: John Kaprielian/Science Source/Photo Researchers, Inc. Page 391 (T): Pete Saloutos/The Stock Market. Page 391 (B): David M. Doody/Tom Stack & Associates. Page 392 (T): U.S. Department of Energy. Page 392 (B): Shone/Liaison Agency, Inc. Page 395: Brian Parker/Tom Stack & Associates. Page 398 (T): James Hanley/Photo Researchers, Inc. Page 398 (B): ML Sinibaldi/The Stock Market. Page 399 (T): Michael Neveux/Corbis. Page 399 (B): Lowell Georgia/Photo Researchers, Inc. Page 400 (T): American Petroleum Institute. Page 400 (B): American Petroleum Institute.

Chapter 15

Page 408: Norbert Wu/Norbert Wu Productions. Page 413: Biophoto Associates/Photo Researchers, Inc. Page 414: Don W. Fawcett/VU/Visuals Unlimited. Page 417: Richard Megna/Fundamental Photographs. Page 428: A. Barrington Brown/Photo Researchers, Inc. Page 429: © Kenneth Eward/BioGrafx, 2000. Page 435: Leonard Lessin/Peter Arnold, Inc. Page 437: M. Baret/RHPHU/Photo Researchers, Inc. Page 444: MIT Biology Hypertextbook. Page 445: Roslin Institute.

Chapter 16

Page 446: PhotoDisc, Inc. Page 448 (T): Richard Megna/Fundamental Photographs. Page 448 (B): Richard Megna/Fundamental Photographs. Page 449: Robert Mathena/Fundamental Photographs. Page 450: Richard Megna/Fundamental Photographs. . Page 451 (L): Cabisco/Visuals Unlimited. Page 451 (R): Cabisco/Visuals Unlimited. Page 454: © The Procter & Gamble Company. Used by permission. Page 455: Dagmar Fabricius/Stock Boston. Page 456: Biophoto Associates/Photo Researchers, Inc. Page 458 (T): Science Photo Library/Photo Researchers, Inc. Page 458 (B): National Medical Slide/Custom Medical Stock Photo, Inc. Page 463: M.I.T. Museum and Historical Collections. Page 473: Richard Megna/Fundamental Photographs. Page 477 (T): Runk/Schoenberger/Grant Heilman Photography, Inc. Page 477 (B): William Warren/Corbis. Page 478: Grant Heilman/Grant Heilman Photography, Inc. Page 479 (T): UPI/Corbis. Page 479 (B): Nathan Benn/Stock Boston. Page 480: National Archives. Page 481: Richard Megna/Fundamental Photographs. Page 484: U.S. Department of Agriculture. Page 485: Richard Megna/Fundamental Photographs. Page 487: U.S. Department of Agriculture. Page 492: United Nations.

Chapter 17

Page 498: Matt Meadows/Peter Arnold, Inc. Page 499: Richard Megna/Fundamental Photographs. Page 501: © Kenneth Eward/BioGrafx, 2000. Page 502: © Kenneth Eward/BioGrafx, 2000. Page 503: Richard Megna/Fundamental Photographs. Page 505 (T/L): Viviane Holbrooke/United Nations. Page 505 (T/R): Richard Megna/Fundamental Photographs. Page 505 (B): David M. Phillips/Visuals Unlimited. Page 507: Richard Megna/Fundamental Photographs. Page 511: Richard Megna/Fundamental Photographs. Page 515: Richard Megna/Fundamental Photographs. Page 519: Richard Megna/Fundamental Photographs.

Chapter 18

Page 528: Steve Mason/PhotoDisc, Inc. Page 529 (T): Al Tielemans/Duomo Photography Incorporated. Page 529 (M): Steven E. Sutton/Duomo Photography Incorporated. Page 529 (B): Paul J. Sutton/Duomo Photography Incorporated. Page 533: AP/Wide World Photos. Page 534: Steve Helber/AP/Wide World Photos. Page 537: Frank Siteman/Stock Boston. Page 539: Eric Grave/Phototake NYC. Page 540 (T): David Madison/Duomo Photography Incorporated. Page 540 (B): Paul J. Sutton/Duomo Photography Incorporated. Page 541: David Madison/Duomo Photography Incorporated. Page 542: David Madison/Duomo Photography Incorporated. Page 545: Goretex. Page 546: Bob Daemmrich/Stock Boston. Page 550: Will & Deni McIntyre/Photo Researchers, Inc. Page 551: Oli Tennent/Stone.

Chapter 19

Page 552: Mauritius, GMBH/Phototake NYC. Page 553: John Hill. Page 559 (T): Dr. Jeremy Burgess/Science Photo Library/Photo Researchers, Inc. Page 559 (B): Stamp from the private collection of Professor C. M. Lang, photography by Gary J. Shulfer, University of Wisconsin, Stevens Point. "Hungary (Scott #2699);" Scott Standard Postage Stamp Catalogue, Scott Pub. Co., Sidney, Ohio. Page 568: Equity Management Inc. (EMI). Page 570: Carl Djerassi. Page 571: Searle Word Headquarters. Page 573: Hank Morgan/Photo Researchers, Inc. Page 578: North Wind Picture Archives. Page 579: Dr. Morley Read/Science Photo Library/Photo Researchers, Inc. Page 584 (T/L): Ed Degginger/Color-Pic, Inc. Page 584 (T/R): VU/Cabisco/Visuals Unlimited. Page 584 (B): Library of Congress. Page 585 (T): Library of Congress. Page 585 (B): The Granger Collection. Page 593 (L): Scott Camazine/Photo Researchers, Inc. Page 593 (R): William D. Adams/FPG International LLC.

Chapter 20

Page 600: Andrew G. Wood/Photo Researchers, Inc. Page 601: The Metropolitan Museum of Art, Catharine Lorillard Wolfe Collection, Wolfe Fund, 1931. Page 617: AP/Wide World Photos. Page 619: Courtesy of UKAEA.

Index

Symbols and Names for Some Simple (Monatomic) Ions

Group	Element	Name of Ion	Symbol for Ion
1A	Hydrogen	Hydrogen ion*	H^+
	Lithium	Lithium ion	Li^+
	Sodium	Sodium ion	Na^+
	Potassium	Potassium ion	K^+
2A	Magnesium	Magnesium ion	Mg^{2+}
	Calcium	Calcium ion	Ca^{2+}
3A	Aluminum	Aluminum ion	Al^{3+}
5A	Nitrogen	Nitride ion	N^{3-}
6A	Oxygen	Oxide ion	O^{2-}
	Sulfur	Sulfide ion	S^{2-}
7A	Fluorine	Fluoride ion	F^-
	Chlorine	Chloride ion	Cl^-
	Bromine	Bromide ion	Br^-
	Iodine	Iodide ion	I^-
1B	Copper	Copper(I) ion (cuprous ion)	Cu^+
		Copper(II) ion (cupric ion)	Cu^{2+}
	Silver	Silver ion	Ag^+
2B	Zinc	Zinc ion	Zn^{2+}
8B	Iron	Iron(II) ion (ferrous ion)	Fe^{2+}
		Iron(III) ion (ferric ion)	Fe^{3+}

*Does not exist independently in aqueous solution.

Some Common Polyatomic Ions

Charge	Name	Formula
1+	Ammonium ion	NH_4^+
	Hydronium ion	H_3O^+
1−	Hydrogen carbonate (bicarbonate) ion	HCO_3^-
	Hydrogen sulfate (bisulfate) ion	HSO_4^-
	Acetate ion	$CH_3CO_2^-$ (or $C_2H_3O_2^-$)
	Nitrite ion	NO_2^-
	Nitrate ion	NO_3^-
	Cyanide ion	CN^-
	Hydroxide ion	OH^-
	Dihydrogen phosphate ion	$H_2PO_4^-$
	Permanganate ion	MnO_4^-
2−	Carbonate ion	CO_3^{2-}
	Sulfate ion	SO_4^{2-}
	Chromate ion	CrO_4^{2-}
	Monohydrogen phosphate ion	HPO_4^{2-}
	Oxalate ion	$C_2O_4^{2-}$
	Dichromate ion	$Cr_2O_7^{2-}$
3−	Phosphate ion	PO_4^{3-}